DEC 01 1998

EURO-CERAMICS

Volume 2: Properties of Ceramics

The Proceedings of the First European Ceramic Society Conference (ECerS '89) held on 18–23 June 1989 at the Maastricht Exhibition & Congress Centre, The Netherlands.

Volume 1: Processing of Ceramics

Synthesis
Forming
Densification
Phases and Reactions

Volume 2: Properties of Ceramics

Characterisation
Electronic Ceramics
Superconducting Ceramics
Traditional Ceramics

Volume 3: Engineering Ceramics

Bioceramics
Standardisation
Fracture
Deformation and Reinforcement
Composites
Surface-Related Phenomena
Applications

EURO-CERAMICS

Volume 2

Properties of Ceramics

including Electronic, Superconducting
and Traditional Ceramics

Edited by

G. de WITH

Philips Research Laboratories, Eindhoven, The Netherlands

R. A. TERPSTRA

Netherlands Energy Research Foundation, ECN,
Petten, The Netherlands

and

R. METSELAAR

Centre for Technical Ceramics,
Eindhoven University of Technology, The Netherlands

ELSEVIER APPLIED SCIENCE
LONDON and NEW YORK

ELSEVIER SCIENCE PUBLISHERS LTD
Crown House, Linton Road, Barking, Essex IG11 8JU, England

Sole Distributor in the USA and Canada
ELSEVIER SCIENCE PUBLISHING CO., INC.
655 Avenue of the Americas, New York, NY 10010, USA

WITH 92 TABLES AND 414 ILLUSTRATIONS

British Library Cataloguing in Publication Data

Euro-Ceramics.
1. Ceramics
I. With, G. de II. Terpstra, R. A. III. Metselaar, R.
666

ISBN 1-85166-426-2 (set)
ISBN 1-85166-430-0 (Vol. 1)
ISBN 1-85166-431-9 (Vol. 2)
ISBN 1-85166-432-7 (Vol. 3)

Library of Congress CIP data applied for

Printed in Northern Ireland by The Universities Press (Belfast) Ltd.

FOREWORD

In September 1987, during the 14th 'Science of Ceramics' conference, the European Ceramic Society was founded. The primary goal of this new society is to improve and facilitate communications both between European countries and between Europe and the rest of the world. Eleven countries have now joined the new society: Belgium, Federal Republic of Germany, France, Italy, The Netherlands, Norway, Portugal, Spain, Sweden, the United Kingdom and Yugoslavia.

To achieve its goal, the Society has decided to organize a set of conferences. This proceedings volume is a record of the first, ECerS '89, which took place in Maastricht, The Netherlands, in June 1989. The aim was to bring together engineers, scientists, technicians and all others with an interest in ceramics. The meeting covered the whole field of ceramics, from traditional clay-based materials to advanced materials with applications in the fields of engineering, electronics, magnetism and medicine. The 30 years tradition of the 'Science of Ceramics' meetings has been incorporated into the ECerS conferences under the heading of Basic Science. In 1991 the next conference of the European Ceramic Society, ECerS '91, will be held in the Federal Republic of Germany, probably near Munich. Directly after this conference the exhibition 'Ceramitec' will be organized. This will further enlarge the scope of the meeting, hopefully in a way that is appreciated by the European ceramic community.

It has been decided to present at each ECerS conference the A. L. Stuijts Memorial Award. As a research scientist at Philips Research Laboratories, Professor Stuijts contributed considerably to the development of magneto- and electro-ceramics, in particular to a better understanding of the relationship between the science and technology of these materials. To commemorate his contributions to the scientific community, an award with his name will be presented to a ceramist noted for his or her outstanding contributions to ceramic science or technology. The awardee will present a memorial lecture. During ECerS '89 the first Stuijts Memorial Award was presented to Professor R. J. Brook, who gave his Memorial Lecture entitled 'The Mastery of Microstructure'. Hopefully with this first memorial lecture a tradition has been started that will last for many years.

As mentioned by the editors, the space available to the authors in these proceedings is limited. I am pleased to announce that the ECerS now has its own *Journal of the European Ceramic Society*. This journal offers the possibility of publishing full-length papers of high scientific quality.

Organizing a conference is an elaborate task which is impossible without the help of many persons. This was true for ECerS '89 as well. It is impossible to mention all who contributed. An exception must be made for all collaborators of the Maastricht Exhibition & Congress Centre, MECC. Their enthusiastic cooperation has contributed considerably to the success of the conference.

I hope that the new series of conferences will be as successful as its predecessor, 'Science of Ceramics', and will fulfil all the needs of ceramists in Europe and the rest of the world.

R. Metselaar
President of the European Ceramic Society, 1987–1989,
Centre for Technical Ceramics, Eindhoven, The Netherlands.

PREFACE

This proceedings volume contains, besides the text of six key-note lectures and an extended abstract text of the first Stuijts Memorial Lecture, a representative selection of the more than 450 papers presented during ECerS '89.

The division between the various topics was approximately as follows: basic science 40%, engineering ceramics including bioceramics 30%, electronic ceramics including high T_c superconductors 20%, and traditional ceramics 10%. As was to be expected, the division is almost the same as in the American Ceramic Society Annual Meetings during the last few years. The fact that proper processing is the key to good properties is reflected here by the large number of contributions dealing with this topic. Moreover, there was a considerable number of papers discussing the processing and properties of composite materials.

Because of the large number of contributions, the number of pages for each manuscript had to be limited. This may seem a handicap to authors. The usefulness of proceedings is, in our opinion, mainly found in an overview of activities going on, rapidly published after the conference. Detailed theories or extensive descriptions of experimental work are out of place in such proceedings. Therefore, apart from practical reasons, in our view a limited number of pages is sufficient. For reasons of speed only limited reviewing was carried out. This procedure limited itself mainly to format and obvious errors. Opinions expressed remain, however, entirely the responsibility of the authors.

The editing of the proceedings would not have been possible without the help of many individuals. Without mentioning them all personally, this is the proper place to thank all those who contributed. One exception, however, must be made: Miss E. Ramaekers performed many secretarial duties, and without her help the editing of these proceedings could not have been completed in so timely a manner.

G. de With
Philips Research Laboratories and Centre for Technical Ceramics, Eindhoven, The Netherlands

R. A. Terpstra
Netherlands Energy Research Foundation, ECN, Petten, The Netherlands

R. Metselaar
Centre for Technical Ceramics, Eindhoven, The Netherlands

Honorary Committee ECerS '89

J. Kremers	Queen's Commissioner of Limburg
Ph. J. I. M. Houben	Mayor of Maastricht
R. W. de Korte	Minister of Economic Affairs
W. J. Deetman	Minister of Education and Science
R. E. Selman	Board of Management DSM
S. van Houten	Board of Management Nederlandse Philips Bedrijven BV
M. C. van Veen	Board of Management Hoogovens Groep BV
G. Kaptein	Board of Management NV Koninklijke Sphinx
G. J. M. Teeuwen	Board of Management Redland Bouwprodukten BV

Advisory Committee ECerS '89

A. McDonald (*Australia*)
F. Cambier (*Belgium*)
P. H. Duvigneaud (*Belgium*)
D. Yen (*China*)
F. Aiazzi (*Italy*)
D. T. Beruto (*Italy*)
G. Ziegler (*FRG*)
G. König (*FRG*)
P. Boch (*France*)
W. Simonet (*France*)
S. Saito (*Japan*)
A. J. Burggraaf (*The Netherlands*)
R. Pampuch (*Poland*)
J. L. Baptista (*Portugal*)
A. E. Benloch (*Spain*)
S. de Aza Pendas (*Spain*)
R. Pompe (*Sweden*)
G. J. Gittens (*UK*)
D. P. Thompson (*UK*)
D. E. Niesz (*USA*)
D. Kolar (*Yugoslavia*)

Scientific Programme Committee ECerS '89

G. de With (*Chairman*)
R. A. Terpstra (*Secretary*)
R. Metselaar
R. Wernicke
R. Nelissen
K. de Groot
K. Keizer
H. J. Timmers

Council of the European Ceramic Society

Belgium:	P. Fierens F. J. Cambier
FRG:	H. Hausner (*President Elect*) M. Blumenberg
France:	R. Cauville D. Broussaud
Italy:	G. N. Babini C. Palmonari
The Netherlands:	J. T. van Konijnenburg R. Metselaar (*President*)
Norway:	Th. Møller J. F. Haslund
Portugal:	A. F. Faria Frasco J. L. Baptista
Spain:	A. G. Verduch M. A. Delgado Méndez
Sweden:	R. Carlsson R. Ahlgren
United Kingdom:	G. J. Gittens D. Taylor
Yugoslavia:	D. Kolar

Sponsors of ECerS '89

City of Maastricht
Commission of the European Communities
Dutch Ministry of Economic Affairs
Elsevier Science Publishers Ltd
NV Gouda Vuurvast
KLM Royal Dutch Airlines
Koninklijke Mosa BV
NV Koninklijke Porceleyne Fles Ao 1653
NV Koninklijke Sphinx
Netherlands Energy Research Foundation (ECN)
D. van Oord & Co. Clay Tiles BV
Philips International BV
Philips Research Laboratories
Redland Bouwprodukten
Strategic Analysis Inc. SAI (Brussels)
Teewen BV
Tichelaar's Koninklijke Makkumer Aardwerk en Tegelfabriek BV
TNO Institute for Applied Physics

CONTENTS

Volume II: PROPERTIES OF CERAMICS including

Electronic, Superconducting and Traditional Ceramics

II.2 ELECTRONIC CERAMICS

THE STUIJTS AWARD

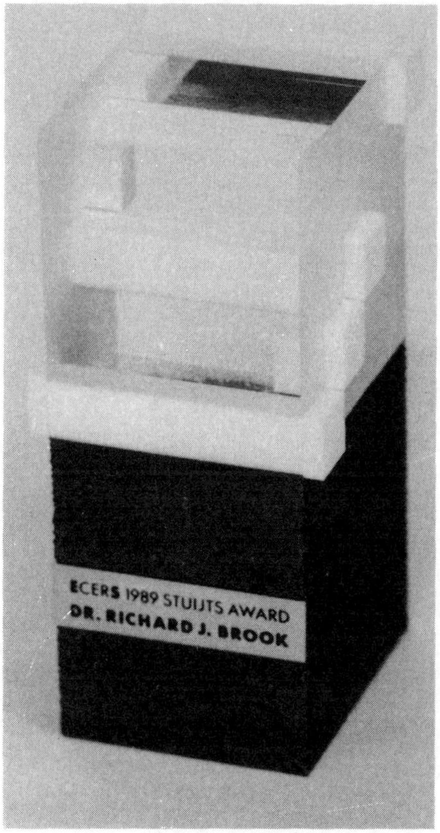

At the founding of the European Ceramic Society it was decided to present at each ECerS conference the A. L. Stuijts Memorial Award. As a research scientist at Philips Research Laboratories, Eindhoven, The Netherlands, Professor A. L. Stuijts (1922–1982) contributed considerably to the development of magneto- and electro-ceramic materials, in particular to a better understanding of the relationship between the microstructure and processing of these materials. He was also a Professor Extraordinary at the Eindhoven University of Technology. In 1976 he was elected Fellow of the American Ceramic Society and in 1977 he delivered the Sosman Memorial Lecture for this society. To commemorate his contributions to the ceramic community, during each conference of the European Ceramic Society an award bearing his name will be presented to a ceramist noted for outstanding contributions to ceramic science or technology. Professor R. J. Brook was the first to receive this award during ECerS '89 in Maastricht, The Netherlands.

Richard J. Brook (1938–) was born in Leeds, UK, and started his career in ceramics at the Massachusetts Institute of Technology, Boston, USA, where he obtained his PhD for research on ferrite thin films in 1966. From 1966 to 1970 he was Assistant Professor of Materials Science at the University of Southern Carolina where he did research on the development of microstructures in ceramics and on electrical properties. After 4 years at the AERE, Harwell, UK, as principal scientist and leader of the Electronic Ceramics Group, where he mainly researched electronic ceramics, Richard Brook joined the University of Leeds as Professor and Head of the Department of Ceramics. Apart from researching ceramic fabrication, solid electrolytes and ceramics processing, he has taken part in many university and external committees. Since 1988 Professor Brook has been one of the directors at the Max Planck Institute for Metals Research, Stuttgart, FRG.

The Stuijts award was presented by Dr F. M. A. Carpay, Executive Director, Philips Research Laboratories, after which Professor Brook delivered the Memorial Lecture entitled 'The Mastery of Microstructure'.

Stuijts Memorial Lecture 1989
"The Mastery of Microstructure"

R.J. Brook
Max Planck Institute for Metals Research
Stuttgart, FRG

In the light of his experience in the successful development of ceramic microstructures to match specific electromagnetic applications, Professor Stuijts proposed a number of themes in 1972 as being ripe for progress in the general task of exploiting ceramics for advanced engineering use. Three of these topics are reviewed with emphasis on the work that has been done and on current attitudes. The three themes are powder preparation, the importance of technological factors, and the need for valid generalised pictures of complex processes.

In powder preparation, the dramatic influence of chemical preparation methods on the quality of powders available for ceramic fabrication is one of the clearest developments of recent years. Major advances in processing convenience and in product reliability have resulted. In respect to technological factors, considerations of cost and of the need for reproducilbility are now seen as critical in determining the rate of adoption of ceramics into new applications; introduction of ceramic components into heat engines which has long been seen as a target in respect to the full and active use of ceramics on the large volume scale is largely determined by such aspects. The very great complexity of ceramics processing science and the resulting risk that detailed explanations of phenomena cannot then be extended for more general use has made the search for precise atomistic interpretation of micro-structural change less persuasive than it once was.

Three sectors that have proved fruitful in the intervening years in tackling the issues that were identified as being of concern are the use of mapping techniques, the use of computer modelling and the use of detailed microstructural characterisation notably in the form of electron microscopy.

These methods have allowed the development of a systematic approach to ceramic processing along the lines that Professor Stuijts was seeking. The approach has shown striking success in the development of the set of high strength and toughness ceramics based on the transformation toughening mechanism; it has also proved effective in the work on high temperature materials based on silicon nitride and the sialons.

The path to the mastery of microstructure that was advocated by Professor Stuijts has been found to be a productive one. The early view that detailed knowledge of precise atomistic mechanisms might one day underlie our approach to ceramic processing has if anything been seen as excessively and unnecessarily ambitious; in its place, a pragmatic pattern of development based on good understanding of the possible contributions of generalised mechanisms has developed.

A framework for the successful processing of ceramics is now in place; the need, however, for imagination in the exploitation of this framework remains as true today as at the time of its recognition by Professor Stuijts.

BORON CARBIDE - A COMPREHENSIVE REVIEW

FRANCOIS THEVENOT
Ecole Nationale Supérieure des Mines de Saint-Etienne
158, cours Fauriel 42023 SAINT-ETIENNE Cédex 2, France

ABSTRACT

Boron carbide, which has a high melting point, outstanding hardness, good mechanical properties, low specific weight, great resistance to chemical agents and high neutron absorption cross section (10BxC, x>4) is currently used in high-technology industries : fast-breeder, light weight armors, and high-temperature thermoelectric conversion.
The contents of this review are - 1.Introduction - 2. Preparations : Industrial preparative routes, powders, sintering (additives, pressureless, hot pressing, HIP) ; Laboratory methods of synthesis : CVD, PVD, plasma, crystal growth. - 3.Analytical characterization - 4. Phase diagram : a peritexy near pure boron, and a wide phase homogeneity range (B_4C-$B_{10.5}C$) - 5. - Rhombohedral crystal structure - A comprehensive model of the whole solid solution may be proposed - 6.Chemical properties - 7. Physical properties : density, mechanical (strength, hardness, toughness..), thermo-electrical properties.- 8 Main industrial applications - 9. Conclusion

1 - INTRODUCTION

In the group of the most important non metallic hard materials (alumina, silicon carbide, silicon nitride, diamond or cubic boron nitride), boron carbide occupies a specific place.

A boron carbide compound was discovered in 1858 [1], then JOLY in 1883 and MOISSAN in 1894 prepared and identified, respectively the compounds B_3C and B_6C The "stoechiometric formula" B_4C was only assigned in 1934 [2]. Then a lot of diverse formulae was proposed by Russian authors ; they have not been confirmed ; in fact, there is a wide phase homogeneity range $B_{4.0}C$-$B_{10.5}C$ (cf. § 4). After 1950, numerous studies were carried out, especially concerning structures and properties. Due to the limited size of this review, I will emphasize on some of the most important or recent references and indicate the more appropriate reviews [1,3-8].

The different topics given in the abstract will be described here .

2 - PREPARATIVE ROUTES

2.1 - Industrial preparation of boron carbide powders [1,3,6,8,9]

2.1.1.- Reduction of boron anhydride (or acid) with carbon : The overall reaction, which is strongly exothermic, is written :

$$2\,B_2O_3 + 7C \longrightarrow B_4C + 6\,CO \qquad (a)$$

The process takes place in two stages :

$$B_2O_3 + 3\,CO \longrightarrow 2B + 3CO_2 \qquad (b)$$

$$4\,B + C \longrightarrow B_4C \qquad (c)$$

Brickets of the B_2O_3-C mixture are placed in an electric-arc furnace. The central zone reaches 2473-2773 K and gives a carbide of composition near $B_{4.3}C$, containing a few % free graphite. The melted carbide is then crushed and milled to produce the grain size appropriate for final use. The pollution introduced during milling is eliminated by acid leaching. The outer zones of the furnace are less hot (2473-1473 K) and contain unreacted products, which must be recycled.

A graphite-tube furnace may also be used at 1973-2073 K, under a protective atmosphere ; a stoechiometric, fine-grained boron carbide (0.5-5 µm) is obtained. But the yield is lower than in the case of arc-melting.

2.1.2 - Reduction of boron anhydride with Mg in the presence of carbon black. The reaction is :

$$2\,B_2O_3 + 6\,Mg + C \longrightarrow B_4C + 6\,Mg\,O \qquad (d)$$

A furnace may be used at 1273-1473K or directly after initiating by a point ignition, the reaction (d) is strongly exothermic. To eliminate magnesia, borides ($Mg\,B_2$ ...) and unreacted Mg metal, the final product is washed with H_2SO_4, or HCl, then with hot water. A stoechiometric carbide with low granularity (0.1-5 µm) is directly obtained. It is possible to prepare graphite-free material ; in other cases, it may contain 2 % graphite.

A soluble carbohydrate that decomposes to carbon may be used with the carbon black [10].

2.2 - Sintering of boron carbide powders

We shall present the different sintering methods of boron carbide powders to prepare dense materials :

(1) hot pressing (and hot isostatic pressing)

(2) pressureless sintering

   (a) using metallic, inorganic, etc..., additives ;

   (b) adding carbon obtained by pyrolysis of an organic precursor : phenolic resin (alone or with polycarbosilane) .

Two recent reviews on sintering have been prepared [9,11] so we shall just cite the more recent references.

2-2-1 - Hot pressing of boron carbide :

2-2-1-1 - Hot pressing of stoechiometric $B_{4.0}C$ (20 at %C) : Hot pressing (HP) is used industrially to prepare different simple shapes. Boron carbide is hard and can be machined only using diamond (or BNc) rectification. To obtain dense products under vacuum or inert atmosphere :

(1) fine and pure powders (< 2 μm). (2) high temperature (2373-2473 K), (3) pressures of 30-40 MPa, and (4) 15-45 min of pressing in graphite dies should be used.

The density, porosity and microstructure depend on the sintering parameters [12].

Densification by sintering during HP results from three successive mechanisms : (1) particle rearrangement, where the enclosed porosity is constant and low ; (2) plastic flow, leading to the closing of open porosity ; or (3) decreases by volume diffusion of the closed porosity by pore elimination at the end of the HP.

2-2-1-2- Hot pressing of boron-rich carbides (2273-2373 K, 20-40 MPa, vacuum or inert atmosphere, 30-60 min) : The limits of the rhomboedral boron carbide phase are : B_4C (20 at%C) to $B_{10.5}C$ (8.8 at%C) (cf. § 4). HP of these compounds is a delicate operation. Extensive carbon diffusion takes place from the mold toward the sample and the sintering of pure boron produces only boron carbide [13]. The diffusion diagram in the B-C system has been established [14-15]. Refractory metallic foils (Ta, Mo, W etc.) can be used for protection. By using a boron nitride barrier (HP or deposited onto graphite mold and punches) it is possible to prepare dense and pure boron and boron-rich phases (between $B_{10.5}C$ and B_4C) by HP in graphite dies [14-15]. The models of densification kinetics of boron carbides were determined [14,16,17]. Carbon-rich boron carbides have the lowest viscosity and are more easily densified than the boron-rich materials [14,17].

2-2-1-3 - Hot pressing with additions [9] : The use of dopants has different purposes : (1) lowering the densification temperature (complete densification at 2023-2173 K) ; (2) increasing the oxidation and thermal-shock resistance ; (3) hindering grain growth, thus improving mechanical properties. But the problem of purity remains : the material is not convenient for nuclear applications.

Doping agents include : (1) the pure elements : Mg, Al, V, Cr, Fe, Co, Ni, Cu - and recently Si and Ti [18,19] ; (2) different compounds : glass, BN, MgO, Al_2O_3, sodium silicate + $Mg(NO_3)_2$ and Fe_2O_3, ethyl-silicate, MgF_2 or AlF_3.

2-2-2 - Hot isostatic pressing of boron carbide : This new promising technique, using glass encapsulation, without any additive, gives high density samples at a low temperature : 1700°C [20]. Hot isostatic pressing (HIP) (1200-1750°C, 700-3000 bar) of B_4C in the presence of excess B, C, N using molds of Ti or Zr, avoids the problem of graphite diffusion : owing to the formation of compounds of m.p. higher than that of the mold material between the mold and boron carbide, leaks were prevented [21,22].

Pressureless sintered B_4C materials (d = 93-97 %) doped with 1 and 3 wt% C (cf. § 2.2.6) "post-HIP ped", without encapsulation at 2000°C under an Ar gas pressure of 200 MPa, for 2h, to final densities > 99 %. Residual porosity was eliminated : mechanical properties and wear resistance were improved. There is a more frequent presence of twins in "HIP ped" samples than in HP ; the SEM micrographs show almost 100 % transgranular fracture [23].

2.2.3 - Microwave sintering of boron carbide [24,25] :Boron carbide is sintered to 0.95 d_{th} in less than 12 min without sintering aids by heating to 2000°C with 2.45 GHz microwave radiation. A mean grain size of 20 µm resulted, twins and microcracking were present. Energy usage for microwave sintering was found to be 18 % less than for inductive H.P.

2.2.4. - Explosion [26,27]

2.2.5 - Pressureless sintering of boron carbide

2.2.5.1 - *Pressureless sintering without additive* [9,11] : Boron carbide possesses strong bonding, low plasticity, high resistance to grain boundary sliding and low superficial tension in the solid state, all factors that hinder sintering. Sintering parameters under an inert atmosphere are known for a wide phase homogeneity range, $B_{8.3}$-$B_{4.1}C$ [28]. The conditions for dense products are : (1) boron carbide powders are preferred (reaction sintering gives porous material) ; (2) the grain size must be as low as possible : < 3 µm (powders with sizes larger than 8 µm cannot be sintered), (3) higher temperature (2523-2553 K), near the boron carbide m.p. (2573-2623 K) are needed.

During sintering recrystallization starts at 2073 K and at > 2273 K grains grow rapidly. The resulting twinning can be removed by high-temperature annealing [29,30]. The impurity-induced activation of the sintering of industrial B_4C arises from the formation of low-melting borides [31] ; the impurities evaporate during sintering.

2-2-5-2 - *Pressureless sintering with inorganic and metallic additions* [9,11] : A lot of inorganic additives has been proposed.

For sintering at < 2073 K : Cr, Co, Ni, glass and alumina with final density < 78 %.

For higher temperature (2423-2523 K) : Si, Al, Mg, TiB_2, CrB_2, Al, SiC, Be_2C and more recently, SiC + Al [32], B + C [33], B + Si (up to 20 at% Si) [18,19], W_2B_5 [34], TiB_2 + C [35], Al or TiB_2 or AlF_3 [36] additions have improved the densification ; but an exaggerated grain growth, a low mechanical strength, and a high impurity content, prove that the additives process is generally not satisfactory.

2.2.6 - Pressureless sintering with amorphous carbon additions : The addition of free graphite can be made to obtain fine grained compounds near the theoretical density [11,33,37-41] The carbon addition may be better produced by in-situ pyrolysis of a Novolaque-type phenol-formaldehyde resin (≈ 9 wt %). The residual free carbon content is 2-2.5 wt%. This method is now possible in industry [11,33,37-41].

A promising new method is the use of two organic precursors e.g. polycarbosilane with a small amount of phenolic resin, giving CSi and C by in-situ pyrolysis ; the resulting boron carbide ceramics have a high density (> 92 %) and contain no free carbon and a small amount of SiC (≈ 5 wt %) [11,33,42].

2.3 - Laboratory preparative routes (recent reviews [9,43])

2.3.1 - Powders - Some recent encouraging new routes. : A hydrogen - Ar plasma using a BCl_3-CH_4-H_2 mixture, produces boron carbide powders in excellent yield, with a wide stoechiometry range (B/C : 15.8 - 3.9) and a fine spherical granularity (20-30 nm) [44-46]. Ultrafine (34 nm) B_4C powders are prepared by the CO_2 laser-driven pyrolysis of the same BCl_3-CH_4(C_2H_4)H_2 mixture. The sintering of the later powders should be promising, unfortunatly they are not still commercially available. [47,48].

An amorphous B-C-Cl precursor precipate is obtained by reaction $TiCl_4$ with BCl_3, in n-heptane in the presence of sodium. After pyrolysis, very fine, agglomerated, irregular particles of B_4C mixed with 10 % carbon excess, are obtained [49,50].

The pyrolysis of boron-containing organic precursors may produce B_4C, pure or with BN, SiC, SiO_2 ... generally with low yields [51].

<u>2.3.2 - Thin solid films obtained by C.V.D. :</u> Films are deposited on different substrates, at high temperature (1273-2073K), using different reactions :

$$4\,BCl_3(BBr_3) + CH_4 + H_2 \longrightarrow B_4C + 12\,HCl \quad (e)$$
$$BCl_3 + C\,Cl_4 + 8H_2 \longrightarrow B_4C + 6\,HCl \quad (f)$$

Calculations and models are used in order to predict the conditions and the nature of B-C deposits [52-56] : boron carbide rhombohedral type phase, tetragonal $B_{50}C_2$, ß boron, graphite, either alone or co-deposited [57].

The mixture BBr_3 [58] (or B_2H_6 [59,60])-CH_4-H_2 was studied into a microwave plasma (PACVD) at a moderate temperature (400-600°C) ; very hard (5000 kg/mm^2) amorphous boron-carbon films were deposited, and the maximum microhardness was obtained for 38 at C% (outside the phase homogeneity range) [58].

<u>2.3.3 - Fibers</u> :They are obtained by C.V.D. of i) boron on a carbon fiber, followed by reaction ii) B_4C, using the reaction (e), on a boron fiber. These fibers reinforce light metals matrices (Al, Ti ...).

<u>2.3.4 - Crystal growth</u> : Single crystals are obtained using i) C.V.D. reactions (e), or better (f), or ii) zone melting of a boron carbide rod. Whiskers and platelets, with different stoechiometries, are prepared according to reaction (f).

3 - ANALYTICAL CHARACTERIZATION

3.1 - Destructive methods for the total C and B content determination [61]

Boron carbide is totally oxidized by fusion with alkaline carbonate, the resulting product being dissolved by HCl. The total boron content is determined either by atomic absorption or emission spectroscopies or as orthoboric acid H_3BO_3, complexed by mannitol, titrated by soda using potentiometry. After oxidation by oxygen at 1600°C, the total carbon is determined as CO_2 by conductimetry or gas chromatography. Several methods are available to characterize free boron, metallic impurities ... [61].

3.2 - Non destructive methods for quantitative analysis

Operating on polished dense materials, the total boron and carbon contents are determined by microanalysis, the characteristic parameters of which have been studied. The accuracy for carbon is 2-4 % and 2 % for boron, the balance (B+C) being in the range 100-102 % [14,62,63].

Using a generalized method of additions of graphite, it is possible to determine free graphite, present in boron carbide powders, using quantitative XRD and the ratio

IC (002)/IB_xC (021) [14,63]. Very fine graphite particles (0.1 μm) (so called "subtle" carbon) could be excluded by this method [64,65].

A new X-ray diffraction method is optimized for a quantitative analysis of both graphite and boron nitride in the presence of textured, powdered or massive, boron carbide (or boron carbide-silicon carbide composites) products.[33,66].

3.3 - Auger electron spectroscopy

The shapes of B-KVV and C-KVV Auger lines observed on ion bombardment cleaned surfaces of B_9C and B_4C were relatively insensitive to the bulk stoechiometry of the samples. This indicates that the local chemical environments surrounding B and C atoms, respectively, on the surface of the ion bombardment cleaned sample do not change appreciably in going from B_9 to B_4C. Microbeam techniques used to study a fracture surface of B_9C material reveal both C-rich (the C-VVV line was graphitic in shape) and B-rich regions.[67].

Measured by Auger Electron spectroscopy, the B concentration of the surface of a B_4C single crystal, heated in the range 600-1300°C, decreased with the heating time and became a steady-state value [68].

3.4.- High resolution imaging microstructures

High resolution transmission electron microscopy used to examine hot pressed boron carbide shows a high density of variable width twins [69]. The same technique was used to study the radiations effects in boron carbide.[70]

4 - PHASE DIAGRAM

A lot of controversial phase diagrams has been proposed in the period 1955-1960 [14,71]. Later on ELLIOTT [72] and KIEFFER [73] assumed a wide phase homogeneity range for boron carbide (9-20 at %C) and a eutectic between B_4C and C lies at 26 at% C and 2400°C.[73].

In order to investigate precisely the whole phase diagram, we decided to study samples prepared by melting and hot-pressing, i.e. in a good thermodynamic equilibrium..

Various mixtures of boron and carbon were electron beam melted. Samples were cut and diamond polished ; their surface was characterized by electron microprobe analysis (cf 3-2), metallographic techniques, XRD (cf.§ 5) and Knoop microhardness (cf.§ 7-5-1) measurements. The metallographic investigations, together with the microanalysis, showed that the boron carbide phase exists in the homogeneity range 9-20 at % C (with a relative accuracy of 3-5 %). In order to confirm these limits we studied the crystal lattice dimensions (cf.§ 5) and the microhardness (cf.§7.5.1) of boron carbide in the above homogeneity range. Boron carbide melts congruently and forms a eutectic between carbon-rich compounds and graphite. For the boron-rich compounds we assumed the existence of a peritectic transformation with a solid solution of carbon (< 1 at %C) in the ß rhombohedral boron structure [14,71,74].

Furthermore, samples were prepared by hot pressing (1800-2200°C, 32,5 MPa, 30-60 min, under argon) a mixture of boron and boron carbide powders. Representing the lattice parameters variation versus C content, the breaks in the lines confirmed the same homogeneity range (8.8-20.0 at %C, i.e. $B_{10.4}C$-B_4C) [14,75]. Therefore we proposed a phase diagram (fig.1).

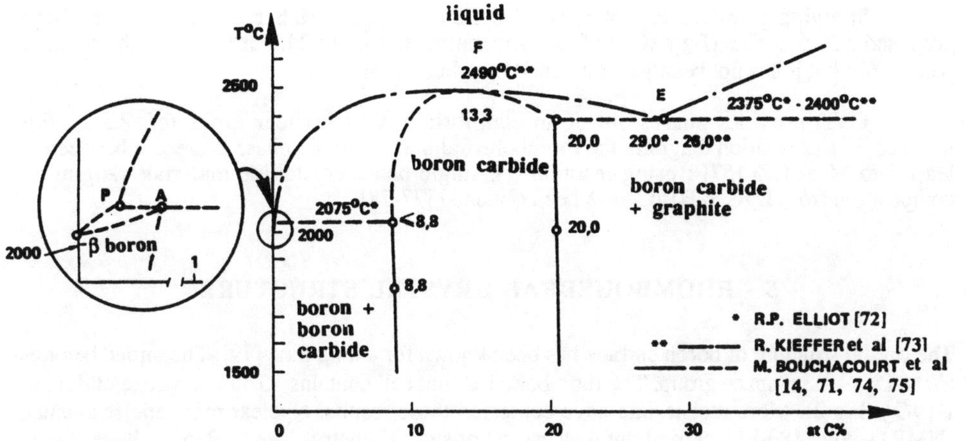

Figure 1. Boron-carbon phase diagram (o : hot-pressed samples)
P : peritexy ; F : congruent melting ; E : eutectic

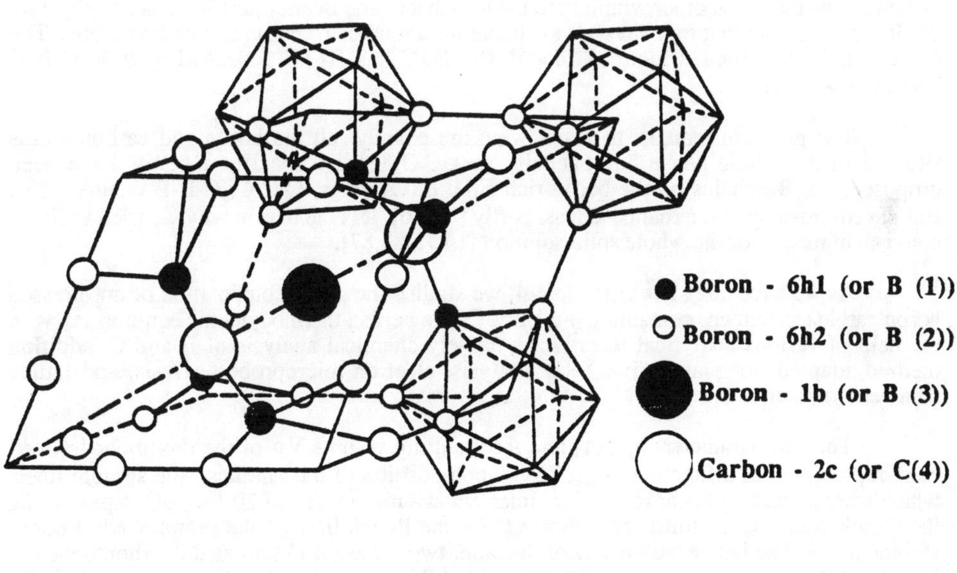

Figure 2. Rhombohedral crystalline strcutrure of boron carbide.

Studying powders or hot pressed pellets, a carbon richer phase limit has been proposed : 21.6 at %C ($B_{3.63}C$) at low temperature and up to 24.3 at %C near the melting point [76] ; but it has not been yet confirmed by other researchers.

CVD provides phases out of an equilibrium, with broader limits (cf. 2.3.2). For instance, in a deposition diagram, the rhombohedral boron carbide phase is present between at least 4 to 16 at C% [57] ; using reaction (f), single phase crystalline materials ranging in composition from B_2C to $B_{17}C$ have been obtained [77-78].

5 - RHOMBOHEDRAL CRYSTAL STRUCTURE

The crystal structure of boron carbide has been known for a long time [79]. The lattice belongs to the $D3d^5$-$R\bar{3}m$ space group. The rhombohedral unit cell contains 15 atoms corresponding to $B_{12}C_3$ (Fig.2). More recent data have contested these results. Nuclear magnetic resonance (NMR) studies [80-81], carried out on a crystal of stoechiometry close to $B_{12}C_3$, have shown that the central position in the C-C-C chain (b site) was partially occupied by boron (60 % in ref.80). Using IR absorption spectroscopy BECHER and THEVENOT have confirmed the existence of the C-B-C chain in compounds such as B_4C and $B_{5.52}C$ [81].

The boron carbides are composed of twelve-atom icosahedral clusters which are linked by direct covalent bonds and through three-atom intericosahedral chains. The boron carbides are known to exist as a single phase with carbon concentrations from about 8.8 to 20 at.%. This range of carbon concentrations is made possible by the substitution of boron and carbon atoms for one another within both the icosahedra and intericosahedral chains [82] (Fig. 2) . It is pointed out that four sites are available for a total of 15 boron and carbon atoms. The most widely accepted structural model for B_4C has $B_{11}C$ icosahedra with C-B-C intericosahedral chains.

The problem remains to know were are exactly all the boron and carbon atoms situated in the whole phase homogeneity range. Otherwise specific subunits have been proposed, i.e. B_4 chains on the boron rich side, B_4C_2 rings, C-C-C, C-B-B chains...[83]. But the controversy still remains, at least partly [83-86]. It is the reason why we tried to find a convenient model for the whole solid solution [14,75,83,87].

As we have already written in § 4, we studied the phase composition of hot-pressed boron carbide specimens containing 0-60 at %C, in a perfect thermodynamic equilibrium, with the help of various analytical techniques, namely chemical analysis of B and C, addition method adapted for quantitative XRD analysis, electron microprobe analysis, and lattice parameter measurements.

The cell parameters a_H, c_H and the resulting volume V_R of the rhombohedral cell were represented as a function of the global composition of the samples. The straight lines, which were plotted by linear regression, intersect at a mean value of 20.1 at %C, representing the C-rich composition limit and 8.8 at %C for the B-rich limit of the rhombohedral boron carbide phase. The lattice parameters of the latter two limits, of C-saturated ß rhombohedral boron and B-satured graphite were determined [14,75].

The previous studies (lattice parameters, analysis of B and C) together with accurate measurements of the density of compounds in the phase homogeneity range allowed us to determine the number n of each atom contained in the unit cell. The value of n_T-varies linearly with the carbon content [14,87] :

$$n_T = 15.47 - 0.019 \text{ [C] at \% } (r = 0.994)$$

The gradients of the plots of the parameters a_H, c_H, a_R and α_R as a function of the C content are discontinuous at the composition 13.33 at %C ($B_{13}C_2$). This experimental result confirms the particular role of this composition within the solid solution range. The volume of the rhombohedral unit cell decreases linearly with the carbon content.[14,87].

For the carbon-rich limiting composition (20.0 at C%) we obtained 15.08 atoms per unit cell, i.e. 12.06 B atoms and 3.02 C atoms. The lattice corresponding to the boron-rich limiting composition (8.8 at %C) consists of 15.30 atoms per unit cell, i.e. 13.95 B atoms and 1.35 C atoms. The unit cell corresponding to the composition 13.3 at %C (so called "$B_{13}C_2$") contains 15.21 atoms i.e. 13.18 B atoms and 2.03 C atoms [14,87].

Finally studies of the variation in the interatomic distances with carbon content, together with ^{13}C NMR and IR absorption spectroscopy, both characteristics of the different interatomic bonds, show the complexity of the solid solution mechanism. We have found two domains, separated by the 13.3 at.% C composition. In the carbon-rich compositions, up to 20 at.% C, we find a partial substitution of a boron atom by carbon in the B_{12} icosahedra and in the centre of the C-B-C chain. In the boron-rich compositions, up to 8.7 at.% C, we find a substitution in the C-B-C chain by boron atoms. Furthermore, over the whole domain, there are interstitial atoms in the lattice (0 for 20 at.% C, 0.33 atom for 8.7 at.% C).This model appears to be the more coherent than has yet been proposed [14,83].

[C] = 20 at.% C	—> [C] = 13.3 at.%C	—> [C] = 8.7 at.% C
$B_{11.77}(C_{0.23})[C(B_{0.23}C_{0.77})C]$	—> $B_{12}(CBC)B_{0.18}C_{0.03}$	—> $B_{12}(CBC)_{0.66}B_{1.33}$
n = 15	—> n = 15.23	—> n = 15.33

Considering the replacement of C atoms within B_4C, ($B_{11}C$) C-B-C, with B atoms, it was recently concluded that entropic and energetic considerations both favored this replacement within the intericosahedral chains, C-B-C —> C-B-B. Once the carbon concentration is so low that the vast majority of the chains are C-B-B, near $B_{13}C_2$, subsequent substitution of C with B atoms occur within the icosahedra, $B_{11}C$ —> B_{12}. Maxima of the free energy occur at the most ordered compositions B_4C, $B_{13}C_2$, ($B_{14}C$?) [82]. But the identity of the chains have to be proved by spectroscopy and the model just considers 15 atoms, whereas we proved there are up to 15.33 atoms in the unit cell.

6 - CHEMICAL PROPERTIES [6,8]

Boron carbide is one of the most stable compounds. Fine powders are slowly oxidized in wet air ; oxygen and water contents increase with time and species like "B_2O_3", HBO_3 or H_3BO_3 are formed at the surface.[88].

Oxidation of hot pressed samples in oxygen starts at 600°C and results in the formation of a thin transparent B_2O_3 film, that becomes cracked after cooling. Up to 1200°C the oxidation process is limited by the diffusion of reagents through the oxide layer.[89-91].

Boron carbide is not attacked by cold chemical reagents ; it is oxidized by hot oxidizing acids (HNO_3, H_2SO_4, $HClO_4$ ...) [92].and fused salts.(cf § 3-1). Chorine attacks B_4C at 600°C, and bromine at 800°C, giving boron trihalides ; B_4C reacts with metal oxides at elevated temperatures to give carbon monoxide and metal borides [93]. Being activated by halides, B_4C is used for the boronizing of steels and alloys [94-96].

7 - PHYSICAL PROPERTIES

7-1 - Density [14,87]
The density of boron carbide increases linearly with carbon content within the homogeneity range of the phase according to the relationship :

$d(g.cm^{-3}) = 2.4224 + 0.00489$ [C] at % ($r = 0.998$)
with 8.8 at % $\leq$ [C] $\leq$ 20.0 at %.

The density measured for B_4C is 2.52 $g.cm^{-3}$, $B_{13}C_2$: 2.488 $g.cm^{-3}$, $B_{10.4}C$: 2.465 $g.cm^{-3}$.

7.2 - Coefficient of thermal expansion

$$\alpha = 3.016 \times 10^{-6} + 4.30 \times 10^{-9}\, t - 9.18 \times 10^{-13}\, t^2 \; (t°C) \; [97]$$

average value $\alpha = 5.73 \times 10\text{-}6 \; K^{-1}$ (300-1970K) [98]

or 4-8 x 10^{-6} K^{-1} (25-800°C) [99].

7.3. Specific thermal capacity of B_4C [100]

$$C_p \; (cal.mol^{-1} \, K^{-1}) = 22.99 + 5.40 \times 10^{-3} \, T \, (K) - 10.72 \times 10^5 \, T^{-2} \, (K).$$

7.4. - Thermoelectrical properties
"Nuclear electric propulsion is under consideration in the USA for future outer planet missions. Previous results from a space craft system study showed that an optimum hot junction temperature is in the range of 1500K for advanced nuclear reactor technology combined with thermoelectric conversion "[101]. A lot of work has been carried out on that subject by SANDIA, JET Propulsion Laboratory under contract with NASA in the USA, starting from 1981 [102-104 ; literature review : 105].

This subject was studied in our laboratory before 1980 and published a bit later [14,105,106]. Because there were many contradictions in the literature, mainly due to uncertain sintering routes of samples [105], the aim of our work was to determine the dependence of the figure of merit Z [via the electrical (σ) and thermal (λ) conductivities and Seebeck coefficient (S), according to the relation $Z = S^2 \sigma/\lambda$], with temperature and carbon content within the phase homogeneity range of boron carbide. Perfectly charactized hot-pressed (§ 2-2) samples were studied [14,105,106].

A lower than theoretical density in boron carbides reduces the thermal conductivity λ; a monotonic decline of λ with increasing temperature is noticed with B_4C ; with large B/C ratios the temperature dependence is essentially nil [14,105,107]. Heat conduction mechanism occurs by phonon diffusion [105,108]. The influence of porosity and grain size on the thermal conductivity of B_4C has been studied [108].

The electrical conductivity of boron carbides increases with temperature [105]. The conduction mechanism is explained by the small polaron hopping [109]. The hopping activation energy E_h increases slightly when the C content increases ; there is perhaps a break in the slope for 13.3 at% C [105].

Boron carbide is a p-type semiconductor ; the Seebeck coefficient S remains positive over the whole phase homogeneity range, increases slightly with temperature and reaches values of 200-300 $\mu V\ K^{-1}$ at 1250 K.

Thermal (λ) and electrical (σ) conductivities and Seebeck coefficient (S) variations vs C content exhibit a break for 13.3 at% C ; this illustrates the particular role played by this compound in the phase homogeneity range (cf. § 5) ; it was not observed by other authors. For all the carbides the figure of merit Z increases with temperature. Furthermore for a given temperature, the 13.3 at% C composition possesses the highest S value, 0.85 x 10^{-3} K^{-1} at 1250 K [105], compared with 0.5 x 10^{-3} K^{-1} at 1273 K obtained by WOOD for B_9C [104]. Our Z value obtained for the $B_{13}C_2$ compound exceeds that of the best p-type Si-Ge alloys, generally used at these high temperatures [104] ; this value underlines the promise of boron carbide for use in high temperature thermoelectric generators.

7.5 - Hardness and wear resistance

<u>7-5-1 - Hardness :</u> boron carbide is among the hardest materials (diamond, BN_c).

Hardness measurements are difficult ; the preparation of samples and conditions of measure are uncertain or unknown, therefore values are scattered, and difficult to be compared [14,71,74,110]. The influence of carbon content is controversial ; for instance ALLEN [111] found an increase of hardness with C content, whereas LIPP and SCHWETZ did not find any influence [110].

Therefore we studied the conditions of KNOOP microhardness HK determination on planar electron - gun - melted samples [14,71,74,112].

Mechanical polishing induces a surface strain, therefore HK decreases when the time of electrolytic etching is increasing ; HK reaches a plateau after 10 sec. HK decreases when the load P increases. The variation of log P versus log L (L length of the print), allows us to determine n, in the Meyer's law $P = aL^n$; after mechanical or electrolytic polishing, this variation is represented by a broken line with two slopes : i.e. (i) n = 1.01 and 1.75 after mechanical polishing (ii) n = 1.20 and 1.80 after electrolytic etching.

The Knoop microhardness increases linearly with the C content, in the phase homogeneity range. For instance, after mechanical polishing, $HK_{200g} = 2910 \pm 90$ kg mm^{-2} (29.1 GPa) for 10.6 at C%, and reaches 3770 ± 80 kg mm^{-2} for 20 atC % (B_4C, with the minimum volume cell (§ 5) and the maximum density (§ 7-1)). After electrolytic etching HK drops down by 25 % : $HK_{200g} = 2840 \pm 60$ kg.mm^{-2} for B_4C.

HK_{200} = 25.5 ± 2.4 GPa for pressureless sintered samples and 29.0 ± 1.5 for hot pressed [33,38].

Vickers hardness HV similarly increases with the C content in CVD samples [113,114]. Extreme hardness values were obtained by microwave plasma deposition (§ 2-3-2). There is an evolution of HV with the deposition temperature of CVD ; a minimum was observed for 1373K, which corresponds to the microstructure with a maximum grain size [113].

Hardness decreases in the presence of free graphite in electron beam melted [71], sintered [38] or CVD [114] samples or Al-Si-C phase in hot pressed boron carbide [115].

Rebound measurements of the hardness of boron carbide indicate no decrease with temperature up to 1300°C. However, static indentation measurements show a continuous decrease of the hardness with temperature [116].

7-5-2 - Wear : Because of its high hardness and strength, boron carbide is inferior in abrasive resistance (arbitrary units = 0.4-0.422), only to diamond (0.613) (cp with SiC : 0.314) [8]. The abrasion mechanisms have been studied [117]. With increasing temperature the friction resistance increased up to 1100K, and it decreased afterwards ; the friction coefficient decreased continuously during heating [118,119] ; B_2O_3, BO(OH) and BO_3H_3 were formed on the friction interface [117,118]. The erosion by particle impact (SiC) is better for B_4C, than for alumina, zirconia, or silicon carbide-based systems [120].

7-5-3 - Machining : Diamond tools or polishing pastes must be used. Electric spark machining is convenient and not expensive [121,122]. In particular there is no surface stress : the strength is constant with increasing etching depth [122].

7.6 - Strength

Almost all the mechanical properties measured on hot-pressed (or sintered) boron carbide samples, with close to theoretical densities, differ and depend on specific impurity contents (especially Al, Si, C, used as dopants) and distribution, porosity, clusters of diffusion pores, grain size ... Therefore measurements are hardly comparable [115].

Strength depends on the hot-pressing temperature and the stoechiometry of the carbide ; small boron additions eliminate free graphite, thus improve strength [13] ; strength would increase with the carbon content of the homogeneous carbide phase [123]. Notice that 3 pt bending tests give higher values than 4 pt [89,115]. Hot-pressed (σ_F = 300 - 500 MPa), or post-HIP treated [23] samples, have higher characteristics than sintered ones (σ_F = 150-350 MPa) [11,23,33,37-39,89,115,121-123].

Strength is highly variable as Weibull modulus is low (5) ; for all etching depths, the Weibull modulus and strength remain constant, thus the flaws are inherent to the materials [122].

Influence of porosity and grain size : Transverse strength σ (compressive, or impact) decreases when porosity P, or grain size D, increase ; several models were proposed [13,121,123] :

$$\mathrm{Ln}\,\sigma = 20.337 - 0.367\,\mathrm{Ln}\,D - 4.974\,P \quad [13]$$

in which σ is in MPa, D in µm, P fractional porosity

$$\sigma = \sigma_0\, e^{-b\omega/\sqrt{DG}}$$

where ω is the total porosity, DG the grain size [123].

The influence of porosity and strength of dense material decreased when the carbon content increased [123]. Strength decreased by a factor of 6-7, with increasing porosity from 2 to 46 % and grain size from 5 to 140 μm [121].

Fracture surface of hot-pressed boron carbide subjected to TRS and impact tests revealed an intercrystalline character at porosities in excess of 15 %, and a mainly transcrystalline character at porosities below 15 %. Rupture was of the brittle cleavage type (river patterns). There were also many twins intersected by cleavage steps [121].

Behavior at high temperature : Under nitrogen atmosphere there was little or no decrease in strength [99] from its room temperature value of 380 MPa up to 1500 K (340 MPa) ; the fracture was fully transgranular at all testing temperatures [122].

In air, there is a gradual decrease in strength between 600 and 1000°C, due to superficial oxidation into B_2O_3 [124]. Above 1200°C, catastrophic B_4C oxidation started, which resulted in a complete degradation of strength [89]. Si and Al dopants improved the oxidation resistance, thus retaining a higher strength at high temperature (< 1400°C) [124].

7.7 - Toughness K_{1c}

K1c is measured by indentation, SENB... methods (2.9 - 3.7 MPa √m) [33,37-39,89,115,122].
A value has been calculated = 1.3 MPa √m [123].

Behavior at high temperature : Under nitrogen atmosphere K1c remained constant at ≈ 3.7 MPa m 1/2 up to 1500 K [122] or decreased from 1.77 at 25°C to 1.31 MPa √m at 1200°C [125].

In air, K_{1c} is not greatly changed up to 1200°C ; here, a crack propagates from a previously made notch, a larger number of defects due to oxidation on the sample surface obviously has no influence on the K_{1c} value [89].

CVD : K_{1c} increases with the carbon content in the phase homogeneity range (B/C : 6-4) ; it decreases in the presence of free graphite [114].

7.9 - Young's modulus E

The relationship $\nu = (E/2G) - 1$ where ν is Poisson's ratio and G shear modulus, is given.
E values are in the range = 360 - 460 GPa [23,33,37,38,89,115,122]. E slightly decreases when temperature (up to 2000°C) and porosity P increase [13,99,123,125,126] :

$E = 460\,[(1 - P)/(1 + 2.999\,P)]$ GPa [125] and

$\ln E = 26.833 - 5.462\,P$ (same expression for B_4C and boron rich-carbides) [13].

E depends on stoechiometry, it decreases with the boron content [123].

E is related to the composition of CVD materials (maximum value : E = 475 GPa for $B_{13}C_2$) [113].

7.10 - Poisson modulus ν

According to the literature, the values of γ split into two groups : 0.14 and 0.18 [122]. Other values were proposed : ν = 0.17 [37] ; 0.178 [122] ; 0.16-0.18 [23] ; 0.21 [126].

7.11 - Shear modulus

158 - 188 GPa [23,37].

7.12 - Thermal shock resistance

Experimental thermal shock resistance (determined by decrease of Young's modulus) was 210° for pressureless sintered samples and 260° for hot-pressed [33,38]. The maximal thermal shock resistance of B_4C pellets (used in nuclear reactors) can be predicted by $\Delta T_{C\ Max}$ = 6.5/R (R is the pellet radius, in m) [108] or experimentally deduced [99].

7.13 - Fracture energy γ

Various values have been proposed : 1.6 [123], 4.5 [125], 12.4 [122] ; 18.9 [13] $J.m^{-2}$.

The dependence of γ with temperature is given [122] . The fracture surface energy at 500°C appeared to be independent of porosity [125], but was inversely related to temperature [122,125]. A regression analysis showed that γ decreased with increasing porosity and average grain size ; variable B content, open and closed porosity, were not significant on the γ value [13].

8 - MAIN INDUSTRIAL APPLICATIONS

Details may be found in diverse reviews [3-8]. According to the previously described properties, the applications may be divided as follows.

8-1 - Uses based on hardness

The major industrial use of boron carbide is as abrasive grit or powder. Particle sizes are available from 1 µm to 10 mm, used as polishing, lapping and grinding media for hard materials such as cemented carbides, technical ceramics ... Boron carbide is far less expensive than diamond.

A second category is wear resistant components made of hot-pressed sintered pieces. Boron carbide sand-blasting nozzles are characterized by minimum wear, even with silicon carbide or corundum grit. Post-HIP sintered components have the best wear resistance [23,127]. Boron carbide ceramic-nozzles are used for water-jet cutting.

Other wear applications : sintered B_4C wheel dressing sticks to produce new cutting edges ; mortars and pestles.

Light-weight armor plates have been used against piercing bullets for the protection of helicopters, or as breastplates for the protection of personnel [128].

8.2 - Chemical uses (cf. § 6)
Boron carbide powders, activated by fluorides (or halides), are used to diffuse boron at the surface of steels ; the resulting Fe_2B thin layer (10-200 µm) is very hard (HV 0.1 = 2400) and wear resistant [94,129]. In certain cases, the formation of a brittle two phase iron borides ($FeB-Fe_2B$) layer, is unconvenient ; the aim is to produce a single Fe_2B phase [95,96].

8.3 - Electrical application (cf. § 7-4) - Boron carbide/graphite thermocouple
The thermocouple consists of a graphite tube, a B_4C bar, and a BN sleeve between them. It can be used to 2200°C, under inert gas or vacuum atmosphere ; there is a linear relationship between its voltage and the temperature at 600-2200°C [130-132].

8.4 - Nuclear applications [4,97,133-135]

8-4-1 - Principles : The main part (95 %) of nuclear power is now produced in reactors controlled by two kinds of absorbing materials : boron carbide ("B_4C") or a ternary alloy Ag-In-Cd. Boron carbide is a neutron absorber widely used because of its high B content, its good chemical inertness and high refractoryness. The neutronic absorption of B_4C is due to ^{10}B, which participates according to the following capture reaction :

$$^{10}{}_5B + {}^1{}_0n \longrightarrow {}^4{}_2He + {}^7{}_3Li + 2.4\ MeV$$

The cross section of this reaction varies from 3850 barns for thermal neutrons to a few barns for fast neutrons (the law following $1/\sqrt{E}$, E energy of neutrons), and allows the use of boron carbide with the natural isotopic content (19.8 % of ^{10}B - the rest being ^{11}B) or enriched up to 90 at% ^{10}B for reactors guiding.

The absorbant nuclear material depends on the specific reactor.

8-4-2 : The main present reactors [134,135] : - Pressurized Water Reactors (PWR). They are the most widely used in the world, and used by EDF in France. Two kinds of reactors now exist : 900 MW, where the Ag-In-Cd alloy is only used, and the last generation 1300 MW in which B_4C is partly substituted to the alloy.

- Fast Breeder Reactors (F.B.R.) : More recent than PWR, their standardization is not completly achieved, but now boron carbide is recognized as neutron absorber. In fact ^{10}B is one of the nuclides which reveals the best behavior under fast neutrons : a large cross section (1 barn in a standard spectrum of a 0.7 MeV average energy) and no resonance in the whole spectrum.

According to the design of the reactor core (number of bars), "natural" boron is used (19.8 % ^{10}B), or enriched : on the one hand 48 % for Phenix, and on the other hand 90 % in Rapsodie (stopped in 1983) or Superphenix. Enrichment (obtained by chemical exchange reactions, or by using resins) is a tedious and very expensive route : the price of 1 g carbide with 90 % ^{10}B is near 4-5 Ecus (or US dollars), whereas "natural" boron carbide costs 15-75 Ecus/kg [6].

8-4-3 - Control rod in PWR [136] : "The fissile material (U_{235}) is placed inside the reactor core in the form of fuel assemblies. These assemblies measure from 4 to 4.8 m in length... Each fuel assembly consists of 264 zirconium tubes or rods, which contain uranium dioxide fuel pellets. The rods are secured in a square framework or "skeleton" by means of 8 to 10 grids, two nozzles and 24 guide thimbles. The control rods slide inside the guide thimbles to conduct the rate of nuclear reaction" [136]. The rod cluster control assembly is constituted by 24 stainless steel cylinders, 4 m in length, in which calibrated boron carbide pellets (diameter 7-8 mm, height 15 mm, weight ≈ 1.4 g) are piled, under argon atmosphere.

8-4-4 - Fabrication of boron carbide pellets : Boron carbide powders are sintered (cf. § 2.2).

- absorber for PWR . "Natural" boron carbide (19.8 % ^{10}B) is used ; pellets are pressureless sintered, the density is ≈ 70 % d_{th}. In France, FBFC produces 2.000 m of pellets/year, i.e. ≈ 190 kg, corresponding to one reactor core.

- absorber for F.B.R. (Superphenix and Phenix) . The diameter of pellets is 20 mm in the Main Control System, and 45 mm in the complementary system. Pellets are hot-pressed to reach 96 % d_{th}.

8-4-5 - Behavior under neutron irradiation : The degree of irradiation of boron carbide is expressed either by a relative value (ratio of consumed ^{10}B atoms to initial number of total atoms) or an absolute value by the capture density (number of consumed ^{10}B atoms - equivalent to the number of He or Li, appeared during irradiation - pro volume unit). The life of boron carbide in the control rod does not reach the theoretical limit fixed by the total consumption of ^{10}B, because of the destruction of the material due to the large volume of He produced. For an irradiation corresponding to 10^{22} captures/cm^3 of B_4C (typical value in F.B.R.), 0.12 g of lithium and 380 cm^3 helium are produced in 1 cm^3 absorber. Under the temperature influence, a part (50-90 %) of helium diffuses outside the material, whereas a part accumulates, thus producing swelling and microcracking, which can break up a pellet into a fine powder.These processes are rather insensitive to the irradiation parameters (temperature, flux and neutron spectrum) [70]. The combination of swelling (increase of 15 vol% for 10^{22} captures/cm^3) and the desagregation of the material induces, for heavy irradiations (> 10^{22} capt./cm^3), a strong mechanical interaction between the absorber and the stainless steel tube which may become broken. This tube is also fragilized under fast neutrons irradiation and by diffusion of both boron and carbon (especially free) coming from the absorber itself ; in Superphenix, boron carbide is directly cooled by metallic sodium, inside the steel tube.

The mean lifetime of an absorber in PWR is 15 years, whereas it should be one year in the fast breeder, for a maximum density of captures respectively of 0.5 10^{22} cm^{-3} and 1.2 10^{22} cm^{-3}.

Recently Transmission Electron Microscopy of samples irradiated by the fast neutrons of a reactor ... was studied to understand the respective roles of helium and point defects in relation with swelling and microcracking [70].

9 - CONCLUSION

Actually, boron carbide is a fascinating technical ceramic. Annual production in the non-communist countries is about 500 t, which is rather small compared to the 500.000 t SiC production [6].

Among the industrial uses, nuclear applications are the most valuable and promising. The price of ^{10}B enriched boron carbide is very high (cf. § 8.4.2) but the development of fast neutron reactors needs boron carbide. Problems due to swelling under irradiation remain, they have to be solved. Research is being carried out. In Europe, France and Germany, seem to be leaders.

Powders are usually prepared either by magnesiothermy or arc melting. Perhaps new reactive powders, from plasma or laser techniques, could be interesting for low temperature pressureless sintering. Organic boron precursors are not yet still developed, but should be useful as sintering additives [43].

New forms for boron carbide are just appearing : platelets and whiskers (cf. § 2-3-4).

Several composites systems are now studied : cermets with Al [137,138], composites B_4C-C, B_4C-SiC [11,33,139,140], B_4C-whiskers SiC [141], B_4C-TiB_2 [142] ... Specific applications for the latter have to be developed.

Acknowledgements : Thanks are due to M. WACHE (FBFC, Romans, F), B. KRYGER (CEN Saclay), G. OSSENA (COGEMA, F), for information on nuclear applications, K.A. SCHWETZ (ESK, FRG) and the Firm QUARTZ & SILICE (Nemours, F) for other industrial applications.

I appreciate the valuable contributions of my former students M. BOUCHACOURT [14] and M. BOUGOIN [33] to the increase of knowledge about boron carbide. I do not forget the care of M.C. MATHAIS for typing the manuscript.

REFERENCES

1. Thévenot, F. and Bouchacourt M., Le carbure de bore : matériau industriel performant, 1ère partie : le point des connaissances physico-chimiques, Ind. céramique, 1979, **732** (10), 655-61.

2. Rigdway, R.R., Boron carbide. A new crystalline abrasive and wear-resisting product, Trans. Am. Electrochem. Soc., 1934, **66**, 117-33.

3. Lipp, A., Boron carbide. Production properties, application, Techn. Rundschau, 1965, **14-28-33** ; 1966, **7** (47 pp).

4. Beauvy, M. and Thévenot, F., Le carbure de bore : matériau industriel performant, 2ème partie : applications industrielles du carbure de bore, Ind. céramique, 1979, **734** (12), 811-4.

5. Schwetz, K.A., Reinmuth, K. and Lipp, A., Herstellung und industrielle Anwendung refraktäre Borverbindungen, Radex Rundschau, 1981, **3**, 568-85.

6. Schwetz, K.A. and Lipp, A., Boron carbide, boron nitride, and metal borides, Ullmann's Encycl. Indust. Chem., 1985, **A4**, 295-307.

7. Thévenot, F., Quelques applications actuelles des borures, Silicates indus., 1986, **51**, (1-2), 17-22.

8. Makarenko, G.N., Borides of the IVb group, in Boron and refractory borides, ed. V.I. Matkovich, Springer Verlag, Berlin, 1977, pp. 310-30.

9. Thévenot, F., Formation of Carbon-Boron Bonds, in Inorganic Reactions and Methods, ed. J.J. Zuckerman, A.P. Hagen, VCH Publishers, New York, 1989, **10**, pp. 2-11.

10. Gray, E.G., Boron Carbide, Eur. Pat. Appl.,11, 524, 28 May 1980 ; Appl. 1978.

11. Thévenot, F., Sintering of boron carbide and boron carbide-silicon carbide two-phase materials and their properties, J. nucl. Mater., 1988, **152**, 154-62.

12. Angers, R. and Beauvy, M., Hot pressing of boron carbide, Ceram. Int., 1984, **10**, 49-55.

13. Champagne, B. and Angers, R., Mechanical properties of hot-pressed B-B_4C materials, J. Am. Ceram. Soc., 1979, **62**, 149-53.

14. Bouchacourt, M., Etudes sur la phase carbure de bore - Corrélations propriétés-composition, Thèse d'état, INPG, Ecole des Mines de Saint-Etienne (F), 1982.

15. Bouchacourt, M., Brodhag, C. and Thévenot, F., The hot pressing of boron and boron rich compounds : B_6O, $B_{10.5}C$-B_4C, Sci. Ceram. , 1981, **11**, 231-6.

16. Brodhag, C., Bouchacourt, M. and Thévenot, F., La cinétique de compression à chaud de céramiques spéciales, Silicates indus., 1981, **4-5**, 91-101.

17. Brodhag, C., Bouchacourt, M. and Thévenot, F., Comparison of the hot pressing kinetics of boron, boron suboxide B6O and boron carbides, in Materials Science Monographs , 16, Ceramic powders, ed. P. Vincenzini, Elsevier, Amsterdam, 1983, pp. 881-90.

18. Telle, R. and Petzow, G., Reaction sintering of boron carbide (B4C) with silicon and titanium, Horiz. Powd. Metall., Proc. Int. Powd. Metall. Conf. Exhib., 1986, **2**, 1155-8.

19. Telle, R. and Petzow, G., Mechanisms in the liquid phase sintering of boron carbide with silicon based melts, Mater. Sci. Monogr., 1987, **38A**, 961-73.

20. Larker, H.T., Hermansson, L. and Adlerborn, J., Hot isostatic pressing and its applicability to silicon carbide and boron carbide, Mater. Sci. Monogr. . 1987, **38A**, 795-803.

21. Adlerborn, J., Isaksson, S. and Larker H., Hot-pressing of parts from boride, carbide, or nitride powders, Ger. Offen, 2, 344, 648, 1974 .

22. ASEA, Manufacture of boron carbide articles, J.P. 62, 207, 762, 1987 ; Appl. 1986.

23. Schwetz, K.A., Grellner, W. and Lipp, A., Mechanical properties of HIP treated sintered boron carbide, Inst. Phys. Conf., 1986, Ser. n° 75, Chap. 5, 413-26.

24. Holcombe, C.E., New microwave coupler material, Ceram. Bull., 1983, **62,** 1388.

25. Katz, J.D., Blake, R.D., Petrovic, J.J. and Sheinberg, H., Microwave sintering of boron carbide, Met. Powd. Rep., 1988, **43**, 835-7.

26. Prümmer, R., Explosive Werkstoffbearbeitung, Keram. Z., 1985, **37**, 243-5.

27. Bairamashvili, I.A., Kalandadze, G.I., Dzhobava, D.S. and Eristavi, A.M., Sintering of boron carbide powders after static and explosive treatment, Soobshch. Akad. Nauk. Gruz. SSR, 1985, **119**, 161-3.

28. Grabchuk, B.L. and Kislyi, P.S., Sintering of boron carbide, Porosh. Met., 1974, **8**, 11-6 ; 1975, **7**, 27-31.

29. Kuzenkova, M.A., Kislyi, P.S., Grabchuk, B.L. and Bodnaruk, N.I., The structure and properties of sintered boron carbide, J. less comm. Met., 1979, **67**, 217-23.

30. ibid Powd. Met. int., 1986, **12**, 11-3.

31. Grabchuk, B.L. and Kislyi, P.S., Sintering of technical boron carbide, Porosh. Met., 1974, **8**, 11-6.

32. Weaver, G.Q., Sintered high density boron carbide, USP, 4,320,204, 1982.

33. Bougoin, M., Frittage sans charge du carbure de bore et de composites carbure de bore-carbure de silicium. Propriétés mécaniques, Thesis, Ecole des Mines de Saint-Etienne (F), 1985.

34. Zakhariev, Z. and Radev, D., Properties of polycrystalline boron carbide sintered in the presence of tungsten boride (W2B5) without pressing, J. Mater. Sci. Lett., 1988, **7**, 695-6.

35. Oh, J.H., Orr, K.K., Lee, C.K., Kim, D.K., Lee, J.K., and Kim, C.H., Sintering of boron carbide, J. Korean Ceram. Soc., 1985, **22**, 60-6.

36. Kanno, Y., Kawase, K. and Nakano, K., Additive effect on sintering of boron carbide, J. Ceram. Soc. Jap., 1987, **95**, 1137-40.

37 Schwetz, K.A. and Grellner, W., The influence of carbon on the microstructure and mechanical properties of sintered boron carbide, J. less comm. Met., 1981, **82**, 37-47.

38. Bougoin, M., Thévenot, F., Dubois, J. and Fantozzi, G., Synthèse et caractérisation de céramiques denses en carbure de bore, J. less comm. Met., 1985, **114**, 257-71.

39. Thévenot, F. and Bougoin, M., Pressureless sintering of boron carbide phase, in Boron-Rich Solids, AIP Conf. Proc. 140, ed. D. Emin, T. Aselage, C.L. Beckel, I.A. Howard and C. Wood, Amer. Inst. Phys., New-York, 1986, pp. 51-8.

40. Prochazska, S. Dole, S.L. and Hejna, C.I., Abnormal grain growth and microcracking in boron carbide, J. Am. Ceram. Soc., 1985, **68**, C-235-6.

41. Dole, S.L. and Prochazska, S., Densification and microstructure development in boron carbide, Ceram. Eng. Sci. Proc., 1985, **6**, 1151-60.

42. Bougoin, M. and Thévenot, F., Pressureless sintering of boron carbide with an addition of polycarbosilane, J. Mater. Sci., 1987, **22**, 109-14.

43. Thévenot, F., Laboratory methods for the preparation of boron carbides, in NATO Adv. Res. Workshop (ARW), The Physics and Chemistry of Carbides, Nitrides and Borides, Manchester, U.K., Sept. 1989 ; to be publ. by Kleuwer, 10 pp.

44. Mac Kinnon, I.M. and Wickens, A.J., The preparation of boron carbide using a radio frequency plasma, Chem. Ind. (London), 1973, 800.

45. Mac Kinnon, I.M. and Reuben, B.G., The synthesis of boron carbide in a r.f. plasma, J. Electrochem. Soc., 1975, **122**, 806-11.

46. Mac Kinnon, I.M., Hamblyn, S.M.L. and Reuben, B.G., Mechanism of formation of boron and boron carbide by reduction of boron trichloride in an r.f. plasma, I.C.P. Inf. Newsl., 1977, **3**, 311-2.

47. Knudsen, A.K. and Langhoff, C.A., Submicron-sized boron carbide powders, PCT Int. Appl. WO 86 04,524, 1986, US Appl.1985.

48. Knudsen, A.K., Laser-driven synthesis and densification of ultrafine boron carbide powders, Adv. Ceram., 1987, **21**, 237-47.

49. Ritter, J.J., A low temperature chemical process for precursors to boride and carbide ceramic powders, Adv. Ceram., 1987, **21**, 21-31.

50. Ritter, J.J., Refractory borides and carbides, USP 4,606,902, 1986.

51. Walker, B.E. Jr, Rice, R.W., Becher, P.F., Bender, B.A. and Coblenz, W.S., Preparation and properties of monolithic and composite ceramics produced by polymer pyrolysis, Ceram. Bull., 1983, **62**, 916-23.

52. Bernard, C., Deniel, Y., Jacquot, A., Vay, P. and Ducarroir, M., Détermination des équilibres chimiques complexes dans les systèmes polyphasés. I méthode de traitement, J. less comm. Met., 1975, **40**, 165-71.

53. Ducarroir, M. and Bernard, C., Thermodynamic domains of the various solid deposits in the B-C-H-Cl vapor system, Proc. Chemical Vapor Deposition Fifth Intern. Conf., 1975, pp. 72-83.

54. Ducarroir, M. and Bernard C., Thermodynamic domains of the various solid deposits in the B-C-H-Cl vapor system, J. Electrochem. Soc., 1976, **123**, 136-40.

55. Vandenbulcke L., Theoretical and experimental studies on the chemical vapor deposition of boron carbide, Ind. Eng. Chem. Prod. Res. Dev., 1985, **24**, 568-75.

56. Rebenne, H. and Pollard, R., Theoretical analysis of chemical vapor deposition of ceramics in an impinging jet reactor, J. Am. Ceram. Soc., 1987, **70**, 907-18.

57. Lartigue, S. and Male, G. , Contribution to the study of tetragonal compounds in the boron carbon system, J. Mater. Sci. lett., 1988, **7**, 153-6.

58. Cholet, V., Herbin, P. and Vandenbulcke, L., CVD of boron carbide from BBr_3-CH_4-H_2 mixtures into a microwave plasma, Proc. 9th intern. symp. Boron, Borides and related Compounds, Duisburg (FRG), Sept. 1987, pp. 423-4.

59. Veprek, S. and Jurcik-Rajman, M., Plasma induced deposition of amorphous boron carbide, Symp. Proc.-Int. Symp. Plasma Chem. 7th, 1985, 90-4.

60. Toyoda, H., Sugai, H., Isozumi, T. and Okuda, T., Formation of pure-boron and boron-carbon alloys by a toroidal glow discharge, in ibid [58], pp. 409-10.

61. Thévenot, F. and Cueilleron, J., Analytical problems in boron and refractory borides, Analusis, 1977, **5**, 105-21.

62. Ruste, J., Bouchacourt, M. and Thévenot, F., Etudes sur le carbure de bore, III microanalyse électronique quantitative, J. less comm. Met., 1978, **59**, 131-8.

63. Bouchacourt, M., Thévenot, F. and Ruste, J., Analyses par microsonde électronique et méthodes chimiques destructrices appliquées à l'étude du système bore-carbone, J. Microsc. Spectros. Electron., 1979, **4**, 143-6.

64. Zuppiroli, L., Kormann, R.. and Lesueur, D., Etude fondamentale sur le carbure de bore, CEA-R-5237, Saclay (F), 1983, 74 pp.

65. Bandyopadhyay, A.K., Beuneu, F., Zuppiroli, L. and Beauvy, M., The role of free carbon in the transport and magnetic properties of boron carbide, J. Phys. Chem. Sol., 1984, **45**, 207-14.

66. Bougoin, M., Fillit, R., Thévenot, F. and Bruyas, H., Determination of free graphite in textured samples of boron carbide and boron carbide-silicon carbide composites, J. less comm. Met., 1986, **117**, 215-23.

67. Madden, H.H., Nelson, G.C. and Wallace, W.O., Auger electron spectroscopy studies of boron carbide, in ibid [39], pp, 121-9.

68. Inoue, S., Fukuda, S., Hino, T. and Yamashina, T.,, Change of surface composition of boron carbide (B_4C) single crystal due to heat treatment, J. Vac. Sci. Technol. A, 1987, **5** (4,Pt2), 1279-82.

69. Mackinnon, I.D.R., Aselage, T. and Van Deusen, S.B., High resolution imaging of boron carbide microstructures, in ibid [39], pp. 114-20.

70. Stotto, T., Etude par microscopie électronique des effets d'irradiation dans le carbure de bore, Thesis, Univ. Paris Sud, 1986 ; CEA-R-5382, 1987, 109 pp.

71. Bouchacourt, M. and Thévenot, F., Etudes sur le carbure de bore. III Domaine d'existence de la phase carbure de bore, J. less comm. Met., 1978, **59**, 139-52.

72. Elliott, R.P., The boron carbon system, Contrat n° AT (11-1), 578, Project Agreement n° 4, Armour Res. Found. Rep. ARF 220012 (Final Tech. Rep.), June 7, 1961.

73. Kieffer, R., Gugel, E., Leimer, G. and Ettmayer, P., Untersuchungen zum System Bor-Kohlenstoff, Ber. DKG, 1971, **48**, 385-9.

74. Bouchacourt, M. and Thévenot, F., The melting of boron carbide and the homogeneity range of the boron carbide phase, J. less comm. Met., 1979, **67**, 327-31.

75. Bouchacourt, M. and Thévenot, F., Analytical investigations in the B-C system, J. less comm. Met., 1981, **82**, 219-26.

76. Beauvy, M., Stoichiometric limits of carbon-rich boron carbide phases, J. less comm. Met., 1983, **90**, 169-75.

77. Kevill, D.N., Rissmann, T.J., Brewe, D. and Wood, C., Growth of crystals of several boron carbon compositions by chemical vapor deposition, J. cryst. growth, 1986, **74**, 210-16.

78. Kevill, D.N., Rissmann, T.J., Brewe, D. and Wood, C., Preparation of boron-carbon compounds, including crystalline B_2C material by chemical vapor deposition, J. less comm. Met., 1986, **117**, 421-25.

79. Clark, H.K., and Hoard, J.L., The crystal structure of boron carbide, J. Am. Chem. Soc., 1943, **65**, 2115-9.

80. Silver, A.H. and Bray, P.J., Nuclear magnetic resonance of boron carbide, J. Chem. Phys., 1959, **31**, 247-53.

81. Becher, H.J. and Thévenot, F., Infrarotspektroskopische Untersuchung des Borcarbids und seiner isotypen Derivate $B_{12}O_2$, $B_{12}P_2$ und $B_{12}As_2$, Z. Anorg. allg. Chem., 1974, **410**, 274-86.

82. Emin, D., Structure and single-phase regime in boron carbides, Phys. Rev. B, 1988, **38**, 6041-55.

83. Conard, J., Bouchacourt, M., Thévenot, F. and Hermann, G., ^{13}C and ^{11}B N.M.R. investigations in the boron carbide phase homogeneity range : a model of solid solution, J. less comm. Met., 1986, **117**, 51-60.

84. Bray, R.J., NMR studies of borates and borides, in ibid [39], pp. 142-67.

85. Alexander, M.N., NMR studies of the structure of boron carbides, in ibid [39], pp. 168-76.

86. Duncan, T.M., The distribution of carbon in Boron Carbides : ^{13}C NMR studies, in ibid [39], pp. 177-88.

87. Bouchacourt, M. and Thévenot, F., The properties and structure of the boron carbide phase, J. less comm. Met., 1981, **82**, 227-35.

88. Matje, P. and Schwetz, K.A., Surface oxygen pick up in submicron SiC and B_4C sintering powders at room temperature, 2nd intern. conf. on ceramic powd. process. sci. Berchtesgaden (FRG), oct. 1988 ; to be pub.

89. Gogotsi, G.A., Groushevsky, Y.L., Dashevskaya, O.B., Gogotsi; Y.G. and Lavrenko, V.A., Complex investigation of hot pressed boron carbide, J. less comm. Met., 1986, **117**, 225-30.

90. Lavrenko, V.A. and Gogotsi, Y.G., Influence of oxidation on the composition and structure of the surface layer of hot-pressed boron carbide, Oxid. Met., 1988, **29**, 193-202.

91. Efimenko, L.N., Lifshits, E.V., Ostapenko, I.T., Snezhko I.A. and Shevyakova, E.P., Oxidation of hot-pressed boron carbide, Porosh. Met., 1987, **4**, 56-60.

92. Makarenko, G.N. and Popova, O.I., Comparative characteristics of phases in the boron-carbon and boron-silicon systems, J. less comm. Met., 1986, **117**, 209-14.

93. Pastor, H. and Thévenot, F., Applications industrielles des composés du bore, Informations Chimie, 1978, **178**, 151-73.

94. Graf von Matuschka, A., Borieren, Carl Hanser Verlag, München, Wien, 1977, 91 p.

95. Goeuriot, P., Thévenot F. and Driver, J., Surface treatment of steels : Borudif, a new boriding process, Thin sol. films, 1981, **78**, 67-76.

96. Goeuriot, P., Fillit, R., Thévenot, F., Driver, J. and Bruyas, H., The influence of alloying element additions on the boriding of steels, Mater. Sci. Eng., 1982, **55,** 9-19.

97. Murgatroyd, R.A. and Kelly, B.T., Technology and assessment of neutron absorbing materials, Atomic Energy Rev., 1977, **15**, 1-74.

98. Tsagareishvili, G.V., Thermal expansion of boron and boron carbide, J. less comm. Met., 1986, **117**, 159-61.

99. Hollenberg, G.W., Thermally induced stresses and fractures in boron carbide pellets, Ceram. Bull., 1980, **59**, 538-41, 48.

100. Stull, D.R. and Prophet, H., Janaf thermochemical tables, 2nd ed. NBS US Gov. Print. Office, Washington, 1971.

101. Lockwood, A., Elwell, R. and Wood, C., Advanced high temperature thermoelectrics for space power, in Proc. Intersoc. Energy Convers. Engineering Conf., 1981, 16th, **2**, 1985-90.

102. Wood, C., Transport properties of boron carbide, in ibid [39], pp. 206-15.

103. Türkes, P.R.H., Swartz, E.T. and Pohl, R.O., in ibid [39]., pp. 346-61.

104. Wood, C., Boron carbides as high temperature thermoelectric materials, in ibid [39], pp. 362-72.

105. Bouchacourt, M. and Thévenot, F., The correlation between the thermoelectric properties and stoichiometry in the boron carbide phase B_4C-$B_{10.5}C$, J. Mater. Sci., 1985, **20**, 1237-47.

106. Thévenot, F. and Bouchacourt, M. Propriétés thermoélectriques de la phase carbure de bore, Silic. Indus., 1984, **49**, 145-9.

107. Moss, M. Thermal conductivity of boron carbides, Mater. Res. Soc. Symp. Proc., 1987, **97**, 77-82.

108. Beauvy, M., Thermal shock resistance of hot-pressed boron carbide, Sc. Ceram., 1981, **11**, 385-90.

109. Emin, D., Electronic transport in boron carbides, in ibid [39], pp. 189-205.

110. Lipp A. and Schwetz, K., Härte und Härtebestimmung von nichtmetallischen Hartstoffen, Ber. Deutsch Keram. Ges., 1975, **52**, 335-8.

111. Allen, R.D., The solid solution series, boron-boron carbide, J. Am. Chem. Soc., 1953, **75**, 3582-3.

112. Bouchacourt, M., Thévenot, F. and Ruste, J., Etudes sur le carbure de bore, I. Métallographie et microdureté Knoop du carbure de bore, J. less comm. Met., 1978, **59**, 119-30.

113. Rey, J., Male, G., Kapsa, P. and Loubet, J.L., Boron carbide coatings : correlation between mechanical properties and L.P.C.V.D. parameters values, Proc. 7th eur. conf. CVD, Perpignan (F), june 1989.

114. Niihara, K., Nakahira, A. and Hirai, T., The effect of stoichiometry on mechanical properties of boron carbide, J. Am. Ceram. Soc., 1984, **67**, C13-4.

115. Gogotsi, G.A., Firstov, S.A., Vasil'ev, A.D., Gogotsi, Y.G. and Kovylyaev, V.V., Mechanical properties and special features of the structure of materials based on boron carbide, Porosh. Met., 1987, **7**, 84-90.

116. De With, G., Note on the temperature dependence of the hardness of boron carbide, J. less comm. Met., 1983, **95**, 133-8.

117. Pilyankevich, A.N., Britun, V.F., Tkachenko, Y.G. and Yulyugin, V.K., Structure of boron carbide surface layers after abrasion at 20-1400°C, Porosh. Met., 1986, **8**, 93-7.

118. Tkachenko, Y.G., Yurchenko, D.Z., Yulyugin, V.K., Molyar, V.N., Murzin, L.M. and Lugovskaya, E.S., High temperature friction and certain properties of hot-pressed boron carbide, Porosh. Met., 1984, **12**, 41-3.

119. Tkachenko, Y.G., Yurchenko, D.Z., and Yulyugin, V.K., Some features of the deformation and fracture of carbides, borides, and nitrides under friction in the temperature range 20-1400°C, J. less comm. Met., 1986, **117**, 271-5.

120. Kingon, A.I., Stone, N.A. and Carr, N.S., Performance of some zirconia based ceramics under conditions of particle erosion, Mater. Sci. Monogr., 1987, **38C**, 2657-66.

121. Osipov, A.D., Ostapenko, I.T., Slezov, V.V., Tarasov, R.V., Podtykan, V.P. and Kartsev, N.F., Effect of porosity and grain size on the mechanical properties of hot-pressed boron carbide, Porosh. Met., 1982, **1**, 63-7.

122. De With, G., High temperature fracture of boron carbide : experiments and simple theoretical models, J. Mater. Sci., 1984, **19**, 457-66.

123. Beauvy, M., Propriétés mécaniques du carbure de bore fritté, Rev. int. htes Tempér. Réfr., 1982, **19**, 301-10.

124. Gogotsi, G.A., Gogotsi, Y.G. and Ostrovoj, D.Y., Mechanical behaviour of hot-pressed boron carbide in various atmospheres, J. Mater. Sci. lett., 1988, **7**, 814-6.

125. Hollenberg, G.W. and Walther, G., The elastic modulus and fracture of boron carbide, J. Am. Ceram. Soc., 1980, **63**, 610-3.

126. Ramana Murthy, S., Elastic properties of boron carbide, J. Mater. Sci. lett., 1985, **4**, 603-5.

127. Knoch, H., Carbide-hochverschleißfeste keramische Sonderwerkstoffe, Tech. Mitt., 1987, **80**, 221-6.

128. Wilkins, M.L., Use of boron compounds in lightweight armor, in ibid [8], pp. 633-48.

129. Epik, A.P., Boride coatings, in ibid [8], pp. 597-612.

130. Hunold, K., Boron carbide/graphite thermocouple for high temperatures, Chem. Tech. (Heidelberg), 1985, **14**, 82, 84.

131. Hunold, K., A thermocouple for high temperatures, Adv. Mater. & Processes, 1986, **9**, 4-5.

132. Kanno, Y., Temperature control of inert gas furnace by tetraboron carbide/carbon thermocouple, Yogyo Kyokaishi, 1986, **94**, 449-51.

133. Reinmuth, K., Lipp, A., Knoch, H. and Schwetz, K.A., Borhaltige keramische neutronenabsorberwerkstoffe, J. nucl. Mater., 1984, **124**, 175-84.

134. Colin, M., Matériaux absorbants neutroniques pour le pilotage des réacteurs nucléaires. In Techniques de l'ingénieur - Génie nucléaire - Structure des réacteurs nucléaires, Paris, 1989, **2**, **B3720**, pp. 1-8.

135. Kryger, B., personal communication.

136. FBFC, Nuclear fuel fabrication, Document, Romans (F), June 1986.

137. Halverson, D.C., Pyzik, A.J. and Aksay, I.A., Processing and microstructural characterization of B_4C-Al cermets, Ceram. Eng. Sci. Proc., 1985, **6**, 736-44.

138. Pyzik, A. and Cohen, A., The role of rearrangement in a densification of ceramic metal composites under hydrostatic pressure, in ibid [88].

139. Bougoin, M., Thévenot, F., Dubois, J. and Fantozzi, G., Synthèse et propriétés thermomécaniques de céramiques denses composites carbure de bore - carbure de silicium, J. less comm. Met., 1987, **132**, 209-28.

140. Dubois, J., Fantozzi, G., Bougoin, M. and Thévenot, F., Microstructure et propriétés mécaniques de matériaux, de types B_xC/SiC frittés, sans, charge, J. Phys. Colloque C1, suppl. n° 2, 1986, **47**, C1-751-5.

141. Wei, G.C., Silicon carbide whiskers-reinforced ceramic composites and method for making same, USP, 4 543 345, 1985.

142. Telle, R., Boride-eine neue Hartstoffgeneration ?, Chemie unserer Zeit, 1988, **22**, 93-9.

CHARACTERIZATION OF GRAIN BOUNDARIES IN SILICON CARBIDE

E. LAURENT-PINSON, S. LAY, G. NOUET, J. VICENS

LERMAT - ISMRa, URA n°1317
14032 CAEN Cedex, FRANCE

ABSTRACT

Grain boundaries have been orientated in α-SiC hexagonal ceramic. About 10% can be described by a 3-dimensional coincidence relation. Due to structural reasons, the boundary plane is often the (0001) plane for one grain and a high indice plane for the other one. In particular cases, bidimensional coincidence can be expected in the boundary plane.

INTRODUCTION

Study of grain boundaries in hexagonal materials were undertaken partly due to the fact that a great number of ceramics developed for thermomechanical application have a hexagonal structure. The aim of this work is to understand the way grains are joined together and if the 3-dimensional coincidence concept used to cubic symmetry and extended to the hexagonal one (1)(2)(3) is adapted to these materials. Observations were previously performed in several ceramics : WC (4) (5) Si_3N_4 (6) and SiC (7). The results presented here will be focussed on SiC.

MATERIALS

The material used in this study is a α-SiC hot pressed powder (Elektroschmelzwerk). It has been sintered at about 2000°C with amounts of 0.3 wt% or 1.5 wt% aluminium based additives (Al, AlN, Al_4C_3, Al_2O_3) as densification aids. As a consequence, a glassy phase (aluminosilicate) is formed and retained mainly at the triple junctions of the grain boundaries (8).

Some specimens of α-SiC-Al have been annealed and deformed at high temperatures (1300°C-1700°C) under vacuum by three point bending. Samples for electron microscopy studies (Jeol 120 CX) have been prepared in the as-sintered and in the deformed material.

Structural considerations

Silicon carbide exhibits a range of ordered structures (polytypism). All silicon carbide structures are made up of a single basic unit, a plane of tetrahedra SiC_4 or C_4Si. Successive layers can arrange in different ways giving rise to cubic (C), hexagonal (H) or rhombohedral (R) unit cells. The most common polytypes observed are the three following 4H, 6H, (often reported as the hexagonal forms of α-SiC), and 3C (β-SiC cubic form).

Chemical nature of grain boundaries

Due to the small amount of additive phase in these materials, grain boundaries appear free from any phase except in some triple junctions but grain boundary segregation can take place. Studies by S.T.E.M were made to examine sintered and hot pressed aluminium doped SiC. Grain boundary segregation of aluminium was found in both materials (9). By using high resolution Auger electron spectroscopy, pressureless sintered SiC materials were studied and grain boundaries analyzed at fractured intergranular regions. This method has revealed the existence of thin Al films at grain boundaries with a thickness less than 1 nm (10).

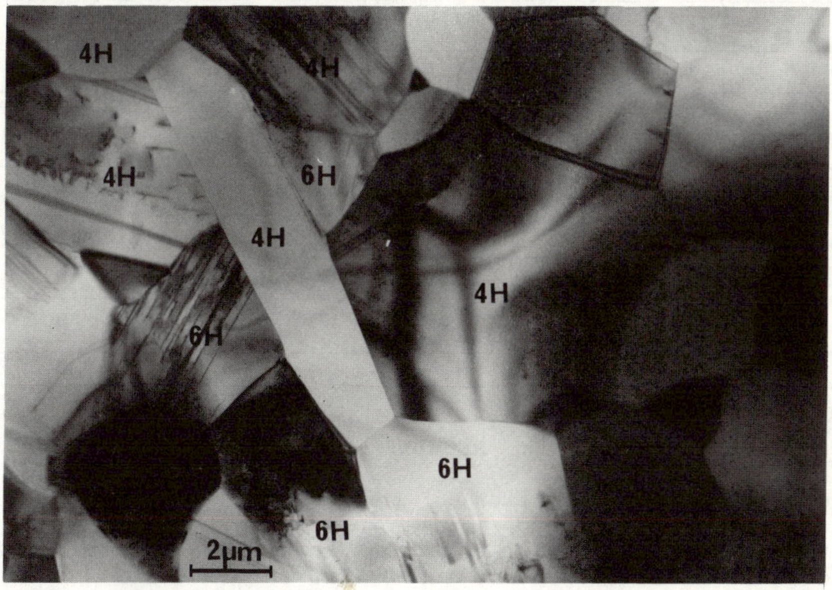

Figure 1. Microstructure of the as-sintered SiC-0.3% Al.

Microstructure

The sintering process leads to a very dense state and a mean grain diameter D_{SiC} = 4 μm. An electron transmission micrograph (Fig. 1) reveals more precisely the shape of the carbide grains. Some of them tend to adopt an elongated shape developing large facets parallel to the {0001} basal plane of the hexagonal structure. Electron diffraction studies have effectively shown that the carbide grains crystallized into the two main hexagonal polytypes 4H and 6H. For example some carbide grains have been indicated on the figure 1 with their corresponding crystallographic structure. Many defects can be seen inside the grains. They are due to either stresses induced during the preparation or to a modification in polytype structure inside a same grain. We note that the additive phase can be visible only inside some triple grain junctions. It is normally a glassy phase which can be observed as crystallized in annealed then deformed specimens.

TABLE

Near coincidence grain boundaries observed in silicon carbide

Σ	Polytype	$\frac{\mu}{\nu}$	rotation axis	rotation angle (°)	grain boundary plane	deviation (°)
19	4H	ind.	$\langle 0001\rangle$	13.17	$\{0001\}_{1,2}$	4.5
19	4H-6H	"	"	"	"	"
9	4H	21/2	$\langle 11\bar{2}0\rangle$	56.25	curved	12
22	"	32/3	$\langle 10\bar{1}0\rangle$	62.96	$\{0001\}_1$	8
27	"	21/2	$\langle 7\bar{2}\bar{5}0\rangle$	87.88	"	4.6
76	"	32/3	$\langle 11\bar{2}0\rangle$	26.53	"	1
6	6H	24/1	"	70.05	curved	8
22	"	"	"	50.48	$\{0001\}_1$ $\{10\bar{1}5\}_2$	5.8
27	"	"	$\langle 10\bar{1}0\rangle$	67.11	curved	5.3
38	"	"	$\langle 1\bar{1}\bar{2}0\rangle$	26.53		6.7
22	4H-6H	96/1	$\langle 10\bar{1}0\rangle$	67.96	$\{0001\}_1$ $\{11\bar{2}6\}_2$	8.8
28	"	96/1	"	44.92		2.9

RESULTS

Geometry of grain boundaries

About 60 experimental orientations have been determined for boundaries between two 4H grains, two 6H grains or 4H and 6H grains. These data have been compared to theoretical coincidence orientations corresponding to the following μ/ν values :

$$\text{4H polytype } c/a = 3.271 : \text{(i) } \mu/\nu = 21/2 \ (c/a = 3.240)$$
$$\text{(ii) } \mu/\nu = 32/3 \ (c/a = 3.266)$$
$$\text{6H polytype } c/a = 4.907 : \mu/\nu = 24/1 \ (c/a = 4.899)$$

Coincidence descriptions between two different polytypes (4H and 6H) have been found by using a common cell for the 4H and 6H structures. This new cell is defined by both a and c parameters :

$$a = a_{4H} = a_{6H} \text{ and } c = 3c_{4H} = 2c_{6H}$$

In this case the μ/ν ratio is equal to 96/1.

The table gives the comparison between experimental and theoretical coincidence orientations. The table reports only examples where a Σ value could be given even if the deviation from the coincidence orientation is rather high in some cases.

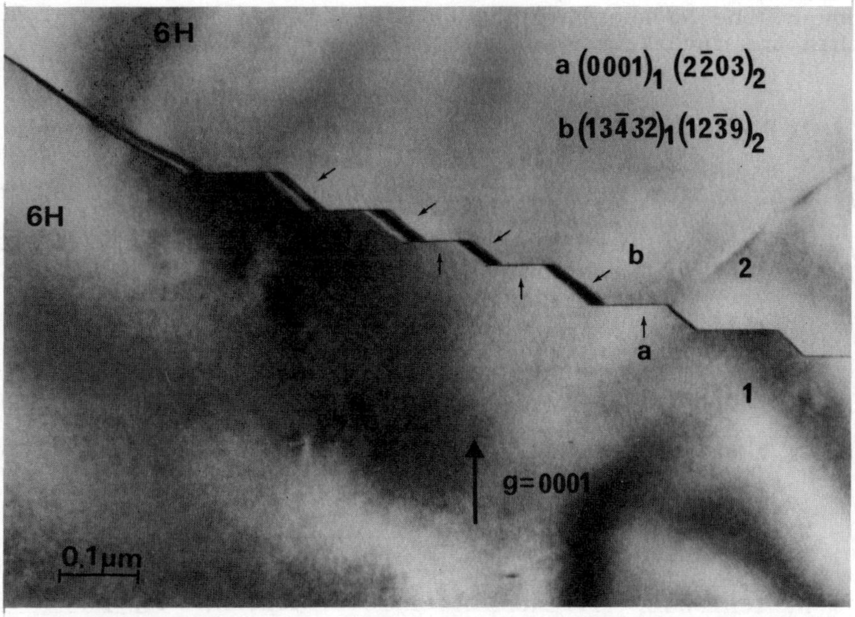

Figure 2. Faceting of the grain boundary in annealed, deformed SiC-Al.

DISCUSSION

The results show that the experimental orientations deviate greatly from the exact coincidence orientations. However in many cases the boundary plane is found parallel to the densest plane (0001) which is the main facet formed in this material after sintering. Two examples (Σ = 19) reported in the table correspond to a twist orientation with a grain boundary plane parallel to the (0001) plane. The other examples are asymmetric tilt grain boundaries with a boundary plane parallel to the {0001} plane for one crystal and parallel to higher indice planes for the other crystal. In the annealed, deformed specimens the grain boundaries are widely facetted with a main facet parallel to the (0001) plane for one crystal (Fig. 2). Due to the particular structures of SiC and the simple atomic configurations of some planes, good matching between planes at the interface are expected to be formed for instance : {0001} with $\{10\bar{1}4\}$ $\{10\bar{1}2\}$ (6H). The last configuration has been already observed (11). This bidimensional approach developed for SiC, was not used for WC for which many grain boundaries can be described by a near coincidence relationship.

CONCLUSION

It seems that in the SiC ceramic, besides the 3-dimensional coincidence approach, it is necessary to take into account the boundary plane nature, for which a good matching within the interfaces can occur for some configurations.

REFERENCES

(1) D.H. WARRINGTON, J. de Phys., 1975, 36-C4, 87

(2) G.L. BLERIS, G. NOUET, S. HAGEGE, P. DELAVIGNETTE, Acta Cryst. 1982, A38, 550

(3) H. GRIMMER, D.H. WARRINGTON, Acta Cryst., 1987, A43, 232

(4) S. LAY, G. NOUET, J.L. CHERMANT, F. OSTERSTOCK, J. VICENS, JIMIS-4, 1985, MInakami Spa, Japan

(5) S. LAY, G. NOUET, J. VICENS, F. Microsc. Spectrosc. Electron., 1986, 11, 179

(6) S. LAY, G. NOUET, J. VICENS, to be published.

(7) E. LAURENT-PINSON, G. NOUET, J. VICENS, Inst. phys. conf. Ser. N°93, Vol.2, EUREM 88, York, England, 1988.

(8) G. NOUET, S. HAGEGE, R. MOUSSA, F. OSTERSTOCK, J. VICENS, J.L. CHERMANT, Rev. Int. Hautes Temper. Refract., 1982, 19, 335

(9) Y. TAJIMA, W.D. KINGERY, J. Mat. Sci. 1982, 17, 2289

(10) R. HAMMINGER, D. GRATHWOHL, F. THUMMLER, J. Mat. Sci., 1983, 18, 3154

(11) H. ICHINOSE, Y. INOMATA and Y. ISHIDA, Ceramic Microstructure'88 edited by J.A. PASK et A.G. EVANS, 1986, 21, 255.

THE ALTERATION OF THE GRAIN BOUNDARY STRUCTURE IN YAG-CERAMICS DUE TO ANNEALING

L. SCHUH*, R. METSELAAR, G. de WITH
Centre for Technical Ceramics, Technical University
5600 Eindhoven, The Netherlands
*now with the Asea Brown Boveri Corporate Research
6900 Heidelberg (Federal Republic of Germany)

ABSTRACT

Annealing conventionally sintered YAG for a long time at high temperature (350 K below T_{sinter}) results in a marked increase of the grain boundary thickness. Impedance measurements showed that the grain boundary conductivity in annealed samples is a function of the oxygen pressure.

INTRODUCTION

Yttrium aluminium garnet ($Y_3Al_5O_{12}$ or YAG) is one of several ceramic materials that can be made translucent [1, 2]. This makes YAG a candidate material to produce Na-discharge lamp envelopes. De With and coworkers [2, 3] have shown that the mechanical and optical properties as well as the Na corrosion resistance are better for YAG than for the presently used translucent α-Al_2O_3. The wall temperature of a burning lamp is 1470-1520 K. Na penetrates along the grain boundaries of Al_2O_3 [4]. YAG has a much higher resistance against such a corrosive attack. It is in no way clear, however, if the grain boundaries have a long time thermal stability far below sintering temperature. We investigated how the grain boundary properties in YAG are altered by a long time anneal at 1670 K.

MATERIALS AND METHODS

YAG powders were prepared using the wet-chemical route (sulphate process) [3]. Doping of the powders was done by adding a proper amount of the acetates. After calcination (at 1570 K) and milling pellets were isostatically pressed and sintered at

2020 K in vacuum for 4 h. The content of aliovalent ions in the sintered sample was analysed spectrochemically to be 150 wt. ppm Mg, 90 wt. ppm Ca and 12 wt. ppm Na. The relative density was 99%.
We investigated the high temperature conductivity by ac-dispersive methods. After polishing disk-shaped samples (1 mm thick and 10 mm in diameter) were provided with painted Pt-electrodes. The electrical conductivity, σ, was measured at various oxygen partial pressures, pO_2, using a Solartron 1174 Frequency Response Analyser. The conductivity cell is described in ref. [5]. Usually a voltage of 100 mV rms was applied across the sample. We used N_2/O_2 and CO/CO_2-mixtures to establish controlled pO_2's. From complex impedance plots bulk- and grain boundary conductivities (σ_b and σ_{gb}) were separated for every single measuring point. One sequence of measurements was performed on as-sintered material. Afterwards the electrodes were removed and the sample was annealed 240 h at 1670 K in air. The sample was polished again, new Pt-electrodes were mounted and the pO2-dependence of the conductivity was determined again.

RESULTS

Curve I in fig. 1 gives σ as $f(pO_2)$ for the as-sintered sample. This curve is typical for acceptor-doped YAG; the defect-chemical interpretation is given in ref. [8]. We should note here that at high and intermediate pO_2 the conductivity decreases following a $pO_2^{3/16}$-dependence. In the defect model this regime is governed by the neutrality condition $[h^{\cdot}]=3[V_{Al}''']$.

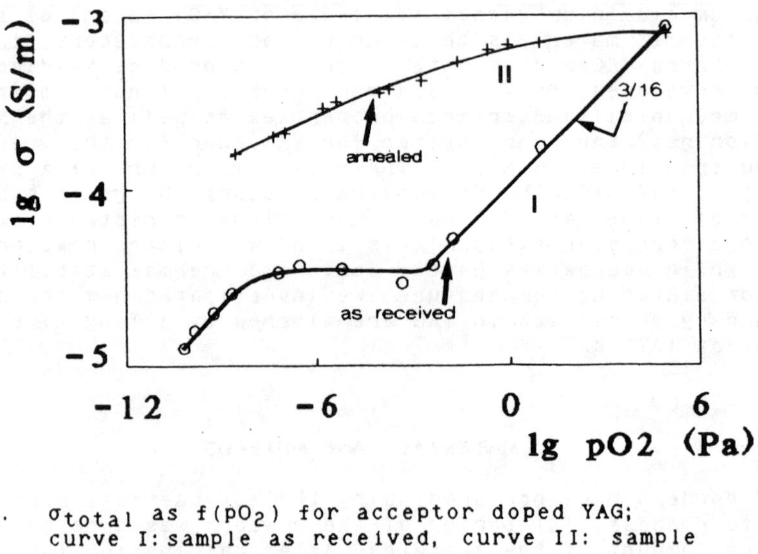

FIGURE 1. σ_{total} as $f(pO_2)$ for acceptor doped YAG; curve I: sample as received, curve II: sample annealed 240h at 1670 K.

Fig. 2 shows for the as-sintered sample two complex impedance plots recorded in pure O_2 and in a CO/CO_2-mixture maintaining $pO2 = 10^{-10}$ Pa. In both diagrams a wide bulk-arc is observed, while the grain boundary dispersion is negligible. The bulk-arc intersects the Z'-axis at zero-resistance.

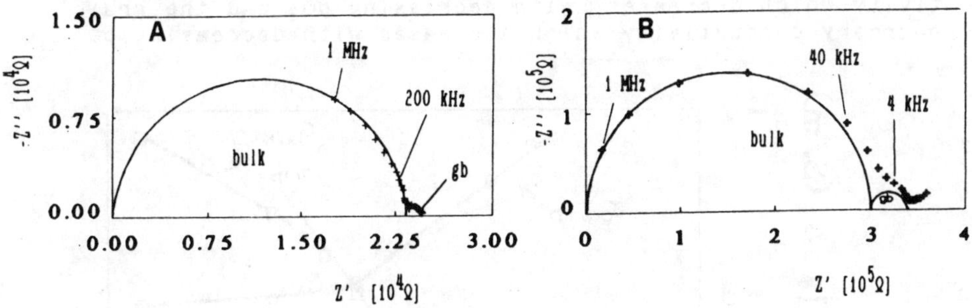

FIGURE 2. Z'-Z''-plots for the as sintered sample (4 h, 2020 K in vacuum) measured in (A) $pO_2 = 10^5$ Pa and (B) $pO_2 = 10^{-10}$ Pa.

Curve II in fig. 1 gives the pO_2-dependence of the conductivity after the sample was annealed 240 h at 1670 K in air. The conduction behaviour has radically changed. The corresponding Z'-Z''-plots in pure O_2 and CO/CO_2 maintaining $pO2 = 10^{-10}$ Pa are given (fig. 3).

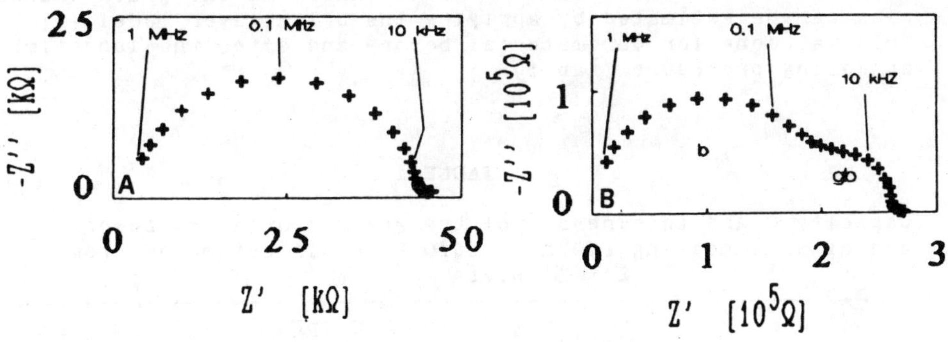

FIGURE 3. Z'-Z''plots of the sample after annealing 240 h at 1670 K in air (A) measured in $pO_2 = 10^5$ Pa and (B) $pO_2 = 10^{-10}$ Pa

After annealing there is a strong and pO_2-dependent grain boundary dispersion. We separated σ_b and σ_{gb} by graphically fitting the arcs in the Z'-Z''-plots recorded at different pO_2's. The oxygen pressure dependencies of σ_b and σ_{gb} are shown in fig. 4. It is obvious that the conductivity of an annealed specimen is the superposition of the bulk conductivity which decreases with decreasing pO_2 and the grain boundary conductivity which increases with decreasing pO_2.

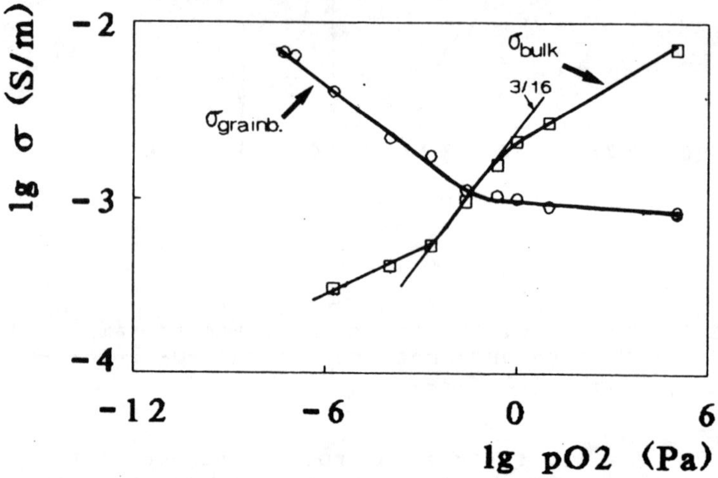

FIGURE 4. The separated grain boundary and bulk conductivity as f(pO_2) in an annealed sample.

From the Z'-Z''-plots an average thickness of the grain boundaries can be estimated by applying the brick layer model [7]. This was done for our material before and after the long time annealing procedure (tab. 1).

TABLE 1

Capacity C and thickness δ of the grain boundaries before and after annealing 240 h at 1670 K in air estimated from Z''-Z'-plots

	C (pF)	δ (nm)
before annealing (date recorded at 1470 K)	1000	1
after annealing (data recorded at 1570 K)	60	11

DISCUSSION

The total conductivity of as-sintered acceptor doped YAG changes proportional $pO_2^{3/16}$ (fig. 1, curve I). From complex impedance measurements (fig. 2) we see that there is only a small pO_2-independent grain boundary dispersion. A grain boundary thickness of 1 nm can be derived from the Z'-Z'' plots.

The total conductivity of the annealed sample is nearly independent from oxygen pressure (fig. 1, curve II). The impedance diagrams indicate that σ_{gb} has become strongly pO_2-dependent. By analysing the Z'-Z''-plots taken at different oxygen pressures the pO_2-dependence of the σ_b and σ_{gb} were separated (fig. 4). The bulk conductivity isotherm is comparable to curve I in fig. 1. At intermediate pO_2's the 3/16-slope is measured again, so the defect chemistry of the grain interior should probably not be altered. The grain boundary thickness can be estimated to be about 11 nm. These results show that the grain boundary properties in sintered materials are altered due to a long time anneal. The thickness of the boundary increases due to a segregation of presumably Ca-aluminate. In a crystalline material the solubility of the impurities is strongly temperature dependent. Cooling down from T_{sinter} with usually 10 K/min leads to a frozen-in high temperature state. When such a material is annealed at intermediate temperatures a segregation of a second phase at the grain boundaries takes place. This process is very slow even at 1670 K.

CONCLUSION

The structure of grain boundaries in YAG slowly alters if the material is kept at high temparature. By analysing impedance plots we found the grain boundary conductivity to increase with decreasing pO_2. The thickness of the grain boundaries increase by a factor of 10 while annealing as received YAG 240 h at 1670 K. This formation of a boundary layer can lead to an impairment of the corrosion resistance.

REFERENCES

1. Gazza, G. E. and Dutta, S. K., U. S. patent 4029755 (1977).
2. With, G. de and Dijk, H. J. A. van, Mat. Res. Bull., 1984, 19, 1669-1674.
3. With, G. de and Parren, J. E. D., Solid State Ionics, 1984, 16, 87-94.
4. With, G. de, Philips J. Res., 1987, 42, 119-130.
5. Hoefsloot, A. M., Thijssen, P. H. F. and Metselaar, R., Silicates Ind., 1985, 3-4, 35-39.
6. Schuh, L., Metselaar, R. and With, G. de, accepted for publication in J. Appl. Phys.
7. Verkerk, M. J., Middelhuis, B. J. and Burggraaf, A. J., Solid State Ionics, 1982, 6, 159-170.

MICROSTRUCTURE OF GRAIN BOUNDARIES IN CERAMICS

HERBERT K SCHMID
Division of Materials Science and Technology, CSIR, P O Box 395,
Pretoria 0001, South Africa

ABSTRACT

The microstructure of grain boundaries and the composition of intergranular phases in various ceramic systems were studied by means of high resolution and analytical transmission electron microscopy (TEM). TEM studies on CeO_2 stabilized ZrO_2 systems (Ce-TZP) revealed the presence of vitreous and crystalline intergranular phases in small pockets at multiple grain junctions. Microanalysis results indicated the segregation of dopants as well as Si and Al impurities at grain boundaries.

In $Si_3N_4/ZrO_2(+CeO_2)$ composite ceramics the presence of a crystalline Zr-Si-O-N phase and a Ce rich vitreous intergranular phase was observed. Although grain boundaries in general appeared to be wetted by the glassy phase, the existence of special low energy boundaries, free of an intergranular phase, was detected.

INTRODUCTION

Fine-grained tetragonal ZrO_2 polycrystals (TZP) have become of interest for advanced structural applications due to their exceptionally good combination of mechanical properties [1]. Recently, the existence of a CeO_2 stabilized 100% tetragonal ZrO_2 solid solution (Ce-TZP) was reported [2], which also showed very promising properties regarding flexural strength and fracture toughness. However, some limitations occur for high temperature applications due to the presence of intergranular glassy phases that result from liquid phase sintering with oxide additions.

It was suggested that in Si_3N_4/ZrO_2 composites the strength and hardness of Si_3N_4 could be combined with the fracture toughness of TZP [3]. Although fully dense materials could be produced without further sintering aids [4], problems may arise from inter-reactions of the components [4,5].

In the present work the microstructure of two ceramic systems was studied by means of high resolution and analytical TEM. The aims of these investigations were (i) to study the influence of La_2O_3 additives on the microstructure of Ce-TZP and (ii) to characterize the microstructure of sintered Si_3N_4/ZrO_2 $(+CeO_2)$ composite ceramic systems.

MATERIALS AND METHODS

Fully tetragonal Ce-TZP ceramics composed of 90 mol% ZrO_2 and 10 mol% (CeO_2 + La_2O_3) stabilizer were prepared via a chemical route and sintered at 1 430 °C in air. Si_3N_4/10 mol.% Ce-TZP composites were prepared by sintering at 1 700 °C in N_2 atmosphere. Electron transparent specimens for TEM observations were prepared by ion-beam thinning and examined in a Philips EM 420 analytical STEM equipped with EDAX energy dispersive X-ray spectrometer (EDS) and GATAN electron energy-loss spectrometer (EELS) attachments.

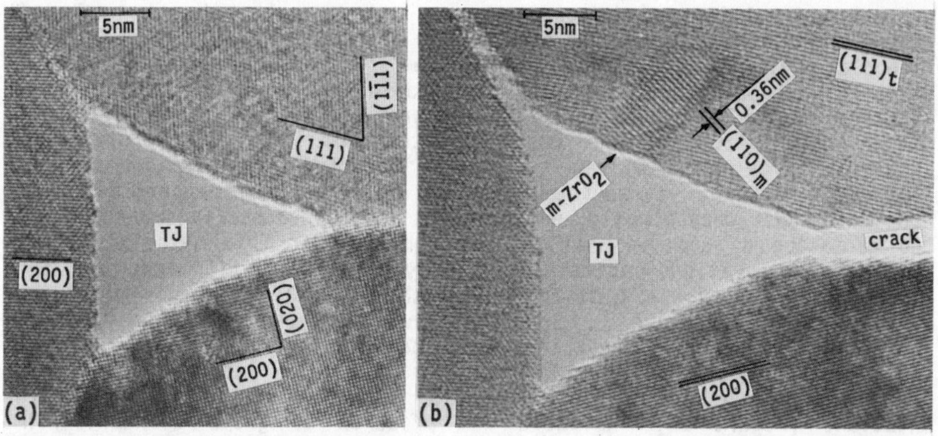

Figure 1. Microstructure of Ce-TZP (L comp.). (a) Intergranular glassy phase in pocket at triple grain junction (TJ); solid-liquid interfaces are facetted. (b) Microcrack caused by partial t→ m phase transformation.

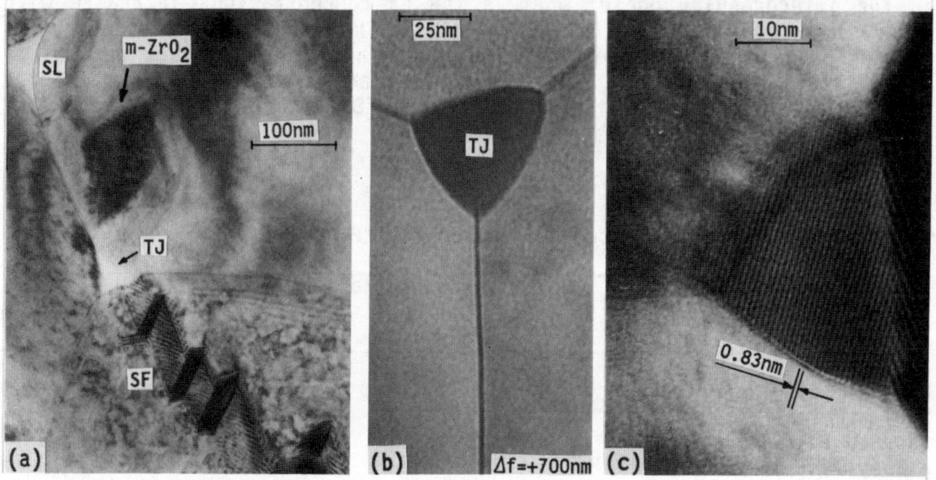

Figure 2. Microstructure of Ce-TZP (H comp.). (a) Partial dislocations associated with stacking faults (SF); SL interfaces are curved. (b) CTEM image and (c) lattice image of crystalline phase at triple grain junction.

RESULTS AND DISCUSSION

Ce-TZP Ceramics

Ce-TZP ceramics containing small fractions of La_2O_3 additives (L composition) showed very good mechanical properties regarding hardness, flexural strength and fracture toughness, whereas in compositions with higher La_2O_3 concentrations (H compositions) a degradation of these properties was observed. The TEM micrographs in Figure 1 show the microstructure typically observed in composition (L). A glassy phase is present in small pockets ($d \leq 10$ nm) at triple grain junctions (TJ). The solid-liquid interfaces (SL) are faceted, indicating an equilibrium state between crystalline matrix and vitreous intergranular phase. The SL interfaces adopt orientations parallel to low energy {111} and {200} lattice planes (Figure 1a). Occasionally, tetragonal → monoclinic transformations of parts of TZP grains were observed in the irradiated sample volume resulting in the formation of intergranular microcracks (Figure 1b).

In compositions (H) the observed pockets of intergranular phase were generally larger ($d \geq 20$ nm) and the SL interfaces were smoothly curved (Figure 2). Both, vitreous as well as crystalline intergranular phases were detected. Occasionally, dislocations split into partials resulting in the formation of stacking faults, were observed in this material (Figure 2a). The observations indicate relatively high dislocation energies, thus plastic deformation via a dislocation mechanism may be very limited even at elevated temperatures. The low dihedral angles formed by the SL interfaces at grain boundaries indicate the penetration of vitreous intergranular phase along grain boundaries. The lateral spacings of Fresnel fringes in out-of-focus images of grain boundaries in edge-on orientations (Figure 2b) indicated the presence of grain boundary films ≈ 0.8 nm thick [6].

The results of the EDS microanalysis are summarized in Table 1. The analysis of Ce-TZP grains indicated the presence of Si and Al impurities within the matrix as well as a pronounced segregation of these elements at grain boundaries. The results suggest a partial substitution of Ce by Si in the intergranular phase, whereas the La concentrations increased with increasing size of second phase pockets. The highest La concentrations ranging between 40 and 70 wt% were observed in the SL interface regions of the vitreous phase and in the crystalline second phase [7]. It appears that in accordance with literature reports [8] the maximum solubility of La_2O_3 in ZrO_2 is in the order of 0.5 mol%.

TABLE 1
EDS microanalysis in Ce-TZP ceramics

Phase	Relative concentrations [wt% cations]					
	Zr	Ce	La	Si	Al	Ca
nominal (L)	84.8	14.4	0.8	-	-	-
ZrO_2 matrix	80	14	0.7	3.5	0.6	-
faceted TJ (L)	74	13	3	6	3	0.5
glassy TJ (H)	61	7	21	6	4	0.5
crystal. TJ (H)	10	6	70	12	-	2

Si_3N_4/ZrO_2(+ CeO_2) Composite Ceramics

Sintered Ce-TZP dispersed Si_3N_4 composits showed reasonably high hardness and flexural strength, but the fracture toughness was found to be rather low ($K_{IC} \approx 4$ MPa $m^{½}$). TEM observations revealed the presence of a vitreous intergranular phase as well as a Zr rich crystalline phase. Both curved as well as faceted SL interfaces preferentially parallel to {10$\bar{1}$0} lattice planes were observed (Figure 3). Most grain boundaries appeared to be wetted by a film of intergranular phase ≈ 1.5 nm thick. Occasionally, special high coincidence grain boundaries free of intergranular phase were observed. The micrographs in Figure 3b,c show a dewetted $\Sigma = 4/[10\bar{1}0]$, $\theta \approx 41.4°$ twist boundary; slight deviations from the exact coincidence orientation are accommodated by sets of edge and screw dislocations.

The results of a combined EDS and EELS microanalysis in Si_3N_4/Ce-TZP composites are listed in Table 2. Surprisingly, the Ce concentration in the Zr rich phase was below the detection limit, whereas high Ce concentrations were measured in the glassy phase. The N:O anion ratios as measured by EELS analysis indicated a slight oxidation of the Si_3N_4 matrix, whereas the O content of the Zr rich phase was rather low. EDS and EELS data would be consistent with compositions close to $Si_2N_2O_4 \cdot 6ZrN$ and $Ce_2O_3 \cdot Si_3N_4$ for the crystalline and intergranular glassy phase, respectively.

TABLE 2
EDS and EELS microanalysis in Si_3N_4/Ce-TZP composits

Phase	EDS [wt% cations]					EELS [atom ratio]
	Si	Ce	Zr	Ca	Ti	N:O
nominal	84.8	2.3	12.9	-	-	94.7: 5.3
Si_3N_4 matrix	98.5	0.6	0.6	0.3	-	94 : 6
Zr rich phase	5	-	95	-	-	90 :10
glassy phase	23	66	5	1	5	55 :45

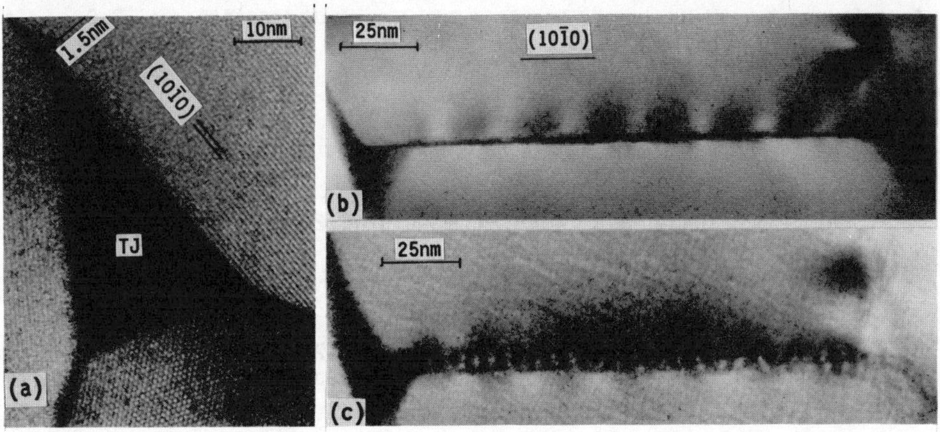

Figure 3. Grain boundaries in Si_3N_4/Ce-TZP composite. (a) Intergranular glassy phase at triple grain junction. (b) Lattice image and (c) CTEM image of dislocaion structure of dewetted (10$\bar{1}$0) twist boundary.

CONCLUSIONS

Ce-TZP ceramics containing small fractions of La_2O_3 additives show excellent material properties. The solubility limit of La_2O_3 in the Ce-TZP solid solution is thought to be in the order of 0.4 - 0.5 mol% at room temperature. La_2O_3 concentrations in excess of the solubility limit result in the formation of excessive amounts of La rich glassy and crystalline intergranular phases and degradation of the mechanical properties. From the in-situ observations of t → m phase transformation of parts of Ce-TZP grains, it is concluded that the t-phase is metastable and a continuous stress will be required to complete the transformation. Hence the t- and m- phases are in near equilibrium and the transformation should be reversible in accordance with recent literature reports [9].

The TEM observations on a Si_3N_4/10 mol% Ce-TZP composite enable the following conclusions to be made:

(i) During sintering under N_2 atmosphere Ce^{+4} is reduced to Ce^{+3} resulting in the complete separation of Ce_2O_3 and ZrO_2.

(ii) Ce_2O_3 together with the excess O reacts with Si_3N_4 resulting in the formation of an essentially Ce-Si-O-N glassy intergranular phase.

(iii) The Ce depleted ZrO_2 transforms to m-ZrO_2, or is reduced to ZrN forming a Zr-Si-O-N oxidation prone crystalline phase.

REFERENCES

1. Rühle, M., Claussen, N. and Heuer, H., Microstructural studies of Y_2O_3 - containing tetragonal ZrO_2 polycrystals (Y - TZP). Advances in Ceramics, 1984, **12**, 352-70.

2. Meriani, S., Metastable tetragonal CeO_2 - ZrO_2 solid solution. J. de Physique (Paris), 1986, C1, 485-89.

3. Claussen, N. and Jahn, J., Mechanical properties of sintered and hot-pressed Si_3N_4-ZrO_2 composites. J. Am. Ceram. Soc., 1978, **61**, 94-95.

4. Terao, K., Miyamoto, Y. and Koizumi, M., Characteristics of ZrO_2 - dispersed Si_3N_4 without additives fabricated by hot isostatic pressing, J. Am. Ceram. Soc., 1988, **71**, C 167-69.

5. Lange, F.F., Falk, L.K.L., and Davis, B.I., Structural ceramics based on Si_3N_4 - ZrO_2 (+ Y_2O_3) compositions, J. Mat. Res., 1987, **2**, 66-76.

6. Schmid, H.K. and Sonnenberg, N., Structure of interfaces in Ce-TZP, Inst. Phys. Conf. Ser., 1988, No. **93**, Vol. 2, 541-42.

7. Schmid, H.K., Microanalysis of grain boundary phases in Ce-TZP, ibid. 539-40.

8. Bastide, B., Odier, P. and Coutures, J.P., Phase equilibrium and martensitic transformation in lanthana - doped zirconia, J. Am. Ceram. Soc., 1988, **71**, 449-53.

9. Hannink, R.H.J. and Swain, M.V., Metastability of the martensitic transformation in a 12 mol% ceria-zirconia alloy : I, Deformation and fracture observations, J. Am. Ceram. Soc., 1989, **72**, 90-98.

INDUCTIVELY COUPLED PLASMA-ATOMIC EMISSION SPECTROMETRY: AN EXCELLENT TOOL FOR THE TOTAL CHEMICAL ANALYSIS OF TITANATE-BASED ELECTROCERAMICS.

J.C. FARIÑAS and M.F. BARBA
Instituto de Cerámica y Vidrio (CSIC)
Departamento de Análisis Químico
28500 Arganda del Rey, Madrid, Spain

ABSTRACT

The application of Inductively Coupled Plasma-Atomic Emission Spectrometry (ICP-AES) for the chemical analysis of titanate-based electroceramics ($BaTiO_3$, Ca-PT, Gd-PT, La-PT, PZT and PLZT) was studied. Acid digestions and fusion procedures for sample dissolution are proposed. The detection limits of possible impurities and dopants were determined as well as the precision of the macroconstituents (Ba, Ca, Gd, La, Pb, Ti and Zr). Using this technique high sensitivity and good precision were attained.

INTRODUCTION

The processing of electroceramic materials introduces impurities which degradate their dielectrical, piezoelectrical and/or semiconducting properties (1). Besides this, the molar ratio of the macroconstituents is considered an important parameter (2) regarding the final behaviour of such kind of materials.

In spite of that only few works on the chemical analysis of electroceramic materials have been reported in the available literature: several japanese researchers (3,4,5) and the present authors (6) have applied the ICP-AES technique to the analysis of $BaTiO_3$-based materials.

Because of its high sensitivity for almost all the elements as well as the linearity of the calibrations over a very wide concentration range, this technique seems to be an appropriate tool for the analysis of major, minor and trace elements in these such of materials.

The aim of the present investigation is to show the high performance of ICP-AES in the chemical analysis of barium titanate, lead titanate and lead titanate-zirconate based materials.

MATERIALS AND METHODS

Instrumentation

A Jobin-Yvon model JY-38 VHR sequential spectrometer with an ICP source was used. This instrument is equipped with a high resolution monochromator and an auxiliary one which was used to determine the alkali elements as well as for measurements with internal standard. The ICP-AES instrument and the operating conditions were described in a previous publication (6).

Materials

The analytical study was performed on $BaTiO_3$,Ca-PT, Gd-PT, La-PT, PZT and PLZT selected materials.

Apart from macroconstituents, the following impurities and dopants were considered: Al, Ba, Ca, Co, Fe, Mg, Nb, Si, alkali metals and rare earth elements (REE).

Sample preparation

The samples were crushed and milled down to a grain size < 30 µm in a tungsten carbide mortar. Four decomposition methods were developed to bring the samples into solution:

Method I: 0.2000 g of sample were treated in a glass beaker with 10 ml HCl and boiled until total dissolution. The solution was transferred to a 100 ml volumetric flask and diluted to volume with distilled water.

Method II: 0.2000 g of sample were treated in a PTFE-lined pressure vessel (7) with 5 ml HCl. Total dissolution requires 14 h at 160ºC. The solution was diluted to 100 ml with distilled water.

Method III: 0.2000 g of sample were treated with 2 drops of water, 2 ml HF and 0.4 ml H_2SO_4 in a PTFE-lined pressure vessel at 170ºC for 14 h. The contents were transferred into a platinum dish and heated in a sand bath until dryness. 10 ml HCl were added and heated. The final solution was diluted to 100 ml with distilled water.

Method IV: 0.2000 g of sample were mixed with 2.00 g of $Li_2B_4O_7$ in a porcelain crucible, and then transferred to a graphite crucible. The mixture was fused in a muffle furnace at 1100ºC for 20 min. The melt was then carefully and quickly poured into a 400 ml beaker containing 100 ml of 1+24 (V/V) HNO_3. The contents were stirred with a PTFE-covered magnetic bar on top of a magnetic stirrer until the melt dissolved. The solution was diluted to 200 ml with 1+24 (V/V) HNO_3.

Yttrium as the internal standard was added to the solutions with a final concentration of 50 $\mu g ml^{-1}$ for the determination of the macroconstituents.

Standards preparation and calibration

The calibration curves for the minor and trace elements analyzed were constructed with a blank solution and a standard of appropriate concentration. Both solutions as well as the standards used in the detection limits determination were prepared matrix-matched with the decomposition reactives of the sample, and with the macroconstituents.

RESULTS AND DISCUSSION

Decomposition methods

Before applying the ICP-AES technique the samples have to be dissolved. Table 1 shows the dissolution procedures, as previously described, used for each selected material.

The $Li_2B_4O_7$ fusion which takes place in a short period of time is valid for all the studied materials. The acid digestion methods have the advantage to introduce a lower amount of impurities and salt concentration to the solution, although the more refractory materials require the use of HF. This acid prevents the determination of SiO_2 because it is lossed as volatile fluoride.

TABLE 1
Dissolution methods for the different electroceramics

Material	Dissolution methods
$BaTiO_3$	II and IV
Ca-PT	I and IV
Gd-PT	II and IV
La-PT	III and IV
PZT	III and IV
PLZT	III and IV

Spectral interferences

The transition metals and the REE, when are excited by a very energetic source as the ICP, show emission spectra very rich in lines. This fact implies the existence of many spectral interferences, specially when these mentioned elements are at a macrocomponent level. This is the case of Gd, La, Ti and Zr in the materials chosen in this work. Because of this, it is necessary to carry out an experimental study of the spectral interferences to select an interference-free analytical line for each element in each material. After a careful study of the above mentioned problem, the valid lines to perform the chemical analysis of these electroceramics are reported in Table 2.

Some of these lines present partial overlappings which can be easily corrigible applying a interelement correction coefficient (8).

TABLE 2
Analytical lines (nm)

Element	$BaTiO_3$	Ca-PT	Gd-PT	La-PT	PZT	PLZT
Al	396.152	396.152	396.152	396.152	394.401	394.401
Ba	455.403	455.403	455.403	455.403	493.409	493.409
Ca	393.366	393.366	393.366	393.366	393.366	393.366
Ce	-	-	413.765	413.765	-	413.765
Co	238.892	237.862	237.862	237.862	230.786	230.786
Fe	238.204	238.204	239.562	238.204	238.204	238.204
Gd	-	-	342.247	342.247	-	342.247
K	766.490	766.490	766.490	766.490	766.490	766.490
La	-	-	333.749	408.672	-	408.672

TABLE 2 (cont.)

Element	$BaTiO_3$	Ca-PT	Gd-PT	La-PT	PZT	PLZT
Mg	279.553	279.553	279.553	279.553	279.553	279.553
Na	589.592	589.592	589.592	589.592	589.592	589.592
Nb	309.418	–	–	–	309.418	–
Nd	–	–	415.608	430.358	–	406.109
Pb	220.353	220.353	220.353	220.353	220.353	220.353
Si	251.611	251.611	251.611	251.611	251.611	251.611
Sm	–	–	442.434	359.260	–	359.260
Ti	334.941	334.941	334.941	334.941	334.941	334.941
Zr	343.823	343.823	343.823	343.823	339.198	339.198

Detection limits

The 3 σ detection limits (DL) of minor and trace elements were determined considering the four decomposition methods.

Table 3 shows, as an example, the DL values of PZT and PLZT ceramics considering the two decomposition methods.In the six tested materials, the DL values obtained with the acid attacks were slightly lower than those achieved with the fusion method, except for alkali metals. In this particular case, due to the ionization effect produced by the flux, the lowest DL values were attained when $Li_2B_4O_7$ fusion was used.

TABLE 3

Detection limits of minor and trace elements in PZT and PLZT materials

Oxide	Dissolution method III in solution ($\mu g ml^{-1}$)	III in sample ($\mu g g^{-1}$)	IV in solution ($\mu g ml^{-1}$)	IV in sample ($\mu g g^{-1}$)
Al_2O_3	0.042	21	0.053	53
BaO	0.006	3	0.009	9
CaO	0.0008	0.4	0.0012	1.2
CeO_2	0.056	28	0.069	69
CoO	0.0054	2.7	0.0066	6.6
Fe_2O_3	0.0062	3.1	0.0063	6.3
Gd_2O_3	0.018	9	0.025	25
K_2O	0.064	32	0.051	51
MgO	0.0004	0.2	0.0004	0.4
Na_2O	0.16	80	0.064	64
Nb_2O_5	0.0096	4.8	0.013	13
Nd_2O_3	0.098	49	0.11	110
SiO_2	–	–	0.048	48
Sm_2O_3	0.064	32	0.075	75

Chemical analysis

In Table 4 are reported the analytical results obtained for impurities and dopants in Ca-PT and PLZT materials. The data of the other materials are not showed because the limited scope of this work. In all the materials, as can be observed in Table 4 for the mentioned electroceramics, there is a good agreement between the results from the fusion procedure and those obtained with the acid digestion.

TABLE 4
Analytical results ($\mu g g^{-1}$) for determination of impurities and dopants in Ca-PT and PLZT materials

	Ca-PT		PLZT	
	Dissolution method			
Oxide	I	IV	III	IV
Al_2O_3	2482 ± 184	2613 ± 152	4327 ± 123	4561 ± 148
BaO	172 ± 14	163 ± 18	58 ± 10	67 ± 11
CaO	–	–	73 ± 14	81 ± 17
CeO_2	–	–	436 ± 48	412 ± 43
CoO	11 ± 5	13 ± 7	<2.7	<6.6
Fe_2O_3	246 ± 27	231 ± 23	131 ± 16	119 ± 13
Gd_2O_3	–	–	23 ± 9	<25
K_2O	321 ± 49	345 ± 77	216 ± 32	233 ± 59
MgO	534 ± 32	502 ± 59	883 ± 64	896 ± 81
Na_2O	1328 ± 141	1476 ± 169	1241 ± 147	1416 ± 177
Nb_2O_5	–	–	5340 ± 268	5474 ± 228
Nd_2O_3	–	–	32 ± 8	36 ± 6
SiO_2	395 ± 53	424 ± 35	–	584 ± 51
Sm_2O_3	–	–	<32	<75
ZrO_2	456 ± 46	427 ± 36	–	–

The results of the macroconstituents determination appear in Table 5, where it can be observed the good precision attained.

TABLE 5
Analytical results (wt.%) for determination of macroconstituents in Gd-PT and La-PT materials

	Gd-PT		La-PT	
	Dissolution method			
Oxide	II	IV	III	IV
Gd_2O_3	17.22 ± 0.07	17.28 ± 0.09	–	–
La_2O_3	–	–	15.86 ± 0.12	15.93 ± 0.11
PbO	59.85 ± 0.19	59.99 ± 0.23	61.08 ± 0.25	61.27 ± 0.28
TiO_2	21.38 ± 0.08	21.31 ± 0.06	21.69 ± 0.13	21.56 ± 0.10

In summary, the electroceramic materials can be simple and quickly analized, obtaining good accuracy and precision, when the ICP-AES technique and the methodology presented in this work are used.

REFERENCES

1. Sauer, H.A. and Fisher, J.R., Processing of Positive Temperature Coefficient Thermistors. J. Am. Ceram. Soc., 1960, **43**, 297-301.

2. Hutchins, G.A., Maher, G.H. and Ross, S.D., Control of the Ba:Ti Ratio of $BaTiO_3$ at a Value of Exactly 1 via Conversion to $BaO.TiO_2.3C_6H_8O_7.3H_2O$. Am. Ceram. Soc. Bull., 1987, **66**, 681-4.

3. Sato, M., Hara, H. and Goshi, Y., Determination of silicon in barium titanate by inductively coupled plasma emission spectrometry. Bunseki Kagaku, 1985, **34**, T31-5.

4. Yamamoto, Y., Azuma, T., Nishimura, Y. and Ueda, K., Simultaneous determination of impurities in barium titanate by ICP-AES. Bunseki Kagaku, 1986, **35**, 631-5.

5. Morikawa, H. and Ishizuka, T., Determination of Impurities in Barium Titanate by Inductively Coupled Plasma Atomic Emission Spectrometry. Analyst, 1987, **112**, 999-1002.

6. Fariñas, J.C. and Barba, M.F., Chemical analysis by inductively coupled plasma atomic emission spectrometry of semiconducting ceramics of barium titanate doped with various metal oxides. Mikrochim. Acta, submitted for publication.

7. Bernas, B., A New Method for Decomposition and Comprehensive Analysis of Silicates by Atomic Absorption Spectrometry. Anal. Chem., 1968, **40**, 1682-6.

8. Fariñas, J.C., Estudio analítico por espectrometría de emisión de plasma de acoplamiento inductivo de materiales cerámicos avanzados con matrices de circona y titanato. Thesis, Universidad Complutense, Madrid, in preparation.

THIN FILM ALUMINA-FERRITE SUBSTRATE INTERACTIONS AND MICROSTRUCTURAL CHARACTERISTICS

LYNNE M. GIGNAC and SUBHASH H. RISBUD
Department of Materials Science and Engineering
University of Arizona
Tucson, Arizona 85721, USA

ABSTRACT

Chemical and microstructural analyses were performed to investigate interactions between rf-sputtered thin alumina films and nickel zinc ferrite substrates. The etchability of deposited alumina films in phosphoric acid was variable and related to the character of the film-substrate interface. The role of deposition conditions and microstructure-phase formation due to ferrite alumina reactions were studied by a number of analytical techniques including transmission electron microscopy (TEM), Rutherford Backscattering Spectrometry (RBS), and Auger Electron Spectroscopy (AES).

INTRODUCTION

Alumina (Al_2O_3) thin films synthesized at temperatures below 500°C by a host of deposition techniques usually have an amorphous structure [1-3]. When the substrate is heated to greater than 500°C during deposition or when the as-deposited films are subsequently annealed, metastable polymorphs of alumina referred to as γ-Al_2O_3 or the stable hexagonal α-Al_2O_3 have been reported [1-4]. Amorphous alumina films are soluble in hot H_3PO_4, HF and buffered HF but when the deposition parameters lead to crystalline (stable or metastable) alumina phase(s), the etch rate of the films is negligible [2-4].

The present work is an investigation of the effect of the various sputtering parameters on the production of alumina films which were partially or completely insoluble in H_3PO_4. The film samples produced in these experiments were examined using various microanalytical techniques to identify the microstructural aspects of readily etchable and chemically resistant films. Optical microscopy, scanning electron microscopy (SEM), x-ray diffraction (XRD), x-ray photoelectron spectroscopy (XPS), Auger electron spectroscopy (AES), Rutherford backscattering spectroscopy (RBS), and transmission/scanning transmission/high resolution electron microscopy (TEM/STEM/HREM) were among the techniques used.

MATERIALS AND METHODS

Aluminum oxide deposition on ferrite substrates was performed in a Perkin-Elmer 2400-SL radio frequency (RF) sputter deposition system. The target was 35 cm in diameter and composed of hot pressed aluminum oxide. The alumina target and ferrite substrates were water cooled and films were deposited at a rate of ≈25 nm per minute.

Three types of films were deposited and analyzed: chemically unetchable, partially etchable, and standard easily etchable films. The unetchable films were so classified, because, after thorough etching, the majority of the film remained on the ferrite. The partially etchable films were almost completely removed by the etching procedure, but a thin layer of alumina still remained on the ferrite surface. The easily etchable films were completely stripped off by the acid. The as-deposited etchable and unetchable thin films were examined using a General Electric XRD-5 x-ray diffractometer with monochromatic CuKα radiation at a scanning rate of 2°min^{-1}. Since the films were ~0.9 μm thick, a detectable x-ray signal from the film was obtained.

After the films were etched, the samples were inspected using a Zeiss ICM-405 optical microscope and an ISI Super IIIA SEM. The area on the sample where the film was exposed to the complete etching procedure was targeted. For SEM analysis, a thin layer of carbon was evaporated on the sample surface to make the sample conducting. Energy dispersive x-ray spectra (EDS) were obtained using a PGT system III spectrometer.

Interactions between alumina films and the ferrite substrate were studied by cross section TEM analyses. Two film surfaces were glued together using an adhesive. Pressure was applied to the glued structure using a machinist clamp, and the glue was cured at 170°C for 2 h. After curing, 300 to 500 μm thick cross-section slices of the film structure were sectioned using a Buehler Isomet slow speed diamond saw. 3 mm diameter discs were cut and TEM samples prepared as discussed by Gignac and Risbud [5].

RESULTS AND DISCUSSION

Categorization of Films

The alumina thin films produced from the alumina deposition experiments were divided into categories according to their chemical etch characteristics. The area on the samples which was subjected to the most severe etching process (70% H_3PO_4 at 70°C for 15 minutes) was inspected with an optical microscope, and, according to remanent film structures found in this region, the films were classified. The categories included: easily etchable (EE) films, partially etchable (PE) films, and unetchable (UE) films. If the film was completely removed during the etching procedure, the film was classified as EE. The EE film samples were produced during the alumina film processing experiment, and these films had a film quality rating equal to 0. For PE films, a film layer or remanent film structures were visible on the surface of the ferrite after etching. The PE films were also produced during the alumina deposition experiment and these films had a film quality rating of 1 or 2. The UE films were created by depositing in a water vapor rich environment or with an Ar/O_2 sputtering gas, and these films could not be removed by the standard etching procedure.

Microstructural Characteristics

The three types of film samples were inspected with an optical microscope and a SEM after the samples had been etched in H_3PO_4 at 70°C for 15 minutes. Micrographs were taken of the top surface of films. The EE film was completely removed by the harsh etching procedure. Remanent alumina film structures of the partially etchable (PE) film were apparent on the top surface of the ferrite and, in particular, at the substrate edge. The film edges showed three regions which included: (i) alumina that appeared barely attacked by the acid, (ii) alumina partially etched by the acid, and (iii) alumina almost completely removed. The size and morphology of these remanent film structures closely resembled the structure of the ferrite (substrate) crystal grains.

For UE films, more of the original alumina film remained on the substrate surface after etching than it did for the PE films. Similar to the PE films, the UE films exhibited a non-uniform etching character, and the film structures remaining on the substrate after etching appeared crystalline. The structure of the film again seemed to be influenced by the structure of the substrate crystals underneath the deposited film.

Evidence for epitaxial film growth was obtained from the high resolution transmission electron microscopy (HRTEM) analysis of the films. Fig. 1 shows a HRTEM image of a cross-section of a PE film. The substrate diffraction pattern was oriented for a [242] zone axis, and the (202) and (111) planes were visible in the HRTEM image of the substrate. The diffraction pattern in Fig. 1 shows the two sets of diffraction spots. The diffraction pattern was indexed and reflections were attributed to the ferrite substrate and the $\gamma(\eta)$-Al_2O_3 phase in the alumina film.

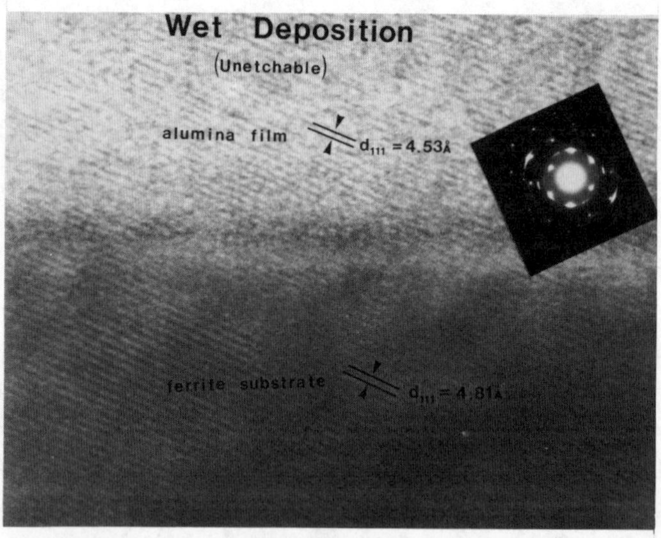

Figure 1. HRTEM of a cross-section of a UE film with a diffraction pattern taken at the film/substrate interface.

SEM-Energy Dispersive X-ray Analysis

When analyzing the samples by SEM, an energy dispersive x-ray spectrometer was used to measure the characteristic x-ray yield in an electron irradiated as-deposited and post-etched UE film. The sole elements detected in both spectra were Al, Ar, Fe, Ni, and Zn which comprised the elements of the substrate and the film. Significant amounts of argon were recorded in the as-deposited films due to Ar entrapment during deposition. Stronger Fe, Ni, and Zn signal and diminished Al and Ar signals were recorded in the post-etched spectrum because less film existed on the ferrite after etching. Since no impurities were detected, the mechanism for formation of the PE and UE films probably does not involve nucleation on impurities. Since the EDS only detects elements down to sodium ($Z > 11$), the presence of low atomic number impurities is still a possibility.

Dot maps of the PE alumina film structures showed a backscattered electron image (BSE) of the top view of a post-etched ferrite surface with remanent alumina film. In the Fe dot map, a high concentration of iron was recorded in the area where the alumina film structure remained. This occurred because the electron beam penetrated through the alumina film into the ferrite. Iron incorporation or diffusion into the film could not be concluded from the iron dot map, but the dot maps did confirm that the film remaining on the ferrite did contain alumina.

REFERENCES

1. Thornton, J.A. and Chin, J., Structure and Treatment Characteristics of Sputter-Deposited Alumina, Am. Ceramic Soc. Bull., 1977, **56**, 504-508.

2. Freiser, R.G., Phase Change in Thin Reactively Sputtered Alumina Films, J. Electrochem. Soc., 1968, **113**, 357-360.

3. Iida, K. and Tsujide, T., Physical and Chemical Properties of Aluminum Oxide Film Deposited by $AlCl_3$-CO_2-H_2 System, Japan J. Appl. Phys., 1972, **11**, 840-849.

4. Ferrieu, E. and Pruniaux, B., Preliminary Investigations of Reactively Evaporated Aluminum Oxide Films on Silicon, J. Electrochem Soc., 1969, **116**, 1008-1013.

5. Gignac, L.M. and Risbud, S.H., Microanalyses of Chemically Etched Thin Film Alumina-Ferrite Interfaces, J. Materials Science, 1989, **24**, 1169-1176.

CHEMICAL AND PHYSICAL PARAMETERS INFLUENCE ON ALUMINIUM NITRIDE BASED CERAMICS THERMAL DIFFUSIVITY

MICHELET J.P., HAFIDI A., LECOMPTE J.P.
JARRIGE J., BILLY M., MARTIN C., MEXMAIN J.
Laboratoire de Céramiques Nouvelles - CNRS U.A. 320
Université de Limoges - FRANCE

ABSTRACT

Ceramics have been elaborated through pressureless sintering, hot pressing and hot isostatic pressing with and without additives (CaO, Y_2O_3,...). The measurements were carried out in the temperature range 300-1000 K using a laser-flash diffusivity technique, the samples surface being coated with colloidal graphite during the test to avoid laser shine-through. Discussion will lead to microstructure action and the sintering aid's influence on the characteristics of obtained ceramics.

INTRODUCTION

High power and high speed systems need high electrical insulating and high thermal conductive materials as high performance ceramic substrate and packaging material.
Nowadays beryllium oxide and silicon carbide with thermal conductivity much higher than alumina are attractive candidates as substrate material. But because of beryllium oxide toxicity, silicium carbide dielectric constant and high production coasts, their use seems to be restricted.
Among these materials aluminium nitride is promising because it has 320 w/m.K theorical conductivity (1). The measures on monocristal give 285 w/m.K (2) instead of 50 to 260 on polycristal (3-6). This is the reason why we tried to densifie aluminium nitride to investigate the role of sintering technics additives, porosity and temperature on the thermal diffusivity.

EXPERIMENTS

1) Samples preparations

Aluminium nitride powder* was prepared by the reduction process of alumina with carbon in nitrogen atmosphere. Specifications of nitride powder is shown table 1.

TABLE 1
Aluminium nitride powder caracteristics

	STARCK		OUR RESULTS
Area B.E.T. m^2/g	3 - 6		4,7
Particle size μm	1,3 - 2,5		0,5 and 1,7 Bimodal
Composition %			
Al	67,9		65,47
N	30	min.	32,5
O	2,0	max.	1,95
C	0,1	max.	0,075

In the first step sintering additives such as calcia or yttria were added to aluminium nitride.
Three sintering methods have been applied to the mixed powder

- Pressureless sintering (P.S.)
  After uniaxial pressing with free die at 150 MPa, the compact bodies were sintered in an aluminium nitride crucible at 1800°C during 5 hours in nitrogen atmosphere
- Hot pressing (H.P.)
  The powder was hot pressed into 8 mm diameter compacts in a boride nitride coated die under a pressure of 25 MPa during one hour between 1700-1800°C in nitrogen flowing
- Hot Isostatic Pressing (HIP)
  it was carried out on four hot pressed compacts (80% density) capsuled in quartz. The pressure and the temperature are gently applied to respectively 170 MPa and 1700°C during one hour.

Thermal diffusivity measurement

Thermal diffusivity was measured by a laser flash method

---

* *Starck Co, Berlin grade B*

(7) using a neodynium-doped glass laser operating at 1,06 µm with pulse length of 400 ms, and a liquid nitrogen cooled In-Sb detector. The pulse energy was 70 J.
The specimen discs of 8 mm diameters were cut and polished to about 4 mm thick.
Both specimen surfaces were coated with carbon to assure complete absorption of the flash, and avoid laser shine-through .

The termal diffusivity "a" has been deduced from the PARKER equation (8) :

$$a = 0,1388 \frac{e^2}{t_{1/2}}$$

where $t_{1/2}$ (s) is the time required for the back surface to reach half of the maximum temperature value and e(cm) is the thickness of the sample.
Thermic losses are expressed from the HECKMAN model (9).

RESULTS AND DISCUSSION

. Porosity

The porosity dependance of diffusity for hot-pressed AlN bodies was measured (figure 1).
It increases with decreasing porosity until near null porosity. For a fully dense sample, diffusivity depends on the microstructural factors such as secondary phases, impurities...

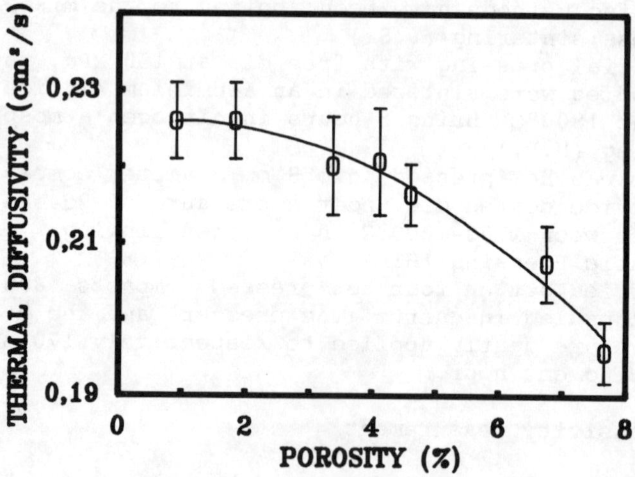

Figure 1. Effect of porosity on thermal diffusivity for HP AlN without additive ceramics obtained.

. **Density and thermal diffusivity**

TABLE 2
Effect of additives on AlN based ceramics density for the three sintering technics.

Sintering Technics	PS		HP		HIP	
AlN samples	d	a (cm^2/s)	d	a (cm^2/s)	d	a (cm^2/s)
Without additive			3,23	0,23	3,26	0,25
1% wt CaO	3,211	0,25	3,17	0,22	3,23	0,23
2% wt Y_2O_3	3,257	0,28	3,26	0,22	3,28	0,23
8% wt Y_2O_3	3,303	0,37	3,36	0,33	3,38	0,33
2% wt CaO-Y_2O_3	3,209	0,31				

. **Diffusivity in function of temperature**

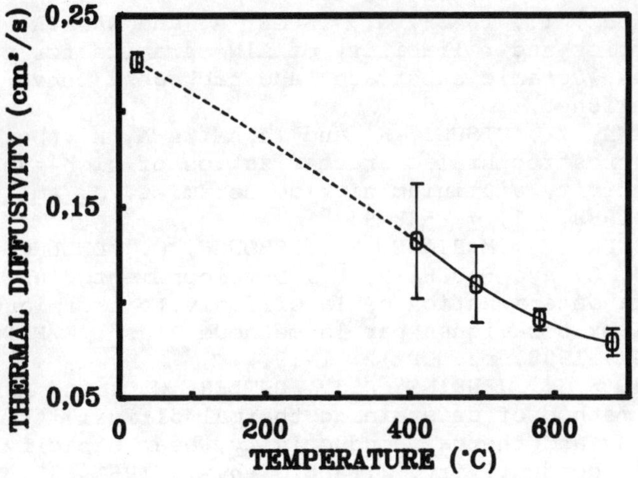

Figure 2. Temperature dependance of aluminium nitride on thermal diffusivity.

CONCLUSIONS

Theses results permit to show that :

1) a low density decreased thermal diffusivity.

2) the additives of CaO or/and Y_2O_3 produced aluminates compounds in sintered AlN ceramics. These compounds permit to obtain fully dense AlN based ceramics and contribute to little thermal diffusivity increase according to the results with porosity.

3) A maximal thermal diffusivity has been obtained with a particular amont of second phase with yttria.

REFERENCES

1. SLACK, G.A., Nonmetallic crystals with high thermal conductivity. J. phys. Chem. Solids, 1973, 34, 321-35.
2. SLACK, G.A., TANZILLI, R.A., POHL, R.O. and WANDERSANDE, J.W., The intrinsic thermal conductivity of AlN. J. phys. Chem. Solids, 1987, 48, 7, 641-47.
3. WERDECKER, W. and ALDINGER, F., Aluminium nitride an alternative ceramic substrate for high power applications in microcircuits. IEEE Trans. Comp. Hyb. Man. Tech., 1984, CHTM-7, 4, 399-404.
4. JARRIGE, J., MEXMAIN, J., MICHELET, J.P., GUYADER, J., LAURENT, Y. and LEBODO, H., Frittage et propriétés du nitrure d'aluminium préparé par réaction NH_3/Al_2O_3. Rev. Chim. min., 1987, 24, 6, 687-96.
5. TAKAHASHI, T., IWASE, N., TSUGE, A. and NAGATA, M., Properties and reliability of AlN ceramics for power devices. Ceramic substrates and packages, Denver, 1987, to be published.
6. KUROKAWA, Y., UTSUMI, K. and TAKAMIZAWA, H., Development and microstructural characterization of high-thermal conductivity aluminium nitride ceramics. J. Am. Ceram. Soc., 1988, 71, 7, 588-94.
7. GITZHOFER, F., MARIAGE, E., ARBOUCH, O., DELLUC, G., MARTIN, C. and FAUCHAIS, P., Développement d'un logiciel pour la détermination de la diffusivité thermique de matériaux céramiques par la méthode flash. S.F.T., Limoges, 1988, pp. CPM12, 1-19.
8. PARKER, W.J., JENKINS, R.J., BUTLER, C.P. and ABBOT, G.L., Flash method of determining thermal diffusivity, heat capacity and thermal conductivity, heat capacity and thermal conductivity. J. Appl. Phys., 1961, 32, 9, 1679-84.
9. HECKMAN, R.C., Finate pulse, time and heat. Loss effects in pulse thermal diffusivity measurements. J. Appl. Phys., 1973, 44, 3, 1455-60.

INFLUENCE OF SOME OXIDES ON THE THERMAL CHARACTERISTICS OF STOICHIOMETRIC AL_2TiO_5

P.P. Demaestri, A. Giachello, P.C. Martinengo
Fiat Research Center - Orbassano (ITALY)

C. Majani
ENEA CRE - Casaccia (ITALY)

ABSTRACT

Aluminium titanate is considered an engineering material expecially where thermal insulation is required.
This is due to its special combination of properties, low thermal expansion coefficient, that it make possible to apply components like portliners into cylinder engine heads directly in aluminium and cast iron casting process.
The aim of this work is to understand the influence of the addition of some oxide like SiO_2, $Zr0_2$ and MgO on the behaviour of the material at high temperature expecially respect to the thermal decomposition of Al_2Ti0_5 at 1000° and 1100°C, which are very critical conditions also in actual engine applications.

INTRODUCTION

Beta aluminium titanate has very low value of thermal conductivity coupled with low thermal expansion coefficient, high melting point, elevate thermal shock resistance and it can be considered for structural components in high temperature applications. Problems arise from the lack of mechanical properties due to the microcracked structure which takes place during cooling period from sintering heat treatment because of the anisotropy of thermal expansion of the Al_2Ti0_5 crystal, and from the thermal decomposition that it suffers and may limit engineering uses (1). Lang, Fillmore and Maxwell (2) have indicated the range of thermal instability of Al_2Ti0_5 between 750° and 1300°C where decomposition occurs into corundum and rutile.
It appears of practical rilevance to find out compounds which can stabilize Al_2Ti0_5 in a wide interval of temperature.
Many authors suggest that by substituting Mg, Si, Zr ions

into the Al_2TiO_5 cell a large effect of stabilization in this way is reached (3). Present paper deals about some work done on this subject considering commercial raw materials of moderate purity, for practical purpose, in view of possible engine applications where components for thermal insulation are needed.

MATERIALS AND METHODS

Starting powders used in these experiments were:

- α - Al_2O_3 with main impurities as Na_2O 0,1%, CaO_2 0,04%, SiO_2 0,02% and with a specific surface area of 1,2 m^2/g

- TiO_2 as anatase contains CaO 0,03%, SiO_2 0,08%, Fe_2O_3 0,16%, SiO_2 0,08% and a specific surface area of 0,10 m^2/g

- MgO, ZrO_2, SiO_2 additives were of reagent grade with specific surface area of 0,51; 26,7; 152,6 m^2/g respectively.
  Because in a previous work it was found that both densification and reactions from constituents are highly enhanced if prereacted Al_2TiO_5 is present, the basic formulation adopted in this research consists in a mixture of 50 wt% Al_2TiO_5 prereacted powder and 50 wt% equimolar Al_2O_3 - TiO_2 powders (named AT). For comparison pourpose three others mixtures were prepared from the AT base material (all by wet alumina ball-jar milling, 48 hours) with 2 wt% MgO addition (designated as ATM); 2 wt% ZrO_2 (as ATZ) and finally with 8 wt% SiO_2 (as ATS).
  Powders obtained were uni-axially pressed, without any binders, at 100 MPa into bars 4x5x30 mm in dimensions.
  All samples were air heat treated in the same sintering cycle which consisted in holding 3 hours at 1450°C.
  Measurements were done on "as sintered" specimen and precisely: density through mercury displacement; open porosity with mercury intrusion porosimeter; flexural strength with 3 point bending, 25 mm span, at an applied load of 0,5 mm/min; mean grain size calculated with the intercept method; thermal expansion with a dilatometer heated and cooled at constant rate of 5°C/min; and quantitative crystalline phases determination using an internal standard with a Ni filtered CuK radiation XRay diffractometer.
  Thermal decomposition were studied ageing in air at 1000°C and 1100°C up to 200 hours and from time to time checking the phases present in the powdered materials.

RESULTS

All data collected from the four sintered ceramics analyzed are summarized in Table 1. The corresponding SEM fracture surfaces are reported for each material in Fig. 1 and Fig. 2.

TABLE 1

Characteristics of sintered Al_2TiO_5 ceramics

CHARACTERISTICS \ MATERIALS	AT	ATM 2wt% MgO	ATZ 2wt% ZrO_2	ATS 8wt% SiO_2
APPARENT DENSITY g/cm^3	3.31	3.24	3.37	3.21
CRYSTALLOGRAPHIC PHASES (XRD)	β-Al_2TiO_5	β-Al_2TiO_5	β-Al_2TiO_5 α-Al_2O_3 tr. TiO_2 tr.	β-Al_2TiO_5 α-Al_2O_3 tr. TiO_2 tr.
OPEN POROSITY mm^3/g	14.5	18.4	10.1	13.0
MEAN GRAIN SIZE μm	4.0	7.1	4.9	3.4
FLEXURAL STRENGTH 3-POINTS MPa	17.6	37.3	23.8	24 8
COEFF. THERMAL EXPANSION α 20-800 °C 10^{-6} °C^{-1}	3.4	2.7	2.6	3.4
α AFTER THERMAL DECOMPOSITION AT 1000 °C 20-800 °C 10^{-6} °C^{-1}				
100 h	4.5	3.6	3.2	4.2
200 h	5.1	3.7	3.8	4.7

a

b

Figure 1. SEM fracture surfaces of materials AT (a) and ATM (b).

c d

Figure 2. SEM fracture surface of materials ATZ (c) and ATS (d).

The two diagrams in Fig. 3 relate to the thermal decomposition behaviour of materials at the two temperatures considered.

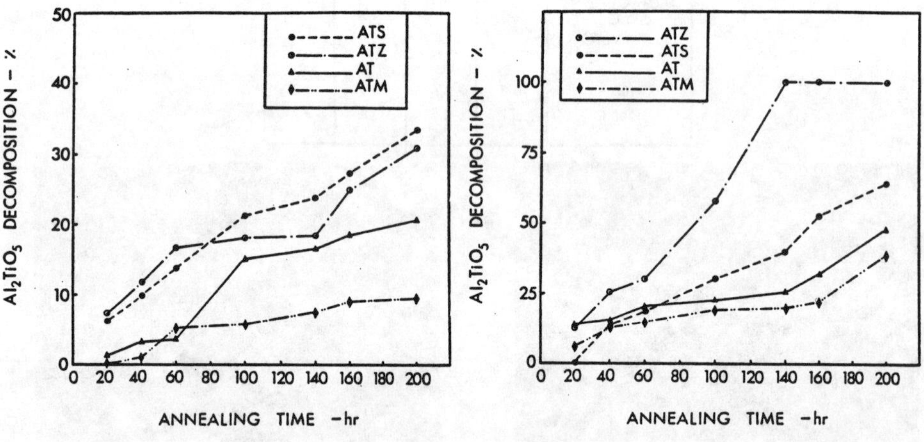

a b

Figure 3. Time dependence of the thermal decomposition of materials at 1000°C (a) and 1100°C (b).

The dramatic change in microstructure is observed in the sequence in Fig. 4 for the ATZ ceramic for 50, 100 and 200 hours respectively of ageing time at 1100°C.

a

c

b

d

Figure 4. SEM fracture surfaces of ATZ aged at 1100°C for 50 hr (a), 100 hr (b) and 200 hr (c, d).

One trend of thermal expansion is visible in Fig. 5 for AT as sintered material and aged for several length of time at 1000°C.

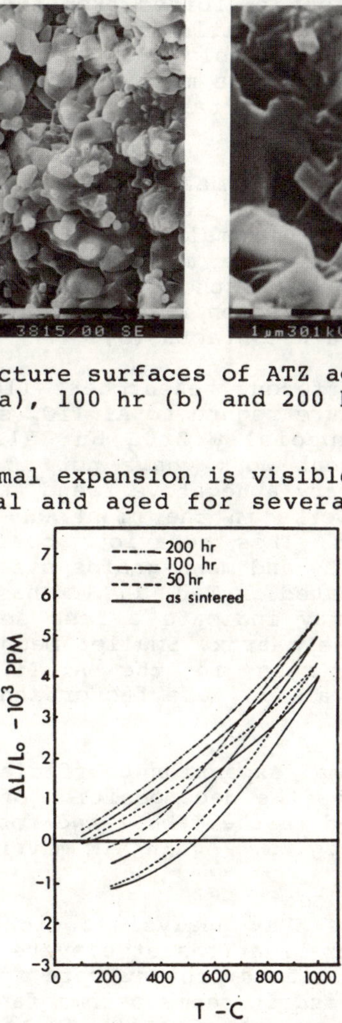

Figure 5. Thermal expansion of materials AT as sintered and aged for several lengths of time at 1000°C.

COMMENTS

Some comments on results obtained so far could be done:

- Impurities in the raw material play a role regard to the reaction kinetics from constituents and in the final characteristics of the sintered material which must not be neglected (4), nevertheless high conversion rate and densification was reached in the reaction sintering cycle (3 hours at 1450°C).

- Also if the mean particles size in all cases under examination of the starting milled powder mixtures range from 1,5 to 2 μm, during sintering at high temperature an extensive grains growth occur not inhibited from the additives expecially for the magnesia addition material (ATM), where mean grains size is 7 μm. Probably ZrO_2 and SiO_2 have a lower diffusion coefficient in the matrix and as grains boundary secondary phase exerts a certain delay effect in grain growing.

- Despite to the quite big mean grains size ATM material shows the highest flexural strength also considering the lowest coefficient of thermal expansion and relative hysteresis correlated to the microcracked structure. This fact suggests what some authors indicate that mechanical properties is mainly based on friction and toothing effects more than grain - grain contacts (5).

- As reported from different researchers 1100°C represent a very severe temperature regard to Al_2TiO_5 stability (3) and in this respect expecially ZrO_2 but also SiO_2, in the amount used in present work, have not at all stabilizing effect, on the contrary a negative trend.
  Only MgO addition works in the right way and merits more deeply investigation. This behaviour could be correlated with the open porosity and mean grains size analyzed in the structures, as indicated in Tab. 1, ATM has for both values high figures which may indicate a less severe compression stress present in the matrix. Smaller mean grains size and high compression stress in the Al_2TiO_5 ceramics, as reported by Kato et al (1), are factors that favor thermal decomposition.

- As obvious the thermal expansion coefficient follows generally an increment as the decomposition of Al_2TiO_5 proceed and data are reported in the Tab. 1 and for AT ceramic also in the diagramm of Fig. 5. Again ATM material demonstrated better results.

- SEM and quantitative XRay analysis indicate during ageing of materials under examinations at elevate temperature, decomposition proceed with different rate for materials with different additives and it seems become faster when rounded Ti-rich particles segregate from the matrix. Finally crystals of rutile with Al rich agglomerats are present (see Fig. 4 c and d) accompanied with an increase of total porosity practically without changing the overall dimension of the sample as observed for the ATZ materials where total decomposition occurs at 1100°C after 100 hours.

CONCLUSIONS

50 wt% prereacted Al_2TO_5 powder plus 50 wt% equimolar Al_2O_3 - TiO_2 powders mistures of commercial materials of moderate purity were sintered at 1450°C, 3 hours with a suitable density and high conversion. In orther to evaluate properties expecially respect to Al_2TiO_5 stability at high temperature, additions of 2 wt% MgO; 2 wt% ZrO_2; 8 wt% SiO_2 to the above cited base material were done.
Results indicate:

. general positive trend for MgO addition is checked as better thermal stability, higher mechanical strength, low coefficient of thermal expansion, also if improvement can be done respect to grain size and microcracked structure

. negative behaviour regard to stability exerts SiO_2 and ZrO_2 addition where Al_2TiO_5 decomposition was deteriorated. Enhanced properties respect to base material were experienced but a lower value if compared with MgO addition

. no single addition seems to control grain growth mechanism of Al_2TiO_5 during sintering heat treatment.

REFERENCES

1. Kato,E.,Daimon,K., and Takahashi,J.,Decomposition temperature of β- Al_2TiO_5, J. Am. Ceram. Soc.,1980,63,355-56.

2. Lang,S.M.,Fillmore,C.L., and Maxwell,L.H.,ibid,1952,48,298

3. Ishitsuka,M.,Sato,T.,Endo,T., and Shimada,M., Synthesis and thermal stability of aluminium titanate solid solutions, J. Am. Ceram. Soc.,1987,70,69-71.

4. Frendenberg,B. and Mocellin,A., Aliminium titanate formation by solid - state reaction of fine Al_2O_3 and TiO_2 powders, J. Am. Ceram. Soc.,1987,70,33-38.

5. Hennicke,H.W. and Lingenberg,W., Dependence of microstructure and physical properties of materials on the basis of aluminium titanate, Proc. of Cer. Mat. and Components for Engines,1986, Travemunde,619-623.

THERMOMECHANICAL PROPERTIES OF ALUMINIUM TITANATE CERAMIC BETWEEN 20 AND 1000°C

B. Freudenberg*, H.-A. Lindner*, E. Gugel*, P. Thometzek°

* Cremer Forschungsinstitut GmbH&CoKG, Rödental, FRG
° Bayer AG, Krefeld-Uerdingen, FRG

Abstract

Thermomechanical property data are provided for a slip-cast grade of aluminium titanate ceramic. Several properties show hysteresis effects under thermal or mechanical cyclic load, a phenomenon attributed to the opening and closing of microcracks. Based on these data, the shrinkage stress acting on the port liner tube in a cast iron matrix is calculated. Experiment and finite element analysis show that aluminium titanate ceramic can withstand the compressive stresses.

1. INTRODUCTION

Low thermal expansion coefficient, good thermal insulation and low Young's modulus are characteristics of aluminium titanate ceramics. These properties, if properly adjusted, allow for the insert-casting of ceramic port liners into the cylinder head (aluminium or cast iron), where they serve as a thermal insulation of the exhaust gas. The cast-in-place ceramic liner must resist not only the thermal shock of the metal casting, but also the shrinkage stress exerted by the contracting metal and thermal cycling in service life. Stress analyses of the different port liner life cycles are required to provide guidelines for an improved geometric design and for further adjustment of the material properties.

The present paper contributes thermomechanical property data for a specific aluminium titanate ceramic, the experimental grade A 1402 developed at the Cremer Forschungsinstitut and based on Bayer raw materials. These data are used for a shrinkage stress analysis of an aluminium titanate tube in a cast iron cylinder.

2. EXPERIMENTAL PROCEDURES

The quality A 1402 is a slip-cast reaction-sintered material, apparent density 3.1 g/cm³, consisting of mainly two crystalline phases: stabilized aluminium titanate and mullite. By controlled variation of processing parameters, the physical properties can be adjusted over a wide range: bending strength 30 - 50 MPa, bending Young's modulus 12 - 23 GPa and thermal expansion coefficient (TEC) (RT - 1000°C) 1.2 - 2.0 x 10^{-6} 1/K. An intermediate strength quality was chosen in view of the port liner application.

The 4-point-bending experiments were performed on unground bars (50 x 4.5 x 3.5 mm³), time to fracture about 10 seconds. The surface previously in contact with the plaster-mold was always on the compressive side. The high-temperature bending strength measurements in air used a special sample holder giving slightly lower strength values due to frictional effects (see figure 4 below). The specimens were fractured between 20 and 30 minutes after exposure to temperature.

The compression samples, cubes (8 mm) or cylinders (Ø/h 8/20 mm) were tested between a rigid plate and a self-aligning hemisphere, time to fracture about 25 seconds. The load was applied normal to the growth direction of the slip-cast samples.

The adiabatic Young's modulus was measured by the resonance frequency method as a function of temperature in vacuum, using ground bending specimens. The mean heating and cooling rate was 4°C/min.

The thermal expansion was determined on bending specimen as well (L = 40 mm), heating and cooling rate 5°C/min in air.

The thermal diffusivities were measured on disks (Ø/h 10/1.5 mm) by the laser flash method. The sample was heated and cooled at an average rate of about 5°C/min in vacuum.

3. THERMOMECHANICAL PROPERTIES

ROOM-TEMPERATURE PROPERTIES

The stress-strain behaviour and strength have been investigated for different types of loading. The bending test gives a strength of σ_{flF} = 40.8 $\pm$ 1.1 MPa (mean value $\pm$ standard deviation), n = 14. The narrow strength distribution is shown in figure 1, together with the maximum likelyhood fit of the Weibull distribution. The fit provides a Weibull modulus of $\hat{m}$ = 39, the 90 % confidence interval ranging from $\hat{m}_l$ = 25 to $\hat{m}_u$ = 52.

The stress-strain curve obtained from the bending test is of sigmoidal shape. The self-alignment of the experimental set-up contributes an extra strain at lower stresses. The intermediate part of the stress-strain curve is almost linear and may justify the representation of the slope by a constant Young's modulus of E_{fl} = 16.9 ± 1.5 GPa. The calculation of this modulus as well as that of the flexural strength assumes a (questionable) linear elastic behaviour over the whole positive and negative strain range, encountered in the bending specimen. At roughly half the flexural strength, an apparent softening of the material is observed, attributed to the propagation of microcracks.

The compressive strength measured on cubes reached σ_{coF} = -517 ± 22 MPa, which corresponds to a common ratio $\sigma_{coF}/\sigma_{flF} \approx 13$. The stress-strain behaviour under cyclic compressive load is shown in figure 2 (cylindrical samples, maximum stress ≈125 MPa). The upward curvature indicates an increasing rigidity due to crack closure [1]. The average compressive Young's modulus for $|\varepsilon|$ from 0 to 1 ‰ is $\bar{E}_{co}$ = 11.9 GPa, and it rises to 13.7 GPa and 17.4 GPa for 0 - 3 ‰ and 0 - 7 ‰, respectively. At the beginning of the unloading cycle, the rigidity of the sample is even higher. It is thought that some of the recently closed cracks remain welded together. They reopen only upon further unloading. After completing the first compressive cycle, the length of the sample is reduced by $\Delta\varepsilon$ = -0.8 ‰. It is seen from figure 2 that further cycling gives a reduced hysteresis area and a decelerated sample length reduction.

The adiabatic Young's modulus is E_{re} = 35 GPa as determined by the resonance frequency method (≈7 kHz), that is at very low strains and high strain rates. A significantly higher Young's modulus measured by the resonance frequency method than by the quasi-static load-deflection method

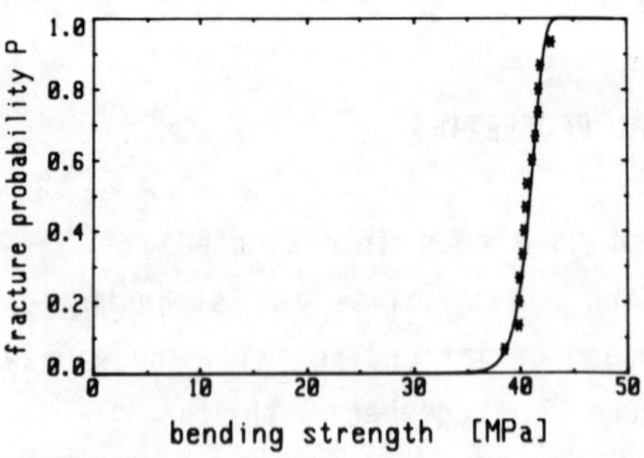

Fig. 1: Weibull plot of the bending strength, measured at room-temperature

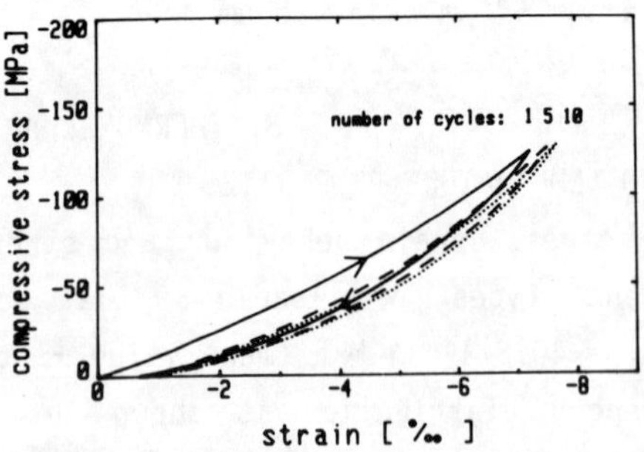

Fig. 2: Stress-strain curve for cyclic compressive load

is unusual for dense ceramics, but represents a common feature of microcracked rocks [2]. This phenomenon may be explained by several strain dependent and/or time-consuming processes that contribute to increased strains in the load-deflection experiment, e. g. sliding of closed crack flanks and slow crack growth. The contribution of time-consuming processes to a length change in aluminium titanate has already been notified in 1952 by Buessem et al. [3].

For a quantitative analysis of thermal- or frozen-in-stresses, the quasistatic load-deformation experiment yields the relevant data. Despite the definitively non-linear behaviour of aluminium titanate ceramics, the approximation by an average Young's modulus might be justified if the strain range for averaging is properly chosen. The quasi-static Poisson ratio has been determined by wire strain gauges applied to bend and compression samples, giving $\nu = 0.135 \pm 0.015$.

SHORT-TIME EXPOSURE TO INCREASED TEMPERATURES

Repeated measurements of the thermal expansion curve (figure 3 shows a single run) give a mean TEC from 20 to 1000°C of $(1.5 \pm 0.3) \times 10^{-6}$ 1/K. This value is much lower than the theoretical mean TEC for dense Al_2TiO_5-ceramic of 10×10^{-6} 1/K (dashed line in figure 3). It is the pronounced thermal expansion anisotropy of the individual Al_2TiO_5-grains that give rise to important stresses on a microscopic scale when cooling the sample after sintering. These local stresses are the driving force for extensive microcracking. When the cold sample is heated up again, most of the solid volume expands into the microcracks, while the macroscopic dimensions remain almost constant. The higher the temperature, the more cracks are closed, the steeper the thermal expansion curve. However, even at 1000°C the slope is far below the theoretical value, suggesting that an important fraction of the microcracks is still open. At the beginning of the cooling cycle, a steeper slope is observed (comparable to figure 2). The closed cracks have sintered together, giving a stiffer skeleton with an increased incremental TEC. A certain supercooling is needed in order to provide the necessary driving force for crack growth.

Progressive crack-healing with increasing temperature has its effect also on bending strength, figure 4. From room-temperature to 1200°C, an about threefold strength increase is observed. Above that temperature strength is reduced by softening of the matrix.

The adiabatic Young's modulus was measured as a function of temperature by the resonance method, figure 5. While the heating part of the cycle follows the general trend, the initial part of the cooling cycle does not. The modulus further increases when cooling from 1000 to 900°C. This reflects an inherent feature of the Young's modulus - temperature relationship, that is to grow with decreasing temperature, given an unchanged microstructure. Below 900°C the onsetting microcracking overrides this effect. At room-temperature, the sample returns to the initial Young's modulus. Because of the additional processes involved, it is not possible to precisely predict the quasi-static Young's modulus as a function of temperature from this resonance frequency measurements.
The thermal diffusivity, figure 6, initially decreases with rising temperature as expected for phonon transport. At temperatures above 500°C the tendency reverses, mostly due to microcrack closure. A further contribution stems from radiative transport. On the cooling cycle the thermal diffusivity remains almost constant down to 700°C and then falls off again.

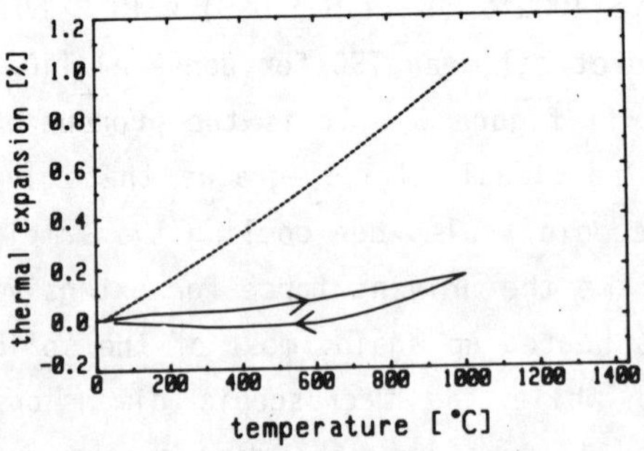

Fig. 3: Thermal expansion curve

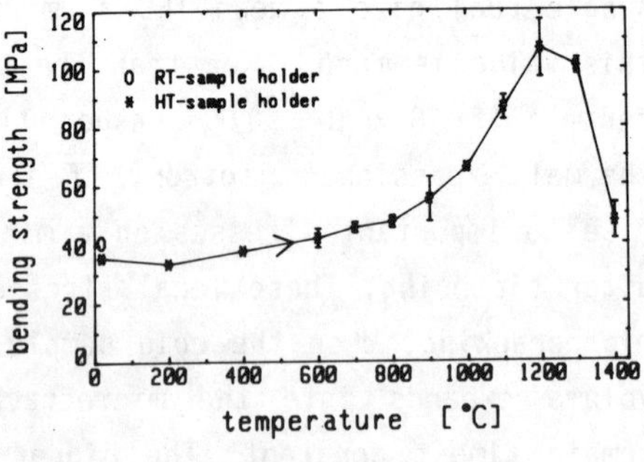

Fig. 4: Bending strength as a function of temperature

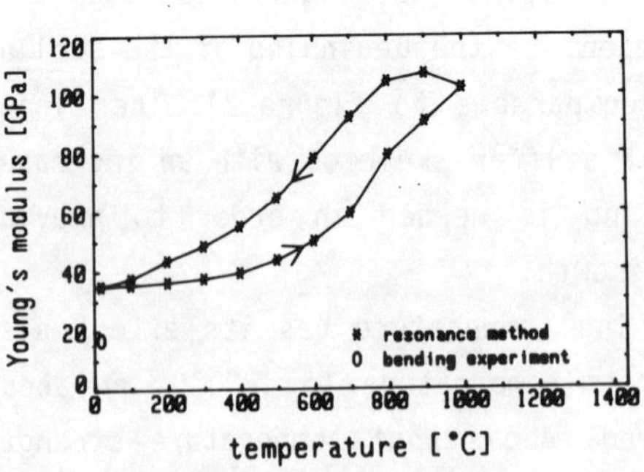

Fig. 5: Adiabatic Young's modulus as a function of temperature

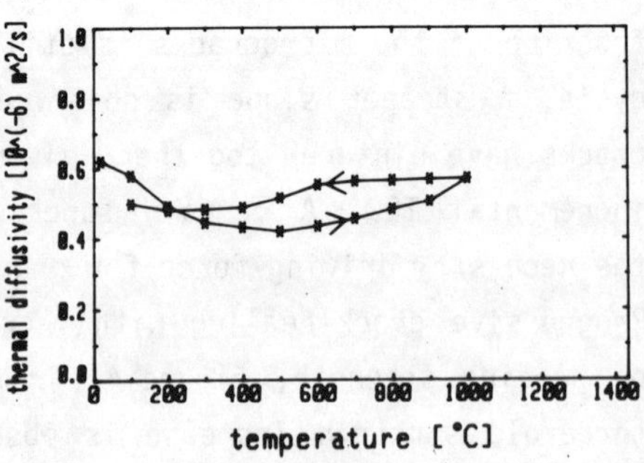

Fig. 6: Thermal diffusivity as a function of temperature

AGING AT INCREASED TEMPERATURES

In a complementary set of experiments, the effect of aging at increased temperatures on thermomechanical properties has been measured. The room-temperature bending strength (figure 7), Young's modulus and TEC were measured after holding the sample 100 hours at constant temperatures between 600 and 1000°C in air. A small but significant increase in strength, Young's modulus and TEC is observed at 1000°C, and there is microstructural evidence for the beginning of decomposition, contrarily to the 900°C sample.

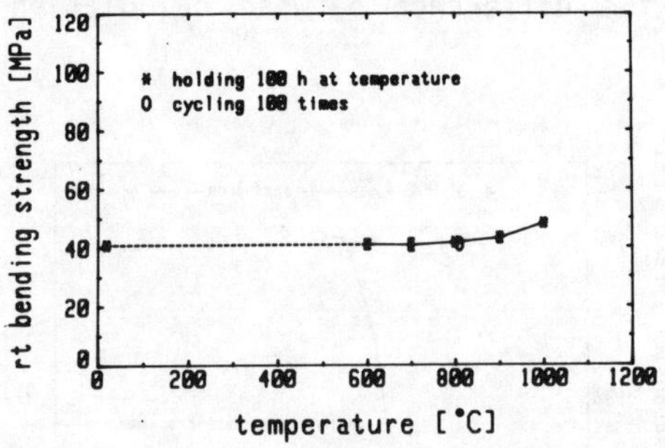

Fig. 7: Room-temperature bending strength after 100 h aging at temperature

The port liner service life conditions in view, the standard bars have been cycled 100 times between room-temperature and 800°C by introduction into the hot furnace and air-quenching. Strength (figure 7), Young's modulus and TEC data correspond to the untreated room-temperature samples within experimental error.

4. ANALYSIS OF SHRINKAGE STRESSES

A two-dimensional plane strain linear elastic finite element model is used to calculate the stress distribution in a doublelayer tube after cooling to room-temperature. The outer cast iron and the inner aluminium titanate layer are represented by 8 curved elements each, one element comprising 8 nodes. It is assumed that due to plastic flow in the cast iron at high temperatures, stresses build up only below 800°C. The dimensions are given in figure 10, together with the material property data. For the aluminium titanate an average Young's modulus is chosen to correspond to the strains encountered in practice. The calculation indicates a radial shrinkage pressure of σ_{rmax} = -28 MPa, acting on the cold ceramic tube, figure 8. The resulting circumferential compressive stress has a maximum located at the inner surface, $|\sigma_{\Theta max}|$ = 186 MPa, figure 9. This represents a stress well below the measured compressive strength of $|\sigma_{coF}|$ = 517 ± 22 MPa. The circumferential stress is expected to show a discontinuity at the ceramic/metal interface, a feature smeared out in figure 9 by the finite element size. The calculated radial displacement of the ceramic tube inner surface is $\Delta R_{calc} = -0.21$ mm.

Casting experiments with a similar geometry confirm that aluminium titanate tubes do indeed survive the thermal shock and the shrinkage stresses exerted by the cast iron, figure 11. The measured radial shrinkage of the ceramic tube is $\Delta R_{meas} = -0.17 \pm 0.02$ mm ($n = 5$). This value corresponds to about 80 % of the calculated displacement, suggesting that the actual stresses are somewhat lower than the calculated ones. In view of the simplifying assumptions (isothermal cooling, sharp plastic-elastic transition, linear elastic behaviour etc.) the difference between calculation and experiment is acceptable.

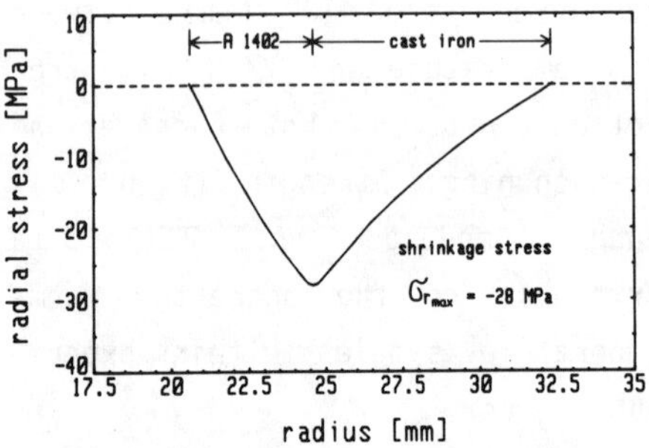

Fig. 8: Radial stress distribution at room-temperature

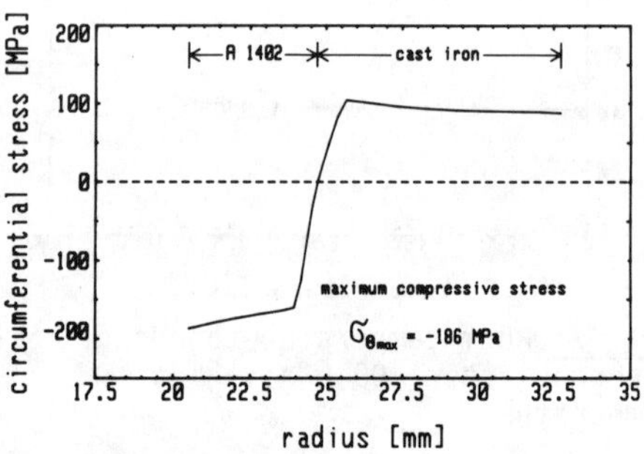

Fig. 9: Circumferential stress distribution at room-temperature

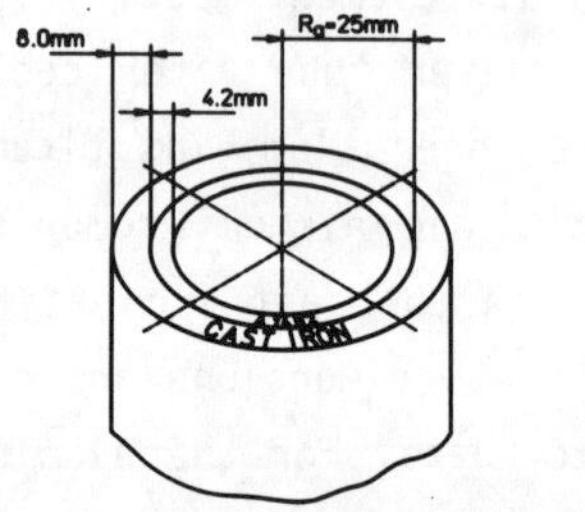

	Ceramic A 1402	cast iron
Young's modulus [GPa]	18	110
Poisson ratio [-]	0.135	0.3
Thermal expansion coefficient [10^{-6} 1/K]	1.3	13

Fig. 10: Dimensions and property data used in the stress analysis

Fig. 11: Aluminium titanate tube in cast iron

5. CONCLUSIONS

The thermomechanical properties of the investigated aluminium titanate ceramic show several effects not encountered with dense ceramics, e. g. a pronounced nonlinear stress-strain relationship at room-temperature, as well as hysteresis effects under mechanical and thermal load. These phenomena are explained by the opening and closing of microcracks. Thus, the property data used for a stress analysis have to be properly chosen, that means strain range and history in experiment and application should correspond.

The shrinkage stresses were calculated for a ceramic port liner tube enclosed in a cast iron cylinder. The calculated maximum compressive stress is situated well below the measured compressive strength. The survival of insert-cast pieces corroborates these finding.

Acknowledgement: The help of R. Wagner (DLR, Cologne, FRG) with adiabatic Young's modulus measurements, W. Neumann (ÖFZ, Seibersdorf, A) with thermal diffusivity measurements and H. G. Trost (MAN Technologie, Munich, FRG) with the metal casting experiments is gratefully acknowledged.

REFERENCES

1. Miller, B. et al., Development of a Ceramic Exhaust Port Liner. In Automotive Ceramics: Recent Developments, Inter. Congr. 29.2.-4.3.1988, Detroit, USA, SAE publication 880677

2. Gebrande, H., Elastizität und Inelastizität. In Landolt-Börnstein Vol. NS, V, 1b, ed. K. H. Hellwege, Springer-Verlag Berlin, 1982, p. 85 cont.

3. Buessem, W. R. et al., Thermal Expansion Hysteresis of Aluminium Titanate, Ceram. Age 60 (1952), p. 38-40

THERMAL EXPANSION STUDY OF ALKALI-SUBSTITUTED CORDIERITES

DANIELE MERCURIO* , PHILIPPE THOMAS*,JEAN-PIERRE MERCURIO*,BERNARD FRIT* and GEORGES ROULT**

* Laboratoire de Chimie Minérale Structurale, (U.A.-C.N.R.S. n° 320) - Faculté des Sciences de Limoges, 123, Avenue A. Thomas 87060 Limoges Cedex - France

** D.R.F.G.-SPH/S, C.E.N. Grenoble, 85 X, 38041 Grenoble Cedex - France

ABSTRACT

Structural studies of alkali-modified cordierite with $M_xMg_2Al_{4+x}Si_{5-x}O_{18}$ formula (M = K, Cs ; x = 0.25 ; 0.50 ; 1.00) were performed between 20 and 825°C using a time-of-flight powder diffraction technique on sintered samples obtained by glass-crystallization. All the crystal structures are related to the pure hexagonal cordierite one : full Al-Si disorder is observed and alkali atoms are located within the hexagonal channels.

The evolution of the relative thermal expansion is analyzed in terms of both nature and concentration of alkali atoms. A very low value of the mean thermal expansion coefficient ($\bar{\alpha} \simeq 0.4\ 10^{-6}\ K^{-1}$ between 20 and 825°C), resulting from an opposite evolution of the a and c parameters, has been obtained for a fully disordered $K_{0.5}Mg_2Al_{4.5}Si_{4.5}O_{18}$ material.

INTRODUCTION

Cordierite $Mg_2Al_4Si_5O_{18}$ occurs in two polymorphic forms : the high temperature, disordered, hexagonal form (sp. gr. P6/mmc) isostructural with beryl and the low temperature, ordered, orthorhombic form (sp. gr. Cmcc) (1,2). Numerous studies have been carried out in order to elucidate the mechanism of the thermal expansion of both polymorphs (3). Nevertheless, little attention has been devoted so far to alkali-substituted cordierites (4).

This paper deals with the study of the influence of potassium and cesium substitution on both lattice dimensions and thermal expansion of these materials.

EXPERIMENTAL

Glassy pieces with compositions $M_xMg_2Al_{4+x}Si_{5-x}O_{18}$ (M = K, Cs ; x = 0.25 ; 0.50 and 1.00) were obtained by melting and water-quenching. After ball-milling for two hours the powders were heated in air (1000°C/2h + 1160°C/2 h) in order to insure complete crystallization. X-ray fluorescence analysis of the crystallized samples did not show any significant deviation from the nominal compositions.

The structural study was carried out between 20 and 825°C using a time-of-flight powder neutron diffraction technique. The diffraction patterns, were fitted to the calculated ones using a full profile analysis program (modified Rietveld method) (5,6).

RESULTS AND DISCUSSION

A projection along the c axis of the room temperature structure of substituted cordierite is shown in fig. 1. The polyhedra arrangement, typical of the hexagonal pure cordierite framework, consists of 6-membered rings of $T(2)O_4$ tetrahedra, centered at the origin, lying in the xOy plane, stacked one above the other and successively rotated ≃ 30° relative to each other (fig. 2a). Rings are linked together laterally and vertically by $T(1)O_4$ tetrahedra and MgO_6 octahedra which, by sharing edges, form a bidimensional network of large overlapping 12-membered rings $[T(1)O_4\text{-}MgO_6]_6$ parallel to xOy.

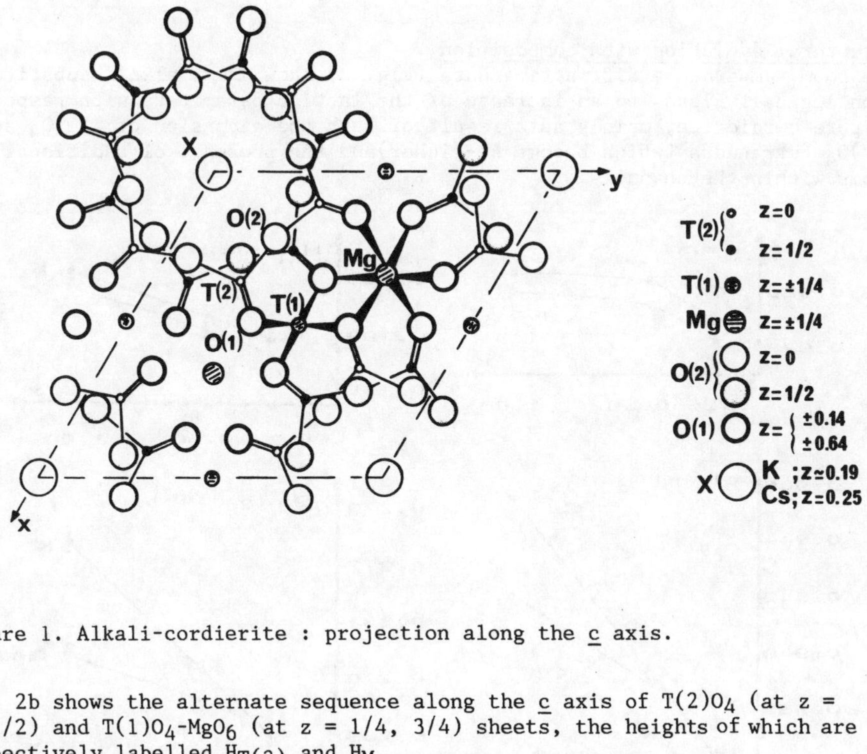

Figure 1. Alkali-cordierite : projection along the c axis.

Fig. 2b shows the alternate sequence along the c axis of $T(2)O_4$ (at z = 0, 1/2) and $T(1)O_4\text{-}MgO_6$ (at z = 1/4, 3/4) sheets, the heights of which are respectively labelled $H_{T(2)}$ and H_M.

The ring stacking produces large channels, parallel to the c axis, in which alkali atoms are statistically distributed among the 4e $(0,\bar{0},z)$ positions.

Ideal $[T(1)O_4\text{-}MgO_6]_6$ rings would not be large enough to fit around the 6-membered $T(2)O_4$ rings. They are therefore severely distorted : the $T(1)O_4$ tetrahedra are elongated along one of their $\bar{4}$ axis (// xOy) and the MgO_6 octahedra are stretched isotropically along the same plane.

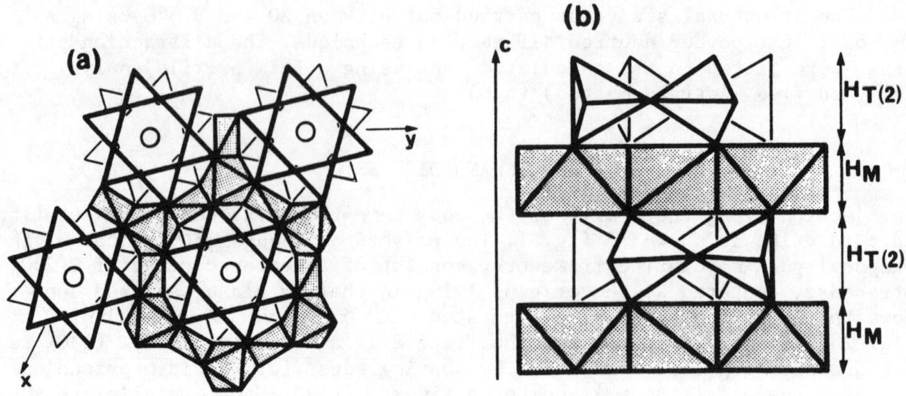

Figure 2. Alkali-cordierite : polyhedra arrangement.

Structural evolution with composition

The room temperature diffraction data (Fig. 3) show that alkali substitution logically leads to an increase of the lattice parameters with respect to pure cordierite, mainly as a result of both the expansion of $T(1)O_4$ and $T(2)O_4$ tetrahedra (which become Al-richer)and the presence of additional atoms within the channels.

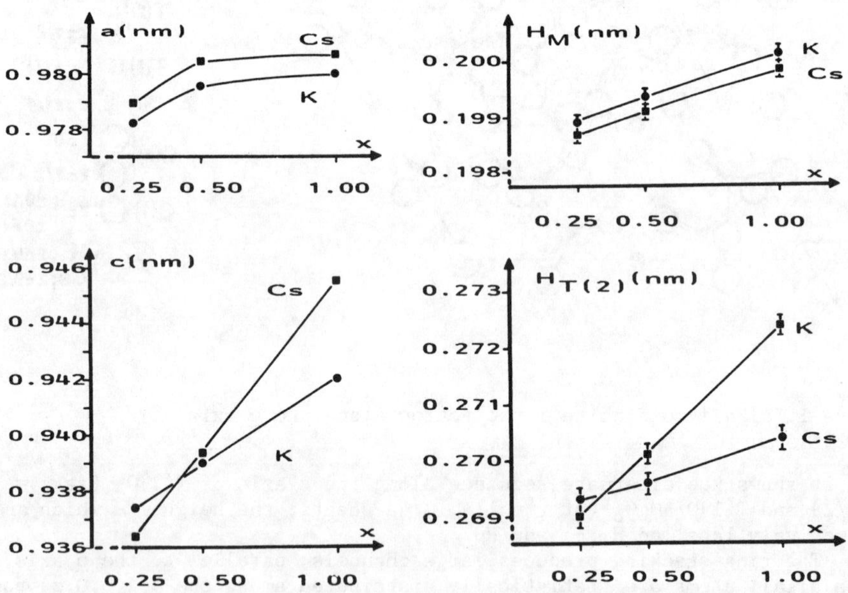

Figure 3. Lattice parameters and sheet heights versus composition.

However two features have to be pointed out.

1°) Whatever the alkali inserted the c parameter increases more rapidly with alkali content than the a parameter does.

As clearly shown by Fig. 4 the alkali atoms inserted within the channels push the T(2) cations away from the centre of these channels. The $[T(1)\text{-}O_4\text{-}MgO_6]$ sheets, already strongly stretched along xOy can no longer accommodate the resulting movements of O(1) atoms whose relaxation therefore mainly occurs along the c axis (H_M and $H_{T(2)}$ increase).

2°) The c parameter increase is far more important for the Cs-substituted cordierites than for the K-substituted cordierites.

Contrary to K^+ cations, Cs^+ cations are located midway to successive $[T(2)O_4]_6$ rings ($z \simeq 0.25$). O(1) atoms participate therefore more strongly to their coordination and $[T(2)O_4]_6$ rings are nearer from each other : actually H_M is systematically lower than for K-cordierites (Fig. 3). As a consequence the $[T(1)_4\text{-}MgO_6]$ sheets are more highly stressed and the only way to accommodate the increasing compression when alkali content increases is to increase strongly their heights ; vertical O(1) movement is therefore increased with respect to the K-substituted cordierites. Moreover such a phenomenon is surely intensified by the stronger $Cs^+\text{-}Cs^+$ repulsive interactions along the channels.

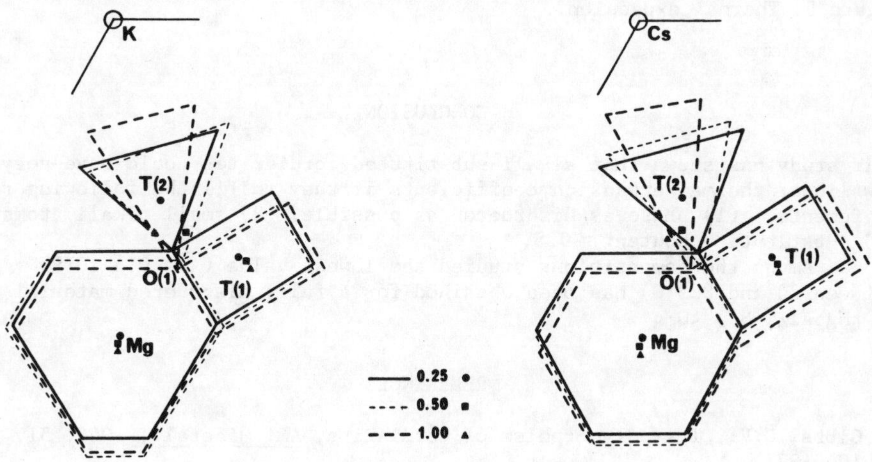

Figure 4. Polyhedra displacements versus composition.

<u>Thermal expansion</u>

It is well known that hexagonal -fully disordered- cordierites have a linear thermal expansion coefficient about two times smaller than the orthorhombic -fully ordered- ones. This phenomenon results from the opposite variation of the <u>a</u> and <u>c</u> hexagonal parameters with increasing temperature : the unit cell slightly expands in the xOy plane and contracts in the Oz direction.

As the temperature increases, if the alkali ion is not too large (K^+), the K-O(2) bond lengths with one of the $[T(2)O_4]_6$ rings increase and the alkali ion moves towards the mean position ($z = 0.25$) between two rings. Until this mean position is not reached, this motion does not inhibit the decrease of the <u>c</u> parameter. On the contrary, if the alkali ion is large (Cs^+) its initial position corresponds to $z \simeq 0.25$: any temperature increase leads the <u>c</u> parameter to increase. The <u>a</u>,<u>c</u> thermal compensation does not occur and therefore the thermal expansion is larger, as shown in Fig. 5.

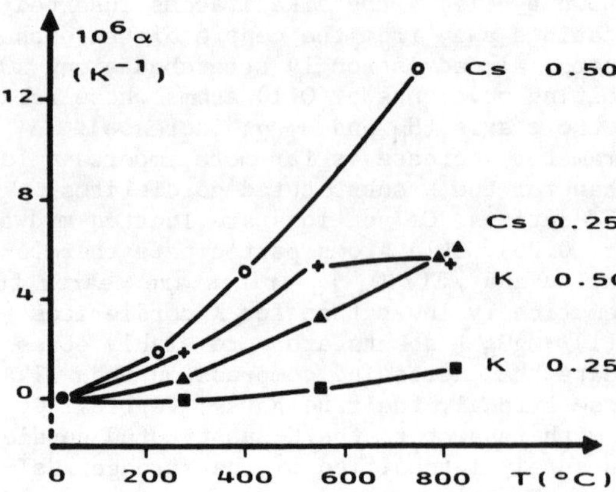

Figure 5. Thermal expansion.

CONCLUSION

This study has shown that alkali-substituted cordierites could have very low linear thermal expansion coefficients if they fulfil the following requirements : (1) phase as disordered as possible, (2) small alkali atoms, (3) substitution content ≃ 0.5.

Among the compositions studied the lowest value ($\bar{\alpha} \simeq 0.4\ 10^{-6}K^{-1}$ between 20 and 825°C) has been obtained for a fully disordered material : $K_{0.5}Mg_2Al_{4.5}Si_{4.5}O_{18}$.

REFERENCES

1. Gibbs, G.V., The polymorphism of cordierite, Am. Mineral., 1966, 51, 1069-87.

2. Maegher, E.P. and Gibbs, G.V., The polymorphism of cordierite, Can. Mineral., 1977, 15, 43-49.

3. Haas, J., Predecki, P., Faber Jr., J. and Hitterman, R.L., Thermal expansion behavior of pure and doped cordierite by time-of-flight neutron diffraction, Adv. X-ray Anal., 1986, 29, 395-405.

4. Mercurio, J.P., Mercurio, D., Frit, B., Kim, H.Y. and Roult, G., High temperature crystal chemistry of a low-thermal expansion material : K-substituted cordierite, in High Tech Ceramics, ed. P. Vincenzini, Elsevier Science Publishers, Amsterdam, 1987, pp. 361-66.

5. Roult, G. and Buevoz, J.L., A time-of-flight spectrometer for elastic neutron scattering on powdered samples, Rev. Phys. Appl., 1977, 12, 581-90.

6. Rietveld, H.M., A profile refinement method for nuclear and magnetic structures, J. Appl. Crystallogr., 1969, 2, 65.

THERMAL COEFFICIENTS OF HIP SINTERED SILICON NITRIDE

KOJI WATARI and KOZO ISHIZAKI
School of Mechanical Engineering
Nagaoka University of Technology
Nagaoka 940-21, Japan

ABSTRACT

The specific heat and thermal diffusivity of HIPped Si_3N_4 with Y_2O_3 and Al_2O_3 additives were measured at different temperatures ranging from 100 to 1300K, by using laser flash technique. The glassy phase quantity could be estimated by the difference of the measured specific heat and the calculated lattice specific heat at low temperatures. HIPped Si_3N_4 with higher β-phase fraction, larger particle size of raw Si_3N_4 and richer Y_2O_3 additive showed higher thermal diffusivity at low temperatures and a larger temperature dependence.

INTRODUCTION

Using Si_3N_4 at high temperatures, it is important to understand thermal coefficients at various temperatures and grain boundary glassy phase quantities. Because the thermal coefficients and the glassy quantity of a sintered Si_3N_4 govern respectively the thermal shock and mechanical resistance.

In this research, the authors measured the specific heat and thermal diffusivity of capsule-HIP sintered Si_3N_4 with Y_2O_3 and Al_2O_3 additives at different temperatures ranging from 100K to 1300K, and discussed about the factor influenced on its thermal coefficients. A method to quantify grain boundary glassy phase is also discussed by using abnormal specific heat of amorphous phase at cryogenic temperatures.

EXPERIMENTAL PROCEDURES

Two types of raw Si_3N_4 powders were used in this study; one (powder L) had an average particle size and specific surface area of 1.4μm and $2m^2/g$, and the other (powder S) 0.3μm and

11.6m^2/g. The raw powders L and S had the same quantities of impurities except for oxygen (0.52 and 1.48wt%). The sintering aids were Y_2O_3 (particle size. 0.9μm) and Al_2O_3 (0.4μm). The total concentration of the sintering aids was fixed at a constant level of 6mol%, but their ratio was changed (a total of 7 different molar proportions, Y_2O_3/Al_2O_3 = 6/0, 5/1, 4/2, 3/3, 2/4, 1/5 and 0/6). The mixed powders were sintered by a capsule-HIP method at temperatures between 1773K and 2023K under 60MPa for 1h in an argon gas. The detailed pre-treatments and sintering conditions were described elsewhere [1].

The HIPped bodies were evaluated by the density and β-phase fraction. Thermal diffusivity and specific heat at temperatures between 100K and 1300K were measured by a laser flash method. The equipments used to evaluate thermal coefficients as well as the detailed procedures were reported elsewhere [2].

RESULTS AND DISCUSSION

Specific Heat

Figure 1 shows the temperature dependence of the measured specific heat of Si_3N_4 sintered at 1973K. The specific heat of the specimens is a sensitive function of the temperature, in particular in the low temperature region.

It is important to obtain the Debye temperature (θ_D) in order to evaluate the physical properties and thermodynamical functions of Si_3N_4. Watari et al. estimated that the sintered Si_3N_4 had Debye temperature of around 1100K, which is calculated by longitudinal and transverse sound velocities [2]. The calculated isochoric specific heat Cv and the experimental isobaric specific heat Cp against T/θ_D were plotted in Figure 2.

As the temperature decreases, the lattice specific heat decreases sharply, in accordance with the Debye T^3 law, which can be applied to estimate the value of Cv within 2% of error for $T \leq \theta_D/10$ [3]. On the other hand, the difference of Cp and Cv tends to zero as the temperature decreases. Therefore the difference between Cp and Cv at low temperatures may be considered the effect of the glassy phase contained in the sintered Si_3N_4.

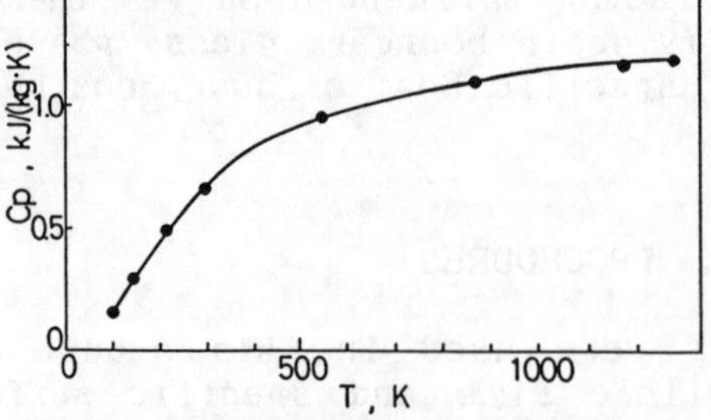

Figure 1. Isobaric specific heat, Cp of HIPped Si_3N_4 (powder S) with 3mol% Al_2O_3 and 3mol% Y_2O_3 vs. temperature, T. HIPped at 1973K under 60MPa for 1h.

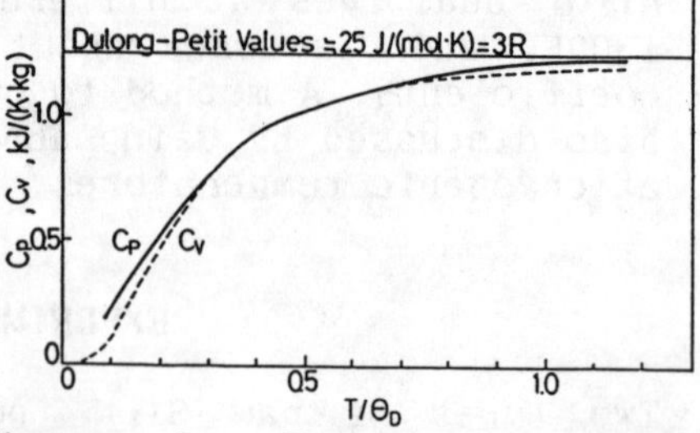

Figure 2. Temperature, T/θ_D dependence of measured isobaric specific heat, Cp and isochoric specific heat, Cv calculated from the Debye theory for HIPped Si_3N_4, using the modified θ_D=1100K. The value of Dulong-Petit rule is also shown.

Sintered Si_3N_4 consists of the major Si_3N_4 phase and minor glassy phase. The glassy phase influences the mechanical and thermal properties of sintered bodies. It is important to know the properties and the amount of glassy phase in the sintered Si_3N_4, but there is not a precise method to measure the amount of glassy phase in sintered bodies. In this research, let us first consider the method to estimate the amount of glassy phase in the sintered bodies.

The specific heat, Cp of sintered Si_3N_4 may be represented by the following equation at a cryogenic temperature.

$$Cp=(C_{SN}\cdot X \quad + \quad Cg\cdot Y)/100 \qquad (1)$$

where, C_{SN} and Cg are specific heats of the pure Si_3N_4 crystal and glass, and X and Y (=100-X) are mol per cent of the pure crystal Si_3N_4 and glassy phase. C_{SN} was approximated as Cv, because Cv is the value of the perfect crystal, which is calculated by the Debye T^3 law. The mol percent of Si_3N_4 phase and glassy phase were calculated from the actual amount of Si_3N_4, Y_2O_3, Al_2O_3 and SiO_2 in mixed powders [4]. Then the amount of glassy phase of HIPped Si_3N_4 (powder S) with 3mol% Y_2O_3 and 3mol% Al_2O_3 was 11.7 mol%. The glass specific heat, Cg can be estimated by using Equation (1), and is shown in Figure 3. The glass specific heat, Cg indicates higher value, compare to that of Cp and C_{SN}.

The relationships between the difference of specific heat and temperature is shown in Figure 4. Cp-Cv means an abnormal specific heat, and is shown below 300K in Figure 4. For estimating the amount of glassy phase, Cp must be measured at low temperatures.

It is known that sintered Si_3N_4 has the oxynitride glass such as Y-Al-Si-O-N in the grain boundary phase. Debye temperature of Si_3N_4 is, therefore, different from that of oxynitride glass in the grain boundary phase. In this research, quisi Debye temperature of oxynitride glasses is estimated from the sound velocity of oxynitride glasses obtained by Messier [5], assuming that oxynitride glasses are crystalline. Table 1 shows the quisi Debye temperature of the oxynitride and other glasses.

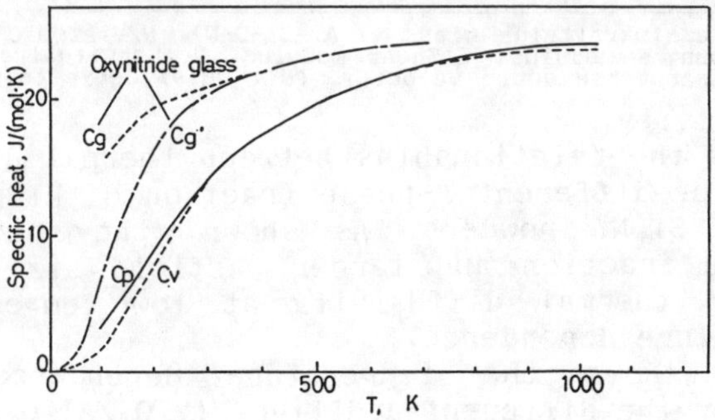

Figure 3. Specific heat vs. Temperature. T. Cg means the specific heat of the glass obtained from Equation (1). Cg' means the isochoric specific heat calculated from the Debye theory for oxynitride glass, using the quisi θ_D'=600K.

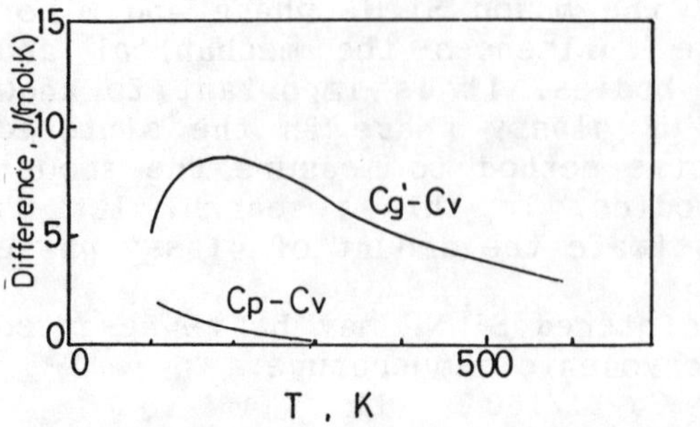

Figure 4. Temperature, T dependence of Cp-Cv and Cg'-Cv. Cp-Cv means the abnormal specific heat of the glass phase. Cg'-Cv indicates the max value at 200K, and the influence of glass can be known at temperature of 200K (= $0.2\theta_D$).

The method to calculate Debye temperature was reported elsewhere [2]. Assuming that quisi Debye temperature of oxynitride glass is 600K, the isochoric specific heat, Cg' of the oxynitride glass is estimated, and then is shown in Figure 3. Also Figure 4 shows Cg'-Cp against temperature. The difference between Cv and Cg' indicates the max value around 200K, and the influence of glass contained in the sintered Si_3N_4 can be known at a temperature of 200K. The amount of glassy phase in the sintered Si_3N_4 can be estimated when the isobaric specific heat Cp measures at 200K. Also there exist the glassy phase in the grain boundary of HIPped Si_3N_4, as the value of Cp-Cv in Figure 4 increased, temperature diminished.

Table 1. Calculation for Quisi Debye Temperature (θ_D') of Oxynitride and Other Glasses.

Specimens	N/V (atm/m³) 10^{29}	E (GPa)	G (GPa)	V_ℓ (m/s) 10^3	V_t (m/s) 10^3	θ_D (K)
N-Glass (6.3 at%N)	0.7862	141	54	7.0	3.8	537
N-Glass (13.3 at%N)	0.8392	157	60	7.5	3.9	569
N-Glass (15.0 at%N)	0.8290	186	72	7.8	4.2	614
Vitreous Silica	0.6624	76	33	--	--	515
$[SiO_2]_n$	0.6615	--	--	6.0	3.8	499
UO_2	0.7333	193	75	--	--	363

N-Glass:Oxynitride glass (Y-Al-Si-O-N), N/V:Atomic density, E:Young's modulus, G:Shear modulus, V_ℓ:Longitudinal sound velocity, V_t:Transverse sound velocity, θ_D': Quisi Debye temperature.

Thermal Diffusivity

Figure 5 shows the relationships between thermal diffusivity and temperature, for different β-phase fraction of HIPped bodies and different raw Si_3N_4 powders. As shown, the specimens with a higher β-phase fraction and larger particle size of raw Si_3N_4 show a higher thermal diffusivity at low temperatures and a larger temperature dependence.

Figure 6 shows the temperature dependence of thermal diffusivity for the different additives (Y_2O_3/Al_2O_3) composition. The specimens containing a higher Y_2O_3 composition had a higher thermal diffusivity at any temperature and larger temperature dependency of thermal diffusivity. The HIPped Si_3N_4 with higher thermal diffusivity at low temperatures were those with

(1) higher β-phase fraction, (2) larger particle size of raw Si_3N_4, and (3) richer Y_2O_3 additive. The detailed discussions of temperature dependence for thermal diffusivity and conductivity in this research were described elsewhere [2].

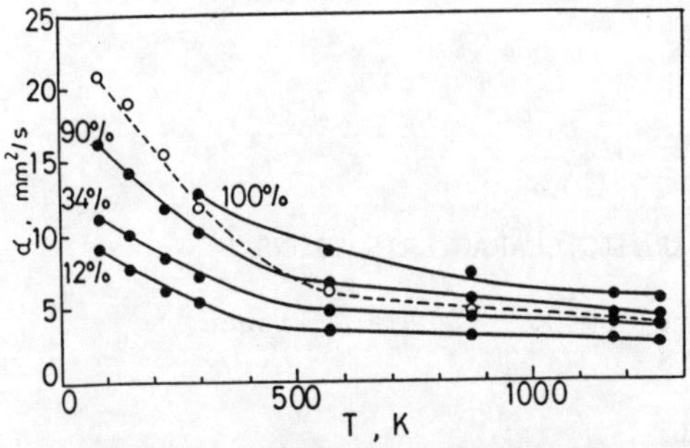

Figure 5. Thermal diffusivity, α vs. temperature, T for different β-phase fractions of HIPped Si_3N_4 (powder S) with 3mol% Y_2O_3 and 3mol% Al_2O_3. HIPping temperatures were 1773, 1823, 1973 and 2023K for β-phase fraction 12, 34, 90, 100% respectively. Open marks indicate the Si_3N_4 (powder L, with 3mol% Y_2O_3 and 3mol% Al_2O_3) sintered at 1973K under 60MPa for 1h.

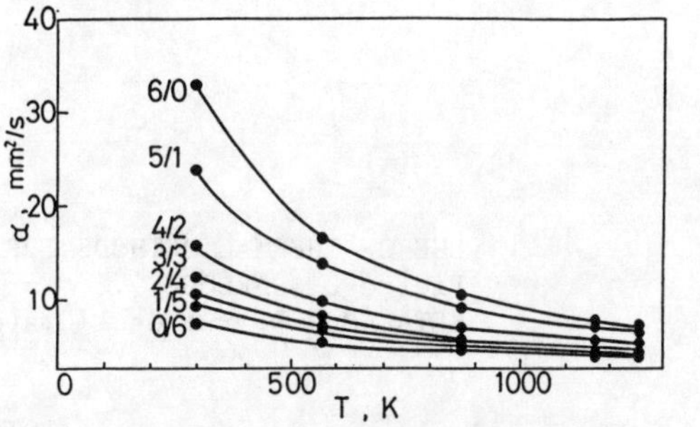

Figure 6. Thermal diffusivity, α vs. temperature, T for different additives (Y_2O_3/Al_2O_3) composition in mol. HIPped at 2023K under 60MPa for 1h using powder S. Relative density and β-phase fraction of HIPped Si_3N_4 were 100% each.

CONCLUSIONS

The specific heat and thermal diffusivity of capsule-HIPped Si_3N_4 with Y_2O_3 and Al_2O_3 6mol% in total as additives were measured at different temperatures ranging from 100 to 1300K. From the present work, the following was revealed.

(1)The isobaric specific heat Cp agreed with the calculated specific heat Cv by the Debye theory, assuming that silicon and nitrogen atoms are lattice dynamically equivalent.

(2)The specific heat of the glass present in the Si_3N_4 grain boundaries was estimated by the authors. Using that result, the glassy phase quantity could be estimated within 5% error.

(3)As the quisi Debye temperature of the oxynitride glass is around 600K, the difference between the Si_3N_4 phase and glassy phase appeared at 200K proving that a glassy phase exist.

(4)The specimens with higher thermal diffusivity at low temperatures were those with higher β-phase fraction, larger particle size of raw Si_3N_4 and richer Y_2O_3 additive. For these specimens the thermal diffusivity showed a lager temperature dependence.

REFERENCES

1. Takata, H., Martin, C. and Ishizaki, K., J. Ceram. Soc. Jpn. Inter. Ed., 1988, 96, 874-881. 2. Watari, K., Seki, Y. and Ishizaki, K., J. Ceram. Soc. Jpn. Inter. Ed., 1989, 97, 170-78. 3. Rosenberg, H.M., Low Temperature Solid State Physics, Oxford Univ. Press, London, 1963, pp.3-7. 4. Tanaka, K., and Ishizaki, K., Sintering of Advanced Materials -Applications of Hot Isostatic Pressing-, Uchida Rokakuho, Tokyo, 1987, pp.16. 5.Messier, D.R. and Broz, A., J. Am. Ceram. Soc., 1982,65, C-123.

USE OF THERMAL METHODS FOR CERAMIC CHARACTERISATIONS

Pierre LE PARLOUER,
SETARAM, 7, rue de l'Oratoire 69300 - CALUIRE, France

ABSTRACT

The thermal analysis methods provide a very interesting tool for the simulation of the heat transformation of ceramics, of their use at high temperature, for the determination of thermal parameters (specific heat, dilatation,...)

INTRODUCTION

The elaboration of ceramics is connected with thermal treatment, like heating, firing, sintering ... Their applications at high temperature and even very high temperature are more and more investigated in many different areas (space and aircraft industries, nuclear research,...)

In both cases, the thermal analysis methods provide a very interesting tool for the simulation of their transformation under heat in various gas conditions, of their limit of use at high temperature, for the determination of thermal parameters (specific heat, heat of transition) and dimensional parameters (expansion coefficient).

Among the thermal methods adapted for such characterisations, thermogravimetry with simultaneous differential thermal analysis is the most popular and also the most efficient. The performances of such a TG-DTA instrument are again upgrated when coupling an evolved gas analysis technique (GC, MS, FT-IR). High temperature calorimetry is also very useful for the determination of specific heat for ceramics. If the mechanical behaviour of the materials has to be known, high temperature dilatometry has to be used for measuring expansion coefficient, but also for simulating powder sintering.

Thermal methods available for ceramic characterisations

In the field of ceramic characterisations, thermogravimetry is very well known and largely used. Thermogravimetry measures the mass variation of a sample when it is heated, cooled or kept at constant temperature. Built around mainly a sensitive microbalance and a furnace, the thermogravimetric instrument is well adapted to measure mass losses due to dehydration, dehydroxylation, decomposition, reaction, or mass gains due to gas adsorption, and to know the corresponding temperatures of transformation. Most of these phenomena occur during the firing of ceramics, the heating of raw materials, or even during the sintering process after the binder decomposition.

Thermogravimetry allows to easily simulate these thermal treatment operations, and to optimise the manufacturing conditions. Thermogravimetry provides only an information on the mass evolution of the sample. The combination with differential thermal analysis enables to have more data especially on the thermal effects associated with the mass variations and more, to detect phase changes (melting, crystallisation, solidification) in the sample. Such a combined system can be operated from - 150°C to 2400°C. All types of ceramics (oxide and non oxide ceramics, glass, graphite, cement) are investigated with such an instrument. The thermal behavior of various ceramics at high and very high temperature is easily evaluated.

It is possible again to upgrade the performances of such a TG-DTA instrument when coupling an evolved gas analyzer. When a mass loss occurs in the sample, gases are emitted. Their analysis by mass spectrometry or FT-IR enables to identify the different components and to understand the reaction or the transformation. During the firing of ceramics or to investigate ceramic thermal stability such a technique is especially powerful to understand the different phenomena that are occuring in the sample.

For measuring different thermal parameters like specific heat, heat of transition, high temperature calorimetry is used. These data are needed for the ceramic engineering, and also for their use in extreme conditions of temperature. When a sample is heated, a dimensional change also occurs due to dilatation. Dilatometry is in this case the method used for measuring expansion coefficients up to very high temperature (2200°C). Though the determination of dilatation coefficients for ceramic is very important, dilatometry is also used for the simulation of the sintering process. From the dilatometric test, many information are get to optimise industrially this operation and to obtain the best mechanical properties for the ceramics.

Examples of ceramic characterizations with the thermal methods

The family of ceramics includes different types of products : traditional ceramics, oxide ceramics, non oxide technical ceramics, graphite and derivates. For each division, thermal methods find various applications in a large range of temperature.

1. Traditional ceramics

The firing of traditional ceramics, like fired clay, faience, sandstone, porcelaine is easily investigated with the TG-DTA technique. For example the firing of kaolin (Figure 1) is simulated with such a method : dehydration at low temperature, followed by dehydroxylation between 450°C and 750°C, corresponding to the formation of metakaolin. The DTA signal at higher temperature (between 950°C and 1000°C) shows an exothermic effect corresponding to the transformation of metakaolin into alumina and mullite.

2. Oxide ceramics

Oxide ceramics include refractory type ceramics, glass and cements. A ceramic is considered as refractory if its melting temperature is higher than 1500°C and has a good mechanical behaviour at high temperatures. These thermal and mechanical properties are precisely characterised using thermal methods on ceramics like alumina, silica, zirconia, uranium oxide, ...

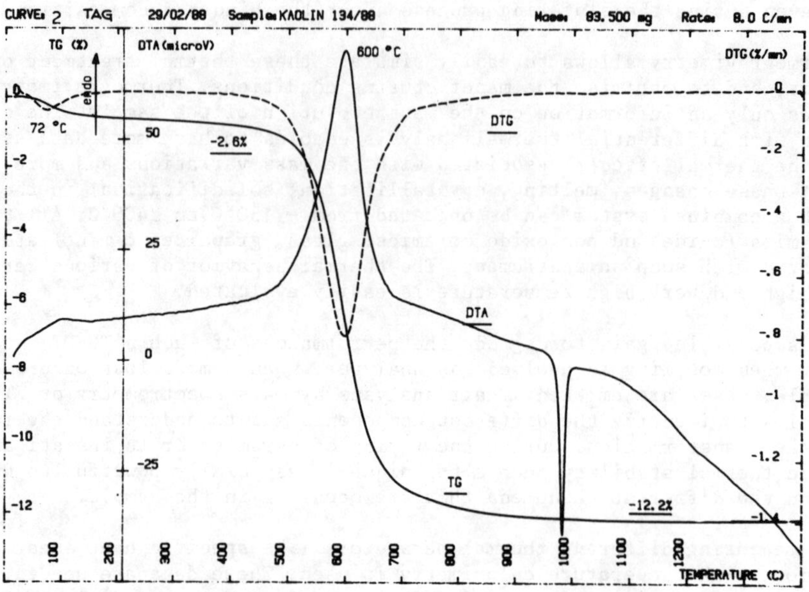

Fig. 1 - Thermal analysis of Kaolin (TG - DTA method)

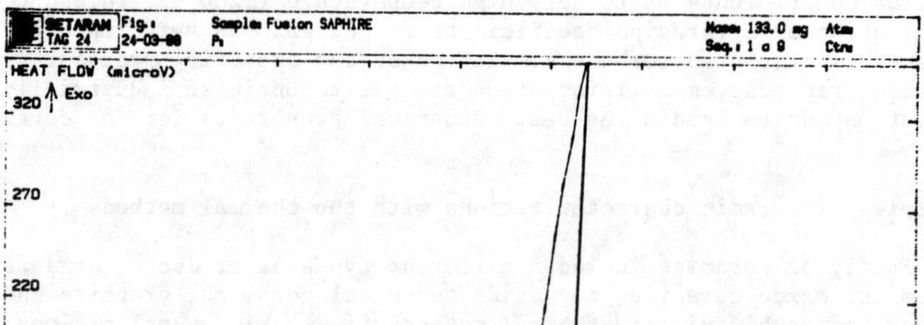

Fig. 2 - Melting and solidification of saphire (DTA method)

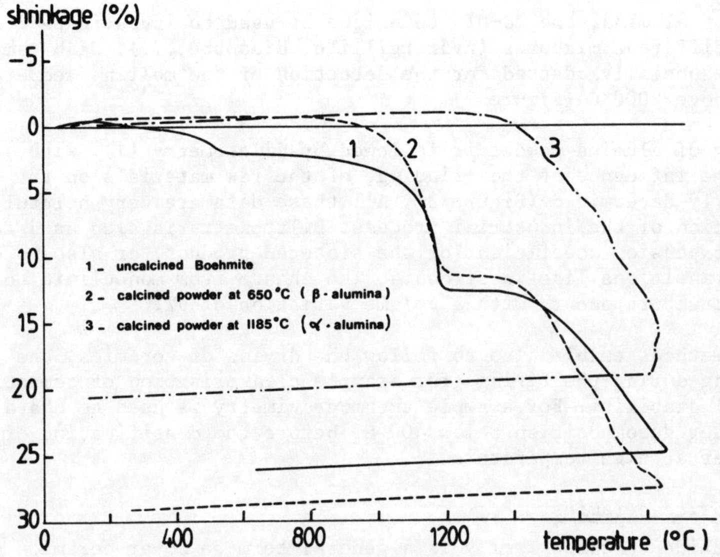

Fig. 3 - Dilatometric investigation of alumina skrinkage

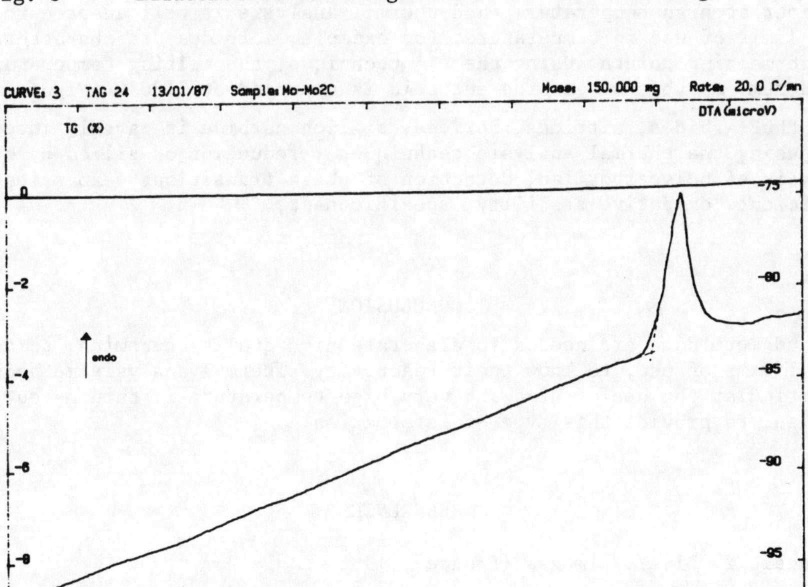

Fig. 4 - Melting of molybdenum-molybdenum carbide eutectic

In the case of alumina, the TG-DTA technique is used to investigate its formation from different mixtures (hydrargillite, diaspore,...). High temperature DTA is especially adapted for the detection of the melting temperature of alumina above 2000°C (Figure 2).

The sintering of alumina powder is followed by dilatometry (1). With such a technique, the influence of the grinding, of the raw materials on the skrinkage is quickly determined (Figure 3). All these data are very helpful for the optimisation of the industrial process. Dilatometry is also used to measure the expansion coefficient of the sintered product, or also to detect phase transitions like in zirconia. The change from monoclinic to tetragonal structure occurs with a volume variation of 9%.

The thermal methods enable also to follow the drying of ceramics, the reactions occuring during the firing (for example clinkerisation of cements), their thermal stability. For example thermogravimetry is used to characterize the binder decomposition below 400°C, before the densification of the ceramic powder at high temperature.

3. Non oxide ceramics

Non oxide ceramics family is a general term to cover ceramics made from carbides, nitrides, metallic borides,... They have especially a good behaviour at high temperature, and thermal analysis is well adapted to fix their limit of use in temperature. For example, carbides are characterised by high melting points. Using the DTA technique, the melting temperature of molybdenum -molybdenum carbide eutectic is measured at 2180°C (Figure 4).

Among the carbides, nitrides, borides, silicon carbide is largely investigated using the thermal analysis techniques : reduction of silica by carbon, pyrolysis of polycarbosilan, detection of phase transitions, expansion coefficient, oxidative stability, specific heat,....

CONCLUSION

More and more data are needed to elaborate high quality ceramics, to define their limits of use, to know their reactivity. Thermal analysis methods, in particular the use of high and very high temperature is turning out very efficient to provide this type of information.

REFERENCES

1. Thesis Y. BIGAY, Limoges (France), 1986

DETERMINATION OF THE THERMAL CONDUCTIVITY OF POLYCRYSTALLINE ALUMINA AND YTTRIUM ALUMINIUM GARNET IN THE TEMPERATURE RANGE 300 -1500 K

C. Wijenberg
Philips Lighting B.V.
P.O. Box 80020, 5600 JM Eindhoven, The Netherlands

ABSTRACT

The thermal conductivity of polycrystalline alumina (PCA) and yttrium-aluminium garnet (YAG) has been determined by means of the Ångström method at temperatures from 300 to 1500 K. Thermal conductivity data have been collected with tube-shaped PCA and YAG samples of dimensions identical to those of ceramic discharge envelopes as applied in high-pressure sodium discharge lamps. The thermal conductivity of PCA and YAG at 1300 K was determined to be 10 ± 1 and 3.0 ± 0.4 J/mKs, resp. At temperatures exceeding 1000 K also the photon conductivity contributes to the overall thermal conductivity.

INTRODUCTION

In order to describe heat fluxes and to predict temperature gradients in ceramic discharge tubes applied in high-pressure sodium (HPS) lamps [1], the thermal conductivity (and its temperature dependency) of the ceramics should be known accurately. Unfortunately, literature data on the thermal conductivity of PCA show a spread which is too large for modelling purposes [2]. Recently, YAG has been proposed as an alternative translucent discharge tube material for HPS lamps [3]. Only a very limited amount of thermal conductivity data for YAG has been reported in the temperature region of interest (900 - 1500 K) [2,4].

The thermal conductivity of translucent ceramics at temperatures exceeding 1000 K is known to depend on the presence of scattering centers for photons such as pores [5]. The pore size distribution and the total pore volume is strongly related to the optical characteristics of the ceramic body [6]. Therefore, measurement of the thermal conductivity should be preferentially done with samples of well-defined transmittance. Since the microstructure of MgO-doped alumina can be a function of the distance from the surface [7] (and thus the overall optical characteristics a function of the dimensions of the ceramic body), the safest approach in collecting thermal conductivity data consists in performing measurements with samples of the same geometry as used in the application of interest.

The described experimental technique for determination of the thermal conductivity (the Ångstrom methöd) allows the usage of tube-shaped samples. Thermal conductivity data have been collected for PCA and YAG samples of well-defined geometry and optical characteristics.

METHODS AND MATERIALS

The Ångström method employs a rod or tube with a heat source generating a sinusoidal temperature variation at one end of the tube. The other side is held at constant temperature T. According to ref. [8] the one-dimensional equation for heat flow in a sample may be written as:

$$d\Theta/dt = D \cdot d^2\Theta/dx^2 \qquad (1)$$

with $D = K/C_p \cdot \rho$ (2)

where Θ = T_1 - T
T = temperature of the surroundings of the sample
T_1 = temperature of the rod at any position and time
D = thermal diffusivity
K = thermal conductivity
C_p = heat capacity at constant pressure
ρ = specific density

Equation (1) can be solved for two places X1 and X2 on the tube. The following solution:

$$D = \pi \cdot f \cdot L^2/\Phi \cdot \ln(d1/d2) \qquad (3)$$

is obtained when the following conditions are obeyed [8]:

			Experimental:
(i)	Θ/T	$\ll 1$	< 0.01
(ii)	$(dD/dT)/D$	$\ll 1$ (K^{-1})	< 0.004 (K^{-1})
(iii)	Bi number	$\ll 1$	< 0.05

where: f = frequency of the temperature waves
L = distance between X1 and X2
Φ = phase difference between the temperature waves at X1 and X2
d1,d2 = amplitudes of the temperature waves at X1 and X2, resp.
Bi = Biot number, $Bi = h \cdot d/K$
h = surface heat transfer coefficient
d = wall thickness

On the ceramic tube (See Table 1 for sample and material characteristics) three coiled wires are attached with a sealing ceramic (5.6 MgO, 38.6 CaO, 8.7 BaO, 45.4 Al_2O_3, and 1.7 B_2O_3 (numbers indicate wt.%)). Figure 1 shows the assembly of the discharge tube and coiled wires (PtRh 10%, wire diameter 0.2 mm, inner coil diameter 0.4 mm, 3 turns/mm, total wire length per coil 250 mm). The heat capacity of Al_2O_3 and YAG was taken to be 1.044 + 1.74E-4 T - 2.79E+4/T^2 [10] and 0.761 + 8.75E-5 T - 1.70E+4/T^2 J/g K respectively. The heat capacity of YAG was calculated from Y_2O_3 and Al_2O_3 data [10] using the addition rule.

TABLE 1
Sample Characteristics

material	outer diam. (mm)	inner diam. (mm)	length (mm)	total transmission (%)	in-line* [1] (%)	relative density (%)	theoretical density (g/cm^3)
PCA	8.9	7.4	100	99	39	> 99.5	3.98
PCA	6.3	4.8	80	99	39	"	
YAG	9.1	7.5	80	91	7	94 98	4.56
YAG	5.1	3.4	48	99	34	> 99.5	

* aperture 0.0015 strad.

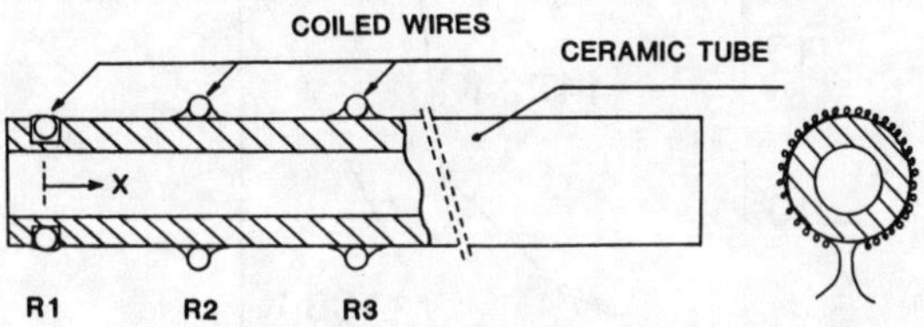

Figure 1. Sample assembly (longitudinal and perpendicular cross-section)

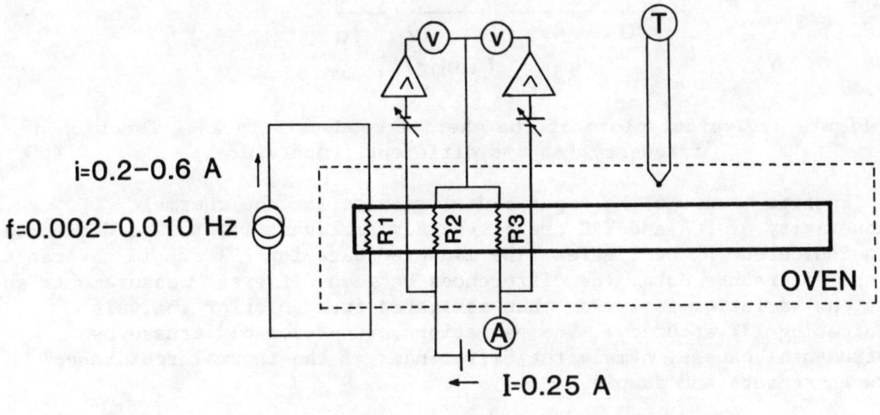

Figure 2. Experimental set-up; electrical parts.

The first coil (R1) is mounted in a groove close to the tube end in order to minimize the thermal resistance between coil and tube and to minimize loss of heat in the -x direction. So-called temperature waves are generated at R1 which is connected to an AC current source (fig. 2). The resistance thermometers R2 and R3 are connected to a DC current stabilizer. The temperature waves injected at R1 generate a sinusoidal variation of the resistances of R2 and R3. As a consequence, a DC voltage with superimposed AC is generated in R2 and R3. When R2 and R3 are of identical electrical resistance and have a similar thermal resistance to the ceramic tube, the voltage phase difference and the amplitude decrement between the coils equals the temperature phase difference and amplitude decrement, respectively. With the latter data the thermal conductivity can be calculated according to equations (2) and (3).

RESULTS AND DISCUSSION

Typical temperature amplitudes measured with R2 en R3 are 0.2-1.5 K and 0.05-0.3 K resp. The sensitivity of the resistance thermometers was found to be 200-400 $\mu V/K$. Figure 3 shows that the thermal conductivity strongly depends on the applied frequency. Frequency-dependent values for K were calculated using equations (2) and (3), and the thermal conductivity was determined by extrapolation to zero frequency. It is also possible to calculate the conductivity from the amplitude ratios at two different frequencies [8]. Results obtained by extrapolation to zero frequency did not differ more than 30 % from results obtained from the amplitude ratios at two different frequencies.

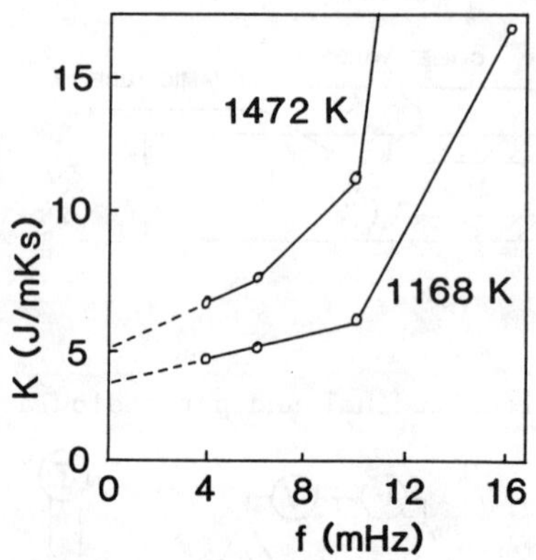

Figure 3. Typical plots of the thermal conductivity as a function of frequency (at two different temperatures)

In figs. 4 and 5 the experimental results for the thermal conductivity of PCA and YAG are given. Also relevant literature data have been indicated in the figures. The measured data for PCA are in the range of the literature data. The differences between different measurements show that the variation is larger than estimated from an error analysis (indicating 13% accuracy). The variations are presumably caused by instrumental causes, namely the differences of the thermal resistance between sensors and sample.

The presently obtained thermal conductivity data for YAG are lower than the literature data. This could be explained by differences in material characteristics and usage of different C_p-data in the calculations. Note that due to parasitic radiation, conduction, and convection, thermal coupling between R1 and R2, R3 can occur. Due to such a coupling systematically too high values for the thermal conductivity are measured. In the present investigation conduction and convection by gas was eliminated by measuring at a pressure < 0.01 Pa. The radiation coupling was minimized by embedding the sample in ZrO_2 powder (average grain size 4 μm). ZrO_2 powder was chosen because it is applicable at high temperatures and the thermal conductivity of loosely packed powder is expected to be less than 1% of the value for alumina [9]. At temperatures exceeding 1000 K a difference appears between translucent and opaque YAG. This is due to different mean free paths for photons.

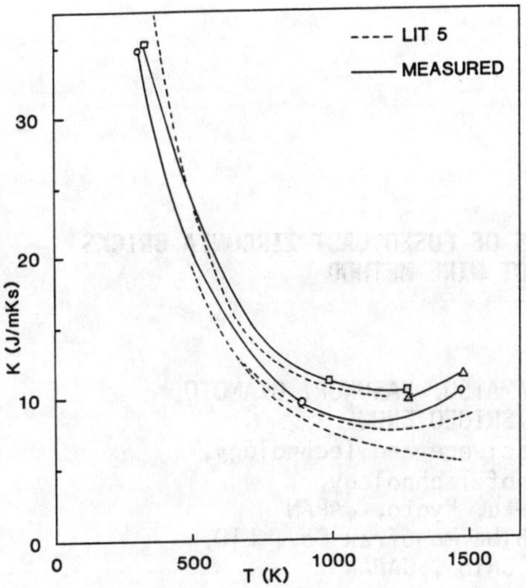

Figure 4. Thermal conductivity of PCA as a function of T.

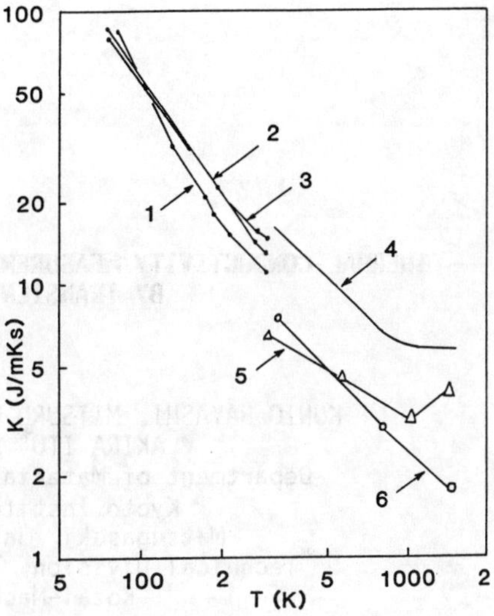

Figure 5. Thermal conductivity of YAG as a function of T.
1 - 4: lit [2,4]; 5: translucent YAG; 6: opaque YAG

ACKNOWLEDGEMENTS

I want to thank my colleagues Mr. T. Duisters for his assistance with the experiments and Dr. A.J.H.M. Kock for his help in preparing this manuscript.

REFERENCES

[1] J.J. de Groot and J.A.J.M. van Vliet, in "The High Pressure Sodium Lamp", Deventer, Kluwer Technische Boeken B.V., 1986.
[2] Y.S. Touloukian e.a., "Thermo Physical Properties of Matter", N.Y., Plenum Press, 1970.
[3] F.J.C.M. Toolenaar and G. de With, Proc.Brit.Ceram.Soc. 37 (1986) 241
[4] Ts.Ts. Dambiev, Depos. Doc. 1981, VINITI 2570-2581.
[5] W.D. Kingery, H.K. Bowen, D.R. Uhlmann, in "Introduction to Ceramics", N.Y., J. Wiley & Sons, 1976.
[6] J.G.J. Peelen and R. Metselaar, J.Appl.Phys. 45 (1974) 216.
[7] R.D. Bagley and D.L. Johnson, Adv.Ceram. 10 (W.D. Kingery, ed.), The Amer.Ceram.Soc., 1984, p. 666.
[8] K.D. Maglic, A. Cezairliyan, V.E. Peletsky, "Compendium of Thermophysical Property Measurement Methods", N.Y., Plenum Press, 1984.
[9] B. Schulz, High Temp.-High Press. 13 (1981) 657.
[10] I. Barin, O. Knacke, "Thermochemical Properties of Inorganic Substances", Berlin, Springer-Verlag, 1973.

THERMAL CONDUCTIVITY MEASUREMENTS OF FUSED CAST ZIRCONIA BRICKS BY TRANSIENT HOT WIRE METHOD

KUNIO HAYASHI, MITSURU WAKAMATSU, YASUNORI OKAMOTO,
AKIRA ITO* AND SHIGEO ENDO*
Department of Materials science and Technology,
Kyoto Institute of Technology,
Matsugasaki, Sakyo-ku, Kyoto, JAPAN
*Technical Division, Toshiba Monofrax Co., LTD.
Kozai-Machi, Chiba, JAPAN

ABSTRACT

The thermal conductivity of the fused cast refractories with various amounts of zirconia and alumina was measured by the transient hot wire method in the range from room temperature and 1000°C. At room temperature, the thermal conductivity of bricks containing large amounts of zirconia was to be higher than that of high alumina bricks in spite of the fact that a alumina crystal has higher thermal conductivity than a zirconia. The thermal conductivity of bricks containing large amounts of alumina decreased slightly with increasing temperature, while that of high zirconia bricks decreased gradually with an increase in temperature.

INTRODUCTION

Fused cast refractories have come into use in place of conventional bonded bricks in glass or metal making plants because of the good resistance to corrosion by molten glass and slag.

Especially, the fused cast refractories with ZrO_2, Al_2O_3 and SiO_2 have the structure of zirconia and alumina eutectic crystals developed in lath form and also a large quantity of zirconia crystals dispersed evenly throughout so that their resistance to molten glass have been increased to the highest level.

In this paper, the thermal conductivity of fused cast refractories with various amounts of zirconia and alumina which is one of the dominating cause of their thermal shock resistance, was measured by the transient hot wire method ranging from room temperature to 1000°C and the relation between their thermal conductivity and composition of refractories was discussed.

EXPERIMENTAL

Preparation of specimens for measuring thermal conductivity

The raw materials were well mixed for preparing four types of specimens and melted in the automatically controlled furnace under suitable conditions for casting, and the fused materials were cast into the molds to obtain the rectangular form as shown in Fig. 1. After the cooling period has been

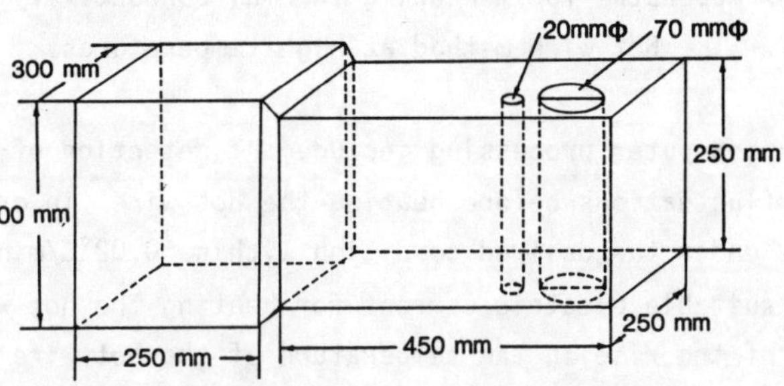

Figure 1. Cast sample configuration for physical and thermal properties.

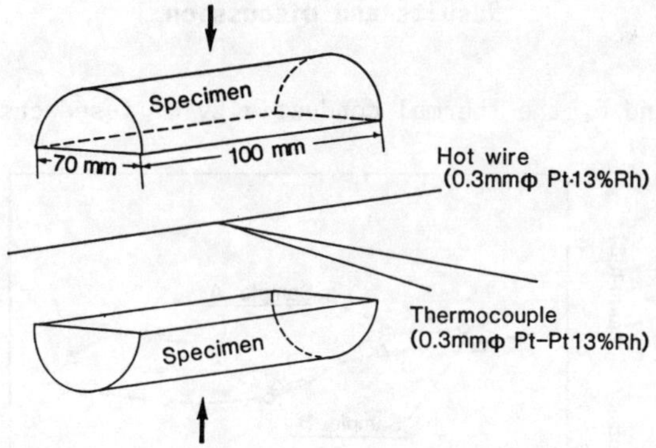

Figure 2. Schematic showing of specimen and setting of probe.

completed in the annealing powder, the semicylindrical specimens(70mmφ x100 mm) for measuring the thermal conductivity shown in Fig. 2 and the cylindrical one (20mmΦ) for physical properties were cut from the each casting. The chemical composition and physical properties are given in Table 1 and the microstructure of samples was shown in Fig. 3.

4. Measurements of thermal conductivity was made up to 1000°C using the apparatus as shown in Fig. 4.

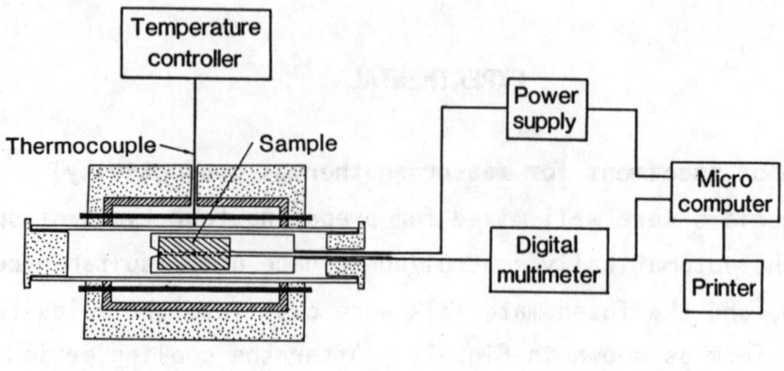

Figure 4. Apparatus for measuring thermal conductivity by transient hot wire method at high temperatures.

The microcomputer processing included: 1)detection of specimen temperature fluctuations before heating the hot wire, in order to make measurements under the desired condition within ±0.02°C/min.[1], 2)the choice of a suitable electric current for heating the hot wire[1], 3)the measurement of the rise in the temperature of the hot wire and 4)the computation of thermal conductivity data[2].

Results and discussion

In Fig. 5 and 6, the thermal conductivity of fused cast samples with

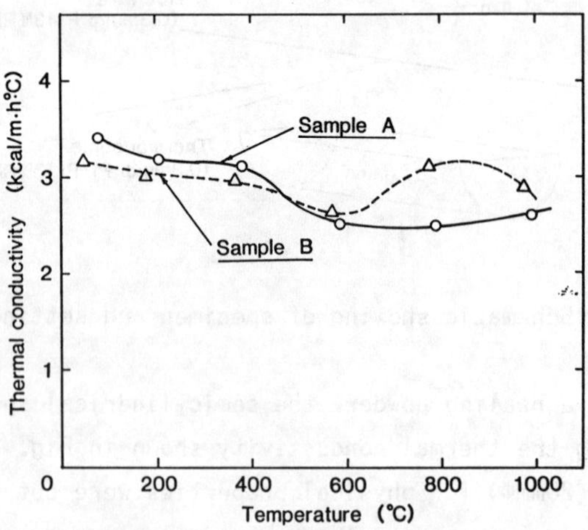

Figure 5. Temperature dependence of thermal conductivity of sample A and B.

TABLE 1

Chemical composition and physical properties of samples used

Sample	Chemical composition (wt%)				Bulk density (g/cm^3)	Apparent porosity (%)	Glass phase (wt%)
	ZrO_2	Al_2O_3	SiO_2	Na_2O			
A	25	57.2	16	1.3	3.67	0.54	20
B	33	50.0	16	1.3	3.76	0.44	21
C	50	36.2	12	1.3	4.17	0.54	15
D	60	26.3	12	1.3	4.33	0.34	9

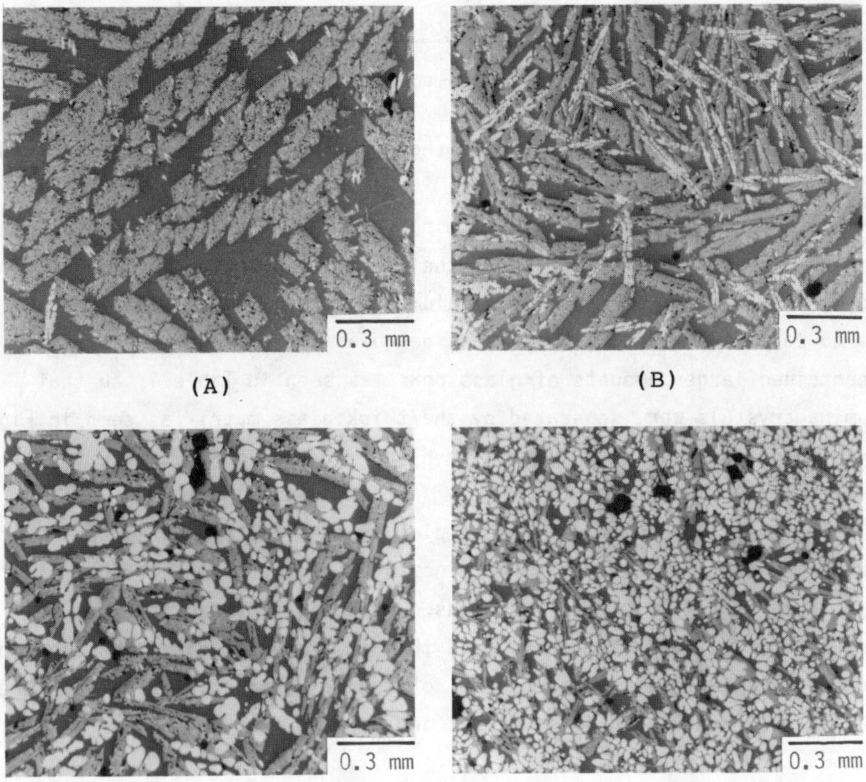

Figure 3. Microstructure of fused cast samples.

Apparatus and procedure for measuring thermal conductivity

The probe consisting of the hot wire (0.3 mmϕ Pt13%Rh wire) on which the hot junction of a 0.3 mmϕ Pt–Pt13%Rh thermocouple to sandwiched two semi-cylindrical specimens as shown in Fig. 2. The sample with the probe was set in a tube furnace controlled by a program controller as shown in Fig.

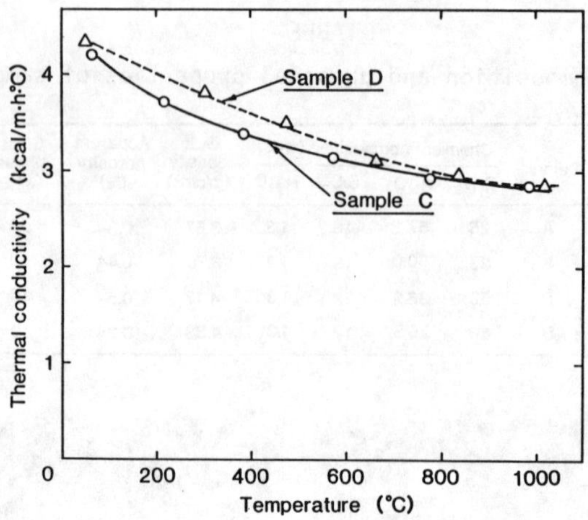

Figure 6. Temperature dependence of thermal conductivity of sample C and D.

various amounts of zirconia and alumina was plotted as a function of temperature. At room temperature, the thermal conductivity of samples A and B containing larger amounts of alumina as compared to samples C and D was to be lower values. It was considered as the reason that samples A and B contained larger amounts of glass phase as seen in Table 1, so that alumina crystals were separated by the thick glass matrix as seen in Fig. 3 in spite of the fact that these have higher thermal conductivity than that of zirconia.

The thermal conductivity of samples A and B containing large amounts of alumina decreased slightly with increasing temperature, while that of high zirconia samples C and D decreased gradually with an increase in temperature. It was also considered that an amount of glass phase having the positive temperature coefficient of thermal conductivity[3] was larger in samples A and B than in samples C and D.

REFERENCES

1. Hayashi, K., Bull. Ceram. Soc. Japan, 1976, **11**, 994-1003.
2. Hayashi, K., Tsujimoto, S., Nishikawa, T. and Imamura, Y., J. Ceram. Soc. Japan, 1986, **6**, 595-600.
3. Kingery, W.D., Bowen, H.K. and Uhlman, D.R., Introduction to Ceramics, John Wiley & Sons, New York, 1976, p.643.

CHARACTERIZATION OF PURE ZrO_2 POWDERS PROCESSED FROM REACTIONS IN MOLTEN SALT MIXTURES

Ravelle-Chapuis R.°, Blanchin M.G.;
Jebrouni M.[+], Durand B.[+],
Département de Physique des Matériaux,U.A. 172,
[+]Laboratoire de Chimie Minérale III,U.A. 116,
Université Lyon I, 43 Bd du 11 Novembre 1918,
69622 VILLEURBANNE Cédex FRANCE.

ABSTRACT

Ultrafine powders of pure zirconia (ZrO_2) were prepared by chemical reaction at low temperatures in molten salt mixtures of nitrates (KNO_3,$NaNO_3$). The reaction product was then crystallized by annealing at increasing temperatures. The structure of the phases present in the powders were identified by X-Ray diffraction spectrometry and electron diffraction. The basic product prepared at 260°C contained both amorphous and monoclinic phases (with traces of tetragonal phase). At increasing calcination temperatures the percentage of tetragonal phase proved to increase and then decrease until the powder became entirely monoclinic at about 1000°C. The powders particles were characterized by conventional transmission electron microscopy (CTEM). Those observations revealed agglomerates and agregates of ultrafine primary particles (mean diameter < 50 nm).

INTRODUCTION

Pure zirconia (ZrO_2) has three polymorphs ,under ambient pressures:

```
            950°C           2370°C      2680°C
MONOCLINIC <---> TETRAGONAL <---> CUBIC <---> MELT     (1)
            1170°C
```

At room temperature, the stable form is monoclinic.
The transformation t --> m is considered to be martensitic in the bulk , and it is accompanied by a volume increase of 3% at 950°C to 4.9% at room temperature. (2)
This transformation causes cracking and subsequent failure of fabricated products.
However, the tetragonal structure exists at room temperature in microcristals, and depends on crystallites size. (3)

EXPERIMENTAL PROCEDURES

Samples preparation

Hydrous zirconium oxychloride ($ZrOCl_2,8H_2O$) was first dessicated under vacuum at 150°C for an hour, in molten salts mixtures ($NaNO_3,KNO_3$), and then heated at 260°C during 1h30mn. The obtained product is called A.

A was then calcinated at different temperatures, in order to observe the transformation t --> m .

The eight samples obtained are called:

A: precursor (prepared at 260°C) E: A calcinated at 700°C
B: A calcinated at 400°C F: ________ 800°C
C: ________ 500°C G: ________ 900°C
D: ________ 600°C H: ________ 1000°C

X-Ray analysis

The phases, crystallite sizes and microstrains of calcined samples were determined by X-Ray diffraction analysis.For the monoclinic-tertagonal ZrO_2 system, the volume fraction of monoclinic phase Vm was calculated from:

$$Vm = PXm / (1 + (P - 1)Xm)$$

Where $Xm = Im(11\bar{1}) + Im(111) / (Im(11\bar{1}) + Im(111) + It(101))$, (Im and It are the integrated intensities of monoclinic & tetragonal peaks), and $P = 1.311 \mp 0.004$. (4).

The integrated intensities and the position of peaks were determined by a fitting program using a pseudo - Voigt function (created by Juan RODRIGUEZ ,ILL Grenoble)

Crystallite sizes were calculated by the analytical approximation method of de KEIJSER. (6).

So we have :

$$D = \frac{\lambda}{B_c{}^f \cos(th)} \quad \text{and} \quad e = \frac{B_g{}^f}{4\tan(th)}$$

Where D is the apparent crystallite size (A), $\lambda = 1.5405981$ Å, e the strain, $B_c{}^f$ and $B_g{}^f$ the integrated breaths, and 2th the position of each peak.

CHARACTERISATION BY CONVENTIONNAL TRANSMISSION ELECTRON MICROSCOPY (CTEM)

The samples were dispersed and deposited directly onto a carbon film on copper grids.

Each sample was examinated: - at low magnification for determination of the agregation state of the powder particles.

- at higher magnification, in order to study grain sizes and shapes ,and the nature of grain boundaries.
- Dark fields images were obtained in order to determine crystallite size for sintered particles .

For each sample ,electronic diffraction patterns were used to determine structure of different phases .

RESULTS

X-RAY analysis

Diffractions profiles of samples A,B,C,D, and E are shown on Fig. 1. The XRD fitted profile of sample H is shown on Fig. 2.

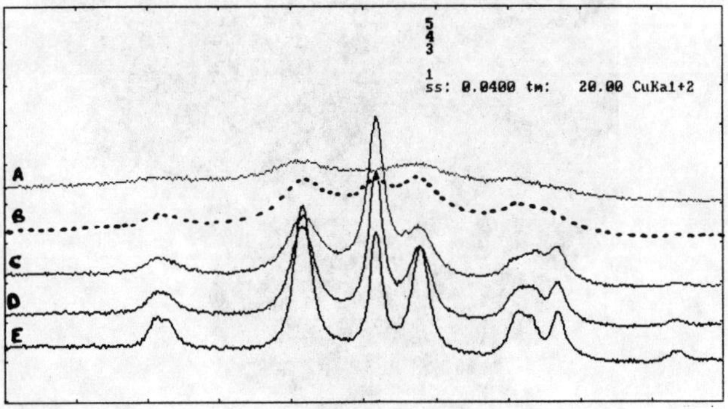

Figure 1. XRD profile of A,B,C,D and E.

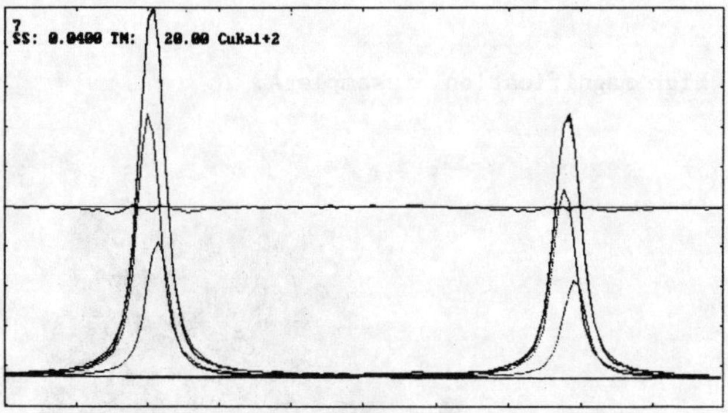

Figure 2. XRD fitted profile of H.

The observed structures are :

A : m + amorphous	E : m + t
B : m + t	F : m + t
C : m + t	H : m
D : m + t	

The relative contents of monoclinic phase and apparent crystallite size are indicated on Table 1.

CTEM

Low magnifications observations show that all powder samples were

formed of similar agglomerates which size between 0.2 and 1.5 μm. At higher magnification ,we can see the grain size :see Figs .3 and 4 ,and Table 1.
The diffractions patterns showed mixtures of tetragonal and monoclinic phases in good agreement with XRD determination results.

Figure 3. High magnification of sample A.

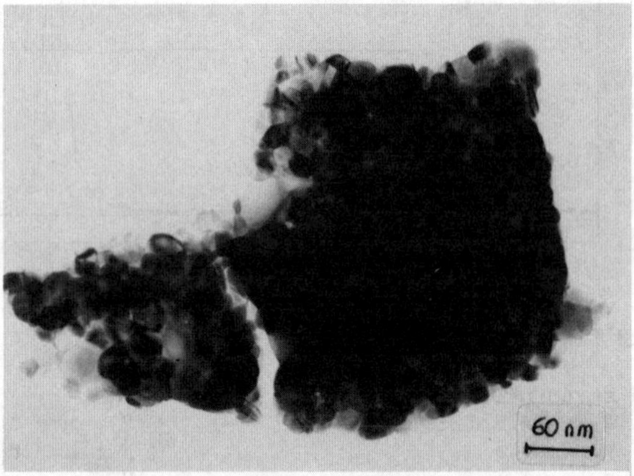

Figure 4. High magnification of sample H.

TABLE 1

Annealing temperature (°C)	Crystallite size(Å) (XRD) Dm(11̄1)	Dm(111)	Dt(101)	(CTEM) D	Vm: % monolinic	% $e*10^4$ Strain
0	28	68	-	55	92	230
400	53	107	48		78	115
500	73	144	110	100	61	96
600	107	148	140		70	78
700	139	203	177	215	80	49
800	215	223	257		91	26
1000	640	587	-	600	100	15

DISCUSSION AND CONCLUSION

At the beginning of calcination ,only amorphous and monoclinic phases with a trace of tetragonal phase were present, and the transformation was from amorphous to tetragonal, which could explain the appearence of tetragonal phase. But, from 500°C, the process t--->m started, so the percentage of tetragonal phase decreased. (7)
For calcination temperature from 400°C up to 700°C, crystallite sizes for both phases were similar. While crystallite size steadily increased, the microstrain decreased (see Table 1). At 800°C, crystallite size of the tetragonal phase was larger than that of the monoclinic phase.At 1000°C ,only monoclinic phase was present .
We can see that crystallite sizes determined by XRD were in good agreement with CTEM observations.

REFERENCES

1. Heuer, A.H. and Hobbs, L.W.,Advanced in ceramics. Vol. 3.Science and technology of Zirconia 1.1981
2. Kriven,W.M. Possible Alternative Transformation Tougheners to Zirconia: Crystallographic Aspects, J. Am. Ceram. Soc.,1988 ,71 (12) 1021-30.
3. Garvie, R.C. Stabilisation of the Tetragonal Structure in Zirconia Microcrystals, J. Phys. Chem. Vol. 82, No. 2, 1978,218-24.
4. Toraya, H. ,Yoshimura, M. and Somiya, S. Calibration Curve For Quantitative Analysis of the Monoclinic-Tetragonal ZrO_2 System by X-Ray Diffraction.Com. Am. Ceram. Soc. 1984 ,C-119,C-121.
5. Enzo, S. ,Fagherazzi,G., Benedetti A., and Polizzi S.,A Profile-Fitting Procedure for Analysis of Broadened X-Ray Diffraction Peaks. J. Appl. Cryst. (1988). 21, 536-42.
6.de Keijser, T.H.,Langford, J.I. ,Mittemeijer, E.J. and Vogels, A.B.P., Use of the Voigt Function in a Single-Line Method for the Analysis of X-ray Diffraction Line Broadening. J. Appl. Cryst.(1982). 15. 308-14.
7. Itoh, T. ,Crystallite growth of ZrO_2 powder, J. Mat. Sci. Lett. ,4 (1985), 1029-32.

HREM OF INTERFACES IN STRUCTURAL CERAMICS

Serge Hagège
ENSCP- 75231 Paris and CECM- 94400 Vitry, France.

ABSTRACT

High resolution electron microscopy of structural ceramics, as aluminium nitride and silicon carbides, gives clear indications on the atomic arrangement at the boundaries and indicates the role of the impurities in the coherency of the interfaces and, by then, the properties.

INTRODUCTION

Transmission electron microscopy under the high resolution mode (HREM) is a very powerfull technique to reveal the true atomic structure of interfaces in materials. Despite severe restrictions in the case of polycrystalline ceramics (orientation of the grains, thickness of the specimen, resolution, interpretation ...) an atomic image can provide new insights on the geometry and the chemistry of an interface and thus the physical and thermomechanical properties of the material. Mechanical and physical properties of structural ceramics are drastically affected by the "history" of the material and consequently by its microstructure. One crucial parameter remains the density and quality of interfaces. As an example, results on a study of interfaces in aluminium nitride (AlN) and silicon carbides (α−SiC and β−SiC) will be presented. Impurities, as oxygen, generate in AlN complicated arrangements of faulted interfaces detrimental to the physical properties. Very pure SiC exhibits coherent twin boundaries and HREM images are straightforwardly analyzed. Large amount of impurities in hot pressed SiC segregate at the boundary to form an amorphous layer.

FAULTED INTERFACES IN ALUMINIUM NITRIDE

Aluminium nitride (wurtzite structure) is very sensitive to the presence of oxygen in its lattice and faulted interfaces are easily produced within the grains. HREM micrographs can demonstrate that the nucleation of these interfaces is related to the impurities and originates in the evolution, during growth, of an simple defect.

Aluminium nitride ceramic materials have a high thermal conductivity and a potential tranluscence, both critically influenced by the impurity content. Our specimen, produced by Toshiba Corporation (Japan), has been hot pressed and contains 1.43 wt% of oxygen. The general microstructure of this material is reviewed in a parallel paper published in this volume [1]

In a large number of grains, when polytypes are not present, impurities induce complex configurations of faulted interfaces (Fig. 1). These configurations are formed by two types of interfaces: one flat, lying in the basal plane, and one curved; the two different surfaces are connected by a dislocation. In the most simple cases the configuration is like a "dome" which is made of a faulted dislocation loop surmounted by a rather spherical faulted surface.

Figure 1. Dome configurations within an AlN 2H grain.

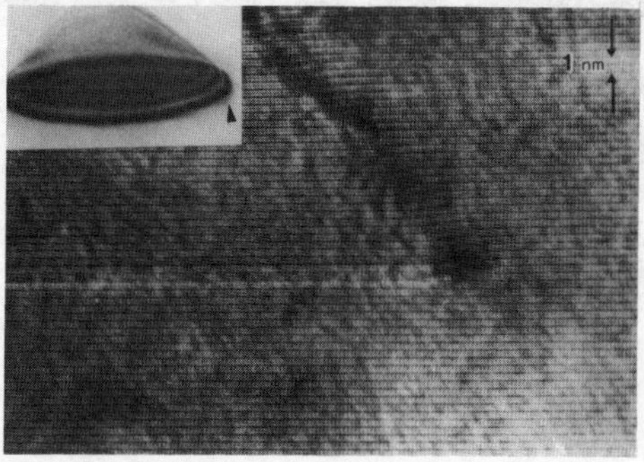

Figure 2. HREM of a part of a simple dome.

The HREM image in figure 2 is the (11$\bar{2}$0) projection of the structure showing the atomic arrangement for the two surfaces with a double stacking fault for the planar part and a shift of the basal planes for the curved part.

It has been demonstrated [2] that the presence of oxygen impurities and metal vacancies is the key point of the nucleation and growth of the dome configuration. This joint effect induces the formation of a double faulted dislocation loop. Inside the loop, the structure is now energetically unstable and a shrinkage of the basal planes occurs in order to give the structure the opportunity to change its coordination from four to six. The size of the dome will increase with a metal vacancy driving force. Complicated configurations are the result of the interaction of a growing dome with dislocations or other defects within one grain.

MULTIPLE TWINS IN β-CVD SILICON CARBIDE

HREM micrographs of CVD β–SiC exhibits particularly well defined multiple twin interfaces which have been analyzed using a coincidence modelisation and a high resolution contrast simulation. The following results demonstrate the high coherency and the fair bond reconstructions of the interfaces in a highly covalent material.

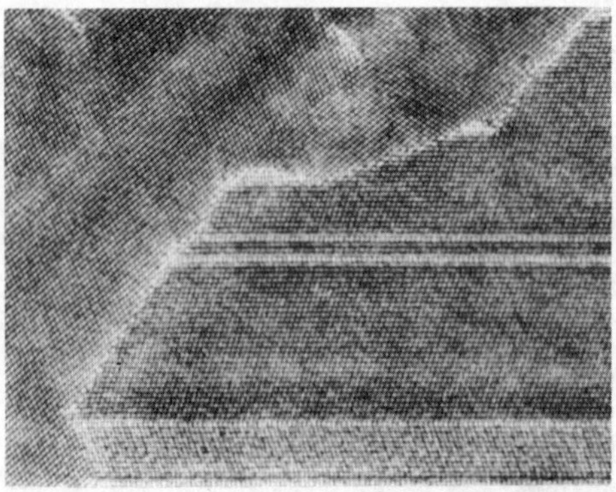

Figure 3. HREM of two Σ3 and one Σ9 boundaries in β-SiC.

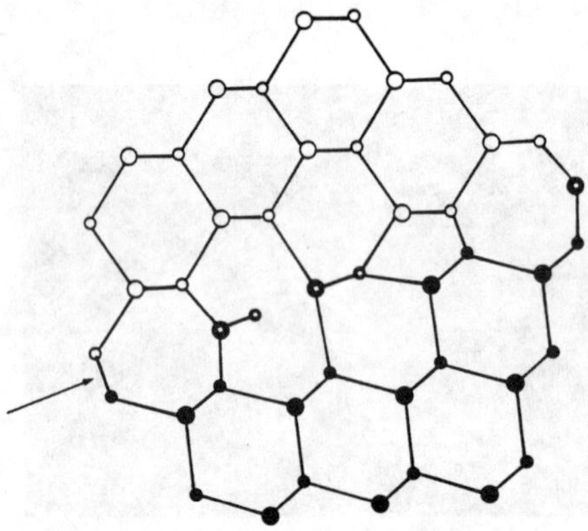

Figure 4. Model of a Σ9 boundary.

The silicon carbide has been prepared at the Institute of Material Research (Sendai, Japan) by chemical vapor deposition (CVD) on a carbon substrat heated at 1500°C. The resulting platelet is polycrystalline with a cubic β structure and an almost perfect texture [3]. The main characteristic resulting from this texture and the high purity of the material is the presence of numerous and well defined stacking faults, twins (Σ3) and subsequently multiple twins (Σ9, Σ27, Σ81 ...)

In figure 3, two Σ3 boundaries interact to give a Σ9 with, at least in its middle part, a well defined periodic contrast (white dots). The boundary plane is in a symmetrical orientation parallel to (221), one of the two twin planes of the Σ9 orientation. Starting within the framework of the coincidence orientation model and the concept of dichromatic pattern (two atoms per unit cell in SiC), the symmetry of a complex, described by a colored space group, can lead to the lowest energy atomic configuration [4]. In the case of Σ9, the choice of the origin of the lattices in between the silicon and carbon atoms and a 0 or +1/2 rigid body translation allows a modelisation of the interface which can be successfully checked against a computer simulation of the HREM image [5]. The models (Fig.4) are made of five and seven atom rings analogous to the ones describing coincidence interfaces in the diamond structure.

INTERFACES IN HOT PRESSED α –SILICON CARBIDE

In order to create bicrystal of α-silicon carbide , studied at the University of Kyushu (Japan), single crystal were first hot pressed (2270K and 40MPa in a vacuum of 2.7mPa) on a polycrystalline α–silicon carbide [6]. This single/poly diffusion bonded interface provide a large number of special grain boundaries, several of which have a particularly simple orientation relationship.

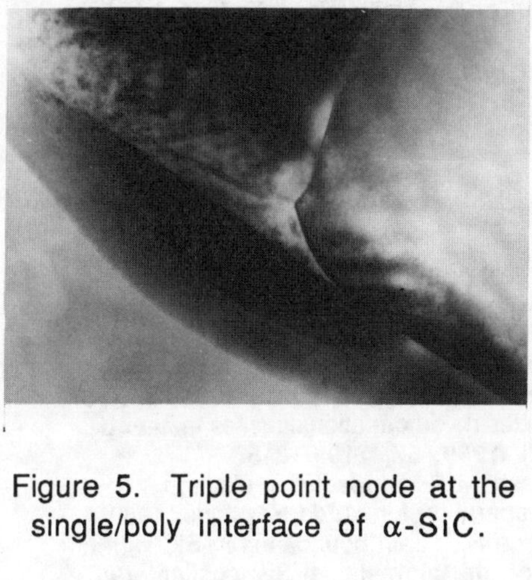

Figure 5. Triple point node at the single/poly interface of α-SiC.

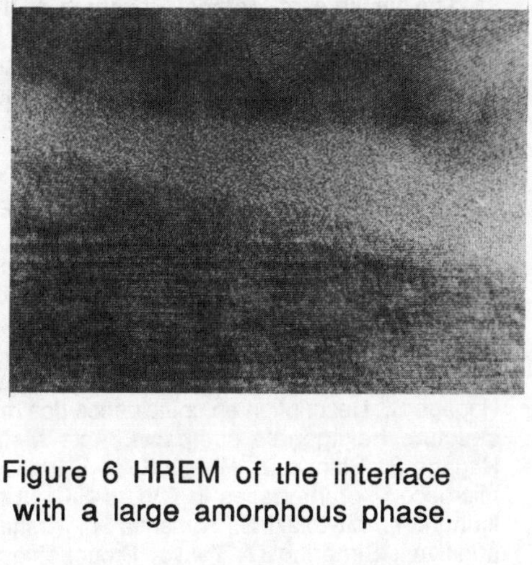

Figure 6 HREM of the interface with a large amorphous phase.

Figure 5 shows the triple point node at the single (lower part) / poly (upper part) interface with a neatly indicated retractation of the triple point inside the polycrystalline part. This retractation is understood as the simplest way for the system to relax by optimizing the dihedral angles between the grains. The single crystal is orientated such as its prismatic plan is normal to the electron beam (note the fringes of the 6H polytype); the upper right grain has the basal plane normal to the electron beam, the upper left grain is randomly oriented. Whatever the orientation of the grains in the polycrystalline part a large (10 nm) amorphous interlayer is present at the single/poly interface (Fig. 6). It is firmly believed that this amorphous phase originates during the diffusion process by migration of the impurities present in the polycrystalline phase. These impurities are mainly the sintering aids present in the material and, during diffusion bonding, grain boundaries are privileged path, leading to an accumulation at the single/poly interface.

Another notable feature of this interface is the $\alpha-\beta$ interface where was before bonding the single α phase. The retraction of the triple point node during diffusion bonding has left behind the thermodynamically stable β phase in a perfect epitaxial orientation leaving a clear trace of the migration of the boundaries during diffusion bonding.

ACKNOWLEDGMENTS

The results presented in this paper have been obtained during the visit of the author at the following research laboratories in Japan:

-The Institute of Industrial Research of the University of Tokyo (Prof. Y. Ishida and Dr. S. Tanaka).

-The Institute of Material Research of the Tohoku University (Prof. M. Hirabayashi, Prof. K. Hiraga and Dr. D. Shindo).

-The Department of Material Research and Technology of the Kyushu University (Prof. H. Yoshinaga, Dr. H. Kurishita and Dr. Y. Ikuhara).

Without their friendly hospitality and efficient cooperation this work would have not been achieved.

REFERENCES

1. Hagège S. and Ishida Y., Structural defects in AlN ceramics using TEM. This volume.
2. Hagège S., Tanaka S. and Ishida Y., Structural analysis of planar defects in wurtzite type aluminium nitride. Journal of the Japanese Institute of Metal, 1988, 52, 1192-98.
3. Hiraga K., Hirabayashi M., Niihara K. and Hirai T., Chemical vapor deposited silicon nitride and silicon carbide. Proc.IXth Internat. Conf. on CVD, Ohio, 1984, 573-582.
4. Hagège S., Description en coïncidence des macles de déformation dans les métaux de structure hexagonale compacte. Acta Metall. 1989, 37, 1199-1215.
5. Hagège S., Shindo D., Hiraga K. and Hirabayashi M., High resolution electron microscopy of interfaces in CVD β-SiC. In preparation for J. de Physique, 1989.
6. Ikuhara Y., Miyazaki H., Kurishita H., Yoshinaga H., Grain boundaries in SiC with no additives. Sintering'87, Tokyo, Proceedings of the International Symposium, 1987.

STRUCTURAL DEFECTS IN ALUMINIUM NITRIDE CERAMICS USING TRANSMISSION ELECTRON MICROSCOPY

S. Hagège* and Y. Ishida+
*ENSCP- 75231 Paris and CECM- 94400 Vitry, France,
+IIS- University of Tokyo, Tokyo 106, Japan.

ABSTRACT

The microstructure of hot pressed, pressureless sintered and CVD aluminium nitride (AlN) ceramic materials of different purities have been analyzed by transmission electron microscopy. The results are critically compared with respect to the nature and level of the impurities and the nature and density of structural defects. Consequences on the properties: mechanical, thermal conductivity ,translucence, are drawn.

INTRODUCTION

Aluminium nitride has a high thermal conductivity with a good electrical resistivity and a rather small thermal expansion coefficient. Thanks to their low reactivity to molten metals, aluminium nitride ceramics are also used in metallurgy as crucibles and evaporator boats. In addition, good quality material can be translucent. These properties, combined to a good mechanical strength, give to the materials a large range of engineering applications, the most accepted one being as substrat for integrated circuits. These properties are very sensitive to the purity of the material and it has been demonstrated that the presence of impurities in a given ceramic material has a strong impact on the microstructure and more particularly on the nature of the interfaces.

In the case of aluminium nitride, oxygen plays a major role on the microstucture by inducing stacking fault in the basal plane of the wurtzite structure of 2H-AlN. Then, depending on the oxygen content and on the nature and the level of metallic impurities in the material, polytypoids and/or dome-like defects can occur [1]. The cumulative effects of these surface defects, induced by the impurities, and the density of dislocations, nucleated during the elaboration, are detrimental to the properties. If the oxygen content is reduced and the material pressureless sintered at a higher temperature, the resulting material can reach very acceptable values of thermal conductivity and be fairly translucent. In the case of CVD material, impurity content is maintained to a minimum value and, by controlling the parameters of the deposition, high quality material, almost defect free, are obtained. Furthermore, subsequent heat treatment on the CVD material (annealing and/or HIP) drastically increase the performance by redistributing the structural defects and the impurities.

HOT PRESSED AlN

The first specimen, produced by Toshiba Corporation (Japan), has been hot pressed for one hour at 1973°K under 220 kg/cm^2. The sample is black colored and therefore not transparent to the light. The thermal conductivity is rather poor (75 W/mK) but the mechanical behaviour is acceptable. Oxygen, with 1.43 wt%, is the main impurity detected during chemical analysis; iron (0.40 wt%) and some other impurities (Fe, Si, C, Ca ....) are also present.

AlN 2H grains are roughly equiaxed with an average size of a few microns (Fig. 1). Every grain has a large concentration of defects most probably introduced in the material by the pressure applied during the manufacturing of the specimen. Polytypoids grains, induced by the large oxygen concentration, are elongated with the basal plane parallel to the longest direction. The 27R structure is the most frequent, as confirmed by X ray analysis. Sometimes, the grains have a complex mixture of unit blocks without forming a definite structure but rather appearing as a compositional modulation [2]. Inclusions of small θ-Al_2O_3 grains are often found inside the 2H grains. Exceptionally large θ-Al_2O_3 grains are also located at grain boundary junction. A $FeAl_2O_4$ spinel phase is infrequently detected in the material but always attached to one side of a polytypoid grain with a perfect epitaxy and a coherent interface. As detailed in a parallel paper [3] faulted interfaces decorate most of the 2H grains.

PRESSURELESS SINTERED AlN

The pressureless sintered samples, produced by Tokuyama Soda (Japan) have been maintained at 2073°K for three hours, the result being a lightly colored beige specimen fairly transparent to the light. Its thermal conductivity has notably improved (140 W/mK). The material is rather pure with only 0.15 wt% of oxygen and almost no other impurities. As a result the microstructure shows a larger grain size with only the 2H phase. Very tiny precipitates (θ-Al_2O_3) were exceptionally found at triple junction with some wetting of the neighbouring grain boundaries. Some faulted interfaces decorate the grains.

CVD AlN

Chemical vapor deposition can provide AlN plates with a good purity at temperature between 1000 and 1500°K. If all the experimental parameters (geometry of the setting, pressure, temperature) are optimized, the material, produced by Toshiba Ceramics (Japan), has very few structural defects and can reach high values of thermal conductivity. These values can be further raised by annealing the specimen. Nevertheless the specimens remain always dark grey, not translucent.

The optimisation of the geometric setting during the deposition has given the best improvement on the properties of the material. First experiments, using a direct deposition (the graphite substrat is directly hit by the incoming species), give poor specimen with a high density of defects (dislocations, stacking fault and other faulted interfaces) and an acceptable thermal conductivity (140-170 W/mK depending on the temperature of the substrat). A modified setting, allowing diffusion of the species in a confined atmosphere around the substrat, decreases indeed drastically the deposition rate but increase the thermal conductivity (200-260 W/mK). The specimen has then a larger grain size (a factor 10 on average) and the only defects are basal stacking faults (Fig.3). The density and the planeity of these faults are the only restrictions to a further improvement of the thermal conductivity, as it has been proved that the best specimens have a low density of stacking faults well located on parallel basal planes. As noted before, annealing improves the results, by lowering the energy of the stacking faults configurations, but the maximum value (260 W/mK) can be reached directly by CVD only, for optimum conditions, demonstrating that an upper value is reached at that level and that an incompressible number of stacking faults is produced during the deposition.

Concerning the transluscency of the material, interesting conclusions can be drawn from the following experiment: A directly deposited plate has been subsequently HIP'ed (2073°K under 2000Kg/cm2 for 2 hrs in nitrogen); the thermal conductivity has not highly improved but the specimen has gained notably in transluscence. TEM experiments show evidences of a recrystallisation (polygonisation, grain boundary migration), segregations of the impurities at triple points and the absence of stacking fault [4]. The same experiment on diffused deposited material did not modify the non-translucence of the specimen.

DISCUSSION

The structure and properties of the ceramics selected for this study can be compared in relation to impurities and structural defects:
Faulted interfaces are associated to a low concentration of oxygen but the existence of metal vacancies is imperative to their nucleation [5]. It is only for higher level that oxygen induces various phases enhancing the mechanical behaviour: polytypoids and spherical precipitates give a bimodal structure and therefore a strengthening of the system.
Translucence has been reached for high purity materials showing very small amount of structural defects (pressureless sintered) but also for HIP'ed CVD material in which a sensible amount of structural defects was still present. It can be asserted that recrystallisation has swept the impurities towards the grain boundaries. Therefore transluscency is mainly related to the amount of impurities in solution in the AlN phase and at a lower level to the density and coherency of planar defects.
Thermal conductivity is directly associated the density of structural defects present in the grain. If the coherency of planar defects (stacking faults or others) is reasonable they have a lesser effect on thermal conductivity. Impurity content has an indirect effect on thermal conductivity because the polytypoid phase and the precipitates, induced by impurities, increase the number of incoherent interfaces.

ACKNOWLEDGEMENTS

This work has been performed while one of the authors (SH) was visiting the Institute of Industrial Science, University of Tokyo (Japan). The specimens were provided by Dr. S. Tanaka (Toshiba Corporation), Dr. M. Kuramoto (Tokuyama Soda) and Dr. K. Hayashi (Toshiba Ceramics); their help and collaboration is gratefully acknowledged.

REFERENCES

1. Hagège S., Ishida Y.and Tanaka S., HVEM and HREM of Interfaces in Aluminium Nitride Ceramics. J. de Physique, 1988, 49-C5, 189-194.
2. Hagège S., Ishida Y.and Tanaka S., TEM Analysis of Impurity Induced Microstructures in Sintered Aluminium Nitride Ceramics. J. Jap Ceramic Society, 1988, 96,1119-26
3. Hagège S., HREM of interfaces in Structural Ceramics. This volume
4. Hagège S.and Ishida Y., to appear in J. de Physique, 1989.
5. Hagège S., Tanaka S. and Ishida Y., Structural analysis of planar defects in wurtzite type aluminium nitride. Journal of the Japanese Institute of Metal, 1988, 52, 1192-98.

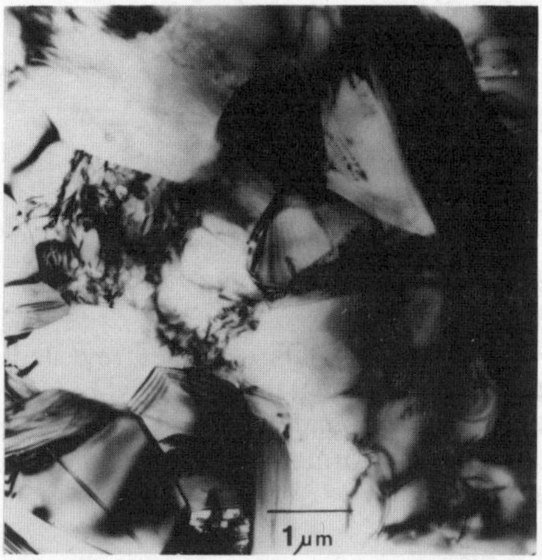

Figure 1. Microstructure of the hot pressed AlN.

Note added: Recent results [6] have confirmed the presence of faulted interfaces in sintered AlN. We maintain our analysis on an oxygen induced, metal vacancy driven mechanism [5]. The possibility of inversion boundaries is rejected after thorough analysis.

[6] Denanot M.F. and Rabier J.,1989, J. Mater. Sci. 24,1594-1598 and Mat. Sci. Eng. A109, 157-160.

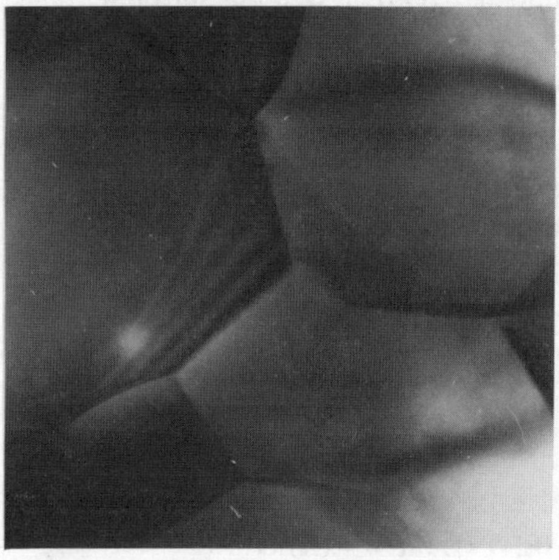

Figure 2. Microstructure of the pressureless sintered AlN.

Figure 3. Microstructure of the CVD AlN.

A TEM STUDY OF Sb-DOPED PZT CERAMICS

MATJAŽ PRISTAVEC, MARIJA KOSEC and VIKTOR KRAŠEVEC
"Jožef Stefan" Institute, University of Ljubljana, Ljubljana, Yugoslavia

ABSTRACT

Transmission electron microscopy of Sb-doped $Pb(Zr,Ti)O_3$ ceramics revealed extended defects and subgrain boundaries within grains. Using the BF-DF imaging criterion, the extended defects were identified as antiphase boundaries (APB'S). It was supposed that extended defects and subgrain boundaries are formed by incorporation of the $Sb^{5+}\,\boxed{Pb}\,Sb^{5+}$ complex into the perovskite lattice.

INTRODUCTION

Materials based on $Pb(Ti,Zr)O_3$ solid solution (PZT) represent the most important group of piezoceramics. Their functional properties generally depend on the Zr/Ti ratio, the dopant ions, as well as the density and microstructure. Most common PZT compositions are located at the morphotropic phase boundary in the vicinty of $Pb(Zr_{0.52},Ti_{0.48})O_3$ where the maximum of piezoelectric activity was observed. The properties of PZT ceramics can be influenced by donors such as La^{3+}, Bi^{3+}, Nb^{5+}, Sb^{5+} etc (1,2).

Since the influence of 5 valente dopants on the microstructure of PZT ceramics is not well known, the present work was undertaken in order to study the microstructural characteristics of these ceramics.

MATERIALS AND EXPERIMENTAL PROCEDURE

PZT specimens doped with 1,2.5,5 and 7.5 at.% of Sb were prepared from high purity oxide powders which were mixed, synthesized at 900°C and subsequently sintered for one hour at different temperatures.

Thin foils for TEM observation were prepared by grinding and subsequent argon-ion erosion. The thinned films were coated with carbon layer. Samples were investigated in a JEOL 2000 FX electron microscope operating at 200kV. EDS analysis were made by a LINK Systems AN 10000.

RESULTS AND DISCUSION

Two distinct microstructural features were observed in doped PZT ceramics with respect to the undoped ones; extended defects (ED) and subgrain boundaries (SGB'S) .

a) Extended defects

Fig. 1a shows a typical view of an ED pinned at the ferroelectric domain boundary. In general the defect surface is irregular and terminates on ferroelectric boundaries or GB'S. Sometimes the defect surface tends to follow a particular orientation e.g. {100} as shown in Fig.1b. Quite often ferroelectric domains terminate at the ED and fix them. The density of an ED was found to be higher in specimens sintered at lower temperatures (1200°C) with respect to those sintered at higher temperatures (1285°C). Exteneded defects were not found in specimens which were hot pressed at 900°C.

Using the BF-DF imaging criterion, the extended defects were identified as antiphase boundaries (APB'S) with the displacement vector parallel to the <110> axis.

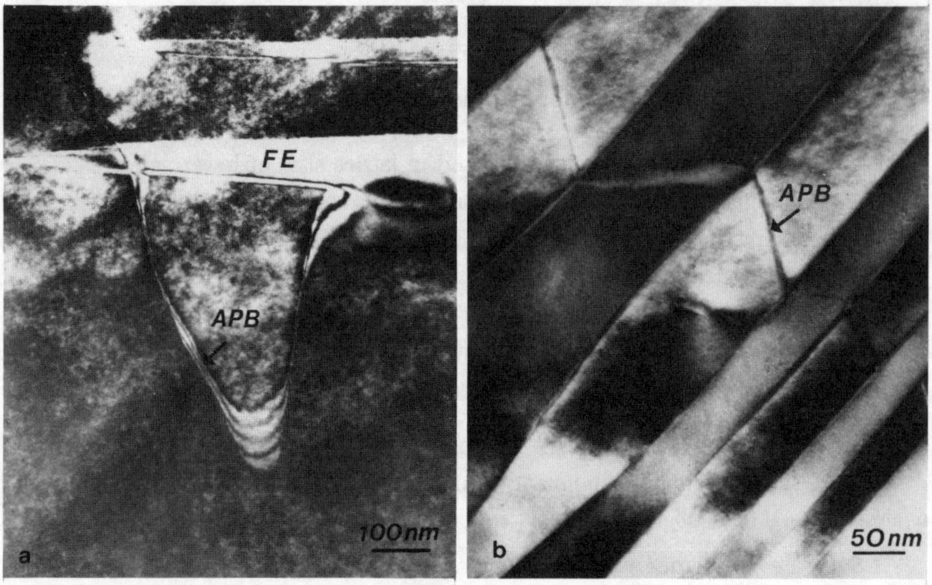

Figure 1a,b: Antiphase boundaries in PZT ceramics, doped with 1 at.% Sb and sintered at 1200°C for one hour.

It was supposed that these extended defects are formed by incorporation of the Sb^{5+} $\boxed{Pb}$ Sb^{5+} complex into the perovskite lattice. The proposed model for an extended defect is shown in Fig.2.

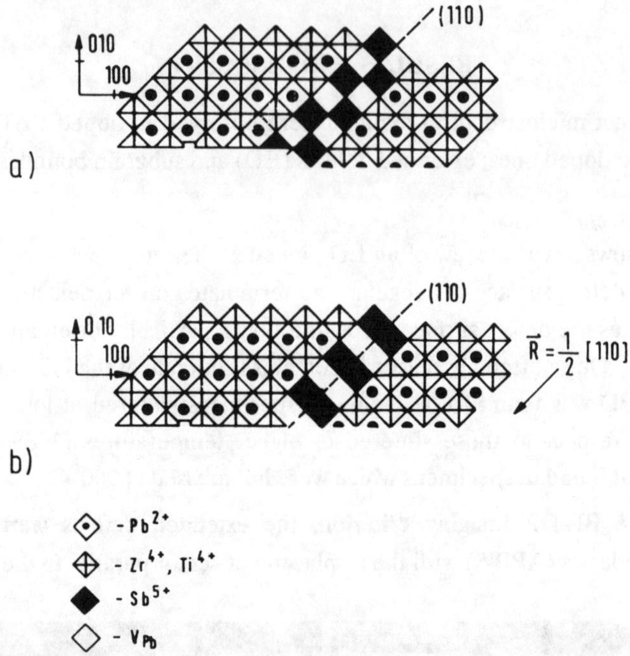

Figure 2. Proposed model for extended defect in Sb-doped PZT ceramics.

Since Sb^{5+} donor ions are incorporated in lattice sites of lower valency (Ti^{4+} or Zr^{4+})(3), the higher valency of the incorporated ions is compensated for by creation of the corresponding number of Pb-vacacies (Fig.2a.). The instability of such configuration is annihilated by shear along the $<110>$ direction in the $\{110\}$ plane producing the defect shown in Fig. 2b.Hence the extended defect in fact, represents a double layer of edge-shared Sb octahedra incorporated between two parts of a crystal which are in an antiphase position with respect to each other.

b) Subgrain boudaries

Fig.3. shows a subgrain boundary crossing a set of ferroelectric domains. EDS analysis showed that subgrain boundary contains more Sb than the rest of the grain.

Subgrain boundaries were observed only in materials doped with more than 5 at.% Sb. In general a larger quantity of dopant reduces the number of extended defects and increases the number of subgrain boundaries.

It is supposed that the subgrain boundary consists of ion complexes similar to those forming extended defects.

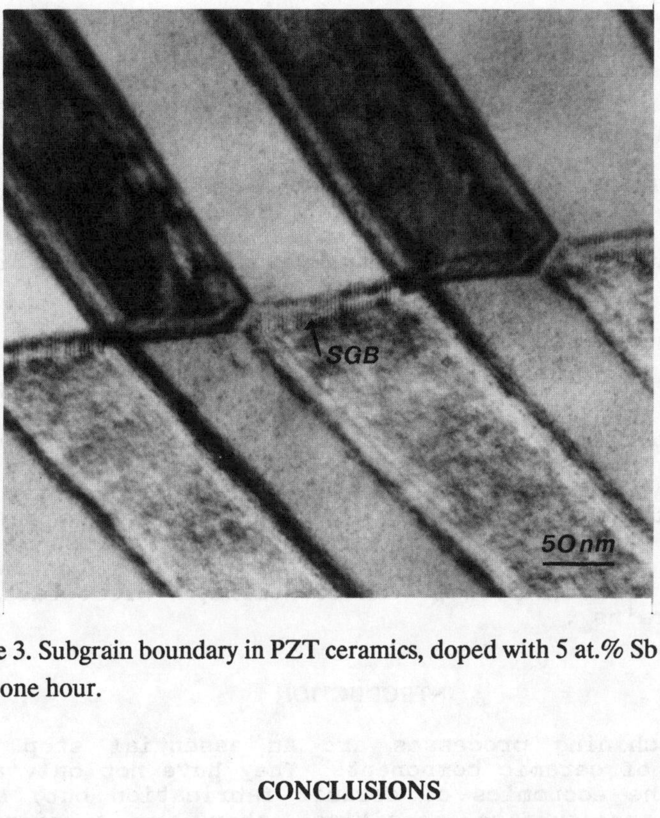

Figure 3. Subgrain boundary in PZT ceramics, doped with 5 at.% Sb and sintered at 1200°C for one hour.

CONCLUSIONS

- In PZT ceramics doped with Sb. extended defects and subgrain boundaries were found.
- A larger quantity of dopant reduces the number of extended defects and increases the number of subgrain boundaries.
- It is supposed that extended defects and subgrain boundaries are formed by incorporation of the $Sb^{5+}\boxed{Pb}Sb^{5+}$ complex into the perovskite lattice.

REFERENCES

1) B. Jaffe, W.R. Cook, H. Jaffe, Piezoelectric Ceramics, Academic Press, London 1971.
2) L. Wu, C.C. Wei, T.S. Wu, C.C. Teng, J. Phys. C.: Solid State Phys., 16 (1983), 2803-2812.
3) V.V. Prisedsky, G.F. Panko, V.V. Klimov, Ferroelectrics, Vol.64, (1985),257-273.

TEM STUDIES OF SSN SURFACES AFTER MACHINING

R.Haushälter , H.Frei, G.Grathwohl
Institut für Keramik im Maschinenbau,
Institut für Werkstoffkunde II,
Universität Karlsruhe, Haid- und Neustr. 7, D-7500 Karlsruhe

ABSTRACT

Sintered Silicon Nitride (SSN) shows different strength values σ_{4P} after different machining treatments. Our aim was the observation of the defects generated by machining in a Transmission Electron Microscope (TEM). To observe these defects in dependence of the depth and to discribe them in a quantitatve way, we used an optimized cross-sectional preparation technique. The predominant defects generated by machining treatment are dislocations, microcracks and disrupted grains .

INTRODUCTION

Abrasive machining processes are an essential step in the fabrication of ceramic components. They have not only a major impact on the economics of ceramic fabrication but, as many properties are surface sensitive, they are a significant factor in the resultant performance of ceramic components. For example, the inherent flaw sensitivity of the strength of brittle ceramics means that the severity of the machining process is often reflected in the final mechanical strength. Thus, it is important to understand the influence of abrasive machining on surface structure and related properties and how these are related to the mechanisms of abrasive machining.

PREPARATION

The only preparation technique for TEM-samples which allow the study of crystal defects from the surface down to a depth of a few µm is a cross-sectional preparation technique. This technique is well established /1,2/ but it was necessary to optimize this technique for ceramic material.
First two bars of SSN are sticked together with a layer of glue as thin as possible. After this, the samples are sliced

perpendicular to the glue layer. These plates with a thickness of about 0.2 mm are ultrasonic cut to discs of 3 mm diameter. After cutting the discs are ground to a thickness of about 100 µm. The best results are obtained by grinding parallel to the layer of glue. The next step is dimpling the sample to a thickness of about 20 µm to minimize time of ion milling to prevent problems due to differential thinning rates of ceramic and glue. Another problem of the eventual disappearance of the glue is the identification of the surface because the glue-sample-interface marks the true surface.
Fig. 1 shows the preparation schematically.

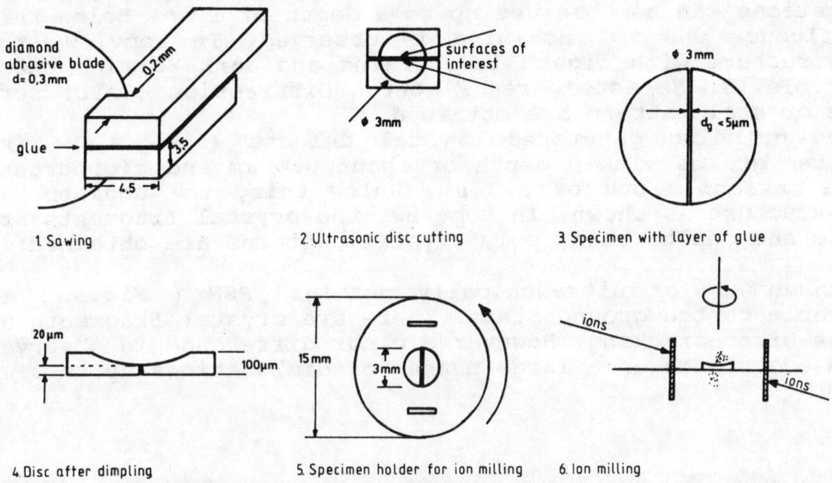

Fig. 1 : cross-section preparation technique

MECHANICAL PROPERTIES

SSN shows different strength values σ_{4P} of machined bending specimens after different mechanical surface treatments. The highest strength value is reached by the fine diamond polished specimen followed by the technical diamond ground and the ultrasonic machined specimen. Fig.2 shows σ_{4P} for the different mechanical treatments.

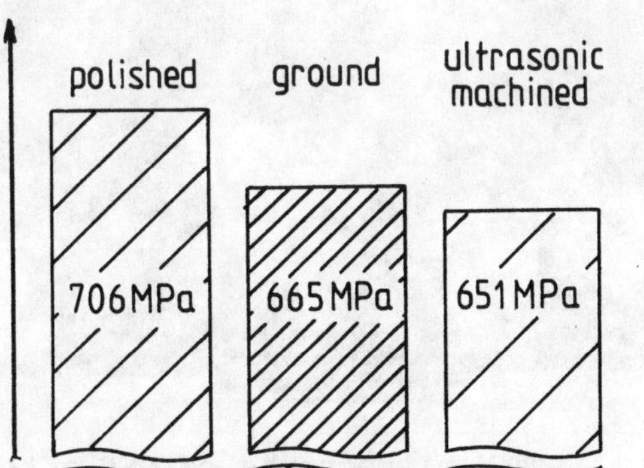

Fig.2 : Strength of SSN (sintered silicon nitride) after different surface finishing.

RESULTS

Dislocations, microcracking and disrupted grains are the predominant defects in the outermost layer below the machined surface of SSN. The polished surface shows the lowest amount of defects of the three machined surfaces. The ground surface shows larger defects and the ultrasonic machined surface has the strongest disturbance due to machining. These results are qualitative comparable to the mechanical properties of the testbars.
Surface polishing of SSN generates a dense dislocation network in a layer of about 0.5 µm in depth (Fig.3). Some dislocations can be observed up to a depth of 1 µm. Below this no influence due to machining is observed. In topview, the microstructure with light Si_3N_4 grains and dark Yb rich phase is visible. In Selected Area Electron Diffraction (SAED) pure single crystal pattern are obtained.
Diamond grinding generates crystal defects (Fig.4) like disrupted grains with a depth of about 0.6 µm and microcracks with a maximum depth of 1.5 µm. Below this, the undisturbed microstructure is shown. In topview, the crystal fragments are visible and due to this, poly-crystal patterns are obtained in SAED.
The subsurface of ultrasonically machined SSN (Fig.5) is comparable to the ground state. There are crystal fragments as well as micro-cracking. However a clear difference is observed in the occurrence of a large number of dislocations up to 3 µm in depth.

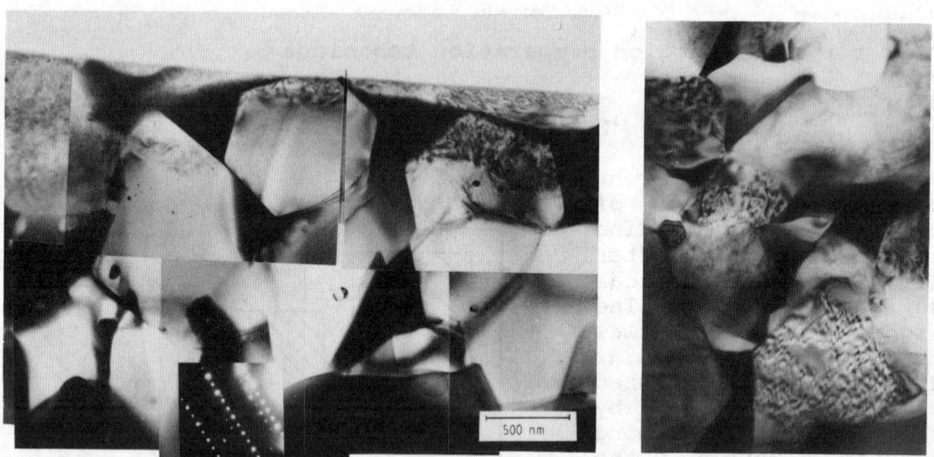

Fig. 3 TEM micrographs of SSN after polishing

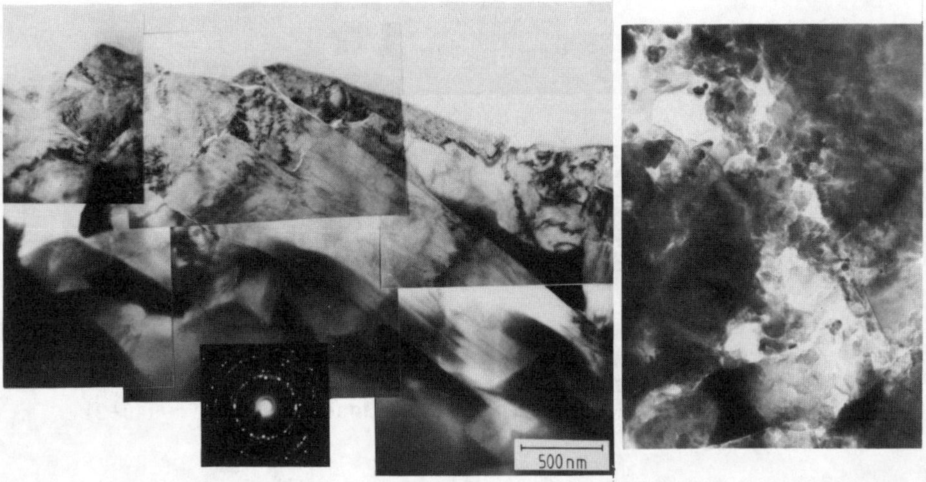

Fig. 4 TEM microcraphs of SSN after grinding

Fig. 5 TEM micrograph of SSN after ultrasonic machining

REFERENCES

1. Braveman,J.C. , Sinclair,R. , Journal of Electron Microscopy Technique 1 (1984) p. 53
2. Shinde,S.L., Jonghe,L.C., Journal of Electron Microscopy Technique 3 (1986) p. 361

CHARACTERIZATION OF SILICON NITRIDE POWDERS BY A TEMPERATURE PROGRAMMED DESORPTION STUDY

MITSURU KAWAMOTO, KOZO ISHIZAKI and CHANEL ISHIZAKI
Nagaoka University of Technology
School of Mechanical Engineering, Nagaoka, Niigata 940-21, Japan

ABSTRACT

Two kinds of silicon nitride powders were characterized by temperature programmed desorption (TPD). The concentration of $[OH^-]$ and $[NH_4^+]$ species desorbed during aqueous soxhlet washing was measured as a function of washing time. The results of the washing studies are discussed together with the TPD spectra of the dried and washed powders at different washing times. The two powders show quite different behaviors.

INTRODUCTION

Silicon nitride has great potential uses as structural material for engines, gas turbines, etc., because of the high strength and good oxidation resistance at high temperatures. To improve these properties, the surface chemistry of the raw powder is of fundamental importance. The present paper attempts to study the surface characteristics of silicon nitride powders using aqueous soxhlet washing and temperature programmed desorption (TPD)[1, 2].

MATERIALS AND METHODS

Powders

Silicon nitride powders were obtained from two different manufactures. The powders are referred to as powder A and powder B. Powder A was produced by precipitation of silicon diimide from liquid silicon tetrachloride and ammonia with subsequent purification of the precipitate and calcination of the imide silicon nitride. Powder B was made by carbothermal reduction of silica and subsequent nitridation by nitrogen gas. Table 1 shows the characteristics of the powders taken from their manufacturing specifications except for the specific surface area, particle size and impurities of O and C, which were measured in our laboratory.

TABLE 1
Characteristics of the silicon nitride powders

		A	B
Crystallinity α/β	(%)	97/3	98/2
O	(wt%)	1.60	2.70
C	(wt%)	0.10	1.02
Fe	(wt%)	<0.01	0.010
Al	(wt%)	<0.005	0.105
Ca	(wt%)	<0.005	0.010
Cl	(wt%)	<0.01	-
Particle size	(nm)	330	760
Specific surface area	(m^2/g)	11.6	4.9

Both raw powders A and B were dried in air at 100°C for two hours. They will be referred here as dried A and B. Also the dried powders A and B were subjected to aqueous washing in soxhlet by 120 ml of deionized water. To avoid flow out of the powder during washing, each powder was pressed by an uniaxial pressing at 250kg/cm^2 and then enveloped in a stainless mesh(400 mesh/inch). During washing, [OH^-] and [NH_4^+] concentrations were measured at room temperature by using a pH meter and a spectroscope (UV-365, Shimazu Ltd.) respectively. The [OH^-] concentrations were calculated from the pH values. The [NH_4^+] concentrations were obtained by Nessler's reagent reaction and the spectroscope. When the concentration of [OH^-] reached equilibrium, the water was changed to a new water. Each water is referred to as 1st, 2nd or 3rd water.

TPD Procedure
An ultra high vacuum system (10^{-10}Torr) and a furnace designed and built by the authors were used for the TPD experiments. The temperature of the furnace located in the UHV system was increased lineally by a PID temperature controller. 15.0 mg of each powder were weighed, and placed in the furnace. During the linear temperature increase, desorbed gases were detected by a quadrupole mass spectrometer (SELECTORR, SPECTROMASS Ltd.). The temperature and partial pressure of the desorbed gases were input into the computer memory.

RESULTS AND DISCUSSION

[OH^-] and [NH_4^+] desorption during aqueous washing
Figures 1 and 2 show respectively the [OH^-] and [NH_4^+] concentration changes of powders A and B during aqueous washing. An obvious difference is observed between the behaviors of the two powders. In the case of powder A, during the 1st water washing there is a gradual increase in the [OH^-] as well as [NH_4^+] concentrations whose values reach a maximum at around 12 hours of washing time. The concentration then drops and reaches equilibrium at about 20 hours of washing. In the case of powder B on the other hand, the increase in concentration of either [OH^-] and [NH_4^+] is small and gradually reaches equilibrium without passing through a maximum like in the case of powder A. It is also observed that the amount of ammonia desorbed by powder A is larger than the one desorbed by powder B (Figures 1 and 2). This topic will be further discuss in connection with the TPD results.

The behavior of the 2nd water of powder A is similar to that of the 1st water of powder B, and the 3rd water washing of powder A shows similar behavior as the 2nd water of powder B, both powders reaching equilibrium [OH^-] and [NH_4^+] water concentrations at around $2x10^{-8}$ mol/l and $1x10^{-4}$ mol/l respectively.

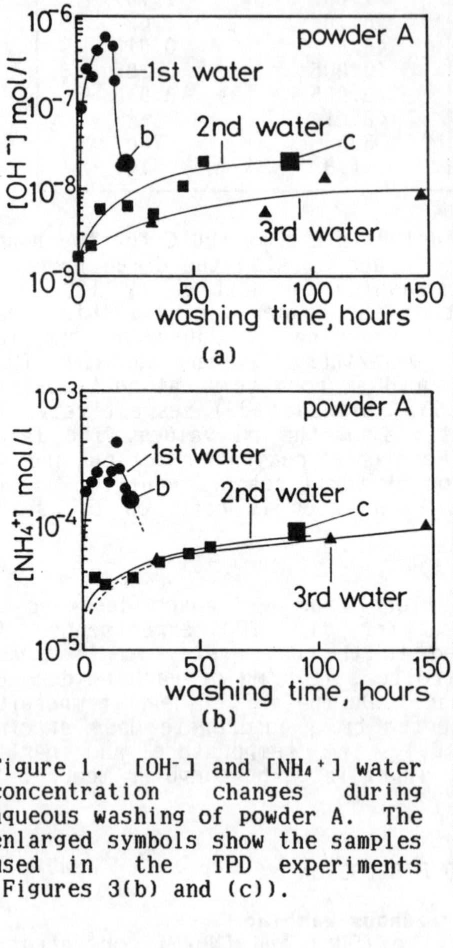

Figure 1. [OH^-] and [NH_4^+] water concentration changes during aqueous washing of powder A. The enlarged symbols show the samples used in the TPD experiments (Figures 3(b) and (c)).

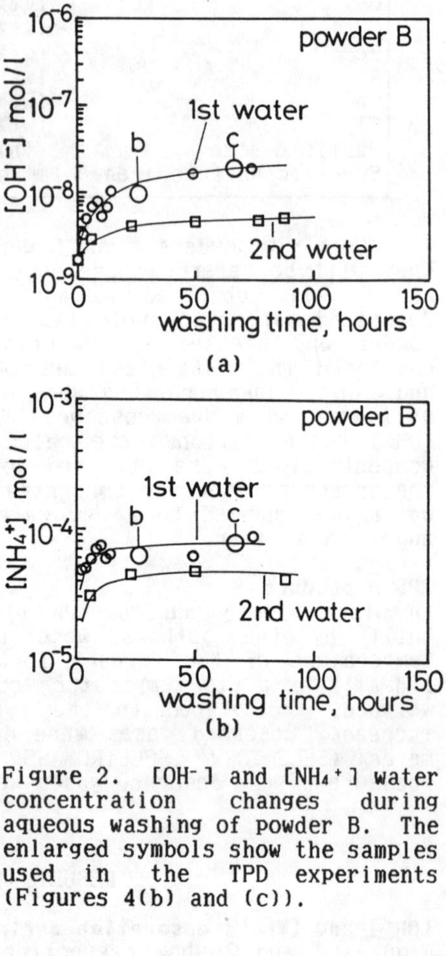

Figure 2. [OH^-] and [NH_4^+] water concentration changes during aqueous washing of powder B. The enlarged symbols show the samples used in the TPD experiments (Figures 4(b) and (c)).

TPD Spectra analyses

Figures 3 and 4 show TPD spectra of the dried powders and the powders after various washing times, which are normalized on a surface area basis. All the TPD experiments were performed at a rate of 50K/min. Water, ammonia and hydrogen are the major desorbed species for both powders. As observed in figures 3(a) and 4(a), the dried powders present quite different desorption patterns. Up to 400K powder A shows two main peaks for H_2O, NH_3 and H_2 with corresponding maximum desorption temperatures. In the case of powder B the amount of desorbed species is larger and broader with no apparent correspondence in the maximum desorption temperatures. Furthermore water is desorbed along all the temperature range.

After washing, the desorption patterns of both powders are again quite different among themselves and in comparison with the respective dried powders. In the case of powder A, the H_2O, NH_3 and H_2 peaks

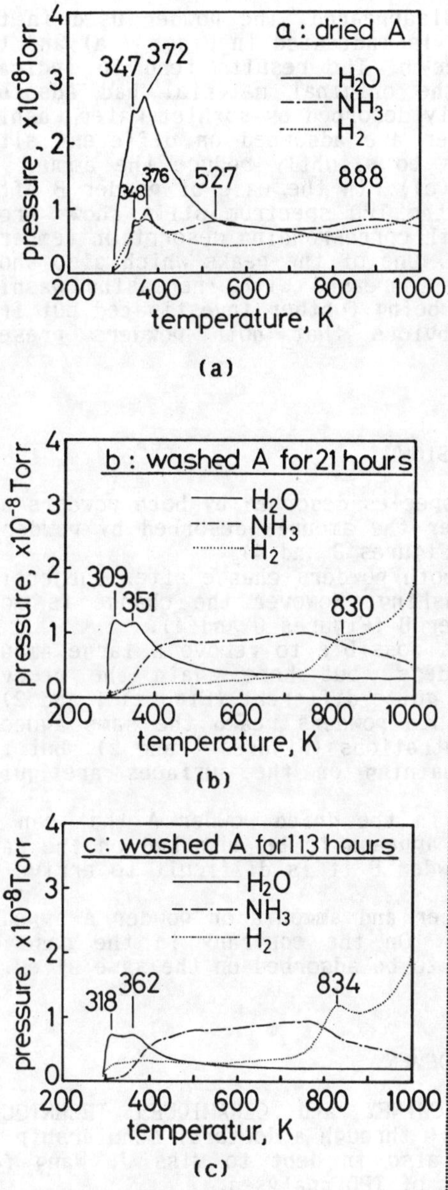

Figure 3 TPD spectra of powder A. (a)Dried powder, (b)the powder after 21 hours aqueous washing(point **b** in Fig. 1), and (c)the powder after 113 hours aqueous washing (point **c** in Fig. 1).

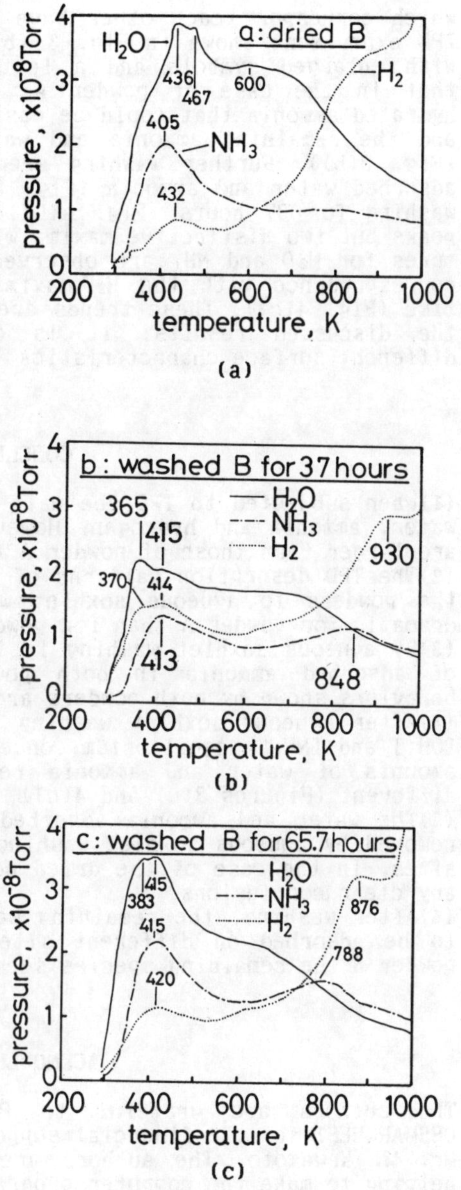

Figure 4 TPD spectra of powder B. (a)Dried powder, (b)the powder after 37 hours aqueous washing(point **b** in Fig. 2), and (c)the powder after 65.7 hours aqueous washing (point **c** in Fig. 2).

which correspond each other have disappeared. The powder used in the TPD experiment shown in Fig. 3 (b) is indicated in Figs. 1(a) and (b) with enlarged symbols and a letter b. The results tend to indicate that in the case of powder A, the original material had adsorbed hydrated ammonia that could be easily desorbed by soxhlet water washing and the remaining ammonia and water are adsorbed on different sites (Fig. 3(b)). Further washing seems to slightly reduce the amount of adsorbed water and ammonia (Fig. 3(c)). In the case of powder B after washing for 37 hours (Fig. 4(b)) the TPD spectrum still shows broad peaks but two distinctive maxima with corresponding desorption temperatures for H_2O and NH_3 are observed. One of the peaks which also shows correspondence with the H_2 maxima increased with increasing washing time (Fig. 4(c)). These trends are being further investigated but from the discussed results. It is obvious that both powders present different surface characteristics.

CONCLUSIONS

(1)When subjected to TPD the main species desorbed by both powders are water, ammonia and hydrogen. However the amounts desorbed by powder B are larger than those of powder A (Figures 3 and 4).
(2)The TPD desorption patterns of both powders change after subjecting the powders to aqueous soxhlet washing. However the change is more dramatic for powder A than for powder B (Figures 3 and 4).
(3)By aqueous soxhlet washing it is possible to remove a large amount of adsorbed ammonia in both powders, but here again the removal behaviors shown by both powders are quite different (Figures 1 and 2).
(4)After aqueous soxhlet washing both powders reach the same aqueous $[OH^-]$ and $[NH_4^+]$ equilibrium concentrations (Figures 1 and 2), but the amounts of water and ammonia remaining on the surfaces are quite different (Figures 3(c) and 4(c)).
(5)The water and ammonia adsorbed in the dried powder A that can be removed by aqueous soxhlet washing appears to be adsorbed on the same sites. In the case of the dried powder B it is difficult to arrive to any clear conclusions.
(6)After washing, the remaining water and ammonia on powder A appears to be adsorbed on different sites. On the contrary in the case of powder B the remaining species seem to be adsorbed on the same sites.

ACKNOWLEDGMENT

The authors are grateful to PECHINEY and CERAMIQUES TECHNIQUES DESMARQUEST for the financial support through a doctoral scholarship to Mr. M. Kawamoto. The authors are also in debt to Miss J. Wang for helping to make the computer program of TPD analyses.

REFERENCES

1. Amenomiya and R.J. Cvetanovic, Application of flash-desorption method to catalyst studies. I: Ethylene-alumina system. J. Phys. Chem., 1962, 67, 144-47.
2. P. A. Redhead, Thermal desorption of gases. Vacuum, 1962, 12, 203-11.

ELECTRONIC STRUCTURE AND CHARACTERIZATION OF TINx COATINGS

E. BEAUPREZ, S. HAREL, J.-L. DEREP
E.T.C.A. / C.R.E.A. (Centre de Recherches et d'Etudes d'Arcueil),
16 bis avenue Prieur de la Côte d'Or,
F - 94114 Arcueil Cedex, France.

N. MONCOFFRE, J. TOUSSET
Institut de Physique Nucléaire de Lyon, Université Claude Bernard Lyon-I,
43 Bd du 11 Novembre 1918,
F - 69622 Villeurbanne Cedex, France.

F. TEYSSANDIER
C.N.R.S. / I.M.P. (Institut des Matériaux et des Procédés), Université,
Avenue de Villeneuve,
F - 66025 Perpignan Cedex, France.

J.-M. MARIOT, C.F. HAGUE
Laboratoire de Chimie Physique (Unité Associée au C.N.R.S.),
Université Pierre et Marie Curie Paris-VI,
11 rue Pierre et Marie Curie,
F - 75231 Paris Cedex 05, France.

ABSTRACT

Nitrogen concentration in TiNx ($0.54 \leqslant x < 1$) CVD coatings has been measured by nuclear reaction spectroscopy (NRS) and x-ray diffraction (XRD). The electronic structure probed by x-ray emission spectroscopy (XES) shows the presence of a strong vacancy-induced peak in the Ti K emission band. Combining the NRS and XRD data with the XES data has enabled us to show that there is a linear relationship between the intensity of the vacancy peak and the nitrogen concentration for $0.7 \leqslant x < 1$. For smaller apparent nitrogen concentrations the intensity of the peak falls off. Using transmission electron microscopy (TEM) we are able to trace this deviation from a linear relationship to the presence of extrinsic stacking faults in samples with $x < 0.7$.

INTRODUCTION

The transition-metal carbides and nitrides belong to a class of compounds which combine metallic conduction with very high melting temperatures and extreme hardness normally associated with covalent bonding. Within this class of hard coatings, titanium nitride films are at present the most studied and used. They are used extensively on cutting tools to increase their wear resistance, but also in electronic applications, e.g. as diffusion barriers in contact structures to silicon integrated circuits (1-2). TiN films have also been studied for their optical properties (3-4).

The transition-metal carbides and nitrides mostly have rock-salt structure with nonmetal atoms occupying the interstitial octahedral sites within the transition-metal face-centered-cubic sublattice. They also exist as substoichiometric compounds over a large composition range with disordered vacancies at the non-metal sites (e.g. $0.41 \leqslant x \leqslant 1$ for TiNx, where x=N/Ti), and their physical properties which depend on electronic structure vary widely with composition (5-6). Therefore, to get a better understanding of the correlation between these physical properties and the N/Ti ratio, we have studied the electronic structure of substoichiometric TiNx CVD coatings by means of high resolution x-ray emission spectroscopy.

PREPARATION AND CHARACTERIZATION

The substoichiometric TiNx samples were deposited by CVD (chemical vapour deposition) on molybdenum substrates heated to 1900 K. $TiCl_4$, H_2 and N_2 were used in the gas reaction to produce TiNx coatings (7). The nitrogen concentration in TiNx was estimated from x-ray diffraction measurements assuming the lattice parameter has a linear dependence on the N/Ti ratio over the $0.6 \leqslant x \leqslant 1$ composition range :

$$a_o \text{ (nm)} = 0.41925 + 0.00467 \text{ N/Ti} \qquad \text{(8).}$$

In order to check the applicability of this method, the nitrogen concentration in TiNx coatings was also determined by nuclear reaction spectroscopy (NRS).

Although ^{15}N constitutes only 0.37 % of natural nitrogen, the nitrogen content and distribution over the depth of TiNx coatings were determined by using the 429 keV resonance in the $^{15}N(p,\alpha\gamma)^{12}C$ nuclear reaction, because with protons as incident particles, the nuclear reaction on ^{15}N is 8 times more sensitive than the reaction on ^{14}N ($^{14}N(p,\gamma)^{15}O$ with a resonance energy of 278 keV) in spite of the isotopic ratio largely in favour of ^{14}N. This difference of sensitivity is due to a higher cross section for the reaction on ^{15}N. Other nuclear reactions can also be used and in particular the $^{14}N(d,p)^{15}N$ and $^{14}N(d,\alpha)^{12}C$ reactions (9), but the nuclear reaction chosen gives a very good depth resolution (∿ 2 to 5 nm). Proton beams generated by a 2.5 MeV Van de Graaf accelerator were used to bombard the TiNx coatings. The gamma ray detection was performed by a NaI(Tl) detector. When the incident energy of protons is equal to 429 keV (the resonance energy), the nuclear reaction occurs at the very surface of the target (half width of the resonance equal to 120 eV, i.e., a theoretical depth resolution of the order of 1 nm) :

$$ {}^{15}_{7}N + {}^{1}_{1}p \longrightarrow {}^{12}_{6}C^{*} + {}^{4}_{2}\alpha $$
$$ {}^{12}_{6}C^{*} \longrightarrow {}^{12}_{6}C + \gamma\ (4.43\ \text{MeV}). $$

But depth information can also be obtained by measuring the gamma ray yield as the proton beam energy is increased during bombardment of the coating. This is because the beam loses energy as it penetrates the coating : the protons reach the resonance energy deeper in the sample dependent on the initial proton energy. We have applied this nuclear analysis method to obtain the nitrogen depth profile distribution in our coatings. The nitrogen calibration was performed by comparison with a stoichiometric titanium nitride standard deposited by PVD (physical vapour deposition) on a zirconium substrate.

NRS and XRD determinations of x agree well for $x \geqslant 0.7$ (Table 1). But, the NRS N/Ti value was found to be smaller than the XRD estimation for $x < 0.7$. We investigated the origin of this discrepancy by looking at the microstructure of samples 8 and 9 (Table 1) using transmission electron microscopy. This analysis has revealed the presence of extrinsic stacking faults in both samples (10). The discrepancies between the NRS and XRD data may be explained by the presence of another phase. It follows that NRS provides a better estimate for chemical composition of these coatings because it gives absolute N/Ti ratios whatever the chemical environment of ^{15}N. However, with extrinsic stacking faults essentially constituted by titanium atoms (10) the real composition of the coating cannot be deduced from lattice parameter measurements.

ELECTRONIC STRUCTURE

In previous papers (11-12) we have described the electronic structure of these TiNx CVD coatings. Here, we summarize the most important results.

TABLE 1
N/Ti ratios determined by nuclear reaction spectroscopy (NRS) and x-ray diffraction (XRD)

Sample	N/Ti ratio		Sample	N/Ti ratio	
	NRS	XRD		NRS	XRD
1	0.99	1.00	6	0.70	0.72
2	0.90	0.89	7	0.67	0.70
3	0.88	0.87	8	0.57	0.70
4	0.76	0.76	9	0.54	0.72
5	0.75	0.73			

The Ti K emission band involves transitions from the local Ti 4p-like valence states to the 1s core level. The Ti $K\beta_{1,3}$ ($3p \rightarrow 1s$) line also recorded along with the Ti K emission band serves as a means of normalizing the valence bands to the same number of emitting Ti atoms. We have shown that the p-like valence states at Ti sites are heavily hybridized with the N 2p states in the TiN compound. This results in the 4p sub-band (which lies just below the Fermi energy (E_F) in elemental Ti) being shifted down by about 6 eV in TiN. As the number of nitrogen vacancies increases, i.e. as the N/Ti ratio decreases, a p-like structure builds up between the 6 eV peak and E_F (see figure 1). The intensity of this vacancy-induced structure is very sensitive to composition. By subtracting out the contribution from the tail of the $K\beta_{1,3}$ emission, we have been able to estimate the intensity of the vacancy-induced structure relative to that of the $K\beta_{1,3}$ line (see figure 2). This processing of the XES data shows that there is a linear relationship between the intensity of the vacancy-induced structure and the N/Ti ratio for $0.7 \leqslant x < 1$. For samples containing stacking faults the XES data indeed indicates a lower vacancy concentration than would be deduced from the nitrogen content.

CONCLUSIONS

Metal-site p valence states are very sensitive to the vacancy concentration in transition-metal nitrides. This may be explained by the fact that the nitrogen atoms and vacancies lie at octahedral sites which favor p-like hybridization effects. The linear relationship between the intensity of the vacancy-induced peak in the density of states and the N/Ti ratio observed by high-resolution XES may be used to characterize such coatings.

In the case of low nitrogen content, however, extrinsic stacking faults may arise which reduce the vacancy concentration in the cubic phase (see figure 2), leading to an apparently higher overall N/Ti ratio. Whereas XRD simply gives the lattice parameter of the cubic phase which as we have seen may not be a reliable measure of composition, XES does provide the correct vacancy concentration and NRS the correct nitrogen concentration. It follows that when a very low nitrogen pressure is used during the CVD process leading to low nitrogen content (7) high resolution XES combined with NRS tests the homogeneity of the sample.

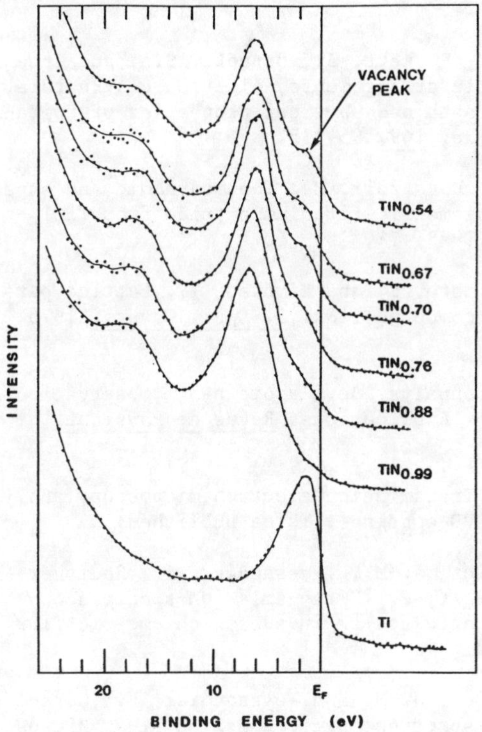

Figure 1. Ti K emission band for Ti and TiNx ($0.54 \leqslant x \leqslant 0.99$). The values of x are those determined from NRS experiments. The spectra have been normalized to the same Ti Kβ$_{1,3}$ emission line intensity.

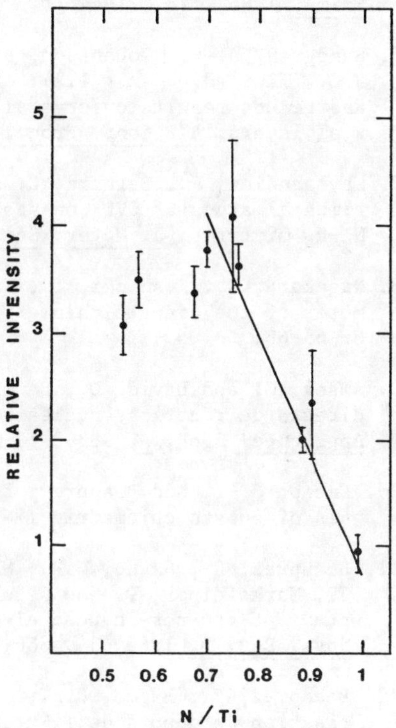

Figure 2. Relative intensity of vacancy-induced states as a function of N/Ti given by nuclear reaction spectroscopy.

REFERENCES

1. Wittmer, M. and Melchior, H., Applications of TiN thin films in silicon device technology. Thin Solid Films, 1982, 93, 397-406.

2. Brat, T., Parikh, N., Tsai, N.S., Sinha, A.K., Poole, J. and Wickersham Jr, C., Characterization of titanium nitride films sputter deposited from a high-purity titanium nitride target. J. Vac. Sci. Technol., 1987, B5, 1741-1747.

3. Rivory, J., Behaghel, J.M., Berthier, S. and Lafait, J., Optical properties of substoichiometric TiNx. Thin Solid Films, 1981, 78, 161-165.

4. Köhler, K., Geserich, H.P. and Christensen, A.N., Optical properties of TiN_{1-x} single crystals. Z. Phys. B, 1986, 62, 319-324.

5. Sundgren, J.-E., Structure and properties of TiN coatings. Thin Solid Films, 1985, 128, 21-44.

6. Sundgren, J.-E., Johansson, B.O., Rockett, A., Barnett, S.A. and Greene, J.E., TiNx ($0.6 < x < 1.2$): Atomic arrangements, electronic structure, and recent results on crystal growth and physical properties of epitaxial layers. AIP Conf. Proc., 1986, 149, 95-115.

7. Teyssandier, F., Bernard, C. and Ducarroir, M., Thermodynamic and experimental study of CVD non-stoichiometric titanium nitride from $TiCl_4$-N_2-H_2 mixtures. J. Mater. Sci., 1988, 23, 135-140.

8. Nagakura, S., Kusunoki, T., Kakimoto, F. and Hirotsu, Y., Lattice parameter of the non-stoichiometric compound TiNx. J. Appl. Cryst., 1975, 8, 65-66.

9. Amsel, G. and David, D., La microanalyse de l'azote par l'observation directe de réactions nucléaires - Applications. Revue de Physique Appliquée, 1969, 4, 383-391.

10. Derep, J.-L. and Beauprez, E., Transmission electron microscopy analysis of substoichiometric TiNx CVD coatings. To be published.

11. Beauprez, E., Hague, C.F., Mariot, J.-M., Teyssandier, F., Redinger, J., Marksteiner, P. and Weinberger, P., X-ray-emission-spectroscopy study of vacancy-induced electronic states in substoichiometric TiNx. Phys. Rev. B, 1986, 34, 886-890.

12. Beauprez, E., Hague, C.F., Mariot, J.-M. and Teyssandier, F., Caractérisation de dépôts de TiNx par spectroscopie d'émission X. J. Microsc. Spectrosc. Electron., 1986, 11, 251-254.

CHARACTERIZATION OF POROUS CERAMIC FUEL CELL COMPONENTS

P.A. NAMMENSMA and S.B. VAN DER MOLEN
Netherlands Energy Research Foundation, ECN
P.O. Box 1, 1755 ZG PETTEN, the Netherlands

ABSTRACT

Since 1986 ECN is involved in molten carbonate fuel cell (MCFC) research. The MCFC consists of an anode of Ni-$LiCrO_2$ and a cathode of NiO separated by an electrolyte-filled matrix of $LiAlO_2$. All components are highly porous. Component fabrication includes powder and slurry preparation, tape casting, drying, heat treatment and roll compaction as required. Starting powders are characterized by a static light scattering technique. The sintering behaviour can be simulated by extensiometry and differential thermal analysis.
Data on pore size distribution and specific surface area are obtained from mercury porosimetry and nitrogen gasadsorption. Such techniques give data that are essential for component development.

INTRODUCTION

In a molten carbonate fuel cell (MCFC) oxygen and hydrogen gas react at about 650°C to produce free electrons. A simple MCFC consists of three components: two electrodes with electrolyte in between. In practice, a matrix material is impregnated with electrolyte and is sandwiched between the electrodes. The components must be porous and their pore sizes have to be matched to each other.
To get a high current density, the ion conduction area of the electrolyte-filled matrix, i.e. the matrix porosity has to be large, however the fluid in the matrix is retained in the pores by capillary forces which are inversely proportional to the pore size, so the pore size cannot be too large. A compromise is found in a pore diameter smaller then 1 µm.
The cathode and anode of the MCFC must be accessible to oxygen gas and hydrogen gas respectively. The pores for these components have to be larger then the pores of the matrix to prevent removal of the electrolyte from the matrix by capillary forces. To ensure leaktightness, a bubble barrier is applied between the anode and the matrix.
The resulting requirements for a well functioning of the MCFC are summarized in the following table [1]. It is evident from the figures in the table that pore sizes of the different components form a sensitive combination. So, to get the right balance, the production process of the fuel cell components must be well controlled.

Table 1
Characteristics for the components of the ECN MCFC

	Material	Porosity %	pore-diameter µm	thickness mm
anode	Ni-10w% Cr	50-60	6-8	0.6
cathode	Ni	75	8-10	0.8
matrix	γ-$LiAlO_2$	65	<1.0	1.0
bubble-barrier	Ni-γ-$LiAlO_2$	50	<1.0	0.2

This control holds for the entire production process as well as for the produced component itself.
In the following the characterization route for fuel cell components development is reviewed. The route is illustrated for one component. In principle the same route holds for manufacturing of the other components of the MCFC.

POWDER CHARACTERIZATION

In the MCFC component development route the powder is characterized by particle size and specific surface area.
The particle size is measured by a static light scattering technique. This technique utilizes the phenomenon of scattering of light by particles. The angle dependent scattering is measured and is mathematically converted to a particle size distribution assuming spherical powder particles and using Mie scattering theory.

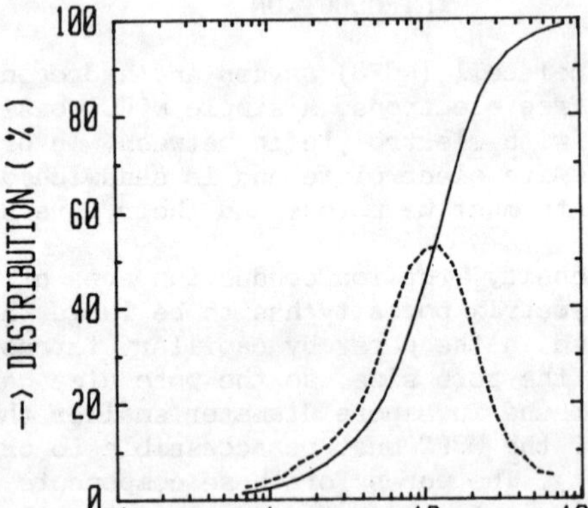

Figure 1. Distribution of lithiumaluminate particle size.

As an example a measured particle size distribution is shown in figure 1. The peak in the particle size distribution lies at 12 µm. The measurement is performed on well dispersed dilute suspensions. To make sure that the measurement gives information on the powder and not on the agglomerated powder, the zeta potential of the powder in suspension is determined. This zeta potential is about -10 mV, sufficiently high to prevent agglomeration. This powder can be used for a matrix that meets the requirements for a cell. The specific surface area of the powder is determined by measuring the adsorption isotherm of nitrogen gas on the powder at liquid nitrogen temperature. In conjunction with the BET model for adsorption the specific surface area of the lithiumaluminate powder from figure 1 is found to be 7.4 m^2/g. The surface area of the powder is one of the parameters determining the porosity of the matrix material.

SUSPENSION CHARACTERIZATION

The powders are mixed with binders, plasticizers and stabilizers to end up with a suspension that can be used for tape casting, the fabrication process used for MCFC component production. The viscosity is used as the main parameter to characterize the suspension. Practical limits are found to be in the range of 800 to 5000 cP.

The suspension is tape cast and dried [2]. Then the heat treatment is performed.

HEAT TREATMENT

The heat treatment can be divided into two steps. The first step is burning out of the binder and the plasticizer. The second step is the sintering process. Both processes can be optimized by simulation techniques. The first step is simulated by simultaneous differential thermal analysis and gravimetry (DTA and TGA). The second step is simulated by extensiometry. Figure 2 shows a simultaneous DTA and TGA experiment on nickel-chromium with binder under nitrogen gas with 25% hydrogen.

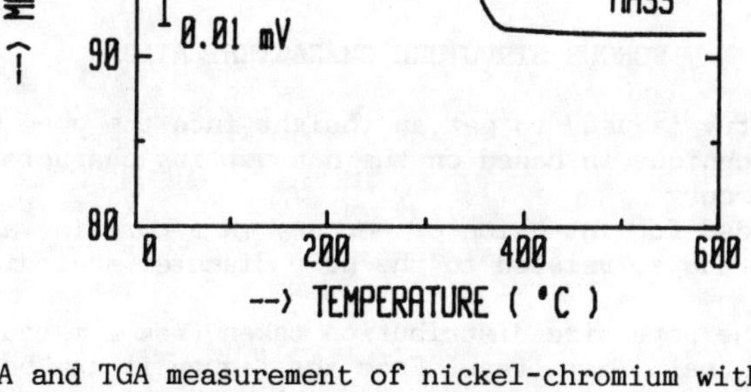

Figure 2. DTA and TGA measurement of nickel-chromium with binder in a nitrogen 25% hydrogen gas atmosphere.

The vertical bar in the figure of 0.01 mV is related to the differential temperature and equals about 1 K.
According to the dT signal there occurs a heat release between about 150 and 450°C. This is due to binder reaction during burn out. The mass change between 250 and 375°C is 8.3%. This appears to be equal to the total amount of the originally binder content. The conclusion from this experiment is that binders in tapes made from this suspension are completely burned out under these conditions between 250 and 375°C.

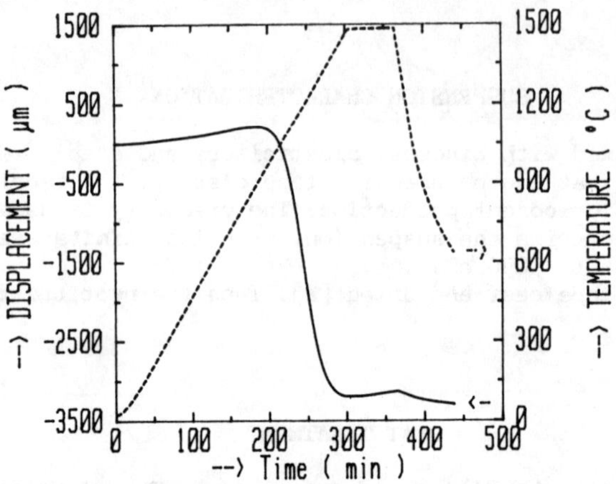

Figure 3. Sinter experiment on a lithiumaluminate sample.

The sintering of a sample of lithiumaluminate with a length of 16.9 mm is shown in figure 3. The temperature curve and the resulting displacement is shown as a function of time.
Temperature is raised by 5 K/min to about 1500°C. At that temperature there is a soak period of 1 hour.
There is no significant shrinkage detected in the first 200 minutes. Between 200 and 300 minutes, the length changes strongly . The measured total shrinkage is about 3400 µm ,which corresponds to a relative shrinkage of 20%. This shrinkage is quite normal for powder sintering. The sintering process starts around 1000°C and ends at 1300°C. Soaking at 1500°C hardly effects the length of the sample.

POROUS STRUCTURE CHARACTERIZATION

Mercury porosimetry is used to get an insight into the pore sizes of components. This technique is based on the non-wetting character of almost all solids by mercury.
The pressure needed for intrusion of mercury in a pore is, assuming a cylindrical pore shape, related to the pore diameter according to the Kelvin equation.
Figure 4 shows the pore size distribution taken from a mercury intrusion curve for a lithiumaluminate tape. From the figure it can be concluded that the open pore volume consists of pores smaller then 1 µm. The total porosity is 65 %. This tape meets the requirements for the matrix as given in table 1.

Figure 4 shows the pore size distribution taken from a mercury intrusion curve for a lithiumaluminate tape. From the figure it can be concluded that the open pore volume consists of pores smaller then 1 µm. The total porosity is 65 %. This tape meets the requirements for the matrix as given in table 1.

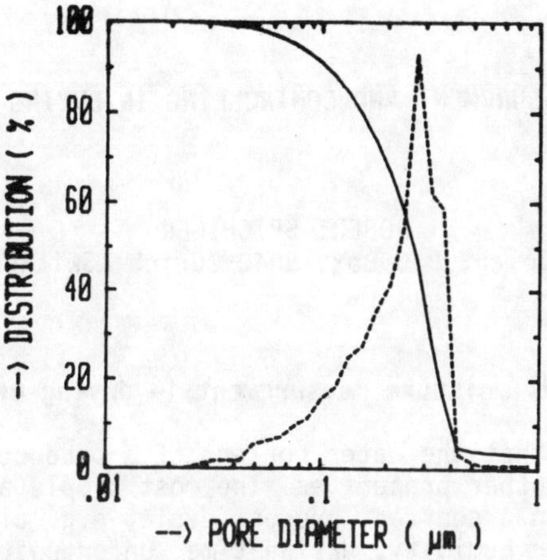

Figure 4. Pore size distribution of a lithiumaluminate tape.

CONCLUSIONS

The characterization of powders, suspensions and tapes as well as the optimization of the heat treatment can be satisfactory covered by the measurement techniques in our laboratory. Control of the measured parameters is indispensable for MCFC component production.

ACKNOWLEDGEMENTS

The project is financially supported by NOVEM on behalf of the Ministry of Economic Affairs.
The help of Herman den Uil for fuel cell discussions and Ruud Nyqvist and Cor van Egmond for characterization is kindly acknowledged.

REFERENCES

1. Boersma, R.J., Van der Molen, S.B., Den Uil, H, De Vries, J.E., Fuel Cell Seminar 1988, Long Beach California, USA.

2 Siewers, E.J., Den Uil, H. and Goris M.J.A.A., Porous MCFC component development, Paper presented at the seminar of Fuel Cell Technology Applications, the Netherlands, 1987.

3. Van der Molen, S.B., De Rooy, R., Siewers, E.J. and den Uil, H., Ceramics for fuel cells, Paper presented at this conference.

HUMIDITY MEASUREMENT AND CONTROLLING IN DRYING PROCESSING

ROBERT SPICHIGER
rotronic ag, P.O.Box, 8040 Zurich, Switzerland

Continuous moisture measurement in drying processes

The aim of drying is that the water content of a product should be reduced without changing its other properties. The most simple aid for process control is to maintain a constant process cycle, e.g. air temperature, air volume, flow rate, air humidity, drying time. Unfortunately this is not possible in industrial practice. The operating conditions change according to the seasons and the weather. Frequently too the state of the product to be dried differs at the beginning of the process. The drying process has consequently to be adapted to these varying conditions. For this reason continuous recording of the product moisture is required. Unfortunately a reliable, direct "water molecule counter" has not yet been devised and the state of the product has to be determined indirectly. The product temperature, conductivity, capacity, inductivity, electromagnetic wave absorption or reflection serve as auxiliaries. All these aids require a product-specific calibration. These therefore represent relative measurements. As during the drying processes hydrophilic materials permit moisture transport from the interior of the product to the surface, the product is not in moisture equilibrium. This means that the product moisture cannot be evaluated with certainty from the surface moisture. Moisture measurement systems deriving their information from the product surface are therefore of highly random usefulness. It is by definition inexpedient to instal continuous water activity or equilibrium moisture measurement systems. Measurement systems which derive their information from the interior of the product require constant granule sizes and constant bulk density. Unfortunately these conditions can often not be fulfilled. Obviously for all these reasons one should stick to the drying medium "air". The volume of water absorbed from the air corresponds exactly to the volume of water released from the product unless other water feed carriers upset this equivalence.

Drying processes

The drying enthalpy - and therefore the necessary energy required for drying a product - is strongly influenced by the water activity. In the case of pure water the evaporation enthalpy amounts to 2300 to 2500 kJ/kg. As the mixing water content falls the evaporation enthalpy rises sharply

because the water to be removed is ever more firmly bonded to the drying ware. With a mixing water content of 0.1% the evaporation enthalpy is of the order of 4000 to 5000 kJ/kg. This explains why it is so expensive to extract the final amount of water from a product.
The rate of drying during evaporation is directly proportional to the evaporation surface and the water vapour pressure differential between the ware to be dried and the ambient temperature. Normally the water vapour pressure increases over a hygroscopic product within approximately the same limits as the vapour pressure over water alone. The water activity of a product is therefore the only temperature-dependent to a limited extent compared with the relative humidity of the air. If the temperature of the product to be dried is increased, this increases its water vapour pressure and hence the water vapour differential between the product and the ambient air. The vapour pressure differential between the product undergoing drying and the surrounding drying air is one of the most important parameters for the drying process. This parameter is influenced by factors such as air temperature and vapour pressure, air velocity, air volume per time unit, the evaporation surface and the sorption properties of the product. The calculation and control of the drying process therefore demands a knowledge of the relationship between water activity and water content (sorption isotherms). In the case of natural products this relationship is not always the same because of their variable chemical composition and hence variable water bonding. One also has to allow for the fact that the sorption isotherms and the water activity are in a defined static equilibrium. This is also the reason for the variations between the theoretical calculations based on sorption isotherms and the measured results in practice. An optimization of the drying process can only be obtained by experiment. Experimental measuring arrangements require reliable transmitters. When reliable data have been determined and the optimized drying parameters are known, the drying process can be automated and electronically controlled.
It is a well-known fact that moisture measurement is considerably more difficult than e.g. the measurement of temperature, pressure, filling level, weight etc. The problem in moisture measurement is that the measuring cell has to be in direct contact with the environment.
A number of different factors are involved in an exact moisture measurement. Of major importance is the long-term stability of the measuring cell. Other criteria are the driftless measurement, the reproducibility of the measured data over the entire measurement range, large temperature application areas of -50 to +200°C, accuracy and resistance to pollution.
The long-term stability and resistance of the measuring cell should not only be designed for laboratory conditions but created for extreme applications in the industrial sphere.
As everyone is aware, soiling occurs in industrial application due to dust, condensation, vibration and thermal shocks. A transmitter is therefore required for industrial process control which has excellent reliability and demands minimum expenditure on maintenance and re-calibration. The reproducibility must be guaranteed for industrial applications over the entire measurement range, i.e. the measuring cell should not drift; it should not have any cyclic effect and naturally no hysteresis; because errors in reproduction cannot be offset.
Another factor is that the size and weight of a sensor have an effect on the measurement which should not be underestimated. It is a well-known fact that a small, very thin-walled sensor can adapt within seconds to changes in environmental conditions.

The Humidity Sensor

Direct Measurement of Relative Humidity

The HYGROMER humidity sensor is a small hygroscopic capacitor that modifies its value as a function of both the water vapour pressure and temperature of the environment. This permits direct measurement of relative humidity after precise calibration of the sensor with NBS references.

Innovative Sensor Design

HYGROMER sensors consist of a thin strip of hygroscopic polymer enclosed between two porous electrodes. This design enables the interchange of water molecules through both surfaces of the sensor. Other capacitive sensors use a non-porous or non-hygroscopic substrate that supports a thin film of polymer and a set of electrodes. Because of this substrate, water molelcules are essentially absorbed through only one surface of the sensor.

Superior Performance

Both the mechanical design of the HYGROMER sensors and the nature of the polymer used in the sensors, provide faster and more uniform absorption of moisture than in conventional sensors. Improvement of sensor performance is significant, especially at high humidity. Even when operated close to condensation for hours, the HYGROMER sensors do not show a sluggish response or "creeping up" of the humidity signal. And, when humidity conditions are reduced, there is no requirement for a recovery period to return the sensor to its original calibration. As a consequence, sensor hysteresis is exceptionally low and this results in excellent repeatability of measurement.

Reliability and Accuracy

Since the early 1980's, HYGROMER sensors have been used with success in thousands of installations either fixed or portable in industrial applications, in research, in meteorology, etc. These applications cover practically every condition within the range of operation specified for the humidity sensor and frequently involve high humidity, dust and contaminated air.

Accuracy of a sensor is a result of such characteristics as linearity, hysteresis, repeatibility, temperature error, etc. Testing done at the National Bureau of Standards in the range of 10 to 95%RH and -40 to +60°C, confirms the basic statements made for linearity, hysteresis, and repeatability of the HYGROMER products. Additional testing done by other laboratories at temperatures up to 150°C for 120 days confirm both the resistance and stability of the sensors.

TABLE 1
Resistance to Contaminants
Maximum Levels Tolerated by the HYGROMER Sensor (max.error: + / - 2.5% RH)

Contaminant	Permanent		Intermittant	
	ppm	g/m3	ppm	g/m3
Gases:				
Carbon Dioxide	no limit		no limit	
Hydrogern	no limit		no limit	
Hydrocarbon Gases	no limit		no limit	
Inert Gases	no limit		no limit	
Nitrous Oxide	no limit		no limit	
Solvents:				
Acetone	3300	8	6600	16
Benzine		150		200
Ethy Acetate	4000	15	8000	30
Ethylmethylketone	3300	8	6600	16
Gasoline		100		150
Toluene	1300	5	3000	12
Xylene	1300	5	3000	12
Alcohols:				
Ethyl Alcohol	3500	6	7000	12
Ethylene Glycol	1200	3	1200	3
Isopropyl Alcohol	4800	12	10000	25
Miscellaneous:				
Acetic Acid	800	2	1200	3
Amonia	5500	4	11000	8
Hydrochloric Acid	300	0.5	5	0.75
Sulphur Dioxide	5	0.013	5	0.013
Sulphur Hydrogen	350	0.5	700	1
Combustion Gases	no measurement		no measurement	
Salt Fog	special filter		special filter	
Amonia Salt Fumes	no measurement		no measurement	

Intermittent: 12 hours exposure and 12 hours regeneration in clean air.

HIGH RESOLUTION SOLID STATE NMR STUDIES OF SOME NEW NITRIDE AND OXYNITRIDE CERAMICS

M.J.LEACH, R.K.HARRIS
Department of Chemistry, University of Durham, Durham DH1 3LE, U.K.
and
D.P.THOMPSON
Department of Mechanical, Materials and Manufacturing Engineering, University of Newcastle upon Tyne, NE1 7RU, U.K.

ABSTRACT

^{29}Si magic angle spinning NMR spectra have been obtained for a variety of crystalline nitride and oxynitride ceramics. Spectra are given for LiSiNO, $LiSi_2N_3$, $MgSiN_2$ and $MgAlSiN_3$, all of which have ordered wurtzite-type structures with the non-metals in 4-fold coordination. Compared with silicon nitride, $LiSi_2N_3$ and $MgSiN_2$ show higher frequency shifts, whereas LiSiNo, which contains SiN_3O tetrahedra shifts to lower frequencies; in $MgAlSiN_3$, the chemical shift is almost the same as for Si_3N_4 itself.

^{29}Si spectra are also given for Ca- and Y- α'-sialons. These show only single peaks at less negative values of chemical shift which depart more from the Si_3N_4 value as the aluminium and interstitial metal content increase.

From the results, correlation of spectra with structural variables such as nearest and next-nearest neighbour coordination and bond angles is discussed.

INTRODUCTION

NMR techniques for the analysis of solid materials have only recently been developed, but interest in the field has intensified, especially in relation to ceramics. These techniques are particularly useful in cases where single crystals are not available and the less precise structural information provided by X-ray powder diffraction techniques requires supplementation. In the last three years several MASNMR studies have been reported for the quaternary sialon system (1-4), silicon carbides (5,6) and glass ceramics (7). More complex Y-Si-O-N oxynitrides have also been explored (8), and extensions of this work into the La-Si-O-N system have been reported both by Dupree et al. (9) and also by the present authors (10). The present paper is an extension of this work to other nitride and oxynitride ceramics, and NMR data are given for both lithium and magnesium silicon nitrides and also for some α'-sialons. All these compounds are either nitrides or derivatives from nitride structures. LiSiNO, $LiSi_2N_3$, $MgSiN_2$ and $MgAlSiN_3$

are ordered orthorhombic variants of the simple wurtzite AlN structure (A), with LiSiNO and $MgSiN_2$ satisfying the unit cell relationships:

$$a = \sqrt{3}a_A \;, \quad b = 2a_A \quad \text{and } c = c_A \qquad \ldots(1)$$

and $LiSi_2N_3$ and $MgAlSiN_3$ satisfying the relationships:

$$a = \sqrt{3}a_A \;, \quad b = 3a_A \quad \text{and } c = c_A \qquad \ldots(2)$$

α'-sialons have structures related to α-Si_3N_4 but with Al substitution for Si mainly compensated by the introduction of large cations (Li, Mg, Ca, Y, and Ln with Z>60) into interstitial sites in the structure, with some additional compensation provided by simultaneous substitution of Si by Al and N by O.

EXPERIMENTAL

Table 1 lists the compounds used in this study. Most of the samples were prepared by firing mixed binary oxide and nitride powders using conditions described in previous papers (11-16). Exceptions to this rule were $MgSiN_2$, which was prepared by nitriding mixtures of Mg and Si, and $MgAlSiN_3$, which was prepared by reacting $MgSiN_2$ and AlN. The product phases were characterized by X-ray powder diffraction to confirm that >90% of the desired phase had been produced.

^{29}Si MASNMR spectra were obtained with a Varian VXR 300 spectrometer operating at a frequency of 59.6 MHz. Samples were packed in ZrO_2 rotors and spun at 3-4 kHz during data acquisition. Chemical shifts are quoted relative to the signal for tetramethyl silane (TMS) ($\delta_{Si} = 0$) via measurements on 3(trimethylsilyl)-1-propane sulphonic acid sodium salt ($\delta_{Si} = 1.18$ ppm). Spectra were acquired using 90° radio frequency pulses (5 μs) and 120 s recycle delays. Line broadening of ca. 100 Hz was applied prior to Fourier transformation.

RESULTS AND DISCUSSION

In ceramic structures, silicon is invariably 4-coordinated by non-metals and ^{29}Si NMR chemical shifts are determined both by the electronegativity of the non-metal (X) (a good indicator of which is the Si - X - Si bond angle as shown in Figure 1) and also by the number of silicon neighbours coordinating it, as expressed by the Q number. Thus in $LiSi_2N_3$, $MgSiN_2$ and $MgAlSiN_3$, silicon is present as SiN_4 units with 3 of the nitrogens coordinated by 3Si + 1Li and one by 2Si + 2Li; in $MgSiN_2$, all the nitrogens are coordinated by 2Mg + 2Si and in $MgAlSiN_3$, Si and Al are disordered (resulting in considerable line broadening) with 3 of the surrounding nitrogens coordinated by 3(Si,Al) + 1Mg, and one nitrogen by 2Mg + 2(Si,Al). The incorporation of Li and Mg in these structures clearly results in higher frequency chemical shifts compared with Si_3N_4. In LiSiNO, silicon occurs as SiN_3O tetrahedra and this results in a lower frequency shift. ^{29}Si NMR spectra for all these compounds are shown in Figures 2 and 3.

Figure 4 shows ^{29}Si spectra for calcium and yttrium α'-sialons of general formulae $CaSi_9Al_3ON_{15}$ and $Y_{0.5}Si_9Al_3N_{14.5}O_{1.5}$. In contrast to α-Si_3N_4,

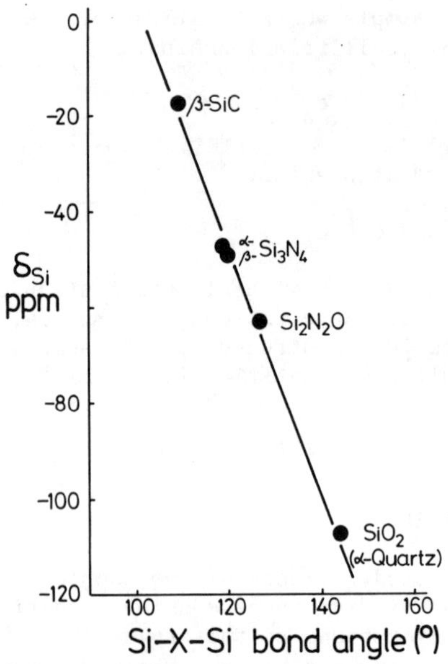

Figure 1. Plot of chemical shift (δ_{Si}) versus Si – X – Si bond angle for silicon-based ceramics.

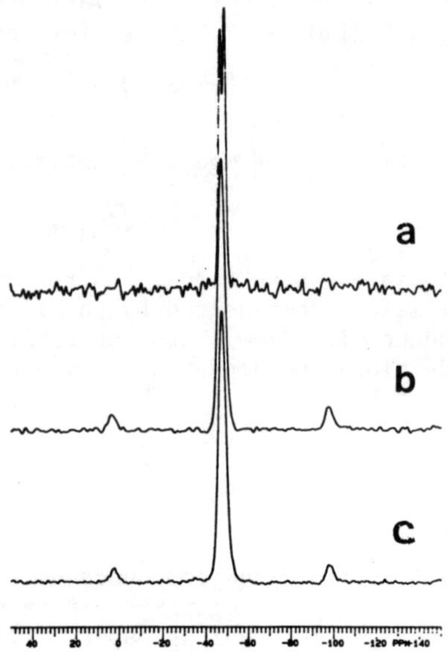

Figure 4. ^{29}Si MASNMR spectra of (a) α-Si_3N_4, (b) Ca α'-sialon, $CaSi_9Al_3N_{15}O$ and (c) Y α'-sialon, $Y_{0.5}Si_9Al_3N_{14.5}O_{1.5}$.

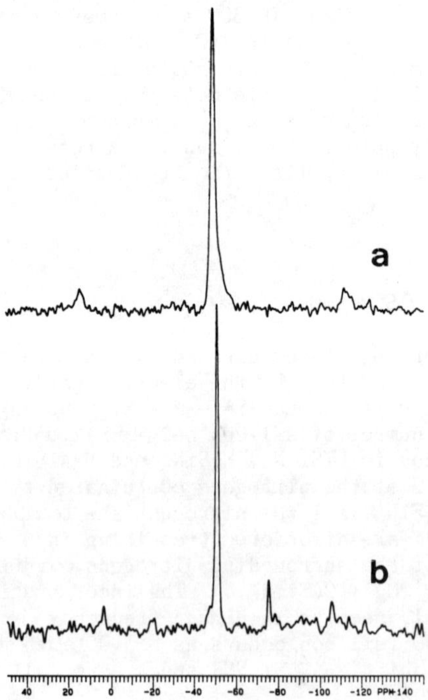

Figure 2. ^{29}Si MASNMR spectra for (a) $LiSi_2N_3$ and (b) LiSiNO.

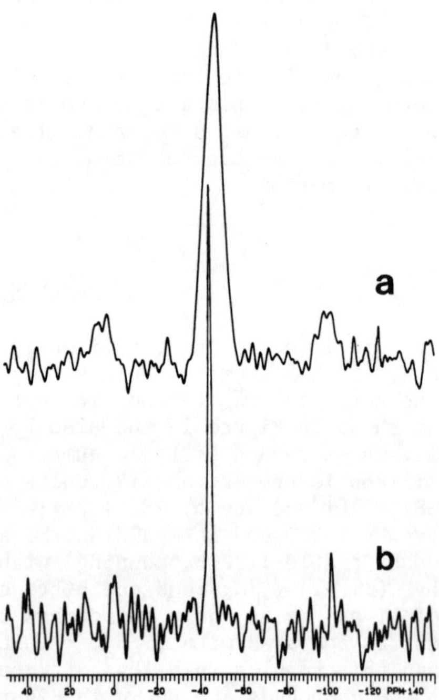

Figure 3. ^{29}Si MASNMR spectra for (a) $MgAlSiN_3$ and (b) $MgSiN_2$.

TABLE 1
^{29}Si chemical shifts for the compounds examined in this study.

Phase	δ_{Si} (ppm)	FWHH(Hz)*
LiSiNO	-51.0	90
$LiSi_2N_3$	-49.3	150
$MgSiN_2$	-44.5	95
$MgAlSiN_3$	-47.9	570
α-Si_3N_4	-47.2,-49.2	120
Ca α'-sialon, $CaSi_9Al_3ON_{15}$	-47.6	240
Y α'-sialon, $Y_{0.5}Si_9Al_3O_{1.5}N_{14.5}$	-47.8	250

* full width at half-height.

the spectra show only a single, slightly broadened peak, and this is typical of other α'-sialons examined in the present work. A plot of chemical shift versus degree of Si : Al substitution shows that the data extrapolate quite closely to the -49 ppm rather than the -47 ppm α-Si_3N_4 peak. This has a similar chemical shift to the single β-Si_3N_4 peak and corresponds to a more uniform SiN_4 tetrahedron. It is interesting that the incorporation of large cations combined with Al substitution in the Si sites appears to "regularize" the more irregular tetrahedron. This seems to be a general trend in sialon ceramics and may be ascribed to the less covalent character of aluminium compared to silicon, which is even more marked when the more covalent nitrogen is replaced by the more ionic oxygen.

CONCLUSIONS

In 1M : 1X wurtzite-type nitrides and oxynitrides, the simultaneous incorporation of other metal cations as, for example, in $LiSi_2N_3$ and $MgSiN_2$, results in ^{29}Si chemical shifts moving to higher frequencies compared with Si_3N_4. In LiSiNO, the SiN_3O structural units result in lower frequency shifts and in $MgAlSiN_3$, the main feature is line broadening due to (Si,Al) disorder.
Chemical shifts for α'-sialons decrease with increasing unit cell dimension regardless of interstitial cation and show only one ^{29}Si peak. The chemical shift versus cell dimension plot extrapolates to the -49 ppm α peak, which corresponds to the more uniform silicon environment, showing that the combined effect of interstitial incorporation plus aluminium substitution makes both the Si sites more similar. The absence of a continuous variation in spectra between α' and α is consistent with preparative observations that there is not a continuous range of solid solution between α-Si_3N_4 and the lowest substitution α'-sialon.

Acknowledgements
M.J.L. is grateful to the U.K. Science and Engineering Research Council for a Research Studentship under the Earmarked Scheme. We also thank Dr G.L. Marshall for useful discussions.

REFERENCES

1. Klinowski, J., Thomas, M.J., Thompson, D.P., Korgul, P. Jack, K.H., Fyfe C.A. and Gobbi, G.C., Polyhedron, 1984, 3, 1267-9.

2. Dupree, R., Lewis, M.H., Leng-Ward, G. and Williams, D.S., J. Mater. Sci. Lett., 1985, 4, 393-5.

3. Turner, G.L., Kirkpatrick, R.J., Risbud, S.H. and Oldfield, E., Am. Ceram. Soc. Bull., 1987, 66, 656-63.

4. Dupree, R., Lewis, M.H. and Smith, M.E., J. Appl. Cryst., 1988, 21, 109-16.

5. Findlay, G.R., Hartman, J.S., Richardson, M.F. and Williams, B.L., J. Chem. Soc. Chem. Commun., 1985, 159-61.

6. Guth, J.R. and Petuskey, W.T., J. Phys. Chem., 1987, 91, 5361-4.

7. Aujla, R.S., Leng-Ward, G., Lewis, M.H., Seymour, E.F.W., Styles, G.A. and West, G.W., Philos. Mag. B, 1986, 54, L51-6.

8. Dupree, R., Lewis, M.H. and Smith, M.E., J. Am. Chem. Soc., 1988, 110, 1083-7.

9. Dupree, R., Lewis, M.H. and Smith, M.E. (to be published).

10. Harris, R.K., Leach, M.J. and Thompson, D.P., accepted for publication in "Chemistry of Materials".

11. Thompson, D.P. in "Applied Crystallography", Eds. Z. Bojarski and T. Bold, Kozubnik, Poland, 1978, 439-53.

12. David, J., Laurent, Y., Charlot, J.-P. and Lang, J., Bull. Soc. Fr. Min. Cryst., 1970, 96, 21-30.

13. David, J., Laurent, Y. and Lang, J., Bull. Soc. Fr. Min. Crist., 1970, 93, 153-6.

14. Perera, D.S., Ph.D. thesis, University of Newcastle upon Tyne, 1976.

15. Hampshire, S., Park, H.K., Thompson, D.P. and Jack, K.H., Nature, 1978, 274, 880-1.

16. Slasor, S. and Thompson, D.P., Proc. Int. Symp. "Non-oxide technical and Engineering Ceramics", Ed. S.Hampshire, publ. Elsevier, 1986, 223-30.

MEASUREMENT OF STRENGTH AND DEFORMATION BEHAVIOUR OF GRANULATED CERAMIC PRESS GRADE POWDER

R. Oberacker, F. Thümmler
Institut für Keramik im Maschinenbau,
Institut für Werkstoffkunde II,
Universität Karlsruhe, Haid-und-Neu-Str. 7, D-7500 Karlsruhe

ABSTRACT

A newly developed system for measurement of single granule load-deformation behaviour and strength is described. Measurements on model speres and silicon nitride press grade powders are presented.

INTRODUCTION

The role of packing inhomogeneities in ceramic powder compacts has been subject of many experimental studies during the last years. Aggreement exists, that inhomogeneous particle arrangement is the source of processing defects in many cases. Especially the dry forming methods bear a high rish for packing inhomogeneities, which is due to a duplex-structure arrangement of the primary powder particles after filling of the die cavity. These structure consists of relatively dense packed agglomerates which themselves are ordered in a much looser random packing. For transforming this structure into a homogeneous random arrangement of primary particles during press compaction, the mechanical behaviour of the agglomerates under the given stress-strain conditions is of high importance. Therefore a method was developed, which should allow for mechanical characterisation of (more or less) geometrically defined agglomerates as they usually can be found in ceramic press grade powders.

THEORETICAL BACKGROUND

Most characterisation methods for agglomerate strength are developed for bulk powders and therefore provide an average property value from a large number of single agglomerates, or in some cases which accord to the principle of weakest link, the lowest strength value of the individual agglomerates. As processing defects result from a few or even one single "hard" agglomerate present in the powder mass, a statistical testing of single granules will be a much better way to detect such sources of defects.

For this purpose, the measurement of crushing strength by the single granule compression test is the only proven method at the present. Agglomerate strength is usually calculated by dividing the crushing load measured in this test by the cross section of the granule [1] which, in the case of spherical granules leads to:

$$\sigma_c = \frac{4F_c}{\pi \cdot D^2} \qquad (1)$$

This is supported by the fact, that agglomerates mostly fail by critical crack propagation through a meridional cross section. For the same reason, a well defined correlation of compressive and tensile strength of agglomerated materials often exists.
Even more interesting seems to be the fact, that not only the fracture load, but also the pre-fracture load-deformation curve measured in such experiments will be a characteristic for the mechanical behaviour of the granule. Under the assumption of spherical particle shape and a continuum nature of the material, two limiting cases can be predicted: Purely elastic behaviour and purely rigid/plastic behaviour of the material. They result in a load-deformation behaviour of the particle in the form:

$$\Delta D = const \cdot F^m \qquad (2)$$

According to the Hertzian therorie the exponent m would be equal to 2/3 in the case of purely elastic behaviour. This gives a progressive load-deformation curve, as the compliance of the particle increases with increasing load, due to the increase in contact area and elastically strained volume. If the deformation becomes plastic beyond a certain stress level, the exponent m increases, due to the degressing stress-strain relationships. For an ideal rigid/plastic material, there is a linear relationship between the compressive load and the area of contact [2]. In a first approximation the area of contact can be assumed to increase linearly with the deformation ΔD of a spherical particle, resulting in the limiting case of 1 for the exponent m. Experimental evidence for similar exponents were also reported for ceramic powder granules [3]. Therefore it is expected, that the exponent m could be used as a measure for the elastic or plastic character of such powders.

STRENGTH AND LOAD-DEFORMATION TESTING SYSTEM

On the basis of some measuring systems reported in literature [3,4,5] the own development was carried out, resulting in an uncomplicated and partially automatised system, which allows for a quick measurement of single granule load-deformation curves on granules with a minimumn diameter of about 100 µm. The system is schematically presented in Fig. 1.The granule (1) is placed between a flexible beam (4) and a micrometer screw, which forms the upper anvil plate. The beam is tilted frictionless by the piezo translator (5) around a solid state bearing, thus moving the granule towards the upper anvil. The load is measured via strain gauges (3), which monitor the strain of the loading beam. The whole load cell is easily interchangeable, so that the measuring range can be chosen according to the required load. Deformation is measured by a noncontacting capacitive or inductive sensor (2).

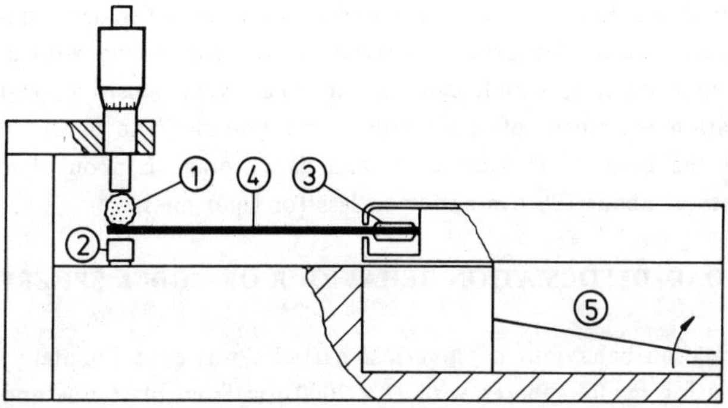

Figure 1a. Granule Strength Tester - Design -

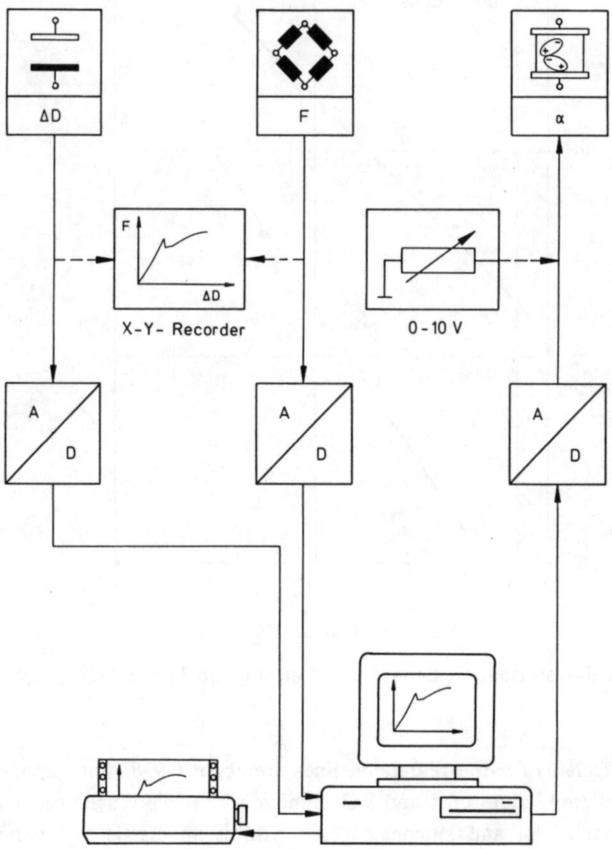

Figure 1b. Granule Strength Tester - Data Aquisition and Control Unit -

The tester is connected to a personal computer which is used for data aquisition and control of the test procedure. The present software allows for testing with a constant tilting rate of the piezo translator, which can be controlled very sensitive, resulting in a minimum deformation increment of < 0.1 μm at the granule. The usual time consumption for charging of the granule, measurement, and data storage is about 2 to 3 minutes for granule diameters of about 100 μm and even less for lager ones.

LOAD-DEFORMATION BEHAVIOUR OF MODEL SPHERES

The load-deformation behaviour of spherical particles was experimentally tested with our strength tester, using model spheres with D = 2000 μm from plasticine and an elastic hard foam material. Fig. 2 shows the load-deformation curves in logarithmic coordinates. The

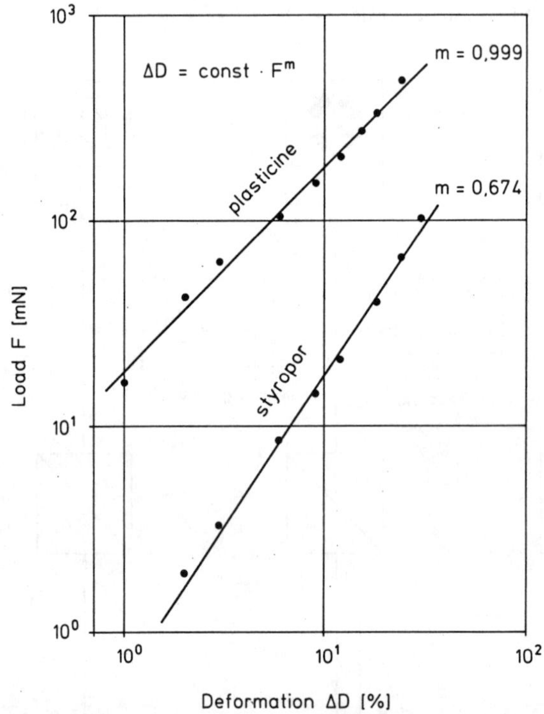

Figure 2. Load-Deformation Behaviour of elastic and Plastic Spheres (D = 2000 μm)

correlation coefficients for the regression lines are about 0.998. The exponents m are very close to the predicted values of 1 and 2/3. Similar results were achieved with 900 ... 1000 μm spheres from styrol and silicone, which exhibit an elastic behaviour at m = 0.63 ...0.75. These results indicate, that equiation (2) indeed can be used for characterisation, at least in the case of ideally spherical particles.

COMPRESSIVE STRENGTH OF PRACTICAL PRESS GRADE GRANULES

Fig. 3 gives an example for measurements on materials of practical relevance. In this case, the crushing load for granulated silicon nitride powder mixtures was estimated for granules classified into narrow size ranges. The powders originate either from a wet mixing process carried out in a tumble mixer (A), or in a planetary ball mill (B). The homogenised slurries were dried under identical conditions.

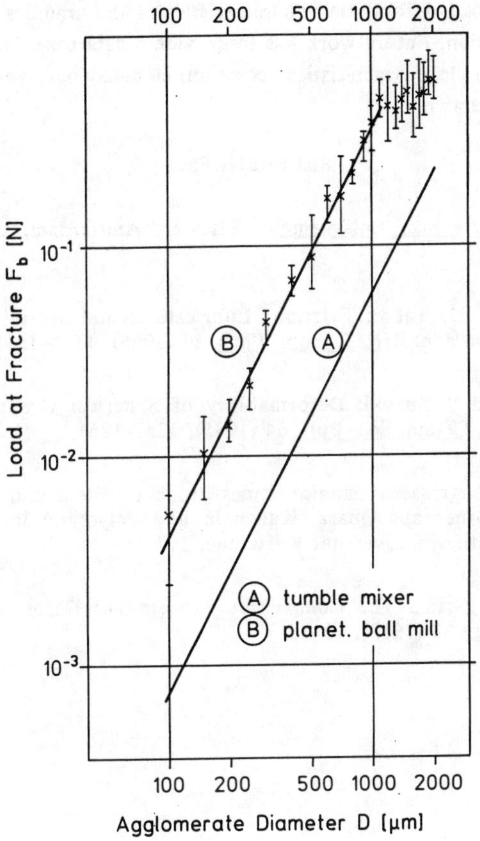

Figure 3. Agglomerate Strength of SSN Powder Mixtures

Strength measurements were carried out on granules with diameters of 100 to 2000 μm. The results in Fig. 3 contain the average, maximum and minimum measured load at fracture for powder (B), and the regression curves for both powders. The slope of these lines, calculated from the interval between 100 and 1000 μm granule diameter, is very close to 2, indicating the validity of equation (1). For the granules with $D > 1000$ μm a square-law equation holds not longer. The slope of the regression curve drops. This indicates a non-uniform distribution of interparticle bonding. Usually the slope of the regression line (in log-log-coordinates) varies between 1 and 2, indicating a concentration of the binding forces at a surface crust in the first, or a homogeneous distribution of the binding forces over the cross section in the latter case [1].

CONCLUSIONS

The strength tester described in this paper allows it to characterise the mechanical behaviour of press grade powders by single granule testing at reasonable efforts. Besides the crushing strength, the load-deformation characteristics of the granules seem to be suitable for further characterisation. Future work has to provide a data base, from which correlations between single granule characteristics, compaction behaviour, and microstructure of green compacts can be drawn.

REFERENCES

1. C.E. Capes: Particle Size Enlargement. Elsevier, Amsterdam, Oxford, New York, 1980

2. F.P. Bowden and D. Tabor: Friction, Lubrication, and Wear: A Survey of Work During the Last Decade. Brit. J. Appl. Phys. 17 (1966), 1521-1531

3. M. Takahashi and S. Suzuki: Deformability of Spherical Granules under Uniaxial Loading. J. Amer. Ceram. Soc. Bull. 64 (1985), 1257-1261

4. K. Steier: Die Korngrössenabhängigkeit des Bruchverhaltens und der Struktur druckbeanspruchter Körner aus Quarz, Kalkstein und Polystyrol im Bereich von 1 bis 1000 µm. Dissertation, Universität Karlsruhe, 1971

5. H. Kuno and J. Okada: The Compaction Process and Deformability of Granules. Powder Technology 33 (1982), 73-79

EFFECT OF ELECTRONIC DEFECTS AND OXYGEN VACANCIES ON MECHANICAL BEHAVIOUR OF OXIDE CERAMICS

K. M. LIANG*, G. ORANGE, G. FANTOZZI
GEMPPM, Bât. 502, INSA Lyon, 69621 Villeurbanne, France.
*Tsinghua University, Beijing, China.
M. GAUTIER, J. P. DURAUD, C. Le GRESSUS
CEA-IRDI-CEN Saclay, 91191 Gif S. Yvette, France.

ABSTRACT

The dependence of mechanical properties of ceramics on microstructure has been well studied: e.g. effect of porosity, grain size, grain boundary, etc. This work deals with the change in fracture properties of yttrium oxide with microscopic defects such as those induced by a non-stoichiometric state or by treatment under an electric field and under ultraviolet radiation.
Oxide ceramics with different stoichiometries (oxygen vacancies) are obtained after heat treatments in vacuo or in air. For an oxygen deficient Y_2O_3 material, a decrease of K_{IC} is observed from 3.5 to 2.3 $MPa{\cdot}m^{1/2}$ whereas no significant dependences are measured with the Vickers hardness. Similar effects are observed after treatment of Y_2O_3 specimens under electric field and ultraviolet radiation: K_{IC} decreases as point defect content is increased in the material. Fracture properties are restored after annealing treatment.
Results are discussed in terms of the evolution of the atomic bonding with the stoichiometry (electronic polarization). The electrostatic energy stored in a brittle material containing charged defects seems to play an important role in the energy balance controlling the crack propagation in ionic/covalent brittle materials.

1. INTRODUCTION

Ceramics have a set of interesting physical and mechanical properties. However the development of these materials is restricted by their brittleness. Nowadays, the fracture toughness of ceramics are being improved (by using finer grain materials, in addition of second phases, by phase transformation toughening, etc.) in order to modify the microstructure. But much less attention has been devoted to the effect of mass-transport and of electron polarization phenomena on the mechanical properties.
Minnear and Bradt[1] studied the fracture characteristics of TiO_{2-x} as a function of stoichiometry. In the case of carbide ceramics (B_4C), results show a close relationship between the B/C ratio with both toughness and hardness[2]. The changes of K_{IC} with that of the microstructure have been extensively studied[3,4]. Rice et al. reported that dense polycrystalline solids with cubic crystal structure showed no significant dependence of fracture energy on grain size. This was verified in the case of Y_2O_3 ceramics by Tani et al.[5]: the grain size dependence of the fracture toughness K_{IC} was almost constant from 0.2 µm to 214 µm. Moreover, no grain size dependence of Vickers microhardness was observed.
However, few studies deal with the influence of point defects such as those induced by non-stoichiometry on the fracture toughness K_{IC} of ceramic oxides as Y_2O_3. In this paper, we report the effect of non-stoichiometry and of electron polarization on the fracture toughness of Y_2O_3 ceramics.

2. EXPERIMENTAL PROCEDURE

2.1. Material Preparation

Dense polycrystalline Y_2O_3 was prepared from a pure powder* (5N) by means of hot isostatic pressing (HIP). In order to obtain a non-porous and fine-grained (~1 μm) material, the powder was first compressed at room temperature with a pressure of 250 MPa. It was subsequently annealed at 850°C using a pressure of 1.3 Pa. Hot isostatic pressing (HIP) was then carried out at 1450°C under 150 MPa for 45 min. Samples were first cut into plates of 1 mm thickness. After polishing, all samples were heated for 4 hours at 1500°C in air (atmospheric pressure), in order to "bleach" the intrinsic oxygen vacancies and the residual stresses resulting from both sintering and polishing procedures. After this treatment all samples had a white color (T1).
Part of the samples was treated at 1700°C in vacuo for two hours. This treatment led to the dissociation of the oxide[6], so a rather large number of oxygen vacancies was produced in this material. These samples submitted to such a heat treatment had a black color (T1 +T2).
Some of the white colored samples (T1) were then put under electric field ($E=10^5$ V/m) and under ultraviolet radiation, respectively, for different hours at room temperature. After the treatment for 120 hours, the samples were then annealed at different temperature (in air) for two hours.
Some of the black colored samples (T1 + T2) were further heated at 1500°C for 2 hours in air in order to restore the white color (T1 + T2 + T3). Other part of the black colored samples were heated at 1000°C in air for different times to observe the change of fracture toughness and hardness.

2.2. Measurements

The determinations of fracture toughness (K_{IC}) and Vickers hardness (H_v) were carried out by using a microindentation hardness test with the applied loads from 1N to 10N. The Vickers hardness was calculated by:

$$H = \frac{P}{\alpha a^2} \qquad (\alpha \approx 2) \qquad (1)$$

and the values of K_{IC} was measured by the indentation formula[7]:

$$14\left[\frac{K_{IC}\phi}{Ha^{1/2}}\right]\left[\frac{H}{E\phi}\right]^{0.4}\left[1 - 8\left(\frac{4\nu - 0.5}{1 + \nu}\right)^4\right] = (c/a)^{(c/18a\ -\ 1.51)} \qquad (2)$$

where K_{IC} is expressed in $MPa{\cdot}m^{1/2}$, H (or H_v) is expressed in GPa, ν is the Poisson's ratio (=0.3), ϕ is the constraint factor (≈3), a is the impression radius and c is the crack length. The value of K_{IC} calculated by this indentation formula is independent of any applied load and any crack profile (Palmqvist or median type).

The depth of tested surface layer changes from 10μm to 80μm according to the applied indentation load. The microstructure of the Y_2O_3 ceramics submitted to these different treatments was characterized using SEM, EXAFS, X-Ray diffraction and X-ray photoemission.

3. RESULTS AND DISCUSSION

3.1 Different Redox Treatments

Results of the materials submitted to different redox treatments are summarized in Table I. A reduction process of the ceramic decreases its toughness by 35% (from 3.5 to 2.3 $MPa{\cdot}m^{1/2}$). This change appears to be a reversible process, as a further oxidation treatment gives back the initial value. However the hardness only appears as a weak variation with the thermal treatments.
The changes of K_{IC} with the thermal treatments corresponding to different redox conditions cannot be attributed to an effect of the grain size (reported in Table I). Firstly, the samples submitted to the heat treatments T1 and T1 + T2 + T3 have the same fracture toughness but they have different grain sizes. Secondly, the samples submitted to the thermal treatments T1 + T2 (oxygen deficient) and T1 + T2 + T3 (oxygen rich) have a similar grain size, but have very different fracture toughnesses. Tani et al.[5] indicated that a difference of grain size from 1 μm to 10 μm cannot lead to a variation in K_{IC} of 30%. These results clearly show the effect of an oxygen deficiency on the fracture toughness of a ceramic oxide.

* Rhône-Poulenc (France)

Previously published results[9,10] show a relationship between an oxygen-deficit of the Y_2O_3 ceramics (non-stoichiometry) and its fracture behavior. A transgranular fracture mode occurs in stoichiometric specimens, whereas an intergranular fracture feature seems to be important in oxygen-deficient ones, where oxygen vacancies are concentrated primarily at the grain boundaries.

Point defects such as oxygen vacancies modify the elastic properties of Y_2O_3 and then its crack propagation behavior. On a microscopic scale, this effect must be related to the change in the nature of the Y-O bond, which tends towards a more covalent character upon reduction, decreasing therefore the bond strength.

The fracture toughness describes the resistance of a brittle material against the critical propagation of a crack. Indeed, most of the theories of brittle fracture involve a critical value for the stress intensity factor K_I (symmetrical opening mode) but not for the applied stress: when K_I is equal to or greater than K_{IC}, the crack is instable and propagates through the material. On the energetic view-point[11], the critical energy value G_{IC} (critical strain energy release rate) which must be dissipated to move the crack is related to K_{IC} and the Young modulus through the formula:

$$G_{IC} = \frac{K_{IC}^2}{E}(1 - \nu^2) \qquad (3)$$

in the case of plane strain. If the elastic coefficients remain constant, a decrease in K_{IC} corresponds to a decrease of G_{IC}: the fracture energy decreases from 57 $J.m^{-2}$ (stoichiometric sample) to 24.5 $J.m^{-2}$ (for an oxygen deficient sample).

Stoichiometry in oxide ceramics seems to affect K_{IC} much more than H. The hardness H is much more concerned to the deformation processes (related to elastic moduli and yield stress). These results show that point defects play an important role in dynamic processes such as crack propagation. As suggested in previous papers[12,13] they clearly set the problem of introducing an additional energy term in the energy balance describing the condition of crack propagation, which would take into account the electrostatic energy stored in a brittle material containing charged point defects.

3.2. Heat Treatment at 1000°C for the Black Colored Samples

The figure 1 shows that the fracture toughness (K_{IC}) and the ratio (K_{IC}/H_v) are function of the heat treatment time at 1000°C in air for the black colored specimens(T1+T2). The fracture toughness and the ratio of K_{IC}/H_v increase with the treatment time. On the other hand, the color of the samples changes from black to white progressively. These changes are consistent with that in the Table I. The ratio of K_{IC}/H_v can be defined[8] as a factor governing the brittle to ductile transition in the contact problem. The inverse quantity H_v/K_{IC} is proposed by Law and Marshall[8] as a convenient index of brittleness: the oxygen deficient samples appear more brittle than the stoichiometric ones. The oxygen deficient defects decrease as the treatment time increases, so the ratio of O/Y and the toughness increase. Minnear and Bradt[1] studied the TiO_{2-x} ceramics and observed that the fracture toughness and the fracture surface energy decrease with increasing x. Their results are consistent with our observations.

3.3. Treatment under Electric Field and Ultraviolet Radiation

Figures 2 and 3 show that the fracture toughness decreases with the treatment time under an electric field and under ultraviolet radiation for the stoichiometric specimens (T1).

When the stoichiometric ceramic, Y_2O_3, is put under an electric field, it will be locally charged, some atoms will be polarized, some bonds of Y-O will be broken when local potentials reach a critical value, so some point defects will appear. The atoms at the grain boundaries are more easily polarized than those inside the grains. Hence, a large number of point defects will be concentrated at the grain boundaries. The fracture toughness will decrease with increasing treatment time and tends to a limit (~2.3$MPa.m^{1/2}$). When the polarized specimens are annealed at low temperature 500°C for two hours, their toughness can return back to their original level (Fig 4). This means that the polarized atoms can restore back to their original equilibrium states with low temperature annealing.

The reason for the fracture toughness decreasing under ultraviolet radiation is the same as that under electric field. When the stoichiometric ceramic samples are put under ultraviolet radiation, some electrons of the material would jump from their equilibrium position. The electrons in the grain boundaries leave their equilibrium position more easily. Therefore, a lot of defects are formed at the grain boundaries in the surface of the material. These defects result in the material becoming more brittle.

4. CONCLUSION

Fracture toughness increases with the ratio of O/Y for non-stoichiometric Y_2O_3 ceramic and it attains a maximum when the material is in stoichiometric state. Hardness seems to be less dependent on stoichiometry.
Electrostatic field and ultraviolet radiation could polarize the atoms and change the electron distribution of the ceramics and make it to become brittle.
The electrostatic energy stored in a ionic-covalent material containing charged defects plays an important role in crack propagation processes.

REFERENCES

1. W. P. Minnear and R. C. Bradt, "Stoichiometry Effects on the Fracture of TiO_{2-x}", J. Am. Ceram. Soc. 63 (1980) 485-490.
2. K. Niihara, A. Nakahira, T. Hirai, "The effect of stoichiometry on mechanical properties of boron carbide", J. Am. Ceram. Soc. 67 (1984) C13, C14.
3. A. G. Evans, "The fracture toughness of ceramics", I. C. F. 4, 1 (1977) 529-556.
4. R. W. Rice, S. W. Freiman, P. F. Becher, "Grain Size Dependence of Fracture Energy in Ceramics", Part I (Experiment), J. Am. Ceram. Soc. 64 (1981) 345-350; Part II (A Model for Non Cubic Materials), J. Am. Ceram. Soc. 64 (1981) 350-354.
5. T. Tani, Y. Miyamoto, M. Koizumi, M. Shimada, "Grain Size Dependences of Vickers Microhardness and Fracture Toughness in Al_2O_3 and Y_2O_3 Ceramics", Ceramic Intern. 12 (1986) 33-37.
6. N. M. Tallan and R. W. West, "Electrical properties and defect structure of Y_2O_3", J. Am. Ceram. Soc. 49 (1966) 401-404.
7. K. M. Liang, G. Orange, G. Fantozzi, "Evaluation by Indentation of Fracture Toughness of Ceramic Materials", J. Mat. Sci., in press(1989).
8. B. R. Lawn, D. B. Marshall, "Hardness, Toughness, and Brittleness: An Indentation Analysis", J. Am. Ceram. Soc., 62 (1979) 347-350.
9. J. P. Duraud, F. Jollet, N. Thromat, M. Gautier, Ph. Maire, C. Le Gressus, E. Dartyge, "Non-stoichiometry of pure and Zr-doped yttria ceramics: an EXAFS and XPS study", submitted to J. Am. Ceram. Soc.
10. F. Jollet, Ph. Maire, M. Gautier, J. P. Duraud, C. Le Gressus, "Influence of Stoichiometry on the Eletrical and Mechanical Behavior of Yttrium Oxide Ceramics", J. Am. Ceram. Soc., 71(1988), in press.
11. J. F. Knott, "Fundamental of Fracture Mechanics", Butterworths (973).
12. C. Le Gressus, Ph. Maire, "Rôle des surfaces et des interfaces dans les phénomènes d'usure, d'adhérences et de claquage", J. Chim. Phys., 2, 295-304 (1987).
13. C. Le Gressus, Ph. Maire, "The role of surfaces et des interfaces in wear, adhesion and breakdown phenomena", Surf. Interf. Anal. 11, 219-228 (1988).

Table I: Fracture toughness (K_{IC}) and Vickers hardness (H_v) of Y_2O_3 ceramics submitted to different redox treatments (applied load from 1N to 10N)

Redox treatment*	Mean grain size	K_{IC} ($MPam^{1/2}$) ± 0.2	H_v (GPa) ± 0.25	$\frac{K_{IC}}{H_v}(\frac{m^{1/2}}{1000})$	$\frac{H_v}{K_{IC}}$
Y_2O_3(HIP) (polished)	/	3	8	0.375	2.67
T1	1 µm	3.5	7.7	0.455	2.2
T1 + T2	10 µm	2.3	7.3	0.315	3.17
T1 + T2 + T3	10 µm	3.4	7.4	0.46	2.17

* T1: 1500°C - 4 hours - in air;
T2: 1700°C - 2 hours - into vacuo (10^{-6} Torr);
T3: 1500°C - 2 hours - in air.

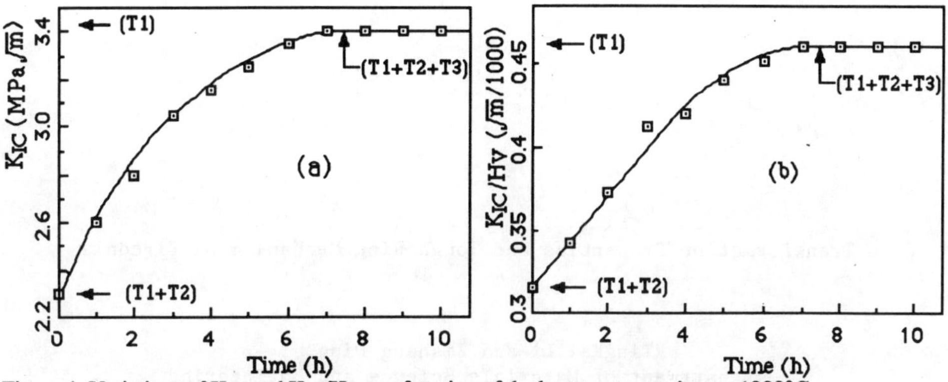

Figure 1: Variations of K_{IC} and K_{IC}/H_v as a function of the heat treatment time at 1000°C for black colored Y_2O_3, non-stoichiometric, submitted to T1 + T2 (applied load: 1N).

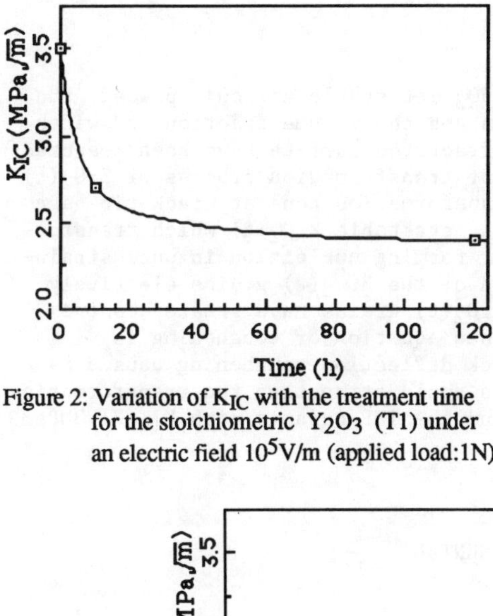

Figure 2: Variation of K_{IC} with the treatment time for the stoichiometric Y_2O_3 (T1) under an electric field 10^5V/m (applied load:1N).

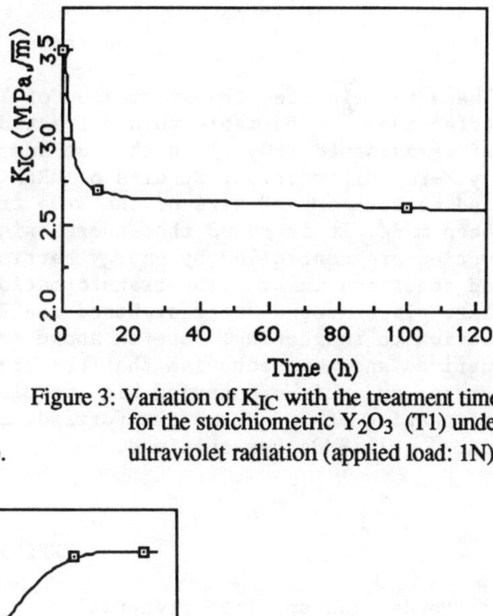

Figure 3: Variation of K_{IC} with the treatment time for the stoichiometric Y_2O_3 (T1) under ultraviolet radiation (applied load: 1N).

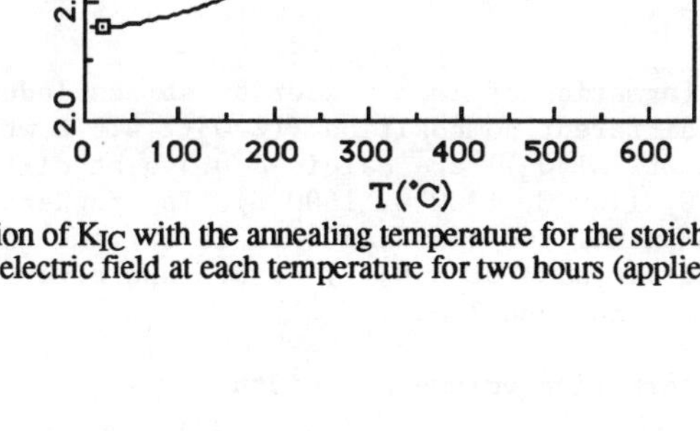

Figure 4: Variation of K_{IC} with the annealing temperature for the stoichiometric Y_2O_3 treated under electric field at each temperature for two hours (applied load: 1N).

Transformation Properties and Toughening Mechanism of Zirconia

TingKai Li and Zishang Ding
Department of Materials Science and Engineering
Zhejiang University, Hangzhou, P.R.China

Abstract

The stress-induced transformation of Y_2O_3 metastable zirconia powder, the effective transformable volume fraction and the volume fraction and width of transformed $ZrO_2(t)$ in the cut and fractured surface have been measured by X-ray diffraction. Studies of HREM of transformation process of $ZrO_2(t)$ and measurement of size of the ZrO_2 transformation zone at crack tip have been made. It is found that there exist metastable $ZrO_2(t)$ which transformation are controlled by energy barrier forming nucleation in unconstrained condition in ZTC. The transformation of the $ZrO_2(t)$ grains electively take place around defects. Hence the $ZrO_2(t)$ grains have greater contribution to toughening. Base on above a new equation of toughening is derived and new mechanism that the crack deflection toughening caused by grains and whiskers having high modulus could strengthen the transformation toughening of ZrO_2 have put forward, and the TZP ceramic with K_{Ic}=21.3MPam$^{\frac{1}{2}}$ and σ_f=1578MPa has obtained.

EXPERIMENTAL

1) Powder and specimen preparation
Powers of α-Al_2O_3 and Y_2O_3 partially stabilised zirconias(YMSZ) for the making of ZTA and TZP ceramics were prepared on the basis of calcining high purity alum ($Al_2(NH_4)_2(SO_4)_4 \cdot 24H_2O$) and coprecipitating mixed solutions of analytic pure $ZrOCl_2 \cdot 8H_2O$ and YCl_3·ZTA and TZP samples were produced by hot-pressing(1600°C, 20MPa).

2) Measurements of transformation of YMSZ powder by stress inducing
Seven YMSZ powders with different composition ((ZrO_2+2,4,6,8 wt% Y_2O_3), (ZrO_2+4wt% Y_2O_3)+5,20,82 wt% Al_2O_3)) are calcined 0.5 h at different temperature (700°C, 900°C, 1100°C, 1300°C, 1600°C). The powders are smashed 50 times by smashing eguipment. The smashing work is 6J/time and 12J/time. The volume fraction and grain size of $ZrO_2(t)$ before and after smashing are measured with X-ray diffraction and TEM.

3) Measurements of transformation volume and width

The effectively transformable volume fraction in ZTA and TZP ceramics and the transformed volume fraction in cut and fractured surfaces of the materials were measured by X-ray diffraction[10]. The width of transformation was estimated by two wavelength method of X-ray diffraction technique (CuKα and MoKα). Calculations for these parameters are as reference [2].

4) Study of TEM on stress induce-transformation
a) The thin film specimen for HREM were prepared by ion thining using conventional means. The ultra high resolution electron microscope JEM200cx with top entry goniometer and LaB_6 cathode was used to image stress induced-transformation of ZrO_2. This TEM routinely provides lattice resolution down to 0.144nm (cross).
b) In situ straining experiments were performed in a high-voltage electron microscope fitted with a double-tilting straining stage. The stress-induced tetragonal(t)-to monoclinic-(m) transformation was studied for ZrO_2(t) inclusions confined in ZTA and TZP.

5) Measurements of Fracture Strength and Toughness
The fracture strength of TZP samples was measured in three point bending using a 20-mm span. The fracture toughness was evaluated by both micro-indentation and chevron notched beam technique.

Results and discussion

1. The properties of stress induced-transformation of ZrO_2(t)
1) There are three types of ZrO_2(t) in ZTC
Figure 1 shows X_b, X_u and X_n, here X_u is volume fractions of ZrO_2(t) of powder calcined at 1600°C temperature that are approximately defined V_u, Hence, the $V_s=X_b-X_u$, V_u is the volume fraction of unconstrained ZrO_2(t) in ZTC. V_s is the volume fraction of constrained ZrO_2(t) in ZTC. X_n is approximately defined V_n that is the volume fraction of untransformable ZrO_2(t) under definite stress condition. Figure 1 shows that there exist three types of ZrO_2(t) in ZTC. a) The metastable ZrO_2(t) by the constraint of the matrix. Transformation of the ZrO_2(t) take place when the ZrO_2(t) particles are released from constrained condition. b) The ZrO_2(t) crystals are also metastable at free condition. In order to make the ZrO_2(t) transform. It is necessary to apply force on or do work. c) Stable ZrO_2(t) crystals.

2) Stress induced-transformation of ZrO_2(t) in powder ZrO_2(t) crystals in powder are approximately under unconstrain condition.
Experiments such as Fig 2,3 and [3]show that a) tne volume fraction of ZrO_2(t) increases with increasing of the content of Y_2O_3 and decreases with the increasing of calcined temperature. b) The grain size of ZrO_2(t) increases with the increasing of calcined temperature and there is no different in grain size before and after smashing. c) The smashing transformation volume fraction has relation to its transformable critical size and the distribution of its particle size d) The transformable stress $_T$ has relation to grain size. stabilizer content and temperature but the constrain of the matrix. e) The larger the applied force, the more the transformed ZrO_2(t).

3) Stress induced-transformation of ZrO_2(t) in ZTC
Experimental results showing the variations of σ_f, K_{IC}, V_e, V_f, V_{ff}, and V_c of ZTA and TZP ceramics with Y_2O_3 content are shown in Fig. 4 It is seen

from these figures that as Y_2O_3 content increases the fracture strength and toughness of the materials passes through a maximum whereas the effective transformable volume fraction of ZrO_2(t) particles in the materials and the volume fraction of transformed particles and the width of the transformation in the named surfaces decrease, and that the transformable volume fraction and the width of the transformation rise with the fracture strength in TZP materials. These results are in agreement with some other experimental observations [4], [5], [2].

4) The properties of ZrO_2(t) transformation at crack tip.
In ZTA, the ZrO_2 having no stabilizer exists as metastable ZrO_2(t) by matrix constraint. ZrO_2(t) grains transformed during crack propagation. The transformation zone is not only located at crack tip. This is because that metastable ZrO_2(t) by matrix constraint could transform when the constraint is removed. In TZP the transformation zone is only at crack tip and uncontinuous along the crack with increasing of stabilizer. This is because that the metastable ZrO_2(t) could transform only when stress is applied on. These phenomena are also found by M.Ruhle at [6]al. It shows that the transformation of the ZrO_2(t) grains electively take place around defacts.

5) Study of HREM on transformation of ZrO_2(t)
By studing of HREM on transformation of ZrO_2(t) it is found that there are dislocations in ZrO_2(t) grains from which the nohomogeneous nucleations form. Fig 5 shows the nohomogeneous nucleation forms inside ZrO_2(t) grain or at boundary. It is known that the formation and movement of dislocations are related to the ZrO_2 transformation as Fig 6. With high-energy boundaries, the associated stress aids the transformation and so assists the movement of dislocations.

2. The transformation toughening mechanism of ZrO_2

From above properties of stress-induced-transformation of ZrO_2 the contributions of ZrO_2 transformation to toughening come from two parts, one is dilational work done by ZrO_2(t) to ZrO_2(m) transformation, another is the work done by the applied stress necessary to initiate the transformation from analysis of fracture mechanics; or one is the change in chemical free energy G^c, another is the transformation energy barrier F associated with the ZrO_2(t) ZrO_2(m) from analysis of thermodynamics. Uing either Griffith, irwin or Evans method the following equations can be derived[2].

$$K_c = [K_o^2 + \frac{2E_c hV_f(G^c - U_{se}f) + 2E_c hV_{f1} F}{1-\nu^2}]^{1/2} \quad (1)$$

and $K_c=0.21E_cV_fe^Th^{1/2}/(1-)+1.26V_fh^{1/2}\ _T(1-2\nu)/(1-\nu)$ (2)

Under conditions where $_T=0$ or F=0, Eq(1),(2) becomes

$$K_c = [K_o^2 + \frac{2E_c hV_f(G^c - U_{se}f)}{1-\nu^2}]^{1/2} \quad (3)$$

which is lange's formula for transformation toughening.

$$K = 0.21E_cV_fe^Th^{1/2}/(1-\nu) \quad (4)$$

which is the formula of Evans.

under definite condition the eq.(1) can transform into eq.(2)[3].

under conditions where $\sigma_T=\sigma_c$, here σ_c is critical transformation stress when $K=K_o+K/2$. eq.(2) becames

$$K=0.63E_cV_fe^Th^{\frac{1}{2}}/(1-\nu) \quad (5)$$

the eq.(5) is similar to Evans' modified formula[7].

3. The crack deflection toughening caused by grains and whiskers could strengthen the transformation toughening of ZrO_2.

Additions of Al_2O_3 and SiC whisker[8] could increase K_{1c} and σ_f of TZP at the same time. This interesting phenomenon is explained by the following mechanism, the crack deflection and branch toughening caused by Al_2O_3 grains could enhance the transformation toughening of ZrO_2[9]. Controlling the content of Al_2O_3 and Y_2O_3 we obtained a TZP ceramic with K_{ic}=21.3MPam$^{\frac{1}{2}}$ and σ_f=1578MPa.

Conclusions

1) There exist three types of ZrO_2(t) in ZTC, the metastable ZrO_2(t) by the constraint of the matrix, metastabe ZrO_2(t) under unconstrained condition and stable ZrO_2(t).

2) In order to make metastable ZrO_2(t) under unconstrained condition which transformation are controlled by energy barrier forming nucleation transform, It is necessary to apply force on or do work. Its transformation electively take place aroud defects at which the stress is concentrated. Hence the ZrO_2(t) grains have greater contribution to toughening.

3. The contributions of ZrO_2 transformation to toughening come from two parts, one is dilational work done by ZrO_2(t) to ZrO_2(m) transformation, another is the work done by the applied stress necessary to initiate the transformation from analysis of fracture mechanics; or one is the change in chemical free energy G^c, another is the transformation energy barrier F associated with the ZrO_2(t) to ZrO_2(m) from analysis of thermodynamics. Based on above a new equation of toughening is derived.

4. The crack deflection toughening caused by grains and whiskers having high modulus could strengthen the transformation toughening of ZrO_2 and TZP ceramics with High strength and toughness are made.

Keywords: Transformation toughening, whisker and grain, structure ceramics and ZrO_2, Al_2O_3, Sic.

Reference

[1] R.C.Garvie, R.H.Hannick, and R.T.Pascoe, Ceramic Steel, Nature 258 (1975) 703

[2] TingKai Li, Suxing Wu and Zishang Ding, Models of Transformation Toughening in ZrO_2-Alloys, Science of Ceramics 14 (1987) 745

[3] TingKai Li, Transformation Toughening Mechanism and Mechanical properties of Zirconia Toughened Ceramics, Ph.D. Thesis (1987)

[4] Kosmac, T., Swain, M.V. and Claussen, N., Mater. Sci.Eng., 71, 65 (1985)

[5] Li,B., Huang, X., Guo, J. and Yan, D., Jour. Inorg.Mater (Chinese) 1, (1986) 129

[6] Ruhle, M., Strecker, A. and Waidelich, D., In Situ Observations of Stress-Induced Phase Transformations in ZrO_2-Containing Ceramics, Advances In Ceramics, Vol.12, (1984) 256

[7] Heuer, A.H., Lange, F.F., Swain, M.V. and Evans, A.G., Jour. Amer. Ceram. Soc., 69(3) (1986);

[8] TingKai Li, Zhijian Shen, Zishang Ding., Unpublished work.

[9] TingKai Li, Zhijian Shen, Zishang Ding., Tetragonal Zirconia Polycrystal Ceramics With High Toughness and Strength, 3rd International Symposium Ceramic Materials £ Component For Engines (1988) USA

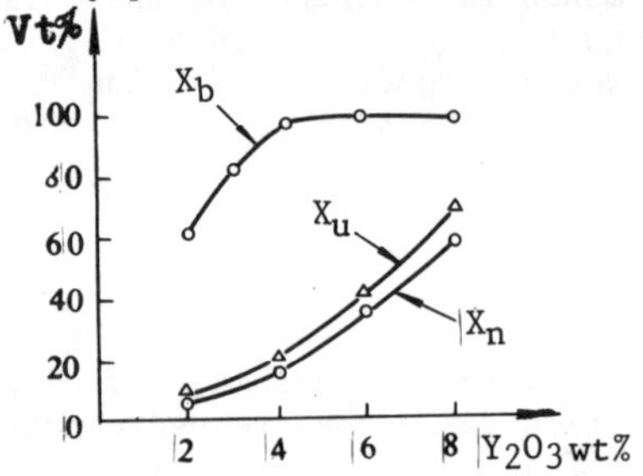

Fig 1. The X_b, X_u, X_n V_x Y_2O_3 Content in TZP

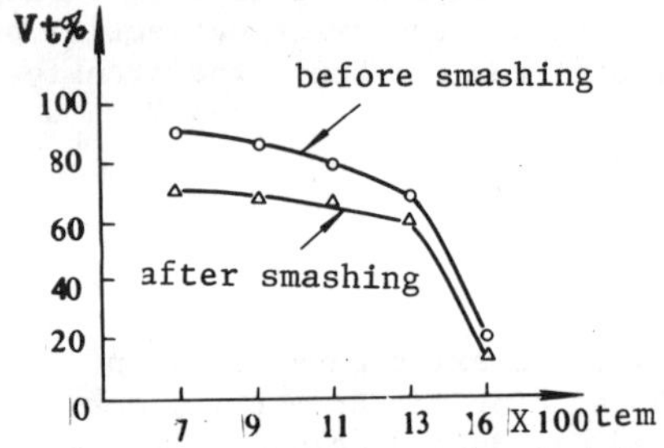

Fig 2 volume fraction of ZrO_2(t) of Y_4MSZ powder V_s temperature

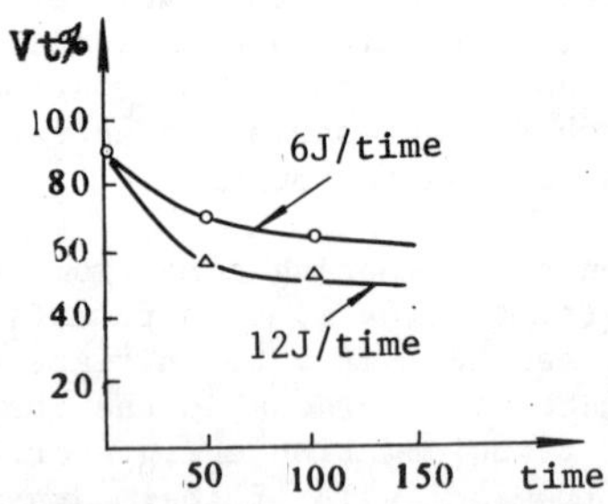

Fig 3 The influences of smashing work and time on V_t of ZrO_2

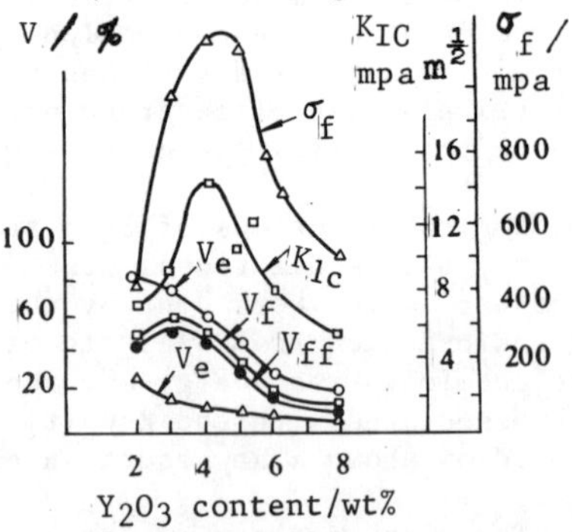

Fig. 4 The influences of Y_2O_3 content on f, K_{IC}, V_e, V_c, V_f, and V_{ff} of TZP ceramics.

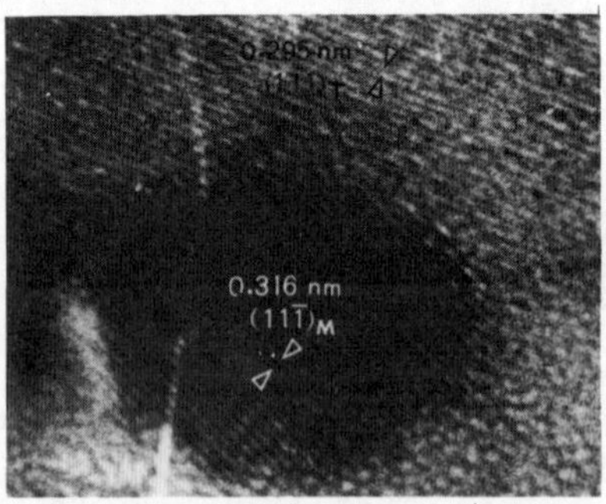

Fig 5. Nohomogeneous nucleation forms inside ZrO_2(t) grains

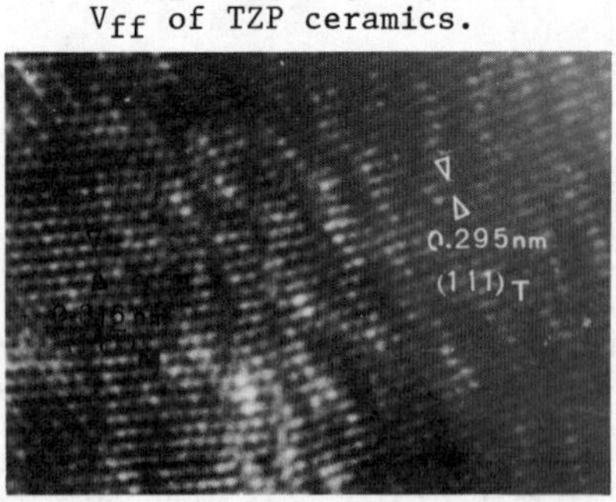

Fig 6 The formation and movement of dislocations are related to ZrO_2 transformation

TETRAGONAL ZIRCONIA POLYCRYSTALS IN THE ZrO_2-TiO_2-CeO_2 SYSTEM

V.C. PANDOLFELLI, W.M. RAINFORTH AND R. STEVENS
Division of Ceramics, School of Materials, Leeds University,
Leeds LS2 9JT, England.

ABSTRACT

Fully dense tetragonal zirconia polycrystalline ceramics have been obtained after sintering a 77.9 ZrO_2-16.5 TiO_2-5.6 CeO_2 mol% composition at 1350°C for 2 hours. The tetragonal phase showed high resistance to stress induced martensitic transformation, which was associated with the presence of short range ordering of the anion lattice. Indications of short range ordering were based on: (a) the presence of superlattice spots for particular directions in the T.E.M. diffraction, (b) the presence of extra peaks on the tetragonal XRD profile, and (c) the high c/a lattice ratio (1.0327 $\pm$ 0.0001) determined for this composition.

INTRODUCTION

The retention of a fully tetragonal zirconia structure at room temperature is most likely at the eutectoid composition. In this paper the composition 77.9 ZrO_2-16.5 TiO_2-5.6 CeO_2 mol%, which retains both the ratio of TiO_2/TiO_2 + ZrO_2 and the CeO_2/CeO_2 + ZrO_2 at their eutectoid level was prepared and fully dense tetragonal polycrystalline ceramics were obtained after sintering at 1350°C for 2 hours. The tetragonal zirconia that formed under such conditions showed a high resistance to the stress assisted transformation to the monoclinic phase. The low transformability in this system is surprising considering that: (a) previous attempts to retain tetragonal zirconia at room temperature in the ZrO_2-TiO_2-CeO_2 system were unsuccessful for compositions containing less than 15 mol% of CeO_2, even with the addition of up to 6 mol% of TiO_2 [1], and (b) fully dense TZP ceramics could not be obtained in the ZrO_2-TiO_2 system [2] or with less than 8 mol% CeO_2 in the ZrO_2-CeO_2 system [3].

A description of the general properties and possible variables that are controlling the stability of this particular tetragonal zirconia composition in the ZrO_2-TiO_2-CeO_2 system is presented below.

MATERIALS AND METHODS

The composition was prepared by the dropwise addition of zirconium n-propoxide and titanium isopropoxide in a continuously stirred solution of distilled water and $Ce(NO_3)_3$-$6H_2O$ kept at pH = 3. In such conditions the Zr and Ti alkoxides precipitate whilst allowing the cerium nitrate to remain in solution. Using additions of NH_4OH, the alkalinity was adjusted to pH = 10 and the Ce^{4+} compound homogeneously precipitated dispersed among the existing small zirconium/titanium agglomerates. After repeated filtration and washing with isopropyl alcohol, the precipitate was calcined at 950°C and then ball milled using zirconia cylinders.

Transmission electron microscopy was carried out on a JEOL 200CX (200 kV) fitted with a Link EDS detector. A Philips PW diffractometer with Ni-filtered CuKα radiation was used for the X-ray analyses. The mechanical measurements were performed on an Instron M1026 testing machine, using a four point bend test rig. The S.E.N.B. technique [4], with samples containing a 100 μm notch width, was employed for the fracture toughness measurements (K_{Ic}). The load-time curve area of a stable fracture from chevron notched (280 μm) samples, was used for the fracture energy (γ_f) measurement [5].

RESULTS AND DISCUSSION

The tetragonal zirconia polycrystalline ceramic obtained in this study was found to be highly resistant to stress assisted transformation to the monoclinic phase. For the sample sintered at 1350°C for 2 hours no surface $t \rightarrow m$ transformation could be detected on the as-ground or fractured surface or even after being immersed in liquid nitrogen. After pulverising the ZrO_2-TiO_2-CeO_2 sample for 10 minutes in a WC shatter box, X-ray diffraction of the resulting powder indicated only 9% by volume of the monoclinic phase. The absence of transformation is considered to be responsible for the low strength and particularly the low toughness values measured in this material (Table 1). The high stability of the tetragonal phase in this system is surprising considering the large grain size (Fig. 1), the levels of CeO_2 and TiO_2 present and given the ready transformability of the tetragonal zirconia in the binary systems.

TABLE 1
General properties of 77.9 ZrO_2-16.5 TiO_2-5.6 CeO_2 mol% composition sintered at 1350°C for 2 hours

Density (g/cm^3)	Grain size (μm)	σ (MPa)	γ_f (J/m^2)	K_{Ic} ($MPa.m^{0.5}$)	Tetragonal phase (Vol%)
5.88	2.7	347*	51*	4.1*	100

*Average of 5 measurements.

When examined in the T.E.M. the tetragonal grains showed a fine scale mottled structure (Fig. 2). The mottling appeared to be associated with the presence of superlattice spots in the diffraction pattern, which were invariably strongest for $\langle 122 \rangle_t$ perpendicular to the foil surface. Some zones systematically failed to give extra spots, for example $\langle 100 \rangle_t$, thus excluding the possibility of their being generated by the monoclinic phase [6,7]. Extensive streaking of the fundamental reflections along $\langle 201 \rangle_t$ was found in those diffraction patterns in which these directions were present. This streaking is suggestive of fine disc shaped precipitates and/or lattice strain along a specific direction.

Figure 1. S.E. micrograph of thermally etched surface of the 77.9 ZrO_2-16.5 TiO_2-5.6 CeO_2 mol% composition sintered at 1350°C for 2 hours.

The results of T.E.M. examination together with the high stability of the tetragonal phase could indicate the possible occurrence of a short range ordering in this composition. Since ordering maximizes the number of unlike neighbours for a given composition of differing atoms subjected to attractive forces, the short range ordering gives rise to a more stable configuration. This would be expected to contribute to the stability of the tetragonal phase, as was found for the 77.9 ZrO_2-16.5 TiO_2-5.6 CeO_2 composition.

Other results associated with the presence of short range ordering in the ZrO_2-TiO_2-CeO_2 composition were obtained by X-ray diffraction. The XRD profile of the tetragonal phase showed additional weak peaks corresponding to 4.928 Å, 2.338 Å and 2.027 Å interplanar spacings, which could not be explained as being due to any other crystalline phase. In addition the c/a tetragonal lattice parameter ratio for the ZrO_2-TiO_2-CeO_2 composition was found to be significantly greater than was shown by the ZrO_2-CeO_2 or the ZrO_2-Y_2O_3 systems (Table 2).

TABLE 2
Tetragonal lattice parameter ratio (c/a) and lattice volume for TZP ceramics

Composition (mol%)	c/a	Vol (mm^3)
97.5 ZrO_2-2.5 Y_2O_3	1.0163 ± 0.0002	0.1344
88 ZrO_2-12 CeO_2	1.0190 ± 0.0001	0.1398
77.9 ZrO_2-16.5 TiO_2-5.6 CeO_2	1.0327 ± 0.0001	0.1389

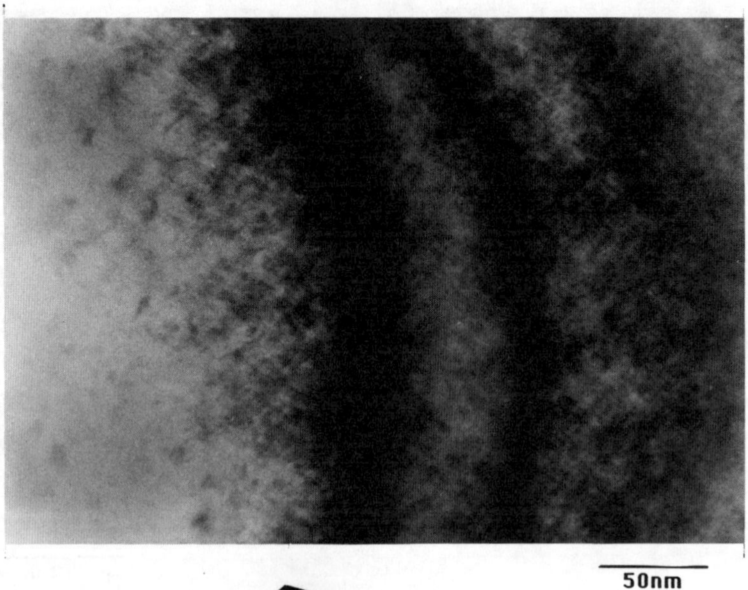

(A)

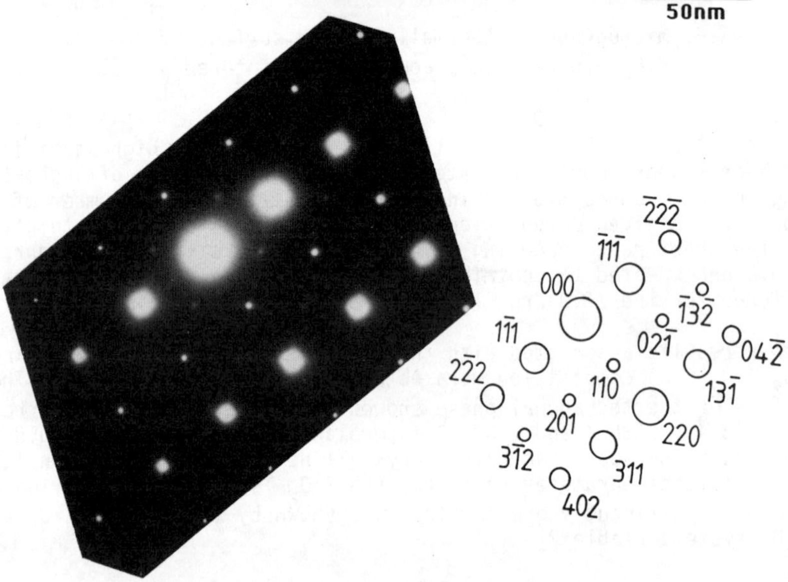

(B)

Figure 2. (A) Bright field micrograph of the tetragonal phase in the sample sintered at 1350°C for 2 hours showing a mottled structure on a very fine scale.

(B) Selected area diffraction pattern for Figure A. B = $\langle 1\bar{1}\bar{2}\rangle$ with superlattice spots at {110} and {201} and streaking along $\langle 201\rangle$.

In order to confirm the presence of any ordered lattice, attempts were made to form a lattice fringe image from the superlattice reflections. These reflections were found to be too faint in regions where the foil was thin enough to give phase contrast and only fringes derived from the tetragonal lattice could be imaged. Further T.E.M. studies will be carried out in this area.

CONCLUSIONS

Tetragonal zirconia polycrystalline ceramics have been obtained in the ZrO_2-TiO_2-CeO_2 system. The high resistance to the stress assisted $t \rightarrow m$ transformation demonstrated by the 77.9 ZrO_2-16.5 TiO_2-5.6 CeO_2 mol% composition is believed to be mainly associated with the presence of short range ordering in this tetragonal zirconia. In the sample sintered at 1350°C for 2 hours, evidence for the presence of short range order was based on (a) the observation of extra spots for particular directions in the T.E.M. diffraction pattern, (b) the presence of additional weak peaks on the XRD profile of the tetragonal zirconia, which could not be explained on the basis of possible crystalline primary phases in this system, and (c) the c/a tetragonal lattice parameter ratio, which is higher than for any TZP ceramic reported in the literature.

ACKNOWLEDGMENTS

V.C. Pandolfelli is grateful to CNPq project no. 2522/84 for sponsoring his studies at Leeds University.

REFERENCES

1. Longo, V. and Podda, L., Ceramurgia, 1974, **4**, 3-10.
2. Pandolfelli, V.C., Nettleship, I. and Stevens, R., to be published in Complex Microstructures, ed. R. Stevens, Proceedings of British Ceramic Society Vol. **42**, 1989.
3. Tsukuma, K., Am. Ceram. Soc. Bull., 1986, **65**, 1386-89.
4. Pabst, R.F. In Fracture Mechanics of Ceramics, Vol. 2, ed. R.C. Bradt, D.P.H. Hasselman and F.F. Lange, Plenum Press, 1974, pp. 555-65.
5. Nakayama, J., J. Am. Ceram. Soc., 1965, **48**, 583-87.
6. Muddle, B.C. and Hannink, R.H.J., J. Am. Ceram. Soc., 1986, **69**, 547-55.
7. Bestgen, H., Chaim, R. and Heuer, A.H., J. Am. Ceram. Soc., 1988, **71**, 826-31.

MICROWAVE PROPERTIES OF ZrO_2-TiO_2-SnO_2 CERAMICS

D M Iddles, A J Bell and A J Moulson[1]

Cookson Group plc, Central Research, Egham, UK
[1]School of Materials, University of Leeds, Leeds, UK

Abstract

Commercially available microwave dielectrics often incorporate sintering aids and dopants to reduce processing temperatures and modify the dielectric properties. However, the mechanism through which these additives influence dielectric loss is not obvious. The role of selected sintering aids and dopants, their effect upon microstructural development and their correlation with dielectric loss at microwave frequencies is investigated. For specimens of density greater than 90% of theoretical, the influences of bulk defect chemistry upon dielectric loss appear to dominate those of the microstructure.

1. Introduction

Microwave communication systems require tight frequency control to maximise information density. Filters are therefore common elements in such circuits. Conventional air cavity resonators are bulky and expensive to manufacture. However, the use of high relative permittivity, low loss dielectrics, in the form of dielectric post resonators, coaxial resonators or as substrates for strip-line resonators, allows significant size reduction[1]. The "Q" of a dielectric resonator, which determines its selectivity and attenuation, is directly related to the dielectric loss of the constituent material, $Q \approx 1/\tan\delta$, where $\tan\delta$ is the dielectric loss factor. Hence, materials with low dielectric loss factors are preferred. In addition, good temperature stability of resonant frequency is also sought. The temperature coefficient of resonant frequency, $\tau_f = 1/f_o.df_o/dT$, where f_o is the resonant frequency and T is temperature, is often quoted.

There are a number of materials which have been found to offer a combination of properties which are suitable for the fabrication of microwave dielectric resonators and substrates. These have properties in the range : $10 < \epsilon_r < 90$, $\tan\delta < 3 \times 10^{-4}$ and $-10 < \tau_f < 10$ MK^{-1}. Zirconium tin titanate (ZTS), a solid solution of ZrO_2, TiO_2 and SnO_2, or $Zr_xTi_ySn_zO_4$ where x+y+z=2, offers an excellent compromise of properties and ease of processing for a wide range of applications. It has been the subject of a number of publications, reporting on its structure, range of existence and dielectric properties[2]. Compositions with ϵ_r = 35, Q = 10000 and τ_f = 0 MK^{-1} are readily identified.

It has been found that ZTS powders do not readily sinter by solid state diffusion, therefore liquid phase sintering aids, such as ZnO, have been invariably added to achieve good densification at temperatures between 1200°C and 1400°C. There are a number of publications and patents which report upon the efficacy of a variety of sintering aids and dopants[2,3 & 4] which are employed to increase densification, control microstructure and enhance Q. This work attempts to clarify the effects of the different types of additives and microstructural features upon Q at microwave frequencies. These include liquid phase sintering aids, (e.g. ZnO), "acceptor" and "donor" dopants (e.g. La^{3+} and Nb^{5+}), porosity, grain boundary phases, and grain size.

2. Experimental

The materials were prepared by conventional solid state synthesis. A single ZrO_2-TiO_2-SnO_2 composition was used throughout, this being $Zr_{0.875}Ti_{0.875}Sn_{0.25}O_4$, subsequently referred to as ZTS. To determine the influence of controlled porosity on Q, additions of small size polymethyl methacrylate (PMMA) spheres were made to ZTS powders. Two sizes of sphere were used, 4 and 90µm. A ZnO-modified material was used in order to avoid the effects of uncontrolled porosity present in samples of "pure" ZTS. Burn out of the spheres and sintering of the pellet were accomplished by following a ramp of 1°C min^{-1}, with a 2 hour hold at 200°C, 340°C, 600°C and 1000°C, with a final soak at 1350°C for 4 hours. The fired densities were measured by the water immersion technique. The microwave properties, ϵ_r and Q, were measured using a modified Hakki and Coleman technique.[5]

To study the effects of different additives and sintered grain size upon dielectric properties, three materials with different additives were fired for various times. The compositions studied were:

ZTS + 1.0 wt% ZnO	A
ZTS + 1.0 wt% ZnO + 0.5 wt% La_2O_3	B
ZTS + 1.0 wt% ZnO + 0.5 wt% Nb_2O_5	C

Pressed pellets of each composition were sintered at 1350°C for up to 128 hours. After firing, the weight loss, density, ϵ_r, Q, τ_f and grain size were all measured.

3. Results

The behaviour of and Q as a function of controlled porosity, or percentage of theoretical density, can be seen in Figure 1. For materials of > 90% theoretical density, Q appears to be independent of porosity within the ± 10% accuracy limits of the measurement technique. However, there is a marked decrease in Q with increasing porosity once the density has fallen below 90%.

The influence of sintering time on weight loss, Q and grain size

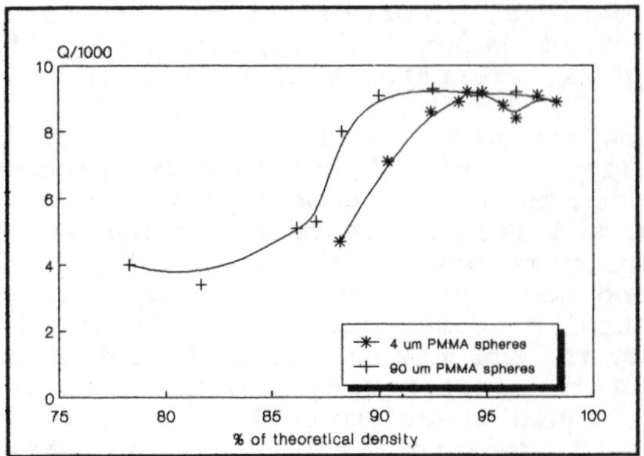

Figure 1 Q as a function of density.

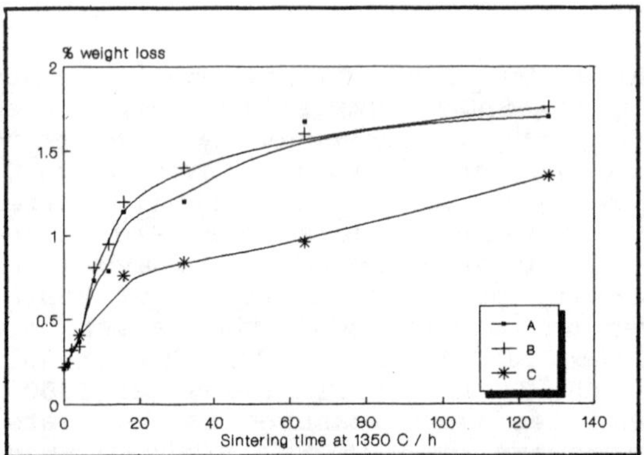

Figure 2 Weight loss as a function of sintering time.

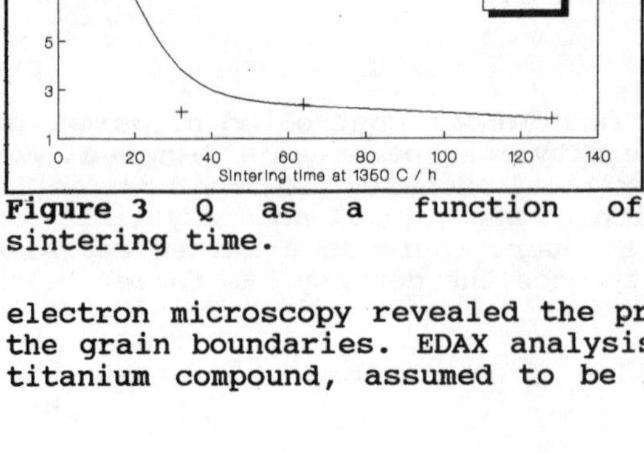

Figure 3 Q as a function of sintering time.

for the three compositions can be seen in Figures 2 to 4. No change, in either phase content or lattice parameters, as a function of sintering time was detected by XRD analysis. Optical microscopy revealed no evidence of major second phases. However, ZnO-rich phases, not resolved by optical microscopy, are presumed to exist at the grain boundaries.Figure 2 illustrates that all compositions lose mass as a function of sintering time. X-ray fluoresence (XRF) analysis of the cation composition demonstrates that up to 90% of the ZnO is volatilised during sintering to 128 hours.

Q measured at 4 GHz, as a function of sintering time, is shown in Figure 3. The Q of compositions A and C are both seen to increase with time, although the value for C, the Nb^{5+}-doped material, is consistently higher than that of composition A sample. The Q of composition B (La^{3+}-doped) initially increases as a function of sintering time to a peak of 10000 but after 10 hours begins to decrease, until after 128 hours at temperature it has a value of only 2000.

Grain size as a function of sintering time is shown in Figure 4. After 128 hours at temperature, the La^{3+}-doped material (composition B) has twice the grain size of the Nb^{5+}-doped compositions (C).

Samples examined by electron microscopy revealed the presence of a second phase at the grain boundaries. EDAX analysis showed this to be a zinc-titanium compound, assumed to be in the form of Zn_2TiO_4. The

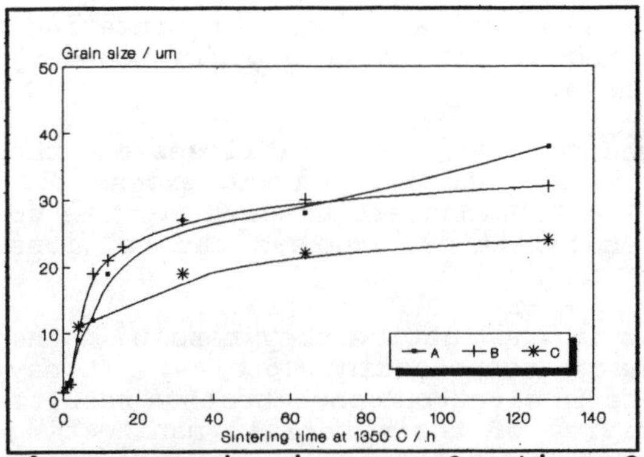

Figure 4 Grain size as a function of sintering time.

presence of La^{3+} ions in the zinc titanate phase was confirmed for composition B, but Nb^{5+} ions were not found to be localised in the grain boundary phase of composition C and were shown to be evenly distributed throughout the matrix.

4. Discussion

It is assumed that the decrease in Q, due to increasing porosity may be attributed to the effects of interfacial polarisation at pore surfaces. However, it might be expected that there should be some dependence upon pore size, with smaller pores producing a more marked degradation of Q. Although this appears to be the case, the effect is perhaps not as marked as might be expected from a consideration of the relative pore surface areas. The fact that Q is virtually independent of porosity for densities greater than 90% of theoretical allows the effects of porosity to be ignored from the following discussion of the influence of sintering aids and dopants upon dielectric properties.

Grain growth as a function of sintering time is a feature of all the compositions studied. As the behaviour of Q, expressed as a function of grain size, is markedly different for the three compositions, it may be assumed that grain size, per se, does not influence Q.

The improved densification achieved through the addition of ZnO is due to liquid phase sintering. The liquid phase, as shown by EDAX and TEM, is in the form of zinc titanate, Zn_2TiO_4. The dielectric properties of this composition have been measured by Wakino[3] as $\epsilon_r = 37$ and tan $\delta > 0.1$ at 1 MHz. The presence of minor amounts of this phase at grain boundaries in ZTS is certain to lead to a reduction in the Q of the ceramic.

It has been shown that on prolonged sintering ZnO volatilizes from the material, suggesting that the amount of second phase decreases as a function of sintering time. This is consistent with the result shown in Figure 7 for composition A, in which Q increases with decreasing ZnO content. With increasing sintering time the composition approaches that of unmodified ZTS and the Q may be assumed to approach the intrinsic value of the pure material.

The La^{3+} ion is some 30% larger than the largest of the principal cations, (Zr^{4+}), and therefore cannot be accommodated as a substituent. The ZTS lattice does contain a number of large interstitial sites and La^{3+} ions may be assumed to occupy these if necessary. However, during the initial stages of sintering, if necessary. However, during the initial stages of sintering, La^{3+} is considered to be located in the Zn-Ti-O grain boundary phase, as shown by the EDAX data.

During sintering for prolonged periods, ZnO volatilizes and the composition of the grain boundary changes, leaving excess TiO_2 and La_2O_3 to enter the bulk. The Ti^{4+} ions are assumed to take up their conventional sites in the lattice, whereas the La^{3+} ions enter as interstitials:

$$La_2O_3 \longrightarrow 2La_i^{\cdot\cdot\cdot} + 3/2O_2 + 6e' \quad (1)$$

Once the La^{3+} ions are present in the lattice they result in the formation of electrons as charge compensating species. It may be postulated that the increase in electron concentration results in a decrease in the resistivity of the material, manifesting itself as a decrease in the dielectric Q at microwave frequencies.

Nb^{5+} ions are of similar ionic radius to the cations of the ZTS lattice and therefore may be assumed to have some solubility as substituents for either Zr, Ti or Sn. EDAX investigations confirm that Nb does not segregate to the grain boundaries and is present throughout the lattice. This distribution remains unaffected by changes in composition of the grain boundary phase as a function of sintering time. The observed changes in Q are therefore similar to those seen for composition A due to the disappearance of the grain boundary phase. That is, Q increases as a function of sintering time. It is apparent, though, that the Q of the Nb doped material is consistently higher than that of the undoped material (composition A). It is thought that Nb^{5+} ions are able to compensate for the effects of small quantities of common impurity ions, such as Fe_2O_3 and Al_2O_3, which may be present in the ZTS. The impurities are thought to substitute onto the cation sites producing oxygen vacancies as lattice defects:

$$Fe_2O_3 \longrightarrow 2Fe_{Zr}' + 3O_o^x + V_o^{\cdot\cdot} \quad (2)$$

The presence of Nb^{5+} ions can reduce the number of oxygen vacancies which result from the presence of impurities, minimising the number of lattice defects:

$$Nb_2O_5 \longrightarrow 2Nb_{Zr}^{\cdot} + 4O_o^x + \tfrac{1}{2}O_2 + 2e' \quad (3)$$

$$Fe_2O_3 + Nb_2O_5 \longrightarrow 2Fe_{Zr}' + 2Nb_{Zr}^{\cdot} + 8O_o^x \quad (4)$$

As discussed by Wakino[6], the intrinsic microwave Q of this class of materials is a function of the lattice vibrational modes. Q may therefore be expected to vary as a consequence of changes to the material which influence the vibrational spectra. A decrease in the oxygen vacancy concentration, might therefore be expected to produce an increase in Q, as demonstrated by the present work.

6. References

1. J.K. Plourde and Chung-Li Ren, IEEE MTT 29 754, (1981)
2. G. Wolfram and H.E. Gobel, Mat.Res.Bull, 16, 1455, (1981)
3. K. Wakino, K. Minai and H. Tamura, J.Am.Ceram.Soc., 67, 278, (1984)
4. U.S. Patent 4,785,375 TAM Ceramics, Inc. (1988)
5. B.W. Hakki and P.D. Coleman, IRE Trans. MTT, 8 402, (1960)
6. K. Wakino, M. Murata and H. Tamura, J.Am.Ceram.Soc., 69, 34,(1986)

ELABORATION AND CHARACTERIZATION OF BORON RICH BORON CARBIDE

Dominique GOSSET, Bernard KRYGER
Centre d'Etudes Nucléaires, DTech/SECS/LMA , Bat. 524
91191 Gif/Yvette cedex , FRANCE

ABSTRACT

Boron carbide is used as a neutron absorber in French nuclear power plants. In order to improve its efficiency, we have elaborated boron rich boron carbides B_nC (compositions from B_4C to $B_{10}C$) . The best material (density > 90%, grain size < 20μm) is obtained by hot-pressing a mixing of B_4C and boron powders protected with a tantalum lining. We have performed cell parameters measurements on pellets. The results we obtain corroborate those previously published. At last, we deduce from thermal properties determinations that heat conductivity of B_8C is three times lower than B_4C's one. In-pile tests of B_8C are in progress in the French PHENIX FBR reactor.

INTRODUCTION

Boron carbide is one of the best neutron absorber thanks to the high neutron absorption cross section of ^{10}B:

$$^{10}B(B_nC) + {}^1n \rightarrow {}^4He + {}^7Li + 2.7\ MeV$$

It is used as control-rods material in French PWR and FBR. The best behavior is obtained with :
- High density pellets (to increase neutron absorption efficiency)
- Low grain size and opened porosity (to allow He release) ;
- High thermal properties (to reduce thermal stresses) .

Nuclear industrial materials have the B_4C (+ free carbon) composition but the boron carbide phase ranges from B_4C to $B_{10}C$. A B_nC boron-rich boron carbide would then have many advantages :
- Higher boron content (up to 15 at.%) or same boron content but lower density and lower helium-induced swelling and thus increased life-time ;
- Lower carbon activity and thus lower cladding carburization.

We have elaborated boron rich boron carbide pellets B_nC (n=4 to 10) ; they have been characterized by X-Ray analysis, microstructure and thermal properties determination.

ELABORATION OF THE PELLETS

Preparation of powders

The boron rich composition is obtained by mixing boron carbide and pure boron powders. A homogeneous mixing is obtained after several steps :

Dispersion of the B_4C powder clusters in liquid nitrogen ;
Mixing of B_4C and boron powders in pure water ;
Drying at low temperature (< 50°C) ;
Sieving (< 100 µm) .

Hot-pressing conditions

Obtaining high density and low grain size B_nC pellets requires high temperature (≃ 2000 °C) and pressure (≃ 80MPa). B_nC has then to be hot-pressed in a graphite die. Due to the high reactivity of B_nC , a lining has to be interposed between the die and the powder (Fig.1) .

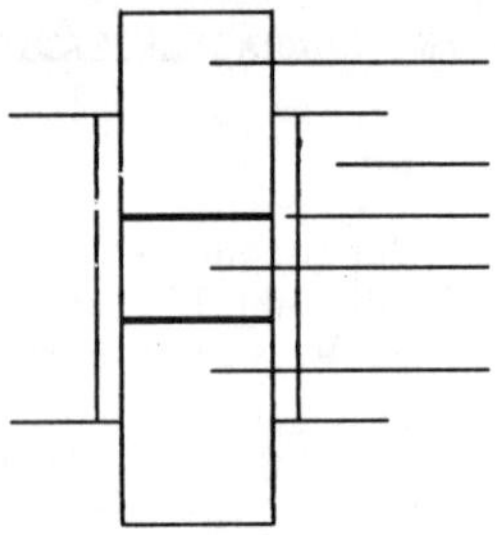

Figure 1. Hot pressing die

We have tested several materials (graphite foils, HIP boron nitride, tungsten and tantalum foils) . The best results have been obtained with HIP BN and Ta foils with the following uses :
HIP BN : low density pellets (e.g. for measurements of cell parameters) ;
Ta foil : high density pellets (e.g. for irradiation tests) .

Hot-pressing is performed in vacuum (≃ 10 Pa) . The furnace has a graphite heater and temperature is raised at 1000 °C/hour. After heating, the die is left to cool naturally (≃ 30 °C/mn) .

Some pellets have been hot-pressed in a two-steps process :
-1st step without pressure to allow the B_4C+B reaction ;
-2nd step with pressure for hot-pressing.

Some results are reported in table 1. We note here that hot-pressing of B_8C requires temperatures 150 °C lower than B_4C's ones in order to obtain the same density.

Table 1 : Hot pressing conditions and properties of B_nC pellets

PEL. n°	hot pressing conditions: 1st step 1800°C,60mn no pressure	pres. MPa	temp . °C	time mn	density g/cm^3	density %	composition B_T wt.%	C_T wt.%	n B_nC
1609	no	61	2000	30	2.49	99	83.1	11.3	8.2
1620	no	81	1950	60	2.13				
1631	yes	95	1950	60	2.42	97	83.2	12.1	7.6
1639	yes	84	1920	60	2.47				
1646	yes	97	1880	30	2.15	87	83.1	13.2	7.0
1659	yes	97	1900	30	2.30	93	84.4	12.4	7.6

X-RAY ANALYSIS

Structure

B_nC cell is rhomboedral. Its basic elements are an icosaedron and a chain whose compositions change from B_4C ($B_{11}C$-CBC) to $B_{10}C$ (B_{12}-$C_{1.4}B_{1.9}$).

Pellets analysis : absorption correction

B_nC has a low absorption coefficient ($\simeq$ 5 cm^{-1}) . Analysing pellets, we then observe distorded peaks. Figure 2 shows the resulting correction we determined.

Analysis method

Pellets are analysed in a vertical diffractometer using CuKβ radiation. Cell parameters are determined by a least squares method using around 20 peaks from 20° to 150° (2Θ) . The calculation takes into account corrections for absorption, sample misalignment and diffractometer alignement.

Results (fig.3)

We confirm Bouchacourt's results [1], with slightly lower values for cell parameters :

- B_nC ranges from around 9 to 20 carbon at.% ;
- $B_{13}C_2$ (13.3 Cat.%) is a special composition for parameters variations ;
- The cell volume varies linearly with composition.

This method allows non-destructive composition determinations of pellets.

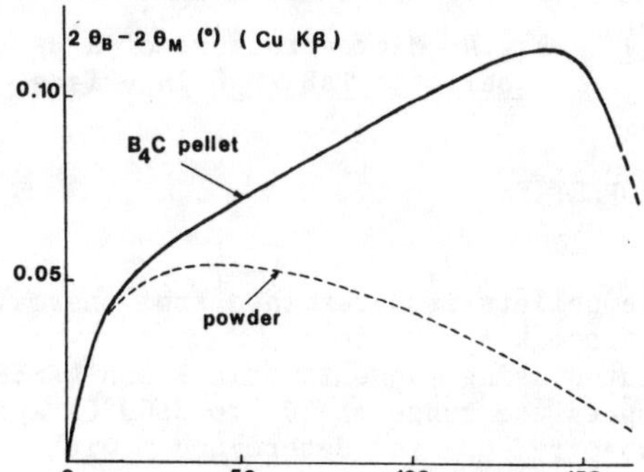

Figure 2. Absorption correction
$2\Theta_B$: Bragg Angle
$2\Theta_M$: Maximum of the peak

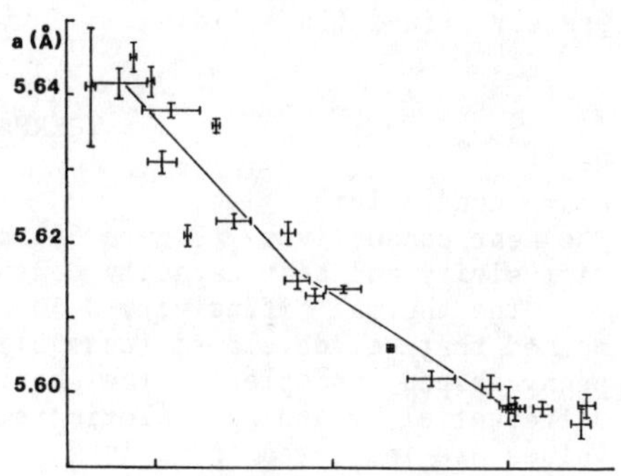

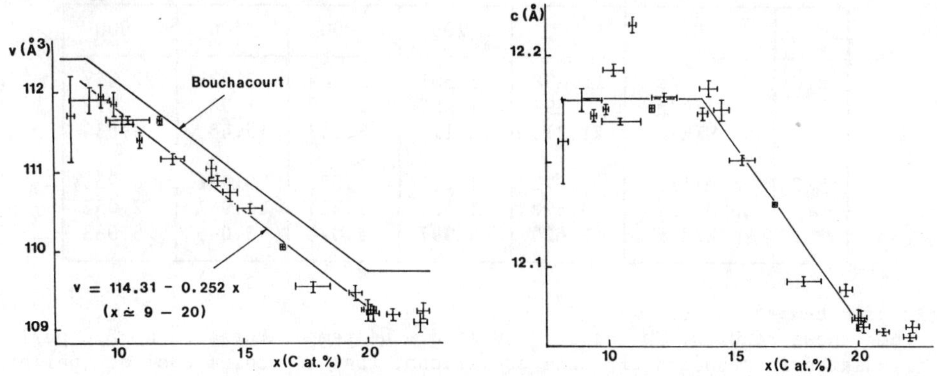

Figure 3. B_nC cell variations vs composition
Left : Rhomboedral cell volume, as compared with Bouchacourt's results [1]
Right : Hexagonal cell parameters

STRUCTURE OF THE PELLETS

Microstructure

The microstructure of the pellets is homogeneous. It reproduces the structure of the powders, that is large recrystallised grains (less than 20 µm) coming from the boron powder and clusters (around 50 µm) of little grains (less than 5 µm) coming from the B_4C powder. For the low density pellets elaborated with a tantalum lining, the porosity is somewhat higher near the surfaces.

Tantalum lining

From microprobe analysis, we observe a very little tantalum diffusion in the pellet : less than 0.2 wt% near the surfaces, non measurable at a 500 µm depth. During hot-pressing, the tantalum reacts with the powder to become TaB_2.

The foil has a good mechanical behavior (no deformation, few cracks) . Still better results are obtained with a pre-carburized (TaC) lining.

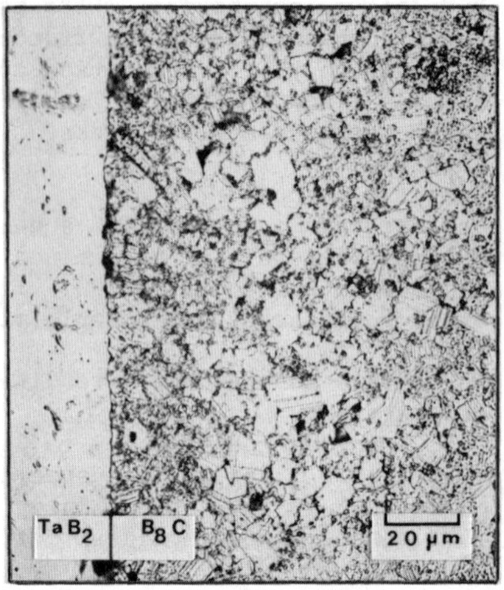

Fig.4. Microstructure of a B_8C pellet : TaB_2/B_8C interface

THERMAL PROPERTIES

Heat conductivity

The heat conductivity (Figure 5.) of the pellets is determined from thermal diffusivity and heat capacity measurements : k = D.Cp.d

The thermal diffusivity D is measured using an automatic Flash-LASER method that we developed (possible temperature range : 20 to 2800°C) and penny-shaped samples. The heat capacity Cp is determined with a differential scanning calorimeter using two heat-flow meters. Typical values are indicated in table 2.

TABLE 2 : Thermal properties of B_nC pellets

	T °C	0	200	400	600	800
B_4C	D mm^2/s	11.498	5.601	3.784	2.938	2.457
	Cp J/K/g	0.699	1.489	1.809	1.983	2.092
	k W/K/m	21.313	22.117	18.157	15.450	13.630
B_8C	D mm^2/s	2.072	1.253	1.058	0.986	0.953
	Cp J/K/g	0.672	1.534	1.824	1.967	2.052
	k W/K/m	3.620	4.997	5.020	5.045	5.083

In-pile temperatures

From those results, we have calculated the temperatures in a B_nC pellet irradiated in standard FBR core conditions, that is sodium cooling (pellets surface temperature ≃ 600 °C) and fast neutron absorption (specific power resulting from the absorption ≃ 300 W/cm^3) . Figure 6 shows the resulting temperatures for a B_4C and a B_8C pellet (diameter = 16 mm) .

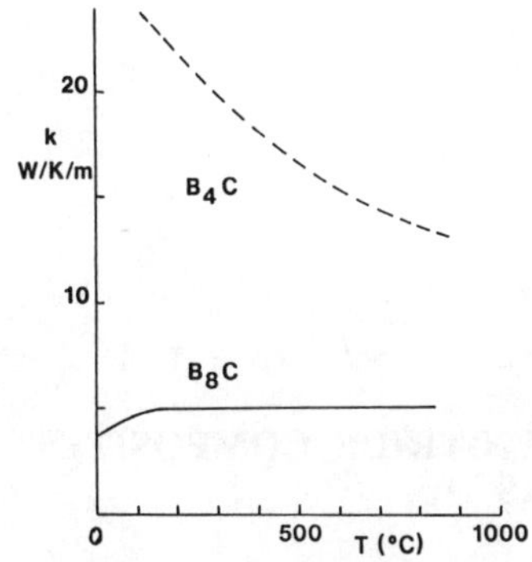

Figure 5. Heat conductivity of boron carbide

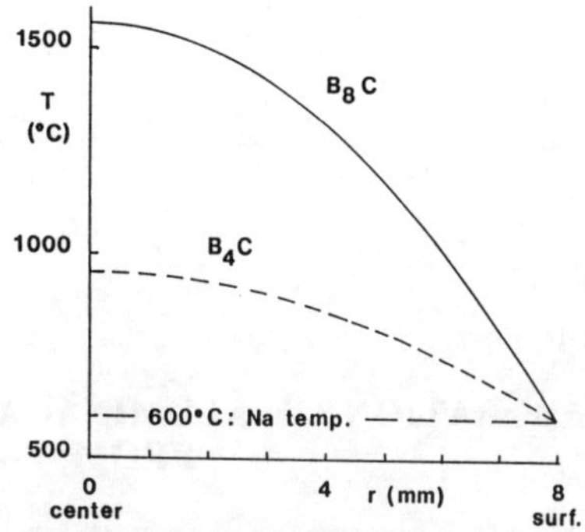

Figure 6. Calculated temperatures in irradiated pellets

CONCLUSIONS

We have elaborated boron rich boron carbide pellets. The best results are obtained by hot-pressing a mixing of B_4C and boron powders in a graphite die protected with a tantalum carbide lining.

The microstructure of the pellets is homogeneous and the lining has good mechanical and chemical behaviors.

Evaluating the X-Ray absorption correction for pellets, we are able to determine precisely their cell parameters. This allows non-destructive composition determinations.

The low thermal conductivity of the boron rich boron carbide (as compared with B_4C's one) leads to high thermal gradients in the irradiated pellets, which induces high thermal stresses and causes pellets cracking. An irradiation experiment is in progress in PHENIX French FBR reactor in order to test the behavior of this material at high temperatures.

REFERENCES

1. M. Bouchacourt, Thesis Grenoble (France) 1982

ELABORATION OF ALUMINA-γALUMINIUM OXYNITRIDE COMPOSITES FROM VARIOUS POWDERS

SAPPEI J. , GOEURIOT D.,THEVENOT F.*
LAURENT Y., GUYADER J., L'HARIDON, P. **
BERRICHE Y., TREHEUX D.***
*Ecole Nationale Supérieure des Mines de Saint-Etienne
158, cours Fauriel F 42023 Saint-Etienne Cédex 2
**Université de Rennes I. Laboratoire de chimie des matériaux
Campus de Beaulieu, Av. du Général Leclerc
F. 35042 Rennes Cédex F
***Ecole Centrale de Lyon
36, route de Dardilly -F 69130 Ecully.

ABSTRACT

Aluminalon (Trade Mark) is an alumina based ceramic composite material, containing up to 30 v/o of γ spinel aluminium oxynitride. The first generation of this material was obtained by reaction bonding of mixed alumina and Starck aluminium nitride.
Starting from alumina , a new nitridation process, at low temperature (about 1200°C), has led to pure (carbonless) and more reactive AlN powder.
From this process, we have elaborated different partially nitrided alumina powders containing γAlON or AlN, and a pure γAlON powder.
This study deals with elaboration of Aluminalon from five different ways. The first three use mixed powders : mixed Al_2O_3 and AlN powders, mixed Al_2O_3 and γAlON powders, and mixed Al_2O_3 and partially nitrided Al_2O_3 containing 8 w/o of nitrogen. The two other ways use partially nitrided alumina powders (about 1 w/o of nitrogen) obtained at 1300°C and 1100°C. The materials were prepared by hot-pressing at temperature about 1650°C. These materials were analysed by scanning electron microscopy and X-ray diffraction. Density, room temperature mechanical properties (toughness, strength, hardness), and tribological properties were measured.
The properties of these new composite materials are compared with those of the first generation.

INTRODUCTION

A new nitridation process based on the reactivity of ammonia with different powders of alumina, has given pure (carbonless) and more reactive nitrided [1] and oxynitrided [2] powders. The products of this chemical reaction (AlN or AlONγ) depend on the temperature, the time and the flux of ammonia.

Starting from powders obtained by this process (pure AlN, pure AlONγ and different partially nitrided aluminas) we have elaborated alumina based ceramic composite materials containing about 20 v/o of γ spinel aluminium oxynitride. These materials were prepared by uniaxial hot-pressing, and then characterized.

The properties of these new composite materials are compared with those of the first generation [3-4] using Starck aluminium nitride obtained by carbothermal nitridation.

EXPERIMENTAL PROCESS

Five different powders were prepared by alumina nitridation :

TABLE 1
Powders Characteristics

powder reference	starting powder	nitridation conditions (NH_3)	%N	X Ray Analysis of the nitrided powder	Specific area m^2/g
A	γAl_2O_3*	T > 1200°C [1]	≈34	pure AlN	10.5 (Øgrain= 0.1 to0.3μm)
B	γAl_2O_3*	1300°C 4h	≈ 8	transition Al_2O_3, αAl_2O_3, AlN	12
C	Boehmite**	1300°C 30 min	≈1	transition Al_2O_3, αAl_2O_3,AlN	------
D	Boehmite**	1100°C 20h	≈1	Transition Al_2O_3,γAlON	83
E	Boehmite**	1100°C-40h then heated at 1700°C-1h-N_2 [5]	≈4	γAlON	2 (Øgrain= 0.1μm with large grains about 3μm)

* A125 Baïkowsky - ** PSB160 Martinswerk-Lonza

Aluminium nitride (A), γ aluminium oxynitride (E) and then partially nitrided alumina (B) were ball-milled with an EXAL α alumina (CR6 BAIKOWSKY) in order to obtain after sintering about 20 vol/o of γ AlON in an alumina matrix, that is to say :
- alumina (a) + 3.12 w/o of AlN (A) : aA
- alumina (a) + 13.3 w/o of B : aB
-alumina (a) + 21.4 w/o of γ AlON (E) : aE

The partially nitrided boehmite powders D and C were used as obtained .
Samples of 12g were uniaxially pressed (40 MPa) at room temperature in a graphite matrix and sintered according to the following cycle :

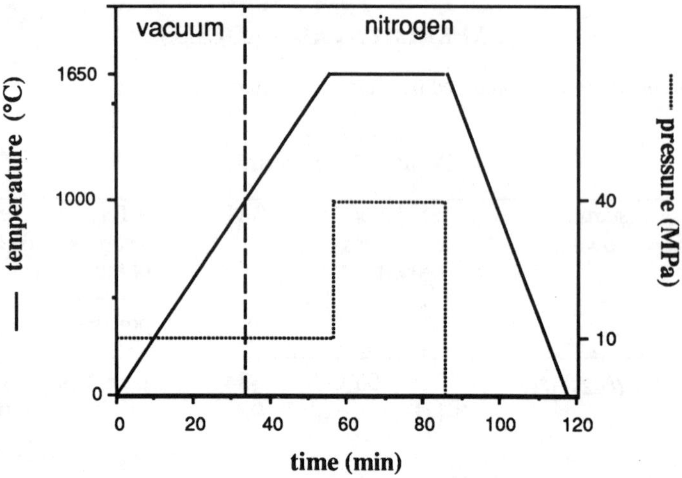

Figure 1. Hot-pressing cycle of the Al_2O_3-γAlON composites.

Samples were analysed by X-rays diffraction to verify the γ AlON ratio. Microstructures were observed by scanning electron microscopy on polished and thermally etched samples.
Once sintered, the samples were cut into bars of 3 x 4 x 17 mm^3 to be tested for mechanical properties. Three point bending tests were performed on polished and chamfered specimens. For microhardness, a Vickers indentor was applied under a load of 10N during 30 seconds on different polished samples. The toughness was measured on SENB bars (ASTM standard).
The wear tests were performed by using steel ring (52100, ϕ 35 mm, 64 HRC)) rotating against a plane ceramic sample (12 x 12 x 5 mm^3).
The rubbing faces were polished and chamfered. The obtained roughness was about 0.02 to 0.04 μm. The sliding speed was about 0.36 m/s. The applied load was 700N. Friction coefficient was recorded during the test. The wear volume was calculated with the width of traces.

RESULTS AND DISCUSSION

The relative green density, after 40 MPa uniaxial pressure, is about 52 % for mixed powders (aA, aE, aB) whereas it was only 23 % for nitrided powders (D, C). This low green density of nitrided powders is due to their grain morphology. These grains are acicular and pile up badly.

Despite these low green densities of nitrided powders, all the samples are well densified by hot-pressing. All relative densities are higher than 99 %.
As shown on fig. 2, the composite microstructures depend on the starting powders or mixtures. The first generation material (Al_2O_3 + AlN Starck : aAs), (Fig. 2.a) was poorly homogeneous despite of an attrition mixing and had a grain size of about 3 to 4 µm with some larger grains upper to 5 µm. Using the new AlN (A) (Fig.2.b), the composite aA is very homogeneous and presents fine grains (≈ 2 µm), in relation with the fineness of the AlN powder (SA ≈ 10.5 m^2/g instead of 6 m^2/g for the AlN Starck). This fine dispersion inhibits the alumina grain coarsening during sintering.
Using the other powders (B,D,E) lead to fine grain composites (Fig. 2.c.d.e), except the boehmite treated at 1300°C (C). In this case, the microstructure (Fig. 2.f) shows alumina grains coarsening. This phenomenon is attributed to the sintering of alumina powder during nitridation at 1300°C. The sintered alumina grains cannot react with ammonia and that leads to an inhomogeneous powder .
In spite of its low specific area (2 m^2/g), it is interesting to mix pure γAlON (E) with alumina. The γAlON powder is easily milled during the mixing with alumina (ball-milling) and the composite (aE) is then homogeneous (Fig. 2.e).
The γAlON presents two kinds of forms :
- γAlON grains similar to Al_2O_3 grains
- an intergranular structure along alumina grain boundaries.
In the composites aAs, aA, aB the intergranular phase is rare ;this kind of morphology is largely developed in the composite D (from Al_2O_3 partially nitrided in γAlON). On the opposite, the composite aE (Al_2O_3 + γAlON mixture) presents no γAlON along grain boundaries.

Table 2 shows the mechanical properties at low temperature (20°C) of our materials and those of the first generation material (aAs).

TABLE 2
Mechanical properties

Samples	flexural strength σ_f (MPa)	toughness K_{1c} (MPa $\sqrt{m}$)	hardness Hv (GPa)
aAs	540 ± 40	3.9	18.2 ± 1.4
aA	660 ± 60	3.85 ± 0.25	20.9 ± 1.3
aB	590 ± 20	-----	20.7 ± 0.6
D	640 ± 60	3.65 ± 0.03	19.8 ± 0.2
aE	700 ± 80	-----	20.3 ± 1.8

These results showed that the use of ammonia nitrided powders enhanced the mechanical properties (flexural strength, hardness) compared with the use of carbothermal aluminium nitride. This can be correlated with the more fine and homogeneous microstructures. The toughnesses are nearly the same. The best flexural strength is obtained with mixed γ aluminium oxynitride and alumina powder (aE).
Figure 3 shows the evolution of the wear volume during the friction tests. All the studied samples obtained from mixed powders aAs, aA, aE nearly show the same wear resistance.

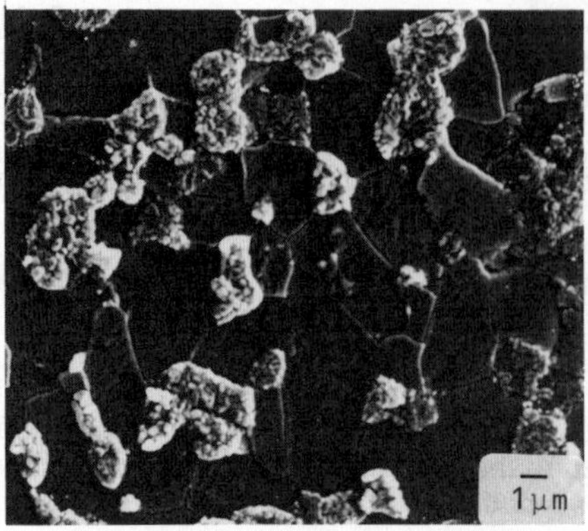

a) **aAs** from Al_2O_3+AlN Starck

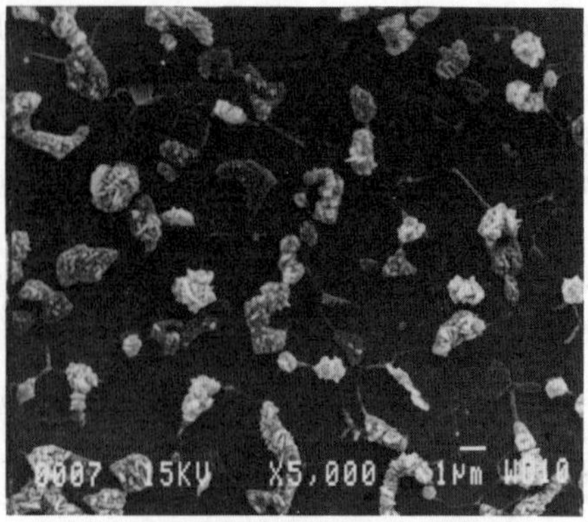

b) **aA** from Al_2O_3+ new AlN

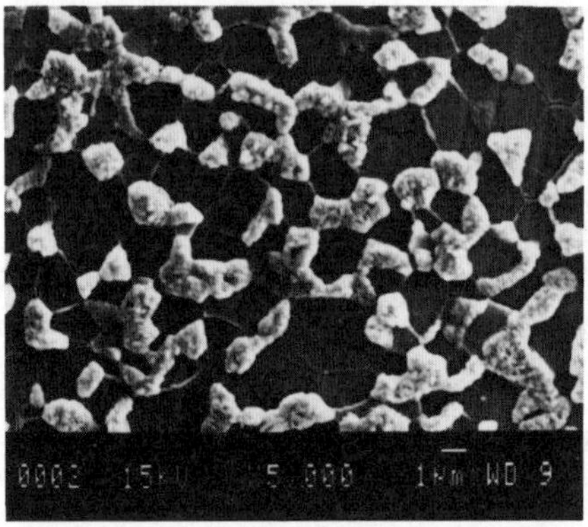

c) **aB** from Al_2O_3+Al_2O_3 partially nitrided in AlN (8w/o of N)

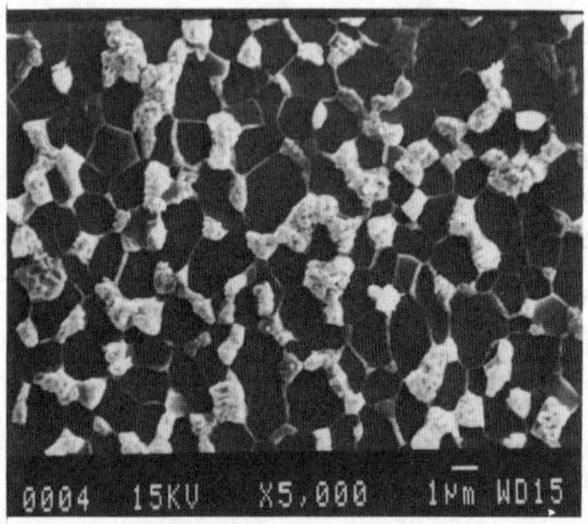

d) **D** from Al_2O_3 partially nitrided in γAlON (1w/o of N)

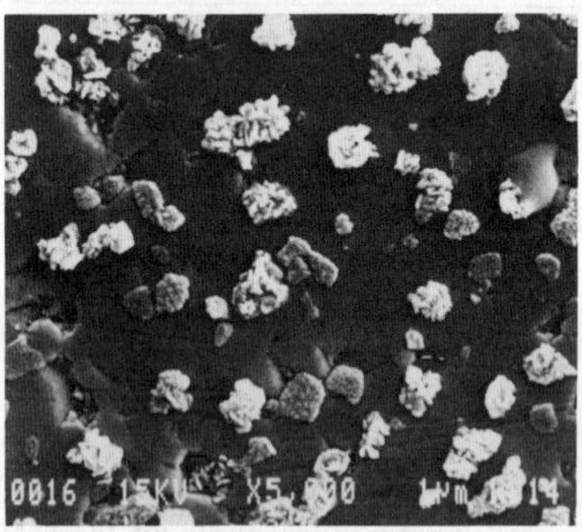

e) **aE** from Al_2O_3+γAlON

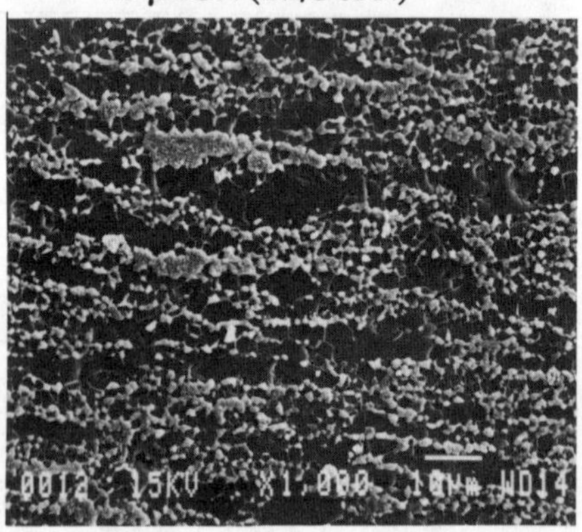

f) **C** from Al_2O_3 partially nitrided in AlN (1w/o of N)

Figure 2. Microstructures of the Al_2O_3-γAlON composites.

The nitrided powder D seems to lead to a better wear resistance material than others. The presence of γAlON in intergranular structure seems to be efficient for the enhancement of this phenomenon.

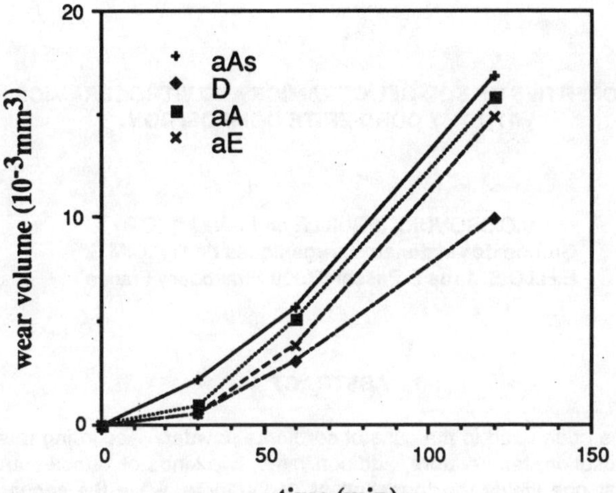

Figure 3. Wear volumes during friction tests for the composites.

CONCLUSION

By using fine and reactive AlN or partially nitrided aluminas, the Al_2O_3-γAlON composites can be obtained with fine and homogeneous microstructures that enhanced mechanical properties, compared with those of the first generation composites.
Alumina partially nitrided in γAlON (D) leads to enhanced wear resistance material, due to the presence of an intergranular γAlON phase.

REFERENCES

1. Guyader, J., Verdier, P.and Laurent, Y., French Patent n° 8415104, 2.10.84.

2. Sappei, J., thesis, ENSM Saint-Etienne, in progress.

3. Goeuriot, D., thesis, University Claude Bernard-Lyon I, 1987.

4. Goeuriot-Launay, D., Goeuriot, P., Thévenot, F, Orange, G, Fantozzi, G., Trabelsi,R. and Treheux, D., Ceramics International, in press.

5. Goeuriot-Launay, D., Guyader, J., Laurent, Y., L'Haridon, P., Sappei, J and Thévenot, F., French, Patent, n° 8901754, 10.02.89.

ACKNOWLEDGMENTS : the authors thank G.ORANGE (G.E.M.P.P.M. LYON INSA LYON F.) for some mechanical tests.

PROPERTIES OF SOL-GEL CERAMICS AND VITROCERAMICS WITH THE CORDIERITE COMPOSITION

J.C.BROUDIC, J.GUILLE and S.VILMINOT
Groupe des Matériaux Inorganiques de l'I.P.C.M.S.
E.H.I.C.S. 1 rue B.Pascal 67008 Strasbourg France

ABSTRACT

Sol-gel synthesis have been used in the case of cordierite powders. According to some elaboration parameters such as dilution, temperature, addition rates, two kinds of samples are obtained. After 1100°C firing, the first one yields the formation of α-cordierite, while the second one yields the formation of a mixture of $MgAl_2O_4$ and μ-cordierite. X-ray diffraction measurements on powders calcinated at different temperatures reveal in both cases the intermediate formation of μ-cordierite Infra-red spectroscopy shows different structures for both kinds of samples. The ceramics of the first type are characterized by a dielectric constant lower than 5 and a thermal expansion coefficient close to 1.10^{-6} K^{-1}.

INTRODUCTION

Its low dielectric constant and a thermal expansion coefficient close to that of silicon make cordierite based ceramics attractive materials for microelectronic packaging. Sol-gel methods usually allow the elaboration of ceramic powders with high purity and high reactivity, yielding a strong decrease of the sintering temperature. Such a method has been used in the case of cordierite (1). This paper will focus on the thermal evolution of cordierite amorphous precursors by means of X-ray diffraction and infra-red spectroscopy. The ceramics obtained after sintering below 1100°C have been characterized by measurements of their dielectric constant, microhardness and thermal expansion coefficient.

EXPERIMENTAL

Cordierite powders have been elaborated using a sol-gel method previously described (2). Aluminum, $Al(NO_3)_3.9H_2O$, and magnesium, $Mg(NO_3)_2.6H_2O$, nitrates in solution in alcohol are added to a solution of tetraethylorthosilicate, $Si(OC_2H_5)_4$ (TEOS), in the same solvent. Diluted ammonia is used as hydrolysis agent, giving rise to a white gel. After drying at 250°C, the xerogel has been characterized according to the nature of the phases obtained after firing at temperatures ranging from 700 to 1100°C. Two kinds of samples have been obtained that will be referred as A and B respectively : A gives rise to α-cordierite after calcination below 1100°C, B does not transform into α-cordierite in the same temperature range.

PHASE CHARACTERIZATIONS

X-ray diffraction

After 2 hours calcination at 700°C, both kinds of samples (A, B) appear to be amorphous by X-ray diffraction. For higher temperatures, whereas A samples give rise to stoichiometric μ-cordierite at 900°C (ASTM 14-249) which transforms into α-cordierite at 1100°C (ASTM 13-293),the B samples yields the formation of $MgAl_2O_4$ and non stoichiometric μ-cordierite with higher SiO_2 content as revealed by refinement of cell parameters (Table 1). For B samples, the α-cordierite formation occurs at 1250°C by solid state reaction between the non stoichiometric μ-cordierite and $MgAl_2O_4$.

TABLE 1

Evolution of the a cell parameter of μ-cordierites with calcination temperature

Calcination temperature (°C)	a (Å) Type A	a (Å) Type B
900°C	5.204	5.172
950°C	5.206	5.169
1000°C	5.204	5.152
1100°C	-	5.124

For stoichiometric μ-cordierite, a = 5.200 Å

A question arising from this study concerns the possible presence of a residual amorphous phase after calcination at 1100°C. A semi-quantitative study has been performed by X-ray diffraction measurements starting from the same quantities of A and B powders obtained at 700°C mixed to the same quantity of α-Al_2O_3 as a reference.

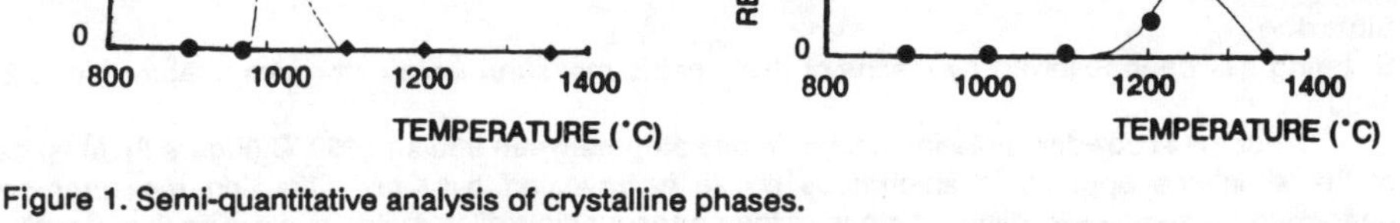

Figure 1. Semi-quantitative analysis of crystalline phases.

Comparison of the diffraction line intensities puts into evidence :

-A powder : The μ to α-cordierite transformation appears to be restricted to the temperature range 970-1050°C (Figure 1.a).The α-cordierite intensities increase after thermal treatment at 1350°C compared to the ones observed at 1100 and 1200°C. As μ-cordierite is no more detected above 1100°C, it seems that this increase could be related to the crystallization of a vitreous phase.

-B powder : As shown on figure 1.b, we first observe a decrease of the μ-cordierite intensities between 900 and 1100°C related to the formation of $MgAl_2O_4$ from the non-stoichiometric μ-cordierite. The μ to α transformation starts above 1100°C and at 1350°C the α-cordierite intensities are bigger than the corresponding ones for the A powder. This observation reinforces the assumption of a vitreous phase in A samples.

Infra-red spectroscopy
Infra-red spectroscopy has been used to get more insight into the structures of the amorphous phases obtained at 700°C.

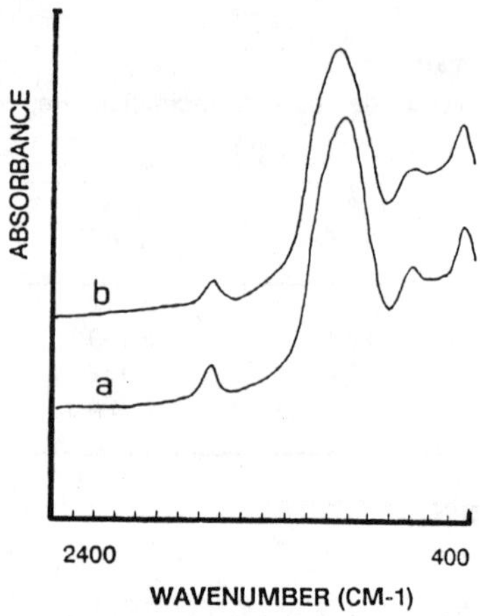

Figure 2. IRFT spectrum of amorphous powders (a) type A (b) type B

As shown in figure 2, differences between both A and B type powders appear as far as stretching T-O-T bands in TO_4 tetrahedra (T = Si or Al) are considered.The observed stretching bands appear at 1020 cm^{-1} and 1080 cm^{-1} for A and B respectively.

Considering that TO_4 tetrahedra have their stretching band around 1085 cm^{-1} and 960 cm^{-1} for T = Si and Al respectively, we can conclude that structural differences exist in the amorphous powders. For A type powders, the observed stretching frequency at 1020 cm^{-1} is related to a structural distribution of both Al and Si within tetrahedrally coordinated sites, structure which can give rise to stoichiometric μ-cordierite. For B type powders, the stretching frequency is very closed to that of pure SiO_4 tetrahedra and can be related to the further formation of a non-stoichiometric μ-cordierite with higher SiO_2 content.

Sintering
Sintering has been followed by means of dilatometric measurements in the temperature range 20-1400°C.

For the A powder, sintering occurs in one step between 850 and 950°C (Figure 3). Most part of the shrinkage appears in amorphous phase as revealed by X-ray diffraction measurements performed on samples at different steps of the corresponding dilatometric curve. The final densities measured on pellets sintered 2 hours at 1100°C are better than 95% of the theoretical density of α-cordierite.

For B powder, sintering occurs within two steps (Figure 4). The first one, between 850 and 950°C is the most important and also happens in amorphous phase. The second one corresponds to the sintering of $MgAl_2O_4$ which crystallization occurs around 970°C. The final volume expansion at 1250°C is due to the reaction between μ-cordierite and $MgAl_2O_4$ yielding the formation of α-cordierite.

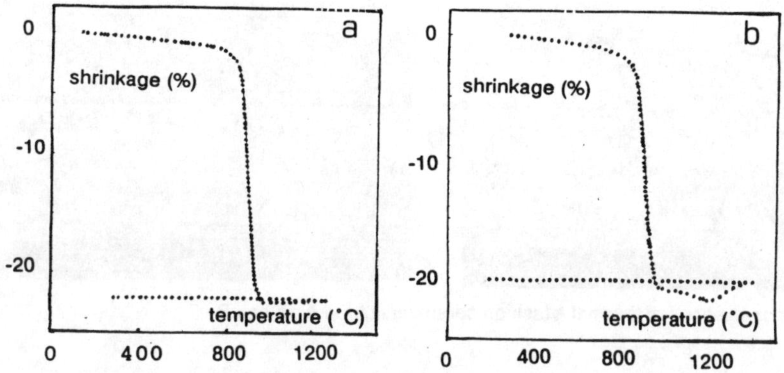

Figure 3. Dilatometric curves of samples (a) type A, (b) type B

CERAMIC CHARACTERIZATIONS

Dielectric constant

Measurements have been performed at 1 MHz freqency on sintered pellets.

At 900°C, all ceramics, exhibiting stoichiometric or non-stoichiometric μ-cordierite structure, have a dielectric constant between 5 and 6.

At 1100°C, the A type ceramics have the lowest dielectric constant around 4.7 ± 0.25. This value seems to be related to the phase structure since all samples exhibiting the α-cordierite structure have the same dielectric constant. So, B type samples for which the conversion into α-cordierite has taken place at 1250°C have the same low dielectric constant, around 4.7, whereas the mixture of non-stoichiometric μ-cordierite and $MgAl_2O_4$ obtained at 1100°C has a dielectric constant about 5.5.

Thermal expansion coefficient

By dilatometric measurements, we have investigated the thermal expansion coefficient of both kinds of ceramics, obtained after calcination at 900, 1100 and 1400°C.

At 900°C, the A type sample exhibits the μ-cordierite structure and its expansion coefficient is between 4 and 5.10^{-6} K^{-1}. At 1100°C, the same sample has been converted into α-cordierite and the expansion coefficient is now lower than 1.10^{-6}. Further calcination at 1400°C does not modify the measured value.

At 900°C, the B type sample exhibits the non-stoichiometric μ-cordierite structure and its thermal expansion coefficient is between 4 and 5.10^{-6} K^{-1}. At 1100°C, the mixture of non-stoichiometric μ-cordierite and $MgAl_2O_4$ has a value around 5.10^{-6} K^{-1}. At 1400°C, the transformation into α-cordierite is achieved and we observe a value lower than 1.10^{-6} K^{-1} characteristics of the α-cordierite.

Thermal attack

We have compared the two kinds of ceramics A and B after a thermal treatment of 2 hours at 1100°C followed by a polish and a subsequent annealing at 1100°C during 16 hours. Different morphologies are revealed. The A ceramics show grains which seem to have coalesced and the memory of the former agglomerates does not appear. The ceramics seem to be very homogeneous at a large scale. The B ceramics show a structure made of a great number of grains of different sizes (10 to 100 nm) It is possible to think that these various sizes are the result of a two steps sintering process.

a b

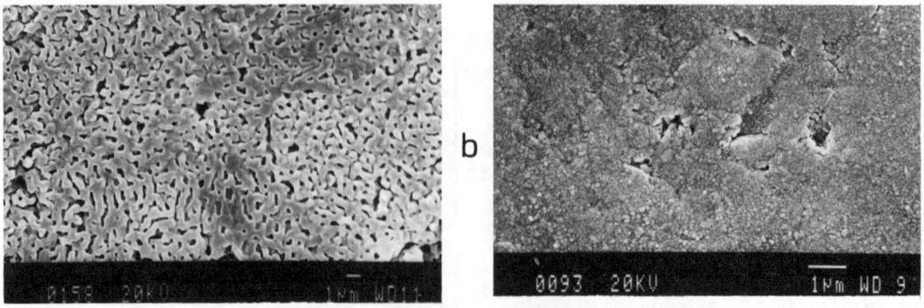

Figure 4. SEM observations of thermal attack on ceramics (a)type A, (b)typeB

Mechanical properties

Both ceramics have been checked using a Vickers indenter and a 4N load. Evaluation of the toughness was obtained on measuring the cracks formed around the impresses and using the relationship given by J. LANKFORD (3). Values are listed in table 2.

TABLE 2
Mechanical properties of A and B ceramics

ceramic type	annealing temperature (°C)	hardness (10^{-9} Pa)	toughness (10^{-6} $MNm^{-3/2}$)
A	900	5.19	1.82
A	1100	5.68	1.78
B	900	8.24	2.78
B	1100	8.95	1.71

It appears that the B ceramics have better mechanical properties than the A ones, despite the toughness at 1100°C which is lower. The correlation between the nature of the cristallographic phase and the mechanical properties is not evident because both μ-cordierite ceramics obtained at 960°C have very different toughness and the modification occuring at 1100°C in the crystallographic phase does not seem to interfere with respect to the hardness.

CONCLUSION

Using a sol-gel process, two kinds of cordierite powders have been elaborated (A and B) which differ by the intermediate formation of stoichiometric (A) or non-stochiometric (B) μ-cordierite. According to their dielectric constant and thermal expansion coefficient, both kinds of samples can be used in electronic packaging and copper cosintering is possible as demonstrated elsewhere (4).

REFERENCES

1. Gense.C., Chowdry.U., Mater.Res.Soc.Symp.Proc.72 (1986) 297-307.
2. Broudic.J.C., Vilminot.S., Bernier.J.C., Mat.Sci.Eng. A109 (1989) 253-259.
3. Lankford.J., J.Mater.Sci.Lett.2 (1982) 493.
4. Oliver-Broudic.V., Strasbourg University These (1988)

The authors wish to thank CRICERAM (38560 JARRIE) for its financial support of this work.

ACOUSTIC EMISSION FROM THERMAL FRACTURE OF CERAMICS

K. J. KONSZTOWICZ,
Atlantic Research Laboratory, National Research Council of Canada,
Halifax, N.S., B3H 3Z1

ABSTRACT

Crack initiation and growth during thermal shock tests of alumina and Mg-PSZ advanced refractory were investigated by the analysis of acoustic emission (AE) amplitudes. The growth of cracks that were detected by AE was assessed by SEM observations. The measurements of strength loss after thermal cycling with various applied temperature differentials correlated with continuous monitoring by acoustic emission and confirmed the effects of microcrack growth on the resistance to thermal shock damage.

KEYWORDS

Crack growth, thermal shock, acoustic emission

INTRODUCTION

Crack growth in ceramics subjected to thermal shock is not believed to occur before the temperature difference reaches its critical value, ΔT crit [1]. With this approach, however, it is difficult to explain the failure of ceramics under thermal shock fatigue conditions with $\Delta T<\Delta T$ crit.

The purpose of this work was to investigate the initial stages of crack formation under thermal shock of lower than critical severity, followed by the microcrack coalescence with increasing ΔT. Further, the effect of microcrack growth on thermal shock fatigue resistance of ceramics is examined. The work is based on the analysis of acoustic emmission (AE) amplitude distributions [2].

METHODS

Materials investigated

Two materials with different response to thermal shock have been chosen: (i) polycrystalline alumina*, providing high thermal shock resistance, and (ii) an advanced magnesia - partially stabilised zirconia fine-grained refractory# with a high resistance to thermal shock damage. Both materials contain a glassy phase, which in alumina fills the space between the grains while in zirconia it forms closed inclusions.

Quenching tests

Samples cut to dimensions of 50x5x5 mm were individually heated in a vertical electric furnace to temperatures ranging between 120 to 1020°C in intervals of 100°C. After stabilization, five to seven samples were quenched into a silicone oil bath at 20°C. An acoustic emission 100-300 kHz transducer was attached to the bath container, and signals were transmitted by a 60dB preamplifier to the main AE system** with a gain of 46dB and a threshold level of 1.6 V. One sample in every batch had one surface polished and etched (in HF, 2 min). These surfaces were

---

* prepared by slip-casting of 95% Al_2O_3, 2.5% talc and 2.5% kaolin mixture in aqueous slurry, and sintering at 1500°C to 98% T.D.

\# produced by Didier Refractories

** Physical Acoustic Corp., Model 3000/3104.

submitted to SEM*** observations for the development of microcracks. Flexure strength was determined in 3-point bending## with a cross-head speed of 0,01mm/min and a roller span of 38mm.

Thermal Shock Fatigue

Samples cut from the same materials to dimensions 22.86 x 2.54 x 2.54 cm were also assessed for thermal shock fatigue behaviour in a standard ribbon test equipment###, providing cycles of rapid heating and cooling [3]. Temperature differences of 400°C, 600°C and 800°C were applied to zirconia refractory, 400°C and 500°C in the case of alumina. For each ΔT 15 samples were placed on the test bench and after the first, third and fifth thermal shock cycle the strength loss was determined as an average of 5 samples in 3-point bending## with a cross head speed 0,5 mm/min and a roller span of 170mm. In one of the bars for each set a steel waveguide was fitted and an AE 150 kHz transducer was then placed on its end, followed by the 60 dB preamplifier. The AE system gain was set at 48 dB and the threshold level was raised to 2.0 V.

RESULTS AND DISCUSSION

Quenching Tests

Strength losses (average values) as a function of the applied quench temperature difference ΔT for the alumina and zirconia are shown in Fig. 1a and b. Fig. 1c and d show the growth of the average number of AE events for each sample set as a function of the applied thermal stress.

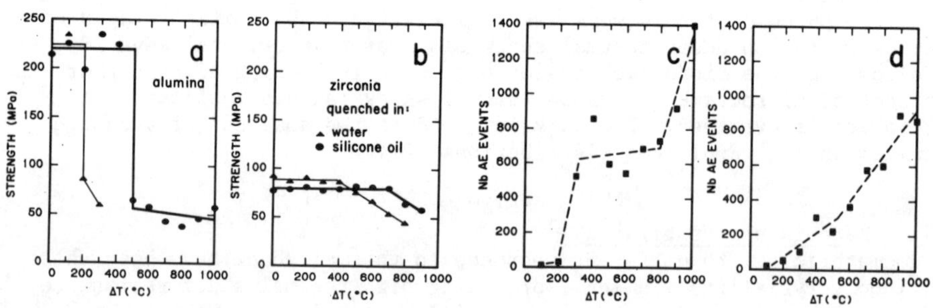

Figure 1. Thermal shock by quenching. Strength loss of a) polycrystalline alumina, and b) advanced zirconia refractory, with corresponding growth of average AE events number for c) alumina, and d) zirconia.

As indicated by the "Amplitude Distribution" plot for zirconia (Fig. 2a), all emissions within the temperature differential of 100°C to 500°C are of a low amplitude character (below ∿53dB). The sources of these AE events are the microcracks developing within the glass phase inclusions (Fig. 3a). The amplitude distribution plot (Fig. 2a) changes its curvature about ΔT=600°C, which is much below the "critical" temperature differential of 800°C leading to the first significant loss of strength . Microscopic observations confirm [4] that at ΔT=600°C the microcracks extending from glass phase inclusions coalesce

***Jeol, model JSM 840.
Instron 1125.
Produced by Harrop.

and cross the crystalline phase forming larger cracks. The observed change in shape of the amplitude distribution plot introduces AE events of higher amplitude (>53dB), thus higher energy is required to propagate the crack through the crystalline phase. The cracks developed in this manner link together at "critical" temperature differential of applied thermal shock and form optically detectable long cracks.

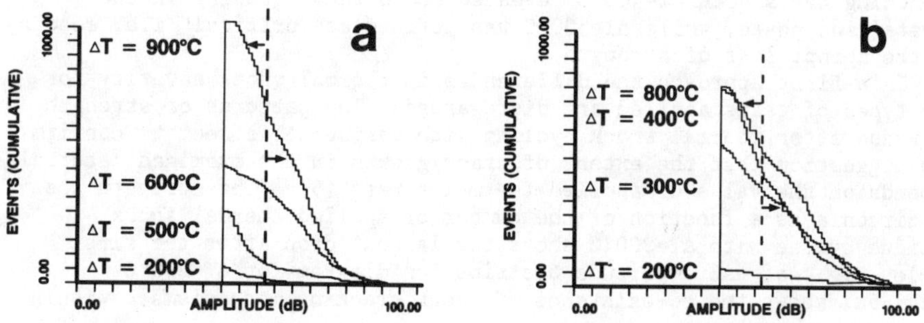

Figure 2. AE amplitude distribution plots, a) zirconia b) alumina. Dashed lines and arrows indicate lower and higher amplitude events.

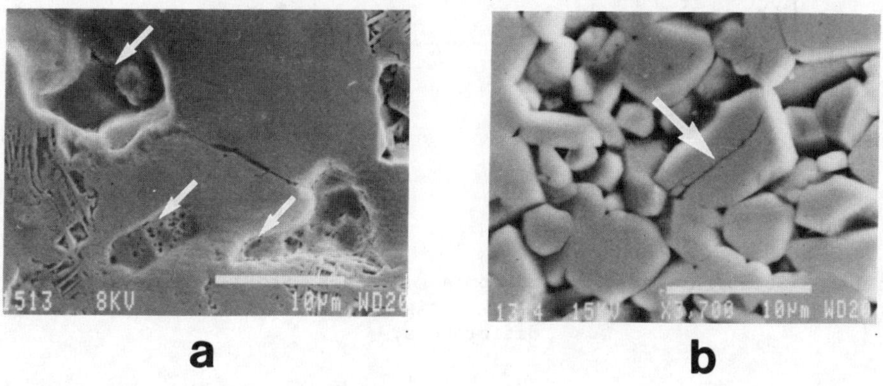

Figure 3. Microcracks at ΔT<ΔT crit.: a) within glass phase inclusions of zirconia (ΔT=400°C, ΔT crit.=800°C) b) within alumina grains (ΔT=400°C, ΔT crit.=500°C)

There is no significant growth in the number of AE events detected for alumina quenched in silicone oil at low temperature differences, 100°C to 200°C (Fig. 2b). Rapid cooling at ΔT=300°C gives a high number of AE events with small participation of higher amplitude events (over about 53 dB). Comparing this with zirconia, it can be suggested that at this level of thermal shock the cracking in alumina occurs mainly within the weaker glass phase. Microscopic observations for the quench temperature difference ΔT=400°C reveal the presence of fine microcracks (Fig. 3b) within the crystalline phase. This feature correlates with the appearance of higher amplitude AE events (Fig. 2b).

When alumina samples are quenched at critical ΔT=500°C, a few (usually two, three) long cracks traversing the surface of the sample, close to the central section, appear under the microscope.

Thermal Shock Fatigue

Three ΔT's of thermal shock were determined in quenching tests as leading to different stages of development of stress-induced microcracks in examined zirconia. The first is ΔT=400°C, at which microcracks were observed within the glass phase. The second, ΔT=600°C, is where the coalescence of microcracks occured. Finally ΔT=800°C resulted in the first strength losses observed. In alumina, the quenching tests with ΔT=400°C revealed microcracks growing in the crystalline phase, while ΔT=500°C was defined as "critical" i.e. leading to the abrupt loss of strength.

In a first approach the differences in thermal shock severity for two types of tests applied are disregarded. The patterns of strength decrease after thermal shock cycling with various ΔT's seem to confirm the suggestion that the extent of crack growth in the examined materials depends on the value of applied thermal stress [5]. The strength loss of zirconia as a function of the number of applied thermal shock fatigue cycles with ΔT=800°C shows the largest drop after the first cycle, Fig. 4a. AE amplitude distribution diagram for ΔT=800°C (Fig. 5a) shows the co-existence of both cracking mechanisms: within

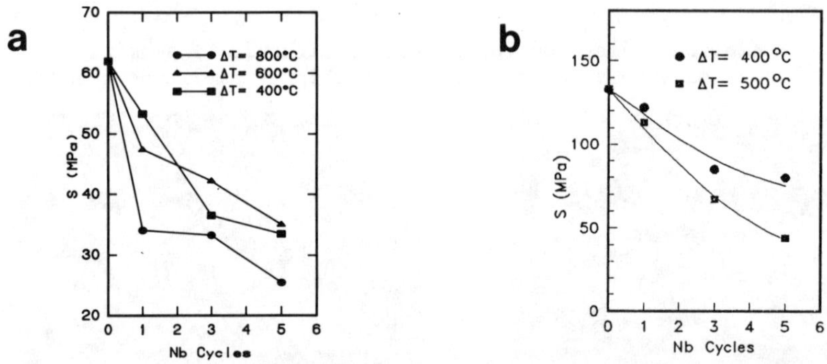

Figure 4. Strength loss of a) zirconia refractory and b) alumina, as a function of number of thermal shock fatigue cycles and temperature differences.

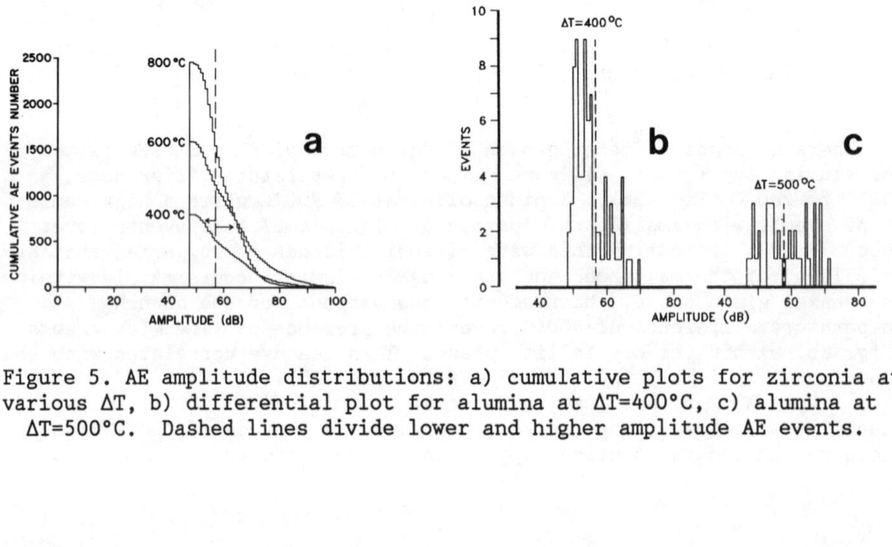

Figure 5. AE amplitude distributions: a) cumulative plots for zirconia at various ΔT, b) differential plot for alumina at ΔT=400°C, c) alumina at ΔT=500°C. Dashed lines divide lower and higher amplitude AE events.

the glass phase (low amplitude events), and within the crystalline phase (higher amplitude events).

The amplitude distribution plot for ΔT=600°C (Fig. 5a) confirms the participation of higher amplitude AE events (crack linking) and as a consequence, the "Strength Loss" curve for ΔT=600°C decreases almost monotonically (Fig. 4a). Since the crack extent in zirconia cycled with ΔT=400°C is limited during the first cycle mostly to the size of glass phase inclusions and probably some neighbouring grains, the strength loss after this first cycle is the smallest (Fig. 4a). The amplitude distribution diagram for these samples (Fig. 5a) confirms also the dominant participation of low amplitude events in the crack growth process. The next cycles of heating and cooling reduce the strength to the same level as cycling with ΔT=600°C, which means that even smallest defects may grow by coalescence.

In the case of alumina, increasing ΔT and the number of thermal shock cycles increases the amount of catastrophic failure (Fig. 4b). There is, however, less damage in thermal shock cycling with ΔT=400°C than with ΔT=500°C. As it has been found in quenching tests, single thermal shock with ΔT=400°C produced crack initiation within glass phase, and also a limited propagation of microcracks in the crystalline phase. Not incidently though, differential amplitude distribution plot for ribbon test of alumina at ΔT=400°C features the domination of AE events of low amplitudes (Fig. 5b). The same plot (Fig. 5c) for ΔT=500°C, which was defined as critical ΔT in quenching tests, shows an important amount of higher amplitude AE events, thus more through-crystalline phase cracks.

SUMMARY AND CONCLUSIONS

* The analysis of acoustic emission amplitudes enables the definition of crack initiation during thermal shock by quenching with temperature differences less than ΔT crit. Crack propagation can be also successfully monitored by AE in more severe thermal shock conditions.
* Experimental evidence indicates that low amplitude AE events originate from microcrack formation in a weaker microstructural element, ie. the glass phase. High amplitude AE events are due to crack propagation through crystalline phase of the ceramic body.
* Thermal shock cycling of ceramic materials in ribbon test produces strength losses in good agreement with the observed stages of crack growth.

REFERENCES

[1]. Bever M.B. (Ed.), Thermal Shock in Ceramics. Encyclopedia of Materials Science and Engineering. 1986 Pergamon, 4953-55.

[2]. Pollock, A.A., Acoustic emission amplitude distributions. International advances in nondestructive testing. Gordon and Breach, Science Publishers Inc., U.S.A., 1981, 215-239.

[3]. Semler, C.E. and Hawisher, T.H., Evaluation of thermal shock resistance of refractories using the ribbon test method. Am. Ceram. Soc. Bull., 1980, 59 [7], 732-38.

[4]. Konsztowicz, K.J., Crack growth and acoustic emission in ceramics during thermal shock. J. Am. Ceram. Soc. To be published.

[5]. Lewis III, D., Rice, R.W., Comparison of static, cyclic and thermal shock fatigue in ceramic composites. Cer. Eng. and Sci. Proc., Sept.-Oct., Vol. 3, 1982, 714-721.

2.192

APPLICATION OF MICROFOCUS RADIOLOGY AND DYE PENETRANTS IN QUALITY AND FRACTURE TOUGHNESS DETERMINATION OF CERAMIC SAMPLES

H. KÖLKER and P. HENZE
ELEKTROSCHMELZWERK KEMPTEN GmbH, West-Germany

ABSTRACT

The x-ray microfocus technique using image processing is applied to detect flaws and density fluctuations in green and sintered ceramic components. Various problems and recent improvements of the method such as detection limit, contrast resolution capability, and beam hardening are addressed and experimentally investigated. Surface cracks are investigated by capillary flaw inspection using liquid dye penetrants. This method is extended to assist fracture toughness determination in ceramics.

OBJECTIVES

Flaws of about 50 - 100 µm currently limit the strength of SiC and Si_3N_4 within the range 400 - 800 MPa. Nondestructive methods are required to detect such flaws.
This paper reviews recent improvements in x-ray microfocus analysis and dye penetrant testing at ESK.

CONCLUSIONS

- X-ray microfocus analysis including image processing detects and counts the pores and metallic inclusions in SiC and Si_3N_4 above a mean diameter of 25 µm.
- For the detection of cracks x-ray analysis is supplemented by capillary flaw inspection with liquid dye penetrants.

METHODS AND RESULTS

ESK has applied the microfocus technique (1-3) since 1984, using both film methods and image processing with integration, different correction routines, and filtering.
When a beam of x-rays passing through a ceramic sample hits a pore or crack in the material it appears brighter on a screen or film than a neighbouring beam. Fig. 1 shows the intensity difference between the two beams as a function of the absorption coefficient µ.

For maximal contrast
(Eq.1) $$\mu = \frac{1}{x}$$ (x = sample thickness)

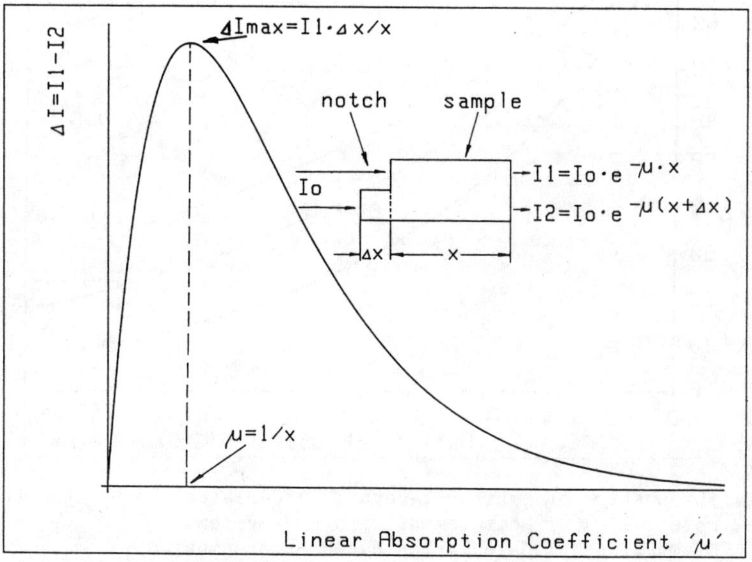

Figure 1. Intensity difference of two neighbouring x-ray beams suffering different absorption in a ceramic sample due to a notch, pore, or crack of length x.

For a given material μ is a function of x-ray wavelength. Consequently, maximal resolution requires a certain wavelength.

At first sight, this is difficult to achieve with the "white" radiation of x-ray tubes.
However, it was experimentally verified in Fig. 2 that the white x-radiation is effectively filtered during its passage through a ceramic sample such that only the shortest wavelength leaves the sample.
It is the cut-off wavelength $\lambda_{min} = 12,4/V$ of the x-ray tube (V = anode voltage).
Consequently, maximal resolution requires a certain anode voltage.

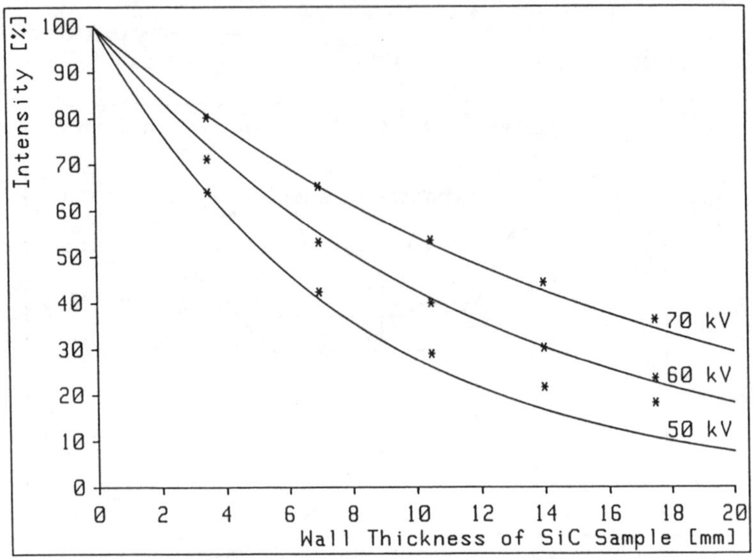

Figure 2. Attenuation of "white" x-rays in SiC-plates.
Dots : Measurement with ESK-System
Drawn lines : Calculation based on assumption that the rays leaving the sample are dominated by the cut-off wavelength min=12.4/V (A) (V=Anode voltage in kV).

This result is applied to detecting notches in the surface of an SiC bending sample (Fig. 3). The resolution is much better for 50 kV than for 100 kV.

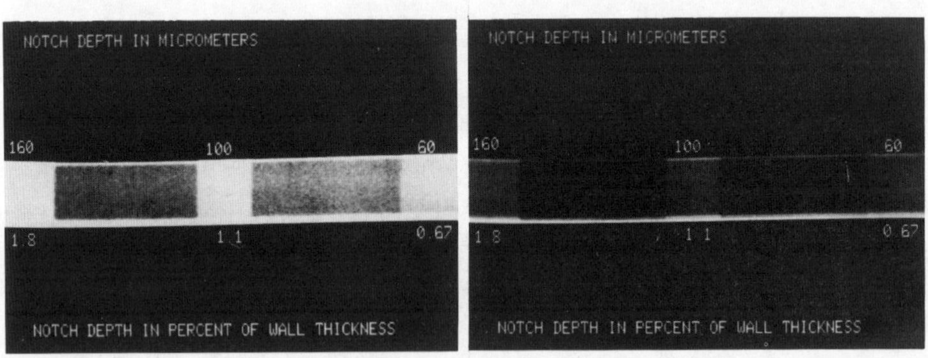

a: Anode voltage = 50 kV (maximal contrast resolution)

b: Anode voltage = 100 kV (poor contrast)

Figure 3. x-ray microfocus image of an SiC plate (wall thickness 9 mm) with tree notches across its surface.

According to Fig. 1

$$\Delta x/x = \Delta I_{max} / I_1$$

The right hand side of this equation is a constant for a given set up (contrast resolution capability of the system), so the flaw depth Δx that can be resolved is a given fraction of the sample thickness x. In the ESK system this fraction is 0.5 % (see Fig. 3).

This sensitivity is high enough to detect the flaws in SiC and Si_3N_4 that are responsible for fracture.
In SiC (SSiC) pores of 25-100 µm were found (Fig.4 a) and in Si_3N_4 (SSN) strongly absorbing inclusions of similar size (Fig. 4 b).

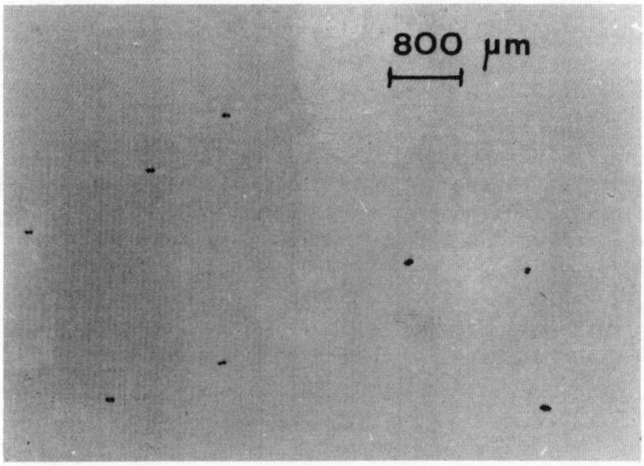

Figure 4 a. x-ray microfocus image of pores in an SiC-plate (5 mm thick)

Figure 4 b. x-ray microfocus image of strongly absorbing inclusions in a Si_3N_4-plate (5 mm thick)

DYE PENETRANT TEST

Cracks that are open at the surface can be filled with liquid dye penetrants all the way down to the bottom, even if the crack opening is less than 0.1 µm. This feature of dye penetrants is applied to detecting cracks in ceramic components and also to marking artificial cracks introduced into bending specimens for the purpose of measuring the fracture toughness.

REFERENCES

(1) W. Grellner, P. Henze, K.A. Schwetz, A. Lipp: "Nondestructive Testing of SiC and Si_3N_4 Turbine Components" in: Proc. 3rd Europ. Conf. Nondestr. Testing, M.R. Caletti (Ed.), Vol. 5 (1985), 78-87

(2) W. Grellner, P. Henze, K.A. Schwetz, A. Lipp: "Nondestructive Testing of Injection Moulded, Green and Sintered Silicon Carbide Turbine Components", in: Technical Ceramics (H. Nosbusch and I.V. Mitchell, Eds.), Elsevier Applied Science/London and N.Y. (1988), 291-297

(3) H. Kölker, P. Henze, K.A. Schwetz, A. Lipp: Proc. 3rd Int. Symp. Ceram. Mat. & Comp. for Engines, Las Vegas 1988

MULTIPLE CAPACITANCE MEASUREMENTS AS A METHOD OF ASSESSING THE QUALITY OF CERAMICS IN THE GREEN STATE

M. R. PIRAMOON and P. M. MARQUIS
School of Metallurgy and Materials
University of Birmingham, P. O. Box 363,
Birmingham B15 2TT.

ABSTRACT

Multiple capacitance measurements on green compacts have been used to assess their quality. It has been shown that the distribution of the measurements can be related to the presence of agglomerates of either binders or ceramic particles, as well as other aspects of the quality of the moulded component. The behaviour of an alumina/wax model system compared to a computer model is discussed and the results related to studies of a silicon nitride injection mix. Finally the potential of the approach as a new form of non destructive testing of green compacts prior to firing is discussed.

INTRODUCTION

Injection moulding is an established commercial method for processing ceramic components [1]. The process involves mixing ceramic powders and organic materials as binders and plasticisers, followed by injection, organic material removal and finally sintering. The mechanical properties of a ceramic product manufactured by this process is mainly controlled by the homogeneity of the mixture in the green state prior to sintering which can be adversely influenced by the presence of ceramic particle agglomerates[2].

The plastics, oil and waxes, which are generally used as binders and plastisizers in the formation of green compacts can also store electrical energy, when subjected to an electric field[3]. The capacity which a volume of material shows for storing electrical energy is determined by the redistribution of electrons and protons (polarization) of each atom making up the substance[4]. Differences in polarization phenomena at different locations on green compacts can be detected in form of capacitance measurements and thus related to changes in the composition and distribution of the components including the presence of particle agglomerates in the compacts.

EXPERIMENTAL

The filler materials used in the study were two different alfa-alumina powders, with size distributions of 5 and 60 microns and an alpha type silicon nitride. A low melting point wax with trade name of Okarin 1865Q was used as a binder.

The blended wax was loaded under identical processing condition with different alumina weight fractions, using a roller type mixer. Identical samples from each blend containing ratios of 5 and 60μm alumina particles between 10 to 100% were press moulded. Samples of silicon nitride/wax mixtures with silicon nitride weight fractions of 24% and 48% were also produced. Capacitance measurements of all the moulded samples were measured at 1KHz using a capacitance measuring bridge

RESULTS AND DISCUSSION

Figure 1 shows how the particle size distribution and alumina weight fraction affected the capacitance measurements of the alumina/wax compacts. It can be seen that capacitance measurements are affected by both the size distribution and weight fraction in the compact. These measurements have been used to construct a computer model to simulate the effect of particle agglomeration on the capacitance measurements of unit volumes of green compacts. The model enables the random generation of particles size distributions in unit volumes of binder and calculates the related weight fraction and total surface area of the particles in the simulated compacts. Particle size distribution and capacitance values for each structure can then be calculated and shown in the form illustrated in Figure 2. Using the model it is possibile to analyse the effect of variations in particle size distribution and composition on the capacitance of unit volumes of the green compacts. The capacitance values obtained from simulated structures have been compared with experimental capacitance measurements from different positions of green compacts. By matching the simulated and experimental variations in capacitance measurements it is therefore possible to establish the types of structure present in the compact. The approach is illustrated for the case of a silicon nitride/wax compact.

In the computer simulation a particle weight fraction of 24% was selected and no limitation on the particle size distribution on the simulated structures was imposed. A compact of 5 micron silicon nitride at the same weight fraction was also prepared. Capacitance values for simulated structures were calculated and experimental multiple capacitance measurements on silicon nitride/wax green compacts were obtained as shown graphically in figure 3. The similarities between the capacitance profiles from both the simulated structure and from the green compacts illustrate that the changes in capacitance measurements at different positions on the compacted sample have mainly been caused by changes in particle size distribution and not composition (ie: weight fraction of silicon nitride in the compact). Grouped multiple capacitance measurements from different silicon nitride/wax samples have also shown similar patterns of behaviour to the simulated structure. Examination of the structures in different samples has shown that there are significant differences between capacitance measurements obtained from homogeneous and agglomerated structures, Figure 4. In general homogeneous samples exhibit a reduction in spread and an increase in the mean capacitance values of the

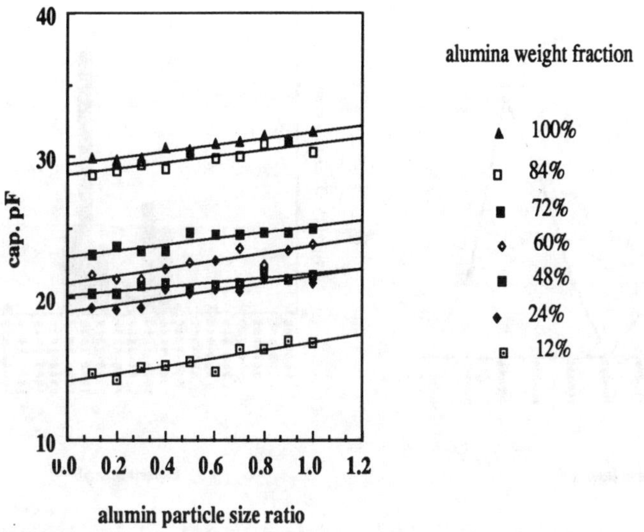

Figure 1 A graph showing variation of capacitance measurements with changing surface area and particle weight fraction in the alumina/wax green compacts

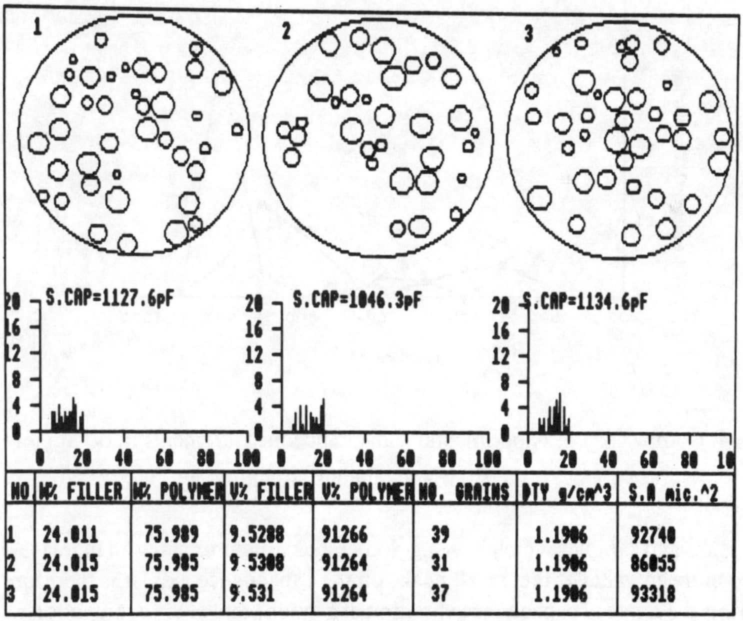

NO	W% FILLER	W% POLYMER	V% FILLER	V% POLYMER	NO. GRAINS	DTY g/cm^3	S.A mic.^2
1	24.011	75.989	9.5200	91266	39	1.1906	92740
2	24.015	75.985	9.5300	91264	31	1.1906	86055
3	24.015	75.985	9.531	91264	37	1.1906	93318

Figure 2 Capacitance measurements for three simulated structures with identical weight fraction and different surface area

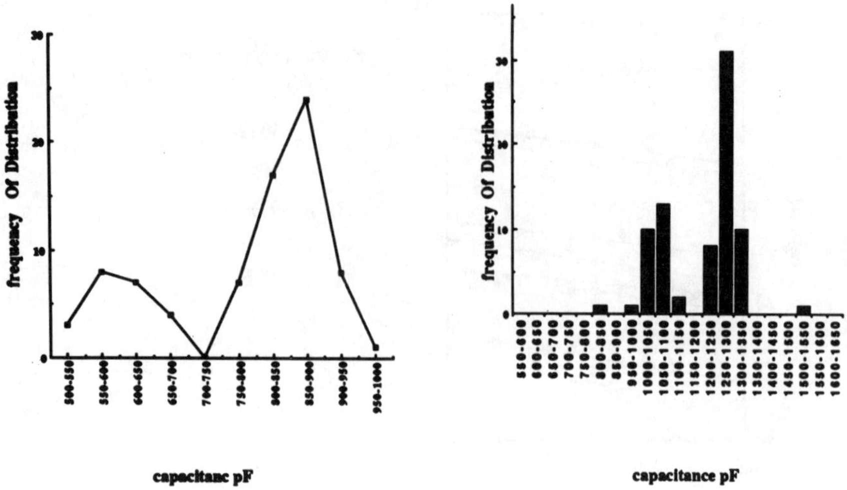

Figure 3 A graph showing comparison between simulated and experimental capacitance values

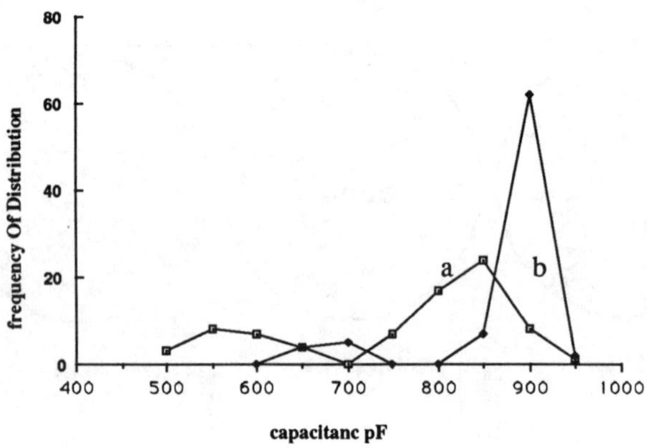

Figure 4 Comparison of experimental capacitance measurements in (a) agglomerated and (b) homogeneous structures

higher peaks and samples which have agglomerated structures show an increase in spread and decrease in mean value of the small peaks on the capacitance axis. It is therefore possible by monitoring the range of capacitance variation of a green compact in conjunction with statistical methods such as process control charts to non destructively assess the performance of a ceramic moulding process. The approach provides a non-destructive method of inspecting both the state of the blends in a compacts and the selection of optimum processing conditions for maximising the homogeneity of the compacts

a b

Figure 5 Typical micrographs of (a) homogeneous and (b) agglomerated structures

CONCLUSION

The work in this paper demonstrates a systematic method for assessing ceramic and polymer mixtures in formed bodies prior to sintering by using multiple capacitance measurements on the green compacts.The experimental results and computer simulations show that changes in capacitance measurements provides quantitative information about particle agglomerates and changes in composition of the blends making up the compacts. This approach can be used as a non destructive inspection technique for monitoring the state of green compacts.

REFERENCE

[1] Richerdson, D. W., Modern Ceramic Engineering , Dekker, Inc, New York, 1982, pp. 202-206.

[2] Roosen, A., Hausner, H., Techniques for Agglomeration Control During Wet-Chemical Powder Synthesis , Adv. Ceram. Mat., 1988, Vol. 3, No.2, 131-37.

[3] Mountvala, A. J., Onoda, G. Y., Onesto, E. J., Dielectric Evaluation of Oxide Powder-Binder Intractions , Amr. Ceram. Soc. Bull., Vol. 50, No.7, 627.

[4] Robert, H. C., Dispersion and Absorption in Dielectrics , J. Chem. Phys., April 1941, Vol. 9, 341-51.

GAMMA DENSOMAT - A NEW QUALITY CONTROL INSTRUMENT IN CERAMICS

GEORG SCHLIEPER, VOLKER ARNHOLD
Sintermetallwerk Krebsöge GmbH, P. O. Box 5100
D-5608 Radevormwald, West Germany

ABSTRACT

The determination of the density is an important part of quality control in ceramics, as many other properties are influenced by porosity. Conventional methods of density determination are often tedious and time consuming, require skilled laboratory personnel and the tested parts are usually destroyed. The Gamma Densomat provides a tool to nondestructively determine local densities in short times with high accuracy. The measurement is based on the absorption of gamma radiation in the tested material. Green compacts and fired components can be tested with equal precision. Applications for this instrument are in the field of technical ceramics, refractories, magnets, grinding and friction materials, cemented carbides, powder metallurgy.

INTRODUCTION

Many properties of ceramic materials are influenced by the porosity. A nonuniform distribution of porosity in ceramic compacts will cause distortion during firing which may lead to the rejecton of the finished product or to a substantial increase in the cost of finishing operations. Excessive porosity in refractory components or grinding tools may reduce their lifetimes dramatically. The magnetic properties of ceramic magnets are closely correlated with porosity and porous ceramics (filters, catalysts) require a close control of porosity in order to guarantee reproducible properties.

For the nondestructive determination of porosities of compacted and sintered ceramic components an instrument was developed which is suitable to be used in a production plant [1].

DENSITY DETERMINATION BY GAMMA RADIATION

The method for the nondestructive determination of the density is based on the absorption of gamma radiation in matter [2]. When penetrating through the tested section, the radiation is weakened (Fig. 1). The absorption law

$$n = n_o \exp(-\mu \rho x) \qquad (1)$$

contains the material constants μ (mass absorption coefficient) and ρ (density). The other factors n_o (intensity of original beam), n (intensity of weakened beam), and x (thickness of test part) are determined by measurement. If the absorption coefficient is known, the density can be calculated as

$$\rho = \frac{1}{\mu x} \ln \frac{n_o}{n} \qquad (2)$$

This equation is valid for dense as well as porous materials. The absorption coefficient depends only on the material composition and the radiation energy, but not on the porosity. This is the reason why this method can be applied for the determination of porosities.

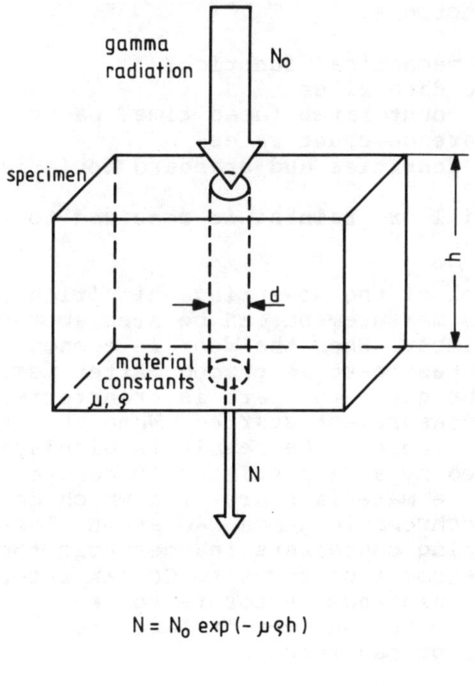

Figure 1. Absorption of gamma radiation in a test part.

THE GAMMA DENSOMAT

An instrument to apply density measurements by gamma radiation in industrial production is the Gamma Densomat (Fig. 2).

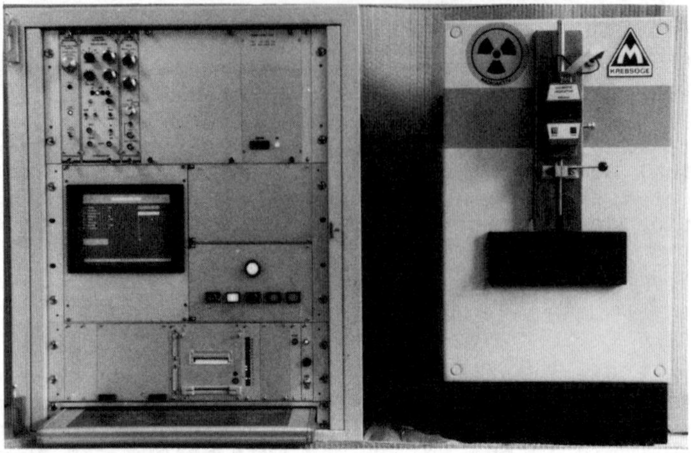

Figure 2. The Gamma Densomat

The control unit on the left contains a computer providing the following functions.

- Control of all mechanical functions
- Handling of the data files
- Corrections of count rates (dead time, background)
- Storage of reference count rates
- Computation of densities and standard deviations

No particular skill or training is required to work with this instrument.

On the front panel of the measuring unit (right) a dial gauge for the thickness measurement can be seen above the door of the measuring chamber. When the door is opened, a tray comes out on which the test part is placed. After measuring the thickness with the gauge the part is transferred into the chamber and the measurement started. When the preselected counting time is elapsed, the result is displayed on the screen and printed by a data printer if desired. As radiation sources radioactive materials are used which decay slowly generating a monochromatic gamma radiation. The sources are mounted in shielding containers inaccessable for the operator. The Gamma Densomat conforms to German safety requirements to prevent unintended exposure to radiation. It can be operated in the production area or laboratory without any additional safety precautions.

ACCURACY OF MEASUREMENT

The intensity of gamma radiation generated by radioactive decay is subject to statistical variations whose magnitude is

$$\Delta N = \sqrt{N} \qquad (3)$$

where N is the number of counted decays and ΔN is the standard deviation of N. This gives rise to statistical errors both in the intensities of the original and the weakened radiation. Additionally absorption coefficients of the tested materials have to be determined once before densities can be measured. These are also subject to a statistical error. For reasons of brevity the total statistical error of density measurements is given here without deduction [3].

$$\frac{\Delta \rho}{\rho} = \frac{0.076}{\sqrt{n_o}} + \frac{x_o}{2x}\left(\frac{0.041}{\sqrt{n_o}} + \frac{1}{\sqrt{n_o t}} \exp\left(\frac{x}{x_o}\right)\right) \qquad (4)$$

The graph of this equation is shown in Fig. 3 for n = 10000 cts/sec and t = 60 sec as a function of x. The standard deviation is a minimum near the optimum thickness x_o. The value of x_o is different for each material and radiation energy.

$$x_o = \frac{2}{\mu \rho} \qquad (5)$$

For accurate density measurements it is necessary to select the radiation energy properly so that for the tested material x_o is not too far from the thickness x.

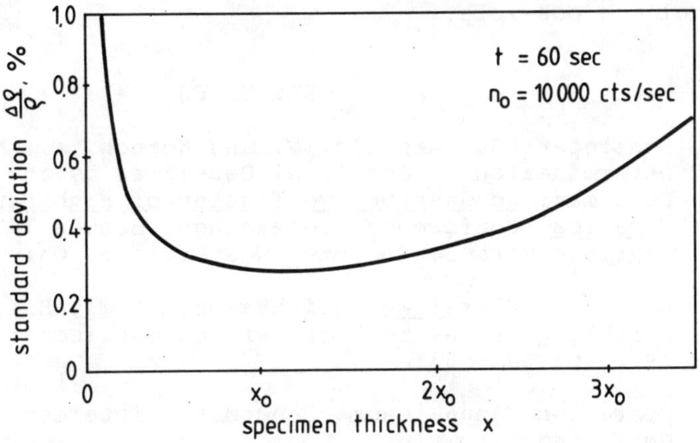

Figure 3. Standard deviation of density measurements as a function of the specimen thickness.

LIMITATIONS

A prerequisite for any absorption measurement is that the thickness of the material is well defined within the volume where the radiation passes through the part. This imposes some limitations to the geometries of parts. The upper and lower surfaces should be as flat and smooth as possible and parallel to each other. Sections with oblique surfaces cannot be tested. Curved surfaces of rings or bushes, however, can be measured with high accuracy as long as the radius of curvature is at least 4 times larger than the aperture diameter.

It is obvious that with a smaller aperture the restrictions for parts geometries become less. Therefore a small aperture size is generally desirable. However, the intensity of the gamma radiation, which is proportional to the cross section of the aperture, must be high enough to obtain the required accuracy.

ADVANTAGES

Density measurements with gamma radiation are useful in all cases where regular readings have to be taken. In quality control of ceramics it is of equal importance to test green compacts and fired components. Advantages are short measuring times (generally 60 seconds or less), safe and simple operation of the instrument (no specially trained personnel), no scrapping of tested parts, and the capability for SPC extension. Idle times of presses can be reduced as the tool-setter can control the density of a compact immediately himself and does not have to rely on the results of a lab. Regular density checks during production runs and even 100 % parts control are possible.

REFERENCES

1. Schlieper, G., Arnhold, V. and Kozuch, A., Non-destructive Determination of Sectional Densities by the Gamma-Densomat. Nondestructive Testing of High-Performance Ceramics, Conference Proceedings, Boston 1987, The American Ceramic Society, Westerville, Ohio, p. 457 - 471

2. Hart, H., Flüssigkeitsdichtemessung mit Hilfe von Kernstrahlung, B. G. Teubner Verlagsgesellschaft Leipzig, 1972, p. 60 - 63.

3. Operation Manual Gamma-Densomat, Sintermetallwerk Krebsöge GmbH, Radevormwald

INFLUENCE OF THE CERAMIC MORPHOLOGY ON THE NATURE OF THE ELECTRICAL PROPERTIES OF $\alpha-Fe_2O_3$

J.P. BONNET, M. ONILLON, C.V. SANTILLI
Laboratoire de Chimie du Solide du CNRS
Université de Bordeaux I
351, cours de la Libération
33405 Talence cedex, France

ABSTRACT

Fine powders of wealky Na-doped $\alpha-Fe_2O_3$ were obtained from precipitation of Fe^{3+} ions in aqueous solution, drying and calcination.
The properties of these powders depend strongly on the ageing of the initial precipitate in the solution.
The ceramics obtained from fresh precursors - strongly agglomerated - are characterized by the presence of extended defects, the order of which increases as the sintering goes on. The p-type conductivity of these ceramics is mainly due to the presence of cationic impurities.
The steady appearance of localized defects occurs during the sintering period of aged precursors, weakly agglomerated. As these defects are associated to donor states, a predominant n-type conductivity is observed in the corresponding ceramics.

INTRODUCTION

The peculiar electrical properties of iron sesquioxide $\alpha-Fe_2O_3$ make it a valuable material for various applications |1,2|.Indeed, as it is generally the case, the macroscopic properties of sintered $\alpha-Fe_2O_3$ - which result from combination of both bulk and intergranular characteristics - depend strongly on the microstructure.

The aim of the present work is to find out to which extent the structural defects and the grain boundaries can act upon the electrical properties of hematite ceramics.

MATERIALS PREPARATION

All hematite powders used as precursors were obtained by firing for 16 h aqueous precipitates of Fe(III) ions, either fresh or aged, as described in our previous paper |3|. Ageing was achieved by holding the precipitate for 140 h at 80°C in the solution from which it was formed (pH=5) Firing temperatures of 400 and 600°C for fresh and aged powders were chosen, respectively, in order to obtain ceramics with identical apparent specific mass.

The hematite powders, shaped as cylinders (0.5 g, 8 mm in diameter) under an uniaxial pressure of 200 MPa, were sintered in air for 1 up to 16 h at 1100°C and then quenched to room temperature in the same atmosphere. In the following, the symbols F and A shall be used to qualify the materials obtained from fresh and aged precipitates, respectively.

The characteristics of the ceramics prepared are summarized in Table I.

TABLE 1
Some physical and microstructural features of the ceramics studied

Sintering conditions			Apparent specific mass	Grain size
Temperature (°C)	Time (h)	Precursor (F = fresh A = aged)	$(g.cm^{-3})$	(μm)
1100	1	F	5.02	3.2
1100	4	F	5.02	5.8
1100	8	F	5.04	7.3
1100	16	F	5.05	8.8
1100	1	A	5.10	2.3
1100	4	A	5.11	4.0
1100	8	A	5.13	5.2
1100	16	A	5.12	6.2

MICROSTRUCTURAL CHARACTERIZATION

The microstructures of the ceramics were described in a previous paper |4|.

The differences in the microstructure of F and A ceramics are mainly due to the evolution of defects and residual microstrains during the sintering stage.

Fig. 1 shows TEM images of F and A ceramics, sintered in air for 1 and 16 h respectively. It appears that both materials contain structural defects. However, depending on the agglomeration state in the precursor, these defects differ both in nature and quantity.

In the A ceramics sintered for 1 h, the defects are ordered according to a double network (Fig. 1-c) which tends to disappear as the heat treatment time is increased, leading to randomly distributed extended defects. The image of such a material, obtained after 16h at 1100°C, is shown on Fig. 1-d.

The evolution of defects in F materials is more complex: many defects are present after 1 h at 1100°C (Fig. 1-a) ; they tend to order during further heat treatment, leading ultimately to the quasi ordered network shown on Fig. 1-b.

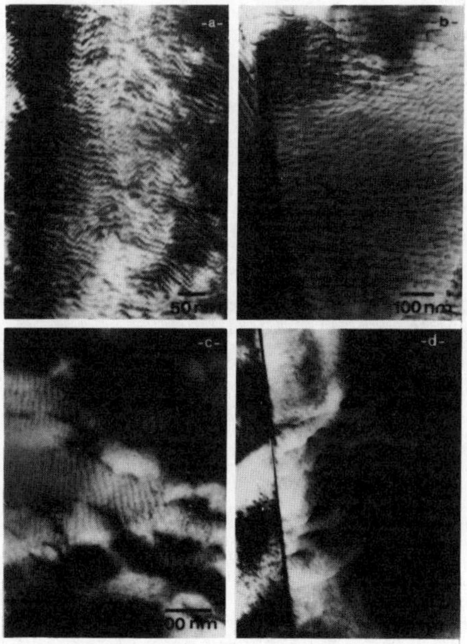

Figure 1 - Structural defects in ceramics sintered in air at 1100°C:
a - F material sintered for 1h, b - F material sintered for 16h
c - A material sintered for 1h, d - A material sintered for 16h.

Fig. 2 shows the variation with the sintering time t_s of the magnitude of the microstrains in the |012| direction (represented by the value of the ε parameter determined using the method described by WAGNER |5|). When the sintering time is prolonged, ε increases continuously in the A-type materials, while the double network makes way for the randomly distributed extended defects. On the contrary, after a quick decrease, ε remains independent of sintering time in the F-type materials.

It is worhtwhile noting that, in either case, the more ordered materials are also those with the weakest microstrains.

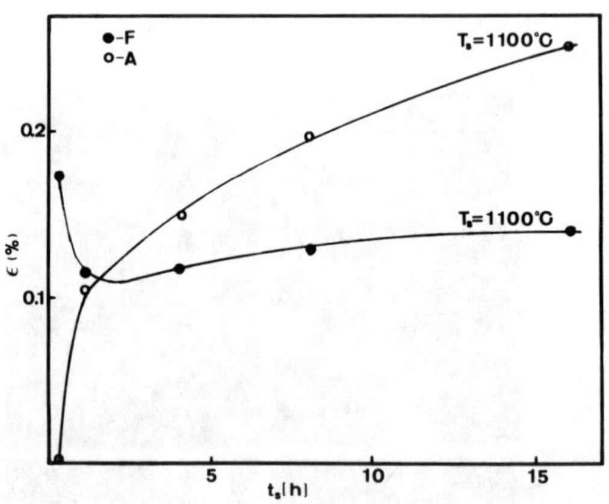

Figure 2 - Influence of the sintering time at 1100°C on the microstrains determined in |012| direction.

ELECTRICAL PROPERTIES

The thermoelectric power, α , of ceramics was measured in a direction parallel to the pressing axis, using the technique described by DORDOR et al. |6|. All measurements were carried out in air, from 400 up to 800°C, since below 400°C several ceramics exhibit non linear electrical properties |7|.

Fig. 3-a and 3-b represent respectively the values of the thermoelectric power coefficient α as a function of temperature for various F and A ceramics. All materials were sintered in air at 1100°C, but for variable holding time t_s. It appears that α is positive for F ceramics and negative for A ceramics.

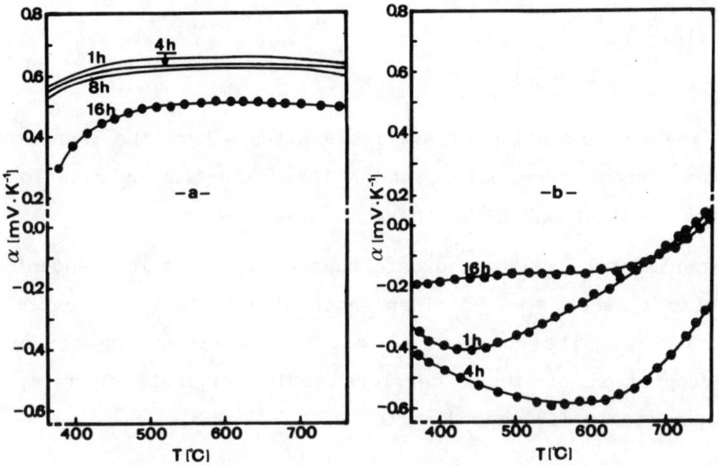

Figure 3 - Evolution of the thermoelectric power coefficient of ceramics, versus temperature, for various sintering times at 1100°C.
a -F-type materials, b -A-type materials

After chemical analysis, F and A ceramics appear to have the same composition (table II). The explanation of their drastically unlike electrical properties shall then be searched into their difference of morphology. We discuss the particular case of every material in the following.

TABLE II
Impurity mass concentrations in A and F ceramics (ppm)

	Cl^-	NH_4^+	Na^+	Other cations
A ceramic	400	100	300	50
F ceramic	350	100	320	50

F materials

The ceramics obtained are p-type semi-conductors (Fig. 3-a), with a positive thermoelectric power coefficient α fairly constant with temperature. Such features are characteristic of a saturated narrow band semi-conductor with α expressed as :

$$\alpha = -\frac{k}{q}\left(\ln \frac{N}{P} + A\right) \tag{1}$$

where N is the concentration of available sites - here the iron atoms-, P the concentration of free holes, and A their kinetic contribution. k is the Boltzmann constant and q the charge of the electron.

Our experimental results indicate that α is almost independent on the sintering time t_s when $t_s < 8h$. Then, according to (1), the concentration of p carriers is quite constant in all the corresponding ceramics. An important proportion of these carriers could originate from impurities such as sodium (table II) according to:

$$Na_2O \rightarrow 2Na_{Fe}^{2'} + O_O + 2V_O + 4h^{\bullet}$$

A materials

The ceramics obtained are n-type conductors. The variation of α with temperature (Fig. 3-b) is similar to that observed by Benjelloun et al. |8| in the case of weakly doped α-Fe_2O_3 single crystals. However, like F ceramics, the A ones shall contain a fair amount of acceptor states since both materials have practically the same impurity content. The appearance of donor states, in amounts large enough to overwhelm the conductivity contribution of the p carriers, could in principle result from a departure from stoichiometry (i.e. x in Fe_2O_{3-x}). In fact, our experimental determinations indicate that, for both A and F ceramics, the departure from stoichiometry is nil within the limits of sensitivity of the thermogravimetric equipment used ($\Delta x \sim 2.10^{-4}$). Then the only mechanism that can be responsible for the predominant n conductivity observed is the appearance of donor states originating from the structural defects which are present in the A ceramics together with microstrains. Until now we could not determine the exact nature of the defects involved.

REFERENCES

1. Nakatani Y., Sakai M. and Matsuoka M. Jpn. J. Appl. Phys., 1983, 22, p.912.

2. Leygraf C., Hendewerk M. and Somorjai G., J. Solid State Chem., 1983, **48**, p.357.

3. Santilli C., Onillon M. and Bonnet J.P., Ceramics International, in press.

4. Santilli C.V., Bonnet J.P.,Rabardel L. and Onillon M., Ceramics International, 1989, **15**, p.99

5. Wagner C.N.J., Local Atomic Arrangements for X-Rays Diffraction, (Ed.), Cohen J.B. and Hilliard J.E., Gondon and Breach, New-York, 1966, p. 219.

6. Dordor P., Marquestaut E. and Villeneuve G., Rev. Phys. Appl., 1980, **15**, p. 1607.

7. Santilli C.V., Onillon M. and Bonnet J.P., to be published.

8. Benjelloun D., Bonnet J.P., Doumerc J.P., Launay J.C., Onillon M. and Hagenmuller P., Mat. Chem and Phys., 1984, **10**, p. 503

GRAIN BOUNDARY EFFECTS IN SEMICONDUCTING $BaTiO_3$

S.B.DESU
Department of Materials Engineering
Virginia Polytechnic Institute and State University
Blacksburg, Virginia, 24061, U.S.A

ABSTRACT

A comprehensive model based on the interrelations between dopant segregation, dopant incorporation, and redox reactions is proposed to account for the effects of donors in lanthanide doped $BaTiO_3$. The model describes the effect of lanthanide concentration on both the room temperature conductivity and grain size of doped $BaTiO_3$. The model also explains the dependence of the critical lanthanide concentration, at which doped $BaTiO_3$ has highest room temperature conductivity, on the size of lanthanide ion. The model incorporates the influence of process parameters through redox reactions that occur at the grain boundaries. Both grain boundary barrier layer effects and positive temperature coefficient resistance can be explained by this model.

INTRODUCTION

Valence doping of $BaTiO_3$ results in a very interesting phenomenon. A small amount (0.2 atom%) of higher valence ions like Nd^{3+} introduced at the Ba^{2+} site or Nb^{5+} at the Ti^{4+} site, produces semiconducting $BaTiO_3$ whose conductivity is as much as 12 orders of magnitude higher than that of pure stoichiometric $BaTiO_3$. This material, however, is semiconducting only within a narrow range of dopant concentrations and is insulating at higher dopant concentrations. Depending on the processing conditions, valence doping can also result in boundary layer devices such as internal boundary layer capacitors (IBLC) or positive temperature coefficient resistors (PTCR). Despite numerous attempts [1–12], no comprehensive theory is available which can account for all the observed phenomenon. In contrast with earlier models [1–12], a comprehensive model is presented based on dopant segregation induced heterogeneous defect distribution; which interrelates the electrical properties, interfacial segregation, and microstructure of donor doped $BaTiO_3$.

EXPERIMENTAL METHOD AND OBSERVATIONS

Coprecipitated powders with [Ba+dopant]/Ti ratios of 1.000 ± 0.002 were used in this investigation. Precise details of the experimental method are given elsewhere [13]. In common with earlier results [1–9], the room temperature electrical conductivity of polycrystalline lanthanide doped $BaTiO_3$ at low dopant concentrations, increased with increasing donor concentration, reached a maximum at a critical donor concentration $[D]_c$, and then decreased with higher donor concentrations (Figure 1–2). Figures 1–3 show that the value of $[D]_c$ increased while the concentration range (in which low conductivity is observed) broadened, with decreasing radius misfit (between radii of donor dopant and Ba^{2+} ions), decreasing partial pressure, increasing cooling rate, and increasing sintering temperature.

Besides room temperature conductivity, the grain size was also a function of dopant (Nd) concentration. For samples sintered at 1380°C for 2 hours, the grain size of highly conductive Nd doped $BaTiO_3$ samples varied from 25 to 40 μm, whereas for resistive samples it was around 1 to 2 μm. The samples which had high room temperature conductivity and Nd concentrations around 0.124 (grain sizes > 20 μm), showed a PTCR effect (sharp increase in electrical resistance with increasing temperature around Curie point).

DISCUSSION

In an earlier paper [14], it was shown that the high conductivity of Nd–doped $BaTiO_3$ was related to the presence of Ti^{3+} ions. It was also shown that the Ti^{3+} concentration followed room temperature conductivity, increasing below $[D]_c$, and decreasing above $[D]_c$ [14]. At sintering temperatures, barium titanate is reduced forming, depending on the conditions, either individual Ti^{3+} ions or Magnelli phases such as Ti_xO_{2x-1} which contain trivalent titanium ions. These reduction products react with Nd_2O_3 forming compounds containing Ti^{3+} ions such as $NdTiO_3$. Since these compounds are isostructural with $BaTiO_3$, they are easily soluble and form conducting solid solutions $\{Ba_{1-x}Nd_x[(Ti^{4+})_{1-x}(Ti^{3+})_x]O_3\}$. This type of Nd incorporation is called electronic incorporation whereby the conductivity increases due to increasing Ti^{3+} ion concentration, with increasing Nd concentration.

According to the above scheme, the number of dopant atoms that are incorporated by electronic compensation depends on the amount of Ti^{3+} formed during sintering and/or annealing conditions (temperature, oxygen partial pressure, etc.). One would thus expect with increasing sintering temperature and reducing sintering atmosphere, an increased degree of reduction, and in turn increased donor concentration which is electronically incorporated. This will result in an increase in $[D]_c$ and broadening of the low conductivity range, which is in accordance with the observations in Figure 1. If the concentration of Nd (donors) is higher than that of Ti^{3+} at the high–temperature treatment conditions, the Nd will be incorporated through a vacancy compensation mechanism forming solid solutions of the type $Ba_{1-x}Nd_x[(Ti^{4+})_{1-0\cdot 25x}(V_{ti}'''')_{0\cdot 25x}]O_3$ [12,13 & 15]. In vacancy compensation, the donors do not contribute to electronic conductivity and thus the conductivity of Nd–doped $BaTiO_3$ decreases.

As the samples are cooled from the sintering/annealing temperatures, the point defect equilibrium shifts in the same direction as does an increase in partial pressure of oxygen. In other words, the samples will oxidize during cooling. This involves reestablishment of defect equilibria by diffusion of point defects. In general, during cooling, grain boundaries follow the falling temperatures in that they successively reestablish the defect equilibria valid for each temperature more readily than the grain interiors, and thus result in a heterogeneous defect distribution in the sample. Since the diffusion constant of the defects decreases rapidly with falling temperatures, the defect structure corresponding to the effective quenching temperature will be "frozen–in" at room temperature.

Moreover, during cooling, the donor concentration at which the vacancy compensation is initiated will also decrease. Depending on the effective quench temperature, oxygen partial pressure, and concentration; the dopant incorporation may shift from the electronic to the vacancy mechanism. This shift in dopant incorporation can be represented as:

$$Ba_{1-x}Nd_x[(Ti^{4+})_{1-x}(Ti^{3+})_x]O_3 + O_2 \rightleftarrows$$

$$Ba_{1-x}Nd_x[(Ti^{4+})_{1-0\cdot25x}(V_{ti}'''')_{0\cdot25x}]O_3 + 0.25x\ TiO_2 \quad (1).$$

As previously reported [13], there is a pronounced tendency for dopants to segregate to the grain boundaries in $BaTiO_3$, thus enhancing the donor concentration at the grain boundaries in comparison to that of the grain bulk. For an average Nd concentration of 0.326 atom%, grain size of 2 μm, a segregation layer of 20 nm, (experimental values for samples sintered at 1380°C for 2 hours in air and cooled 100°C/hour) and for a reasonable segregation coefficient of 20; we obtain Nd concentrations of 0.208 and 4.16 atom% for grain bulk and grain boundary regions, respectively. Therefore with increasing dopant concentration, the shift in dopant incorporation for grain boundaries occurs much earlier than the bulk.

An estimate (based on defect calculations [13]) of donor concentration at which a shift in compensation takes place at an effective quench temperature of 500°C in air results in a value of 3 to 4 atom%. This is a reasonable result since it does not contradict the fact that single crystals of $BaTiO_3$ with donor concentrations close to 2 atom% are semiconductive [16]. For a polycrystalline donor doped $BaTiO_3$ samples, due to dopant segregation, the grain boundary region may be in vacancy compensation while the grain bulk may be in electronic compensation (see the case above). This difference in compensation is generally controlled by the kinetics of defect diffusion. The apparent diffusion constant of a defect (D_{app}) in polycrystalline materials can be written in terms of its grain boundary diffusion constant (D_g) and lattice diffusion constant (D_l) as

$$D_{app} \cong D_l + \delta D_g / L \qquad (2)$$

where δ and L are the thickness of the grain boundary layer and mean grain radius, respectively. Therefore, the formation of resistive layers by vacancy compensation on the surface of the donor doped $BaTiO_3$ grains, not only depends on the cooling rate, oxygen partial pressure, along with factors enhancing the defect diffusion constants (such as the presence of liquid phases), but also depends on the grain size.

Besides affecting electrical properties of grain bulk and boundary, the donor segregation to the grain boundary also effects the grain boundary mobility, and thus the average grain size of donor doped $BaTiO_3$. Since the enhanced concentration of dopants at the boundary impedes its mobility, exaggerated grain growth and average grain size will decrease with increasing average donor concentration. In fact, the present investigation as well as earlier reports [4,7 & 9] have shown that highly resistive donor doped $BaTiO_3$ has very small grain sizes on the order of 1–2 μm.

Based on this model, the experimental results can be easily understood. If the samples are quenched from the sintering temperature (Figure 2), a shift to vacancy compensation is kinetically prohibited and therefore the Nd concentration range in which low conductivity is observed increases. Further, low temperature annealing (1100°C) of these quenched samples should restore the defect structure corresponding to the effective quench temperature. This explains why the samples with 0.326 atom% Nd became insulative, whereas the conductivity of samples with 0.124 atom% Nd did not change with annealing time (Figure 4).

If the grain boundary donor segregation is key to the high resistivity of donor doped samples, one would expect that under the same processing conditions, the donors with large segregation coefficients should have lower $[D]_c$ values. As the ionic radius of the donor occupying the Ba^{2+} site decreases (radius misfit increases), the tendency for segregation increases. Therefore, the $[D]_c$ should decrease. This is observed in Figure 3. However for dopants like Tm^{3+}, low conductivity was not observed at any concentration. This is in agreement with our model.

Based on the proposed model, the PTCR effect can also be explained. The formation of highly resistive grain boundary layers surrounding the semiconductive grains, results in a potential barrier due to the equilibriation of Fermi levels. Therefore, space charges will develop at the vicinity of the grain boundary due to the flow of electrons from the semiconducting grain interior to the highly resistive grain boundary. The space charge layer could be shielded by the spontaneous polarization in ferroelectric regime (thus, decreasing the barrier height) if its thickness is much smaller than the grain size. Therefore, the material should show high conductivity below the Curie point. Above the Curie point, no shielding is present and the conductivity decreases drastically. The grain boundary potential barriers, which are responsible for the PTCR effect in our model, are formed by the high resistance grain boundary layers in contrast to surface states in Heywang's model [8].

SUMMARY

The low temperature electrical properties of Nd–doped $BaTiO_3$ are determined by Nd segregation to the grain boundary. Nd incorporation is achieved either by electronic compensation at low concentrations or by vacancy compensation at high concentrations. Furthermore, segregation of Nd to the grain boundaries inhibits the grain growth. At small concentrations, the Nd incorporation by electronic compensation explains the high conductivity. As the average dopant concentration increases, the Nd concentration at the grain boundary rapidly increases due to segregation. This has two important effects, 1) the dopant incorporation at grain boundary shifts from electronic to vacancy compensation and results in the formation of high resistance layers, and 2) grain size decreases rapidly due to the significant dopant drag (at high concentrations) on the grain boundary mobility. Thus, for high dopant concentrations, the Nd–doped $BaTiO_3$ becomes insulating. The proposed model based on the interrelationships between dopant segregation, defect structure, defect diffusional kinetics, and microstructure successfully explains the diverse phenomena associated Nd–doped $BaTiO_3$.

ACKNOWLEDGEMENTS

The author wishes to thank D.A Payne for illuminating discussions, and K.Bartelmehs for editing this manuscript.

REFERENCES

1) Sauer, H.A. and Fisher, J.R., "Processing Of Positive Temperature Coefficient Thermistors," J.Am.Ceram.Soc., 1960, 43, 307–9.
2) Tennery, V.J. and Cook, R.L., "Investigation of Rare–Earth Doped Barium Titanate," J.Am.Ceram.Soc., 1961, 44, 187–93.
3) Goodman, G., "Electrical Conduction Anomaly in Samarium–Doped Barium Titanate," J.Am.Ceram.Soc., 1963, 46, 48–54.
4) Ueda, I. and Ikegami, S., "Oxidation Phenomena in Semiconducting $BaTiO_3$," J.Phys.Soc.Jpn., 1963, 20, 546–52.
5) Saburi, O. and Wakino, K., "Processing Techniques and Applications of Positive Coefficient Thermistors," IEEE Trans.Comp.Parts, 1963, CP–10, 53–66.
6) Jonker, G.H., "Some Aspects of Semiconducting Barium Titanate," Solid–State Electron., 1964, 7, 895–903.
7) Swiggard, E.M. and Clabaugh, W.S., "Preparation of Barium Titanate

Semiconductors Containing Controlled Amounts of Neodymium," Am.Ceram.Bull., 1966, 45, 777–81.
8) Heywang, W., "Semiconducting Barium Titanate," J.Mat.Sci., 1971, 6, 1214–24.
9) Murakami, T., Miyashita, T., Nakahara, M. and Sekine, E., "Effect of Rare–Earth Ions on Electrical Conductivity of $BaTiO_3$ Ceramics," J.Am.Ceram.Soc., 73, 6, 294–97.
10) Daniels, J., Hardtl, K.H., Hennings, D. and Wernicke, R., "Defect Chemistry and Electrical Conductivity of Doped Barium Titanate Ceramics," Philips Res.Rep., 1976, 31, 487–559.
11) Heydrick, H. and Krauer, U., "Grain Boundary Effects in Ferroelectric Barium Titanate," Ferroelectrics, 1981, 31, 151–56.
12) Lewis, G.V. and Catlow, R.A., "Defect Studies of Doped and Undoped Barium Titanate using Computer Simulation Techniques," 1986, 47, 89–97.
13) Desu, S.B., "Interfacial Segregation in Perovskites and Applications to Boundary Layer Devices," Ph.D Thesis, University of Illinois, Urbana, USA, 1982.
14) Desu, S.B., "The Nature of Defects in Semiconducting $BaTiO_3$," Adv.Ceram., to be published.
15) Shaikh, A.S. and Vest, R.W., Defect Structure and Dielectric Properties of Nd_2O_3–Modified $BaTiO_3$," J.Am.Ceram.Soc., 1986, 69, 689–94.
16) Bursian, E.V., Girshberg, Ya.G. and Starov, E.N., "Small Polarons in Conducting Barium Titanate Crystals," Phys.Stat.Sol., 1971, 46b, 529–33.

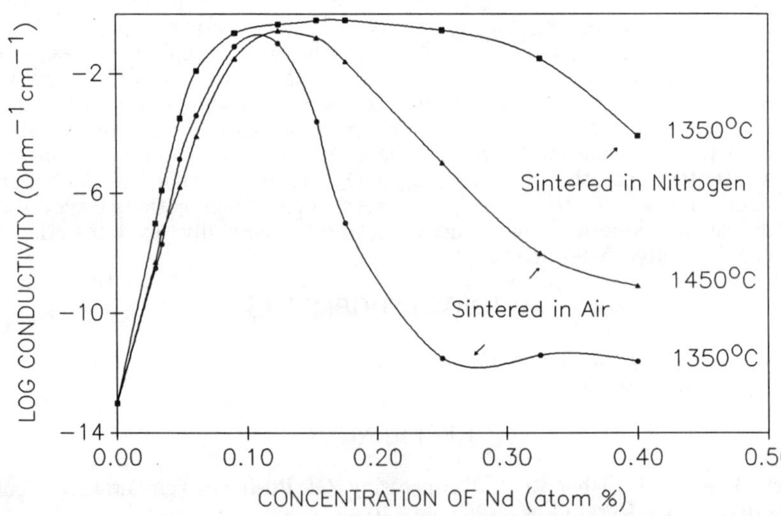

Figure 1. Effect of Sintering Temperature, and Atmosphere on the Relation and Room Temperature Conductivity of Polycrystalline Nd–doped $BaTiO_3$ Samples Cooled at 100°C/hour.

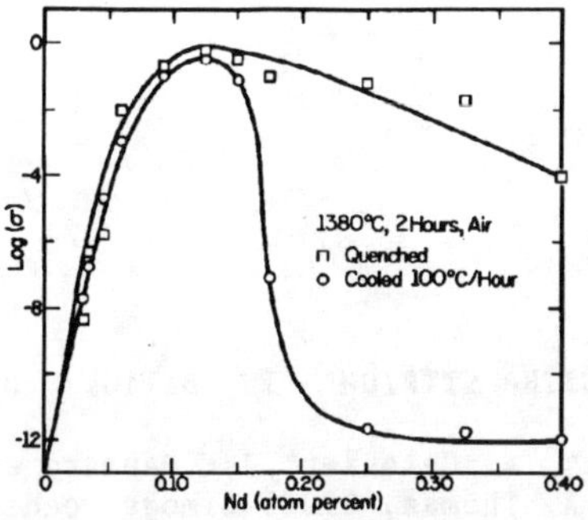

Figure 2. Effect of Cooling Rate on the Relation between Nd-concentration and Room Temperature Conductivity of Polycrystalline Nd-doped $BaTiO_3$.

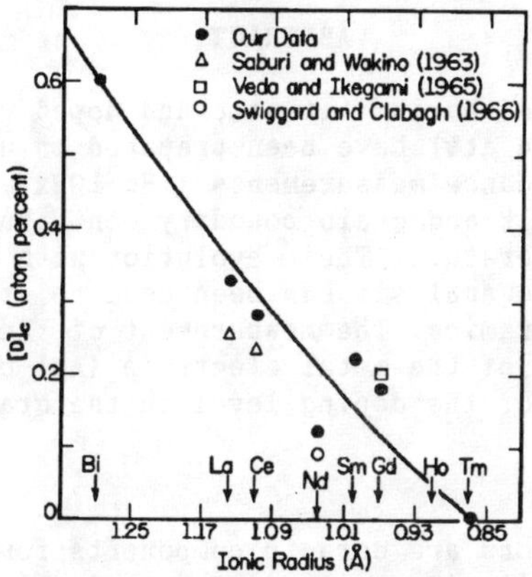

Figure 3. Relation Between Ionic Size of Donor and Critical Donor Concentration $[D]_c$ at which Conductivity Maximum is Observed. Samples were Sintered at 1380°C for 2 hours in Air and Cooled 100°C/hour.

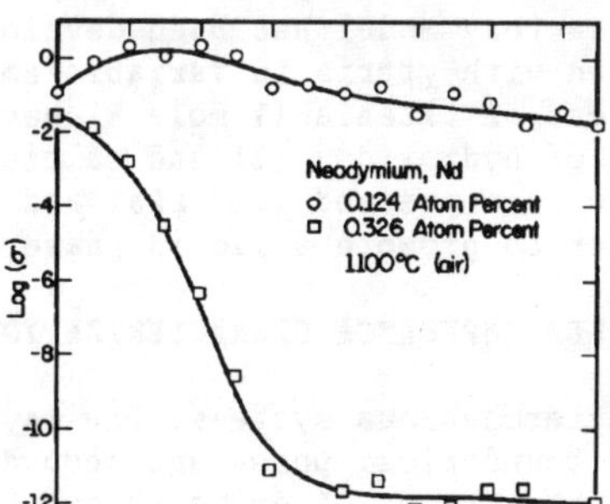

Figure 4. Relation Between Heat Treatment Time and Room Temperature Conductivity of Polycrystalline Nd-doped $BaTiO_3$ Samples which were Sintered at 1380°C for 2 hours in Air and Quenched to Room Temperature.

EFFECT OF DOPING WITH YTTRIUM IN $BaTiO_3$ BASED P.T.C. THERMISTORS

P.Blanchart, J. Machet*, A. Celerier*,J.F.Baumard and P.Abelard
UA CNRS 320 ENSCI Av. A. Thomas, 87065 Limoges cedex France.
* Laboratoire de Physique ionique, Faculté des Sciences,
Av. A. Thomas, 87060 Limoges cedex France

ABSTRACT

PTC thermistors made of barium titanate and doped with yttria in variable amount (0.1-0.4 at%) have been prepared on using a chemical route. The complex impedance measurements (1Hz-1GHz) have permitted the determination of the bulk and grain boundary contributions to the total resistance at room temperature. Their evolution with the dopant content is discussed. Microprobe analysis has been used to study the repartition of the dopant in the ceramics. The measurement of the capacitance -voltage characteristic of the metal electrode (Ag)-ceramic interface gives also an estimate of the doping level in the grains.

$BaTiO_3$ based thermistors are ceramic components for electronics. Doped with yttrium, $BaTiO_3$ is semiconducting and this explains the low resistance of these devices at room temperature. Above 120°C, which is also the temperature of the ferro-paraelectric transition, the resistance increases rapidly by several orders of magnitude. It has been clearly established that the PTC effect is a grain boundary property. The high resistance above 120°C is due to the presence of potential barriers at the interfaces. A satisfactory model has been developed by Heywang [1].

Powders of $BaTiO_3$ doped with yttria in variable amount (0.1 to 0.4 mole %) and with an excess of titania (1 mole %) have been prepared on using a coprecipitation of hydroxides [2] and calcined at 1150°C for two hours. Disk samples have been pressed (125 MPa) and sintered during two hours at 1350°C, in order to promote a liquid phase sintering mechanism.

COMPLEX IMPEDANCE CHARACTERIZATION.

PTC thermistors are heterogeneous systems. One may distinguish grains of various sizes, grain boundaries, pores and second phases. Furthermore, the device includes two electrodes. A dc measurement of the resistance gives us only an effective resistivity, which is insufficient to a proper understanding. In this respect, complex impedance measurements have proved to be useful.

An alternating signal of variable frequency is applied to the device. The real and imaginary parts of the impedance are determined separately and reported in the complex impedance plane. Such a spectrum is presented in Fig.1 for a PTC sample with sputtered Ni-Cr electrodes, which are ohmic and of a low resistance. It has been obtained on using two different apparatus, the Solartron 1170 at low frequencies (1 Hz-1 MHz) and the Hewlett Packard 4191 at high frequencies (1 MHz-1 GHz). The graph can be viewed as the superposition of arcs of circle, which reminds us the spectrum of an electrical circuit made of several resistor-capacitor cells mounted in series. Using a least square analysis of the data and an appropriate model described in Ref. [3], the contributions of the grains and grain boundaries to the total resistance can be evaluated. It appears that the latter is the most important one. According to the Heywang model, it is due to residual potential barriers.

Evolution of Rb and Rjg with the dopant content.

The variations of the bulk and grain boundary resistivities are depicted in Fig.2. They can be described as U-shape curves and this has been described already in several publications. However, this is the first time that both contributions have been measured separately and prove to follow the same behaviour. It can be seen also in Fig.2 that the grain boundary resistivity is more affected by variations of the dopant content. This is in qualitativ agreement with the Heywang model which implies that the bulk conductivity σ_b is proportional to the dopant content :

$$\sigma_b = qN_D\mu_e \qquad (2)$$

where μ_e is the electronic mobility i.e. 0.5 cm²/V.s at room temperature, while the grain boundary resistivity σ_{gb} depends exponentionally on it:

$$Vi = \frac{Q^2}{2qN_D\varepsilon_o\varepsilon} \quad \text{and} \quad \sigma_{gb} \equiv \exp\left(\frac{qVi}{k_BT}\right) \qquad (3)$$

Vi is the potential barrier, Q the charge at the interface and N_D the dopant content in the adjacent grains. However, if one tries to derive an estimation of N_D from the measured value of the bulk resistivity using Eq.(2), one finds that Nd is one order of magnitude lower than the nominal content of the original powder.

Microprobe analysis (CAMECA) was used to ascertain the dopant concentration in the grains. The mean composition and standard deviation of twelwe points in each sample are reported in Table I.

Table I: Results of microprobe analysis

ppm at in powders	ppm at. measured	ppm at. standard deviation
1000	380	227
2000	620	142
4000	830	102

These measurements confirm that only part of the dopant is incorporated into the grains during the sintering process. The standard deviation decreases with the nominal content. This is certainly due to an improvement of the experimental uncertainty but also without doubt to a more uniform distribution of yttrium.

It is not surprising in these conditions to find very high concentrations of yttrium, up to several percents, in the secondary phases. The analysis shows that they are not homogeneous. Part of them have a composition close to $Ba_6Ti_{17}O_{40}$ in agreement with previous studies while yttrium concentrates in small pockets of variable composition. We conclude that this element is not soluble in the compound $Ba_6Ti_{17}O_{40}$ and that these secondary phases are in a metastable state, may be partly amorphous.

Interpretation.

The lowest resistivity (5 Ω.Cm.) was obtained for a 0.2 at% doping level. This corresponds to a concentration of conduction electrons of 160 ppm, i.e. lower than the real concentration of the dopant in the grains by a factor of 4. This can be attributed to a compensation mechanism by acceptors. The origin is twofold. According to a previous study at high temperature [4], common impurities in the starting powders are most often acceptors. This explains why samples with a nominal content below 0.1 at% are insulators. The impurity content in our samples may be evaluated to be 300 ppm about. The rest of the acceptors are cationic vacancies present in the bulk material. During the liquid phase sintering, the dopant can be incorporated into the grains in two ways, in a reduced form:

$$2Y_2O_3 + 4TiO_2 \longrightarrow 4Y_{Ba}^{o} + 4e' + 4Ti_{Ti}^{x} + 12O_o^{x} + O_2(gas) \quad (4)$$

i.e. with a free energy gain

$$\Delta G1 = \Delta G1^{o} + 8k_BT.Log(N_D) + k_BT.Log(pO_2) - 2\mu(Y_2O_3) - 4\mu(TiO_2)$$

if the electroneutrality relationship $[e']=[YBa^{o}]$ is taken into account, and in an oxidized form. In the latter case, the dopant ions are compensated for by cationic vacancies :

$$2Y_2O_3 + 3TiO_2 \longrightarrow 4Y_{Ba}^{o} + V_{Ti}^{4'} + 3Ti_{Ti}^{x} + 12O_o^{x} \quad (5)$$

$$\Delta G2 = \Delta G2^{o} + 5k_BT.Log(N_D) - 2\mu(Y_2O_3) - 3\mu(TiO_2)$$

or

$$2Y_2O_3 + 6TiO_2 \longrightarrow 4Y_{Ba}^{o} + 2V_{Ba}^{2'} + 6Ti_{Ti}^{x} + 18O_o^{x} \quad (5')$$

$$\Delta G3 = \Delta G3^{o} + 6k_BT.Log(N_D) - 2\mu(Y_2O_3) - 6\mu(TiO_2)$$

The dopant concentration in the grains, N_D and the compensating mechanism depend on the yttria and titania chemical potentials $\mu()$ in the liquid phase and therefore on the nominal composition, in particular the Ba/Ti ratio. The latter affects also the defect chemistry of barium titanate but previous studies have clearly proved that the solubility of titania in barium titanate is low, about 100 ppm [6], i.e. lower than the impurity content. An oxygen rich atmosphere and a high concentration of the dopant favour reaction (5) and (5'). Increasing the yttrium content, conduction electrons become minority defects and the material is again an insulator. The concentration of cationic vacancies depends on the cooling procedure, because the thermodynamic equilibrium is not maintained up to room temperature [5]. It is most probable that a strict, $[e']=[Y^{o}]$, electroneutrality relationship is not realized.

CAPACITANCE - VOLTAGE MEASUREMENTS.

Semiconducting samples were covered on one side by a sputtered Cr-Ni electrode, giving a good ohmic contact. Evaporation of silver under vacuum (5.10^{-6} atm.), followed by an oxydation post treatement at 400°C produced a rectifying contact. In the complex impedance plane, see Fig. 3, it is associated to a third semicircle observed at low frequencies. Selecting an appropriate frequency (1000 Hz.), we have made capacitance-voltage measurements. A plot of $1/C^2$ versus voltage, see Fig. 4, shows a straight line, indicating the presence of a potential barrier. The intersection whith the X axis, -Vo, depend on the preparation conditions. This is usually attributed to the presence of a thin intermediate layer due to contamination(s). Taking into account a capacitor in series with the depletion layer capacitor, it is possible to improve the fit and decrease Vo. The slope of the staight line is inversely proportionnal to N_D which can be evaluated. It depends also on the permittivity of the material ε. At room temperature and for a small applied electric field, ε is equal to 1500. Then N_D is equal to 6 $10^{17}/Cm^3$ i.e. lower than the previous estimate by a factor of two. However, a very hight electric field exists in the depletion layer and the domain wall contribution to ε is strongly decreased [7]. We think that it could be the cause of the discrepancy.

REFERENCES

1) W.Heywang, "Resistivity Anomaly in Doped Barium Titanate", J. Am. Ceram. Soc., 47, 10, 484-490 (1964).
2) F. Chaput, Thèse Université de Limoges, France 1989.
3) Ph. Blanchart, To be published.
4) N.H. Chan and D.M. Smyth, "Defect chemistry of donor doped $BaTiO_3$", J. Am. Ceram. Soc., 67, 4, (1984).
5) N.H.Chan, R.K.Sharma, and D.M.Smyth., "Solubility of TiO_2 in $BaTiO_3$", J. Am. Ceram. Soc. 64, 556 (1981).
6) J. Daniels, R. Wernicke, Philips Research Reports, Part 4, 31, 487-559, 1976
7) A. von Hippel, Reviews of Modern Physics, 22, 3, 221-237, 1950.

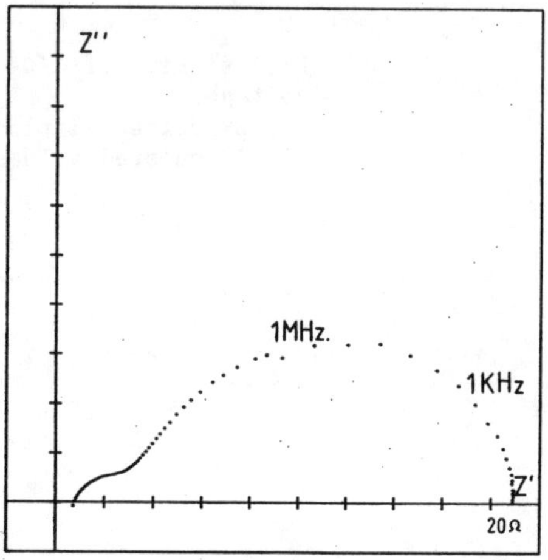

Fig. 1: Complex impedance spectrum.

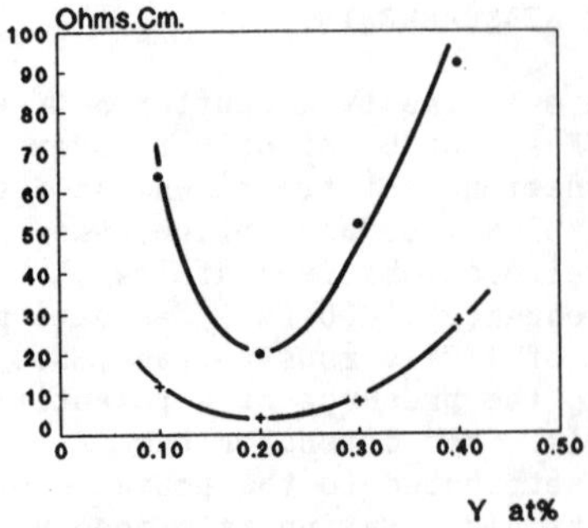

Fig. 2: (+) Bulk and (▲) grain boundaries resistivities against nominal yttrium concentration.

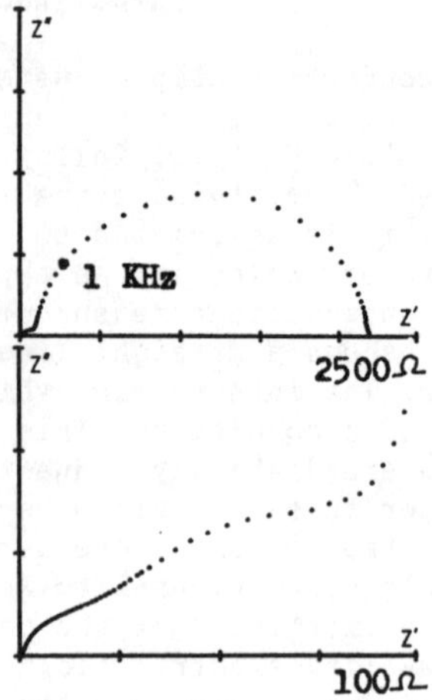

Fig. 3: Complex impedance spectrum showing a semicircle associated to a contact metal (Ag)-ceramic.

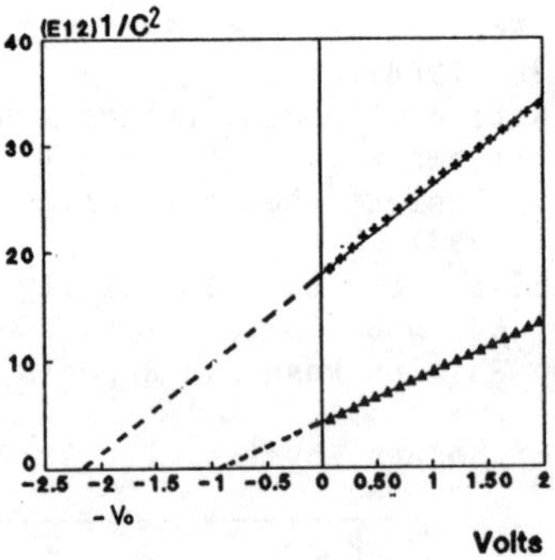

Fig. 4: Plot of $1/C^2$ versus voltage.
(+) Experimental plot.
(▲) Calculated values.

EFFECT OF MgO ADDITION ON A ZnO VARISTOR COMPOSITION

A. SMITH, Y. BOYAT, J. F. BAUMARD and P. ABELARD
Ecole Nationale Supérieure de Céramique Industrielle
C.N.R.S. UA 320, 47 à 73 Av A. Thomas
87065 Limoges Cedex, France

ABSTRACT

The present paper examines the effect of MgO addition on the electrical characteristics of a ZnO simple varistor composition. In the first part we describe the role of MgO addition in ZnO by itself. It gives an either conductive or resistive ceramic depending on the MgO content. When MgO is added to a basic varistor composition including oxides such as Bi_2O_3, MnO or Co_2O_3, it acts as a grain growth inhibitor. This leads to a significant increase in the breakdown voltage per unit length which is of consequence for the overall varistor characteristic.

INTRODUCTION

Zinc oxide varistors are ZnO-based ceramic semiconductor devices with non linear current-voltage characteristics (1-5). Their electrical behaviour can be described by the relation $I = KV^{\alpha}$. The exponent α, usually specified at a current density of 1 mA/cm^2, is 25 to 50 in typical ZnO varistor and is used as a figure of merit. Varistors are usually produced by conventional ceramic processing routes and consist of conductive ZnO grains surrounded by electrically insulating barriers. These barriers derive from trap states at the boundaries induced by additive elements such as bismuth, cobalt or manganese. The device can be considered as insulating up to the breakdown field above which it becomes conductive. The breakdown field specified at a current density of 1 mA/cm^2 is of the order of 2 to 3 V/barrier. To achieve a given breakdown voltage, two alternatives are possible :

(i) for fixed grain size, one can change the varistor thickness ;

(ii) for fixed varistor thickness, the grain size can be varied to increase the number of barriers.

In the present paper we examine the second alternative. Magnesium oxide is a well known grain growth inhibitor and densification aid for the sintering of alumina (6). Since alumina and zinc oxide have similar phase diagrams with respect to magnesium oxide (7-8), we have chosen to add magnesium oxide to the ZnO ceramic, individually or mixed with other oxides. The electrical behaviour has been examined.

MATERIALS PREPARATION

The ceramic powder was prepared through a conventional processing route : mixing of the different oxides, attrition milling in alcohol and drying of the powder by alcohol evaporation. The powder was isostatically pressed into disks and sintered for 1h at different temperatures depending on the content and the chemical nature of the dopants. The sintered samples had a diameter and a thickness of about 8 mm and 0.3 to 1 mm, respectively. Gold electrodes were deposited by sputtering under argon on the faces of the disks and give a small contact resistance of the order of 0.3 Ω. Two types of electrical characterization were carried out :

(a) V-I measurements through the application of a linear potential sweep (the input voltage ranged between 0 and 600V) ;

(b) impedance measurements in the 1 MHz - 1 GHz frequency range with the Hewlett Packard 4191A impedance analyzer. This technique measures the complex impedance of a ceramic sample as a function of the applied frequency. The bulk, the grain boundaries and the electrodes can be distinguished in terms of electrical response. In a first approximation the data points due to each contribution are spread over arcs of a circle in the impedance plane. In the particular case of ZnO the 1 MHz - 1 GHz frequency range is suitable for the study of the grain boundary electrical behaviour (9). For some compositions with very resistive grain boundaries impedance measurements were also made in the 100 KHz - 1 Hz range . For this latter domain, a frequency generator was coupled with a lock-in phase sensitive detector.

Two types of samples were prepared :

(a) ZnO mixed with a variable quantity of MgO only in order to cover two areas in the ZnO - MgO phase diagram : the domain where MgO is soluble in ZnO and the two phase region (i.e. simultaneous presence of zincite and periclase). The sintering temperature was 1400°C. The aim of this first type of compositions was to investigate the particular role of MgO on ZnO electrical behaviour.

(b) ZnO doped with a fixed amount of MnO, Co_2O_3 and Bi_2O_3 (i.e. 1 mol% with respect to ZnO for each oxide) and a variable quantity of MgO. The sintering temperature was 1100°C.

RESULTS AND DISCUSSION

The results can be divided into two categories which correspond to the two types of prepared compositions.

Effect of increasing percentages of MgO on ZnO electrical behaviour

For a MgO content ranging from 0 to 9.6 mol% the impedance does not exhibit a grain boundary semi-circle (Fig. 1). The only contribution to the frequency response of the sample is the grain interior. The shape of the spectrum can be explained by inductive effects at high frequency inherent to the particular geometry of the measurement cell (9). The measured value of the impedance at 1 MHz is real and equal within experimental error to the DC resistance of the sample. It does not change greatly for MgO ranging from 0 to 9.6 mol%. Therefore the magnesium ions replace the zinc cations in some of the zinc sites. Conduction takes place through a connected network of zinc oxide grains with negligible intervening resistive grain boundaries.

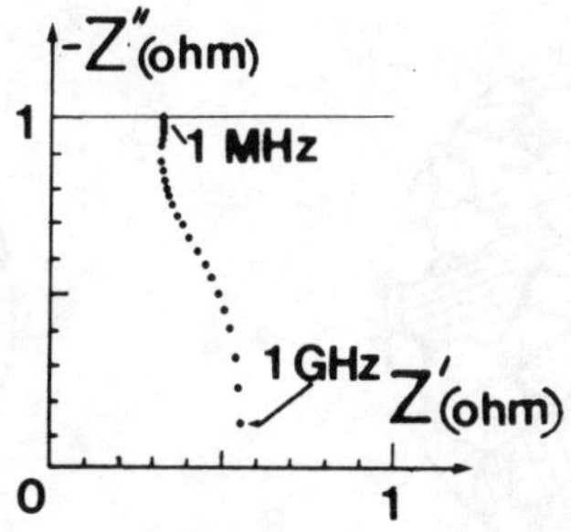

Figure 1. Impedance spectrum for ZnO doped with 9.6 mol% MgO.

By increasing the amount of MgO (respectively 28.8, 44.7 and 57.4 mol%) the impedance spectra show arcs of a circle which become larger as the periclase phase content is increased (Fig. 2). For greater MgO contents (respectively 60, 62.5 and 65 mol%) the samples are highly resistive ($R > 10^{14}\Omega$). The impedance spectra only have the beginning of an arc of a circle. For all these compositions, since the sample response to an AC signal changes with the amount of periclase phase, it can be deduced that the arcs of a circle in the impedance spectra are due to the periclase phase. In any case, the interpretation of an impedance spectrum for a two phase system is more complicated than for a one phase material. At low frequency the current does not flow in a straight line from one electrode to the other. It has to go through tortuous and constricted pathes in between the insulating periclase grains (Fig. 3a). At high frequency the impedance of the sample, appproximatly equal to 1/Cw is small. Effectively routes which were too resistive at low frequencies are short circuited by the associated capacitance. The alternating current then flows evenly through the entire ceramic (Fig. 3b).

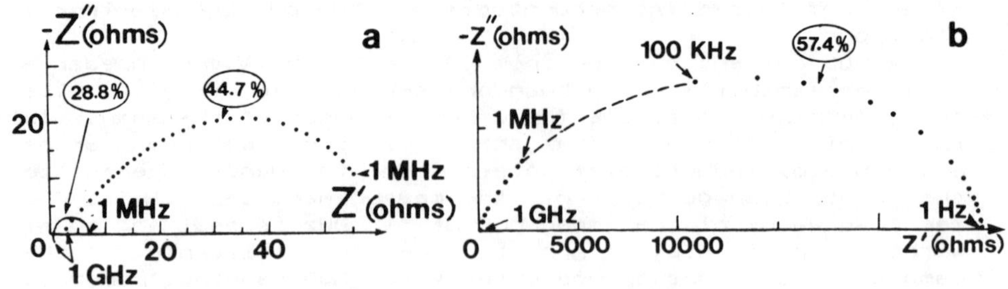

Figure 2. (a) Impedance spectra for ZnO doped with 28.8 mol% and 44.7 mol% MgO. (b) Impedance spectrum for ZnO doped with 57.4 mol% MgO.

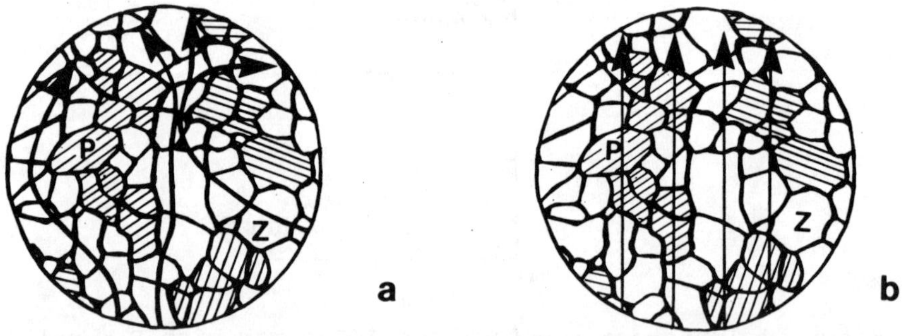

Figure 3. Schematic representation of the current lines through a two phase ceramic (Z : Zincite ; P : Periclase). (a) Low frequency situation. (b) High frequency situation.

We have investigated the electrical behaviour of all the compositions during the application of a linear potential sweep. All these compositions had an ohmic behaviour.

Compositions with a varistor effect

ZnO doped with a fixed amount of MnO, Co_2O_3 and Bi_2O_3 and a variable quantity of MgO shows a varistor effect. The corresponding average grain sizes are given in Table 1. Up to 42.8 mol% of magnesium oxide the samples tested electrically by V-I measurements were about 1 mm thick. Beyond this content the thickness ranged from 0.3 to 0.6 mm because of measurement practicability. Values of the non linearity coefficients and the breakdown fields measured at 1mA/cm² are given in Table 1. The V-I characteristics are presented in Fig. 4.

TABLE 1
Average grain sizes, non linearity coefficients and breakdown voltages (measured at $1 mA/cm^2$) for the different compositions.

Composition Number	MgO content (mol%)	Average grain size (μm)	α	Breakdown Field (V/mm)
0	0	21	15	84
1	1	12	9.5	125
2	3.9	11.4	11	127
3	9.6	5.8	13	305
4	13.2	5.6	17	310
5	42.8	2.1	11.5	490
6	44.6	2.6	11.5	720
7	46.4	2.8	10	1000

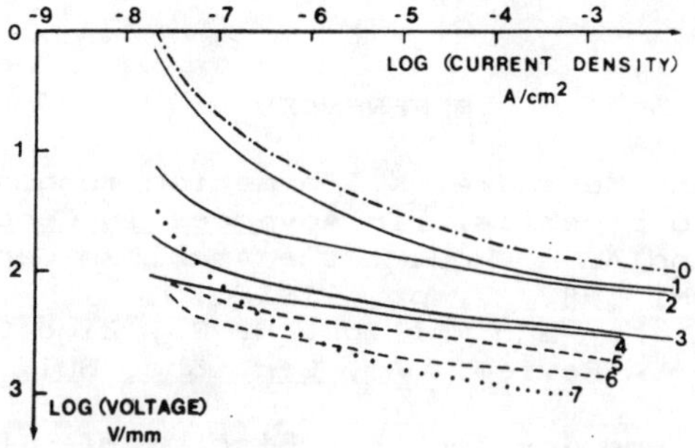

Figure 4. V-I characteristics obtained on samples with compositions presented in Table 1.

The results show :

(i) the addition of magnesium oxide yields a small grain ceramic. Magnesium oxide acts as a grain growth inhibitor.

(ii) The breakdown voltage per unit length is increased in comparison with a varistor composition without magnesium oxide. Moreover, it increases with the magnesium oxide content.

(iii) The leakage current decreases when the magnesium oxide content is increased up to 13.2 mol%. For greater amounts the leakage current either keeps on decreasing or increases depending on the tested samples. A possible explanation is that a non uniform distribution of the dopants within the ceramic sample due to the processing route yields regions of different conductivities. Since the tested samples are 0.3 to 0.6 mm thick it is possible for more conductive regions to occur right across the entire ceramic thickness. This could increase the leakage current.

The present data show that the addition of magnesium oxide (up to about 13.2 mol%) to a varistor composition improves its electrical characteristics. Its exact role with respect to potential barrier formation at the grain boundaries is difficult to evaluate and cannot be separated from the role of the other dopants.

CONCLUSION

Adding MgO to ZnO gives a ceramic which progressively goes from a conducting to an insulating state. The ceramic does not show a non linear behaviour. When the ZnO is doped with a fixed amount of MnO, Co_2O_3 and Bi_2O_3, the addition of MgO inhibits the grain growth. A direct consequence is the increase of the breakdown voltage per unit length. Lastly the leakage current can be decreased.

ACKNOWLEDGMENTS

Part of the work was done under reseach contract #150 with Thomson LCR in collaboration with Drs Ganne and Gaucher.

REFERENCES

1. Inida, M. and Matsuoka. K, Formation mechanism of nonohmic ZnO ceramics, in Advances in Ceramics, ed. by M. F. Yan and A. H. Heuer, the American Ceramic Society, Columbus, OH, 1982, **7**, pp. 91-106.
2. Levinson, L. M. and Philipp, H. R., Zinc Oxide Varistors - A Review, Am. Cer. Soc. Bull.,1986, **4**,pp. 639-45.
3. Blatter, G. and Greuter, F., Electrical Properties of Grain Boundaries in the Presence of Deep Bulk Traps, in Polycrystalline Semiconductors, ed. by G. Haberke Springer-Verlag, Berlin, 1985, pp. 118-37.
4. Einziger, R., Grain Boundary Phenomena in ZnO Varistors, in Grain Boundaries in Semiconductors, ed. by G. E. Pike, C. H. Seager and H. J. Leamy, Elsevier Science Publishing Company Inc., New-York, 1982, pp. 343-55.
5. Pike, G. E., Semiconductor grain-boundary admittance : Theory, Phys. Rev. B, 1984, **2**, pp. 795-802.
6. Coble, R. L., Transparent Alumina and Method of Preparation, U. S. Patent 3 026 210, March 20, 1962.
7. Segnit, E. R. and Holland, A. E., The system ZnO-MgO-SiO_2, J. Am. Ceram. Soc., 1965,**8** , pp.409-12.
8. Alper, A. M., McNelly, R. N., Ribbe, P. G. and Doman,R. C., J. Am. Ceram. Soc., 1962, **6**, p. 264.
9. Smith, A., Baumard, J. F., Abelard, P., and Denanot, M. F., J. Appl. Phys. , 1989, **65**, pp. 5119-125.

NTC THERMISTORS OF HIGH RELIABILITY

A. FELTZ, F. SCHIRRMEISTER, J. TÖPFER and M. STEINBRÜCK
Department of Chemistry
Friedrich Schiller University
Steiger 3, Jena, 6900, GDR

ABSTRACT

Two different methods of preparation of thermistor ceramics based on $NiMn_2O_4$ are compared pertinent to the density of grain structure and the electrical data. An additive of a PbO/GeO_2 glass for liquid phase sintering is involved. Furthermore preparation and resistivity behavior of $Mg_{2-x}Ti_{1+x}O_4$ mixed crystals are reported.

INTRODUCTION

Two main roots appear to exist in the developement of NTC thermistor ceramics. The first starting from Fe_3O_4 involves solid solutions with insolating or low conducting spinels such as $MgCr_2O_4$, $ZnCr_2O_4$ or Zn_2TiO_4 [1]or with $MgFe_2O_4$, $MnFe_2O_4$ or $ZnFe_2O_4$ [2]. The second concerns the spinel systems $Ni_xMn_{3-x}O_4$ and $Co_xMn_{3-x}O_4$ or combinations of both, often modified by addition of copper oxide or $Cu_xMn_{3-x}O_4$, respectively [3-6]. Obviously, spinel type transition metal oxides imply structural suppositions yielding comparatively low values of the resistively R at relatively high values of the activation energy, which is an advantage with respect to the sensitivity $\varepsilon = (dR/R)/dT = - B/T^2$. The activation energy E_σ is expressed by $B = E_\sigma/k$ in the equation with R_r as the resistivity at the reference temperature.

$$R = R_r \exp\left[B\left(\frac{1}{T} - \frac{1}{T_r}\right)\right]$$

Taking into account cylindrical size of the specimens with a thickness $0.13 \leq L \leq 6$ mm, L equal to the diameter D = 2r, and resistivity variation in the range $10^0 \leq R \leq 10^6 \Omega$

at 300 K, the specific resistivity ϱ may be altered in the limits $0.5 \leq \varrho \leq 10^4\ \Omega$ cm. Using for the pre-exponential factor $\varrho_0 \approx 3\cdot 10^{-3}\Omega$ cm as an approach deduced from conductivity measurements of oxide semiconductors these data result in $0.13 \leq E_\sigma \leq 0.39$ eV for the activation energy or 1500-B-4500 K. Thereby allowing thermistors with sensitivities $0.017 \leq \varepsilon \leq 0.05$ K^{-1} (at T = 300 K) to be obtained. Considering that resistivities can be easily measured with an accuracy $\Delta R/R \leq 10^{-4}$ the resolution of temperature registration achieved should be better than $\Delta T = \pm\ 0.01$ K.

On the other hand, catalogue data qualifying thermistors commonly show a tolerance up to several centi degree. Obviously, the excellent intrinsic properties of the materials remain covered because of variations in freezing the different equilibria of cation distribution or defect formation affecting the conductivity, of inhomogeneous density distribution in the bulk and of fluctuations in the grain structure between specimens of the same batch and between different charges. Indeed, preparation of NTC thermistors of high reliability is a challenge to solid state chemistry. In the latter time thermistors interchangeable with an accuracy $\Delta T = \pm\ 0.1$ K between room temperature and about 100 °C have become known [4].

It is the aim of this paper to study the reproducibility of the electrical properties of ceramic samples based on the $Ni_xMn_{3-x}O_4$ system, especially $NiMn_2O_4$ as an inverse spinel, $Ni^{II}_{0.35}Mn^{II}_{0.65}[Ni^{II}_{0.65}Mn^{IV}_{0.65}Mn^{III}_{0.70}]O_4$ [7], whose cation distribution is expected to depend on temperature. Furthermore mixed crystals $Mg[Mg_{1-x}Ti^{IV}_{2x}]O_4$ have been involved, whose cation distribution is assumed to be widely fixed.

PREPARATION

According to the phase diagram [8] $NiMn_2O_4$ exists as a stable compound in the range $720 \leq T \leq 900$ °C in air and in oxygen up to 1150 °C. Two preparation roots have been investigated:

Calcination of $NiCO_3/2\ MnCO_3$ mixtures at 650 °C, (A), leaving behind a mixture consisting predominantly of NiO and α-Mn_2O_3. Predensification by granulometric handling and compacting with pressure of about 150 M Pa yield cylindric samples (D = 4 mm) with a green density ϱ_0 of about 55 %, who were fired at 1100 °C in oxygen (40 h). For studying the influence of liquid phase sintering on grain growth and densification 1 mole-% $(PbO)_{1.88}(GeO_2)_{1.12}$ glass was substituted for $NiMn_2O_4$.

Decomposition of mixed crystals $NiMn_2(C_2O_4)_3 \cdot 6\,H_2O$ in air at 350 °C, (B), allows to prepare a cubic defect spinel phase $Ni^{II}_{0.95}Mn^{III}_{1.54}Mn^{IV}_{0.37}\square_{0.14}O_{4.00}$ which is metastable and has to be understood as an intermediate state between $NiMn_2O_4$ and $Ni^{II}_{0.8}Mn^{IV}_{1.6}\square_{0.6}O_4$ [9]. Following the phase diagram [8] $NiMnO_3$ and α-Mn_2O_3 is formed even at room temperature on leaving the material in air:

$$Ni^{II}_{0.95}Mn^{III}_{1.54}Mn^{IV}_{0.37}\square_{0.14}O_4 + 0.15O_2 \longrightarrow 0.95\,NiMnO_3 + 0.48\,Mn_2O_3$$

In the system $Co_xMn_{3-x}O_4$ the formation of a cubic defect spinel phase has been observed, too [10]. After annealing and pre-densification the pellets formed from the defect spinel powder retain an increased sintering activity allowing to obtain ceramics of higher density and improved homogeneity of the grain structure.

Formation of $Mg[Mg_{1-x}Ti^{IV}_{1-x}Ti^{III}_{2x}]O_4$ ceramic samples demands a closed system with extremely low oxygen partial pressure at temperatures up to 1450 °C [11]

$$(1-x)Mg_2TiO_4 + xTi_2O_3 + xMgO \longrightarrow Mg[Mg_{1-x}Ti^{III}_{1-x}Ti^{III}_{2x}]O_4$$

To obtain homogeneous samples the procedure has to be repeated after grinding. Small additions of Boron act as an oxygen getter forming B_2O_3 which supports densification by liquid phase sintering.

All the samples have been contacted with Ag past fired at 300 °C (1/2 h) yielding ohmic contacts which has been proved by variation of the thickness.

MEASUREMENT OF RESISTIVITY AND TEMPERATURE

In order to prove the interchangeability of thermistor samples, e.g. with a tolerance $\Delta T = \pm\,0.1$ K in an interval such as $T_1 \leq T \leq T_2$ (T_1 = 273 K, T_2 = 343 K), the temperature has to be measured absolutely with an accuracy of $\Delta T_e = \pm\,0.01$ K, which has been performed with a calibrated Pt-100 resistor. At a given temperature 5 samples have been simultaneously measured utilizing a constant current source (100 μA), and a high resolution digital voltmeter. The total error is composed by 4 parts (reference temperature T_r for instance 298K).

$$\left|\frac{\Delta R(T)}{R(T)}\right|_{tot} = \left|\frac{\Delta R(T)}{R(T)}\right|_m + \left|\frac{B}{T^2}\Delta T_e\right|_m + \left|\frac{\Delta R(T_r)}{R(T_r)}\right|_s + \left|\frac{(T_1-T_2)\,\Delta B}{T_1 T_2}\right|_s$$

The first term, the error of the resistivity measured may be kept $\leq 10^{-4}$, and the error of the temperature measurement with $\Delta T_e = \pm\, 0.01$ is also sufficiently small. Variation of the sample data resulting from processing is collected in the two last terms. $\Delta T \leq \pm\, 0.1$ K requires $|\Delta R(T_r)/R(T_r)|_s <$ $\pm\, 0.01$, which commonly needs adjusting by geometrical levelling. Subsequently, interchangeability of thermistors is predominantly controlled by $\Delta B \simeq B\,\Delta T(T_1^2 + T_2^2)/(T_2-T_1)T_1T_2$, i.e. $\Delta T = \pm\, 0.1$ K is consistent with $\Delta B \leq \pm\, 13$ K for B = 3920 K.

RESULTS

TABLE 1

Properties of thermistor ceramic samples $NiMn_2O_4$ A_o and B_o without and A_1 and B_1 with additive for liquid phase sintering and of $Mg_{1-x}Ti_{1+x}O_4$ ceramic samples

$NiMn_2O_4$	ρ_{rel}/%	z	ρ(298 K)/Ωcm		Ω cm	B/K	ΔB/K
A_o (1)	73.9	30	5612	±	198	3818	± 4
A_o (2)		10	5343	±	69	3818	± 2
B_o (1)	86.7	30	2058	±	155	3750	± 6
B_o (2)	90.3	30	1848	±	209	3778	± 12
A_1 (1)	77	30	2935	±	210	3839	± 5
A_1 (2)		10	4042	±	127	3842	± 6
B_1 (1)	97.7	29	1729	±	105	3800	± 8
B_1 (2)	97.4	30	1688	±	68	3806	± 4
$Mg_{2-x}Ti_{1+x}O_4$			ρ(293 K)/Ω cm				
x = 0.02	93.1	4	49500	±	1470	3680	± 11
	92.9	4	37700	±	10000	3511	± 7
x = 0.05	89.2	4	34500	±	8000	3705	± 11
after fatigue 150 °C (150h)	89.2	4	37700	±	1000	3776	± 9

The results are collected in table 1 indicating an excellent reproducibility of the B values for specimens in combination with high density prepared by liquid phase sintering from oxalate mixed crystals who act as a precursor state [11]. z means the number of single samples measured. Because of uncertainties of the effective contact area and thickness the specific resistivity data show rather more extended deviations. In every case the two charges have been seperately prepared. Aging has been carried out in a cyclic regime: heating and cooling with 1 K/min and keeping at the minimum and maximum temperature of - 40 °C and + 85 °C. For the samples B_1 and B_2 20 cycles did not indicate any changes of the B values. Only the resistivity showed a small drift of about 2 %.

Obviously because of the high sensitivity to oxidizing impurities and the more complicated preparation process the reliability achieved for the $Mg_{2-x}Ti_{1+x}O_4$ specimens is not so good.

REFERENCES

1. Verwey, E.J.W., Haaijnian, P.W. and Romejin, F.C., J. chem. Phys. 15 (1947) 181
2. Jahnson, D.W., Robbins, M., Vogel, E.M. and Lambrecht, V.G., Ceram. Bull. 62 (1983) 597
3. Brit. Pat. 540.844 "Resistance Materials and Methods of Making the Same", October 31st (1941)
4. Walch, H., Siemens Zeitschrift 47 (1973) 66
5. Meguro, T., Sasamoto, T., Yokoyama, T., Yamada, S., Abe, Y. and Torikai, N., J. Ceram. Soc. Jpn. Inter. Ed. 95 (1987) 307
6. Metz, R., Caffin, J.P., Legros, R. and Roosset, A., J.Mat. Sc. 24 (1989) 83
7. Macklen, E.D., J. Phys. Chem. Solids 47 (1986) 1073
8. Wickham, D.G., J. Inorg. Nucl. Chem. 26 (1964) 1369
9. Feltz, A. and Töpfer, J., Z. anorg. allg. Chem., in press
10. Feltz, A. and Ottlinger, M., Z. Chem., in press
11. Feltz, A. and Steinbrück, M., Z. anorg. allg. Chem., in press
12. Feltz, A., Töpfer, J., Schirrmeister, F., Blayer, A., Bühling, D., Kuzelka, K.-H. and Röter, G., Patentschrift DD 266 670 A1 (1989)

PROPERTIES OF Ca-DOPED $LaMnO_3$

M. FANDEL, J. HÖSCHELE, W. SCHÄFER AND R. SCHMIDBERGER

DORNIER GMBH 7990 FRIEDRICHSHAFEN FRG

INTRODUCTION

The mechanism of electronic conductivity in perovskite type oxides depends on the chemical composition of the basic compound ABO_3, the concentration of dopants and in certain cases on the temperature. Metallic behaviour as well as p-type semiconduction appear for a lot of similar compounds which are discussed as electrodes and interconnection materials for Solide Oxide Fuel Cell (SOFC) applications |1,2,3|. According to the model of Goodenough |4|, the process of electron itineration is assumed to be based on localized level hopping or polaron conductivity, giving rise to p-type semiconductor behaviour |5,6|. Mechanical properties like thermal expansion are also known to be dependent on the amount of dopants (2).

SAMPLE PREPARATION

Mixed oxide powders were prepared by the Reaction-Spray-Process (7). This process is based on thermal decomposition of aqueous solutions in a hot reaction chamber (Fig. 1).

The components desired in the powder are dissolved in water and sprayed into the reaction chamber by using a nozzle. Thus each solution droplet results in a powder particle by rapid water evaporation. Due to the rapid evaporation, the homogeneous distribution of the dissolved components is finally transferred to the powder particles.

The overall reaction can be summarized schematically as follows:

$$La(NO_3)_3 + Ca(NO_3)_2 + Mn(CH_3COO)_2 \rightarrow La_{1-x}Ca_xMnO_3 + NO + CO_2 + H_2O.$$

The substitution of La by Ca ranged from 0 to 50 % (x = 0 to 0.5).

The powders were grinded to a medium particle size less than 2 microns, axially pressed (100 MPa) and sintered at a temperature of 1250°C in air or oxygen. The resulting densities were found to be more than 95 % of theoretical density.

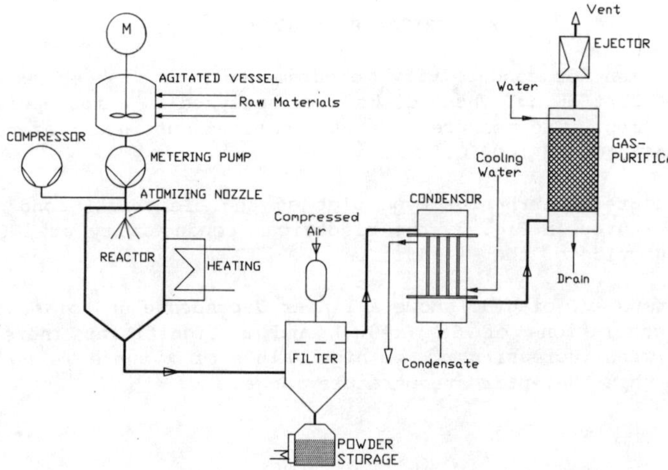

Fig. 1 Scheme of the reaction spray equipment for the synthesis of mixed oxide powders

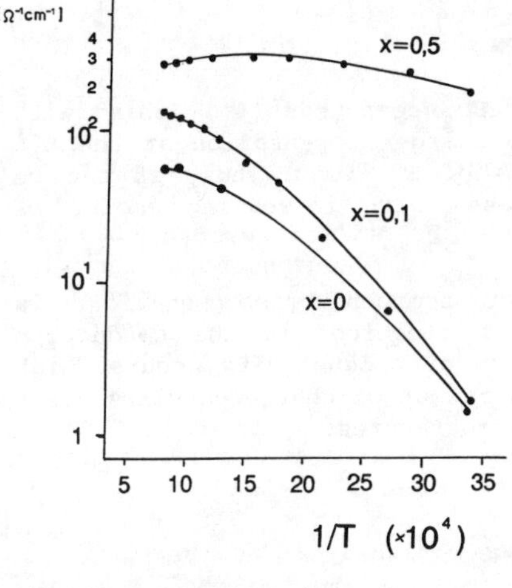

Fig. 2 Arrhenius plot of the electrical conductivity of Ca-doped $LaMnO_3$

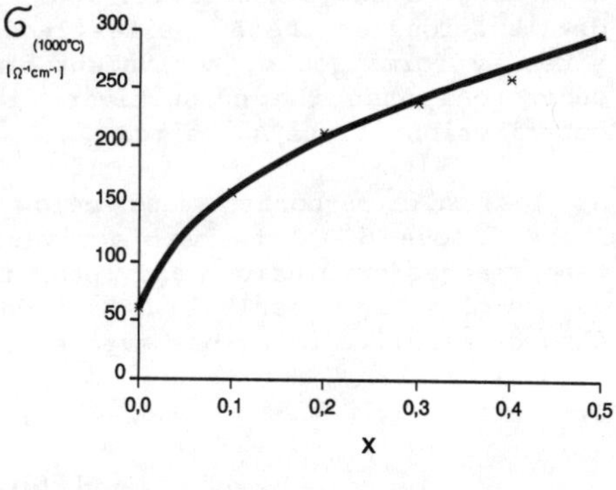

Fig. 3 High temperature conductivity (1000°C) of $La_{1-x}Ca_xMnO_3$ as a function of x

ELECTRICAL PROPERTIES

The test specimen for conductivity measurements were disk-shaped having diameters of about 25 mm. These disks were analyzed in a four-point van-der-Pauw equipment. The measurements were carried out in air at temperatures between 25°C and 1000°C.

Fig. 2 represents an Arrhenius-type plot of the electrical conductivity versus temperature. In Fig. 3, the electrical conductivity at 1000°C is shown as a function of the Ca addition.

The log σ versus 1/T diagram shows a linear dependence up to about 600°C for low concentrations of Ca (x=0.1) and a significant increase of conductivity with increasing x. For high values of x there is no linear dependence within the entire temperature range.

THERMAL EXPANSION

Thermal expansion of the materials having different calcium content was investigated using bar-shaped test specimen ($50x4x4mm^3$) in a commercial high temperature dilatometer (Netzsch 414/1).

Fig. 4 shows the thermal expansion coefficient averaged between room temperature and 1000°C as a function of the calcium content. Thermal expansion increases with increasing calcium content. The variation is between $10.1 \times 10^{-6} K^{-1}$ and $11.4 \times 10^{-6} K^{-1}$.

DISCUSSION

According to the conductivity model for ABO_3 compounds |4|, doping with divalent ions at the A^{3+} sites leads to charge compensation at the B^{3+} sites by formation of equivalent amounts of B^{4+} ions. Thus, it can be understood that the conductivity increases with increasing amount of substitution at the A^{3+} sites.

At low calcium content and below 600°C semiconduction behaviour is found. Above 600°C thermal activation of electrons in the conduction band reaches saturation, and metallic type of conductivity occurs. High values of x also result in a high concentration of charge carriers, leading to metallic behaviour even at low temperatures.

APPLICATIONS

The possibility of influencing the thermal expansion coefficient by varying the composition of the mixed oxides is important for their use as electrodes in SOFC's |2| or other high temperature electrochemical cells. Matching the thermal coefficients of different materials for these cells is one of the most important aspects in materials development for high temperature electrochemistry.

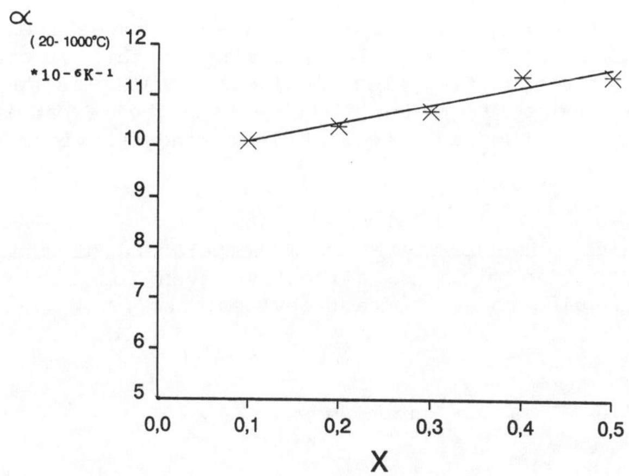

Fig. 4 Averaged thermal expansion coefficient as a function of the Ca-content

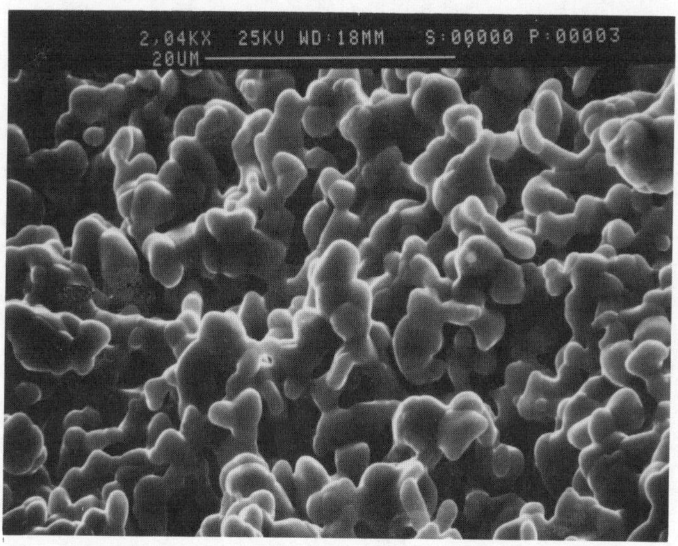

Fig. 5 Porous oxygen electrode of a solid oxide fuel cell prepared from Ca-doped $LaMnO_3$

Fig. 5 shows a porous oxygen electrode of a solid oxide fuel cell prepared from Ca-substituted $LaMnO_3$. The porosity of this electrode layer is about 50 % and the resulting electrical conductivity is approximately 25 % of the bulk conductivity. Due to a well matched expansion coefficient, this electrode shows an excellent long time behaviour even under temperature cycling.

The relatively high conductivity at room temperature of the high calcium-level mixed oxides might be attractive even for low temperature applications, e.g. electrodes, contact devices, etc..

REFERENCES

|1| H. Ohbayashi et al, Japanese Journal of Applied Physics Vol. 13, No. 1, 1974

|2| W.Schäfer,R.Schmidberger, High Tech Ceramics, Elsevier Science Publishes B.V., Amsterdam 1987

|3| D.E.Meadowcroft, Int.Conf.on Strontium Containing Compounds, Halifax, Canada, 1973

|4| J.Goodenough, J.Phys.Chem.Solids, Pergamon Press, Vol.6

|5| D.P.Karim, A.T.Aldred; Physical Review B.Vol.20,No.6, 1979

|6| O. Yamamoto, Y.Takeda et al., Solid State Ionics 22 (1987) North-Holland, Amsterdam

|7| R.Röttenbacher,R.Schmidberger, Ceramic Powders, Elsevier Science Publishers B.V., Amsterdam 1983

INFLUENCE OF MICROSTRUCTURE ON THE ELECTRICAL AND MECHANICAL BEHAVIOUR OF SELECTED COMPOSITIONS IN THE Zr-Ce-Y-O SYSTEM.

N. Khan, C. Leach and B.C.H. Steele.
Department of Materials,
Imperial College of Science, Technology and Medicine,
Prince Consort Road, LONDON. SW7 2BP.

Abstract

The microstructures, mechanical and electrical properties of nominally single phase cubic or tetragonal ternary zirconia-ceria-yttria ceramics are described. Two series are considered; containing 2mol% yttria and 8mol% yttria respectively, but with variable amounts of ceria. Increasing the ceria content was found to increase grain growth during sintering, increase the lattice parameters, decrease modulus of rupture but increase fracture toughness, and decrease slightly the bulk conductivity. Strong cerium segregation to interfaces was noted and may account for the stabilisation of ternary compositions in humid environments.

INTRODUCTION

Doped zirconia ceramics have long found favour in engineering and elecrical applications where hard, wear resistant, inert materials or oxygen-ion conducting ceramics are required [1]. In both these areas, yttria doped zirconia yielding at room temperature a polycrystalline tetragonal (Y-TZP) and/or cubic (Y-CSZ) material is a popular choice. However, there are certain deficiencies associated with Y-TZP which render it inappropriate for some applications. These include destabilisation in humid environments [2], chemical unmixing during long term high temperature anneals [3], and crack growth during cyclic fatigue loading, in addition to a relatively low value of fracture toughness[4]. We believe that the addition of up to 10mol.% of ceria to yttria doped zirconia to form a ternary oxide with similar phase chemistry to the binary compositions may reduce some of these problems without introducing further processing difficulties.

In this contribution some features of the electrical, chemical, mechanical and microstructural behaviour of these ternary ceramics will be briefly described and discussed. Two series are chosen for illustrative purposes; one containing 2mol.% yttria and the other 8mol.% yttria, each with variable amounts of ceria. The compositions are illustrated in fig.1 on a 1500C ternary section through the phase diagram [5]. This phase diagram has been generated from X-ray phase identification in normally sintered samples and defines a 'kinetics modulated' phase assemblage appropriate to sintering time scales but which may differ slightly from the equilibrium situation.

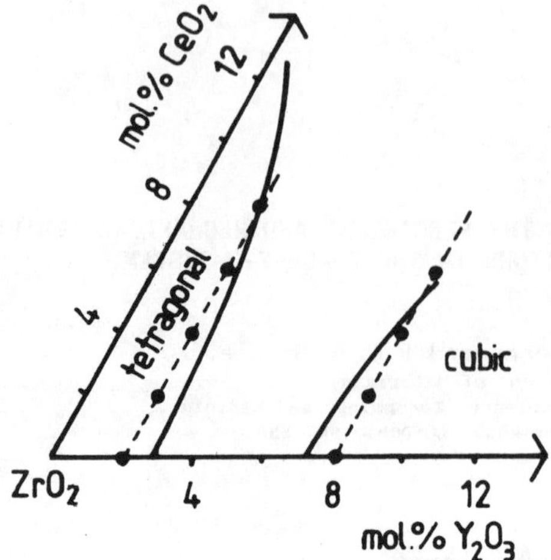

Fig.1. 1500C section through the zirconia rich corner of the ZrO_2-CeO_2-Y_2O_3 'kinetics modulated' phase diagram showing the ceramic compositions selected for study.

EXPERIMENTAL

The ceramics were prepared by mixing, in a ball mill, appropriate quantities of various commercial powders (supplied by Tosoh Inc. Japan), formed into compacts which were isostatically pressed at 300MPa, and sintered in air at 1550C for 4 hours. X-ray diffraction was carried out using a Phillips diffractometer and CuK_a radiation. Test bars were ground to 50um surface roughness prior to 3 point bending. (Test bar dimensions were 3x4x40mm, knife edge span was 25mm, cross-head speed $0.2mm.min^{-1}$, notch depth for fracture toughness measurements 0.5mm) Test discs were sputtered with platinum electrodes for A.C. impedance analysis in a Hewlett Packard HP 4192 LF frequency analyser.

RESULTS

I: Microstructure.

X-ray diffraction showed the samples to be predominantly (>97%) single phase, either tetragonal or cubic and with steadily increasing lattice parameters in sympathy with an increase in the mole fraction of ceria (see table.1). Representative SEM micrographs are presented in fig.2. and show an increase in grain growth during sintering associated with increasing the ceria content of the ceramic (see also table.1). It is noted that this increased grain size does not result in the development of room temperature monoclinic zirconia in the sintered ceramic, demonstraing the increased critical grain size in agreement with published work [6].

ESCA analysis was used to investigate interfaces in these ceramics. By adjusting the sintering conditions it was possible to prepare a ceramic with grain boundary fracture, thus permitting direct analysis. In the case of binary yttria-zirconia ceramics, yttrium is seen to segregate strongly to the surface resulting in a 'saturated' surface concentration of ca. 26% yttrium, as reported elsewhere in the literature [7]. However, the addition of ceria resulted in a swamping of this surface segregation effect and in compositions containing more than ca. 4mol.% ceria no yttrium was detected on the surface, only cerium (fig.3). Thus the character of grain boundary interfaces in the ternary phase field may be expected to be different from that in yttria-doped zirconia. In particular, recent results [8] demonstrate that the rate of destabilisation of TZP's in humid environments can be reduced

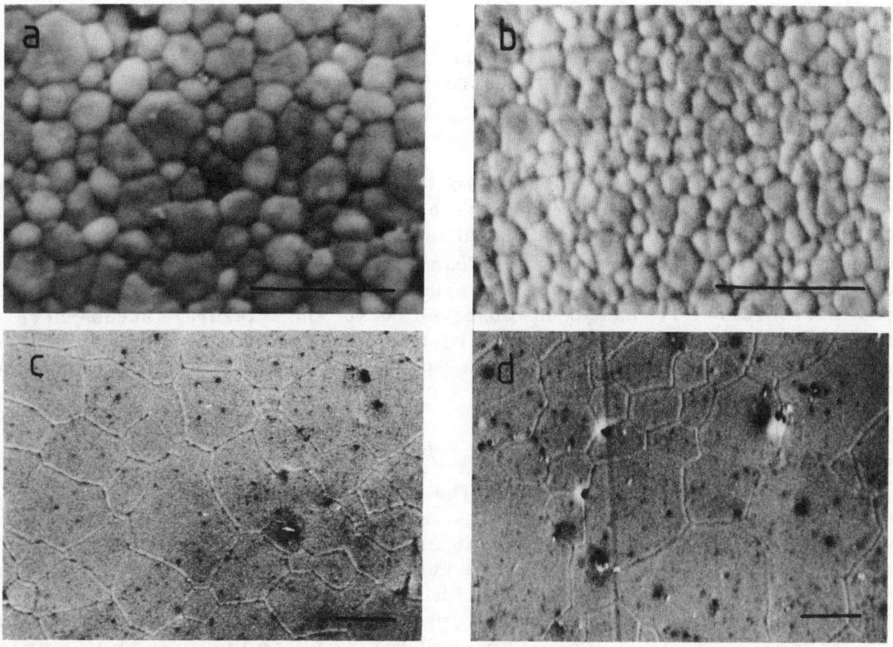

Fig.2 Scanning electron micrographs of representative ternary compositions. Ce2-Y2 (a), Ce6-Y2 (b), Ce2-Y8 (c) and Ce6-Y8 (d). Scale bars =5μm

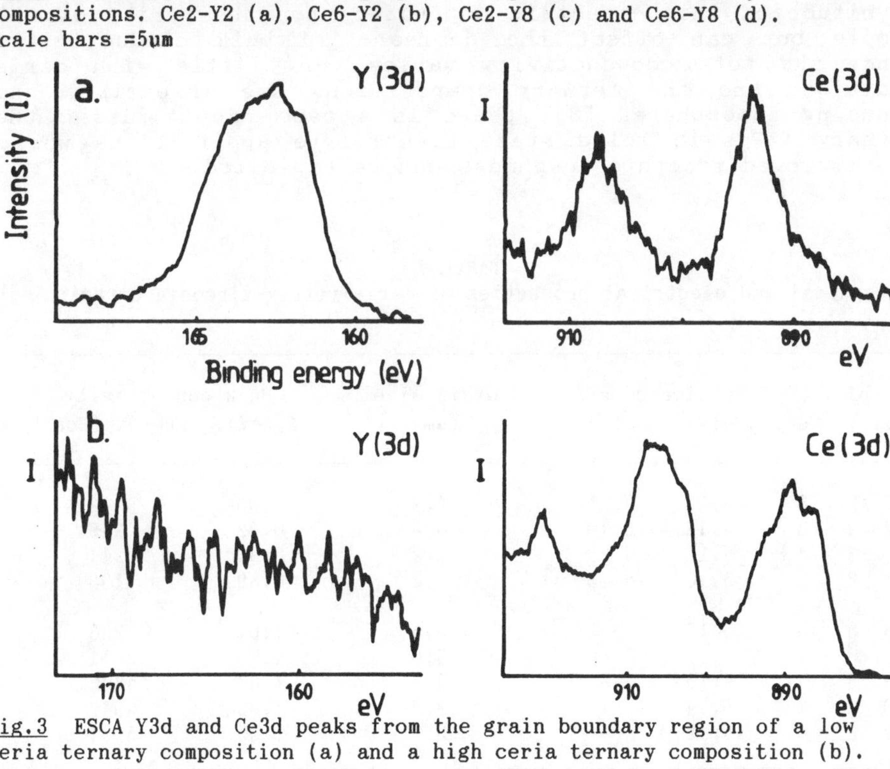

Fig.3 ESCA Y3d and Ce3d peaks from the grain boundary region of a low ceria ternary composition (a) and a high ceria ternary composition (b).

significantly by the addition of ceria to the point where the reaction is inhibited completely at 10mol.% addition. It is suggested that this effect is directly related to the strong surface segregatioon of cerium in these ternary compounds.

II: Mechanical properties.
Room temperature moduli of rupture in the tetragonal series were found to be in the range 900-400MPa, with the higher values recorded in the ceria poor compositions. Fracture toughnesses for the ternary tetragonal compositions are in the range 12-18MPam$^{1/2}$, the higher values recorded for the ceria rich compositions being significantly higher than for yttria doped zirconias. It has been observed that the flexural stress-strain curves of Ce-TZP commonly show a plastic strain prior to failure [9] indicating R-curve behaviour. The possibility that similar fracture processes occur in the ternary compositions is currently being investigated.

III: A.C.Impedance spectroscopy.
Bulk resistivities of samples from the two series are listed in table.1. The constant yttria content within the two series has resulted in constant intrinsic vacancy concentrations. Little variation in the spectra is observed in either series apart from a small decrease in conductivity with increasing ceria content. However, the addition of ceria permits a larger grain size in the ceramic, which has the effect of reducing the grain boundary resistivity, both in the TZP and the CSZ series. The magnitude of this effect is dependent upon the purity of the sample but can offset the decrease in bulk conductivity. Since the total conductivity varies very little with ceria additions and the ternary compositions are insensitive to reducing atmospheres [8], there is a case for utilising the ternary TZP's in solid state electrolyte applications where the improved fracture toughness can be exploited.

TABLE.1
Physical and electrical properties of ceria-yttria-zirconia ceramics.

Sample. Zr	Y	Ce	Lattice const's a(A)	c(A)	Grain size (um)	Bulk conductivity E_A(eV)	Pre-exp(Scm^{-1})
90	2	8	5.13	5.21	1.2	0.93	242
92	2	6	5.12	5.19	0.6	0.92	245
94	2	4	5.11	5.20	1.2	0.91	244
96	2	2	5.10	5.13	1.2	0.89	193
86	8	6	5.17	--	5.6	1.07	4220
88	8	4	5.15	--	5.0	1.05	3611
89	8	3	5.15	--	3.2	1.07	5806
90	8	2	5.14	--	3.2	1.06	5049
92	8	0	5.12	--	3.0	1.05	5661

REFERENCES

1. Section IV of: Advances in Ceramics, vol.12, Science and Technology of Zirconia II (eds. Claussen,N., Ruhle,M., Heuer,A.) The American Ceramic Society. (1984), 465-552.
2. Lange,F.F., Dunlop,G.L., Davis,B.I., J.Amer.Ceram. Soc., 69(1986), 237-240.
3. Stubican,V.S., Hellmann,J.R., in Advances in Ceramics, vol.3, Science and Technology of Zirconia (eds. Hobbs,L., Heuer,A.) The American Ceramic Society. (1981), 25-36.
4. Dauskardt,R.H., Yu,W., Ritchie,R.O., J.Amer.Ceram.Soc. 70(10), (1987), C248-C252.
5. Khan,N., Leach,C.A., British Ceramic Proceedings, 42(1989), 133-138.
6. Duh,J.G., Dai,H.T., Chiou,B.S., J.Amer.Ceram.Soc. 71(1988), 813-819.
7. Winnubst,A.J.A., Verkerk,M.J., Burggraaf.A.J., in Advances in Ceramics, vol.7, Additives and Interfaces in Electronic Ceramics (eds. Yan,M.F., Heuer,A.) The American Ceramic Society. (1983), 177-183.
8. Leach,C.A., Khan,N., submitted to J.Mater.Sci.
9. Tsukuma, K., Shimada, M., J.Mater.Sci. 20(1985), 1178-1184.

CONCENTRATION POLARIZATION EFFECTS IN ZIRCONIA BASED ELECTROCHEMICAL PUMPS

F.M.B. Marques, J.A.L. Baptista, J.R. Frade, D.M. Sá, A.P.M. Carvalho
Universidade de Aveiro, 3800 Aveiro, Portugal

ABSTRACT

Calcia stabilized zirconia (CSZ) based cells, working as oxygen electrochemical pumps were used to investigate cathodic polarization. When oxygen is pumped from a gas with low oxygen content, limiting current densities are observed, indicating mass transport rate determining processes. These results are discussed assuming parallel paths for oxygen supply to the electrochemical reaction sites, their importance depending on working conditions and electrode morphology.

INTRODUCTION

Polarization losses are responsible for decreased efficiency of electrochemical pumps, high temperature fuel cells, and steam electrolyzers. In such devices activation and concentration polarization may become relevant. Their efficiency can be improved if the effects of working conditions and materials characteristics on polarization are known. As part of this objective an attempt was made to identify the effect of temperature, gas flow rate, and oxygen partial pressure on concentration polarization in CSZ electrochemical pumps.

At 1000°C zirconia based electrolytes as CSZ are pure ionic conductors for oxygen pressures ranging from ambient pressure to less than 10^{-15} Pa [1]. In such conditions, a voltage V applied to the electrolyte results in electrochemical transport of oxygen through the material. This is the basis for their use as oxygen pumps. V is related to cell current I_o by:

$$V = V_o + \eta + I_o R_o, \qquad (1)$$

where V_o is the open circuit electromotive force, η the sum of activation and concentration overpotentials in cathode and anode, and R_o the cell resistance. V_o is expressed by:

$$V_o = [RT/(4F)].\ln (P_1/P_2) \quad (2)$$

where R, T and F have their usual meaning and P_1 and P_2 stand for boundary oxygen pressures in contact with the electrolyte. In the experimental work to be described cathodic polarization is predominant, as previously demonstrated [2].

The flux N of molecular oxygen and the ionic current density J are interrelated (J = 4.F.N). Oxygen supply to the electrochemical reaction sites requires oxygen transport from the gas phase. For low oxygen pressures in an inert gas, oxygen flux by diffusion can be described by:

$$N = [D/(RT)].[(P-P_i)/\delta_t] \quad (3)$$

where D is the diffusion coefficient, P and P_i the oxygen pressures in the bulk gas and at the electrolyte interface, and δ_t is the boundary layer thickness. δ_t/D conveniently describes the resistance to oxygen transport. Deposited Pt electrodes show different morphologies depending on deposition techniques, thermal history, and previous working currents. For non-uniform electrolyte surface coverage, large empty areas are surrounded by dense or porous Pt deposits. Electrochemical reaction sites lie on triple phase boundaries of dense deposits. For porous deposits, access to the electrolyte occurs also through the electrode small pores. For this reason, the resistance to gas phase transport actually includes different contributions. The boundary layer thickness outside the electrode will depend on gas flow rate. For oxygen molecules crossing the electrode layer, parallel paths include constriction resistances, normal molecular diffusion processes in large pores, and Knudsen type diffusion for small pores. Transport of oxygen atoms adsorbed on Pt will compete with these processes. Electrode morphology and working conditions will determine the relative importance of these processes.

The driving force for oxygen transport in gas phase is the oxygen pressure gradient. Insufficient supply of oxygen results in local drop in oxygen pressure at the gas electrolyte interface, causing an increase in the driving force that tends rapidly to P/δ_t. Further increase is prevented by shortage of oxygen supply. The maximum flux of oxygen in gas phase and the corresponding limiting current density J_L will be proportional to oxygen pressure in the gas. Temperature causes a small increase in gas phase diffusion coefficients ($D \alpha T^{3/2}$). As $N \alpha D/T$ only a slight temperature dependence is expected for the limiting current density. Concentration overpotential η_c is related to P and P_i by the Nernst relation, or J and J_L by [2]:

$$\eta_c = - [RT/(4F)].\ln (1-J/J_L) \quad (4)$$

Concentration overpotential becomes increasingly important as the current density approaches J_L. Identical results are obtained for constriction resistances to oxygen transport. When adsorption or diffusion of oxygen atoms are rate determining, due to the activated nature of these processes, a different temperature dependence of limiting current densities is found.

The Butler Volmer equation describes the relation between activation overpotential and current. Ohmic behavior is observed for low η, whereas for large overpotentials, Tafel type behavior is observed [3]. Activation polarization should not be ignored even when concentration polarization is large. However, limiting currents are identified when mass transport becomes rate determining [2,3], depending on parameters whose role can be discussed without evaluation of activation overpotential.

EXPERIMENTAL RESULTS AND DISCUSSION

The setup used was identical to the one described by Etsell and Flengas [2], and consisted of a CSZ tube (Degussa) closed at one end, concentric with an impervious alumina tube, both placed in the furnace. Different gases (air or nitrogen with small oxygen additions) could be forced over the cathode, while air was kept in contact with the anode. Electrodes were obtained from Pt paste (Engelhard Nr. 6926) fired in air at 1000°C for several days for morphological stabilization. Two CSZ tubes were used in the experiments, for periods in excess of two months. SEM observations showed that about 40% of the electrolyte surface was not covered by Pt. The remaining part was covered by porous (tube #1) or dense (tube #2) Pt deposits. In both cases electrode thickness was less than 3 µm.

Figure 1(a), for tube #1, shows results obtained for current density (J) versus voltage ($V-V_o$), when oxygen is being pumped from an oxygen + nitrogen gas mixture. The open circuit electromotive force V_o was deduced from the applied voltage V, so that all curves can be superimposed at the origin. These results demonstrate non-ohmic behaviour, and that current density approaches a limiting value at intermediate voltages as already reported in the literature [2]. At all temperatures, for higher voltages, ionic current increases due to the reduction of vestiges of carbon dioxide in the gas phase, as shown only for T=750°C. The onset of electronic conductivity is identifiable only at higher overpotentials. The existence of a limiting current value is related to the presence of a mass transport rate determining step [2-4]. Also shown by data, the limiting current density is almost independent of temperature, except for T<800°C. For higher temperatures, results agree with negligible temperature dependence of a gas phase diffusion controlled supply of oxygen. For T<800°C the kinetics of cathodic reaction is probably different as discussed later.

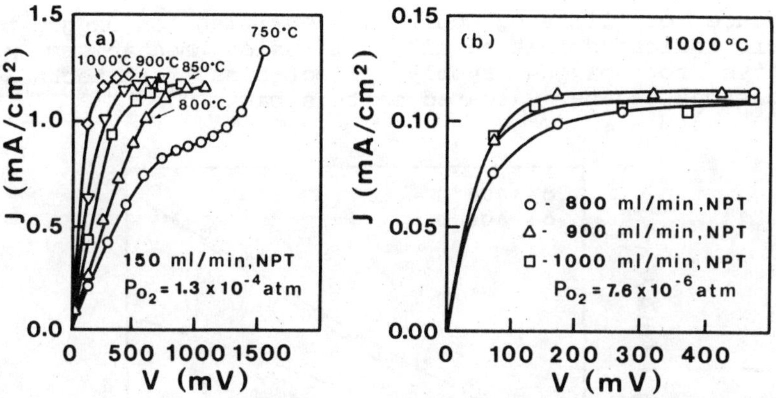

Figure 1. Current density versus voltage, corrected for the open circuit emf, for the cell (-)Po_2,Pt|CSZ|Pt,air(+), for different temperatures (a) and gas flow rates (b).

Figure 1(b), for tube #2, shows that limiting current densities are nearly independent of gas flow rate. As large electrolyte uncovered areas are present, surrounded by dense Pt deposits, a constriction resistance to oxygen transport to the three phase boundary line seems adequate to explain the present results. Adsorption or surface diffusion of oxygen atoms could also account for that tendency.

Figure 2 demonstrates the effect of oxygen partial pressure on the limiting current density, for temperatures of 800 and 1000°C, and gas flow rates from 150 to 650 ml/min NPT. The slope for these data varies from 0.76 for low pressures (correlation factor f=0.95, tube #2) to 1.14 for the highest oxygen pressures (f=0.99, tube #1). If besides data scattering (see also ref. [3]), the change in slope has physical meaning, it seems reasonable to assume that two parallel processes are effective for low oxygen pressures, as suggested in Figure 2; a 0.5 power of oxygen pressure dependence describing diffusion of oxygen atoms over Pt, and the linear pressure dependence corresponding to diffusion of molecular oxygen in gas phase. The slope larger than 1 may be explained by the poor definition of the current density plateau for high oxygen contents in the gas phase. The importance of an activated diffusion process involving adsorbed atoms of oxygen also agrees with temperature dependence of limiting current density, observed in Figure 1(a) when temperature increases from 750 to 800°C.

CONCLUSIONS

The plateau observed in voltage-current data for an oxygen electrochemical pump is due to insufficient supply of oxygen.

Dependence of limiting current densities on oxygen partial pressure indicates that parallel diffusion mechanisms may be effective for oxygen supply. Electrodes for technological applications can be evaluated on this basis.

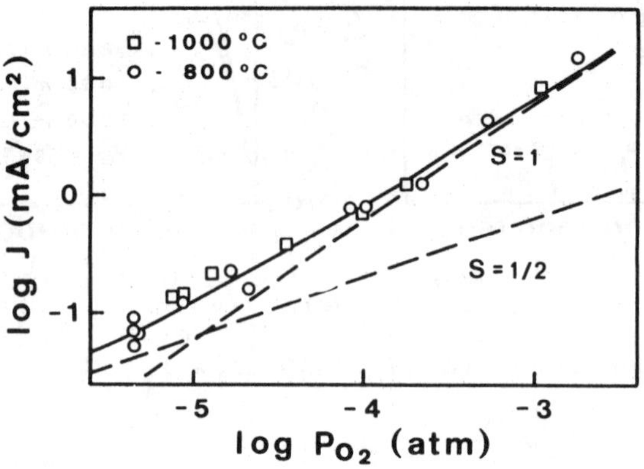

Figure 2. Limiting current density versus oxygen partial pressure in the gas. Slopes 1 and 1/2 refer to possible contributions of parallel mechanisms for oxygen supply.

ACKNOWLEDGEMENT

Work performed under contract (EN3E-0168P) with the Commission of the European Communities (Non-Nuclear Energy Programme).

REFERENCES

1. Etsell, T. H. and Flengas, S. N., N-Type Conductivity in Stabilized Zirconia Solid Electrolytes, J. Electrochem. Soc., 1972, 119, [1], 1-7.

2. Etsell, T. H. and Flengas, S. N., Overpotential Behavior of Stabilized Zirconia Solid Electrolyte Fuel Cells, ibid., 1971, 118, [12], 1890-1900.

3. Wang, D. Y. and Nowick, A. S., Cathodic and Anodic Polarization Phenomena at Platinum Electrodes with Doped CeO_2 as Electrolyte. I. Steady-State Overpotential, ibid., 1979, 126, [7], 1155-1165.

4. Verkerk, M. J., Hammink, M. W. J., Burggraaf, A. J., Oxygen Transfer on Substituted ZrO_2, Bi_2O_3, and CeO_2 Electrolytes with Platinum Electrodes. I. Electrode Resistance dy D-C Polarization, ibid., 1983, 130, [1], 70-78.

EFFECT OF STRONTIUM ADDITION ON THE ELECTRICAL AND THERMAL PROPERTIES OF $LaMnO_3$

J. PALMA, P. DURAN, J.R. JURADO and C. PASCUAL
Electroceramics Department
Instituto de Cerámica y Vidrio (CSIC)
Arganda del Rey, 28500 Madrid, SPAIN.

ABSTRACT

Strontium doped lanthanum manganite is a good electronic conductor and an excellent candidate to be used as electrode in SOFCs. Compositions of strontium-doped lanthanum manganite, $La_{1-x}Sr_xMnO_3$ ($0 \leq x \leq 0.5$), were prepared by using Mn_2O_3, La_2O_3 and $SrCO_3$ as starting materials and by the citrate method. Phase evolution, thermal expansion coefficients, and electrical conductivities were characterized as a function of Sr/La ratio.

INTRODUCTION

The development of SOFC's for the energy production still involves the solution of many problems concerned to materials that are used in their components and the interfaces between them, considering the high temperature and severe environments of the cell operation. In that connected to the material used as cathode, which is responsible of the largest polarization losses in the system, high electronic conductivity, good adherence, and chemical and thermal compatibility with the electrolyte are required.

After high cost and poor long time catalytic noble metals were rejected as cathode materials, high conductive ceramic oxides, specially those with perovskite structure like manganites, cobaltites etc., seem to be the most promising cathode materials.

In this work the influence of strontium on electrical conductivity and thermal expansion coefficient of lanthanum manganite has been considered.

MATERIALS AND METHODS

Compositions of $La_{1-x}Sr_xMnO_3$ ($0.0 \leq x \leq 0.5$) were prepared by the citrate method (1). Precursors were heated slowly up to 200°C, and then calcined for two hours at 1000°C. On the other hand, the same compositions were prepared by solid state reaction of Mn_2O_3, La_2O_3 and $SrCO_3$ fired at 1300°C for 10 hours.

After perovskite phase obtention, powders were uniaxially pressed at 100 MPa in bars of 30 x 5 x 5 mm for conductivity measurements. Samples for dilatometric tests were isostatically pressed at 200 MPa. Sintering was done at temperatures ranging from 1350°C to 1550°C for two hours depending on compositions and final density required. Phases were identified by X-ray diffraction.

Conductivity measurements were made on bars suspended in the homogeneous zone of a furnace by means of Pt wires that worked as electrical contacts. Measurements were carried out from room temperature to 900°C using the AC four-probe method in a frequency range of 10^{-1} to 10^5 Hz with an electrochemical interface SOLARTRON 1286 and a frequency response analyzer SOLARTRON 1255.

Dilatometric measurements were done in a ADAMEL-LHOMARGY DI-24 dilatometer with a constant heating rate of 2°C/min.

RESULTS AND DISCUSSION

Synthesis of the perovskite phase

Perovskite precursor powders prepared by the citrate method kept amorphous up to 600°C. Over this temperature the characteristic diffraction peaks of the perovskite phase were appearing. Well defined peaks, similar to those of samples prepared from oxides, were observed at 1000°C. Compositions with x lower than 0.3 presented a single perovskite phase with hexagonal symmetry corresponding to an oxygen non-stoichiometric compound $La_{1-x}Sr_xMnO_{3+\delta}$. However, compositions of $x \geq 0.3$ show manganese oxide and an orthorrombic perovskite $SrMnO_3$ as secondary phases. The last structure also appeared in calcined powders prepared from oxides, but only for $x \geq 0.4$. After sintering every secondary phase had disappeared except for x=0.5 in which strontium manganite traces still remained. Increased Sr content conducts to a more symmetric hexagonal phase and a lower crystallization which was observed by XRD.

Thermal expansion

The multilayer structure and the conformation process of monolithic fuel cells limit the mismatch of the thermal expansion coefficients between electrodes and electrolyte (2).

Sr substitution do not change manganite expansion behavior in a significant way. The smallest expansion takes place for x=0.2, and becomes larger for the rest. In figure 1, thermal expansion, $\Delta l/l^o$, of Sr doped manganite and a commercial solid electrolyte Y-TZP are plotted vs temperature. The observed mismatch is slightly superior to the usually admitted difference, 500 ppm, but other factors as elasticity modulus and porosity in the real systems must be accounted.

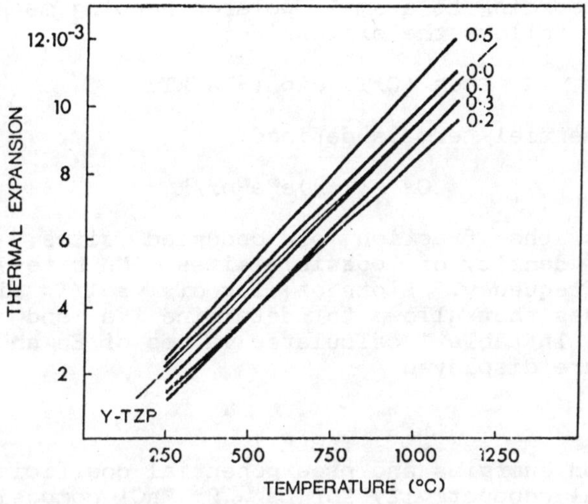

Figure 1. Thermal expansion of $La_{1-x}Sr_xMnO_3$ vs temperature.

Electrical conductivity

Conductivity, as log σ, of the samples prepared from oxides vs the inverse of temperature is plotted in figure 2. Samples prepared by the citrate method present a similar behavior but lower conductivity.

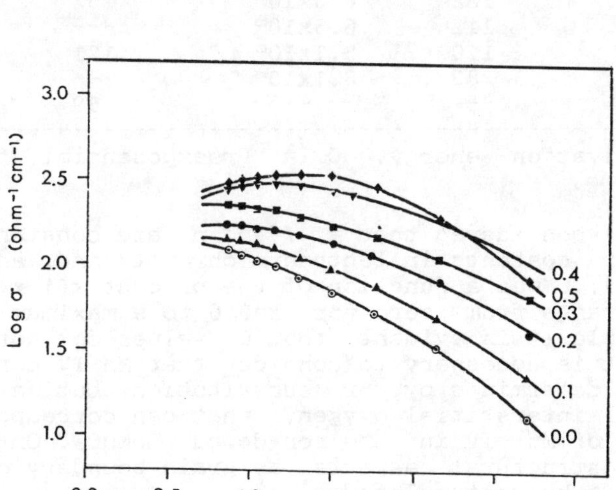

Figure 2. Electrical conductivity of $La_{1-x}Sr_xMnO_3$ vs inverse of temperature.

The presence of Mn IV in Sr doped $LaMnO_3$ suggests a semiconductor behavior where holes, associated to Mn IV, are responsible of conduction (3).

Sr doped manganites acting as semiconductors can be described according to a small polaron hopping mechanism, thus conductivity follows the expression:

$$\sigma= (C/T) \exp(-Ea/kT) \qquad [1]$$

the preexponential term is defined:

$$C= x(1-x)e^2a^2nf/k \qquad [2]$$

where x is the fraction of occupied sites, a the jump distance, n density of possible sites (Mn sites) and f the relaxation frequency. Plots of log σT vs 1/T give a set of straight lines that allows to determine Ea and C for each composition. In table 1 calculated values of Ea and C by least square fit are displayed.

TABLE 1
Activation energies and preexponential coefficients of electrical conductivity for $La_{1-x}Sr_xMnO_3$ compositions.

	OXIDES		CITRATES	
COMPOSITION $La_{1-x}Sr_xMnO_3$	Ea (meV)	C (K/ohm cm)	Ea (meV)	C (K/ohm cm)
0.0	148	$6.2x10^5$	168	$5.1x10^5$
0.1	132	$6.3x10^5$	152	$6.6x10^5$
0.2	112	$6.3x10^5$	---	---
0.3	110	$8.1x10^5$	121	$7.8x10^5$
0.4	83	$8.1x10^5$	---	---
0.5	---	---	89	$4.8x10^5$

Ea = Activation energy, C = Preexponential factor.

Weber (4) considered that a, f and n are constant for the different Sr contents in lanthanum chromite and, according to equation [2], C was a function of the product x(1-x). C values must then range from zero for x=0.0 to a maximum for x=0.5. From the table it is evident that C values do not meet this behavior. It is necessary to consider that Mn IV concentration is not only determined by Sr substitution but also by the presence of interstitial oxygen, that can correspond to more than a 10 % of Mn IV in the non-doped $LaMnO_3$. On the other hand, microstructural aspects as grain boundary density and porosity must be contemplated.

It can be noticed the continuous remarkable decreasing of activation energy as Sr content is increased. According to Goodenough (5), as Sr content rises there is a transition from a low spin (S=2) electronic configuration in Mn III to a high spin (S=3/2) one in Mn IV. That means a band broadening and, therefore, holes become less localized at Mn IV sites, and jump energy decreases.

Figure 2 shows that compositions with x<0.4 have a semiconductor behavior in the whole range of studied temperatures, however 0.4 and 0.5 compositions present a semiconductor to metallic transition at 600° and 450°C respectively. When both Mn IV content and temperature are high enough, broadening of the σ^* band reaches a critical value and carriers become itinerant as in metallic conductors.

CONCLUSIONS

Sr substitution increases electrical conductivity of lanthanum manganite, but not affects remarkably its thermal expansion nor its crystallographic structure.

These compounds present a narrow band semiconductor behavior that at temperatures high enough becomes metallic by broadening of the band. Activation energies and transformation temperatures clearly depend on Sr content.

Electrical properties of these materials are strongly influenced by processing.

REFERENCES

1. Baythoun, M.S.G. and Sale, F.R., Production of strontium-substitued lanthanum manganite perovskite powder by the amorphous citrate process. J. Mat. Sci., 1982, 17, 2757-69.

2. Majumdar, S., Claar, T. and Flandermeyer, B., Stress and fracture behavior of monolithic fuel cell tapes. J. Am. Ceram. Soc., 1986, 69, 628-33.

3. Subba Rao, G.V., Wanklyn, B.M. and Rao, C.N.R., Electrical transport in rare earth ortho-chromites, -manganites and -ferrites. J. Phys. Chem. Solids, 1971, 32, 345-58.

4. Weber, W.J., Griffin, C.W. and Bates, J.L., Effects of cation substitution on electrical and thermal transport properties of $YCrO_3$ and $LaCrO_3$. J. Am. Ceram. Soc., 1987, 70, 265-70.

5. Goodenough, J.B., Covalency criterion for localized vs collective electrons in oxides with the perovskite structure. J. Appl. Phys., 1966, 37, 1415-22.

CHARACTERIZATION AND ELECTRICAL PROPERTIES OF PSZ-Al_2O_3 BASED COMPOSITES

M.T. Hernández, P. Recio, J.R. Jurado and P. Durán
Electroceramics Department
Instituto de Cerámica y Vidrio, C.S.I.C.
Arganda del Rey, 28500 Madrid, Spain

ABSTRACT

Using a sol-gel coprecipitated method from nitrates, partially stabilized zirconia composites have been prepared. The amount of yttria was remained constant (4.5 mol% Y_2O_3) and the alumina dispersed particles amount was ranged between 0 to 30 wt% Al_2O_3.

The electrical conductivity studies performed by an ac impedance spectroscopy method showed a conductivity maximun as Al_2O_3 content was 10 wt%. A hydrothermal annealing experiments indicated that Al_2O_3 strongly reduce the tetragonal-monoclinic tranformation and that improvement was independent on the alumina contents into the composites. The effects of small additions of ceria is also discussed.

INTRODUCTION

Electrically speaking PSZ (partially stabilized zirconia) ceramics have paid little attention in literature. PSZ have been always considered as a structural ceramics; but today it can be an alternative material to FSZ (fully stabilized zirconia) ceramics as a solid electrolyte component. However, like TZP (tetragonal zirconia polycrystalline) material, PSZ suffer the known tetragonal-monoclinic transformation at low temperature (200 ºC) annealing experiments.

It seems to be that Al_2O_3 introduction as dispersed phase has a beneficial effect both on the mechanical properties, the electrical conductivity of the TZP materials (1) and, on ageing effect (2). On the other hand, the beneficial effect of alumina dispersed particles into stabilized zirconia solid electrolytes could be due to the "Al_2O_3-scavenger effect" for SiO_2 in ZrO_2 observed by Butler and Drennan (3).

In the present work the effect of alumina dispersed particles and small additions of ceria on the electrical and ageing behaviour of a PSZ containing 4.5 m% of Y_2O_3 was examined.

MATERIALS AND METHODS

The PSZ-Al_2O_3 composites prepared in this work were obtained employing a method described elsewhere (4). Al_2O_3-dispersed particles addition were in the range of 5-30 wt% of Al_2O_3. PSZ-Al_2O_3 composites with a small amount of ceria (1-2 mol% CeO_2) were prepared by coprecipitating method. Calcined powders were isopressed at 200 MPa and then sintered at 1350º and 1600 ºC for two hours. The thermal stability of the tetragonal phase present in these materials have been examined by means of ageing test in water-vapour at 180 ºC for at least 100 hours and the resulting phases was studied by DRX. Electrical conductivity data was obtained by using ac impedance complex plane analysis in the frequency range of 10 Hz to 10^7 Hz and at temperatures between 200º to 800 ºC.

RESULTS AND DISCUSSION

The SEM observation indicated that PSZ show a bimodal grain distribution, where the tetragonal phase exhibits an average grain size of 3 nm and the cubic ones was around 10-15 nm.

A PSZ-free of Al_2O_3 show a bad ageing behaviour, and after 4 hours of annealing treatment in water-vapour produce a 25% of monoclinic phase. The SEM picture (Fig. 1) depicted a big grains of monoclinic phase along with the appearance of larg cracks in the transformation zone. After 16 hours the material has a 80% of monoclinic phase. When a 10 wt% Al_2O_3 is added a strong reduction of monoclinic phase was observed after 100 hours of annealing treatment (Fig. 2). An increasing amount of Al_2O_3 do not significantly improve the ageing behaviour.

There are various interpretations about the formation of monoclinic phase during ageing experiments. Lange et al. (5) and Schubert (6) considered that the formation of $Y(OH)_3$ is the responsible of the ageing

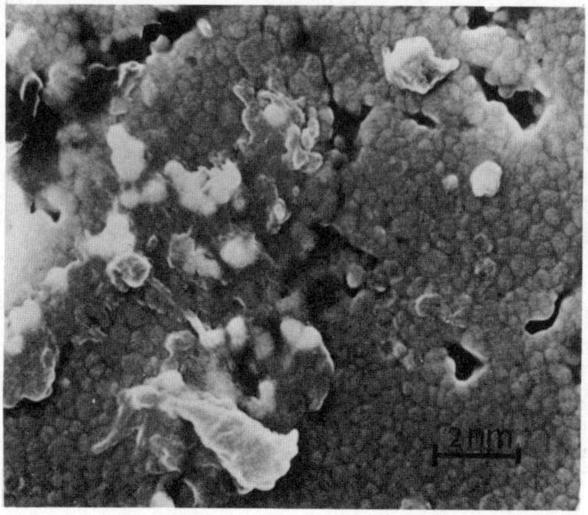

Fig. 1 SEM of Y-PSZ after 4 hours under hydrothermal treatment at 170ºC

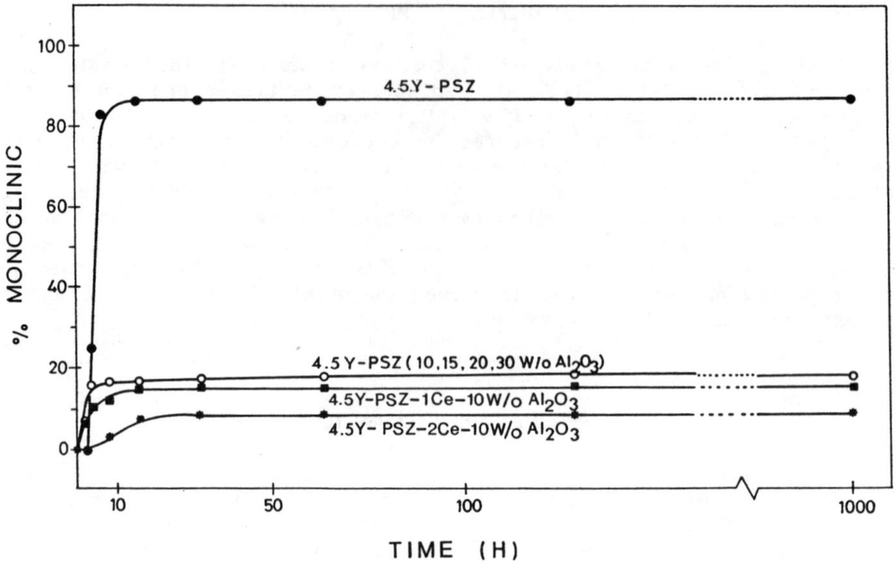

Fig. 2. Monoclinic phase content of PSZ-Al_2O_3 composites vs annealing time.

mechanism. However Yoshimura (8) indicated that the anionic diffussion governed the mechanism hydrothermal conditions. In a first step there is a formation of Zr-OH and Y-OH bonds on the sample surface which creating surfaces stress. In a second step because of OH^- diffusion in the surface a stress accumulation creates strong local deffects which causing the formation of micro and macrocracks nucleating monoclinic phase. A similar explanation in the case of CeO_2 additions was given elsewhere (9).

In partially stabilized zirconia studied in this work, the addition of Al_2O_3 and CeO_2 prevents noticeably the tetragonal transformability, due to alumina acts as second phase increasing the matrix rigidity. Alumina has a constraining effect which improving the mechanical properties of the material, it seems to be that posseses a positive effects when material undergoes hydrothermal test of ageing. In this case alumina acts probably as an inhibitor of active points onto surface of the composite.

The electrical conductivity of composites sintered at 1350ºC does not change when alumina is added, however composites sintered at 1600 ºC show a better sintering behaviour of alumina. Electrical conductivity of both phases goes through a maximum for the composite with a 10 wt% of Al_2O_3. Similar results were reported by Rajendran et al (9). They have observed, in the Y-TZP materials, a disminishing of resistitity of grain-boundary surface area at 10 wt% of Al_2O_3. They interpreted that effect as due to the removing of the vitreous phase located into the boundaries where the ion migration must diffuse. The decreasing in conductivity when the alumina dispersed particle concentration is augmented above 10 wt% is explained as

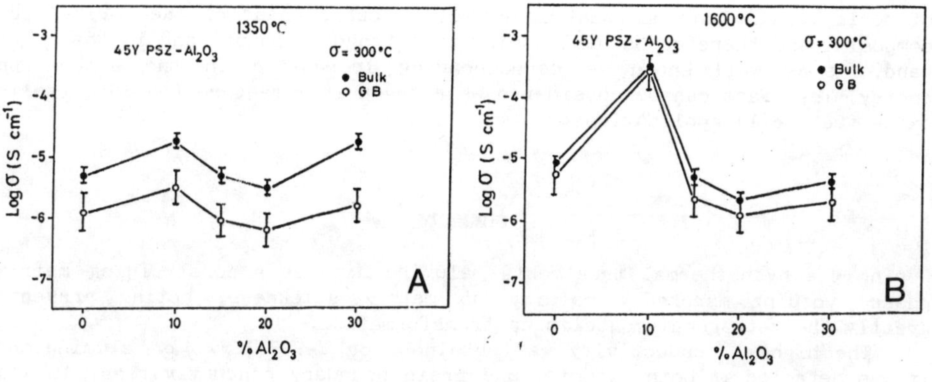

Fig. 3 Bulk and grain boundary electrical conductivities as a function of alumina content. (A) 1300ºC and (B) 1600 ºC sintering temperature

due to the alumina particles start to coagulate which would restrict the anion diffusion path. Both the grain boundary and the grain interior resistivities are strongly reduced. It indicates that the scavenger effect of alumina removing silica impurities from grain interior too, and therefore, the anion path diffusion are cleaned and charge carrier motion is easier. Nevertheless the electrical conductivity values are so high, for the composite 10 wt% Al_2O_3, that one can reasonably think that the scavenger effect is not the only mechanism which can govern those high conductivities. Another important fact is in relation with the effect of small addition of ceria into Y-PSZ-Al_2O_3 composites. On one hand, ceria introduction improving the reduction of tetragonal transformability and allow to increase the conductivity at high temperatures since the addition

TABLE I
Bulk, grain boundary conductivities of Y-PSZ-CeO_2-Al_2O_3 composites.

MATERIAL	Sintering T (ºC)	σ_{bulk} 300ºC	σ_{gb} 300ºC	Working T(ºC) ($\sigma=10^{-2}$)
4.5Y-PSZ-1Ce-10%Al	1350	5.16 10^{-6}	9.05 10^{-7}	750
4.5Y-PSZ-1Ce-10%Al	1600	6.21 10^{-6}	6.21 10^{-6}	700
4.5Y-PSZ-2Ce-10%Al	1350	2.28 10^{-6}	5.90 10^{-7}	700
4.5Y-PSZ-2Ce-10%Al	1600	9.11 10^{-6}	1.06 10^{-5}	650

of ceria causing an important increment in cubic phase of the Y-PSZ-Al_2O_3 composite and therefore a diminishing in tetragonal phase, and in the other hand, it is well known, a corresponding increasing in the activation energy. That fact can be considered as a favourable feature for SOFC (solid oxide fuel cell) applications.

SUMMARY

During the hydrothermal treatments, alumina acts as a constraining matrix phase avoiding marckedly raising up matrix stickness, acting prevents greatly the tetragonal-monoclinica transformation.

The highest conductivity was obtained for Y-PSZ-10%wt of alumina,and it was detected in both interior and grain boundary conductivities. 10 wt% Al_2O_3 content is also the optimum concentration to clean the Y-PSZ grains, grain boundaries ,and triple points for these materials.

Addition of 1-2 mol CeO_2 causes an important increment in activation energy and reduces the tetragonal-monoclinic trasformation. Spite of Y-PSZ-Al_2O_3-CeO_2 could be an atractive candidate for SOFC applications.

REFERENCES

1. SUKUMA, T., UEDA, K. and SHIMADA, M., Strength and Fracture Toughness of Isostatically Hot-Pressed Composites of Al_2O_3-Partially Stabilized Zirconia. J. Am. Ceram. Soc., 1985, **68**, C4.
2. SATO, T., OHTAKI, S., ENDO, T. and SHIMADA, M., Microstructure on the Surface of Yttria-Doped Tetragonal Zirconia Polycrystals (Y-TZP) by Annealing in Humid Conditions. In Advances in Ceramics, V. 24A, ed. S. Somiya, N. Yamamoto and H. Hanagida, The American Ceramic Society, Westerville, Ohio, 1988, pp. 501-509.
3. DRENNAN, J. and BUTLER, E.P., Does Alumina at a Grain Boundary Scavenger in Zirconia?. In Science of Ceramics, V. 12, ed. P. Vicenzini, National Institute Ceramics Technology, Faenza, 1983, pp. 267-272.
4. RECIO, P., PASCUAL, C., MOURE, C., JURADO, J.R., and DURAN, P., Study of Some Processing Aspects in the Y-Tetragonal Zirconia Preparation Using Organic Precursors. In British Ceramic Proc. 38, ed. R.W. Davidge, The Institute of Ceramics, 1986, pp. 127-136.
5. LANGE, F.F, DUNLOP, G.L. and DAVIS, B.I., Degradation During Aging Transformation-Toughned ZrO_2-Y_2O_3 Materials at 250ºC. J. Am. Ceram. Soc., 1986, **69**, 237-245.
6. SCHUBERT and H. PETZOW, G.,. Microstructural Investigations on the Stability of Yttria-Stabilized Tetragonal Zirconia. In Advances in Ceramics, V. 24A, ed. S. Somiya, N. Yamamoto and H. Hanagida, The American Ceramic Society, Westerville, Ohio, 1988, pp. 21-28.
7. YOSHIMURA, M., Phase Stability of Zirconia; Ceram. Bull., 1988, **67**, 1950-1955
8. Hernández, M.T., Jurado J.R. and Durán, P., Influence of CeO_2 content on the electrical properties of Y-TZP ceramics. To be published in Solid State Ionic Periodical.
9. RAJENDRAN, S., DRENNAN, J. and BADWAL, S.P.S., Effect of Alumina Additions on the grain boundary and Volume Resistivity of Tetragonal Zirconia Polycrystals; J. Mat. Sci. Let., 1987, **6**, 1431-1434.

EFFECT OF CeO_2 AND Al_2O_3 ADDITIONS ON THE MECHANICAL AND AGEING BEHAVIOUR OF Y-TZP, TETRAGONAL ZIRCONIA.

M.T. Hernández, F. Capel, J.R. Jurado, P. Durán
Electroceramics Department
Instituto de Cerámica y Vidrio. C.S.I.C.
Arganda del Rey, 28500 Madrid. Spain

ABSTRACT

The addition of CeO_2 in solid solution into a tetragonal matrix strongly inhibit the Tetragonal-Monoclinic transformation during annealing in hydrothermal ageing conditions. On the other hand, the addition of alumina dispersed particles improve the mechanical properties of Y-TZP. In this work, compositions of the ZrO_2-Y_2O_3-CeO_2 system in the ZrO_2-rich region were investigated, as well as Y_2O_3-TZP-CeO_2 composites with addition of 10 wt% of Al_2O_3 were prepared. Fracture toughness, fracture strength as a function of temperature, and thermal shock test were also performed to characterize the composites. Results are correlated with microstructural development and densification path.

INTRODUCTION

Recently an important task in the research of stabilized zirconia materials is to reduce its tetragonal transformability at low temperature especially in hydrothermal conditions (1). It has been reported that with an addition of 7 mol% of CeO_2 the Y-TZP tetragonal phase is retained avoiding the degradation in the most ageing dramatic conditions (2).

On the other hand, when CeO_2 is added a decreasing in density, due to a desmesurate grain growth, was produced. As consequence a diminishing in the mechanical properties was noticed (3).

The present work studies the mechanical properties, fracture toughness, (K_{Ic}), fracture strength, σf, as a function of temperature, thermal shock, and ageing behaviour of tetragonal zirconia containing CeO_2, and composites Y_2O_3-TZP-CeO_2 containing 10 wt% of alumina dispersed particles.

EXPERIMENTAL PROCEDURE

Samples were prepared by the hydroxide coprecipitation method. Sintering temperatures ranged between 1300º and 1350ºC. The fracture toughness was measured by indentation method (100 N load), the Young modulus was determined by an ultrasonic technique, fracture strength as a function of temperature was performed by using the three point flexure method. An INSTRON Universal Machine was employed and the sample span was as high as 50 mm. Thermal shock behaviour was determined in the temperature range of 300º-1000ºC, by heating the samples to the desired temperature for 2h and rapidly cooled in water. The KIC variations was utilized to perform the experiments. The ageing experiments were carried out in air for 50 hours and in water vapor for at least 1000 hours.

RESULTS AND DISCUSSION

The ageing test shown in figures 1-2 indicated that Y-TZP with 3 mol% Y_2O_3 required only an alloying of 2 mol% of CeO_2 to avoid the t-m (tetragonal-monoclinic transformation), while for samples with 2 mol% of Y_2O_3 is necessary alloying at least with 6 mol% of CeO_2 to achieve the thermal stability of the tetragonal phase. That fact can be explained taking into account the ternary phase diagram in the ZrO_2 rich region. The role of CeO_2 to achieve the thermal stability of the Y-TZP material can be accounted for CeO_2 could reinforce the surface sample, preventing the formation of ZrOH groups during the annealing in water vapor. Figures 3-4 show the fracture toughness and relative density as a function of CeO_2 concentration. Both parameters exhibit similar behaviour, and both decreases as CeO_2 concentration increases.

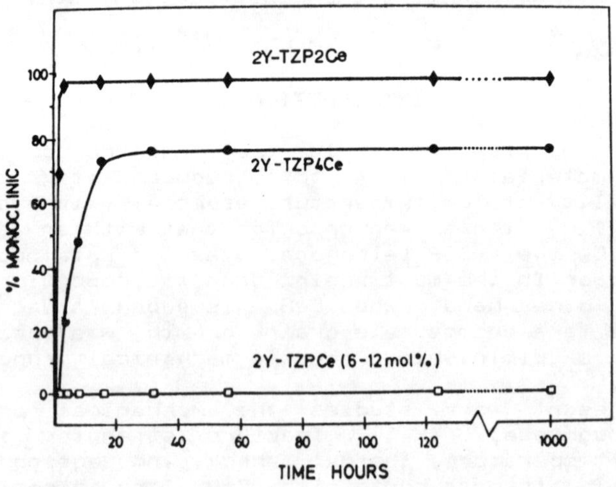

Figure 1. Ageing behaviour of 2Y-TZP-CeO_2 ceramic materials.

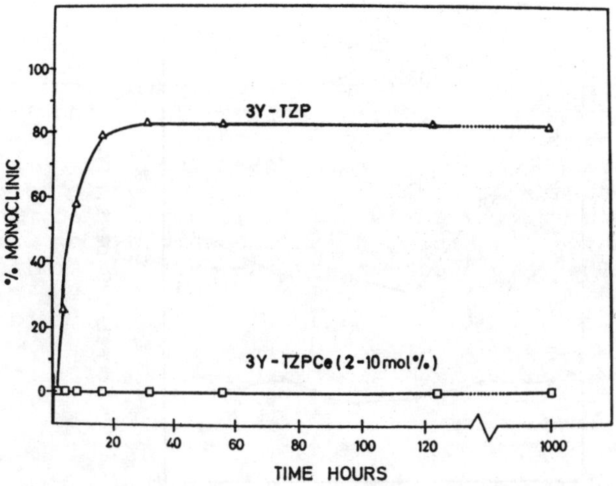

Figure 2. Ageing behaviour of 3Y-TZP-CeO_2 ceramic materials.

That result agreed with those reported by other authors (4). It can be indicated that K_{IC} decreases because of the martensitic transformation as consequence of the increment of thermal stability of tetragonal phase and also of porosity increasing. The dependence of fracture toughness, K_{IC}, on temperature of shocked samples is shown in figure 3. In the temperature region of 300º to 400ºC was present a rapid, but no drastic fracture toughness degradation which is in agreement to the ageing behaviour as above distended.

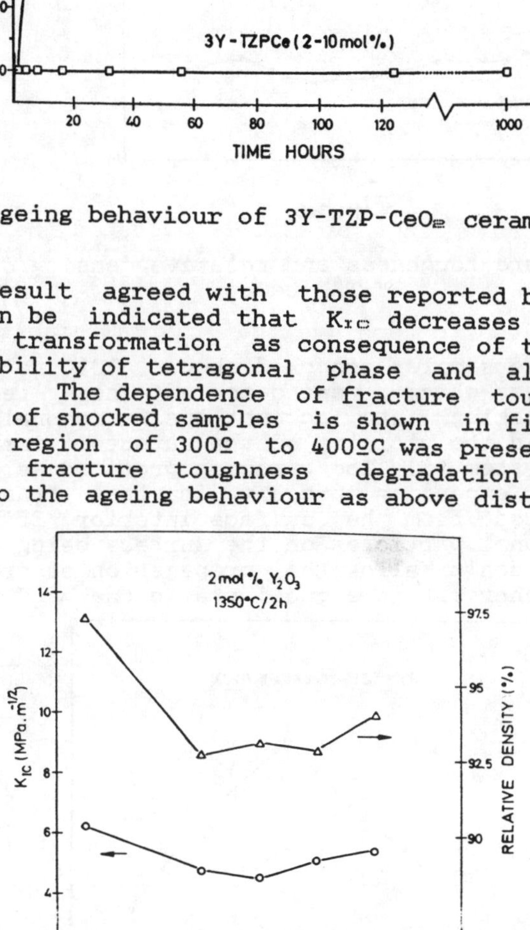

Figure 3. Fracture toughness and relative density of 2Y-TZP-CeO_2.

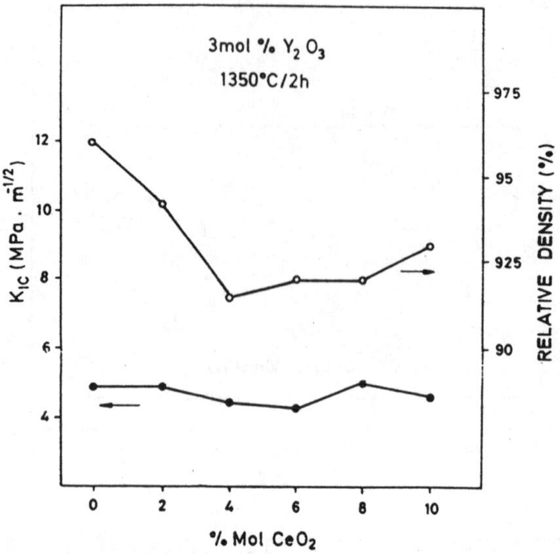

Figure 4. Fracture toughness and relative density of 3Y-TZP-CeO_2.

However, for the samples shocked above 400ºC no significant fracture toughness changes were detected. These results could be explained taking into account the tetragonal-monoclinic transformation in the vicinity of the eutectoid temperature of the ZrO_2-Y_2O_3 system (5). The cracking produced as consequence of the volume expansion present in that transformation, rapidly propagated from the surface interior. If this is so, then the tetragonal particles on the surface being coated by a CeO_2 thin film don't allow the propagation of cracking in a catastrophic manner but in a quasi static one.

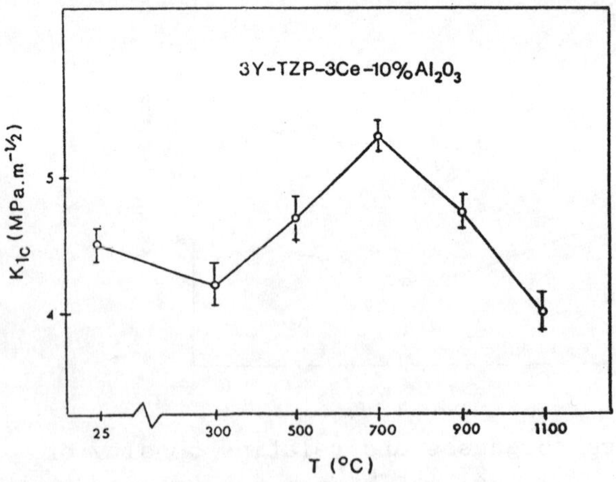

Figure 5. K_{IC} vs thermal shock temperature of 3Y-TZP-3Ce-10 wt% Al_2O_3.

TABLE 1
BENDING STRENGTH (*) OF Y-TZP CERAMICS

Material	25ºC	250ºC	500ºC	800ºC
3Y-TZP	602±5	512±5	309±5	209±5
3Y-TZP-10%Al_2O_3	479±5	489±5	338±5	270±5

(*) σ_f MPa

Table I shows the variation of bending strength as a function of temperature for the two kinds of tetragonal samples. As it can be seen σ_f decreases monotonically as the temperature increases, however in the case of Y-TZP the bending strength decreasing is much more drastic overall in the low temperature region (300ºC), which again coincides with the normal behaviour of the tetragonal zirconia (6) below the eutectoid temperature transformation. The presence of alumina seems to be acting as crack-pinning at least in the low temperature region. At high temperature the bending strength values are very similar to those of a cubic stabilized zirconia. When small amounts of CeO_2 were added a strong bending strength decreasing was produced.

REFERENCES

1. SATO, T. and SHIMADA, M., Transformation of yttria-doped tetragonal ZrO_2 pollycrystal by annealing in water. J. Am. Ceram. Soc., 1985, 68, 365-359.
2. HERNANDEZ, M.T., JURADO, J.R. and DURAN, P., Influence of CeO_2 content on the electrical properties of Y-TZP ceramics. To be published in Solid State Ionic Periodical.
3. DUH, J.G., DAI, H.T. and CHIOU, B.S., Sintering, Microstructure, Hardness, and Fracture Toughness Behaviour of Y_2O_3-CeO_2-ZrO_2. J. Am. Ceram. Soc., 1988, 71, 813-819.
4. SATO, T., FUKUSHIMA, T., ENDO, T. and SHIMADA, M., Thermal Stability and mechanical properties of yttria and ceria doped zirconia/alumina composite. In Science of Ceramics, V. 14, ed. D. Taylor, The Institute of Ceramics, Shelton, Stoke-on-Trent, U.K., 1988, pp. 843-848.
5. PASCUAL, C. and DURAN, P., Subsolidus Phase Equilibria and Ordering in the System ZrO_2-Y_2O_3, J. Am. Ceram. Soc., 1983, 66, 23-27.
6. ORANGE, G. and FANTOZZI, G. Mechanical Properties and Dependence with Temperature of Tetragonal Pollycrystalline Zirconia Materials. Presented at 4th Inter. Conf. on the Fracture Mechanic of Ceramics. Blacksbury, USA, Jun, 1985.

SINTERING AND DIELECTRIC PROPERTIES OF SiO_2-DOPED $BaTiO_3$ CERAMICS

J.F.FERNANDEZ, P.DURAN and C.MOURE
Instituto de Cerámica y Vidrio (CSIC), Electroceramics Department, Arganda del Rey, 28500 Madrid, Spain.

ABSTRACT

Barium titanate powders doped up to 1.0 % weight of SiO_2 were sintered between 1150ºC and 1250ºC for times ranging from 1 to 10 h. The sintering process was followed by a dilatometric method. The microstructure was studied by means of SEM. Dielectric parameters such as the dielectric constants and losses were also determined.

For low amount of SiO_2, the maximum density temperature decreases from 1225ºC to 1175ºC. The end-point sintering temperature lowers to 1200ºC for 0.4% doping. The dielectric constant increases for 0.1% SiO_2 and lowers for doping above 0.4% weight. A two-step sintering process was observed. The maximum density values decreases significatively, and the end-point temperature raises. A second phase, barium titanium silicate, appears. Liquid phase sintering seems to be the dominant mechanism.

INTRODUCTION

Lowering the sintering temperature of the barium titanate based materials allows the use of non precious metals as electrodes for multilayer ceramic capacitors. The lowering is attained by using several compounds which form a liquid phase. This phase aids to the sintering process. Swillan and Gadalla (1) studied the effect of additives such as $PbTiO_3$, CdO, ZnO or CuO which promote densification by reactive liquid phase formation. The use of LiF allows the lowering of the initial sintering temperature up to 800ºC (2) Their main problem is the fluorine volatilization. This phenomenon is reduced by using $BaLiF_3$ instead of LiF (3), but it keep important. Chowdary and Subbarao (4) used low melting point glasses as $74Bi_2O_3$-$26B_2O_3$ to improve the densification behaviour at low temperatures. Some industrial procedures combine silica and

Figure 1. Morphology of $BaTiO_3$ powders.

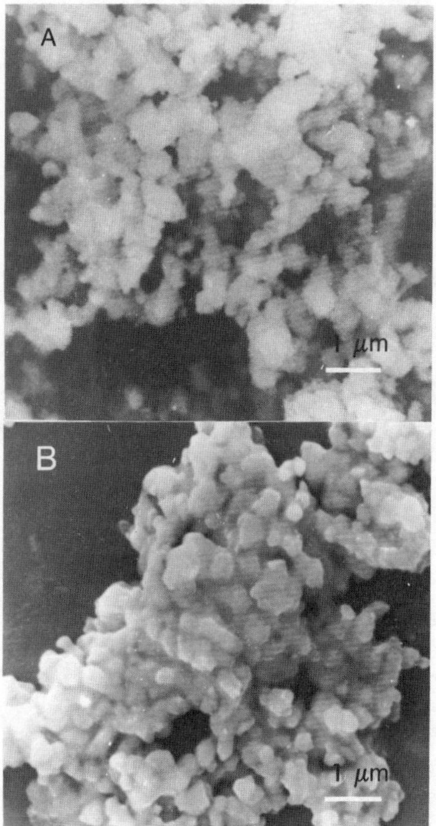

Figure 2. Morphology of $BaTiO_3$ powders with 0.2% (a) and 0.4 % weight silica.

titania or silica and alumina as sintering aids.

In the present work a study of the effect of silica doping on the sintering behaviour and electrical properties of $BaTiO_3$ has been made. Amounts up to 1.0 % weight of SiO_2 were added to pure barium titanate.

EXPERIMENTAL PROCEDURE

The barium titanate (5), shows a mean particle size of 0.75 μm (fig.1). Their specific area is 5 m^2/gr and the stoichiometric ratio Ba/Ti is 1, with 2.78% weight of soluble Ba. The impurity content (measured by ICP and expresed in % weight) are: Al_2O_3 0.16, Fe_2O_3 0.029, P_2O_5 0.13, SiO_2 0.12, ZrO_2 0.094, SrO 0.19, CaO 0.024, Na_2O 0.029 y K_2O 0.005. The colloidal silica added is very pure (99.9%). Their surface area is 390 m^2/gr, and the estimated particle size is 7 nm.

The $BaTiO_3$+SiO_2 mixtures were homogeneized by means of a turbine at 6000 rpm for 4 minutes in isopropylic alcohol. The slurry was dried in an oven at 55ºC for 24 hours. The powder was isopressed at 200 Mpa and sintered in air for several temperatures and times.

RESULTS AND DISCUSSION

Figure 2 shows electron micrographs of dried powders. It can be seen that for silica amounts of 0.4% and above, the whole $BaTiO_3$ particles are recovered by silica gel. The agglomeration state seems higher than the corresponding to the pure $BaTiO_3$. The crystal structure of this last compound is unaffected by silica gel addition, according with XRD analysis.

Fig.3 shows the results of

CRH sintering, performed by means of a high-temperature automatized dilatometer. The shrinkage-temperature and shrinkage rate-temperature curves were obtained. It is possible to see that the silica doping retards the beginning of the shrinkage and anticipates the end. The shrinkage rate curves show that the silica doping lowers the sintering temperature. When the silica amount grows above 0.4% weight, two sintering mechanisms are apparents.

The addition of silica lowers the temperature to which, the maximun shrinkage rate takes place (Fig.4). For the higher

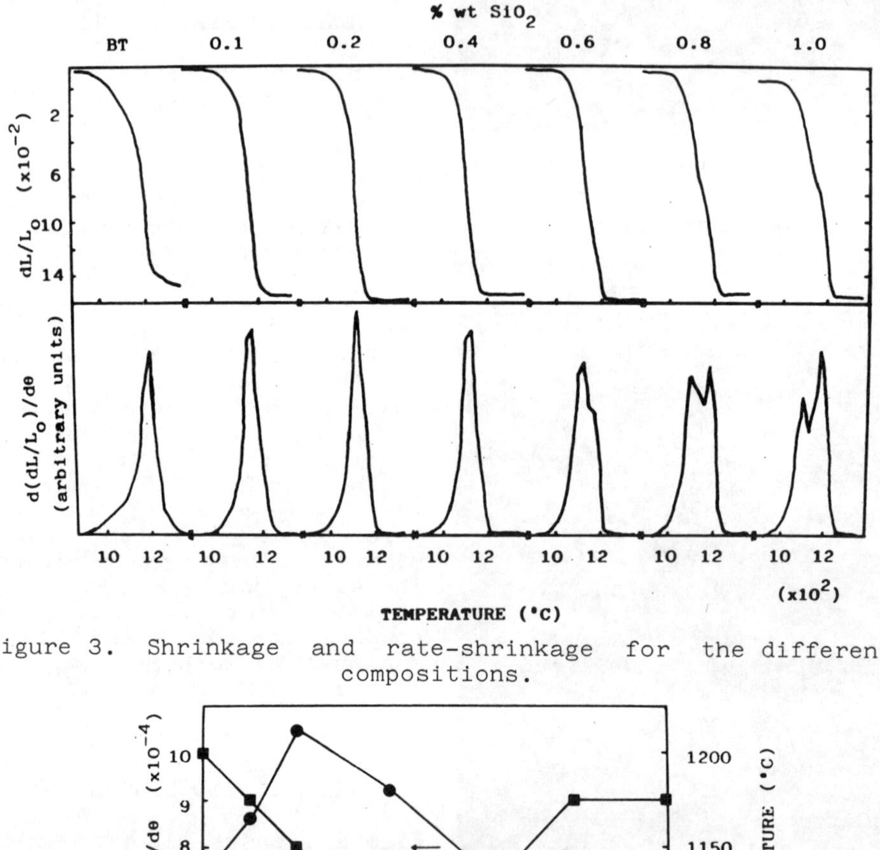

Figure 3. Shrinkage and rate-shrinkage for the different compositions.

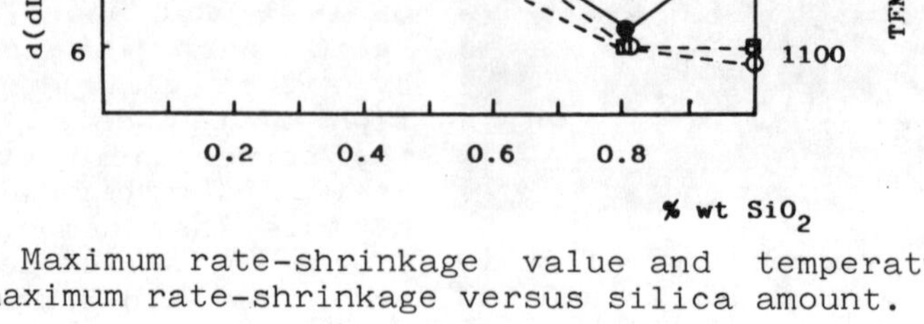

Figure 4. Maximum rate-shrinkage value and temperature for maximum rate-shrinkage versus silica amount.

silica amounts this is valid for the first sintering mechanism. The maximun shrinkage rate appears for 0.2% silica amount and monotonically decreases for higher contents. The second sintering mechanism becomes predominant for silica contents of the 1% weight.

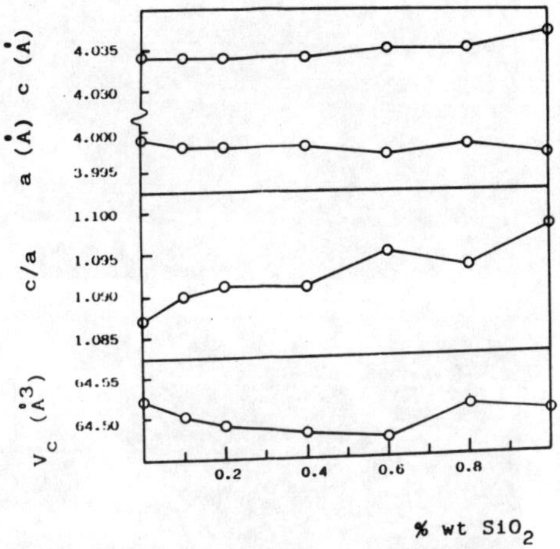

Figure 5. Lattice parameters variation vs silica amount.

Figure 5 shows the lattice parameter variation of the doped $BaTiO_3$ versus the silica amount. The lattice volume slightly decreases when the silica increases. It seems that the silica dissolves, at least partially, into the $BaTiO_3$. At the same time the lattice tetragonality increases. The XRD analysis shows no second phases for 0.1 and 0.2% silica content. Above these amounts, formation of second crystalline phase is observed. This phase has been identified as the barium titanium silicate, $Ba_2TiSi_2O_8$. For silica contents higher than 0.6% the amount of the formed silicate seems to keep constant, with respect to the $BaTiO_3$ present.

Figure 6 shows the density versus temperature curves for 2 h constant time. It can be seen that the temperatures for which the maximun density is attained, are lower for doped samples than the corresponding to pure $BaTiO_3$. The temperature decreasing is as higher as lower the silica doping. On the other hand, the maximun density value, obtained for 0.1% SiO_2, is only slightly smaller than that of pure titanate, and strongly decreases when the silica amount increases. The density variation with temperature, for constant times, was smoother for low amounts of silica. However, it is very pronounced for the highest silica levels.

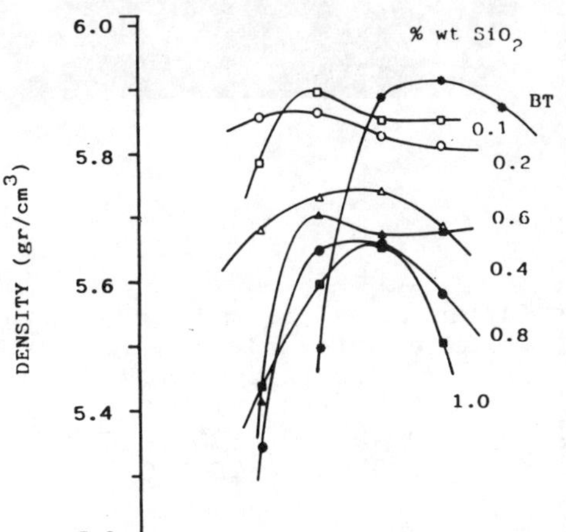

Figure 6. sintering curves versus temperature.

Figure 7 shows SEM micrographs of polished and thermally etched surfaces of sintered samples with the highest density. For doping levels up to 0.2% the particle

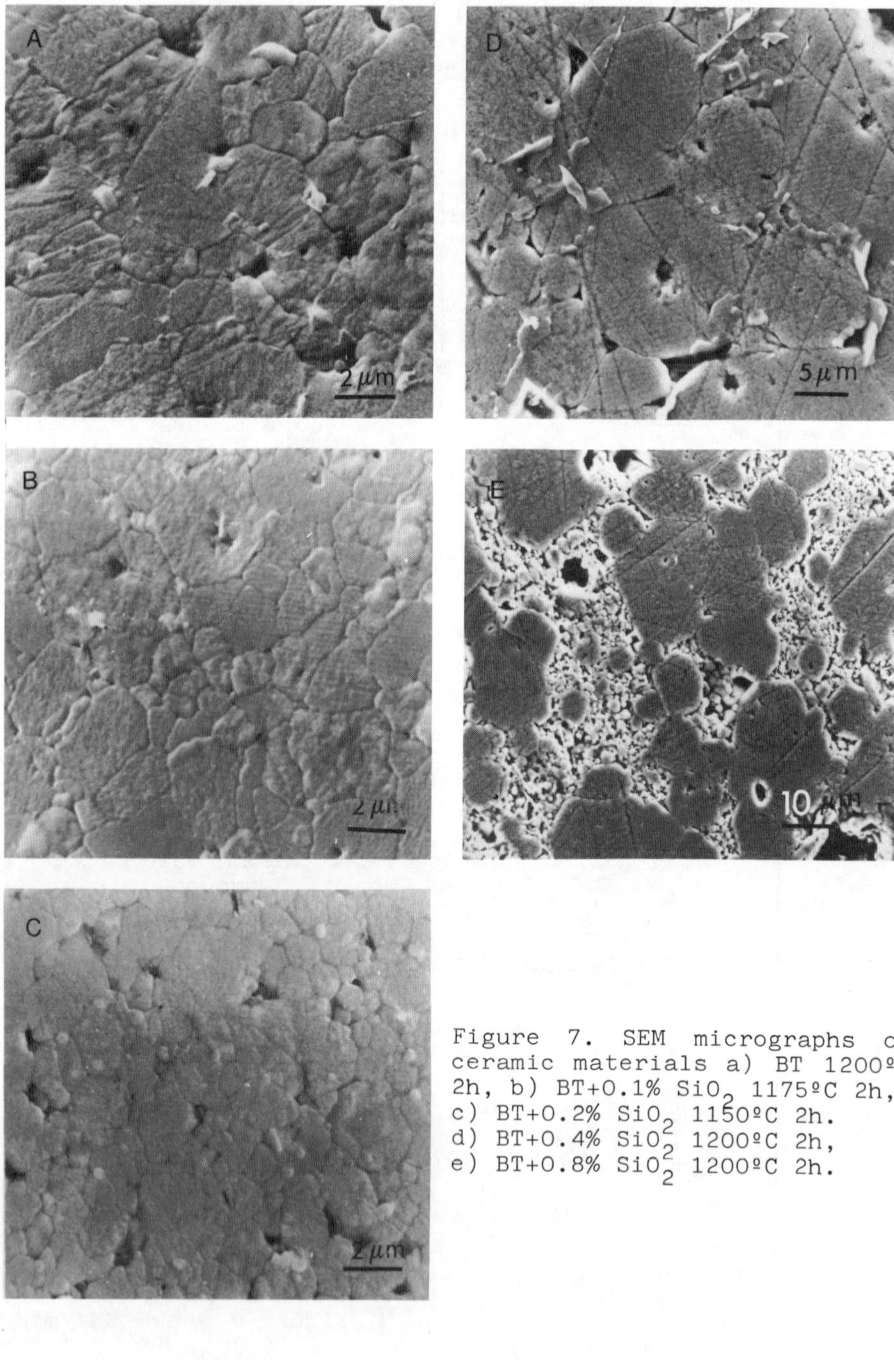

Figure 7. SEM micrographs of ceramic materials a) BT 1200ºC 2h, b) BT+0.1% SiO_2 1175ºC 2h, c) BT+0.2% SiO_2 1150ºC 2h. d) BT+0.4% SiO_2 1200ºC 2h, e) BT+0.8% SiO_2 1200ºC 2h.

size are smaller than the corresponding to the pure titanate (2.1μm and 4,3μm respectively). The porosity is located in the triple points. For silica amounts of 0.4% and higher there is a strong grain growth. For contents of 0.8% a duplex structure is developed. In these samples the silicate phase is located in the grain boundaries. Intragranular porosity becomes apparent.

The addition of silica to the barium titanate causes two different effects according to the doping level. The doping with very low amounts of silica favours the sintering and densification processes by a grain growth control mechanism. The silica, along with the free barium oxide and the impurities, forms a liquid phase. It favours the grain boundary sliding and the pore elimination; therefore the shrinkage rate increases. The amount of that phase is so low that the mass transport is avoided. Residual liquid phase is located in some grain boundaries.

With 0.4% weight of silica, the microstructure is quite different. The grain size reaches values of $\sim$12μm, and secondary phase remains in grain boundaries and triple points. The liquid phase grows, and dissolves the titanate grain boundaries. This fact favours the mass transport and the grain growth. From this liquid phase, barium titanate silicate precipitates and remains located in grain boundaries. Their presence arrests the densification process.

For doping levels above 0.6% value, a duplex structure is observed. It coincides with the double sintering mechanism shown by the shrinkage rate curves above described. The first stage of sintering could be attributed to the rapid growth of the largest grains, in the presence of abundant and no homogeneusly distributed liquid phase, because of the presence of hard aggregates which were no broken during the compaction process. The second step can be caused by a diffusion process between the titanate grains when the liquid phase precipitates the barium titanium silicate, with the consequent lowering of their amount, and the exagerated grain growth collapses.

The figure 8 shows the dielectric constant and losses corresponding to the different samples. The dielectric constant of well densified samples doped with 0.1 and 0.2 % weight of silica, increases with respect to those of the pure titanate of similar density owing to the resulting smaller grain size. On the other hand, for the same sintering temperature samples, the addition of silica decreases the dielectric constant values, because of their lower densities. When the silica amount raises, and the second crystalline phase appears, the dielectric constant sharply decreases. The losses are similar between pure and low-doped titanate, but increases strongly for those samples with glass and crystalline phases in the grain boundaries.

CONCLUSIONS

Small amounts of silica (up to 0.2 % weight) lower the sintering temperature and favour the grain growth control and microstructure homogeneity of the barium titanate. An excess

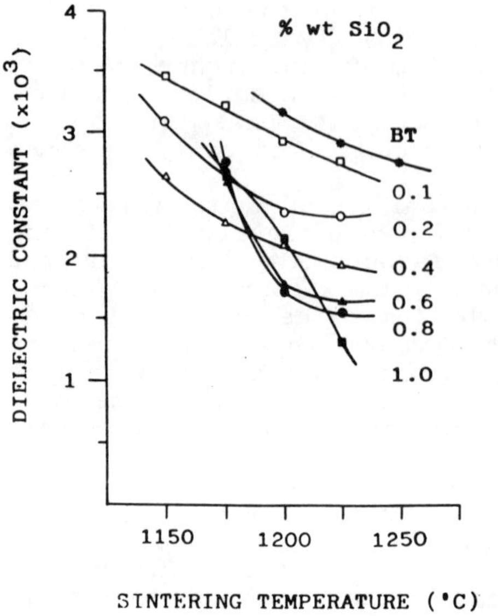

Figure 8. Dielectric properties versus sintering temperature.

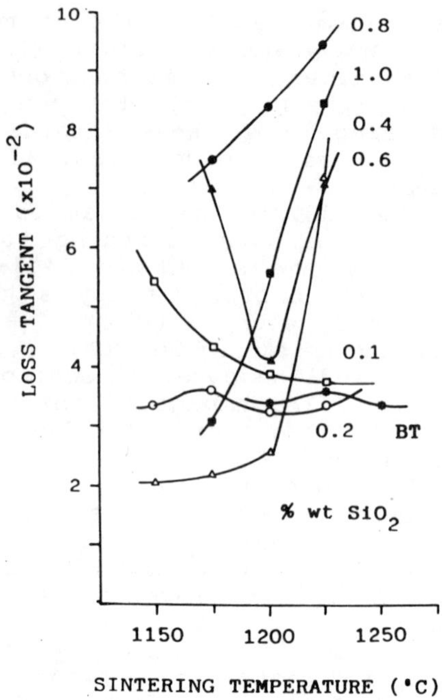

of liquid phase caused by higher silica amounts promotes exaggerated grain growth, lowers the final density and leads to deteriorate the dielectric properties.

ACKNOWLEDGEMENTS

The authors are grateful to RIO RODANO S.A. for finantial support of this work under contract.

REFERENCES

1. Swillan, M.N., Gadalla, A.M., "Effect of Additions on the Sinterability of Barium Titanate" J.Trans.Br.Cer. 1975,74,5,165-69.
2. Walker Jr, B.E., Rice, R.W., Pohanka, R.C., Spaull,J.R., "Densification and Strength of $BaTiO_3$ with LiF and MgO Additives" Am.Ceram.Soc.Bull. 1976,55,3,274-85.
3. Desgardin, G., Mey, I., Raveau, B., Haussonne, J.M., "$BaLiF_3$ a New Sintering Agent for $BaTiO_3$-doped Capacitors" Am.Ceram.Soc.Bull. 1985,64,4,564-69.
4. Chowdary, K.R., Subbarao, E.C., "Liquid Phase Sintrered $BaTiO_3$" Feroelectrics 1981,37,689-72.
5. Fernandez, J.F., Duran, P., Moure, C., "Barium Titanate Processing: Effect of Raw Material on their Synthesis and Sintering Behaviour" In this volume.

SYNTHESIS OF ADVANCED DIELECTRIC MATERIALS

DETLEV HENNINGS
Philips GmbH Forschungslaboratorium Aachen
P.O.B. 1980, 5100 Aachen, F.R.G.

ABSTRACT

High quality ceramic powders for electronic components are nowadays produced to an increasing exent by applying wet chemical preparation techniques. At the miniaturization of temperature stable multilayer capacitors chemically prepared powders are already inevitable. Some of the new preparation routes for $BaTiO_3$, eg. coprecipitation and hydrothermal synthesis are discussed.

INTRODUCTION

In the last five years the outer dimensions of ceramic multilayer capacitors (MLC) have been reduced by a factor of nearly 4. In that time the capacitance per volume must be enhanced by the same factor. The strong increase of the specific capacitance could only be performed by raising the permittivity of the material and reducing the dielectric layer thickness.

Up to now most of the dielectric materials used in ceramic capacitors are based on barium titanate, $BaTiO_3$. In temperature stable dielectrics of the EIA specification "X7R" (maximum deviation of the capacitance from the 25°C reading: ± 15%) very fine (micro heterogeneous) mixtures of ferro-electric $BaTiO_3$ and a paraelectric second phase are applied [1]. The para-electric 2nd-phase becomes ferroelectric, showing a dielectric maximum at the low temperature end of the X7R characteristic (-55°C). The permittivity of ferroelectric $BaTiO_3$ strongly depends on the grain size. Maximum values of the permittivity of $K \gtrsim 6000$ are observed at grain sizes of

0.7 μm [2]. Most of the modern X7R dielectrics are therefore optimized to this grain size. For further increases of the specific capacitance the thickness of the dielectric layer must be reduced.

The utility of thin dielectric layers is limited by the following critical parameters: Dielectric break down, too low insulation resistance, unsufficient life stability, and too high ferroelectric losses.

There are also close relations between the electrical properties and the grain size of $BaTiO_3$. The degradation of $BaTiO_3$-based dielectric materials in DC-fields and the insulation resistance are determined by the defect chemistry as well as the number of grain boundaries between the electrodes [3]. At very low grain sizes (<0.4 μm) the ferroelectric properties of $BaTiO_3$ begin more and more to disappear [2] and are then often characterized as "superparaelectric". The perovskite structure then changes from tegragonal to "pseudocubic". The break-down voltage of such fine-grained materials is considerably higher than that of normal coarse-grained materials. The life stability of thin dielectric layers is inversely proportional to the grain size.

In thin layers of temperature-stable dielectric materials the inter-granular coexistence of ferroelectric $BaTiO_3$ and a paraelectric 2nd phase gives rise to problems with the insulation resistance and life stability. Thin dielectric layers therefore require very fine, regular mixtures of small grains. Hence, there is a strong demand for fine-grained ceramic powders at the manufacture of thin dielectric layers. The powders need a precise control of the grain size and dopant distribution.

Fine-grained and poorly aggregated ceramic powders can no longer be economically prepared in the classical way of mixing and firing oxides and carbonates. New, wet chemical preparation routes have therefore largely replaced the classical routes of powder preparation.

WET CHEMICAL PREPARATION OF BARIUM TITANATE

Aqueous or organic solutions of ions or other compounds can be mixed on a quasi atomic scale. The aim of wet chemical preparation techniques is to transfer this high degree of intermix from the liquid into the solid state. The advantages of wet chemically prepared powders are:

- better homogeneity,
- high purity,
- lower calcination temperatures and
- smaller particle sizes.

Ball milling of chemically prepared powders can be often discontinued. In most cases wet chemical preparation techniques really provide mixtures on a nanometer scale, rather than on an atomic scale.

Wet chemical preparation techniques of $BaTiO_3$ can be subdivided in a few principal routes which are:

- spray-drying and -roasting,
- synthesis from complex precursors,
- coprecipitation,
- sol-gel techniques and
- hydrothermal syntheses.

The advantages and disadvantages of these techniques have been discussed in detail elsewhere [4].

Preparation of highly pure and fine-grained $BaTiO_3$ from barium titanyl oxalate, $BaTiO[C_2O_4]_2.4H_2O$, has now become an industrial route. Sol-gel techniques are increasingly used at the production of highly pure ceramic powders, too. However, in the case of $BaTiO_3$ sol-gel techniques have not yet found larger industrial applications up to now. Hydrothermal synthesis of $BaTiO_3$ seems to be a promising and economic way of preparing submicron and widely monodisperse powders. Even multicomponent perovskites (doped or mixed crystals) can be prepared by this way.

REFERENCES

1. Hennings, D. and Rosenstein, G., J. Am. Ceram. Soc., 1984, **67**, 249.

2. Arlt, G., Hennings, D. and de With, G., J. Appl. Phys., 1985, **58**, 1619.

3. Waser, R., Bajatu, T. and Härdtl, K.H., submitted to J. Am. Ceram. Soc.

4. Hennings, D., British Ceramic Proceedings No. 41 "ELECTROCERAMICS", pp. 1-10, edited by A.J. Moulson, J. Binner, R. Morrel (The Institute of Ceramics, Stoke on Trent, U.K., 1989).

RELAXOR MATERIALS OWING REPRODUCIBLE VERY HIGH DIELECTRIC CONSTANTS, AND THEIR BEHAVIOUR UNDER A D.C. BIAS FIELD

J.M. HAUSSONNE*, M. LE CUN* ; P. BOURDOIS**, G. DESGARDIN**
* CNET Centre Lannion B, B.P. 40, 22301 LANNION Cedex, FRANCE
** I.S.M.Ra., Laboratoire CRISMAT, 14032 CAEN Cedex, FRANCE

ABSTRACT

A review of the literature points out many different results concerning the value of the dielectric constant of relaxor materials : either one particular material is supposed to own many different dielectric constants, or some materials are characterized by surprisingly high constants. The work we present here shows that this is due to the fact that, at the end of the sintering cycle, other phases than the perovskite one can diffuse and segregate as a thin layer on the surface, calculation showing then that the resulting capacitance decreases drastically. Every relaxor material is able to own a very high dielectric constant, until the eventually present secondary phases stay in the bulk material. In this paper, we shall also show the particular behaviour of disc or multilayer relaxor capacitors under a D.C. bias field.

INTRODUCTION

Much work has been devoted to relaxor materials for capacitors, leading to a good knowledge of the synthesis chemistry and of the sintering phenomena [1]. Powders are now supplied, and capacitors are industrially produced.

Unfortunately, some problems remain. We shall present here two of them. The first one deals with the presence in the ceramic of other phases that the pure perovskite one, with as a consequence a large dispersion of the dielectric characteristics of sintered discs. The second one deals with the spectacular evolution of dielectric measurements made on capacitors with a D.C. bias. The results we present here have been obtained on the classical PMN/PFN material with its Curie temperature near 20°C. As previously described [2, 3], we prepared this material by mixing and calcining the starting powders : PbO, $MgCO_3$, Nb_2O_5 and Fe_2O_3. Lithium carbonate is then added, and sintering near 900°C leads to a dense ceramic with a dielectric constant close to 35000.

SECOND PHASES AND DIELECTRIC CHARACTERISTICS

Figure 1 shows a strong dependance of the dielectric constant measured on discs versus the amount of lithium carbonate. There exists an optimum, that is not related to the densification of the material : for low amounts, the ceramic may be not well densed.

On the countrary, for high amounts, it is full densed, although the dielectric constant is far to own high values. X ray analysis of the surface of discs gives the explanation of that. Observations are summerized table 1, colomn 1 of which corresponds to Fig. 1 materials. In the case of high lithium amounts, we noticed the presence on the surface of a second phase, so called "C". The SEM observation of surface (see Fig. 2) shows it with a weak thickness, typically some micrometers. Figure 3 shows how its presence leads to poor dielectric constants : when it is present, the disc is in fact composed of three discs : the relaxor material with a high dielectric constant between two thin materials with a very small dielectric constant. If we assume that the relaxor material owns a dielectric constant close to 35000 and the 1 μm thick C phase a 50 one, calculation shows that the apparent dielectric constant of the disc is only close to 15000. Figure 4 confirms that polishing low dielectric samples until pure perovskite is detected by X ray diffraction on surface leads to much better values.

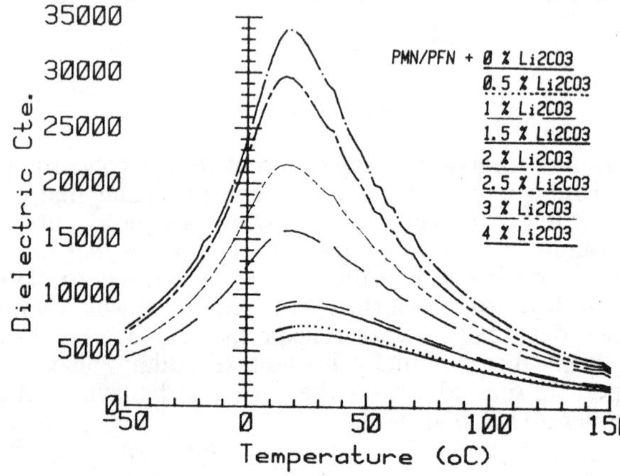

Figure 1. Dielectric characteristics of PMN/PFN + x % Li_2CO_3 sintered 1 hour at 885°, heating 80°/H, corresponding to column 1 of table II.

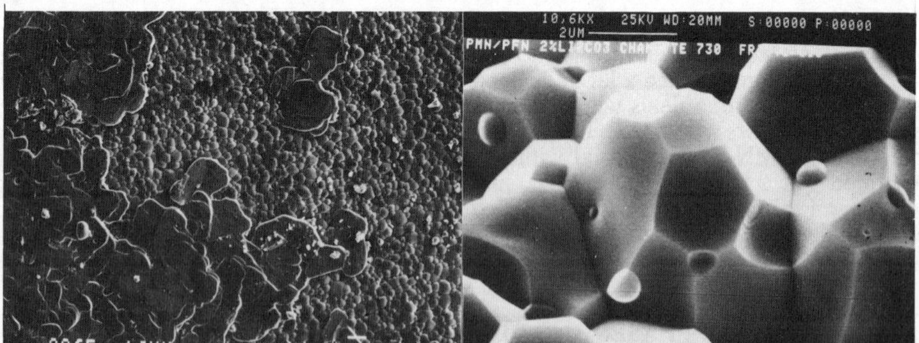

Figure 2. S.E.M. observation of a/ the surface of a disc showing a second face, b/ the fracture of a disc showing a second phase at grains boundaries.

Table I gives together with the first example the evolution of surface phases detected on ceramics made from the same batches, but sintered with different schedules. In every case, the measured dielectric constant has the highest values only when the surface of discs is free of second phase, the optimum of lithium content being to be associated to the definition of the sintering schedule.

TABLE I

Phases Detected by X Ray Diffraction on Surface of Discs

Per. = Perovskite ; Pyr. = Pyrochlore ; C = Phase 'C'
TF = Very Strong ; f = weak ; PD = Poor Densification

% Li2CO3	Sintering 885oC 80 o/h	Sintering 890oC 150 o/h	Sintering 950oC 450 o/h
O	Per. TF, Pyr. f, PD k max. = 6500	Per. TF, Pyr. f, PD k max. = 9000	Per. TF, Pyr. f, PD k max. = 9000
0.5	Per. TF, Pyr. f, PD k max. = 7000	Pure Per., PD k max. = 18500	Pure Per. k max. = 30000
1	Per. TF, Pyr. f, PD k max. = 9500	Pure Per. k max. = 32000	Pure Per. k max. = 29500
1.5	Per. TF, Pyr. f, PD k max. = 9000	Pure Per. k max. = 32000	Per. TF, C f k max. = 27000
2	Pure Per. k max. = 34000	Pure Per. k max. = 26500	Per. TF, C f k max. = 25000
2.5	Per. TF, C f k max. = 30000	Per. TF, C f k max. = 23000	Per. TF, C f k max. = 22500
3	Per. f, C TF k max. = 22000	Per. f, C TF k max. = 23000	Per. f, C TF k max. = 23000
4	Per. f, C TF k max. = 16000	Per. F, C TF k max. = 19000	Per. f; C TF k max. = 20500
5	Per. f, C TF		

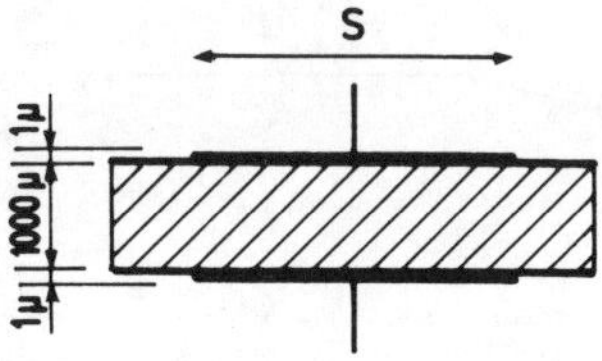

$$C = kS/e = A\varepsilon/e \sim A\varepsilon/e_1$$

$$C_1 = A\varepsilon_1/e_1 \qquad C_2 = A\varepsilon_2/e_2$$

$$1/C = 1/C_1 + 1/C_2 \qquad C = C_1C_2/(C_2 + 2C_1)$$

$$\varepsilon = \frac{\varepsilon_1 \varepsilon_2}{e_2(\varepsilon_2/e_2 + 2\varepsilon_1/e_1)}$$

$$\varepsilon_1 = 35\,000 \qquad \varepsilon_2 = 50$$

$$e_1 = 1000\ \mu \qquad e_2 = 1\ \mu$$

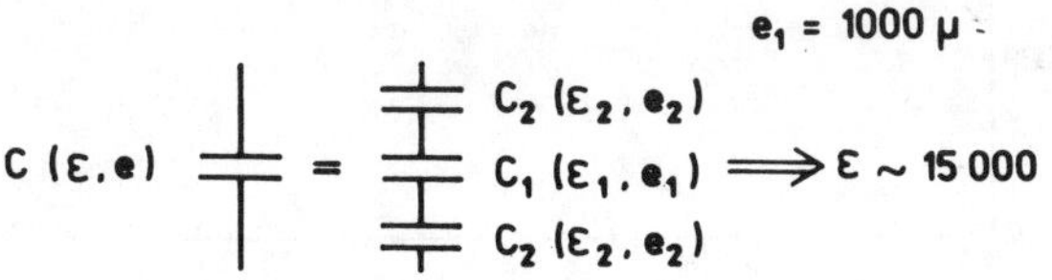

Figure 3. Schedule of a disc with a second phase on surface : principle of capacitance reduction.

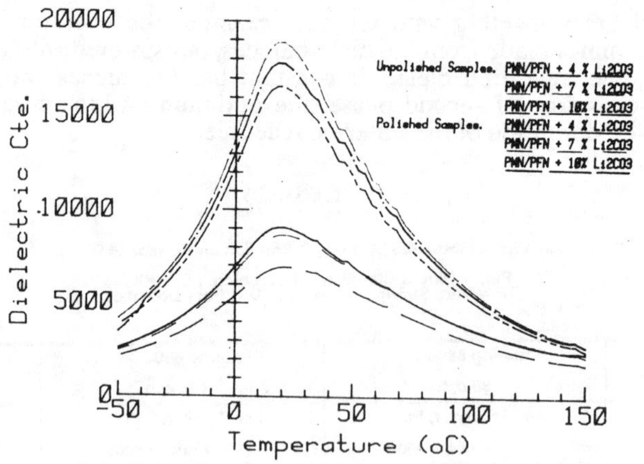

Figure 4. Dielectric characteristic vs temperature of unpolished and polished discs owing a second phase layered on surface.

We have not been able up to now to determine the nature of this second phase unambiguously : its X ray diffraction pattern never corresponds strictly to a known phase. E.D.S. analysis detects only lead. As we notice also the presence of this phase on other perovskites than the PMN/PFN one, we are sure it contains no niobium, iron or magnesium. On the other side, we cannot exclude the presence of lithium. In fact, the X-ray diffraction pattern is not far from the Massicot PbO one, but does not own all the peaks of this material. Both the SEM observation of a second phase at the grain boundaries before phase C appears on surface (see Fig. 2b/), and the fact that the EDS analysis shows that it is lead rich confirm the mechanism leading to a weaker dielectric constant : sintering at a too high temperature, or using too much lithium salt as a sintering aid has as a consequence the diffusion of PbO from grain boundaries where they can be seen by S.E.M. observation up to the surface of discs. That can also allow to imagine an explanation to the fact that some authors use perovskites with an excess of B site cations. In fact, we think that all relaxor materials own dielectric constants as high as 30000 or 35000, but that such values can be obtained only by a perfect control of the material synthesis, and then of its sintering, leading to the control of the localization of the different phases.

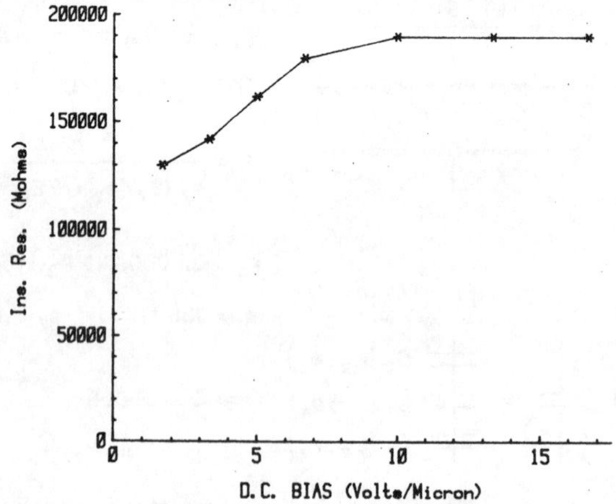

Figure 5. M.L.C. Insulation Resistance versus D.C. bias.

BEHAVIOUR OF THE RELAXORS UNDER A D.C. BIAS

A capacitor is always used with a D.C. or A.C. bias, and is specified for that. The value of the bias it is specified for corresponds to an electric field in the dielectric material, that can be for example near 3 volts/micrometer with 20 μm thick dielectric layers. We have characterized multilayer capacitors we have made with our PMN/PFN material, measuring their capacitance and insulation resistance. MLC had 30 μm thick layers, electrodes being an AgPd alloy with 5 % Pd.

Fig. 5 shows that the insulation resistance begins to increase, and stabilizes at 2.10^{11} Ω up to 20 volts/μm, value corresponding to the break down voltage of the capacitor. This behaviour is satisfying.

Fig. 6 shows the spectacular behaviour of the capacitance value, that falls down dramatically with the electric field. This phenomenum is partially reversible : after a 6 volts/μm polarization at room temperature, a 10 mn relaxation time allows to recover 70 % of the intial value. Heating the capacitor 10 mn at 120°C, temperature where the material owns a paraelectric structure, leads only to 88 % of the initial value.

In order to be sure to avoid any interface effect between the dielectric material and the electrodes, we studied this particular behaviour on discs metallized with the In-Ga eutectic. Discs were 200 μm thick, so we have been obliged to limit the D.C. bias field to 1 volt/μm. Figure 7 a/ shows the ε = f (temperature) characteristic, and figure 7 b/, the relative dependance of capacitance versus, temperature. We see again that dielectric constant (curve 2/) measured with a 1 V/μm D.C. bias is half of the one measured initially (curve 1/). Heating the disc 60 hours at 160°C allows to obtain the curve 3/ close to the intial one, a new measurement with the D.C. bias gives the curve 4/ even lower to the curve 2/. Curves showing the relative variation of capacitance versus temperature show two behaviours : the one corresponding to the initial state or after re-heating the material (curves 1/ and 3/), and the one with a D.C. bias (curves 2/ and 4/). We observe in this last case both the displacement of maximum toward high temperatures, and a slight flattening of the curve. Fig. 7 c/ shows the behaviors of losses : they are smaller with the D.C. bias, particularely in the ferroelectric phase, with a maximum that shifts towards lower temperatures. This is in opposition to the fact that the maxima of dielectric constant corresponding curves shift towards high temperatures !

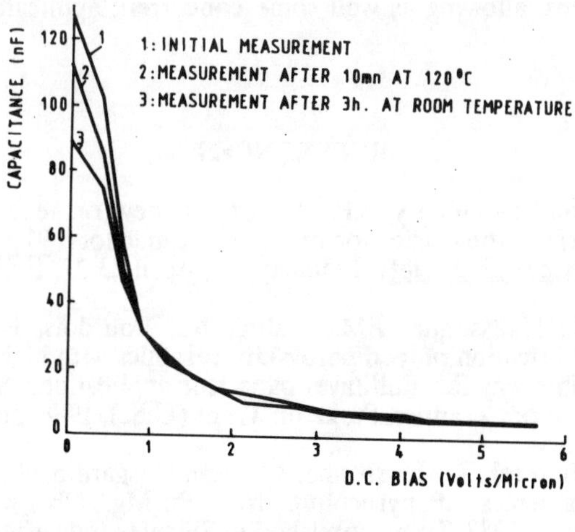

Figure 6. M.L.C. Capacitance versus D.C. bias.

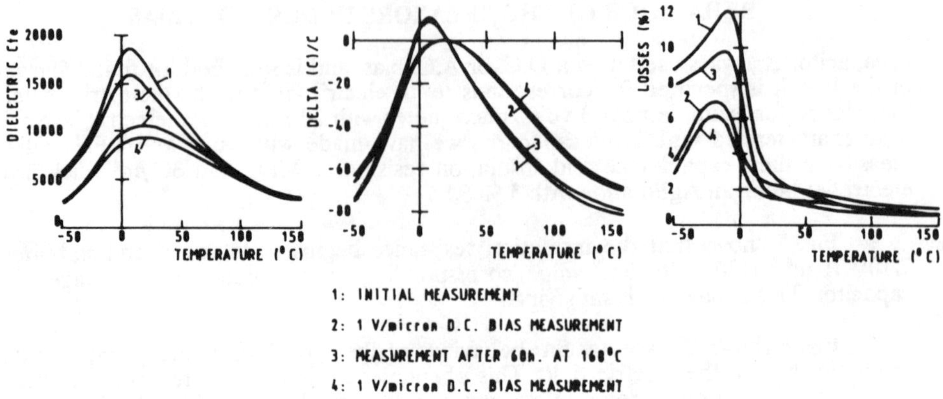

Figure 7. Dielectric constant, capacitance change and losses of discs versus temperature : effect of a D.C. bias.

This particular behaviour seems in contradiction to the one reported by Shrout in his revue paper [1], who described for such D.D. bias a much better compartment : a k = 6500 material had for that bias an increased value, a k = 12000 material only a slight decrease, close to 10 %. Our particular material is a "very high k" material, certainly very different to these previous authors'ones. But, in fact, we have today no explanation to propose.

CONCLUSIONS

Although much work has been done up to now allowing some knowledge concerning the synthesis and the sintering of the relaxor materials, some work remains to be able to control perfectly their technology. Moreover, the electric and dielectric behaviours of the components that are made with them cannot be strictly compared with the one of classical $BaTiO_3$ based components. They have to be understood as new components with new specifications, allowing as well some concurrent applications, and also some new ones.

REFERENCES

1. Shrout, T.R. and Dougherty, J.P., A world review on lead based $Pb(B_1B_2)O_3$ relaxors vs $BaTiO_3$ dielectrics for multilayer capacitors. 91st Annual Meeting of the American Ceramic Society, Indianapolis, April 23-27, 1989

2. Desgardin, G., Haussonne, J.M., Halmi, M., Bourdois, P., Le Cun, M. and Raveau, B., Fabrication of lead perovskite ceramics with high dielectric constants in a reproducible way for multilayer capacitor applications. Science of Ceramics 14, The Institute of Ceramics, Stoke on Trent (U.K.), 1988, pp. 1037-42

3. Bourdois, P., Boterel, F., Haussonne, J.M. and Desgardin, G., Control of the synthesis parameters of pyrochlore free $Pb(Mg_{1/3}Nb_{2/3})O_3$. Electroceramics II, Bruxelles, Sept. 1988. To be published in Silicates Industriels.

IONIC AND ELECTRONIC PROCESSES IN TITANATE CERAMICS UNDER DC VOLTAGE POLARIZATION

RAINER WASER
Electronic Ceramics Group
Philips Research Laboratory Aachen
D-5100 Aachen, F. R. G.

ABSTRACT

Acceptor-doped titanates annealed in oxidizing atmospheres are presented as model materials for the description of various dc voltage-induced polarization and transport processes in dielectrics such as ferroelectric domain wall polarization, dipole orientation, Maxwell-Wagner relaxation, electrode interface polarization, and concentration polarization.

INTRODUCTION

Ceramics based on perovskite-type titanates are of considerable economic importance as dielectric materials for capacitor applications. An understanding of the voltage-induced processes governing these materials is essential for the further development and tailoring of ceramic capacitors. In the present paper, a survey of polarization processes is given which are relevant in the frequency range up to the MHz region:

* ferroelectric domain wall polarization
* dipole orientation
* Maxwell-Wagner relaxation of grains and grain boundaries
* electrode interface polarization
* concentration polarization

The impact of some of these processes on material properties such as the insulation resistance is discussed.
Various experimental methods such as the ac impedance spectroscopy, the potentiostatic step technique (i.e. stimulation by a voltage step), and the galvanostatic step technique (i.e. stimulation by a current step) are commonly employed to investigate polarization and conduction processes in dielectric materials. The present survey emphasizes the potentiostatic step method since it allows to correlate the observations to data obtained by IEC procedures for measuring the insulation resistance and the life stability of multilayer ceramic capacitors (MLCs).

FERROELECTRIC POLARIZATION

After applying a potentiostatic step to a ferroelectric material such as $BaTiO_3$ ceramics, a hyperbolic decay of the polarization and depolarization currents

$$I \propto t^{-\alpha}, \text{ with } \alpha \cong 1$$

is observed over many orders magnitude (Fig. 1). This behaviour which was already reported in Refs. [1, 2] presumably can be attributed to the

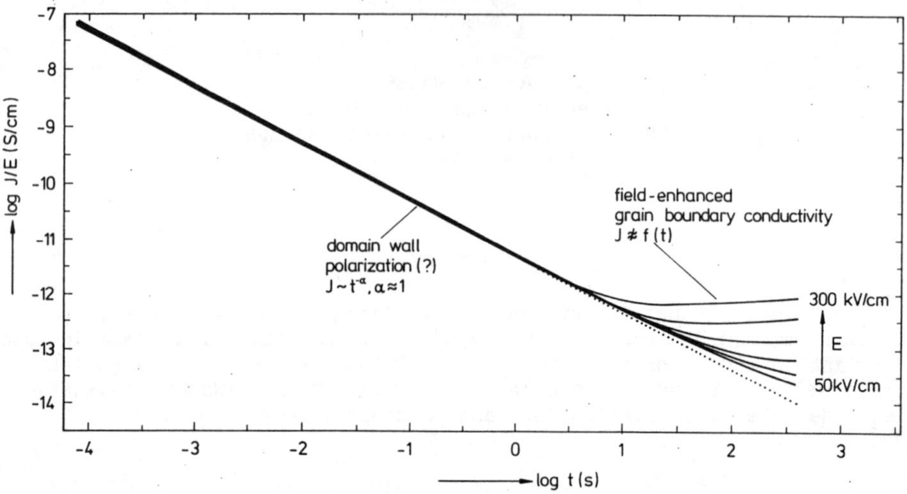

Fig. 1: Current density - field ratio J/E vs. time t at T = 293 K for MLC's based on fine-grained Nb/Co-doped $BaTiO_3$ ceramics.
(———) polarization at various dc field strengths E.
(··········) depolarization at E = 50 kV/cm.

domain wall motion. It does not show a specific relaxation frequency but instead a distribution of the relaxation over a very broad frequency range. This may be caused by a scatter of the activation energy as derived for polarization processes in amorphous insulators such as Al_2O_3 and polyimide which, phenomenologically, show very similar time dependencies [3, 4].
The polarization effects of ferroelectric titanates are studied since their discovery approx. 50 years ago. Recently, the time evolution of these processes attracted interest again due to their importance for the switching times of ferroelectric random access memories (FRAM).

DIPOLE ORIENTATION

The defect structure of electrically insulating titanates is dominated by the almost complete compensation of acceptors and donors. In the case of substitutionally accommodated lower-valent metal ions being the acceptor, oxygen vacancies, $V_O^{\cdot\cdot}$, and - to a much lesser extent - protons, $H_i^{\cdot}$, act as compensating donors. Since these donors are sufficiently mobile at moderate temperatures, there is a trend towards the formation of defect associates due to the Coulomb interaction. EPR and optical spectroscopy

studies on transition metal doped $SrTiO_3$ single crystals revealed various defect associates between the acceptor metal ion and $V_O^{\cdot\cdot}$ (for a review by K. A. Müller see Ref.[5]). IR spectroscopic investigations of Fe-doped $SrTiO_3$ annealed in moistened atmospheres indicated a variety of associate types which involve the proton as the donor part [6].
However, no unambiguous investigation on the dielectric relaxation of defect associates in perovskite-type titanates is reported up to date.

MAXWELL-WAGNER RELAXATION

Applying dc voltages to an inhomogeneously conducting material causes a Maxwell-Wagner polarization effect. The process is frequently observed in ceramic ion conductors (e.g. ZrO_2 [7]) due to the fact that the resistivity across grain boundaries is significantly higher than the resistivity of the bulk of the grains. In alkaline earth titanate ceramics, the Maxwell-Wagner relaxation has been investigated by several research groups using different techniques, e.g. the galvanostatic step [8], the potentiostatic step [9], and the ac impedance analysis [10].

As an example, Fig. 2 shows the time evolution of the current density - field ratio J/E after applying potentiostatic steps of different field strengths E to an acceptor-doped $SrTiO_3$ ceramics. At short time, the J/E characteristics is determined by a field-independent resistive term, R_{grain}, which is attributed to the bulk resistance of the material as confirmed by measurements of single crystals. The material exhibits a Maxwell-Wagner relaxation due to the capacitance of the grain boundaries, C_{gb}. From the relaxation time of approx. 2 s an 'electronical width' of the grain boundaries of approx. 200 nm can be estimated assuming the same dielectric constant in the grain boundary region as in the bulk. At long times, the J/E curves show another resistive term, R_{gb}, which presumably

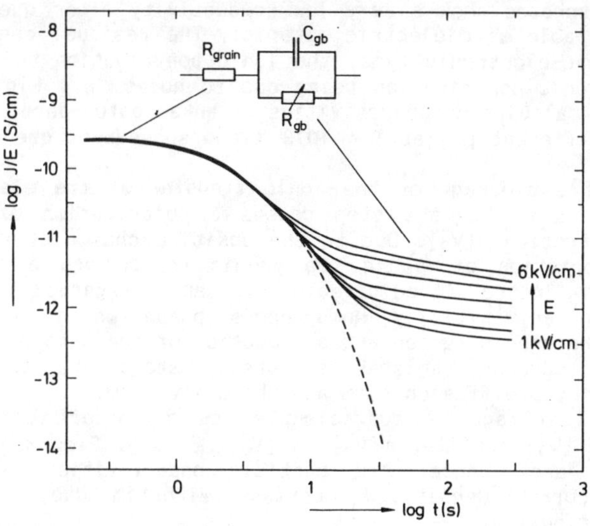

Fig. 2: Current density - field ratio J/E vs. time t during polarization at T = 423 K for 0.1 mol% Ni-doped $SrTiO_3$ ceramics annealed in oxygen. Average grain size: 4 µm.
(------) separated Debye relaxation of R_{grain} and C_{gb}.

can be attributed to the resistance across the grain boundaries. Obviously, R_{gb} is non-ohmic and decreases with increasing fields E.

ELECTRODE INTERFACE POLARIZATION

In general, at any solid electrolyte, the cathodic and/or anodic metal electrode interfaces may block the transfer of electronic or ionic charge carriers. The blockage of electronic carriers is well known for doped Si semiconductors and for donor-doped, semiconducting titanates which - depending on the metal - may form depletion layers, i.e. Schottky contacts. In insulating dielectrics with very low electron and hole concentrations, e.g. acceptor-doped titanates, usually no transfer blockage for electronic carriers is observed.

Mobile ionic defects, on the other hand, usually are (at least partially) blocked at electrode interfaces. Studies on Y-doped ZrO_2 single crystals [7] may serve as a typical example for an oxygen ion conductor. Results reported for single crystalline titanates [11] can be interpreted in the same manner on the condition that the defect structure and defect mobilities are taken into account.

The electrode interface polarization represents an essential precondition for the concentration polarization described below.

CONCENTRATION POLARIZATION

The defect structure of alkaline earth titanates with respect to its dependence on the oxygen partial pressure, pO_2, and on the doping by heterovalent metal ions has been studied intensively in the past two decades, e.g. see Refs. [12-16]. Acceptor-doped titanates annealed in oxidizing atmospheres show a very low conductivity after quenching which makes them suitable as dielectric material. The residual conductivity is of a mixed ionic-electronic type, the ionic contribution being due to $V_O^{\cdot\cdot}$ and the electronic contribution being due to holes, $h^{\cdot}$. Fig. 4 gives an example of the calculated conductivities of an acceptor-doped $SrTiO_3$ after annealing at different pO_2 at T = 1073 K and subsequent quenching to T = 500 K.

In the case of a blockage of the ionic transfer at the electrodes, the application of a dc voltage step causes a polarization of the oxygen vacancy concentration, $[V_O^{\cdot\cdot}]$. Due to the positive charge of $V_O^{\cdot\cdot}$, the field leads to a depletion of $V_O^{\cdot\cdot}$ in the anodic region and a corresponding enrichment of $V_O^{\cdot\cdot}$ in the cathodic region. It can be regarded as a de-mixing process within an initially homogeneous phase which results in an oxidation of the anodic region and a reduction of the cathodic region. The ionic partial current vanishes at steady state due to the mutual compensation of the diffusion term and the drift term.

If the applied voltage is sufficiently low in order to induce small changes in the $[V_O^{\cdot\cdot}]$ profile, $\Delta[V_O^{\cdot\cdot}] << [V_O^{\cdot\cdot}]_{t=0}$ (e.g. Fig. 3a), the defect concentrations and, hence, the partial conductivities remain almost constant. The current density, J, decreases with time due to the decay of the ionic contribution.

At higher voltage stress, $[V_O^{\cdot\cdot}]$ changes by orders of magnitude (e.g. Fig. 3b). This has a tremendous impact on the concentrations of the other defects due the interaction via the local defect equilibria. It leads to an increase of the p-type conductivity in the anodic region and formation

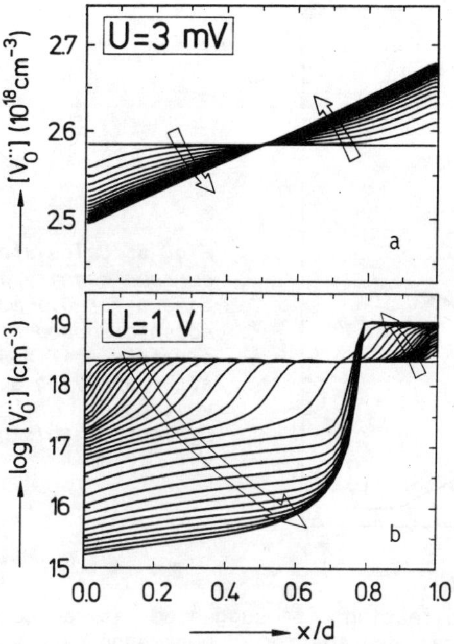

Fig. 3: Calculated time evolution of the oxygen vacancy profiles after applying a dc voltage U to 0.1 mol% acceptor-doped $SrTiO_3$ crystals (thickness d = 0.1 cm) at T = 500 K.

(a) low voltage stress, U = 3 mV.
(b) high voltage stress, U = 1 V.

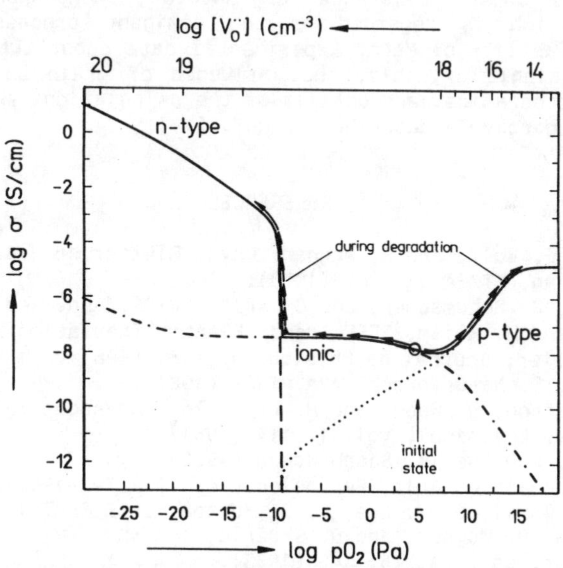

Fig. 4: Conductivity σ for 0.1 mol% acceptor-doped $SrTiO_3$ after annealing at T = 1073 K and subsequent quenching to T = 500 K calculated as a function of the oxygen partial pressure pO_2 during annealing.
Included: an illustration of the effect of a high dc voltage stress.

of a n-type conductivity in the cathodic region as illustrated in Fig. 4 (the upper scale shows the [$V_O^{··}$] which corresponds to the quenched defect equilibria). Therefore, despite the decay of the ionic conductivity, the total current density increases with time (Fig. 5) due to the pronounced rise in the electronic carrier concentrations. The formation of a

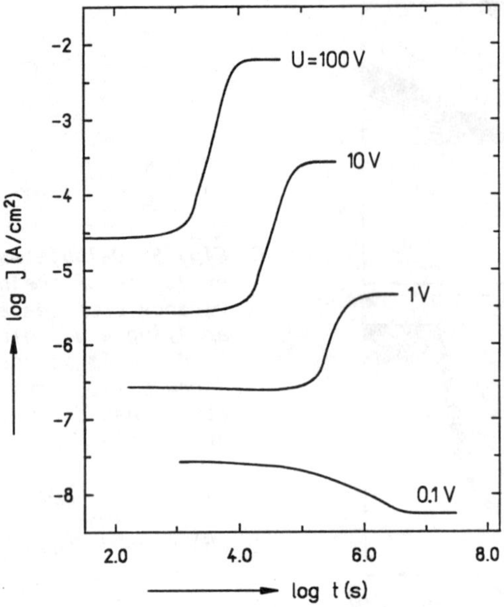

Fig. 5: Calculated current density J vs. time t for 0.1 mol% acceptor-doped $SrTiO_3$ crystals at T = 500 K after applying a dc voltage U. Dielectric thickness: d = 0.1 cm.

pn-junction biased in forward direction is suggested as a consistent explanation of the resistance degradation phenomenon of titanate dielectrics which is regarded as the dominant process limiting the long-term reliability of MLCs. Experimental data about the degradation of doped $SrTiO_3$ single crystals, the influence of grain boundaries on the degradation of ceramics, and details of the calculations sketched in Figs. 3-5 are comprehensively described in Ref. [17].

REFERENCES

[1] B. Wild, F. Wolf, and R. Wagner, Phys. Blätter 16, 164 (1960).
[2] A. Hersping, ETZ-A 91, 265 (1970).
[3] H. Kliem, B. Schumacher, and G. Arlt, CEIDP, Ann. Rep., 168 (1986).
[4] G. Arlt and H. Kliem, IEEE Trans. Electr. Insulation 24, 185 (1989).
[5] K. A. Müller, Journal de Physique 42, 551 (1981).
[6] R. Waser, Z. Naturforsch. 42a, 1357 (1987).
[7] T. G. Tratton, D. Reed, and H. L. Tuller, Advances in Ceramics, (ed. L. M. Levinson), Vol. 1, 114 (1981).
[8] J. Maier, PhD thesis, Saarbrücken (1982).
[9] H. Neumann and G. Arlt, Ferroelectrics 69, 179 (1986).
[10 S. S. Villamil, H. Y. Lee, L. C. Burton, IEEE CHMT-12, 482 (1987).
[11] R. Stumpe, D. Wagner, and D. Bäuerle, phys. stat. sol. (a) 75, 143 (1983).
[12] N.-H. Chan and D. M. Smyth, J. Electrochem. Soc. 123, 1584 (1976).
[13] J. Daniels, K. H. Härdtl, D. Hennings, and R. Wernicke, Philips Res. Repts. 31, 489 (1976).
[14] N.-H. Chan, R. K. Sharma, and D. M. Smyth, J. Am. Ceram. Soc. 64, 556 (1981).
[15] N.-H. Chan, R. K. Sharma, and D. M. Smyth, J. Electrochem. Soc. 128, 1762 (1981).
[16] H.-J. Hagemann and D. Hennings, J. Am. Ceram. Soc. 64, 590 (1981).
[17] R. Waser, T. Baiatu, and K. H. Härdtl, submitted to J. Am. Ceram. Soc.

DIELECTRICS WITH CONTROLLED TEMPERATURE COEFFICIENT OF THE DIELECTRIC CONSTANT FOR CERAMIC MULTILAYER CAPACITORS WITH BASE METAL ELECTRODES

W.Noorlander
Philips Components, Bredeweg 10, 6042 GG Roermond, the Netherlands

ABSTRACT

$SrZrO_3$-$SrTiO_3$ forms a continuous series of solid solutions with a structural change from orthorombic for $SrZrO_3$ to cubic for $SrTiO_3$. The dielectric constant increases from 35 for $SrZrO_3$ to 300 for $SrTiO_3$ and its temperature coefficient decreases from +150 to -3000ppm/^oC
MnO proves to work very well as a sintering aid.
The influence of firing in an oxidizing and a reducing atmosphere on the dielectric properties has been investigated. When fired in air the TC_ϵ proved to be more negative. In the case of Ni inner electrodes there is a more pronounced increase of the equivalent series resistance.

INTRODUCTION

The mechanization of the production of ceramic multilayer capacitors makes the material cost of increasing importance. One of the possibilities to overcome this is the replacement of the noble metals in the inner electrodes and end-terminations by co-fired base metals. For that purpose special ceramic dielectrics have to be developed due to the fact that base metals have to be fired in a non-oxidizing atmosphere. The investigations described in this paper are restricted to temperature stable dielectrics with a controlled temperature coefficient of the dielectric constant.(TC_ϵ)

MATERIALS AND METHODS

Amongst the known dielectrics for class one capacitors the zirconates[1,2,3,4,5] were expected to be the most stable with respect to the influence of the firing atmosphere on the dielectric properties.

Fig.1 shows the results of the crystallographic investigations of Iskhakov and Keler [1]. The phase diagram has been determined at 1600°C in air. The most important part of their work with respect to dielectric materials is the line $SrZrO_3$-$SrTiO_3$, a continuous series of solid solutions with a gradual change in structure from orthorhombic for $SrZrO_3$ to cubic for $SrTiO_3$. The dielectric constant increases from 30 for $SrZrO_3$ to 300 for $SrTiO_3$, the TC_ϵ decreases from P100 to N3000 ppm/°C.

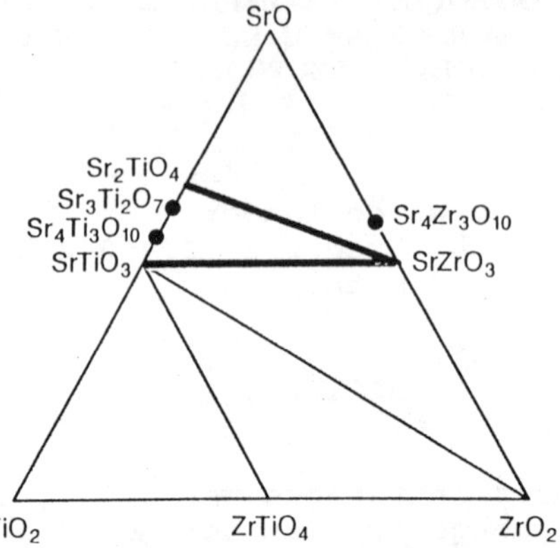

Figure 1. Phase relations in the system SrO-ZrO_2-TiO_2.

The comparison of the physical properties of metals to be used as electrode material learns that nickel is an acceptable material with respect to resistivity, coefficient of thermal expansion and melting point. The influence of the magnetic properties will be discussed later.
The first aim of the investigation was to lower the sintering temperature to a value far below the melting temperature of Ni (1455°C). As was shown by Hagemann and Hünten [4] manganese oxide works very well as a sintering aid for alkaline earth zirconates. The introduction of MnO as a flux and the use of Ni as electrode material restricts the temperature-oxygen partial pressure region in which the samples have to be sintered. The oxygen partial pressure for this investigation was realised by saturating a mixture of N_2 and 6% H_2 with water vapour at 25°C.
To evaluate the influence of MnO on the dielectric properties a series of samples (Multilayer capacitors with 5 active layers, Ni-electrodes and cofired Ni endterminations) was made with 4.5 mole % $SrTiO_3$ in $SrZrO_3$, the amount of MnO varying between 1.5 and 6.5 mole %. They have been fired at 1375°C, 3 hrs soak. The dielectric properties are summarized in table 1. There is hardly any influence of the manganese content on the dielectric properties with exception of the insulation resistance.
In fig.2 the dielectric properties as a function of the $SrTiO_3$ content are compared to the results of Iskhakov and Keler.

TABLE 1
The dielectric properties of $SrZr_{0.995}Ti_{0.045}O_3$ + X% MnO

X (mol %)	ϵ_r	TC_ϵ (ppm/°C)	RC (sec)
1.5	31	+23	17800
2.5	33	+24	15700
3.5	31	+23	13000
4.5	32	+23	13000
6.5	34	+24	2800

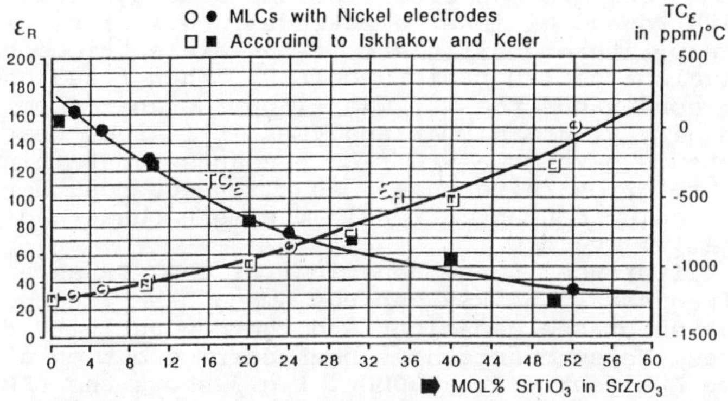

Figure 2. ϵ_r and TC_ϵ at 1 MHz.

A very good agreement exists between the measurements on discs with 3 to 4% by weight $SrO.B_2O_3$ fired in air at 1450°C and on multilayer capacitors with manganese oxide as a flux and fired in a reducing atmosphere at 1350°C. Life test results confirm the expectation that alkaline earth zirconates are stable when fired under reducing conditions.[6]

SINTERING AND POWDER MORPHOLOGY

SEM EDAX analyses show a SrO rich manganese containing second phase and an almost complete wetting of the grains. It is concluded that sintering is accomplished by a reactive liquid phase. This means the powder morphology, the amount of flux and the sintering temperature are interrelated. This is illustrated by the next series of experiments. The sintering temperature range has been determined for samples with the NPO composition and 2.5 mole % MnO as a function of the specific surface area (SSA). The experiments have been carried out in a gradient kiln. The variation in SSA was realised by varying the milling time. The temperature at which the shrinkage is constant (the onset temperature of the sintering temperature range) increases from 1200°C with 2.6 m^2/g to 1360°C with 5.4 m^2/g. To the sample with a SSA of 3.6 m^2/g an increasing amount of flux has been added. With 4.15 mol % MnO the onset temperature of the sintering range decreased from 1320°C to 1200°C.
Firing in air of the NPO composition with 2.5 mol % MnO and a

SSA of 2.6 m^2/g leads to even higher densities (99.5% i.s.o 97.5%) but the onset temperature of the sintering range increases to 1300°C. The colour changes from manganese double plus pink to Mn_3O_4 brown.

MULTYLAYER CAPACITORS WITH Pd AND Ni ELECTRODES COMPARED.

Because the dielectric properties proved to be almost equal when fired in an oxidizing or non oxidizing atmosphere it was possible to compare capacitors of the same construction and value with Pd and Ni inner electrodes. Deviations were found in the temperature coefficient of the dielectric constant and the equivalent series resistance(ESR). When fired under reducing conditions the TC_ϵ was +16ppm/°C and -6ppm/°C when fired in air. With SEM EDAX analyses it is seen that in that case there is next to a SrO rich manganese containing second phase a ZrO_2 rich third phase. As a consequence there is a higher titanium concentration in the main phase and thus a more negative TC_ϵ.
Another difference is the behaviour of the ESR as a function of the frequency. The ESR depends among other things on the construction of the capacitor and the conductivity of the electrodes. Measurements have been carried out on a HP 4191A Impedance Analyser up to 500MHz. Fig.3 shows the ESR as a function of the frequency.

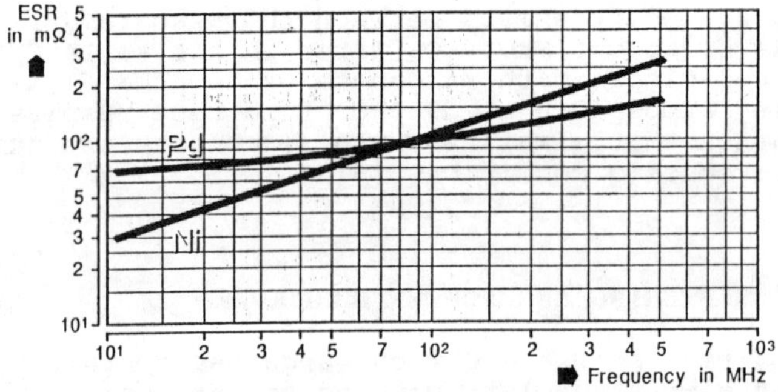

Figure 3. ESR as a function of the frequency, capacitance 120pF, dimensions 1206 with both Ni and Pd innerelectrodes.

The more pronounced increase in the case of Ni is evident, the ESR increases with the square root of the frequency. A possible explanation of the difference in behaviour is the influence of the magnetic properties of Ni on its conductivity,the skin effect. The skin depth δ of a metal plate with infinite dimensions is given by:

$$\delta=\sqrt{\frac{2}{2\pi f\mu_o\mu\sigma}}$$

In which f = the frequency
μ_o = the permeability constant

μ = the magnetic permeability
σ = the electrical conductivity

In the case of Ni the skinn depth at 80 MHz is 4μm and at 1000MHz 2μm and approaches the the electrode thickness. With Pd these thicknesses are resp. 20 and 5 μm.
This might give an explanation for the difference in behaviour between Ni and Pd in multilayer ceramic capacitors at frequencies up to one GHz.

CONCLUSIONS

The system $SrZrO_3$-$SrTiO_3$ with an addition of manganese oxide proved to be suitable as a dielectric material for multilayer capacitors with both nickel and palladium electrodes.
With a proper adaption of the flux content to the powder morphology the material can be sintered in a reducing as well as in an oxidizing atmosphere at temperatures between 1200 and 1350°C.
Dielectrics can be prepared with a temperature coefficient from P100 with a dielectric constant of 30 to a TC_ϵ of at least N1000 with a ϵ_r of 100 with excellent insulating properties.
The increase of the equivalent series resistance as a function of the frequency in the case of Ni is more pronounced but well within the customers requirements.

LITERATURE

1. Iskakov, Kh.Sh. and Keler, E.K., Solid state reactions in the SrO-TiO_2-ZrO_2 system. Russian Journal of Inorganic Chemistry, 1962, 8, 1005-10.

2. Iskakov, Kh.Sh. and Keler, E.K., Conditions of formation and electrical properties of $SrTiO_3$-$SrZrO_3$ solid solutions. Russian Journal of Inorganic Chemistry, 1962, 8, 1010-13

3. Kell, R.C., Greenham, A.C. and Olds, G.C.E., High permittivity temperature-stable ceramic dielectrics with low microwave loss. J. Am. Cer. Soc., 1973, 7, 352-54

4. Hagemann, H.J. and Hünten, S., Method of producing a dielectric having a perowskite structure. U.S. patent 4,260,663 April, 7 1981.

5. Hiroshi Kishi, Minoru Oshio, Shunji Murai, Takeshi Wada and Masami Fukui, Low temperature sintered ceramic capacitor with a temperature compensating capability, and method of manufacture. U.S. patent 4,700,264 and 4,700,265.

6. R. Waser, Long term reliability of multilayer ceramic capacitors. Proceedings of the 2nd international conference on passive components: materials, technology, processing. Paris, Nov. 18-20, 1987

MICROSTRUCTURAL DEVELOPMENT AND MICROWAVE DIELECTRIC PROPERTIES OF ZIRCONIUM TITANATE CERAMICS SINTERED WITH Nd_2O_3

F. AZOUGH AND R. FREER
Materials Science Centre,
University of Manchester/UMIST
Grosvenor Street, Manchester M1 7HS, U.K.

ABSTRACT

Zirconium titanate ceramics have been prepared by the oxide route, using Nd_2O_3 as a sintering aid. The products were ~ 95% theoretical density, and almost single phase. The microstructures were sensitive to heat treatment, changing from a disordered structure after rapid cooling, to a more ordered structure with a superlattice after very slow cooling. The dielectric Q value was ~ 700 at 5 GHz.

INTRODUCTION

Zirconium titanate-based ceramics have long been known as temperature stable dielectrics [1]. The structure of $ZrTiO_4$ has been investigated by several workers [2-4] and shown to be based on the α-PbO_2 structure. There is, however, a phase transition ~1120°C between the high temperature and low temperature forms, which is characterised by a change in the c-axis unit cell parameter. In this study, several structural modifications have been formed by varying the cooling rate after sintering. The products have been investigated scanning and transmission electron microscopy.

EXPERIMENTAL

Materials and Preparation

Starting materials were ZrO_2 (Magnesium Electron MEL 20 grade), TiO_2 (Tiona AG) and Nd_2O_3 (R.E. Acton, UK). Equimolar amounts of ZrO_2 and TiO_2 were mixed together and 2 wt % Nd_2O_3 was added (as a densification aid). The mixture was vibro-milled wet for 8 hours (using ZrO_2 balls), calcined at 1100°C for 4 hours, and vibro-milled again for 12 hours.

Pellets were pressed at 120 MPa to sizes of 10 mm diameter and 15 mm high (for high frequency dielectric studies) and 16 mm diameter and 4 mm thick (for microstructural studies). Sintering was performed at 1400°C for 4 hours, with cooling rates which varied from water quench, air quench, 1800°C/hour to 1°C/hour.

Characterisation of Powders and Products

X-Ray diffraction analysis, using CuK_{α} radiation and a Philips diffractometer with horizontal goniometer, was employed to investigate calcined powders and sintered discs. Product morphologies were examined by optical microscopy and scanning electron microscopy (Philips 505 and 525). Transmission electron microscopy (Philips EM430 and 400) was used for lattice image, electron diffraction and crystallographic studies. High frequency dielectric measurements were performed at 5 GHz with a network analyser and sweep oscillator (post resonator method).

RESULTS AND DISCUSSION

With the exception of the pellets sintered at 1400°C and rapidly quenched into water (which cracked and fragmented), all the specimens developed uniform microstructures and high densities (approximately 95% of the theoretical value).

X-ray diffraction analysis confirmed that the water-quenched products were single phase $ZrTiO_4$, whilst samples cooled at slower rates contained traces of ZrO_2 and TiO_2. Indeed, the X-ray peaks for the minor phases increased in intensity as the cooling rate increased. Unit cell dimensions for $ZrTiO_4$ showed little dependence on cooling rate for the a and b directions but the length of the c-axis decreased significantly as the cooling rate decreased (Figure 1).

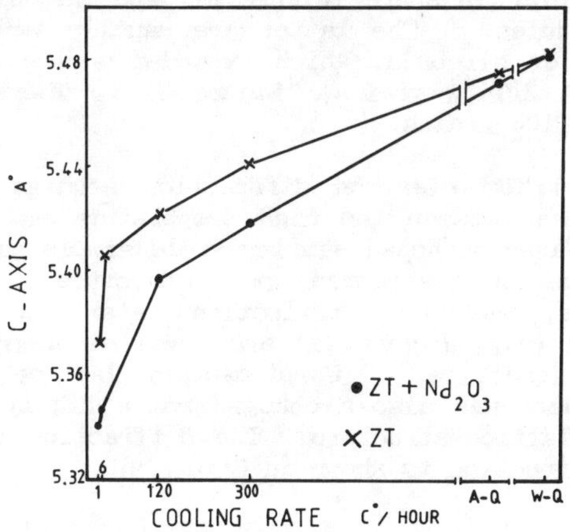

Figure 1. Length of c-axis for $ZrTiO_4$ (sintered at 1400°C) as a function of cooling rate. Data for pure zirconium titanate is included for comparison.

From previous work [5] it is clear that the rapidly quenched $ZrTiO_4$ corresponds to the high-temperature form and the very slowly cooled specimens correspond to the low-temperature form. McHale and Roth [4] observed a similar variation in structural form, and cell parameters, for specimens sintered and equilibrated at temperatures in the range 800-1500°C, and subsequently water quenched.

Figure 2 shows an optical micrograph of a polished and etched specimen which had been subjected to air quenching during fabrication. The grains are typically 5 μm.

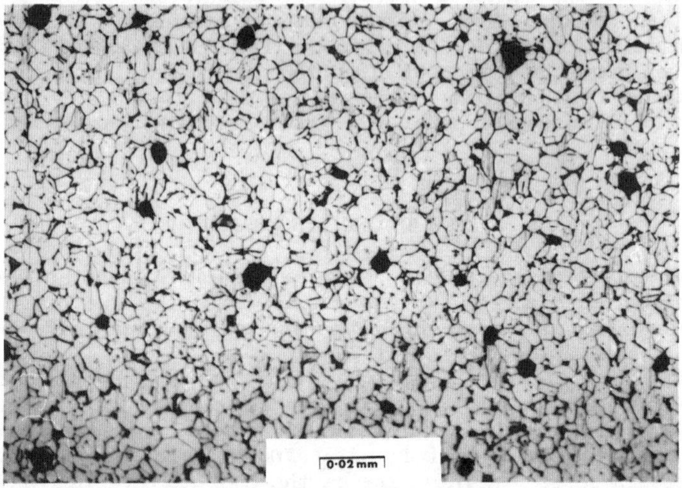

Figure 2. Optical micrograph of air-quenched $ZrTiO_4$ specimen.

Electron microscopy revealed that the slowly cooled $ZrTiO_4$ grains are frequently irregular in shape (Figure 3), and the traces of ZrO_2 are present as small nodules. The latter are usually twinned, suggesting the monoclinic form of zirconia, which expanded during cooling to cause cracking of the host $ZrTiO_4$ grains, (Figure 3). There are no obvious signs of individual TiO_2 grains.

High resolution TEM electron diffraction studies confirmed major structural differences between the high temperature and low temperature forms of $ZrTiO_4$. Water quenched samples yield simple patterns which may be intpreted as a cation disordered, pbcn structure. As the cooling rate becomes slower, satellite reflections start to appear in the diffraction patterns (e.g. Figure 4a) and there is a gradual transition to a more ordered structure. Complementary lattice imaging studies demonstrated that there was also a change from a highly modulated to an incommensurate superlattice structure. The diffraction pattern for this structure in the a direction is shown in Figure 4b.

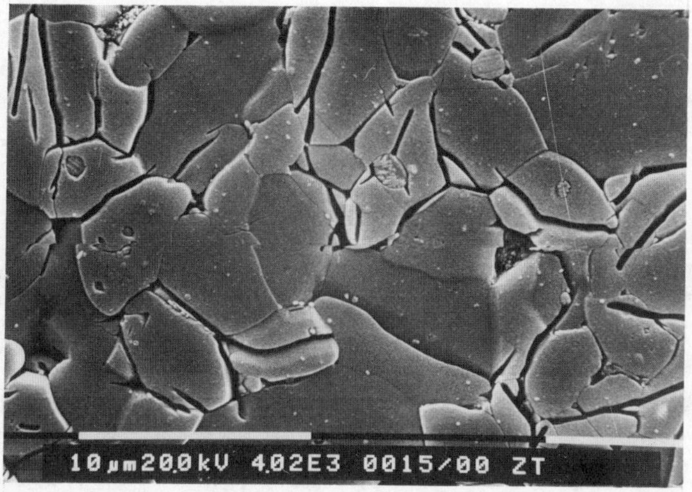

Figure 3. SEM micrograph of $ZrTiO_4$ cooled at 1°C/hour during fabrication.

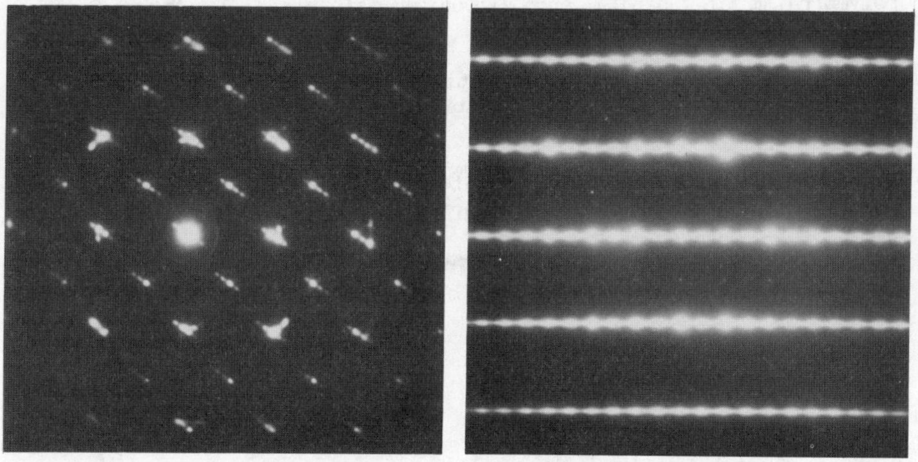

Figure 4. Electron diffraction patterns of $ZrTiO_4$ sintered at 1400°C and cooled by air quenching (a), and cooled at 1°C/hour (b).

At a frequency of 5 GHz, the dielectric constant was found to be 38-43 (depending on cooling rate) and the Q value was approximately 700 for all the samples. The use of Nd_2O_3 as the densification aid is thought to be responsible for the poor loss characteristics as Q values ~ 4000 have been obtained with alternative sintering aids [6].

CONCLUSIONS

The structural state of zirconium titanate depends upon its preparation and heat treatment. Rapid cooling from 1400°C yields a disordered structure. With slower cooling, modulations begin to develop and eventually a superlattice is formed in the a direction.

REFERENCES

1. Rath, W., Keram. Radsch., 1941, 49, 137.

2. Newnham, R.E., Crystal structure of $ZrTiO_4$. J. Amer. Ceram. Soc., 1967, 50, 216.

3. Lynch, R.W. and Morosin, B., Thermal expansion, compressibility and polymorphism in hafnium and zirconium titanates, J. Amer. Ceram. Soc., 1972, 55, 409.

4. McHale, A.E. and Roth, R.S., Investigation of the phase transition in $ZrTiO_4$ and $ZrTiO_4$-SnO_2 solid solutions, Comm. Amer. Ceram. Soc., 1983, C-18.

5. McHale, A.E. and Roth, R.S., Low-temperature phase relationships in the system ZrO_2-TiO_2. J. Amer. Ceram. Soc., 1986, 69, 827.

6. Azough, F. and Freer, R., in preparation.

THE CRYSTALLIZATION OF FINE PLZT POWDERS DERIVED FROM NITRATE SOLUTIONS AT LOW TEMPERATURE

Yoshio Yoshikawa and Kaoru Tsuzuki
College of Engineering, Nihon University, Koriyama, Fukushima 963, Japan

ABSTRACT

Fine, homogeneous lead lanthanum zirconate titanate(PLZT) powder is prepared by hydrolysis of an aqueous nitrate solution with ammonia gas. The powder exhibits high reactivity when calcined. The kinetics of crystallization between 400°C and 600°C are studied isothermally by XRD and thermoanalytically by DTA. The isothermal crystallization kinetics is best fitted to a first-order equation derived for a random nucleation-controlled system on a large number of small particles with an activation energy of 297 KJ/mol. Characterization of the powder by X-ray diffraction line broadening, density, surface area, and SEM observation in the range of 400°C to 500°C show a microstructure consisting of 20-nm subunits agglomerated into 60-nm particles. It seems that a single nucleus generates on the subunit and grows within the subunit. The activation energy for a non-isothermal process by means DTA techniques is found to be in agreement with that obtained from the isothermal process.

INTRODUCTION

For almost 10 years, PLZT electrooptic devices have been used in shutter applications, display and photoferroelectric image storage. Recently, various chemical preparation(CP) methods[1,2] for powder processing have been investigated to obtain a sinterable powder for PLZT ceramics having compositional homogeneity, high purity, fine grain, and low porosity structures. Sintering process using such a sinterable powder would expect (1) a reduction in sintering temperature, (2) better control microstructure and related properties, and (3) sintering methods other than hot-pressing. In one work[3], highly reactive powders were prepared from an inorganic-salt precursor.

Chemically prepared PLZT powder is often finer and, consequently, less convenient than that produced by mixed oxide(MO) method, because on a submicrometer scale, the smaller the particles the greater their tendency to aggregate during calcination. This particle aggregation precludes the dense packing and uniform, small pore size that, in turn, permit sintering to 100% relative density at relatively low temperature. Further research on the kinetics of aggregation and crystallization at low calcination temperatures could lead to major advances in the fabrication of ceramics.

MATERIALS AND METHODS

MATERIALS

The raw materials used to prepare the aqueous nitrate mixture were $Pb(NO_3)_2$, $La(NO_3)_3$, $ZrO(NO_3)_2$, and $TiO(NO_3)_2$. The mixed solution of composition $Pb_{0.92}La_{0.08}(Zr_{0.65}Ti_{0.35})_{0.98}O_3$(PLZT 8/65/35) was stirred in a reaction vessel, cooled to 1°C, and hydrolyzed with ammonia gas. The filtrated oxyhydroxide coprecipitate was then ultrasonically dispersed in distilled water, ultrasonically rinsed with acetone, and dried overnight at 40°C in an air oven.

ISOTHERMAL KINETICS

The as-dried powder was placed in a platinum crucible and isothermally heated in air. Powder X-ray diffraction(XRD) data were obtained for quantitative analysis of the reaction product. The integral intensity of the (110)peak of the pseudo-cubic PLZT phase was determined by scanning at 0.25 °C /min and measuring the area under the curve. The crystallite size was also calculated using the (110)peak from the Sherrer formula,

$$D = 0.9\,\lambda\,/\,\beta\cos(\theta) \qquad (1)$$

where D is the crystallite size, λ the cuKα radiation wavelength, and β corrected half-width.

Thermoanalytical kinetics and powder characterization

The quantitative data necessary to determine kinetic parameters were obtained using differential thermal analysis(DTA) techniques. For non-isothermal experiment, a fixed quantity of 80 mg of dried powder was heated from 400°C to 600°C at rates of 2.5, 5, 10, and 15°C/min. The exothermic peak attributable to the heat of crystallization was recorded in the range ± 10 uV at rate of 10°C/min. The heat of inversion of quartz was used as references for instrument calibration. Agglomerate state and particle size distribution were determined by Scanning electron microscopy (SEM). Powder density was measured using a helium-gas pycnometer, and specific area was determined by a single point BET method.

RESULTS AND DISCUSSION

MICROSTRUCTURE

X-ray diffractometry revealed that the coprecipitated by hydrolysis and the as-dried powder were amorphous. As shown in Fig.1(a), SEM observation revealed that it consisted of spherical agglomerates. After calcination at 500°C, the diameter of the agglomerates was nearly constant at 60 nm(Fig.1(b)). The surface area and density of the as-dried powder, on the other hand, changed from 140 to 35 m^2/g and from 4.2 to 7.9 g/cm^3, respectively. The agglomerates therefore must have a substructure consisting of many smaller oxide units. The equivalent diameter(d_{BET}) of the as-

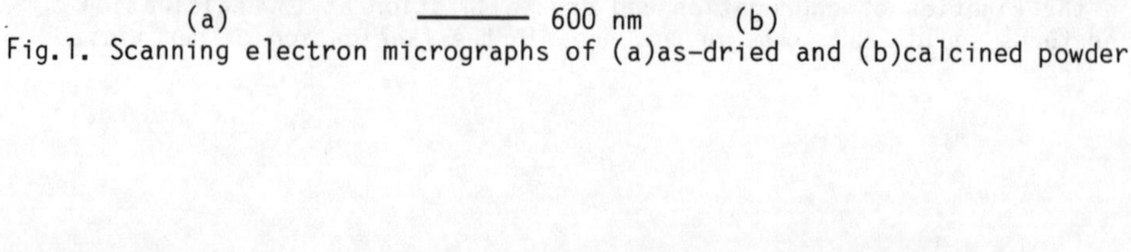

(a) —— 600 nm (b)

Fig.1. Scanning electron micrographs of (a)as-dried and (b)calcined powder.

dried powder, calculated from the surface area(S_{BET}), and the density(ρ), from $d_{BET}=6/(\rho \cdot S_{BET})$, revealed that the agglomerates contained numerous 10-nm-diameter particles, corresponding to the subunits. The values d_{BET} of powders calcined at 400°C and 500°C were also calculated at 16.2 and 21.7 nm, respectively. These calculations indicated an increase in the subunit size with increasing temperature, accompanied by densification and weight loss.

Figure 2 illustrates the TG-DTA curves of as-dried powder. The TG curve shows weight loss below 100°C due to the removal of loosely bonded absorbed water and trapped acetone and dehydration of the complex oxyhydroxide from 100°C to 400°C, resulting in a 13% total weight loss. An XRD analysis of the samples before and after the exothermic peak at 500 °C showed only a pseudo-cubic PLZT crystallization phase. The exotherm must essentially represent crystallization, since little weight change occurred in the relevant temperature range. The heat of crystallization estimated from the DTA peak area is 12.5 KJ/mol.

KINETICS

Figure 3 shows the volume fraction of crystallization, X (measured from the XRD intensity) plotted as a function of time, t, for as-dried powders between 400°C and 500°C. The Avrami transformation kinetic equation[4] was used to evaluate the XRD data. The volume fraction after time is given by

$$X = 1 - \exp(-At^n) \tag{2}$$

where n is a constant dependent on the details of the nucleation and growth mechanisms, and A is related to the rates of nucleation and growth. For isothermal conditions, taking logarithms and rearranging Eq.(2) results in

$$\mathrm{Ln}[-\mathrm{Ln}(1-X)] = \mathrm{Ln}(A) + n\,\mathrm{Ln}(t) \tag{3}$$

where a plot of Ln[-Ln(1-X)] as a function of Ln(t) yields the values of n and A. This plot is illustrated in Fig.4, resulting in a mean constant n value of 1.08. Therefore the first-order equation gives the best fit to the experimental data: A mechanism of random nucleation, with nucleation as the rate-determinig step, therefore may be inferred. It should be noted, however, that Eq.(2) derived by Avrami is based on the three-dimensional growth of randomly distributed nuclei on the large crystals. It is difficult to include the effect of particle size on kinetic behavior in the general Eq.(2). Some aspects of this influence were considered by Mample[5]. When the particles are small enough, effectively, a single

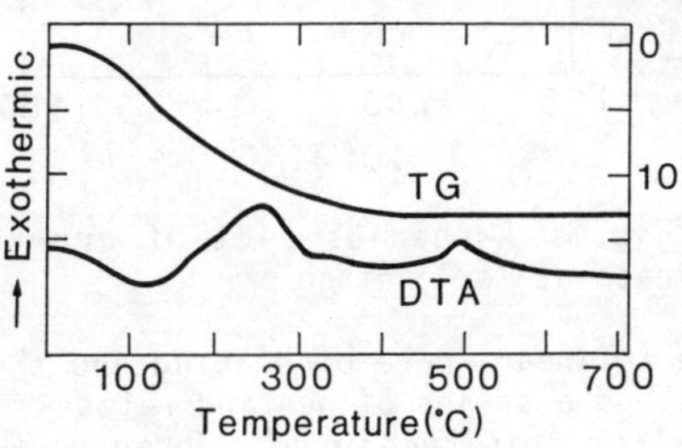

Fig.2. TG-DTA curves of as-dried powder.

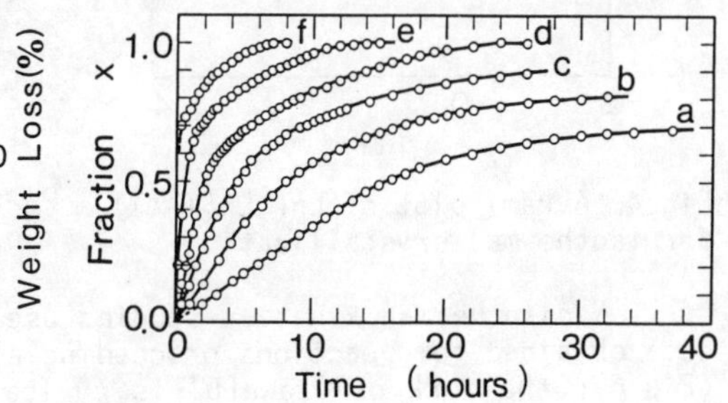

Fig.3. Isothermal crystallization curves at (a)409, (b)417, (c)425, (d)434, (e)450, and (f)483°C.

nucleus is generated on each particle. For this system the rate of crystallization is controlled by nucleation in an assemblage of identical reactant fragments, and the first-order expression is obeyed in random nucleation as follows:

$$dX/dt = k(1-X) \tag{4}$$

where k is a rate constant, and

$$-Ln(1-X) = kt \tag{5}$$

which is a restatement of Eq.(2), with n=1. This kinetic equation is supported by such direct experimental observation as the equivalent diameter(d_{BET}) and the XRD line broadening crystallite size, as described next.

The temperature dependence of k in Eq.(5) apparently demonstrates a simple Arrhenius behavior, i.e.,

$$k = \upsilon \exp(-E/RT) \tag{6}$$

where υ is the frequency factor, E the apparent activation energy of crystallization, and R a gas constant. The plot of the logarithm k values obtained from Eq.(5) vs. the reciprocal of absolute temperature revealed an apparent activation energy of 297 KJ/mol.

The crystallite size, as determined by XRD, increased linearly during the early stages of isothermal heating (0.1<X<0.5) and did not change significantly with further heating (X>0.5). Figure 5 plots the logarithm of growth rate vs the reciprocal of absolute temperature. The slope of the line shows an apparent activation energy of 60 KJ/mol, much smaller than that obtained for nucleation using the first-order equation.

The crystallite size determined by XRD agrees with the equivalent diameter of the primary particle measured from the surface area. Scanning electron microscope observation showed little change in the shapes and sizes of isothermally treated agglomerate particles. A single nucleus in a primary particle therefore apparently develops into a single crystallite.

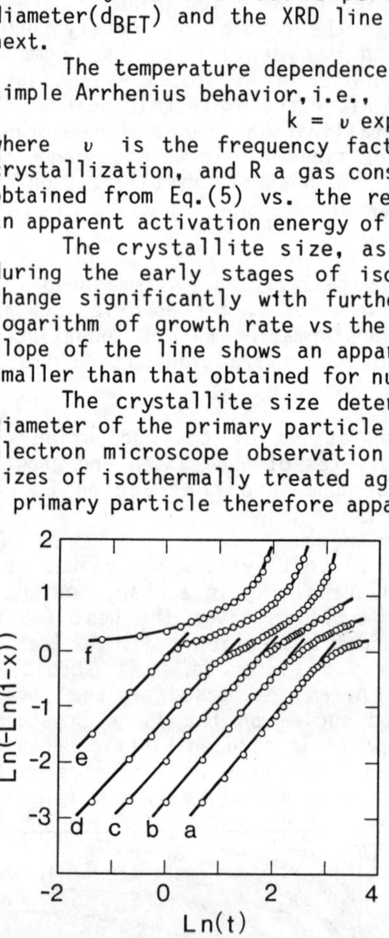

Fig.4. Avrami plot of Ln(-Ln(1-X)) for isothermal crystallization.

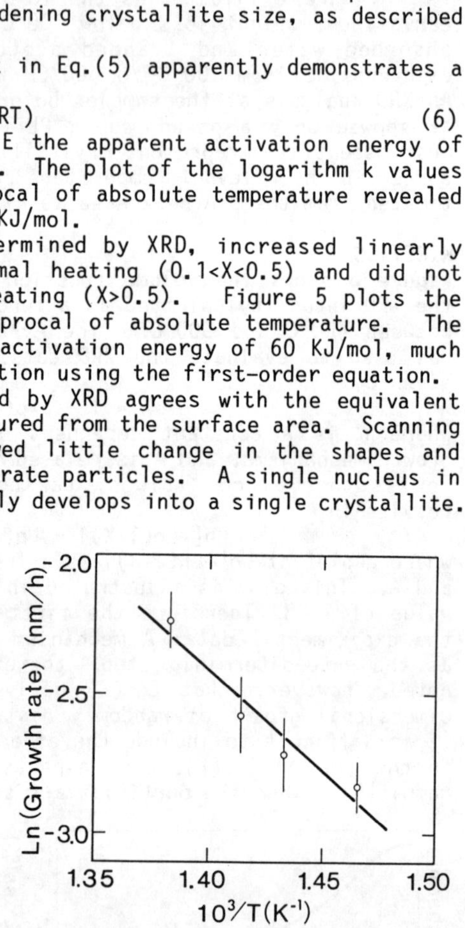

Fig.5. Arrhenius plot of growth rate at early stage.

Nonisothermal kinetics studies use a linear rate of heating and the data obtained for reactions proceeding along a series of heating rates (Fig.6); the work of Ozawa[6] is typical of this technique. Three equations can be written for the first-order model: Eq.(4), Eq.(6), and

$$T = T_0 + B\cdot t \tag{7}$$

where B=(dT/dt) and is the constant heating rate and T_0 is the temperature for the onset of crystallization. An alternative expression is

$$dX/dT = (\nu/B)(1-X)\exp(-E/RT) \quad (8)$$

Integration of the rising-temperature-rate expression can be represented by

$$\int_0^X dX/(1-X) = (\nu/B)\int_{T_0}^T \exp(-E/RT)dT \quad (9)$$

The right side of the expression can be integrated to T if E is known. The value of the integral has been calculated by Doyle[7], who obtained an approximation to the integral as follows:

$$\mathrm{Ln}[\int_{T_0}^T \exp(-E/RT)dT] \doteqdot -5.330 - 1.052(E/RT) \quad (10)$$

Using the approximation in Eq.(9) yields

$$\mathrm{Ln}[-\mathrm{Ln}(1-X)] \doteqdot \mathrm{Ln}(\nu) - \mathrm{Ln}(B) - 1.052(E/RT) + \mathrm{const} \quad (11)$$

A plot of Ln(B) vs $1/T(X=X_i)$ should be linear, where X_i is a specified value of X. Figure 7 illustrates the plot at X_i=0.2, 0.4, and 0.6, resulting in the activation energy of 321, 323 ,and 352 KJ/mol, respectively. The energy does not change from the early to the main stage and also in agreement with that for the isothermal process.

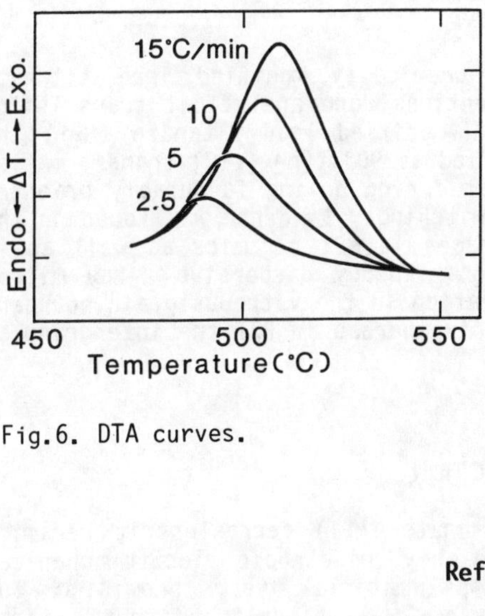

Fig.6. DTA curves.

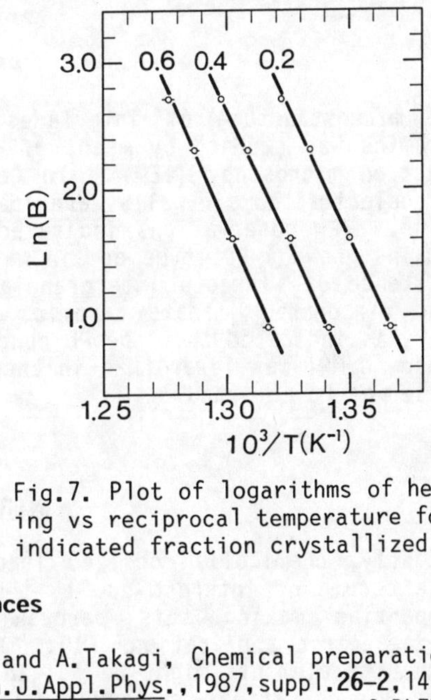

Fig.7. Plot of logarithms of heating vs reciprocal temperature for indicated fraction crystallized.

References

1. Y.Yoshikawa, K.Tsuzuki, T.Kobayashi and A.Takagi, Chemical preparation and Two-Stage Sintering of PLZT. Jpn.J.Appl.Phys., 1987, suppl.**26-2**, 145-8.
2. Y.Yoshikawa, K.Tsuzuki, T.Kobayashi and A.Takagi, Preparation of PLZT Powders from Several Aqueous Solutions. J.Mat.Sci., 1988, **23**, 2729-34.
3. Y.Yoshikawa, K.Tsuzuki, The Effect of Ultrasonic Washing on CP-PLZT Powder Characteristics. Proceedings of the MRS International Meeting on Advanced Materials, 1989, **3**.
4. M.Avrami, Kinetics of phase change. J.Chem.Phys., 1939, **7**, 1103-12, 1940 **8**, 212-24, 1941, **9**, 177-84.
5. W.E.Brown, D.Dollimore and A.K.Galwey, Reactions in the Solid State, in Comprehensive Chemical Kinetics Vol.22, ed. C.H.Bamford and C.F.H.Tipper Elsevier Scientific Publishing Company, 1980, PP. 41-113.
6. T.Ozawa, A New Method of Analyzing Thermogravimetric Data. Bull.Chem. Soc.Japan, 1965, **38**, 1881-6.
7. C.D.Doyle, Kinetic Analysis of Thermogravimetric Data, J.Appl.Poly.Sci., 1961, **5**, 285-92, 1962, **6**, 639-42.

MICROSTRUCTURE OF MODIFIED LEAD TITANATE CERAMICS

HERBERT K SCHMID
Division of Materials Science and Technology, CSIR, P O Box 395,
Pretoria 0001, South Africa

ABSTRACT

The microstructure of interfaces in chemically modified lead titanate ceramics was studied by means of conventional and analytical transmission electron microscopy (TEM). In Ca-Co-W modified lead titanate (MPT) the ferroelectric domain walls were identified as 90° type {110} transformation twins. TEM observations indicated that during poling favourably oriented domains grow by 90° type domain wall switching. Evidence was found on the existence of vitreous intergranular phases in MPT ceramics as well as in lead zirconate titanate ceramics (PZT). Energy dispersive X-ray microanalysis indicated that the Pb concentration in the vitreous grain boundary phase in MPT was lower than in the matrix whereas in PZT the intergranular phase was Pb enriched.

INTRODUCTION

Recently, chemically modified lead titanate (MPT) ferroelectric ceramics have become of interest due to their highly anisotropic electromechanical properties making this perovskite-type material very promising for piezoelectric applications [1]. It is now well established that a large number of elements might be suitable additives for partial substitution of A or B site cations in lead titanate. The largest anisotropy of coupling factors was found in Ca-Co-W modified MPT [1]. Another commercially important piezoelectric material is lead zirconate titanate (PZT) due to its high coupling factor. It is now well known that excess PbO may assist densification of PZT during sintering [2]. The mechanical and piezoelectric properties are influenced by the amount of additives in the composition and by the processing conditions. Unfortunately, limitations occur in these materials due to the presence of residual intergranular phases that result from liquid phase sintering. Such second phases are normally in a vitreous state and form either from sintering aids or from impurities present in the starting materials.

In the present study the microstructure of two piezoelectric ceramic systems were investigated by means of conventional and analytical TEM. Emphasis was put on (i) the microstructural characterization of ferroelectric domain boundaries and grain boundaries, and (ii) the microanalysis of residual intergranular phases in Ca-Co-W doped MPT and in Si doped PZT ceramics.

MATERIALS AND METHODS

Dense Ca-Co-W modified MPT ceramics with nominal composition $Pb_{.7}Ca_{.3}Ti_{.94}(Co_{.5}W_{.5})_{.06}O_3$ plus 1 mol% MnO and PZT ceramics with nominal composition $PbZr_{.52}Ti_{.48}O_3$ were prepared by the mixed oxide route and sintered at 1 120 °C for 5 hours and 1 260 °C for 2 hours, respectively. One mol% PbO was added to MPT and 2.7 mol% to PZT in excess of the stoichiometric compositions in order to compensate for PbO losses during processing. The PZT material was doped with 0.5 wt% SiO_2 in order to study the influence of SiO_2 impurities on the microstructure of PZT ceramics.

Electron transparent specimens were prepared by ion-beam thinning and were examined in a Philips EM 420 analytical STEM, equipped with EDAX energy dispersive X-ray spectrometer (EDS) attachment.

RESULTS AND DISCUSSION

MPT Ceramics

The Ca-Co-W modified MPT ceramics showed a high degree of electromechanical anisotropy; ratios of thickness coupling factor to planar coupling factor $k_t/k_p \approx 12$ were measured in >97% dense materials having average grain sizes between 2.8 and 5 µm [3]. The ferroelectric domain structure as typically observed in MPT after poling in an external electric field (50 kV/cm) is shown in Figure 1. The domain boundaries were identified as 90° type {110}

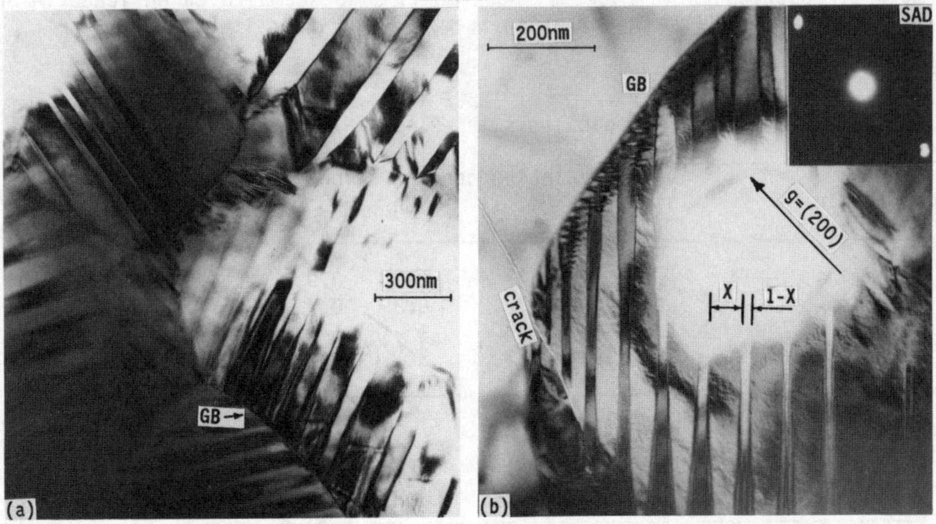

Figure 1. Microstructure of MPT ceramics. (a) Grain boundaries (GB) and ferroelectric domain boundaries. (b) Area contrast between (110) twin variants (dynamical TEM BF image), diffraction spots split along [110].

twin related ferroelectric domain walls. In order to cancel the transformation strains four equal shears are required in this twin system [4] giving rise to diffraction spot splitting across (110) as shown in Figure 1b. The slight deviation from normality of the c-axes in adjacent twin variants gives rise to a distinct areal contrast of domains imaged under dynamical diffraction conditions. This small change in the excitation error s across coherent interfaces results in the formation of characteristic δ-type fringe contrasts in tilted domain boundaries [5]. From theoretical considerations of a martensite type transformation the equilibrium fractions X and (1-X) of the twin variants can be predicted as function of the lattice parameters of parent and product phases [6]. Calculations yielded a theoretical equilibrium domain width ratio X:(1-X)=2:1 whereas in TEM images of poled material the domain width ratio varied between 3.5:1 and 5.3:1. It is suggested that in the poling process the favourably oriented twin variant grew at the expense of the other variant by 90° wall switching.

TEM investigations revealed the presence of vitreous intergranular phases situated in small pockets at triple grain junctions (TJ) and in thin films at grain boundaries (GB) as shown in Figure 2. The lateral spacings of Fresnel fringes in out-of-focus images indicated the presence of intergranular films at grain boundaries about 1 nm thick.

The chemical composition of matrix phase and residual intergranular glass phase was determined by EDS microanalysis, the results are summarized in Table 1. The experimentally determined composition of the matrix agreed well within experimental errors with the nominal composition. The EDS analysis of pockets of intergranular phase indicated a marked segregation of W, Co, Mn and Ca modifiers as well as of Si impurities in grain boundary regions, whereas the Pb concentration was below the concentration in the matrix. An attempt was made to calculate the true concentrations of segregate atoms in thin GB films by deconvolution of the X-ray production functions of matrix and intergranular phase within the sampling volume (heterogeneous case) [7]. The results confirmed the same segregation pattern as observed in TJ pockets, however Ti also seems to be enriched in GB films whereas the Pb concentration seems to be rather low.

TABLE 1
EDS microanalysis in MPT and PZT ceramics

Phase	Relative concentration [wt% cations]							
	Si	Ca	Ti	Mn	Co	W	Pb	Zr
nominal (MPT)	-	5.5	21.4	0.3	0.9	2.6	69.7	-
matrix	-	5.4	22.2	0.3	0.7	2.5	68.9	-
TJ	1	6	21	1	1	4	66	-
GB* phase	2	12	29	2	2	9	44	-
nominal (PZT)	0.3	-	8.3	-	-	-	74.4	17.0
matrix	0.5	-	7	-	-	-	79	14
glassy TJ	5	-	2	-	-	-	88	5

* calculated values (for electron probe size d=3 nm)

PZT Ceramics

The microstructure of grain boundaries, as typically observed in PZT ceramics doped with 0.5 wt% SiO_2 is shown in Figure 2c. A vitreous intergranular phase was detected in pockets at triple grain junctions up to 50 nm in size. The relative concentrations of cations in the PZT matrix and in the glassy phase, as determined by EDS microanalysis, are listed in Table 1. Indications are that sintering with excess PbO not only resulted in the formation of a PZT matrix having a Pb content higher than the stoichiometric composition, but also seems to be the major source for the formation of glassy intergranular phase. The Si dopant showed a marked tendency towards segregation at grain boundaries, its concentration was increased in the glassy phase by at least one order of magnitude over the nominal concentration. Since both Zr and Ti go into the glassy phase at a rate approximately proportional to their respective fractions in the nominal composition, the Zr:Ti ratio in the PZT matrix is thought to be unaffected by the formation of intergranular phase.

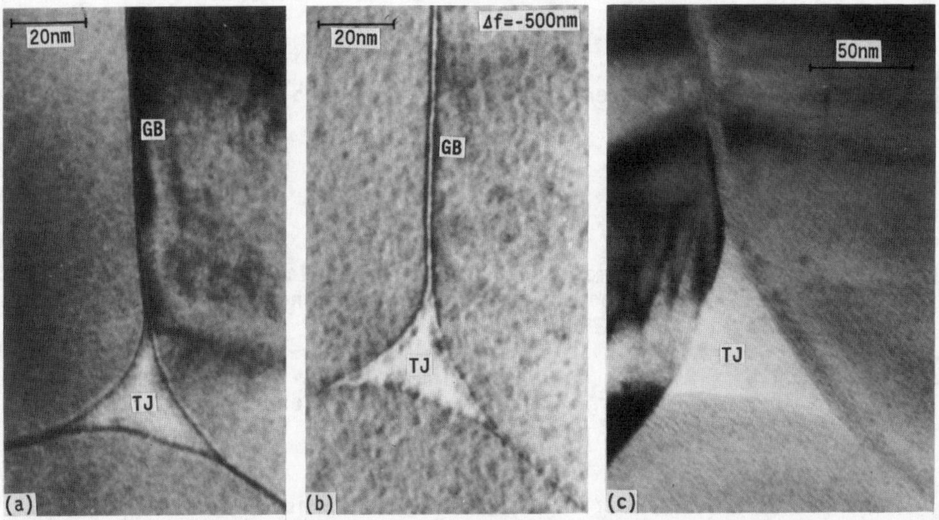

Figure 2. Microstructure of grain boundaries (TEM BF images) . (a) in-focus, and (b) under-focused image of glassy intergranular phase in MPT. (c) Glassy phase in pocket at triple grain junction (TJ) in Si doped PZT.

CONCLUSIONS

The present TEM observations on the microstructure of Ca-Co-W modified MPT and SiO_2 doped PZT ceramics enable the following conclusions to be made:

(i) Strains brought about by the cubic→tetragonal phase transformation are relieved by twinning on {110} planes. Each twin variant is a single ferroelectric domain separated by 90° type domain boundaries. During the poling process favourably oriented domains grow by 90° type domain wall switching.

(ii) Vitreous intergranular phases exist in small pockets at triple grain junctions and in films at grain boundaries approximately 1 nm thick.

(iii) In MPT ceramics Ca, Mn, Co and W modifiers show a marked tendency towards segregation in grain boundary regions. Si impurities, not detectable in the matrix, are concentrated in the glass phase. The Pb content in the glass phase is below the level in the matrix composition. In summary, it can be concluded that the microstructural evolution is controlled by the segregation behaviour of modifiers and Si impurities in this material.

(iv) In PZT ceramics sintering with excess PbO results in Pb concentrations above the stoichiometric composition in the bulk material and in the formation of a Pb rich vitreous intergranular phase. The SiO_2 impurities show strong segregation in grain boundary regions and promote the formation of a lead-silica-zirconia based glass phase. Since at the sintering temperature the PbO vapour pressure over the glassy phase can be assumed to be considerably higher than over the matrix phase, the presence of a Pb rich glass phase will promote the PbO losses from surface regions of PZT samples.

Acknowledgement
The MPT and PZT sample material was kindly supplied by DR de Villiers and Ö Soykan, respectively.

REFERENCES

1. Yamashita, Y., Yokoyama, K., Honda, H. and Takahashi, T., (Pb, Ca) $((Co_{1/2}\ W_{1/2}),\ Ti)O_3$ piezoelectric ceramics and their applications. Jap. J. Appl. Phys., 1981, **20**, Suppl. 20-4, 183-87.

2. Goo, E.K.W., Mishra, R.K. and Thomas, G., Transmission electron microscipy of Pb $(Zr_{0.52}Ti_{0.48})O_3$, J. Am. Ceram. Soc., 1981, **64**, 517-19.

3. De Villiers, D.R. and Schmid, H.K., Piezoelectricity and microstructure of modified lead titanate ceramics. subm. J. Mat. Sci., 1989.

4. Tanner, L.E. and Ashby, M.F., On the relief of ordering strains by twinning. phys. stat. sol., 1969, **33**, 59-68.

5. Gevers, R., Delavignette, P., Blank, H., Van Landuyt, J. and Amelinckx, S., Electron microscope transmission images of coherent domain boundaries. phys. stat. sol., 1964, **5**, 595-633.

6. Wechsler, M.S., Lieberman, D.S. and Read, T.A., On the theory of the formation of martensite. J. Met., Trans AIME, 1953, **197**, 1503-15.

7. Schmid, H.K., X-ray microanalysis with high spatial resolution in the analytical electron microscope, Proc. Electr. Micros. Soc. South. Afr., 1988, **18**, 35-36.

A MODEL FOR COMPOSITIONAL CHANGES IN INTERLEVEL DIELECTRIC BOROPHOSPHOSILICATE GLASS DURING VLSI DEVICE PROCESSING

S.B.DESU
Department of Materials Engineering
Virginia Polytechnic Institute and State University
Blacksburg, Virginia, 24061, U.S.A

ABSTRACT

The compositional changes in interlevel dielectric borophosphosilicate glass (BPSG), due to the PBr_3 gettering process, are analyzed by various analytical techniques such as a X–ray fluroscence, secondary ion mass spectroscopy (SIMS), Auger electron spectroscopy (AES), and transmission electron microscopy (TEM). The PBr_3 gettering process resulted in phosphorus enriched layers on BPSG, whose thickness depended on parameters such as time, temperature, and partial pressure of PBr_3. A model is proposed to account for the formation of these phosphorus enriched layers and its dependence on gettering parameters. BPO_4 crystallites are also formed in some cases, depending on the gettering parameters and the boron composition of BPSG.

INTRODUCTION

Borophosphosilicate glass (BPSG) films are used as an interlevel dielectric in VLSI devices to isolate the source–drain interconnects from the gate level interconnects [1]. The flow of interlevel dielectric BPSG and impurity gettering are two important steps in VLSI device processing [2]. Gettering removes process–induced defects and impurities and thus aids the fabrication of high performance, reliable devices. Flow of interlevel dielectric gives rise to smoother surfaces which facilitate further lithography and processing [3,4]. Recently, a combined PBr_3 getter–flow process was proposed to achieve both the backside phosphorus gettering as well as BPSG flow [5]. This process consists of heat treating the device wafers, after the BPSG deposition and back side etch, in an atmosphere containing both PBr_3 and O_2.

In the present investigation, the flexibility of the getter–flow process was studied by characterizing the compositional changes in BPSG film after gettering. A model is proposed for the observed phosphorus incorporation and the formation of phosphorus enriched layers.

EXPERIMENTAL METHOD

The BPSG films were deposited in a low pressure chemical vapor deposition reactor using SiH_4, PH_3 (5% in N_2), B_2H_6 (2% in N_2), and O_2 as reactants[4]. The glass was deposited (typically 1.3 μm thick) at around 400°C and 0.28 torr on 125 mm diameter Si (100) wafers which were pre–coated with 100 nm of SiO_2. The average concentration of boron and phosphorus in the film is 1 ± 0.1 and 4.9 ± 0.1 wt% respectively. Immediately after glass deposition, the wafers were rinsed in DI water for 15 sec, and then placed in the getter–flow furnace. Gettering–flow temperature, time, and partial pressure of PBr_3 were varied from 800 to 1000°C, 2 to 120 min, and 1 to 10 torr, respectively. The oxygen partial pressure was kept constant at 75 torr. The films were analyzed after deposition, after annealing in 2% O_2 in N_2 at 950°C, and after PBr_3 getter–flow process. Analysis consisted of using techniques such as TEM, SIMS, and AES which are sensitive to compositional variations as a function of depth. Data collected from these techniques were calibrated to results from chemical analysis and electron microprobe studies. Thickness and refractive index of the films were monitored using a Rudolph ellipsometer at a wavelength of 632.8 nm and with an angle of 70 deg. between the polarizer and analyzer. Etch rates of the BPSG films after various processing steps were determined by using a standard 30:1 BHF solution at 20±1°C and measuring thickness before and after etching at intervals of 1 min.

RESULTS AND DISCUSSION

The TEM, SIMS, and AES depth profile data showed that the as deposited and densified films (annealed in 2% O_2 in N_2) were uniform in composition throughout the entire thickness. However, after PBr_3 gettering (at 950°C for 30 min in 7.6 torr of PBr_3), analysis showed a 0.45 μm thick BPSG surface layer which had high phosphorus concentration (up to 13 wt%) [6]. Another indication that the P concentration increased in the outer layer of the BPSG is the change in chemical etch rate (which is known to be sensitive to both phosphorus and boron concentration) observed after the PBr_3 gettering. This change in etch rate as a function of time for different gettering conditions was also investigated to obtain an idea of possible phosphorus concentration [6]. The etch rate for the as deposited film was about 65 nm/min, whereas the value for densified glass is around 30 nm/min. The phosphorus enriched region showed an etch rate of 80 nm/min (for gettering at 950°C for 10 min in 7.6 torr of PBr_3). The etch rate steadily decreased with further removal of the surface layer, whereby finally the etch rate was found to be identical with that of the densified glass.

The thickness of this surface layer (enriched with phosphorus for a fixed BPSG composition) is a function of gettering conditions such as temperature, time, oxygen partial pressure, and partial pressure of PBr_3. Figure 1 shows the dependence of the surface layer thickness on gettering time at different temperatures. It can be seen from the Figure 1 that the thickness is proportional to square root of time for the temperatures investigated, indicating that the rate controlling mechanism may be diffusion. Furthermore, Figure 2 shows an Arrhenius dependence of reduced thickness [thickness/$(\text{time})^{1/2}$] on temperature with an activation energy of 1.37 eV. Figure 3 shows a dependence of the surface layer thickness on the partial pressure of PBr_3 [$(\text{Thickness})^4 \propto PBr_3$ partial pressure]. The surface layer thickness was measured by both SIMS and etch rate measurements.

These results can be understood by the following model which assumes diffusion as the rate controlling step. The interaction of PBr_3 with oxygen produces P_2O_5 according to equilibrium reaction,

$$2\,PBr_3 + 2.5\,O_2 = P_2O_5 + 3\,Br_2 \tag{1}$$

which in turn reacts with BPSG to form a composition similar to the liquidus composition (at the processing conditions) of the B_2O_3–P_2O_5–SiO_2 phase diagram. Because of the sluggish nature of the kinetics in glasses, only the top layer is converted into liquid whereby the rate of liquid film advance may be controlled by diffusion. For a low percentage of conversions, the surface concentration of P_2O_5 formed can be obtained from Equation 1 as:

$$C^s(P_2O_5) \;=\; K_1\,[P(PBr_3)]^{1/2}\,[P(O_2)]^{5/8} \qquad (2)$$

where $P(PBr_3)$ and $P(O_2)$ are partial pressures of PBr_3 and O_2, respectively, and K_1 is constant (which includes the factors for converting concentrations to partial pressures).

Due to the interaction between P_2O_5 and BPSG, the surface layer of thickness (X) is converted into the liquidus composition with an enriched phosphorus concentration, $C^l(P_2O_5)$. Figure 4 depicts the phosphorus concentration profile of the BPSG film. There are two interfaces to consider. The outer surface where the concentration $C^s(P_2O_5)$ is determined by the gas–liquid equilibrium (Equation 2), and the liquid–BPSG interface (at X), whose phosphorus concentration $C^l(P_2O_5)$ is determined by the liquidus composition at the processing conditions. The formation of this liquid is very crucial to the flow process. The rate of advance of this liquid film can be obtained by diffusion kinetics.

If $D(P_2O_5)$ is the diffusion coefficient of P_2O_5, the flux of P_2O_5 (J) across the film is given by

$$J \;=\; -D(P_2O_5)\;d[C(P_2O_5)]/dx \qquad (3)$$

where $d[C(P_2O_5)]/dx$ is the gradient of P_2O_5 across the film which can be approximated as

$$d[C(P_2O_5)]/dx \;\cong\; [C^s(P_2O_5)-C^l(P_2O_5)]/X \qquad (4).$$

Since $C^l(P_2O_5)$ is constant for a given processing condition, from equations 2, 3, and 4 we obtain

$$J \;\cong\; K_2\,D(P_2O_5)\,[P(PBr_3)]^{1/2}\,[P(O_2)]^{5/8}\,/X \qquad (5)$$

where K_2 is a new constant. The rate of advance of the liquidus film is then represented by

$$dx/dt \;=\; J/C^l(P_2O_5) \qquad (6)$$

$$dx/dt = K_2\,D(P_2O_5)\,[P(PBr_3)]^{1/2}\,[P(O_2)]^{5/8}\,/[X\,C^l(P_2O_5)] \quad (7).$$

Integrating Equation 7 we obtain

$$X^2 = 2K_2\,D(P_2O_5)\,[P(PBr_3)]^{1/2}\,[P(O_2)]^{5/8}\,t\,/[C^l(P_2O_5)] \quad (8).$$

Thus, this simple model shows a dependence of the phosphorus enriched surface layer thickness on $P(PBr_3)$, $P(O_2)$, time (t), diffusion coefficient of P_2O_5 $[D(P_2O_5)]$, and the composition of BPSG. $D(P_2O_5)$ depends exponentially on temperature and on the viscosity of the film, which itself depends on the temperature and boron concentration of the BPSG. From a plot of Log $(X/t^{1/2})$ verses 1/T holding other parameters constant, the activation energy (E) for the phosphorus enriched surface layer formation can be obtained. The experimental value obtained is 1.37 eV (Figure 4). This value should not be confused with the activation energy for the diffusion of P_2O_5, since both $C^l(P_2O_5)$ and K_2 are also temperature dependent.

Besides the enrichment of phosphorus in the surface layers during the flow/getter step, depletion of boron and the formation BPO_4 crystallites was also observed [6]. The propensity of BPO_4 crystallite formation increased with increasing boron concentration in the starting BPSG film. This depletion of boron and crystallite formation are accounted for by the following reactions [6]:

$$2\ PBr_3 + O_2 + B_2O_3 = P_2O_5 + 2\ BBr_3 \quad (9)$$

$$BBr_3 + PBr_3 + 2\ O_2 = BPO_4 + 3\ Br_2 \quad (10).$$

SUMMARY

In the flow/PBr_3 getter step of VLSI device processing, significant compositional changes were observed in the interlevel dielectric BPSG. A phosphorus enriched surface layer is formed, the thickness of which is a function of T, t, $P(PBr_3)$, $P(O_2)$, and composition of starting BPSG film. The formation of this phosphorus enriched layer aids the flow of the BPSG, but the flexibility of the process is limited due to the formation of hygroscopic, highly doped phosphorus layers (13 wt%) and the possibility of BPO_4 crystallite formation. The functional dependence of the surface layer is explained by a simple model which assumes diffusion as the rate limiting process.

ACKNOWLEDGEMENTS

The author wishes to thank J.A.Taylor for providing AES and SIMS depth profiles, V.C.Kannan for TEM data, and K.Bartelmehs for editing this manuscript.

REFERENCES

1) Kern, W. and Schnable, G.L., Chemically Vapor–Deposited Borophosphosilicate Glasses for Silicon Device Applications, RCA Rev., 182, 43, 423–457.
2) Sze, S.M., VLSI Technology, McGraw–Hill, New York, 1983.
3) Ravi, K.V., Imperfections and impurity phenomena. in Microelectronic Materials and Processes, ed. R.A. Levy, Kluwer Academic Publishers, London, 1986, pp.769–761.
4) Kern, W., Dielectric Materials, ibid, pp.247–270.
5) Taylor, J.A., private communication.
6) Desu, S.B and Taylor, J.A., Analytical Characterization of the Interaction of PBr_3 and O_2 with BPSG, to be published.

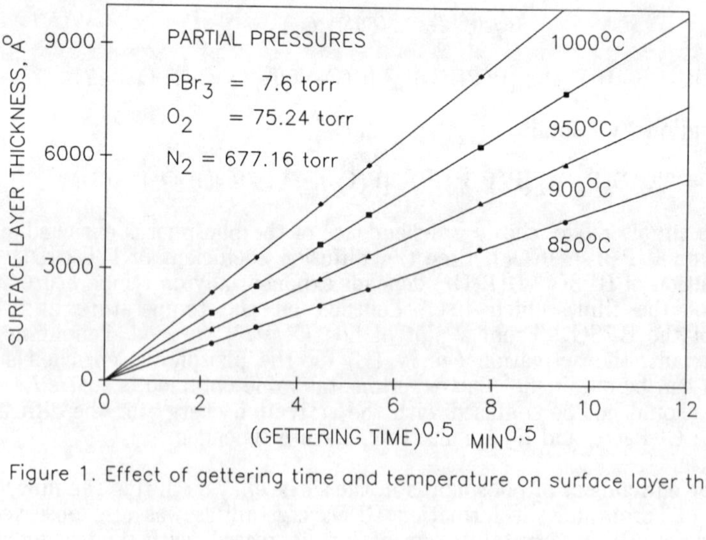

Figure 1. Effect of gettering time and temperature on surface layer thickness

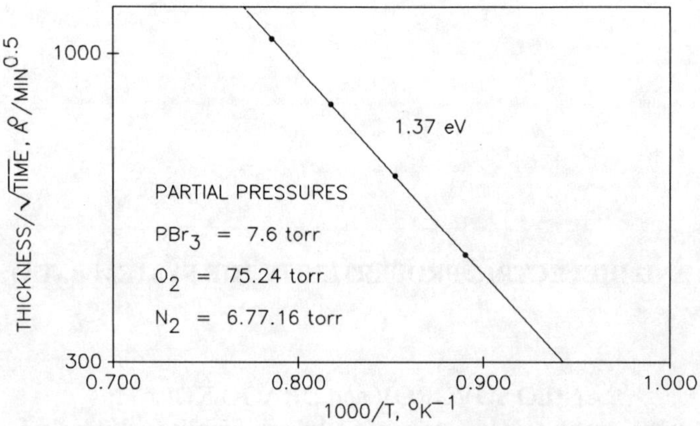

Figure 2. Effect of gettering temperature on reduced surface layer thickness

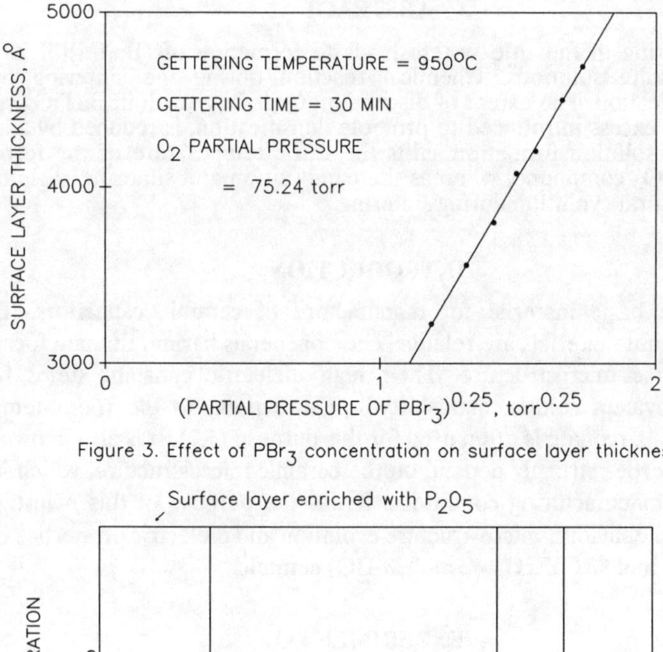

Figure 3. Effect of PBr_3 concentration on surface layer thickness

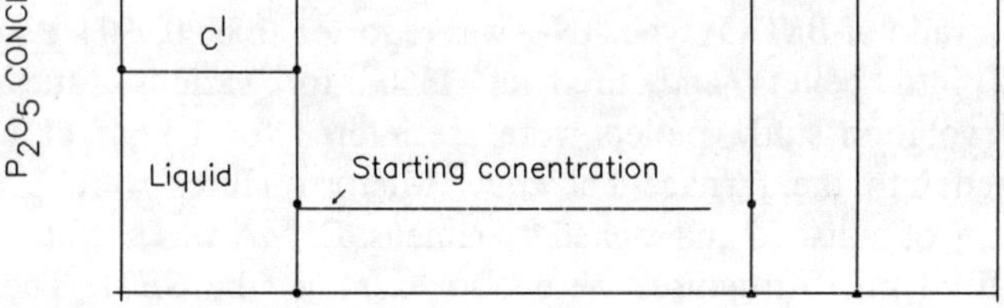

Figure 4. Steady−state model for P_2O_5 diffusion in BPSG

SINTERING AND DIELECTRIC PROPERTIES IN THE SYSTEM $BaTiO_3$-$CaZrO_3$

DANILO SUVOROV and DRAGO KOLAR
"J. Stefan" Institute, University of Ljubljana, Ljubljana, Yugoslavia

ABSTRACT

Reaction sintering in the title system leads to formation of (Ba,Ca)(Ti,Zr)O_3 and (Ca,Ba)TiO_3 solid solutions. Chemical reaction during the sintering operation disturbs densification. The extent of discontinuous grain growth in $BaTiO_3$ caused by the small TiO_2 excess introduced to promote densification, is reduced by addition of $CaZrO_3$. Solid solution formation shifts the Curie temperature of the ferroelectric (Ba,Ca)(Ti,Zr)O_3 compound, whereas the composition and sintering regime control the microstructural evolution during sintering.

INTRODUCTION

$BaTiO_3$ is the basic material for manufacture of ceramic capacitors. The high dielectric constant materials are relatively homogeneous barium titanate formulations that have coarse microstructure. Their high dielectric constant stems from the addition of isovalent solutes that shift the Curie point to the room temperature region. Calcium zirconate is often used for this purpose (1,2). It is also known that the dielectric properties strongly depend on the ceramic microstructure, which in turn is influenced by manufacturing conditions. It was the purpose of this investigation to study sintering behaviour, microstructure evolution and dielectric properties in 82 mol % $BaTiO_3$ - 16 mol % $CaZrO_3$ - 2 mol % TiO_2 ceramic.

EXPERIMENTAL

Fine grained commercial materials (Kyoritsu, Japan) for capacitor production were used (BaO/TiO_2 ratio of $BaTiO_3$ type AN-2 was reported to be 0,997). Powders were mixed, pressed into pellets and fired at 1360° for various times. For the microstructure evolution study, pellets were presintered for 1 hour at 1300°C and thereafter pushed into the furnace hot zone. Microstructures were examined by optical microscopy of polished and etched specimens. EPMA was used to analyse the inclusions. As fired microstructures were also examined by SEM. The extent of chemical reaction was examined by the powder XRD technique. For dielectric measurements, sintered disks ϕ = 10 mm, h = 1 mm were measured at 1 kHz.

RESULTS AND DISCUSSION

Reaction sintering of the $BaTiO_3$ - $CaZrO_3$ ceramics

On heating, $BaTiO_3$ and $CaZrO_3$ react to form barium zirconate phase with some Ca and Ti in solid solution and calcium titanate phase with some dissolved $BaTiO_3$ (3). $BaTiO_3$-$BaZrO_3$ form a solid solution $Ba(Ti_{1-y}Zr_y)O_3$ over the whole composition range. The solubility of Ca in $Ba_{1-x}Ca_xTi_{1-y}Zr_yO_3$ solid solution was reported to be 17,5 mol % at 1300°C (1) and increases to 28 mol % at 1600°C (4).

Reaction betwen $BaTiO_3$ and $CaZrO_3$ was detectable by X-ray analysis already after 1 hour's heating at 1100°C. Reaction is slow and no heat effect could be detected on the DTA curves. This chemical reaction is believed to interfere with the sintering process in $BaTiO_3$-$CaZrO_3$ mixtures, as observed by dilatometric analysis. Fig. 1 shows shrinkage curves for various compositions. $BaTiO_3$ starts sintering at around 1100°C and sinters considerably in the temperature region 1300 - 1350°C; $CaZrO_3$, being more refractory, demands higher sintering temperatures. It begins to shrink at around 1250°C and sinters to high density in the temperature region 1450 - 1500°C. $BaTiO_3$ - 16 mol % $CaZrO_3$ mixture starts to shrink at approximately the same temperature as $BaTiO_3$; however, after a few % shrinkage the kinetics slow down and temperatures of 1550 - 1600°C are needed for densification. Microstructural examinations showed high porosity which is ascribed to the Kirkendal effect due to differences in the diffusion coefficients of various constituents in the $BaTiO_3$ - $CaZrO_3$ ternary mixture. Kirkendall type porosity was demonstrated in $BaTiO_3$ - $CaZrO_3$ diffusion couple fired for 15 hours at 1450°C (Fig. 2).

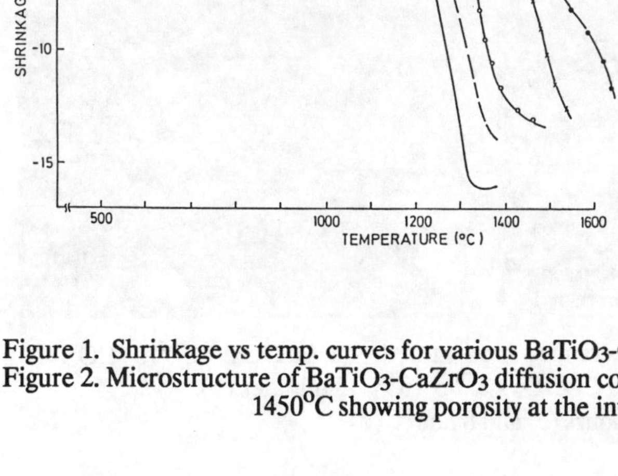

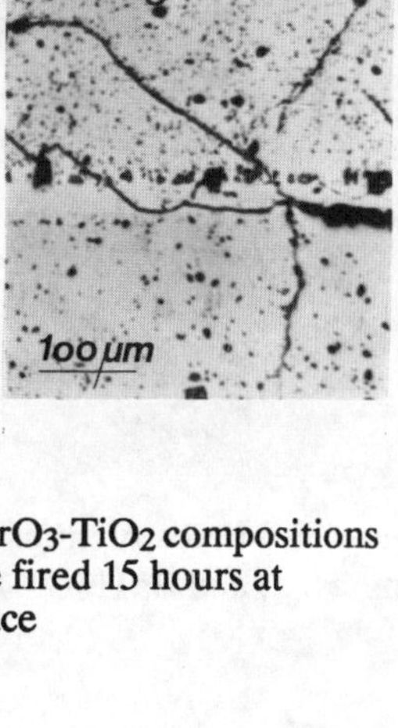

Figure 1. Shrinkage vs temp. curves for various $BaTiO_3$-$CaZrO_3$-TiO_2 compositions
Figure 2. Microstructure of $BaTiO_3$-$CaZrO_3$ diffusion couple fired 15 hours at 1450°C showing porosity at the interface

Sinterability may be improved by avoiding solid state diffusion. It is well known that the addition of a small amount of TiO_2 enhances sinterability in $BaTiO_3$ based systems by forming a low melting $BaTiO_3$-$Ba_6Ti_{17}O_{40}$ eutectic at around 1310°C. The TiO_2 effect is demonstrated in Fig. 1: sinterability of $BaTiO_3$ - 16 mol % $CaZrO_3$ - 2 mol % TiO_2 mixtures is improved so that sintering temperatures of 1350 - 1380°C are sufficient to achieve high density - above 95 % of theoretical value in 1 hour's firing time.

Microstructure evolution

Sintering of $BaTiO_3$ in the presence of aTiO_2- rich eutectic is accompanied by rapid discontinuous grain growth. Fig.3a shows a coarse grained 98 mol % $BaTiO_3$ - 2 mol % TiO_2 structure after 15 minutes sintering at 1360°C. The average intercept length was measured as 40 μm. In contrast, the microstructure of an 82 $^m/o$ $BaTiO_3$ - 16$^m/o$ $CaZrO_3$ - 2 $^m/o$ TiO_2 sample sintered under identical conditions shows much finer grain size, with an average intercept length 3 - 4 μm (Fig. 3b). The microstructure shows fine inclusions of a second phase. Microprobe analysis confirmed that inclusions are rich in Ca. Figs. 3b-d demonstrate grain coarsening with firing times from 15 minutes to 8 hours. The reduction in the grain size of the ferroelectric phase in $BaTiO_3$ - $CaZrO_3$ mixtures was already noted (4).

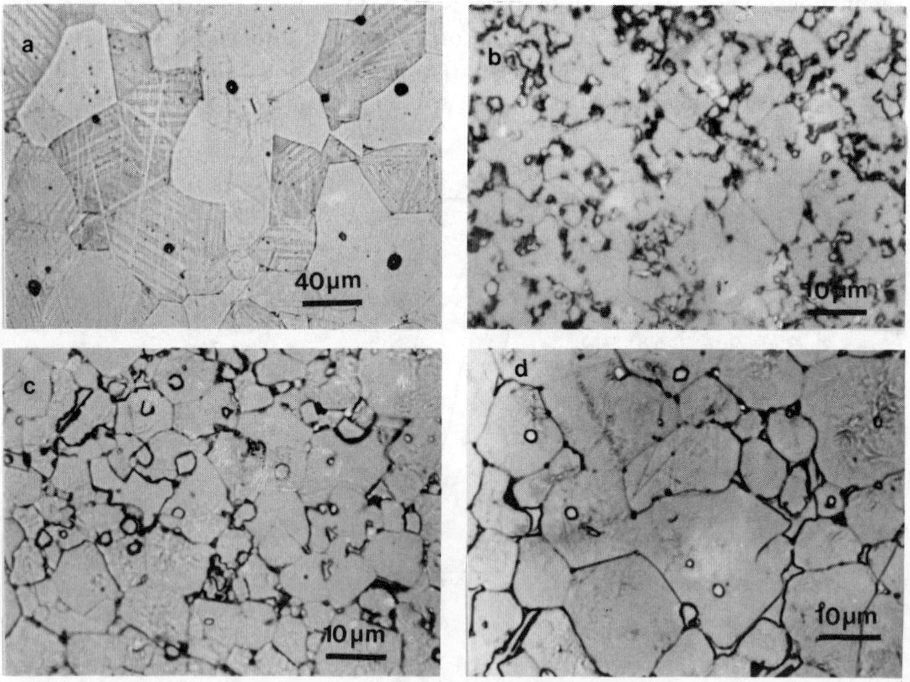

Figure 3. Microstructures of ceramics fired at 1360°C: $BaTiO_3$ - 2 $^m/o$ TiO_2, 15 minutes (a) and BaTiO3 - 16 $^m/o$ $CaZrO_3$ - 2 $^m/o$ TiO_2, 15 minutes (b), 2 hours (c) and 8 hours (d)

Electrical properties

Fig. 4 shows ε vs. T curves for 82 m/o $BaTiO_3$ - 16 m/o $CaTiO_3$ - 2 m/o TiO_2 compositions, fired at 1360°C for 15 min, 2 hours and 8 hours, respectively. Whereas an ε value below the Curie temperature decreases with increasing firing time, the peak at the Curie temperature increases in value and broadens. Changes in the intensity and broadness of the peak accompany coarsening of the microstructure, as demonstrated in Fig. 3. This is in accordance with the known fact that the increase of the grain size of $BaTiO_3$ causes a decrease of ε below the Curie point, while the maximum of ε becomes sharper and higher (5,6). The gradual shift of T_c from 65°C to 50°C with increasing firing time is believed to be due to improved homogenization of the constituents (Table 1.).

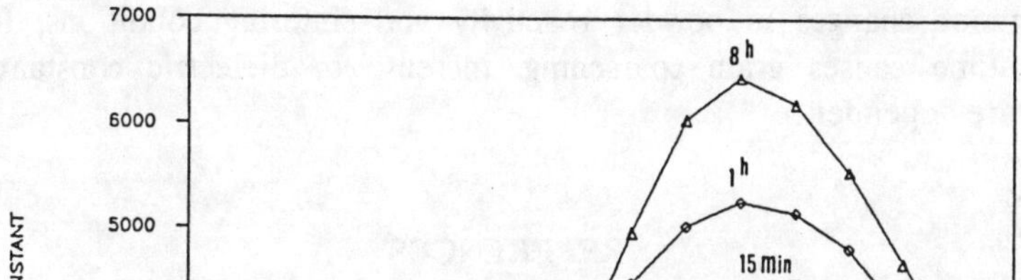

Figure 4. Dielectric constant versus temperature of $BaTiO_3$ - 16 m/o $CaZrO_3$ - 2 m/o TiO_2 composition sintered at 1360°C

TABLE 1

Electrical properties of $BaTiO_3$ - 16 m/o $CaZrO_3$ - 2 m/o TiO_2 composition sintered at 1360°C

SOAK. TIME	Tc (C)	DK at 25°C	DK at Tc	tg δ . 10^4	Δ C/C (%) -25°C	Δ C/C (%) +85°C
15'	65	3934	4455	174	-47.8	+4.6
1 h	55	4500	5251	266	-59.1	-5.9
8 h	50	4762	6441	293	-67.5	-2.7

CONCLUSIONS

Reactive sintering in the system $BaTiO_3$ - $CaZrO_3$ in the composition range close to the immiscibility gap results in a $Ba_{1-y}Ca_yTi_{1-x}Zr_xO_3$ matrix with isolated $Ca_{1-w}Ba_wTiO_3$ inclusions. Chemical reaction hinders densification. Additional porosity appears to be due to the unequal diffusion coefficients of diffusing species. Sintering is improved by the addition of TiO_2, which forms a eutectic at the sintering temperature. Liquid phase sintering is accompanied by discontinuous grain growth; however, far less pronounced then in the system without $CaZrO_3$. The dielectric properties of the investigated compositions confirm the known fact that the modified material is suitable for manufacture of ceramic capacitors according to Z5U specifications. However, discontinuous grain growth makes the system sensitive toward minor changes in powder reactivity and sintering conditions. Increasing sintering time causes grain coarsening, increase in dielectric constant and its temperature dependence.

REFERENCES

1. Jonker, G.H., Kondensatormaterialien mit hoher Dielektrizitatskonstante, Philips Tech. Rundschau 1955, **17**, 127 -135
2. M. Kahn, D.P. Burks, I. Burn and W. A. Schulze, Ceramic capacitor technology, In Electronic Ceramics, ed. L. M. Levinson, M. Dekker, New York, 1988, pp. 191-274
3. McQuarrie, M. and Behnke, F. W., Structural and dielectric studies in the system (Ba,Ca)(Ti,Zr)O_3, J. Am. Cer. Soc. 1954 **37**, 539-543
4. Hennings, D. and Schreinemacher, H., Temperature dependence of the segregation of calcium titanate from solid solutions of (Ba,Ca)(Ti,Zr)O_3 and its effect on the dielectric properties, Mat. Res. Bull., 1977 **22**, 1221-1226
5. Buessem, W. R., Cross, L. E., and Goswami, A. K., Phenomenological theory of high permittivity in fine-grained barium titanate, J. Am. Cer. Soc. 1966, **49**, 33-36
6. Hennings, D., Schnell, A., and Simon, G., Diffuse ferroelectric phase transition in $Ba(Ti_{1-y}Zr_y)O_3$ Ceramics, J. Am. Cer. Soc. 1982, **65**, 539-544

OPTIMIZATION OF PZT POWDERS. QUANTIFICATION BY PHYSICAL METHODS OF GRINDING POLLUTION. EFFECTS ON SINTERING AND CERAMIC PROPERTIES.

R.MAKALA, B.THIERRY, M.DESCAMPS
C.R.I.T.T. Céramiques Fines Techniques 59600 MAUBEUGE FRANCE
G.PALAVIT, P.VAST
Laboratoire de Chimie Appliquée,USTL, 59655 VILLENEUVE D'ASCQ FRANCE

ABSTRACT

After calcination, PZT powders have low physical properties which decrease electrical performance. So to improve them, effects of attrition grinding have been studied through various parameters as grinding time, dry material rate...The evolution was followed by measuring grain size distribution, specific surface.However,attrition milling performance is accompanied by an important effect of contamination. We tried to control this phenomenon by quantitative analysis: an original method was developped by using density measurement with gas pycnometer. Optimized powders obtained by this way allowed us to study firing and sintering stages.Various time and temperature cycles were made to define best conditions of traditionnal sintering to obtain ceramics without open porosity to be densified by H.I.P. So we get densities upper to 99% with improved physical and electric properties.

INTRODUCTION

Lack of homogeneity and reproductibility appears as an important phenomenon in the manufacture process of piezoelectric PZT ceramics (1). This defect comes particularly from the misapreciation of powders initial state. The characteristics of the starting powders particle size, distribution, aggregates,specific surface have a great influence on the reactivity during firing. Densification and microstructural development during ceramic processing are affected by characteristics of the starting PZT powders. So it seems needful to perfectly know the characteristics of raw materials to improve them to an optimal state and to be able to standardize operating mode conditions to be sure to have an homogeneous chemical and physical quality of PZT product. Objectif of this paper is to show the effects of these parameters at different stages of the material manufacturing. Attrition milling has been utilized to improve powders quality.However, attrition milling performance is attended with an important contamination effect,this one has been quantified by analytical method at the different stages of the process.

MATERIAL AND METHODS

For the study PZT composition selected was $Pb(Zr_{0,54},Ti_{0,46})O_3$.Different qualities of raw materials were used to prepare the mixtures. Two grades of ZrO_2 (coarse and fine),one quality of TiO_2 and two sorts of PbO: a coarse quality and the same product after grinding. For the last one, we have studied the effect of the attrition milling on granulometric properties of the powders. Contamination rate of the powder proceeding from zirconia balls used as grinding media is quantified by measuring the powder density at different times of grinding by the mean of gas pycnometer. Excess ZrO_2 found in PbO powder is deducted from the weighing of raw oxides to keep a constant composition. Mixtures were carefully mixed by wet ball-mill mixing in agate jar. After firing, PZT powder was either crushed with mortar or grinding by attrition milling in optimized operating conditions. Powders were compressed isostatically (4 kbars) without a binding agent to a cylindrical shape. Specimens were then covered with packing powder (PZ + PZT) in an alumina crucible and heated to the sintering temperature between 1100 to 1300°C. Density and residual porosity were mesured by mercury porosimeter. Sintering parameters were studied by dilatometer (4). Ceramics were post-densified by HIP to remove the residual porosity (5).

RESULTS

Powders optimization by attrition milling
Generally particle size decrease with attrition milling time. Example of PbO powder size reduction is shown in Fig.1.

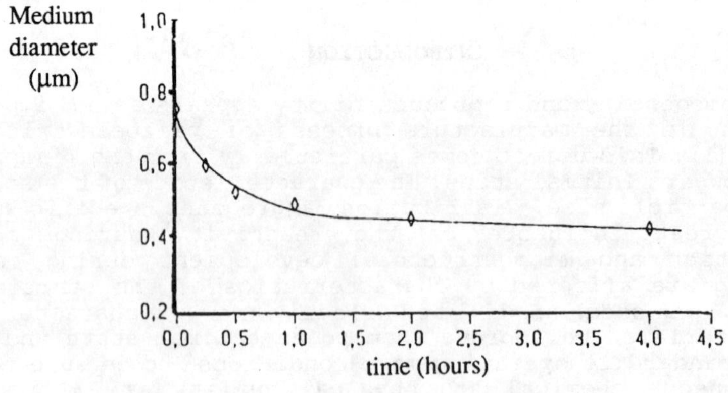

Figure 1. Effect of attrition milling time on size reduction of PbO powder.

Figure 2 demonstrate that attrition milling is a potentially useful method of reducing PbO powder to particle in a range from 0.2 to 1 µm with a medium diameter of .45µm after 2 hours grinding.

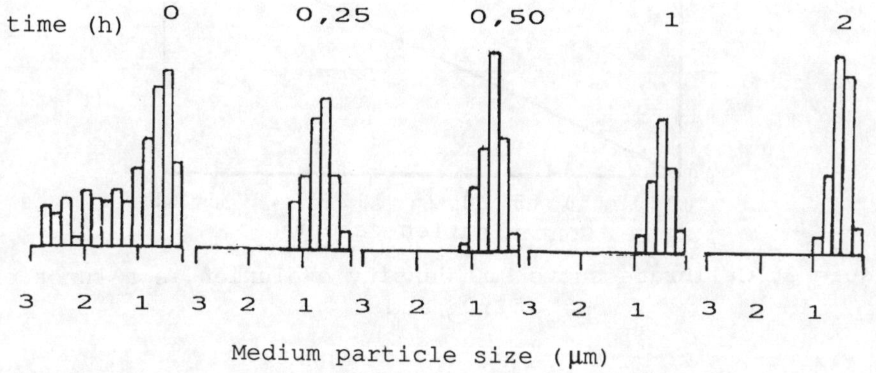

Figure 2.Effect of attrition milling time on size distribution of PbO powder.

Quantification of grinding pollution

A quantitative method was developed, based on powder density determination by gas pycnometer. This apparatus is capable of measuring a very low volume variation. This method consist in measuring powder density evolution after different grinding times. Powder and grinding media have different densities. During attrition grinding, a part of matter constitutive of zirconia balls pollute the powder, involving powder density modification: $D = D_o \pm \varepsilon$This value" ε can be quantified by gas pycnometer.

Procedure

A calibration curve was determinated by measuring density of a raw material sample in which were successively added known quantities of crushed ZrO_2 balls. For each add $|\Delta D|$ is calculated. The different values permit to define a curve $|\Delta D| = f(\%\ add)$. This curve can be assimilated to a straight line a range of 0 to 10% weigh adding. To quantify pollution level, milled powder density is determinated comparatively to this same powder as raw material. That permit to calculate $|\Delta D|$ value. By means of calibration curve, a concentration of polluted agent can be associated to the ΔD value. For instance, PbO grinding pollution: at first is defined a calibrate curve by the means of known adds of crushed ZrO_2 balls in PbO powder (Fig.3). During PbO powder attrition milling, samples were removed periodically and density were determined for each of them. By means of calculate curve it can be defined a curve showing the pollution rate as a function of grinding time (Fig.4)

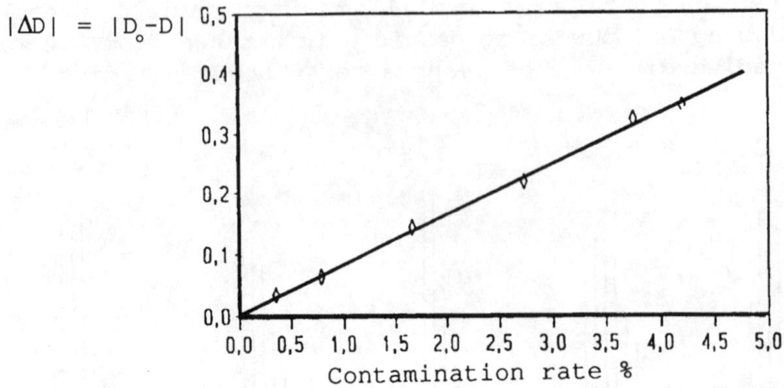

Figure 3. Calibrate curve PbO density evolution as a function of ZrO_2 add.

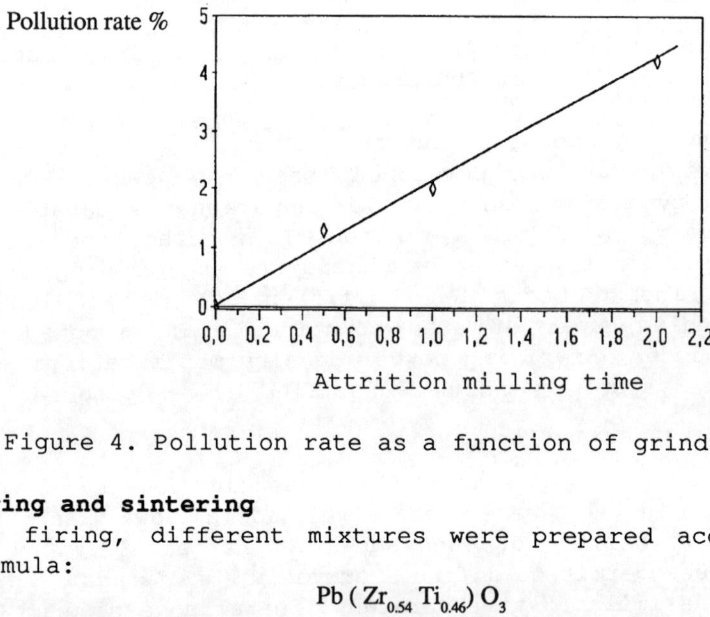

Figure 4. Pollution rate as a function of grinding time.

Firing and sintering

For firing, different mixtures were prepared according with formula:

$$Pb(Zr_{0.54}Ti_{0.46})O_3$$

Raw materials used to prepare the chemical composition are described in table 1. It can be obtained PZT structure at lower firing temperature during a short time (800°C, 6 h). Results show that to have optimized PZT powders, the product should contain about 6% pollution (zirconia). A dilatometric study of the shrinkage shows that the best temperature range is between 1200-1250°C. Curves of figure 5 shows the effect of granulometric quality of PZT powder on shrinkage rate. The best results are obtained with PZT powder ref.B, prepared with optimized raw powders

TABLE 1

	coarse ZrO_2 (A1)	fine ZrO_2 (A2)	TiO_2 (B)	coarse PbO (C)	grinded PbO
medium particle size (µm)	0.5	0.4	0.6	0.8	0.4
distribution size (µm)	0.2-0.4	0.2-0.8	0.2-1.4	0.2-2.8	0.2-1
specific surface (m^2/g)	12	18	6.6	1.5	2.5
chemical purity	>99%	>99%	>99%	>99.5%	97.5% 2.5 ZrO_2

After 4 hours at 1250°C density is higher than 98% of the theorical value with a very low level of opened porosity.

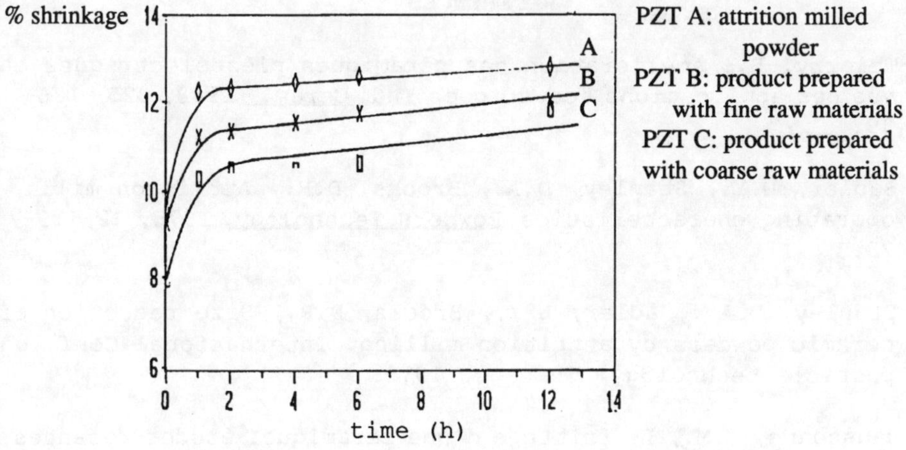

Figure 5. Shrinkage rate at 1250°C as a function of granulometric quality of the PZT powders.

Growing grain rate is very low function of the temperature for ceramics prepared with optimized powders. The conventionnal sintering stage has an objective of preparing the samples without open porosity for HIP system without mold to remove the residual closed porosity. A HIP cycle, 1100°C-200 bars-1h, has been carried out on sintered specimens: we have obtained density higher than 99,5% of theorical value.

CONCLUSIONS

The purpose of this paper is to study the effect of powders characteristics at the different stages of PZT ceramic manufacturing. The following general remarks appear to be useful for optimization of the ceramic product. It is desirable to well know characteristics of raw materials and to improve them to obtain a small particle size, a narrow particle dispersion, a large specific surface to favour firing at low temperature. Attrition milling seem a very good aid to have ideal powders. Results have shown that during a short grinding time, powders characteristics may be improved, but the contamination problem is not negligible (4-6%). So, for raw materials grinding, it may be solved by using grinding media of the same chemical composition: an original method has been developed based on density measurement with gas pycnometer.Uniform and densified microstructures in sintered ceramics can be obtained (>99,5%) with powders having optimized characteristics, particle coarsering must be avoided to prevent grain growth during sintering.

REFERENCES

1. Thierry, B., Amélioration des céramiques piézoelectriques en vue des applications acoustiques.Ind. Ceram., 1988, **825**, 186-89.

2. Sadler, L.Y., Stanley, D.A., Brooks, D.R., Attrition mill operating characteristics.Powders Technology, 1975, **12**, 19-28.

3. Stanley, D.A., Sadler, L.Y., Brooks, D.R., Size reduction of ceramic powders by attrition milling. International Conf. on particle technology, Chicago, 1973.

4. Haussonne, J.M., Le frittage d'une céramique: études récentes. Rev. Int. Hautes Temper. Réfract., France, 1984, **21**, 95-106.

5. Hardtl, K.H., A simplified method for the isostatic hot pressing of ceramics. Philips Tech. Rev., 1975, **2.3**, 65-72.

EFFECTS OF MICROSTRUCTURE ON THE MECHANICAL AND ELECTRICAL PROPERTIES OF Yb–TZP CERAMICS

M. González, C. Moure, J.R. Jurado and P. Durán
Instituto de Cerámica y Vidrio (CSIC)
Electroceramics Departament
Arganda del Rey, 28500 Madrid, Spain

ABSTRACT

Some compositions in the ZrO_2-rich region of the ZrO_2-Yb_2O_3 system were studied. Fine grained YbO_2-containing sintered samples approximately 100% tetragonal zirconia exhibited good mechanical properties (K_{IC}- 6-8 $MPam^{1/2}$). Those samples obtained by a sintering- precipitation process showed a both mechanical and electrical behaviour strongly affected by microstructural development. In order to know its applicability as a high temperature solid electrolyte the electrical conductivity was determined in the 300-800ºC temperature range.

INTRODUCTION

The tetragonal stabilized zirconia is well known because of its good mechanical properties and as an ionic conductor. On the other hand, the tetragonal phase can be retained at room temperature with the addiction of a relatively small amounts of oxides. The added cations (di or trivalent) sustitute the Zr^{+4} ions in the lattice. One of these stabilizer ions is Yb^{+3}, about which few mechanical or electrical behaviour data exist (1-2), particularly in the ZrO_2-rich region of ZrO_2-Yb_2O_3 system.

EXPERIMENTAL PROCEDURE

Samples were prepared by a wet chemical method using $Yb(NO_3)_3$ and $Zr(Butox)_4$ as the starting materials (3). Calcined coprecipitated powders were sintered in the temperature range of 1300º-1500ºC.

Ceramic characterization was carried out by using water inmersion method for bulk densities measurements, XRD and SEM. Vicker's indentation method was employed to obtain fracture toughness data. Dc and ac-impedance spectroscopy in the frequency range of 10^{-1} to 10^6 Hz in a SOLARTRON 1255-1286 model were used for electrical characterization, over a temperature range of 300º to 800ºC.

RESULTS AND DISCUSSION

Densification

Densification evolution as a function of both temperature and Yb_2O_3 contents are shown in FIGURE 1A and 1B. Density goes through a maximum for the sample composition of 3.5 mol% Yb_2O_3,and then decreses for higher amounts of Yb_2O_3 (Fig.1A). It could be explained taking into account the phase equilibrium diagram, since that composition can correspond to a single tetragonal phase ceramic. For the 2.5 mol% Yb_2O_3 an unusual density decreasing at 1450ºC can be clearly seen in FIGURE 1B. It could be due to the presence of monoclinic phase in the cooled sample or, on the other hand, to a grain size above the critical grain size for the martensitic transformation.

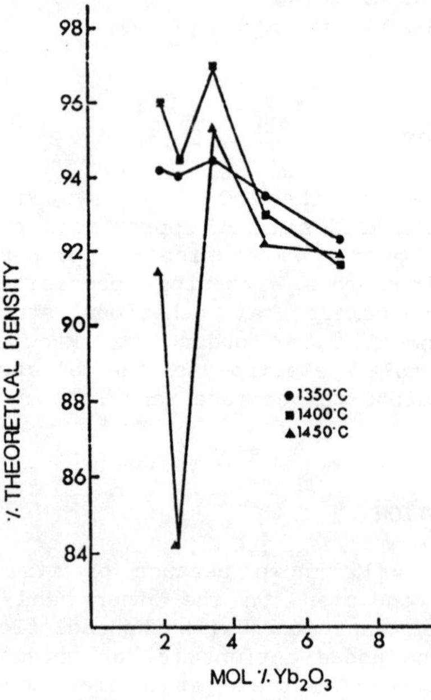

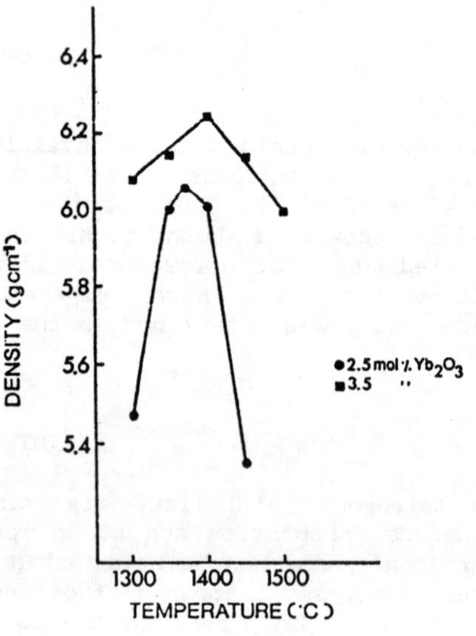

Figure 1. **a)** Density evolution for the sintered samples as a function of Yb_2O_3 content for several temperatures.
b) Comparative behaviours with temperature between 2.5 and 3.5 mol% Yb_2O_3 compositions.

Microstructure

The microstructure observed in FIGURES 2A,B and C indicate a fine-grained tetragonal matrix (at 3.5 mol% Yb_2O_3), with an average grain size about 0.5 m (FIG 2A), in other case (5 mol% Yb_2O_3) isolated cubic grains are noticed with an averange grain size of - 2 m (FIG.2B) and in FIGURE 2C can be observed a fully stabilized cubic phase at 7 mol% Yb_2O_3 composition.

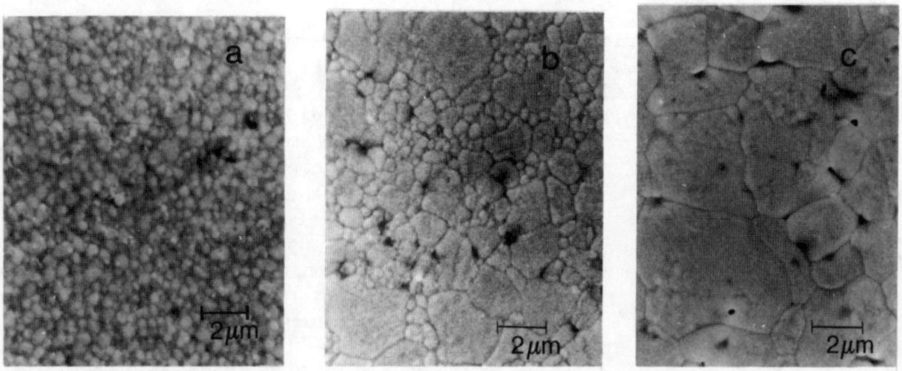

Figure 2. Microstructural SEM development for the following compositions sintered at 1400ºC during 2 hours:
a) 3.5 mol% Yb_2O_3; **b)** 5 mol% Yb_2O_3; **c)** 7 mol% Yb_2O_3

Mechanical Behaviour

Fracture toughness versus Yb_2O_3 contents is described in FIGURE 3. The critical stress intensity factor decreases as Yb_2O_3 amounts increases and that fact agrees well with the phase increasing symmetries. However, in the tetragonal region (2.5-4 mol% Yb_2O_3) at 1350ºC there is a K_{IC} maximum at 3.5 mol% Yb_2O_3. such a behaviour could be justified if it is taking into account the critical grain size and the vinicity of the phase boundary. Concerning to the compositions located at 2 and 2.5 mol% Yb_2O_3 which at 1450ºC exist as tetragonal+monoclinic structure, their density is very low and therefore K_{IC} drops sharply.

The both average grain size and fracture toughness as a function of heat-treatment is shown in FIGURE 4. As time and temperature increases the grain size grows because of the arising cubic phase, which causes a quickly K_{IC} decreasing.

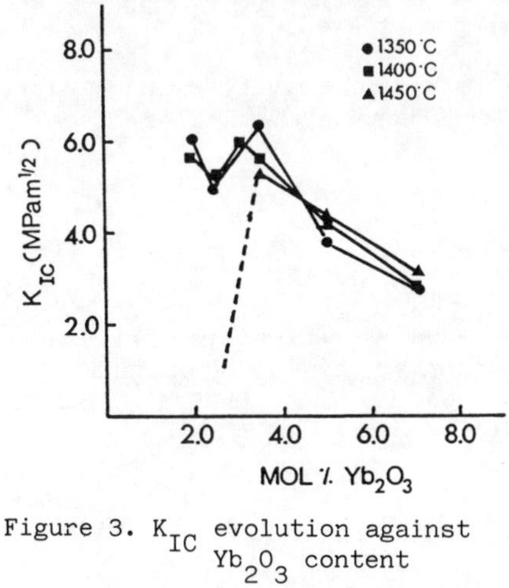

Figure 3. K_{IC} evolution against Yb_2O_3 content

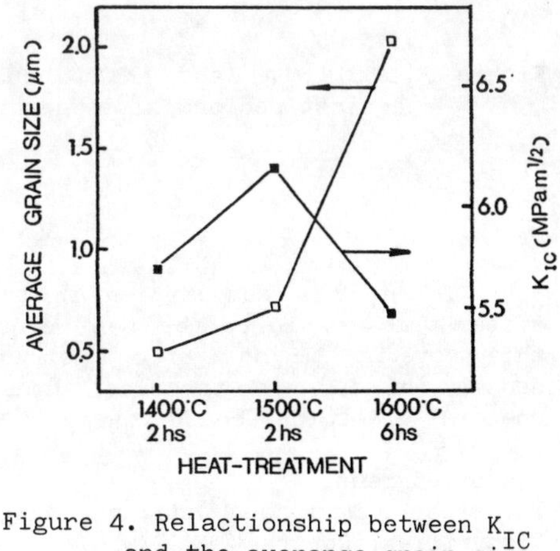

Figure 4. Relactionship between K_{IC} and the averange grain size with various heat-treatments

Ionic Conductivity

Both grain interior and grain boundary ionic conductivity data were calculated from the ac-impedance spectra. FIGURE 5A shows the bulk conductivity versus reciprocal temperature for four selected compositions (2.5, 3.5, 5, and 7 mol% Yb_2O_3). It is observed that the curve slope increases as the oxide concentration raises, and, consecuently, the activation enthalpy for the conduction process. This behaviour is similar to that ocurring in the ZrO_2-Y_2O_3 systems (4,5).

Concerning to the grain boundary conductivity versus reciprocal temperature curves (FIG.5B), it can be seen that the lowest value is obtained for a 3.5 mol% Yb_2O_3 tetragonal single phase ($-6*10^{-3}$ Scm^{-1} at 700ºC), where the average grain size of about 0.5 m is so small that makes a large grain boundary density. However, the 2.5 mol% Yb_2O_3, a monoclinic + tetragonal composition, has an average grain size larger and the grain boundary conductivity is also higher than that for 3.5 mol% Yb_2O_3. The grain boundary conductivities of 5 and 7 mol% Yb_2O_3, which are basically cubic symmetries, is therefore augmented. That fact is well illustrated in FIGURES 6A, B and C, where it can be noticed the arc evolution as a function of heat-treatments induced phases. As temperature raises the phase symmetry changes from tetragonal+monoclinic (A) to tetragonal+cubic (C) and therefore the grain boundary arc is being diminished. On the right side of FIGURE 6, the SEM micrographs illustrate the microstructure of the samples tested.

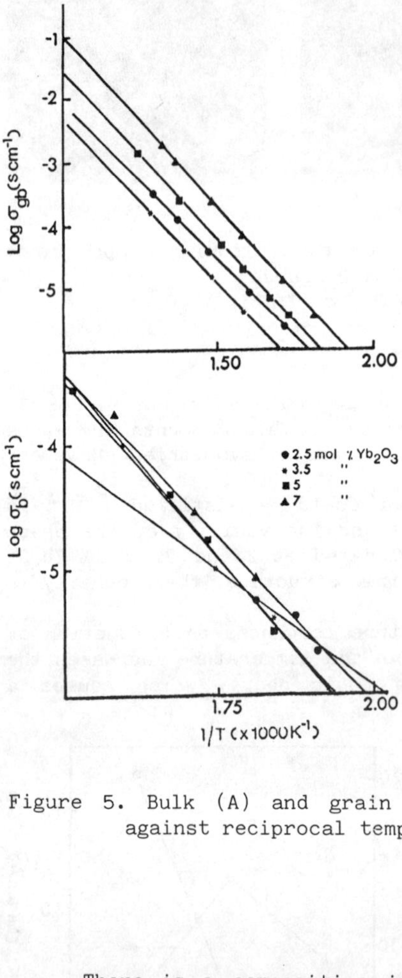

Figure 5. Bulk (A) and grain boundary (B) conductivity contributions against reciprocal temperature for several compositions

CONCLUSION

There is a composition in the ZrO_2-rich region of the ZrO_2-Yb_2O_3 system which exhibits the best electrical and mechanical properties. In this way, a 3.5 mol% Yb_2O_3 tetragonal single phase composition with density of 97% (% theorical density) shows a K_{IC} = 6 $MPam^{1/2}$ and a dc-total electrical conductivity of $1.3x10^{-2}$ $\Omega^{-1}cm^{-1}$ at 800ºC.

Acknowledgment
The present work was under the auspics of the Comission of the European Communities, Contract EN3E/0167/E.

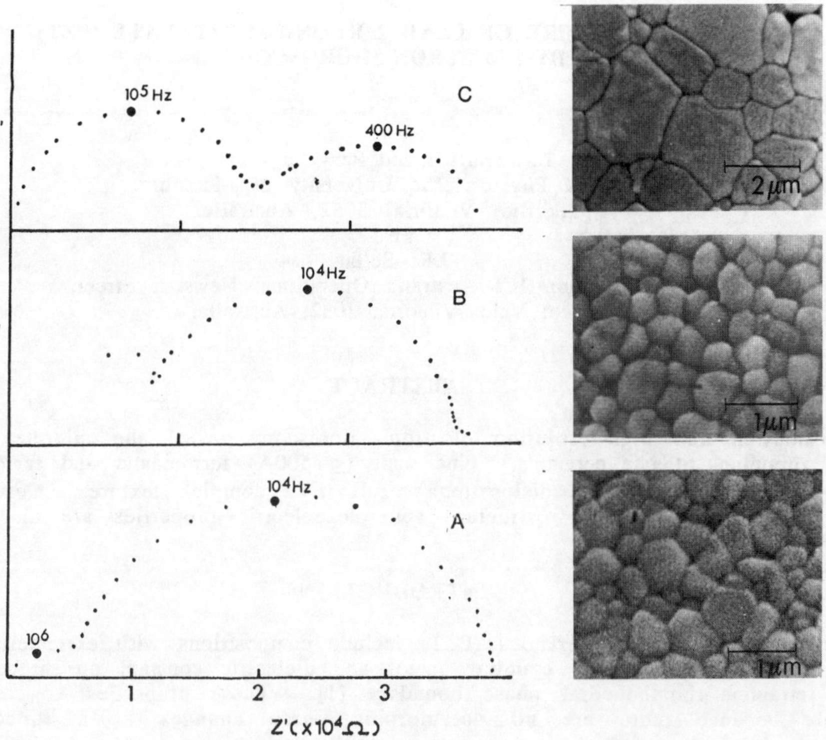

Figure 6. Complex resistivity plots and their respective SEM micrographs for samples containig 2 mol% Yb_2O_3 sintering at: a) 1400ºC-2hs, b) 1500ºC-2hs, c) 1600ºC-6hs

REFERENCES

1. Gogè, M., Letisse, G. and Gouet, M.,Electrochemical Study of Samaria and Ytterbia Stabilized Zirconia in the Low Temperature Range: 400º-800ºC. Sol. Stat. Ion., 1983, **9-10**, 937-44.

2. Yamamoto, O., Takeda, Y., Kanno, R., Kohno, K. and Kamiharai, T., Electrical Conductivity of Polycrystalline Tetragonal Zirconia ZrO_2-M_2O_3 (M= Sc,Y,YB). J. Mater. Sci. Letters, 1989, **8**, 198-200.

3. Durán, P., et al., Y(Er)-Doped Tetragonal Zirconia Polycrystalline Solid Electrolyte. Part 1. Powder processing. J. Mater. Sci., 1989, **24**, 708-16

4. Bonanos, N., Stotwinski, R.K., Steele, B.C.H. and Butler, E.P., High Ionic Conductivity in Polycrystalline Tetragonal Y_2O_3-ZrO_2. J. Mater. Sci. Letters, 1984, **3**, 245-8.

5. Durán, P., Recio, P., Jurado, J.R., Hernandez, M.T. and Moure, C., Y(Er)-Doped Tetragonal Zirconia Polycrystalline Solid Electrolyte. Part 3. Electrical Properties. J. Mater. Sci., 1989, **24**, 717-725.

MICROSTRUCTURE OF LEAD ZIRCONIUM TITANATE (PZT) BY ELECTRON MICROSCOPY

by
L.A. Bursill and Peng Ju Lin
School of Physics, The University of Melbourne,
Parkville, Victoria, 3052, Australia.
and
J.R. Sellar
Research Group, ICI Australia Operations, Newsom Street,
Ascot Vale, Victoria, 3032, Australia.

ABSTRACT

Transmission and high-resolution electron microscopy reveal the microtexture of lead zirconium titanate ceramics. Fine scale ($\leq$ 500Å) ferroelastic and ferroelectric twin domains, as well as dislocations result in a complex texture. Correlations between stoichiometry, microstructure and piezoelectric properties are discussed.

INTRODUCTION

Lead zirconium titanate ceramics (PZT) include compositions with extremely strong piezoelectric effects. The coupling factor and dielectric constant are highest near the tetragonal-rhombohedral phase boundary [1]. These properties are relatively stable because there are no polymorphic phase changes as a function of temperature and the Curie points are typically 360°C, compared with 120°C for $BaTiO_3$.

The aim of the present work was to observe submicron texture for PZT ceramics of commercial interest. Bright- and dark-field transmission electron microscopy are used to characterize the shape and size distribution of ferroelastic and ferroelectric twin domains in tetragonal and rhombohedral phases of PZT. This allows much finer detail to be revealed than was the case for surface replica techniques [2]. Attempts are also being made to apply atomic resolution electron microscopy to investigate domain wall widths and surface roughness, as well as interactions between the twin variants and between the twins and dislocations due to plastic deformation. Preliminary results are reported here for a pure PZT specimen $Pb(Zr_{0.53},Ti_{0.47})O_3$.

It was discovered that PZT containing $NiNbO_3$ or SrO tends to decompose chemically during observation under electron irradiation at high magnifications. Some of these results are included since they concern the more general problem of radiation hardness of PZT ceramics.

EXPERIMENTAL METHODS

Samples of $Pb(Z_{0.53},Ti_{0.47})O_3$, $(Pb_{0.94},Sr_{0.06})(Zr_{0.53},Ti_{0.47})O_3$ and $Pb(Zr_{0.53},Ti_{0.47})O_3$ + 6 mol% $NiNbO_3$ were prepared using Z-TECH EF super grade zirconia powder, together with PbO, TiO, NiO, SrO and Nb_2O_5 powders in appropriate proportions. The powders were wet-ball-milled for 12h, dried, pressed into pellets, calcined at 870°C for 2h, milled for a further 12h, dried and hot-pressed at 660°C for

2h, then sintered at 1220°C for 2h. Silver contacts were applied to polished ceramic discs by firing for 10 min at 700°C. Finally the discs were poled at 120°C for 10 min in oil under a DC field of 2.5 kV/mm.

X-ray powder diffractometry showed that the predominant phases were tetragonal (PZT), tetragonal (PZT + SrO) and rhombohedral (PZT + 6 mol% $NiNbO_3$).

Thin specimens suitable for transmission electron microscopy were prepared by mechanical grinding and dimpling, followed by ion-beam thinning using argon. Additional high-resolution specimens were prepared by crushing the ceramic disc under chloroform in an agate mortar and pestle, and dispersing thin fragments on a holey carbon support film. Thin edges were then carefully aligned with <001>, <011> or <111> directions (pseudocubic perovskite-type indices) parallel to the incident electron beam. A JEOL-4000EX instrument was used, operated of 400 keV, and capable of point resolutions of 1.6Å.

RESULTS

(a) $Pb(Zr_{0.53},Ti_{0.47})O_3$. Low magnification shows grain size of about 7 µm. Complex intersecting fringe contrasts, indicative of multiple polysynthetic twin lamellae give rise to textural detail within each grain on the scale of ≃ 50 nm (Fig. 1a). The enlargement (Fig. 1b) shows lath-like domains, typical of 90° twin lamellae, as well as many fringe contrasts due to extended defects inclined to the projection axis. The latter contrast is typical of 180°-domains [2]. A dark-field image (Fig. 1c) shows a complex dislocation interaction with twin lamellae. The somewhat overwhelming complexity of the domain contrasts, and the fine scale, have prevented a systematic identification of the many combinations of ferroelastic and ferroelectric twin domains possible in this material. Further dark-field analysis, using different low-order reflections, is being attempted using annealed specimens, when the domain texture is expected to be less complex and, at least, occurring on a coarser scale.

a — 1µm b Fig. 1 Domain Texture of PZT c — 20nm

High-resolution images of this specimen showed areas of perfect perovskite-type structure with changes in image detail and symmetry occurring on the scale of several 100Å (Fig. 2a). Such changes are consistent with the existence of coherent ferroelastic and/or ferroelectric domains (see Discussion below).

(b) $Pb(Z_{0.53},Ti_{0.47})O_3$ + 6 mol% $NiNbO_3$. This rhombohedral specimen proved remarkably sensitive to electron irradiation, changing dramatically during observation for several minutes at 400 keV (Figs. 3a-c). The small (150Å diameter), multiply-twinned particles were relatively very stable. They were identified as exsolved zirconia, by comparison of the diffraction patterns after irradiation (Fig. 3d) with data for polycrystalline zirconia powders. The diffraction pattern for the as-prepared PZT + $NiNbO_3$ is shown for comparison in Fig. 3e. The small particles were surprisingly active, fluctuating in orientation and changing their twinned structure and shape during observation. The remaining material became glassy (amorphous) after 10-20 minutes observation.

It has not yet been possible to complete bright- and dark-field studies of the domain structure of this rhombohedral phase. However, some high-resolution images were obtained (Fig. 2b) before significant decomposition occurred in the beam. Thus a fine-scale coherent microdomain texture was revealed by local changes in periodicity and symmetry of the lattice images, with fluctuations on the scale of 50-100Å. This texture was remarkably similar to results obtained previously for silver-ion-exchanged lithium tantalate [3]. The corresponding proton- and Fe-doped lithium niobates showed similar microdomain textures and electron beam sensitivity [4,5]. The papers [3-5] should also be consulted for a discussion of the coherency and structural relationships between various pseudocubic and rhombohedral perovskite phases.

(c) $(Pb_{0.94},Sr_{0.06})(Zr_{0.53},Ti_{0.47})O_3$. This specimen gave results intermediate between the previous two cases. Thus there was significant chemical reactivity under electron irradiation. The final decomposition product has not yet been identified. The high-resolution images were similar to those shown in Fig. 2.

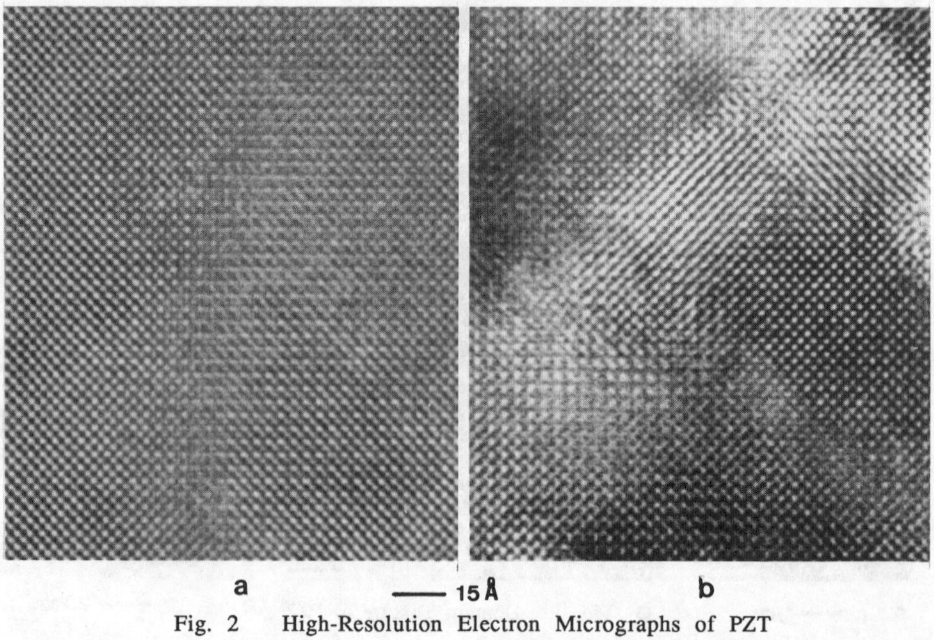

Fig. 2 High-Resolution Electron Micrographs of PZT

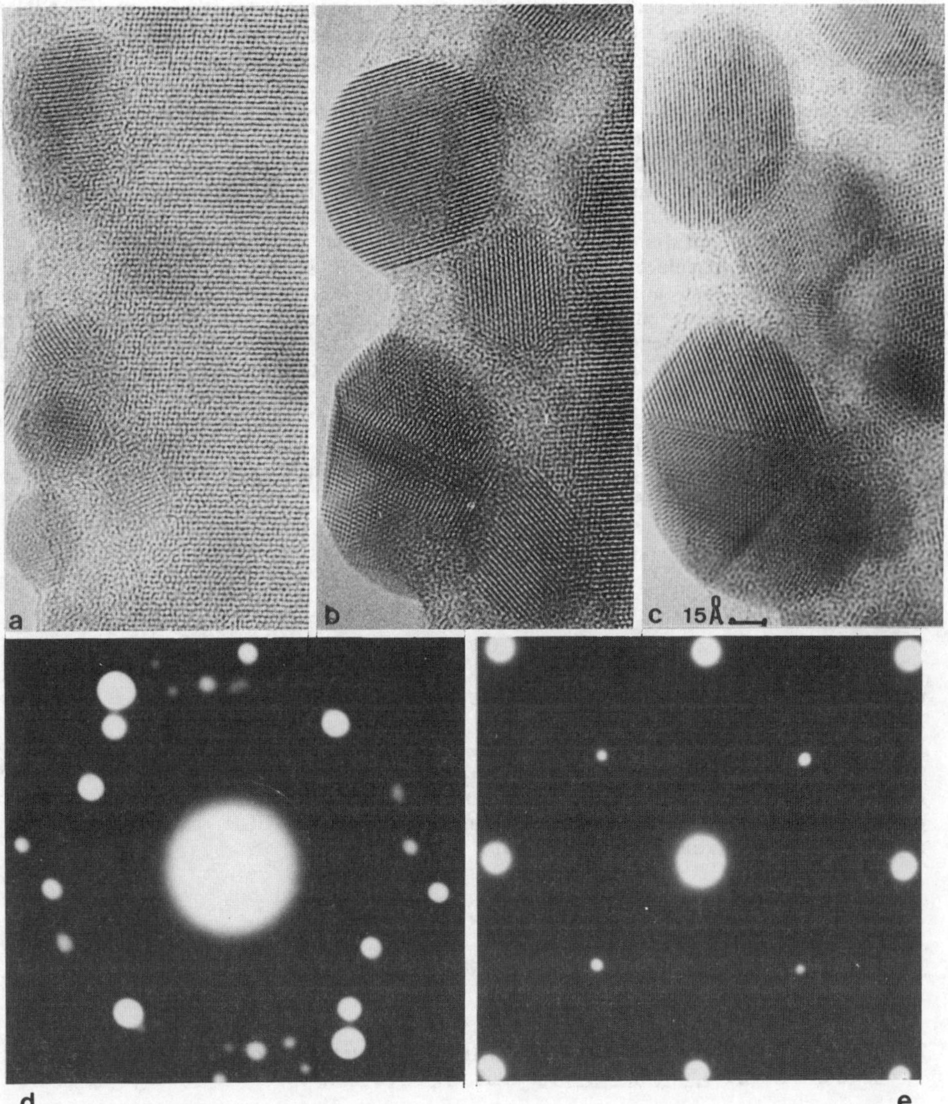

Fig. 3 Electron-Irradiation Effects on PZT + 6 mol% $NiNbO_3$

DISCUSSION

Given the observed microstructures for pure PZT it is apparent why predominantly tetragonal phase preparations should show significant hysteresis effects in poling and depoling fields. It is clear why the combination of activation of a large number of ferroelastic and ferroelectric twinning variants, together with some plastic deformation during processing and poling below and Curie temperature, may be a unique property of the perovskite based crystal structure. Thus movement of ferroelastic domains will allow a mechanical response (shape change) which interacts intimately with the ferroelectric twin domain distribution, essentially via octahedral tilting within the perovskite structure [3-6].

The coherent fine-scale microdomain texture observed for the 6 mol% $NiNbO_3$ specimen should arise due to a combination of multiple twinning of the rhombohedral phase and structural fluctuations (local changes of space group) due to inhomogeneity of $NiNbO_3$ distribution. Given the relatively small pseudocubic distortion in this case, i.e. a slight stretching along one of the eight equivalent <111> perovskite cell direction ($\alpha = 89°40'$), we expect that predominantly rhombohedral phase PZT ceramics may show distinct hysteresis characteristics, compared to tetragonal varieties. This point remains to be quantified.

The present results refer to standard ceramic processing routes, known to give favourable piezoelectric products. However, it seems to us that the existing processing routes are almost certainly not fully optimized. The present results demonstrate that knowledge of the interaction between stoichiometry, processing, microstructure and piezoelectric characteristics is rather primitive. Further microtextural studies are required to allow some of the experimental parameters to be monitored and controlled (e.g. domain size and distribution). The effects of controlled amounts of plastic deformation, for restricting twin boundary movement, should be productive in producing "hard" piezoelectrics. The role of microstructural changes in understanding the ageing characteristics in various environments also remains to be quantified.

CONCLUSIONS

It is established that a rich variety of microstructural information is accessible in PZT, using TEM and HREM imaging techniques. Further analytical studies are continuing. Low irradiation dose techniques will be required to study the microtexture of PZT + additives.

This work was supported financially by the Australian Research Council. Peng Ju Lin is grateful for the award of a National Research Fellowship. We are grateful for the support of D. Cousins and A. Vecchie (Z-TECH) also C. Blake and J. Kariuki (ICI Australia Research Group).

REFERENCES

1. Jaffe, B., Cook, W.R. and Jaffe, H., Piezoelectric Ceramics. Academic Press, NY, 1971, **Ch. 7**.

2. Gerson, R., Variations in Ferroelectric Characteristics of PZT Due to Chemical Modification. J. Appl. Physics, 1960, **31**, 188-194.

3. Peng Ju Lin and Bursill, L.A., Silver-Ion-Exchanged $LiTaO_3$: Intergrowth Defect Structures. Ferroelectrics, 1988, **77**, 69-70.

4. Bursill, L.A. and Peng Ju Lin, Electron-Beam Sensitivity of $LiNbO_3$(Fe,Ti,H). Ferroelectrics, 1988, **77**, 81-89.

5. Yan Yong, Feng Duan, Peng Ju Lin and Bursill, L.A., Electron Microscopic Study of Proton-Exchanged $LiNbO_3$. Ferroelectrics, 1988, **77**, 91-100.

6. Glazer, A.M., Octahedral Tilting in Perovskites. Acta Crystallogr., 1972, **B28**, 3384-89.

EFFECT OF MICROSTRUCTURE ON THE DIELECTRIC PROPERTIES OF LEAD MAGNESIUM NIOBATE CERAMICS

DONG HEON KANG and KI HYUN YOON
Department of Ceramic Engineering, Yonsei University
Seoul, Korea

ABSTRACT

The effects of microstructure on the dielectric properties, such as dielectric constant, K, and dielectric hysteresis characteristics of lead magnesium niobate (PMN) ceramics have been studied as a function of the amount of excess MgO. The dielectric constant and remanent polarization(P_r) increased with increase in grain size, and then decreased with grain growth inhibitation. The diffuseness coefficient (δ) showed a tendency nearly reverse to that for the dielectric constant and remanent polarization change.

INTRODUCTION

Ferroelectric lead magnesium niobate ($Pb(Mg_{1/3}Nb_{2/3})O_3$, PMN) with a perovskite structure has recently attracted interest for use in MLC and electrostrictive devices due to the large K and the low firing temperature. However a limitation to the utilization of PMN in device application has been the lack of a simple, reproducible fabrication technique for pure perovskite PMN ceramics. Earlier studies have reported that the dielectric properties of PMN are generally affected by the pyrochlore phase and also by the microstructure (1,2,3). The present work was undertaken to prepare pyrochlore-free PMN ceramics with both 40 m/o excess PbO and excess MgO in the range of 0 to 70 m/o and to investigate the effect of microstructure, such as grain size, density, and second phase precipitation, on the dielectric properties of PMN ceramics.

MATERIALS AND METHODS

The oxides used to synthesize the ceramic composition were reagent grade PbO, MgO and Nb_2O_5 powders. Compositions investigated in this experiment were the following : stoichiometric composition (PMN-N), specimens with excess MgO composition with 40 m/o excess PbO (PMN-0 to PMN-6), as shown in Table 1. The fabrication process used is the mixed oxide procedure, apart from the addition of excess PbO after the 4 hr calcining step at 800°C. PVA binder (1 w/o) was added to the calcined powders which were then pressed and sintered at 900°C for 4 hrs. Polished and fractured sections were examined by a SEM equipped with an EDX. And etching with 5%HCl+0.5%HF for 10 mins. were used to reveal the grain boundary. The mean grain sizes were determined by the linear intercept method. For dielectric measurements were carried out using a LCR meter and the dielectric hysteresis loops were observed by using the Sawyer-Tower circuit.

RESULTS AND DISCUSSION

The physical and dielectric properties of the PMN series specimens as a function of the amount of excess MgO are reported in Table 1. The PMN series specimens (PMN-0 to PMN-6) processed with 40 m/o excess PbO were fired at 900°C but the PMN-N specimen with no excess PbO added was fired at 1000°C. As tabulated, the density and grain size increased slightly with excess PbO addition due to the liquid phase sintering. Furthermore the formation of pyrochlore phase which mainly caused by the volatilization of PbO during the sintering process was suppressed effectively(4). The XRD patterns in Fig. 1 show the reduction in the amount of the pyrochlore phase (cubic $Pb_3Nb_4O_{13}$, P_3N_2). Thus the increase in K with addition of excess PbO can be attributed to the decrease in the pyrochlore phase and the improvement in sinterability. Also as indicated in Table 1, a complete conversion to the perovskite PMN occurred with the addition of 5 m/o excess MgO (PMN-2) because of the compensation for poor dispersability and reactivity of MgO through the addition of excess MgO.

However, the behavior of the dielectric properties in all of these compositions can not simply explained by the pyrochlore phase, because K decreased with the addition of MgO in excess of 10 m/o even no pyrochlore phase was found to be present. So in this work, the increase in K in the range of 0 - 5 m/o excess MgO can be explained by the increase of grain size,

TABLE 1
Physical and dielectric properties of the PMN series specimens as a function of the amount of excess MgO.

Specimen No.	Excess MgO (m/o)	Density (g/cm^3)	Average grain size (um)	Pyrochlore phase (%)	K_{max} T_C (^oC)	ρ_{25^oC} ($x10^{11}$) ($\Omega \cdot cm$)	δ	P_r (uC/cm^2)
PMN - N	0	6.86	1.4	22	4500(-9)	3.1	81	5.1
PMN - 0	0	7.78	2.0	9	10200(-10)	13	59	5.7
PMN - 1	2	7.80	3.1	3	10800(-6)	5.2	55	6.2
PMN - 2	5	7.82	4.2	0	13200(-4)	2.2	52	6.6
PMN - 3	10	7.82	4.0	0	12500(-4)	2.3	53	6.4
PMN - 4	20	7.79	3.75	0	12010(-5)	2.4	59	5.2
PMN - 5	50	7.65	3.6	0	10780(-4)	2.0	62	5.3
PMN - 6	70	7.50	3.58	0	10060(-5)	1.9	64	5.0

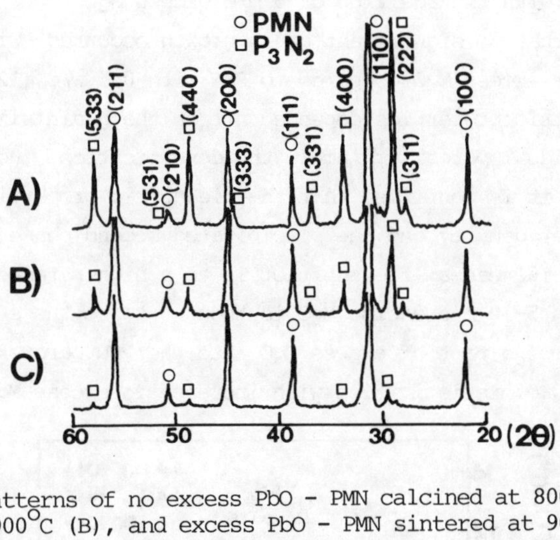

Figure 1. XRD patterns of no excess PbO - PMN calcined at 800^oC(A), sintered at 900^oC (B), and excess PbO - PMN sintered at 900^oC (C).

density and also the removal of the pyrochlore phase. The grain size dependency is related to the relative volume of grain boundary with lower K (pyrochlore, impurities, MgO, PbO, etc.), which is electrically connected in series with the bulk of the grain(1). For greater than 10 m/o excess MgO the decrease in K is believed to be due to the decrease of grain size and density, especially the precipitation of the second phase as indicated in Fig. 2-(C). In the PMN-6 specimen of Fig. 2-(C), the area (b) was identified as the MgO rich second phase and the area (a) as a PMN matrix, which were analyzed by EDX. Fig. 2-(A), (B) show the no excess MgO added PMN-0 and the grain grown PMN-2 specimen which contains 5 m/o excess MgO, respectively. Swartz et al.(1) explained that K increased due to the significant increase

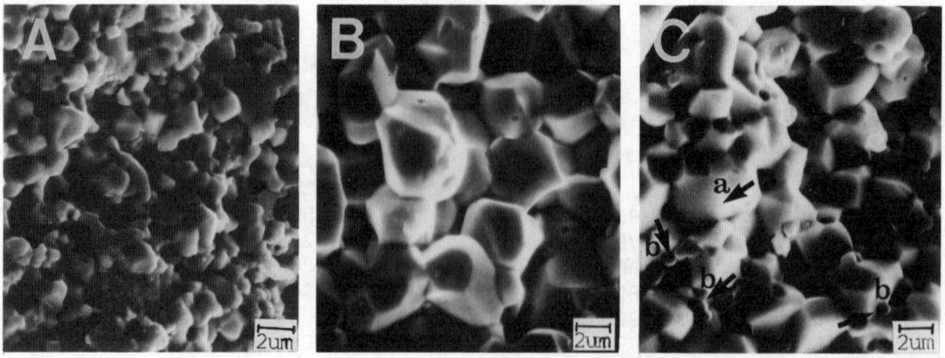

Figure 2. SEM photographs of (A) PMN-0 (B) PMN-2 and (C) PMN-6 (a ; PMN matrix, b ; MgO rich second phase).

in grain size with the addition of 5 m/o excess MgO. On the contrary, Guha (2) reported that no significant grain growth occurred with the addition of 3 w/o (72 m/o) excess MgO compared to PMN with 0.5 w/o (12 m/o) excess MgO, so the dielectric properties depended not on the grain size but on the amount of residual porosity, namely the densification. However, according to this work, it is confirmed that the dielectric properties are affected by the grain size, density and the precipitated second phase. When the amount of excess MgO increases, the shift of T_C to a higher temperature for the specimens with 5-10 m/o excess MgO is thought to be the effect of incorporation of a part of excess MgO into the PMN perovskite cell(4). Also the degraded d.c. resistivities with increase in excess MgO may have been

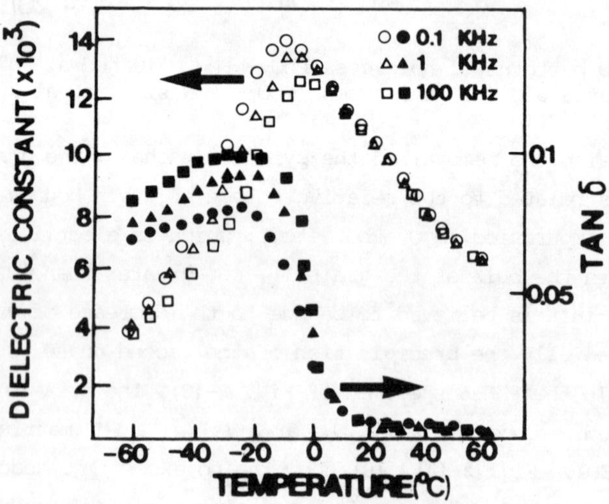

Figure 3. K and Tan δ vs. T for PMN-2 specimen as a function of frequency.

caused by a similar effect(4). But for excess MgO greater than 10 m/o the variations can not be directly correlated to the composition. As already observed in other Pb-perovskite type compounds, K and TANδ vs. T curves in Fig.3 show a broad maxima for K and an increase in T_C with increasing frequency, characteristic of relaxor ferroelectrics, and a diffused phase transition is also apparent. So δ has been calculated and shown in Table 1. The lower K corresponded to the larger δ, and P_r increases up to a composition of 5-10 m/o excess MgO addition and then decrease, which probably can be explained by the effect of grain size(5,6). Also P_r of about 0.4-0.7 uC/cm^2 can still be observed near T_C perhaps owing to the induced polarization of paraelectric phase.

CONCLUSIONS

This work confirmed that the pyrochlore phase alone could not explain the dielectric properties of PMN ceramics. With the addition of up to 5-10 m/o excess MgO, the pyrochlore phase was completely eliminated. At the same time, K and P_r increased with increasing grain size and density. But further addition of excess MgO caused a degraded K and P_r mainly due to increase in concentration of the precipitated MgO rich second phase, and decrease of grain size and density.

ACKNOWLEDGEMENT

This work was supported by the Ministry of Education of Korea.

REFERENCES

1. Swartz, S. L. and Shrout, T. R., Dielectric properties of lead-magnesium niobate ceramics. J. Am. Ceram. Soc., 1984, **67** (5), 311-15.
2. Guha, J. P., Comment on "Dielectric properties of lead-magnesium niobate ceramics". J. Am. Ceram. Soc., 1985, **68** (3), C-86 - C-87.
3. Kang, D. H. and Yoon, K. H., Dielectric properties due to excess PbO and MgO in lead magnesium niobate ceramics. Ferroelectrics, 1988, **87**, 255-64.
4. Lejeune, M. and Boilot, J. P., Low firing dielectrics based on lead magnesium niobate. Mat. Res. Bull., 1985, **20**, 493-99.
5. Haertling, G. H., Hot-pressed lead zirconate-lead titanate ceramics containing bismuth. Am. Ceram. Soc. Bull., 1964, **43**(12), 875-79.
6. Martirena, H. T. and Burfoot, J. C., Grain-size effects on properties of some ferroelectric ceramics. J. Phys. C.: Solid State Phys., 1974, **7**, 3182-92.

PYROELECTRIC PZT CERAMICS

JEAN-PAUL HAHN[1,2], BENEDIKT BARBISCH[6], RENE BEYELER[4], PETR BOHAC[2,3], MARTIN FORSTER[5], GERNOT KOSTORZ[1], ALFRED MARTIN[6]
[1] Angewandte Physik, ETH Zürich, CH 8093 Zürich; [2] formerly AFIF c/o ETH Zürich, CH-8093 Zürich; [3] now at Nichtmetallische Werkstoffe, ETH Zürich, CH-8092 Zürich, [4] AFIF, c/o ETH, CH-8093 Zürich; [5] Cerberus AG, CH- 8708 Männedorf; [6] Schaffner AG, CH- 4708 Luterbach

ABSTRACT

The pyroelectric properties of $Pb(Zr,Ti)O_3$ ceramics were studied as a function of Zr/Ti ratio, stoichiometry and doping. The best figure of merit is obtained for MnO_2 doped ceramics (1.89 mol% MnO_2) and 0.7 mol% lead oxide deficiency. The variation of the Zr/Ti ratio showed a steep rise of the figure of merit on passing the morphotropic phase boundary from the tetragonal to the rhombohedral phase region.

Doping PZT 60/40 ceramics with MnO_2 or U_3O_8 improves the figure of merit in the solubility range. A second phase was detected for U_3O_8 contents ≥ 0.5 mol%; increasing the doping concentration of MnO_2 increased the presence of ZrO_2 particles, and a second phase appears for concentrations > 2 mol%.

INTRODUCTION

Lead zirconate titanate ($Pb(Zr,Ti)O_3$) ceramics exhibit a sharp maximum of the dielectric constant near the morphotropic phase boundary [1]. PbO volatility poses a problem during sintering of PZT ceramics and can be controlled by a bed powder consisting of either $PbZrO_3$ or PZT of the same composition. These powder beds have a great influence on the pyroelectric coefficient and the dielectric constant. The dielectric constant, the electrical resistivity and the pyroelectric coefficient are greatly influenced by doping the ceramic. MnO_2 [2] and U_3O_8 [3] reduce the dielectric constant, U_3O_8 enhances the pyroelectric coefficient [3] as well as the figure of merit F defined as $F = p/(\varepsilon \rho c_p)$ where p denotes the pyroelectric coefficient [$\mu C/m^2K$], ε the dielectric constant, ρ the density [g/cm^3] and c_p the specific heat [J/gK].

MATERIALS AND METHODS

For the investigations on stoichiometry, the starting materials were PbO(Fluka, puriss. p.a. >99%), ZrO_2(Degussa), TiO_2(Riedel-de-Haën, >99.8%), MnO_2(Cerac, >99.9%). For the other investigations $PbZrO_3$(Cerac, >99.7%) and $PbTiO_3$ (Cerac, >99.9%) were used for the basic composition. The oxides added to the basic composition (PZT 60/40), were MnO_2(Cerac) and U_3O_8(Cerac, >99.8%). All starting materials were weighed out in the desired molar amount and mixed in a wet milling process with HDPE containers and ZrO_2 balls (Tosoh).

The mixed powders were dried, crushed and passed through a 100 µm sieve before calcining at 900°C for 6 h. After calcining the baked powders were once more crushed and passed through the sieve before wet-milling them again with ZrO_2 balls for 20 h in a centrifugal ball mill (Fritsch Pulverisette 6).

After drying, the powders were spray-dried with an organic binder and then pressed into discs of 13 mm diameter at a pressure of 1.47 kbar yielding a green density greater than 60% of the theoretical density.

The specimens for the investigations on the stoichiometry were sintered at 1180°C using either $PbZrO_3$ or PZT of the same composition as atmosphere powder. All the other specimens were sintered at 1206°C using powder of the same composition as atmosphere powder. The resulting density was greater than 97% of the theoretical density. The sintered and lapped specimens were electroded with a silver-palladium paste. After contacting, the specimens were poled in a silicon oil-bath at 170°C, under an applied voltage of 2.5 kV/mm for 10 minutes.

One pill of each batch was lapped, polished and chemically etched for examinations for grain size distribution and the occurrence of second phases. The dielectric constant and the resistivity were measured with an RLC-meter at a frequency of 159 Hz. The pyroelectric coefficient was determined by measuring the pyroelectric current generated during heating the pill at a constant rate.

RESULTS

The stoichiometry of the sintered specimens depends on PbO partial pressure. Sintering a stoichiometric composition of PZT 60/40 (1.9 mol% MnO_2) on a $PbZrO_3$ powder bed results in a weight (PbO) gain of about 0.4 wt% while sintering on a PZT bed leads to a weight loss of 0.7 wt%. For compositions with PbO excess excess the powder bed has only a minor influence because the partial pressure of PbO is higher than for $PbZrO_3$ or PZT. These weight changes during sintering must be accounted for in the investigations on the influence of stoichiometry on the pyroelectric coefficient. Making these corrections for PbO loss, we get a sharp maximum in F near -0.7 mol% PbO (Fig. 1), independent of the powder bed used. The figure of merit is increased by the use of $PbZrO_3$ powder bed.

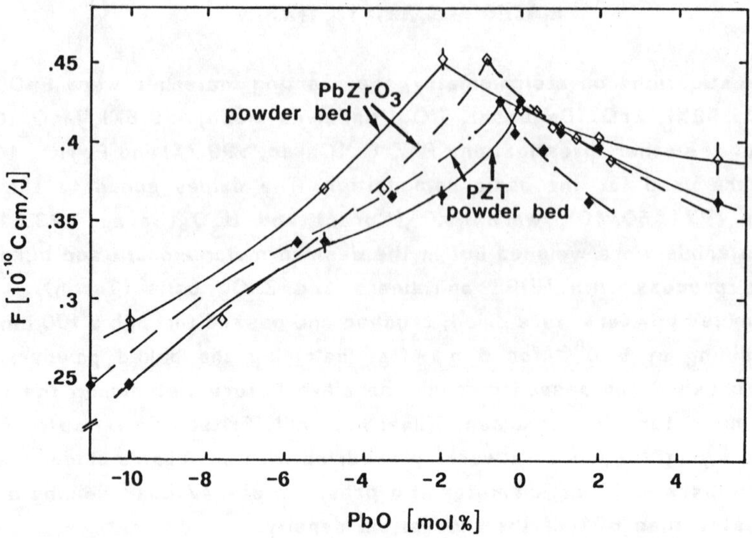

Figure 1. Figure of merit F (---) with and (——) without corrections for PbO loss during sintering.

The figure of merit presents a sharp rise near the morphotropic phase boundary and runs into a plateau for $PbZrO_3$ contents higher than 55 mol%, i.e. for compositions with a rhombohedral structure.

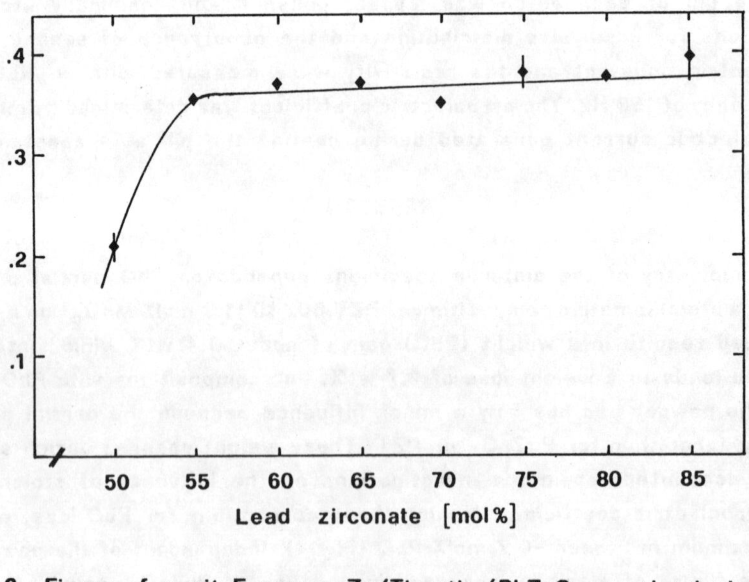

Figure 2. Figure of merit F versus Zr/Ti ratio ($PbZrO_3$ powder bed; 1240°C sintering temperature).

Doping PZT 60/40 ceramics with MnO_2 raises the pyroelectric coefficient from 172 (undoped) to 382 [$\mu C/m^2K$] (doped with 2 mol% MnO_2).Simultaneously the figure of merit increases from 0.14 to 0.31×10^{-10} [Ccm/J].

Doping with U_3O_8 first increases p from 172 to 250 [$\mu C/m^2K$] for 0,1 mol% added. At higher concentrations of U_3O_8 p decreases to 187 [$\mu C/m^2K$] for 2 mol% added.

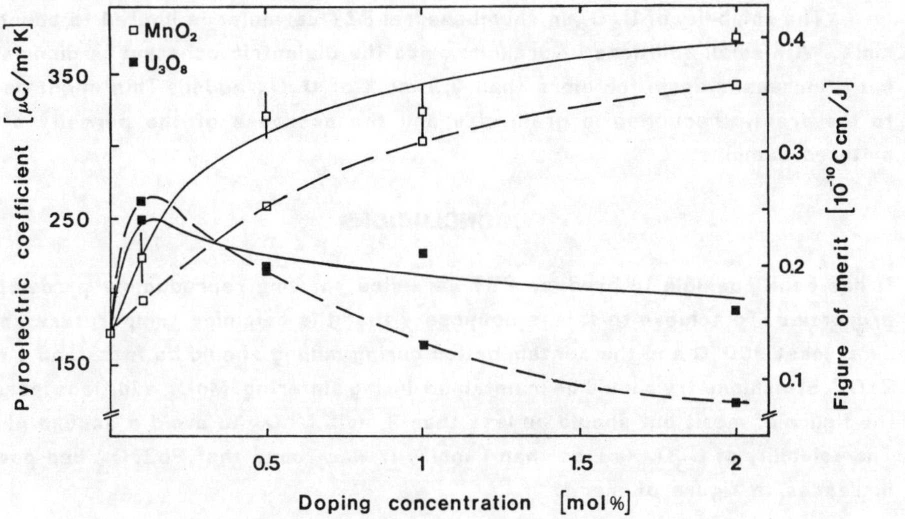

Figure 3. Pyroelectric coefficient (---) and figure of merit (——) as a function of doping content.

DISCUSSION

Stoichiometry

With increasing deviation from stoichiometry the figure of merit of MnO_2 doped ceramics decreases. Wu et al. [4] showed that Mn doping may occupy A- as well as B-sites in the perovskite lattice. The observed shift of the maximum in F near - 0.7 mol% PbO might be caused by partial substitution of Pb by Mn according to the following equation (□ = vacancy)

$$ABO_3 + x\, M_2O_3 \left[A_{1-3x}\, M_{2x}\, \square_x\right]BO_3 + 3x\, AO\nearrow \qquad (1)$$

In eq. (1) it is assumed that MnO_2 has been reduced to Mn_2O_3.

Zr/Ti ratio

Uchida [5] showed that the maximum obtainable polarisation after poling of ferroelectric ceramics is higher for rhombohedral ceramics than for tetragonal ones. This leads to a higher pyroelectric coefficient and a higher figure of merit.

Doping

Doping with MnO_2 increases the domain wall mobility due to the formation of A-site vacancies which are caused by the substitution of Pb^{2+} by Mn^{3+}. The increased mobility facilitates the poling of the doped ceramics and therefore leads to a higher pyroelectric coefficient.

The solubility of U_3O_8 in rhombohedral PZT ceramics is limited to about 0.5 mol%. With small additions of uranium oxide the dielectric constant is diminished, but it increases again for more than 0.5 mol% of U_3O_8 added. This might be due to the drastic reduction in grain size and the decrease of the porosity of the sintered samples.

CONCLUSIONS

It has been possible to produce PZT ceramics showing reproducible pyroelectric properties. To achieve this, it is necessary that the calcining temperature should be at least 900°C and the contamination during milling should be less than 1 mol% ZrO_2. Stoichiometry should be maintained during sintering. MnO_2 additions increase the figure of merit but should be less than 2 mol% MnO_2 to avoid a second phase. The solubility of U_3O_8 is less than 1 mol%. It was found that $PbZrO_3$ bed powder increases th figure of merit.

ACKNOWLEDGEMENTS

The authors wish to thank Prof. L. J. Gauckler for helpful advice and discussions. Thanks are due to the Schweizerischer Nationalfonds zur Förderung der wissenschaftlichen Forschung for financial support.

REFERENCES

1. Jaffe, B., Cook, W.R. Jr., Jaffe, H., Piezoelectric ceramics, Academic Press, New York, 1971, p. 142.
2. Wu, L., Wei, Ch.-Ch., Wu, T.-S., Teng, Ch.-Ch., Dielectric properties of modified PZT ceramics. J. Phys. C: Solid State Phys., **16**,(1983)2803-2812.
3. Hardiman, B., Reeves, C.P., Zeyfang, R.R., Pyroelectric properties of modified lead titanate zirconate ceramics. Ferroelectrics, **12**,(1976),163-165.
4. Wu , L., Wu, T.-S., Wei, Ch.-Ch., Liu, H.-Ch., The DC resistivity of modified PZT ceramics. J. Phys. C: Solid State Phys., **16**,(1983)2823-2832.
5. Uchida, N., Ikeda, T., Electrostriction in perovskite-type ferroelectric ceramics. Jap. J. Appl. Phys, 1967, **10**,1079-1088

STRUCTURAL STUDIES OF PMN CERAMICS

P. Bonneau*, E. Husson* and A. Morell**
*Laboratoire de Chimie-Physique du Solide URA CNRS D0969
Ecole Centrale de Paris. 92 295 Chatenay-Malabry Cedex. France.
**Laboratoire Central de Recherches.Thomson CSF.
Domaine de Corbeville. 91 404 Orsay Cedex. France.

ABSTRACT

The disordered crystal structure of the cubic $PbMg_{1/3}Nb_{2/3}O_3$ was determined at room temperature by X-ray and neutron diffraction. The electron diffraction patterns revealed the existence of a superstructure and of diffuse streaks due to atomic shifts in the (110) planes of the structure. At high temperature the structure becomes ideal cubic perovskite. At low temperature no evolution was observed on the X-ray and neutron patterns, even at 5K except a sensible widening of the basis of some peaks.

INTRODUCTION

Lead magnesium niobate $PbMg_{1/3}Nb_{2/3}O_3$ (PMN) is one of the most interesting relaxor because of its dielectric properties and its low temperature of sintering. At room temperature it has the ABO_3 cubic perovskite structure of space group Pm3m with a_c#4Å. Its Curie temperature is about -7°C (1). According to Shebanov (2) the low-temperature phase is rhombohedral and the ferroelectric transition which is diffuse, occurs in a temperature range of several tens of degrees. The interesting dielectric properties are originated by the diffuse character of the ferroelectric transition; so, we have undertaken a complete structural study by X-ray, neutron and electron diffraction as a function of temperature.

MATERIALS AND METHODS

PMN ceramics have been prepared according to the process described by Swartz et al (3): first, magnesium niobate $MgNb_2O_6$ is formed, by calcination of MgO and Nb_2O_5 at 1050°C; then, perovskite PMN is obtained at 800°C by the reaction of PbO with $MgNb_2O_6$. To avoid the formation of a parasite pyrochlore phase, an excess of 4% mol

MgO was added prior to reaction with Nb_2O_5. Pellets were pressed and sintered at 960°C for 6 hours and O_2 flux in a PbO-rich atmosphere.

X-ray diffraction patterns were recorded at a slow Θ scan (0.05°/mn) on a high sensitivity diffractometer "Microcontrol" using Kα radiation of a copper rotating anode.

Neutron diffraction patterns were collected at the Institut Laue-Langevin using the D1A diffractometer specially adapted for the study of polycrystalline samples.

Selected aera electron diffraction (SAED) patterns were obtained with an electron microscope JEOL 100 CX.

RESULTS

The cubic structure at room temperature

Our recent work (4) on the structural refinement from X-ray diffraction patterns, by the integrated intensities method, has revealed disordered atomic positions in the PMN cubic phase at 297 K : Pb atoms shift of about 0.32 Å along [110] directions leading to twelve local disordered positions around the special position and Nb atoms shift of about 0.18 Å along [110] or [111] directions leading respectively to twelve or eight disordered positions around the special position.

This work has been completed by a neutron diffraction study.in order to find the displacements of oxygen atoms. The refinement calculations by the Rietveld's method, give shifts of about 0.16 Å for these atoms with local disordered positions uniformly distributed on rings in the faces of the cube. Figure 1 gives a model of the disordered structure with exaggerated atomic displacements corresponding to that found by Itoh et al (5) for the cubic $BaTiO_3$.

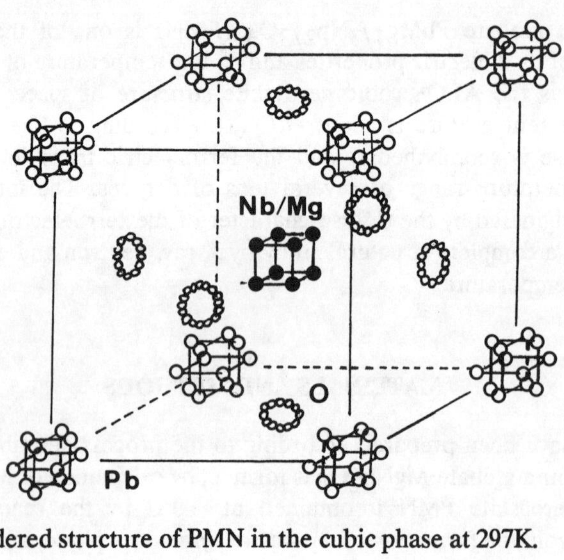

Figure 1: Disordered structure of PMN in the cubic phase at 297K.

The discrepancy factors defined by R_{wp} and R_B are respectively equal to 7.78% and 1.62% which is quite satisfactory:

$$R_{wp} = 100\Sigma_i w_i[y_i(obs) - y_i(calc)]^2 / \Sigma_i w_i[y_i(obs)]^2$$

$$R_B = 100\Sigma_i[y_i(obs) - Sy_i(calc)] / \Sigma_i y_i(obs) .$$

$y_i(obs)$ and $y_i(calc)$ are observed and calculated intensity at the ith step, $w_i = 1/y_i(obs)$ S is a scale factor.

The electron diffraction study has pointed out:
-the existence of superstructure spots of the type (h+1/2,k+1/2,l+1/2) which could correspond to a face centered cubic cell with a parameter $a\#2a_c$. The superstructure, not detected by X-ray diffraction, seems to characterize an order in microdomains; this result is confirmed by high resolution transmission electron microscopy images where we observed ordered microdomains of 40 to 100 Å. The superstructure could be induced by a 1-1 ordering of the Nb and Mg cations on the B site of the perovskite (6).

-the existence of sets of linear diffusion streaks which are due to a planar disorder in the (110) planes of the structure: this result is in good agreement with the shifts of the Pb and Nb/Mg atoms deduced from X-ray and neutron diffraction study. Figure 2 presents SAED patterns obtained for $(100)^*$, $(110)^*$ and $(111)^*$ reciprocal planes.

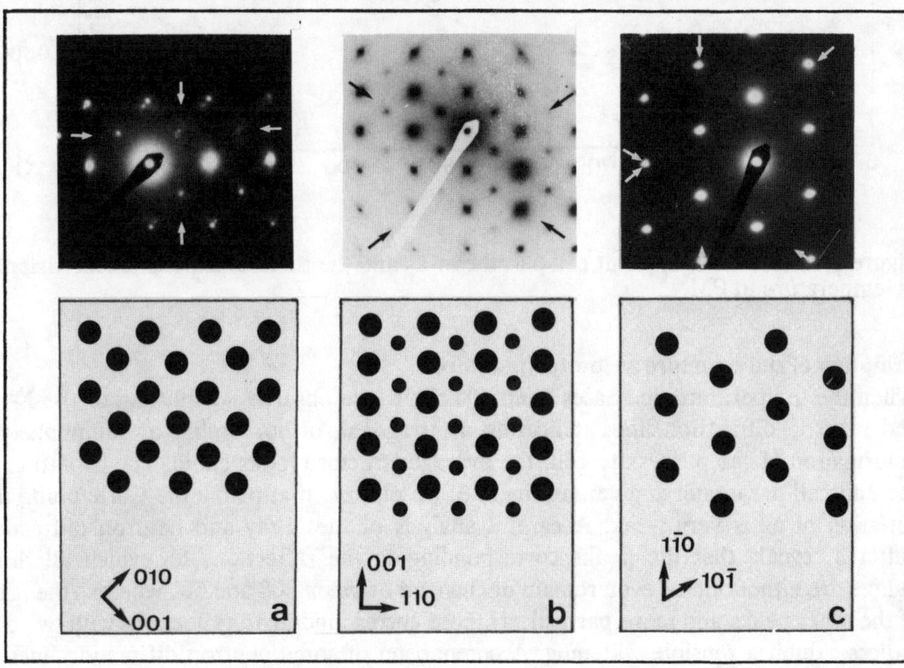

Figure 2: SAED patterns and reciprocal lattice planes: (a) $(100)^*$, (b) $(110)^*$, (c) $(111)^*$.Arrows indicate diffusion streaks.

Evolution of the structure at high temperature

Our X-ray diffraction study between 297 and 1023K allowed us to determine by least square refinement the evolution of the cubic cell parameter and the thermal expansion coefficient α as shown on figure 3.Moreover, structural refinement calculations showed that atomic shifts decrease when the temperature increases: above 600K the Nb atoms are no more located at any disordered positions and above 900K, the Pb atoms are also on the special position of the ideal cubic perovskite structure.

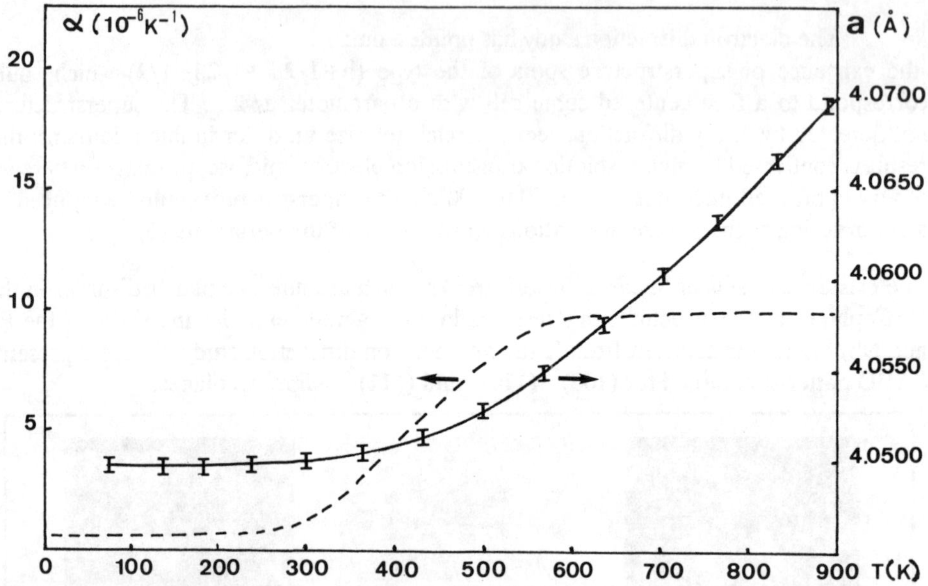

Figure 3: Evolution of the unit cell parameter a_c and the thermal expansion coefficient α vs temperature in PMN

Evolution of the structure at low temperature

When the temperature decreases from 300 to 5K, we observe no splitting of the X-ray and neutron diffraction lines indicating a tetragonal, orthorhombic or rhombohedral deformation of the perovskite cell; the average structure looks cubic. The evolution of the unit cell parameter is given on figure 3; we observe that below the Curie point, the variation of a_c is very weak. A careful analysis of the X-ray and neutron diffraction patterns reveals that the peaks corresponding to the reflections for which all three indices are either odd or even remain unchanged between 308 and 5K, whereas the basis of the other peaks and more particularly those corresponding to reflections with two odd indices exhibit a sensible widening. A comparison of some neutron diffraction lines at 308 and 5K is given on figure 4. The disordered cubic structural model previously described on figure 1 is not valuable for the ferroelectric phase. Structure refinement calculations taking into account correlation displacements of atoms are now in progress.

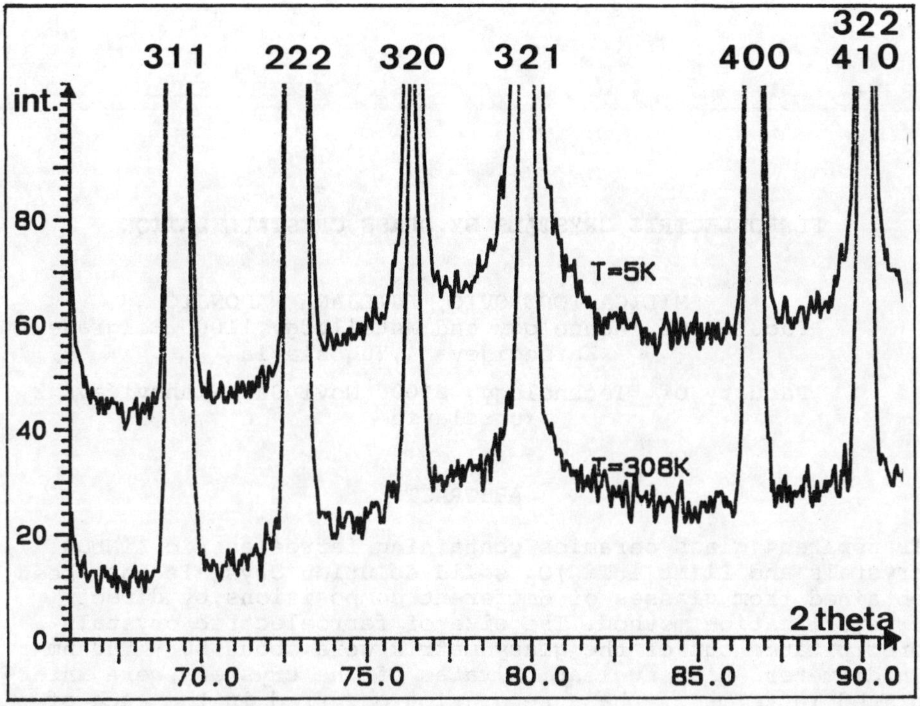

Figure 4:Comparison of some neutron diffraction lines at 5 and 308K.

Acknowledgements:
We are particularly indebted to M. Lavergne for his technical assistance in the high resolution electron microscopy investigation.

REFERENCES

1. M. Lejeune, Thèse de Doctorat, Limoges, 1986.
2. L.A. Shebanov, P. Kapostins and J. Zvirgzds, The structural change of PMN in the diffuse phase transition region, Ferroelectrics, 1984, **56**, pp.1057-1060.
3. S.L. Swartz, T.R. Shrout, W.A. Schulze and L.E. Cross, Dielectric properties of Lead-Magnesium Niobate Ceramics, J. Am. Ceram. Soc., 1984, **67**, p 311-314.
4. P. Bonneau, E. Husson, A. Morell, Structural study of PMN ceramics by X-Ray diffraction between 297 and 1023K, Mat. Res. Bull., 1989, **24**, pp 201-206.
5 K. Itoh, L.Z. Zeng, E.Nakamura and N.Mishim, Crystal structure of $BaTiO_3$ in the cubic phase, Ferroelectrics, 1985, **63**, p.29.
6. E. Husson, M. Chubb, A. Morell, Superstructure in $PbMg_{1/3}Nb_{2/3}O_3$ Ceramics revealed by High Resolution Electron Microscopy, Mat. Res. Bull., 1988, 23, pp 357-361.

FERROELECTRIC CRYSTALS BY GLASS CRYSTALLIZATION

MILICA TODOROVIĆ, LJILJANA RADONJIĆ
Faculty of Technology and Metallurgy,11000 Belgrade, Karnegijeva 4,Yugoslavia

Faculty of Technology, 21000 Novi Sad,Vlahovićeva 2, Yugoslavia

ABSTRACT

Transparent glass ceramics containing ferroelectric $LiNbO_3$ crystals and $Li(Nb_{1-x}Ta_x)O_3$ solid solution crystals have been obtained from glasses of different compositions,by direct crystallization method. The size of ferroelectric crystals precipitated out of the glasy matrix were about 50 - 500 nm in diameter. High nucleation rates of the crystals were interpreted in terms of phase separation occurred in the each of investigated glass systems and of Al_2O_3 presence as a more important parameter than the molar ration B/N.

INTRODUCTION

The formation of ABO_3 type ferroelectrics with perovskite type structure, such are $LiNbO_3$,$LiTaO_3$ and their solid solutions,by the crystallization of glass has been reported by a few authors (1-4). Such ferroelectric glass ceramics have been formed by controlled crystallization of glasses which contain major çuantities of the oxide constituents of a ferroelectric phase in addition to minor amounts of a glass-forming oxide. Lithium niobate and its solid solutions are of interest because of their increasing application in electrooptic and aurface-acoustic devices (5)

The purpose of this work is to define the relationship between the mechanism of glass nucleation in investigated glass systems and the glass composition. The mechanism of crystallization is described first in a simple glass system, $Li_2O-Nb_2O_5-SiO_2-Al_2O_3$ and than in a more complex system,$Li_2O-Nb_2O_5-Ta_2O_5-SiO_2-Al_2O_3$.

EXPERIMENTAL

A 30 batch mixture with the composition given in Table 1 was prepared from reagent grade Li_2CO_3, SiO_2, Nb_2O_5, Ta_2O_5 and Al_2O_3. They were melted in platinum crucible at 1300 - 1450°C, for the time of 1h. The melts were poured into an highly cooled steel plate and pressed into platelets approximately 1 mm thick. Transparent glass samples were very difficult to get and only for some B/N values it was possible, where B/N=$(Nb_2O_5 + Ta_2O_5)/(SiO_2+Al_2O_3)$ molar ratio.

TABLE 1

Glass composition of the $Li(Nb_{1-x}Ta_x)O_3$ systems

Specimen	Composition, mol.% Li_2O	Nb_2O_5	Ta_2O_5	SiO_2	Al_2O_3	B/N	x
L-1	29	28	-	37	6	1.51	0
L-2	43	13	6	28	10	0.50	0.5
L-3	35	9	9	37	10	0.39	0.7

The glass specimens were introduced into the preheated furnace at the appropriate temperature, kept at this temperature for some time, and finally taken out from the furnace and allowed to cool in air. Heat treatment conditions and appearance of the specimens are given in Table 2.

TABLE 2

Heat treatment conditions of the $Li(Nb_{1-x}Ta_x)O_3$ glasses

Specimen	Temperature, °C	Time min	Appearance of the specimens
L-1/1	600	10	Transparent, phase separation
L-1/2	620	10	Transparent, $LiNbO_3$ nuclei
L-1/3	650	10	Transparent, $LiNbO_3$ crystals
L-2/1	670	15	Transparent, phase separation
L-2/2	700	15	Transparent, $Li(Nb_{1-x}Ta_x)O_3$ nuclei
L-2/3	750	15	Transparent, $Li(Nb_{1-x}Ta_x)O_3$ crystals
L-3/1	730	15	Transparent, phase separation
L-3/2	770	15	Transparent, $Li(Nb_{1-x}Ta_x)O_3$ nuclei
L-3/3	800	15	Transparent, $Li(Nb_{1-x}Ta_x)O_3$ crystals

The specimens heat treated were subjected to powder X-ray diffraction (XRD) analysis. Scanning electron microscepy, JEOL 35, was used for microstructure investigation.

RESULTS AND DISCUSSION

Glasses containing a low amount of network formers are very fluid and therefore crystalize rapidly at a temperature close to the liquidus. A rapid cooling rate from the liquid to the glasy state is necessary to prevent crystallization.

XRD analysis of the as-prepared glass specimens did not shown nucleation appearance. Figure 1. show XRD patterns of the specimens L-1, L-2 and L-3,after the heat treatment and the standard values for $LiNbO_3$ and $LiTaO_3$ crystals. It could be seen that in the each of investigated systems,the crystalline phases of $LiNbO_3$ (specimen L-1) and $Li(Nb_{1-x}Ta_x)O_3$ solid solution crystals (specimens L-2 and L-3) were identificated.

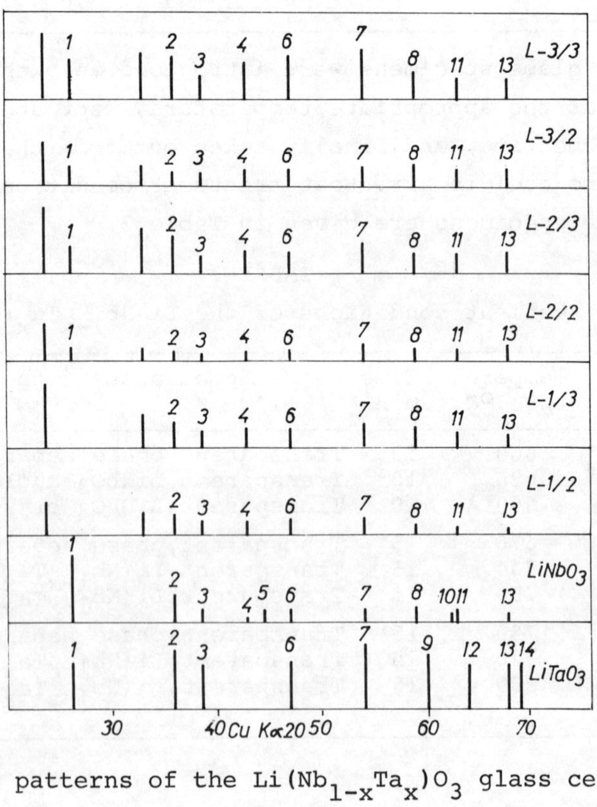

Figure 1. XRD patterns of the $Li(Nb_{1-x}Ta_x)O_3$ glass ceramics

Figure 2. show electron micrographs of the specimens L-1,L-2 and L-3, after heat treatment of the $Li(Nb_{1-x}Ta_x)O_3$ glass systems. According to XRD patterns (Fig.1.) diffraction intensivity is somewhat higher in the system with greater content of Ta_2O_5 oxide (Fig.2,D and F) but it seems that microcrystallinity features are similar.

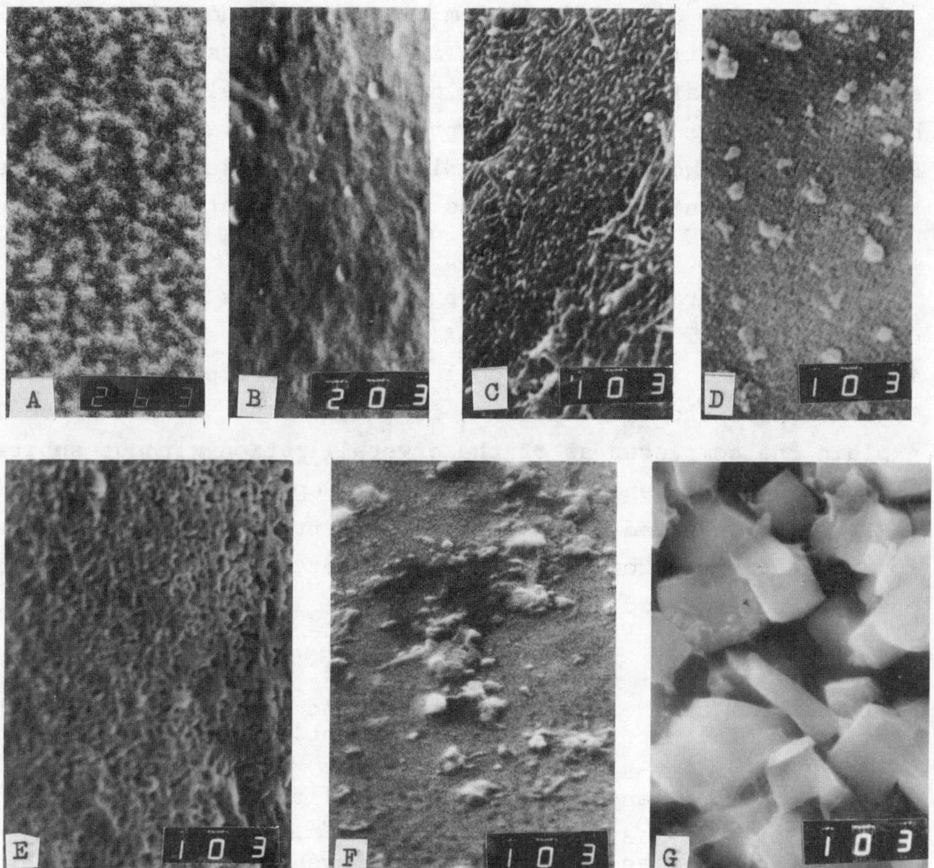

Figure 2. Electron micrographs of $Li(Nb_{1-x}Ta_x)O_3$ system after heat treatment of the specimens:L-1: A)620°C and B) 650°C; L-2: C) 700°C and D) 750°C; L-3:E)770°C, F)800°C and G)900°C

In the system $LiNbO_3$ (L-1) at the low temperature of the heat treatment, 600°C, only phase separation occurred (6). At the higher temperature,620°C,$LiNbO_3$ nuclei appeared (Fig.2,A) and at 650°C crystallization of $LiNbO_3$is extensive(Fig.2,B),on the scale about 100-200 nm.

In the system $Li(Nb_{1-x}Ta_x)O_3$ also at lower heat treatment temperatures phase separation occurred.At the higher temperatures,700°C for the system L-2,x=o.5(Fig.1.and Fig.2.C)and 750°C, for the system L-3,x=0.7(Fig.1.and Fig.2.E)$Li(Nb_{1-x}Ta_x)O_3$nuclei appeared. At the temperatures 750°C,for the system L-2(Figs.1. and 2,D)and 800°C,for the system L-3 (Figs.1.and 2,F),crystallization of the $Li(Nb_{1-x}Ta_x)O_3$ solid-solution crystals is extensive, on the scale about 50-500 nm. Full crystallization could be achieved by increasing heat treatment temperature up to 900°C,as it shown in Fig.2,G.Typical cubic crystals (perovskite type) are present, but the glass ceramics is not transparent.

CONCLUSIONS

Phase separation followed by nucleation is characteristic of the crystallization in the $Li(Nb_{1-x}Ta_x)O_3$ system.Nucleation is attributed to subsequent coalescence of the phase-separated regions until the critical size is reached.This mechanism could explain the completeness of the crystallization without substan tial grain growth,and its rapidity(1). Replacement Nb_2O_5 with Ta_2O_5 oxide resulted in the higher heat treatment temperatures for nucleation and crystal growth,but better ferroelectric properties are expected.

REFERENCES

1. Layton M.M.and Herzog A. Structure and Crystallization of glasses of low network formers content,Glass Technol,1969, 10,50
2. Ito S,Kokubo T.and Tashiro M.Crystallization process of a $LiTaO_3$-SiO_2-Al_2O_3 glass,Bull.Inst.Chem.Res.1978,54,307-11
3. Ito S.Kokubo T.and Tashiro M,Transparency of $LiTaO_3$-SiO_2-Al_2O_3 glass ceramics in relation to their microstructure, J.Mater.Sci. 1978,13,930.-38
4. Herczog A.Microcrystalline $BaTiO_3$ by crystallization from glass,J.Amer.Ceram.Soc.1964,47,107-15
5. Shin-Ichi Hirano and Kazumi Kato, Preparation of Crystalline Films Preferred Orientation by Hydrolysis of Metal Alkoxides Advanced Ceram.Mater.1988,1,501-6
6. Todorović M.and Radonjić Lj.Direct crystallization of an ABO_3 ferroelectric glass ceramics, High Tech.Ceramics,Edited by P.Vincenzini,Elsevier Science Publication B.V.Amsterdam, 1987,2106.

SYNTHESIS, SINTERING AND DIELECTRIC MEASUREMENTS OF PMN-PT CERAMIC MATERIALS

Antoinette MORELL
Thomson-CSF, Laboratoire Central de Recherches
Domaine de Corbeville, 91404 Orsay Cedex, France

ABSTRACT

Solid-solutions between lead-magnesium niobate and lead-titanate are synthesized according to the columbite route. Sintering parameters (temperature, time, atmosphere) are studied. The dielectric characteristics of the sintered pellets are measured as a function of temperature (between - 50 and + 150°C) and frequency (between 1 KHz and 10 MHz). Among all the prepared compositions, many are Z5U ceramic materials. As an example, at 20°C and 1 KHz, one composition has a dielectric constant of 22150 and a loss tangent factor of 1.7 10^{-2}, the sintering taking place at 900°C for 4 hours in air atmosphere. Micrographs of the sintered materials are presented.

INTRODUCTION

Lead-magnesium niobate ($Pb\ Mg_{1/3}\ Nb_{2/3}\ O_3$ or PMN) based materials are characterized by high dielectric constants and their potential for low-sintering temperatures (1, 2, 3). The main feature of the properties of PMN is a broad maximum dielectric peak just below room temperature. The magnitude and the temperature of this maximum respectively decreases and increases as frequency increases. This behavior is typical of relaxor ferroelectrics. It is also known that dielectric properties of PMN can be enhanced by addition of $PbTiO_3$ (PT) (3, 4).

The PMN material family has been reported to be quite difficult to synthesize reproducibly using conventional solid-state reaction methods. The presence of a metastable pyrochlore phase can be detrimental to the dielectric properties. In order to balance the PbO losses during sintering, the addition of an excess of PbO to PMN has been often used.

In this study we do not have used any PbO excess. But in order to avoid pyrochlore formation and to increase the dielectric constant, MgO excess has been introduced, as other authors do (1, 5) during the PMN synthesis. In a first part of this paper we present the optimization of the MgO excess value and in a second part, we study the system (1 - x) PMN - x PT.

EXPERIMENTAL PROCEDURE AND CHARACTERIZATION

The raw materials used throughout this investigation were reagent-grade oxides of lead, magnesium, niobium and titanium.

1° Synthesis of PMN and Characterization

In order to bypass the formation of the intermediate pyrochlore phase, PMN synthesis has been performed according to the process developed by S.L. Swartz et al. (1). It consists in a two steps reaction :

$$MgO + Nb_2O_5 \;\text{----->}\; MgNb_2O_6$$
$$1/3MgNb_2O_6 + PbO \;\text{----->}\; PbMg_{1/3}Nb_{2/3}O_3$$

1. MgO is prereacted with Nb_2O_5 to form the columbite-type phase $MgNb_2O_6$. The excess α of MgO, between 0 and 4 mole % is added prior to reaction with Nb_2O_5. After attrition milling in water of the raw materials and drying, the calcination takes place in alumina crucibles at 1050°C for 48 hours in air atmosphere.

2. PbO is added to the columbite, attrition milled in water, dried and calcined at 800°C for 4 hours in O_2 atmosphere. The reacted powders are then milled in order to brake the agglomerates. To limit PbO loss during sintering the pellets (ϕ = 1,3 mm, t = 2 mm) are embedded in PMN powder and placed in a closed alumina crucible.

Dielectric measurements are made on sintered pellets, copper electroded without any polishing of the surfaces and carried out on an automated system. Capacitance and dissipation factor are measured at various frequencies, between 1 KHz and 10 MHz, and temperatures, between - 50 and + 150°C.

Table 1 gives the dielectric measurements at 1 KHz and 20°C of PMN samples corresponding to various amounts of MgO excess and for two sintering temperatures.

TABLE 1
Dielectric characteristics of PMN samples

		900°C - 6h - O_2	945°C - 6h - O_2
PMN	ε	3630	5820
	tg $.10^{-2}$	1.7	0.6
PMN - 2 % MgO	ε	8650	9060
	tg $.10^{-2}$	0.7	0.9
PMN - 3 % MgO	ε	13800	14570
	tg $.10^{-2}$	0.6	0.8
PMN - 4 % MgO	ε	16500	16400
	tg $.10^{-2}$	0.6	0.7

Dielectric constant is very sensitive to the MgO excess. For values higher than 4 %, the final sintered material is not homogeneous. Therefore, at 900°C for 6 hours in O_2 atmosphere we obtained PMN samples having at 1 KHz and 20°C the following dielectric characteristics : ε = 16500 and tg δ = 0.6 10^{-2}. These results compared to those reported by others authors (5) show that, in our case, the sintering temperature is 300 to 400°C lower. The samples are translucent and have a good ceramic quality with a mean grain size of 4 µm. X-ray diffraction analysis show that only the perovskite phase is present. For the second part of this study concerning the (1-x) PMN - x PT investigation, all compositions have been prepared with a 4 % MgO excess.

2°. Synthesis of (1 - x) PMN - x PT Materials and Characterization

The two steps reaction has also been used. For this serie of experiments, x has been varied from 0 to 10 %. The experimental procedure is the same as it was for PMN alone except for the sintering atmosphere, air or O_2 have been used. Table 2 gives the dielectric measurements (1 KHz, 20°C) of the samples. Figure 1 shows the variations of ε and tgδ as a function of x for the samples sintered at 900°C for 6 hours under air or O_2 flow.

TABLE 2
Dielectric measurements of (1-x) PMN - x PT materials

	(1 - x) PMN - x PT							
x	0 %	2.5 %	3.75 %	5 %	6.25 %	7.5 %	10 %	
ε tg $.10^{-2}$	16500 0.6	20850 1.3	24300 2.5	27100 5.8	20450 6.8	17500 7.7	15950 8.0	900°C 6h-O_2
ε tg $.10^{-2}$	14450 0.7	17900 1.1	21450 1.6	22500 4.5	19150 5.9	17250 6.5	15150 7.1	900°C 6h-air
ε tg $.10^{-2}$	13250 0.9	15850 0.9	19150 2.1	21900 4.2	17650 5.6	16450 5.4	14500 4.6	875°C 4h-air
ε tg $.10^{-2}$	15100 0.6	18100 0.9	22150 1.7	25000 5.1	20700 6.3	18000 7.4	16700 7.7	900°C 4h-air
ε tg $.10^{-2}$	13550 0.7	16250 0.7	21300 2.0	24100 4.1	19450 5.8	17500 4.2	16000 4.0	925°C 4h-air

The main conclusions of these results are :
- ε increases up to a maximum value for x = 5 % and then decreases regularly.
- tg δ increases as x varies from 0 to 10 %. For x = 5 %, tg δ is very high, 5.8 % and 4,5 % respectively under O_2 and air atmosphere.
- regarding to the limit of tgδ values for Z5U capacitor applications (tgδ < 3.10^{-2} at 1 KHz and 20°C) it is concluded that the optimum value for x is around 3.75 %.
- because for industrial applications it is easier to sinter in air atmosphere and in short time we tried other sintering conditions : 875°C, 900°C and 925°C for 4 hours in air.

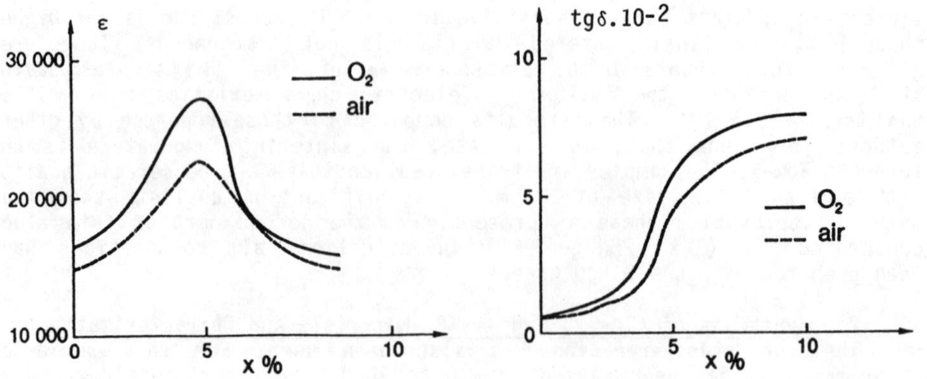

Figure 1. ε and tgδ versus x of (1-x) PMN - x PT ceramics (900°C-6h).

As it is shown in Table 2, the maximum value of the dielectric constant is always obtained on materials sintered at 900°C. ε increases for $0 \leqslant x \leqslant 5$ % and then decreases. A regular increase of the dissipation factor tgδ is observed. In capacitor applications, high ε values and low tgδ are required. Therefore, the best compromise is obtained for the 0,9625 PMN - 0,0375 PT material sintered at 900°C for 4 hours in air. In this case at 20°C and 1 KHz :

$$\varepsilon = 22150 \qquad tg\delta = 1.7\ 10^{-2}$$

As for the PMN synthesis with MgO excess, the sintering temperatures are much lower (300 to 400°C) than those indicated by other authors (4, 5). Figure 2 represent respectively the variations at 1 KHz of ΔC/C and Δ tgδ/tgδ versus temperature for $0 \leqslant x \leqslant 10$ %. An increase of the transition temperature from - 5°C to + 47°C when x increases from 0 to 10 % is observed. A corresponding increase is observed at higher frequencies.

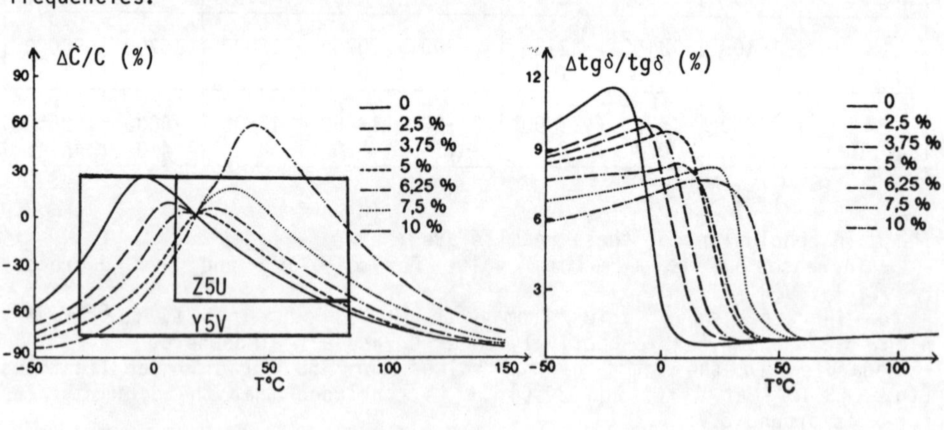

Figure 2. Variations at 1 KHz of ΔC/C (%) and Δtgδ/tgδ (%) of (1-x) PMN - x PT ceramics.

For Z5U or Y5V capacitors, $\Delta C/C$ varies in precise limits and $tg\delta < 3.10^{-2}$ (20°C, 1 KHz). Therefore the limit value of x giving acceptable materials is comprised between 3.75 and 5 %. Figure 3 shows SEM micrograph of the material with x = 3.75 %. The structure is regular and the grain size is about 4 µm. Materials have resistivities of 10^{13} Ω.cm.

Figure 3. SEM micrograph of the 0.9625 PMN - 0.0375 PT ceramic (1cm = 5 µm).

CONCLUSIONS

The results of this investigation are summarized below :

1) PMN-PT ceramics, prepared by the columbite route with a MgO excess of 4 % have been sintered at low temperature between 875°C and 925°C, in air, in a PbO-rich atmosphere.

2) The maximum dielectric constant, 22150 (1 KHz, 20°C) is achieved with the 0.9625 PMN - 0.0375 PT composition sintered at 900°C - 4 h - air. The corresponding dissipation factor is $1.7\ 10^{-2}$.

3) The microstructure is regular with grain size of 4 µm. The resistivity is high (10^{13} Ω.cm).

4) The dielectric compositions correspond to the Z5U or Y5V type capacitors.

We would like to thank DRET for his financial support.

REFERENCES

1. Swartz S.L. and Shrout T.R., Fabrication of perovskite lead magnesium niobate, Mat. Res. Bull. 17, 1245-1250 (1982).

2. Swartz S.L., Shrout T.R., Schulze W.A. and Cross L.E., Dielectric properties of PMN ceramics, J. Am. Cer. Soc. 67 (5), 311-315 (1984).

3. Lejeune M. and Boilot J.P., Low firing dielectrics bases on lead magnesium niobate, Mat. Res. Bull. 20, 493-499 (1985).

4. Guha J.P., Anderson H.U., Microstructural inhomogeneity in sintered PMN-PT based dielectrics, J. Am. Ceram. Soc., 70 (3), C39-C40 (1987).

5. Swartz S.L., Shrout T.R., Schulze W.A., Cross L.E., Dielectric properties of lead magnesium niobate ceramics, J. Am. Ceram. Soc., 67 (5), 311-315 (1984).

SHORT-RANGE ORDER IN PMN CERAMICS

E.Husson*, N. de Mathan*, P.Bonneau* et A.Morell**
*Laboratoire de Chimie-Physique du Solide URA CNRS D0969
Ecole Centrale de Paris. 92295 Chatenay-Malabry Cedex. France
**Laboratoire Central de Recherches. Thomson-CSF.
Domaine de Corbeville. 91404 Orsay Cedex. France.

ABSTRACT

A TEM study of $PbMg_{1/3}Nb_{2/3}O_3$ (PMN) ceramics reveals the existence of a short-range order which could correspond to a 1:1 ordering of Mg and Nb ions on the B-site of the perovskite. The short-range order can be modified by a suitable thermal annealing. The evolution of the dielectric properties is analyzed.

INTRODUCTION

$PbMg_{1/3}Nb_{2/3}O_3$ (PMN) is a ferroelectric relaxor material which has at 20°C the cubic perovskite structure $AB'_xB''_{1-x}O_3$ (a_c # 4 Å). Since PMN exhibits a diffuse phase transition, this material has a high dielectric constant through a large temperature range. Therefore it is an interesting ceramic from an industrial point of view ($tg\,\delta = 0{,}6.10^{-2}$ and $\varepsilon = 15000$ at 20°C and 1 kHz).
Generally, the diffuse transition is considered as induced by composition fluctuations in the ceramics (1). We discuss here the possibility of a local order on the B-site of the PMN perovskite ceramics by electron diffraction and high resolution electron microscopy (HRTEM).Then, we present the modification of this order after thermal annealing of the ceramics.

MATERIAL AND METHODS

PMN ceramics were prepared according to the process described by Swartz and Shrout (2): first, magnesium niobate $MgNb_2O_6$ was formed by calcination of Nb_2O_5 at 1050°C; then perovskite PMN was obtained at 800°C by the reaction of PbO with $MgNb_2O_6$. To avoid the formation of a parasite pyrochlore phase, an excess of 4% MgO was added prior to reaction with Nb_2O_5. Pellets were pressed and sintered at 900°C for 6 hours and O_2 flux in a PbO rich atmosphere. Samples for dielectric measurements were prepared from the sintered pellets (12 mm diameter and 1,3 mm thickness) by evaporating copper electrodes.

X-ray diffraction patterns were performed at a low θ scan on a high sensitivity diffractometer MICROCONTROL using K_{α} radiation of a copper rotating anode.
Electron diffraction patterns and HRTEM images were obtained with a 400 kV microscope from JEOL.
Image simulations were performed with EMS Stadelmann's program using the multislice method (3).

RESULTS AND DISCUSSION

The electron diffraction pattern of the $(110)^*$ reciprocal plane of PMN is given on figure 1. The weak superstructure spots at the (h+1/2, k+1/2, l+1/2) positions correspond to a face-centered cubic cell with a parametrer $a = 2a_c \# 8$ Å. Such a superstructure could be induced by the existence of a local order on the B site of the perovskite, i.e. a 1:1 non stoichiometric ordering analogous to that observed in $PbSc_{1/2}Ta_{1/2}O_3$ ceramics (4). It involves the coexistence of microdomains with Mg-Nb-Mg-Nb cations ordering and microdomains richer in Nb cations. Consequently, strong charge effects in the ceramics inhibit the ordered domains growth.

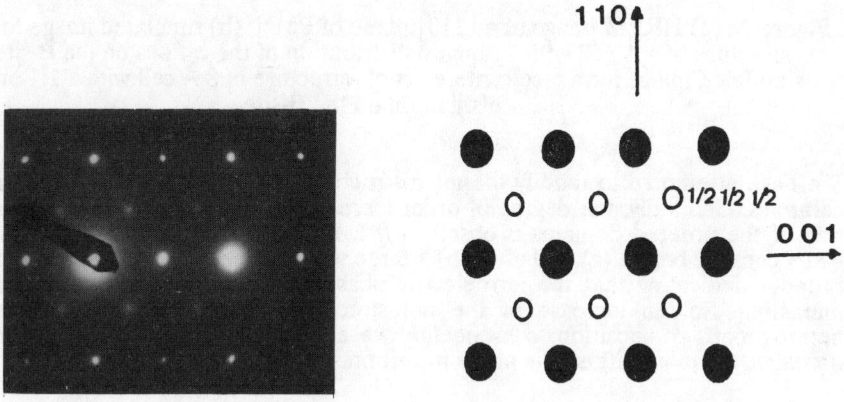

Figure 1. Electron diffraction pattern of a $(110)^*$ plane of PMN

Figure 2(a) shows the HRTEM image corresponding to the diffraction pattern given on figure 1. On some areas, the contrast consists in rows of bright dots spaced by about 4 Å on the [001] direction and by 2.86 Å on the [110] direction. The simulated image (figure 2(b)) shows that this contrast is coherent with a random distribution of the Mg and Nb ions on the B site of PMN (i.e. a statistically homogenous disorder). On some other areas, of about 40-100 Å, we observe the lack or the attenuation of one bright dot over two. These areas are responsible for the superstructure spots observed on electron diffraction patterns. The simulated image (figure 2(c)) confirms that this contrast is coherent with a 1:1 ordered distribution of the cations on the B-site.The two calculated images correspond to a crystal thickness of 95 Å and a defocus of 100 nm.
This superstructure is not revealed by X-ray diffraction because of the small size of the ordered domains.

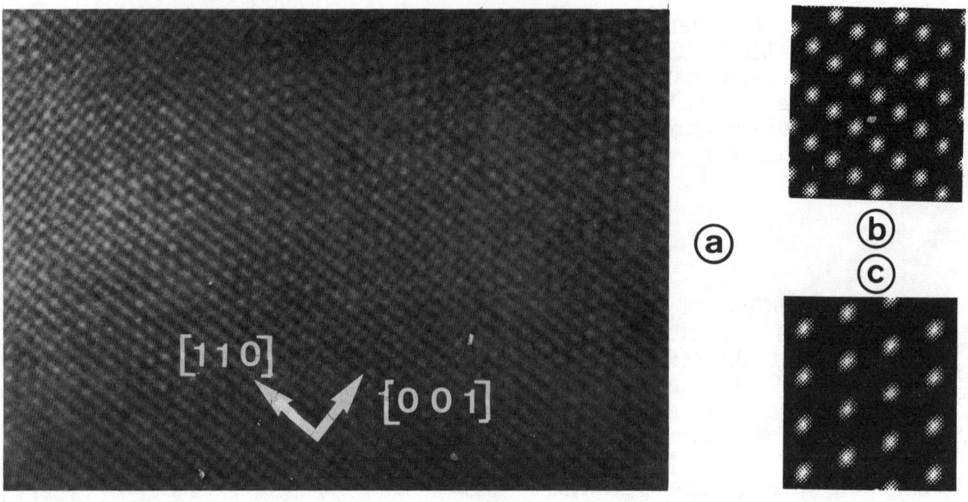

Figure 2 . (a) HREM image of a [110] plane of PMN; (b) simulated image for a cubic structure of 4 Å cell with a random distribution of the cations on the B-site; (c) simulated image for a face-centered cubic structure of 8 Å cell with a 1:1 ordered distibution on the B-site.

We have attempted to modify the microstructure of PMN by thermal annealing of the ceramics. Generally, the degree of order increases and a corresponding increase in the size of the ordered domains is observed (4). Figure 3 shows the curves $\frac{\Delta\varepsilon}{\varepsilon}=f(T)$ for a PMN ceramic before (a) and after (b) a three weeks annealing at 800°C.The curve (b) is broader, indicating that the ferroelectric phase transition becomes more diffuse after annealing. So, an increase of the non-stoichiometric 1:1 order promotes a more heterogenous composition; consequently we can assume that the greater the ordered domains, the more diffuse the phase transition.

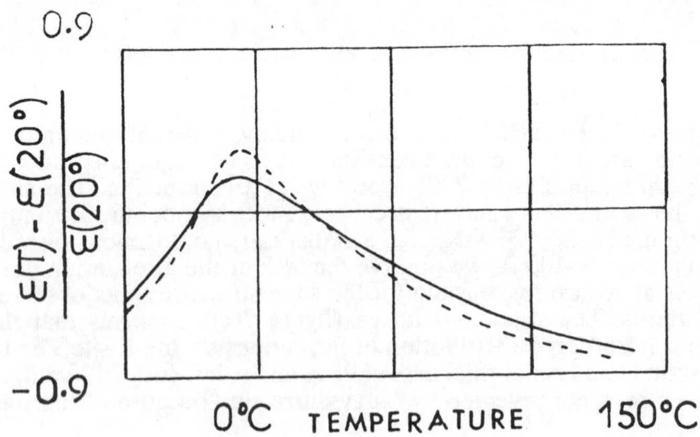

Figure 3. Relative dielectric constant variation with temperature
(a)- - - before annealing
(b)—— after annealing

On figure 4 are reported the hysteresis loops of a ceramic before annealing (a) and after annealing (b). The loop (a) exhibits a usual shape but the loop (b) is doubled. Double loops are commonly observed on antiferroelectrics and are explained by a field induced phase transition from the antiferroelectric phase to the ferroelectric phase (5).

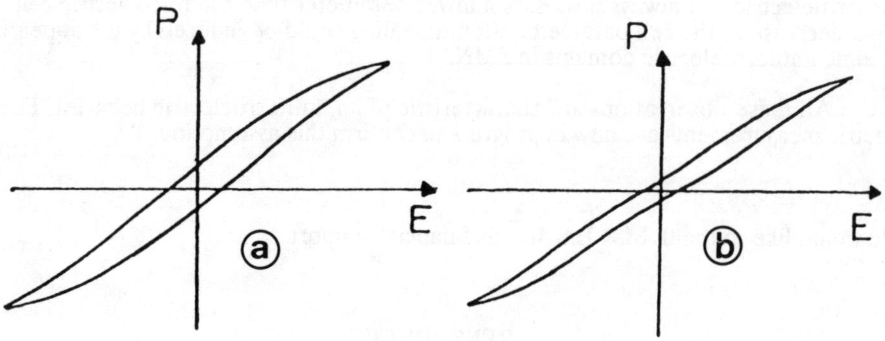

Figure 4.Hysteresis loops at -50°C before annealing (a) and after annealing (b).

Figure 5 presents the $\varepsilon = f(T)$ curves for an annealed sample measured under 0 kV/cm (a) and 4 kV/cm (b) dc poling field. When a dc field is applied, the transition temperature T_c is shifted towards low temperatures. If the same measure is performed on a PMN sample before annealing, no evolution is observed. For a ferroelectric material, one can expect a shift towards the high temperatures. On the contrary, a shift towards the low temperatures is generally ascribed to an antiferroelectric phase (6).

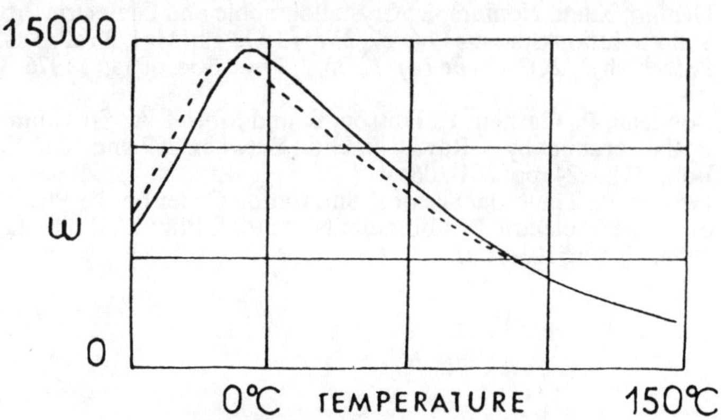

Figure 5.Variation of dielectric constant with temperature
(a) —— under 0 kV/cm
(b) - - - under 4 kV/cm

Structure refinement from X-ray diffraction patterns of PMN reveals especially strong statistical Pb atomic shifts around their cristallographic site (7). Identical structure refinements performed on annealed ceramics show a slight increase of the Pb atomic shift (0.285 Å and 0.295 Å respectively before and after annealing) and a smaller cell parameter (4.0482 Å and 4.0477 Å respectively before and after annealing). PMN and PIN ($PbIn_{1/2}Nb_{1/2}O_3$) materials should be compared. Well ordered PIN is antiferroelectric whereas disordered PIN is ferroelectric. X-ray diffraction study reveals that for PIN, the Pb atom displacement increases in the ordered phase (8). The antiferroelectric cell always possesses a lower parameter than the ferroelectric cell. The slight decrease of the cell parameter after annealing could be induced by the appearance of some antiferroelectric domains in PMN.

All these observations are characteristic of an antiferroelectric behavior. Further electric measurements are now in progress to confirm this assumption.

We would like to thank M.R.E.S. for his financial support.

REFERENCES

1. Smolensky, G.A., Physical Phenomena in Ferroelectrics with Diffuse Phase Transition. J. Phys. Soc. Jap. 1970, 28, supp,.pp.26-37.
2. Swartz, S.L., Shrout, T.R., Schulze, W.A. and Cross, L.E.J. Dielectric Properties of Lead Magnesium Niobate Ceramics, Am. Ceram. Soc., 1984, 67, pp.311-314.
3. Stadelmann, P.A., EMS. A Software Package for Electron Diffraction Analysis and HREM Image Simulation in Materials Science. Ultramicroscopy, 1987, 21, pp.131-146.
4. Setter, N. and Cross, L.E., The Role of B-Site Disorder in Diffuse Phase Transition Behavior of Perovskite Ferroelectrics. 1980, 51, pp.4356-4360.
5. Randall, C., Barber, D., Whatmore, R., Groves, P., A Microstructural Study of and ß Phases in 8.2/70/30 PLZT, 1987, 76, pp. 311-318.
6. Uchino, K.and Nomura, S., Crystallographic and Dielectric Properties in Solid Solution Systems Pb(Fe2/3W1/3)O3-Pb(Mg1/3Ta2/3)O3 and Pb(MgNb)1/2O3-Pb(FeTa)1/2O3, J. Phys. Soc. of Jap., 1976, 41, pp. 542-547.
7. Bonneau, P., Garnier, P., Husson, E. and Morell, A., Structural Study of PMN Ceramics by X-Ray Diffraction Between 297 and 1023 K. Mat Res. Bull., 1989, 24, pp. 201-206.
8. Groves, P., The Influence of B-Site Cation Order on the Phase Transition of Antiferroelectric Lead Indium Niobate, J. Phys. C: Solid Stat Phys., 1986, 19, pp.5103-5120.

STRUCTURE AND PTCR STUDIES IN BARIUM LEAD TITANATE PREPARED BY THE MOLTEN SALT SYNTHESIS METHOD

KI HYUN YOON, MAN HWA LEE AND SANG OK YOON
Department of Ceramic Engineering
Yonsei University, Seoul, Korea

ABSTRACT

The variation in lattice constants and PTCR effect in the barium titanate system prepared by the molten salt synthesis (MSS) method have been investigated. The c/a ratios of $(Ba_{1-x}Pb_x)TiO_3$ $(0<x<0.5)$ prepared by the MSS method are greater than those of the conventional mixed oxides (CMO) method due to a decrease in PbO loss. In Nb doped $(Ba_{0.8}Pb_{0.2})TiO_3$ with increasing the KCl weight ratio the room temperature resistivity decreases. These results can be explained by the charge compensation effect between electrons due to the Nb ion and holes due to the K ion.

INTRODUCTION

When an appropriate amount of a rare earth element is added to ferroelectric $BaTiO_3$, the $BaTiO_3$ is converted to a semiconductor, according to the controlled valency principle, and shows an abrupt increase of resistivity at near the Curie temperature, i.e. the Positive Temperature Coefficient of Resistivity (PTCR) phenomenon. The PTCR effect in $BaTiO_3$ ceramics is strongly influenced by the microstructure of the sintered body which can be altered by changing of starting materials[1]. In our previous report[2] the ceramics produced by molten salt synthesis (MSS) method showed better microstructure and a larger PTCR effect than the ceramics made by the conventional calcining of mixed oxides (CMO) method. Semiconducting ceramics whose Ba sites can be filled with Pb ions have come under special study[3] since they can be used in high temperature heaters. Gol'tsov et al.[4] reported that the PTCR phenomenon occurred at about 500 °C in La_2O_3 doped $PbTiO_3$ ceramics. Therefore, in this experiment barium-lead titanate powders were prepared by the molten salt synthesis and the calcining of mixed oxides methods and the difference in lattice parameters of these two kinds of ceramics were evaluated. Also, for $(Ba_{0.8},Pb_{0.2})TiO_3$, whose Curie temperature is about 200 °C and is widely used in high temperature heaters, the effect of KCl flux amount on $(Ba_{0.8}Pb_{0.2})(Ti_{0.994}Nb_{0.006})O_3$ ceramics were investigated in terms of the change in the density, the microstructure and the resistivity-temperature characteristics.

EXPERIMENTAL PROCEDURE

Barium-lead titanate powders, $(Ba_{1-x}Pb_x)TiO_3$, with x in the range of 0.1 to 0.5 mole were prepared by the MSS and CMO methods. Raw materials used, $BaCO_3$, PbO, TiO_2, and KCl, were all above 99.9% in purity. In the MSS method, the raw materials and the salt (with weight ratio of KCl to raw materials 0.8) were weighed and milled for 20 hrs with zirconia balls in ethyl alcohol. Then the mixtures were dried on a hot plate with stirring and placed in an electrical furnace at 800 °C and allowed to react for various periods of time. After reaction, the salt was washed out with deionized water until no Cl ions were detected using $AgNO_3$ solution[2]. In the CMO method, the oxide mixtures were calcined at 900 °C for 2hrs.

These synthesized powders were granulated with 2 wt% PVA and pressed into pellets 1 cm in diameter by 0.2 cm thickness under uniaxial pressure at 900 kg/cm^2. The pressed pellets were sintered at 1300 °C for 30 min. on a Pt plate in an electrical furnace. The synthesized powders and the sintered specimens were examined by X-ray diffraction analysis using CuKa radiation. The weight ratio of KCl flux used to raw materials was 0.2, 0.4, 0.8, 1.0. The pressed pellets were sintered at 1300 °C for 30 min. In order to measure the resistivity Ni electrode was coated on opposite surfaces of the polished specimens by an electroless plating method. The temperature-resistivity characteristics were measured by the voltage drop method[2]. The microstructures were observed using a metallurgical microscope after etching with an 85% H_3PO_4 solution at 80°C for 15-30 sec.

RESULTS AND DISCUSSION

1. Synthesis and lattice parameters of the $(Ba_{1-x}Pb_x)TiO_3$

Figure 1 shows X-ray diffraction patterns of the powders synthesized by the MSS method at 800 °C. The reaction time of the $(Ba_{1-x}Pb_x)TiO_3$ was 2hrs for x=0.1 mole and 1 hr for x=0.2 mole and above.

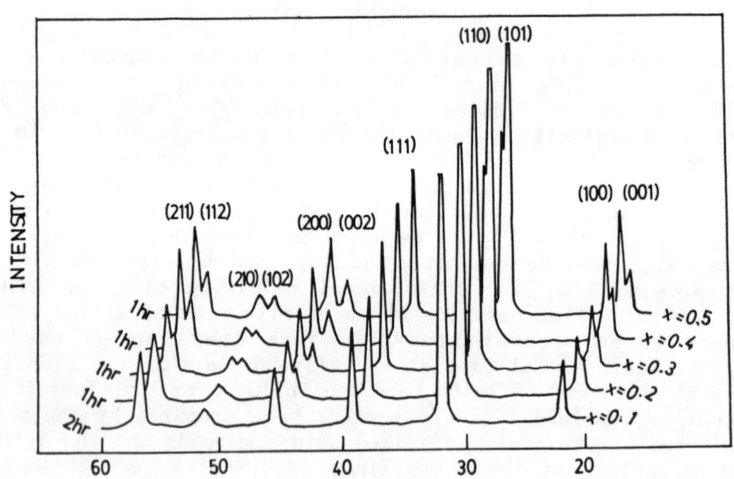

Figure 1. XRD patterns of $(Ba_{1-x}Pb_x)TiO_3$ prepared at 800°C by molten salt synthesis.

As shown in this figure, no other substances were detected except the barium-lead titanates, but with increase of the Pb content, some of the peaks split due to the substitution of Pb ions for Ba ions. so an increase of the tetragonality occurred in the crystalline lattice. The synthesis temperature for the MSS method is lower than that for the conventional CMO method. The reason is that in the MSS method the KCl flux (m.p.=772°C) changes to the melt phase during the synthesis and improves the synthesis reaction due to increase in the interdiffusion between the raw materials[2].

Figure 2. shows lattice parameters and c/a ratios for various composition of the $(Ba_{1-x}Pb_x)TiO_3$ specimens prepared by the MSS and the CMO method sintered at 1300 °C for 30 min.

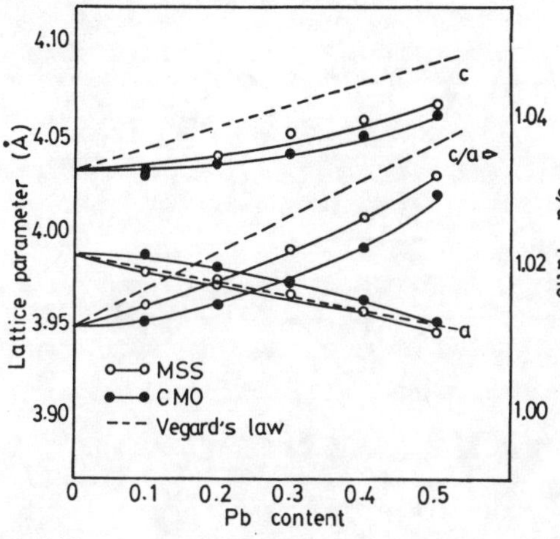

Figure.2 Lattice parameters and c/a ratios of MSS and CMO.

As the amount of Pb ions substituted increases, the length of the a-axis decreases but that of the c-axis and the ratio c/a increases. And the length of the c-axis deviates negatively in comparison with the dotted line based on Vegard's law. According to the report of Shevanov et al.[5], in the $(Ba,Pb)TiO_3$ system the interaction between structural factors in the solid solution is very complicated, that is, the high polarizability and particularity of an unfilled core electron of the Pb ion affects the property of structural components and the substitution of Pb ion for a Ba ion changes the degree of interaction from acoustic to the optical vibration mode. This leads to an increase in the compressibility in the crystalline lattice and results in the negative deviation in the barium-lead titanate solid solution. The results shown in figure 2. are comparable with those of Shevanov et al.[5] and Shirane et al.[6]. However, a slight difference between the MSS and CMO specimens can be seen in the lattice parameters. This difference is conjectured to be due to vaporization loss of the Pb ions during the synthesis in the CMO method, because the barium-lead titanate powder is calcined at 900 °C which is over the melting point of the PbO (m.p.=880 °C). On the other hand, the MSS powder reacted below melting temperature, so it is considered that the sharper increase in the c-axis and decrease in the a-axis occurred in the CMO specimen. And as the content of Pb ions increases in the solid solution, the c/a ratio of the specimen for the two methods increases, but the specimens sintered from the MSS powders show larger tetragonality than those from the CMO powders.

2. Effects of KCl flux on $(Ba_{0.8}Pb_{0.2})TiO_3$

Figure.3 shows the microstructures of 0.6 mole% Nb doped $(Ba_{0.8}Pb_{0.2})TiO_3$

specimens with various KCl flux weight ratios. The grain size of the specimens decreases with increase in the amount of KCl, but the densities of the specimens are only slightly influenced by the KCl content. This can be explained due to the size reduction of the primary particles which are prepared through a molten salt at a lower temperature, and are thus affected by the flux content. The synthesized particles depend on the grain size in the sintered specimens. This agrees well with our previous reports[2].

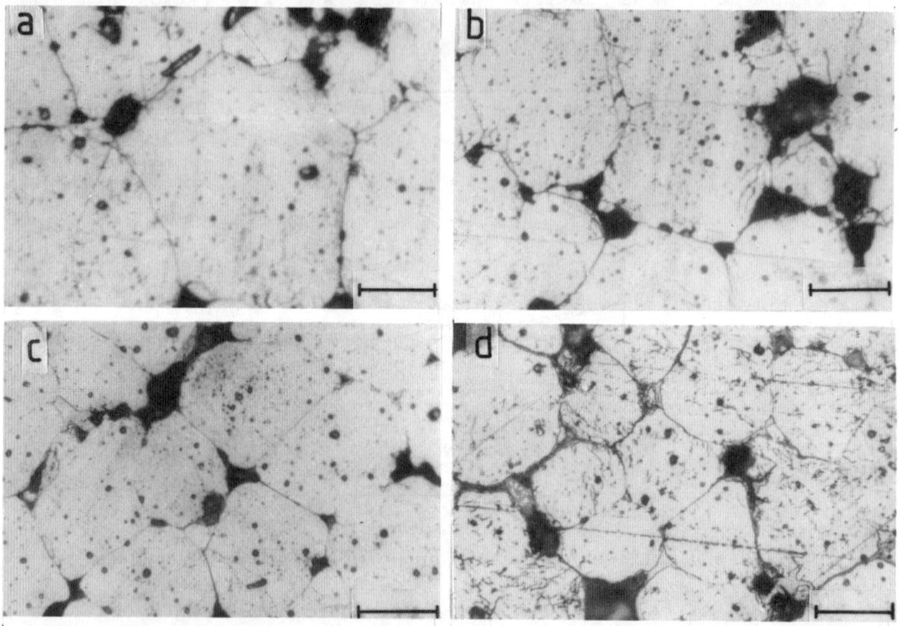

Figure.3 Microstructures of $(Ba_{0.8}Pb_{0.2})(Ti_{0.994}Nb_{0.006})O_3$ with KCl weight ratio of a)0.2 b)0.4 c)0.8 d)1.0 (bar=10 um).

The effect of the KCl weight ratio on the resistivity-temperature characteristics of the 0.6 mole% Nb doped $(Ba_{0.8}Pb_{0.2})TiO_3$ ceramics is shown in figure.4. As the KCl flux content increase, the room temperature resistivities of the specimens decrease, but the PTCR effect increases and then decreases again. According to our report[2], when Sb_2O_3 is doped in $BaTiO_3$, the room temperature resistivity and the PTCR characteristic of the specimens vary greatly with the KCl weight ratio as represented in the following equation; $KCl = K'_{Ba} + h^{\cdot} + Cl_2(g)$.

With increasing KCl content, the charge compensation effect between the holes created by K ions which are substituted at Ba sites and the electrons donated by donor ions increases in n-type semiconducting $(Ba_{0.8}Pb_{0.2})TiO_3$; thus the room temperature resistivity of the specimens increases. However, in Nb_2O_5 doped $(Ba_{0.8}Pb_{0.2})TiO_3$, there is no critical relation to the KCl content due to the difference in substituting sites between K ions and Nb ions. As shown in this figure the decrease in the room temperature resistivity can be explained by the charge compensation effect between conducting electrons and accepting holes due to K ions. When the KCl weight ratio is 0.2, the large resistivity of the specimen cannot be explained, but we think that the holes due to K ions correspond to other holes due to the added Pb ions from equation as follows;

$$PbO = V'_{Pb} + 2h^{\cdot} + Pb(g) + O_O$$

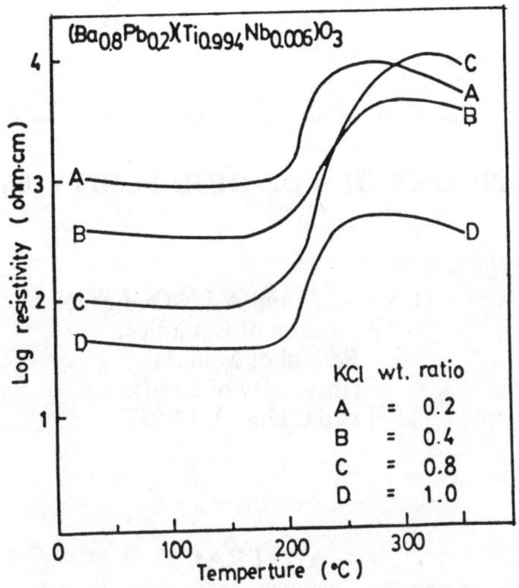

Figure 4. Resistivity-temperature characteristics of $(Ba_{0.8}Pb_{0.2})(Ti_{0.994}Nb_{0.006})O_3$.

And Kuwabara[1] reported that a large PTCR effect could appear in porous body with small grain size and homogeneous size distribution. The variation of the PTCR characteristics in figure 4 can be explained by observation of the microstructure of the specimens shown in the figure 3.

CONCLUSION

1) The c/a ratios of the sintered specimens from the MSS method are larger than those of the CMO method due to decreased PbO loss.
2) The resistivity at the room temperature decreases with increasing the KCl weight ratio. This is can be explained by the charge compensation effect between electrons due to the Nb ions and holes due to the K ions.

ACKNOWLEAGEMENT

This work was supported by the Korea Science and Engineering Foundation.

REFERENCES

1. Kuwabara, M., Effect of microstructure on the PTCR effect in semiconducting barium titanate ceramics, J. Am. Ceram. Soc., 64 (1981), 639.
2. Yoon, K.H., Oh, K.Y., and Yoon, S.O., Influence of synthesis methods on the PTCR effect in semiconductive $BaTiO_3$, Mat. Res. Bull., 21 (1986), 1429.
3. Ichikawa, Y., and Calson, W.G., Yttrium doped ferroelectric solid solutions with positive temperature coefficient of resistivity, Ceram. Bull., 42(1963), 312.
4. Gol'tsov, Y.I., Prokopalo, O.I., and Belova, L.A., Positive temperature coefficient of resistivity in doped lead titanate,Sov. Phys. Solid State, 14(1982), 805.
5. Shevanov, L.A., Fritsberg, V.Y., and Freimane, M.Z., Validity of Vegard's law in ferroelectric solid solutions based on $BaTiO_3$ and $SrTiO_3$, Sov. Phys. Solid State, 23 (1981), 373.
6. Shirane, G. and Suzuki, K., On the phase transition in barium-lead titanate, J. Appl. Phys. Jpn., 6 (1951), 274.

DIELECTRIC PROPERTIES OF HIGH-PURITY $BaTiO_3$ CERAMIC

D.A.HALL and A.J.MOULSON
Division of Ceramics,
School of Materials,
University of Leeds,
Leeds, U.K. LS2 9JT.

ABSTRACT

The microstructural-dependence of the dielectric properties of $BaTiO_3$ ceramics is well-documented. To date, most publications have been confined to the effect of grain size on the real part of permittivity, ε_r'. The present paper describes the ceramic processing necessary to produce fine- and coarse-grained materials and reports the effect of grain size on the temperature- and frequency-dependence of the complex permittivity (both ε_r' and ε_r''). The results are discussed in terms of the existing models.

INTRODUCTION

The effect of grain size on the relative permittivity, ε_r', of polycrystalline $BaTiO_3$ is well established(1). A decrease in the grain size (from >10 μm to ~1 μm) is generally found to cause an increase in the room temperature value of ε_r' (from ~1700 to ~5000) and a broadening of the cubic (paraelectric) to tetragonal (ferroelectric) phase transition.

Various authors have attributed this increase in ε_r' to the effects of internal stress(2,3), to a distribution in the value of the Curie temperature(4), or to an increase in the domain-wall contribution to ε_r'(5). If the increase in ε_r' is directly related to domain structure, then it would be expected that frequency- and temperature-dependent domain-wall relaxation processes would also be influenced by grain size. Therefore, the objective of the present study was to determine the complex permittivity, ε_r^*, as a function of temperature and frequency, with particular interest focussed on the low-temperature (~120 K) region, where Arlt et. al. have observed a relaxation in the 90° domain-wall contribution to ε_r^* for PZT(6).

EXPERIMENTAL METHODS

Green compacts were prepared by isostatically -pressing a high-purity, chemically-derived, powder◊ at 200 MPa. Coarse-grained polycrystalline $BaTiO_3$ was obtained by conventional sintering for 1 hour at 1350°C. The resulting material had a relative density ~95%, as measured by water immersion, and a grain size ~10-50 μm.

Fine-grained materials were prepared by firing at temperatures between 1200 and 1350°C in a fast rise furnace†. Heating and cooling rates were 50°C / minute, with a hold time of 5 minutes. The optimum firing temperature was found to be 1260°C, at which a relative density ~89% was achieved. Under these conditions, the resulting microstructure was reasonably uniform, with a grain size ~0.5 μm. Lower firing temperatures led to reduced final densities, while at higher firing temperatures abnormal grain-growth occurred.

Specimens for dielectric measurements were machined discs having a thickness ~1mm and a diameter ~10mm, with fired-on Pt / Ag electrodes.

Low-temperature (-180 to +150°C) dielectric measurements were made using an automated system based on a Hewlett-Packard impedance analyser§, with a 300-series controller, and an environmental chamber#. Measurements were taken both on heating and cooling at a rate of 2±0.5°C/minute, with an interval of 1°C. Results were obtained in the form of ε_r^* vs. temperature curves at 5 spot frequencies (100 Hz to 1 MHz).

Wide-frequency scans (0.01 Hz to 100 kHz) at selected temperatures above ambient were made using a frequency response analyser¢ (FRA) controlled by an Apple microcomputer. In this case, the specimen was held in a computer-controlled vertical tube furnace. Temperature stability during the frequency scans was ±0.5°C.

The complex permittivity values were calculated from the measured specimen capacitance, loss tangent, and dimensions, without any correction for porosity.

RESULTS AND DISCUSSION

The results for the temperature-dependence of ε_r^* for the coarse- and fine-grained materials are given in figs. 1-3. The most obvious characteristics of these curves are the three peaks in ε_r' associated with the cubic→tetragonal (~120°C), the tetragonal→orthorhombic (~5°C), and the

◊ HPB $BaTiO_3$, Cookson Group plc.

† Vectstar Ltd.

§ HP 4194A, Hewlett-Packard Ltd.

Tenney Junior, Tenney Engineering inc.

¢ Solartron 1174, Schlumberger Electronics Ltd.

orthorhombic→rhombohedral (~-80°C) phase transitions. As reported in the literature(7), the cubic→tetragonal transition is shifted to a slightly lower temperature (123→119°C) and the tetragonal→orthorhombic transition to a higher temperature (5→20°C) for the fine-grained material. In addition, for the fine-grained material the peaks are more diffuse.

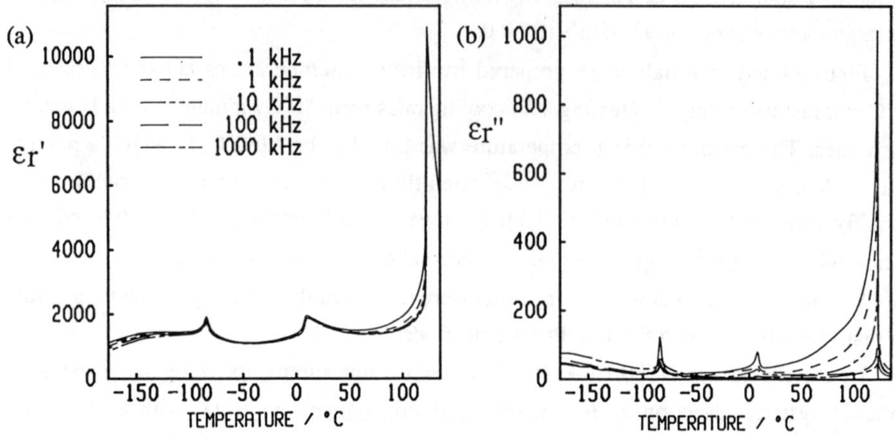

Fig.1. Variation of ε_r* with temperature for coarse-grained $BaTiO_3$ ceramic. Results on cooling.

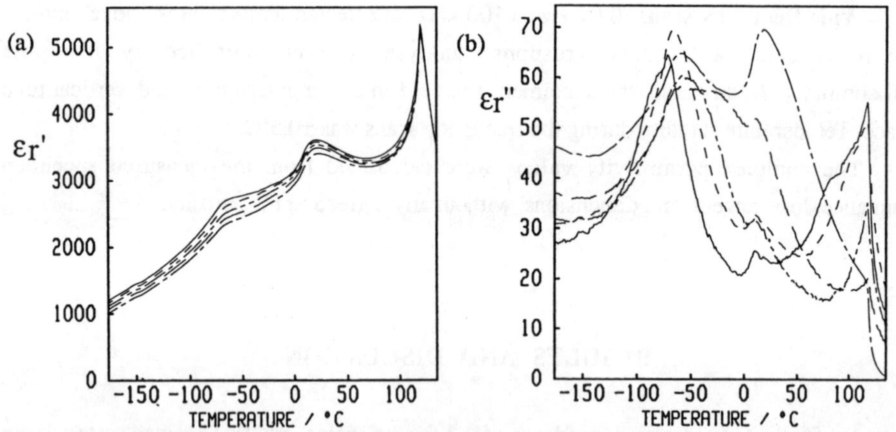

Fig.2. Variation of ε_r* with temperature for fine-grained $BaTiO_3$ ceramic. Results on cooling. Key as for fig.1.

The most striking effect of the reduction in grain size is a general increase in the value of ε_r' over almost the whole of the temperature range studied. An increase in the loss, ε_r'', and a gradual reduction in ε_r' begin to occur for both materials at temperatures below the orthorhombic→rhombohedral transition, fig.3, although the low temperature capability of the equipment was insufficient to fully characterise this effect.

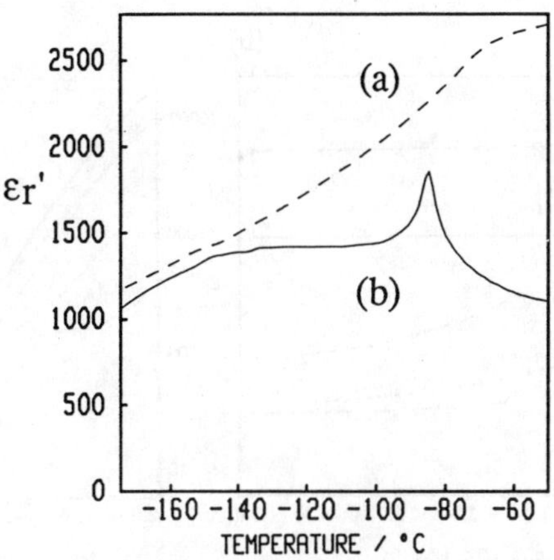

Fig.3. Variation of ε_r with temperature for (a) fine- and (b) coarse-grained $BaTiO_3$ ceramic. f = 1 kHz.

Arlt et. al. observed a relaxation in the dielectric (ε_r*), piezoelectric (d*), and elastic (s*) parameters for tetragonal PZT at ~-150°C(6), which was attributed to 90° domain-wall relaxation. $BaTiO_3$ ceramics were said to exhibit similar behaviour, though from figs.1-3 it would appear that this must occur at a lower temperature. It is clear that the **relative** decrease in ε_r' below the transition is greater for the fine-grained than for the coarse-grained material (~60% compared with ~24%). Hence, there is a strong indication that the reduction in ε_r' in this low-temperature region, assumed to be due to a domain-wall contribution, is greater in the fine-grained material. Further work, involving lower-temperature measurements, will be required to substantiate this.

One aspect of the ε_r* vs. temperature curves for coarse-grained $BaTiO_3$ ceramic, not discussed previously, is the increase in both ε_r' and ε_r" with decreasing frequency at temperatures just below the Curie temperature, T_c. The fact that the frequency-dependent component of ε_r* is confined to the ferroelectric phase suggests that its origins lie in domain-wall movements. In order to investigate this possibility, the frequency-response at selected temperatures, both above and below T_c, was determined using the FRA. The results are shown in fig.4.

The first obvious feature is the increase in ε_r' and ε_r" with decreasing frequency below ~ 1Hz.This is evident at temperatures both above and below the Curie point, becoming more pronounced at the higher temperatures. There are several possible reasons for this, including :

(a) low-frequency dispersion in ε_r*, as discussed by Jonscher(8);

(b) 'barrier-layer' effects.

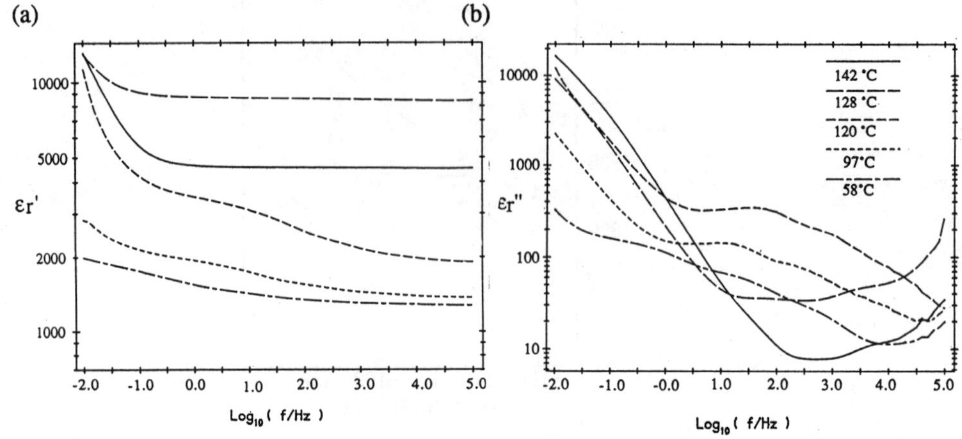

Fig.4. Variation of ε_r* with frequency for coarse-grained $BaTiO_3$ ceramic.

When plotted in the complex impedance plane the results indicated that the specimen impedance was consistent with two elements acting in series, a bulk component dominant at higher frequencies, and a higher-resistance barrier effect dominant at lower frequencies. Further work is required to investigate the cause of this effect and to determine the nature of the boundary layer, for example whether it involves regions of differing resistivity in the volume of the specimen or within each individual grain.

At intermediate frequencies (~0.1 to 100 Hz) a strong relaxation in ε_r' (~2000) and a peak in ε_r'' are apparent for temperatures below T_c. The frequency corresponding to the peak value is temperature-dependent, implying a thermally-activated relaxation process, with a relaxation time τ given by:

$$\tau = \tau_0 \exp(-E / kT) \tag{1}$$

for which the values $E = 0.79$ eV , $\tau_0 = 1.5.10^{-12}$ s are estimated. This well-defined relaxation was not observed in the fine-grained material, which instead showed a much broader dispersion throughout the temperature range studied, as is apparent from fig.2.

It seems likely that the relaxation in ε_r* in the coarse-grained $BaTiO_3$ ceramic is related to the ferroelectric domains, which are reported to be bounded by 90° and mixed 90°/180° domain-walls(9). The fact that the relaxation was not observed in the fine-grained material further supports the view that the increment in ε_r' is associated with the mixed domain-walls, since this structure is only observed for grain sizes above 10μm[5]. The relation to the 90° domain-wall relaxation occurring at lower temperatures, or the high-frequency dispersion described by Poplavko et. al.[10] is not yet clear.

CONCLUSIONS

The complex permittivity, ε_r*, has been determined as a function of temperature and frequency for fine- and coarse-grained $BaTiO_3$ ceramic. Low-temperature measurements indicate that the reduction in ε_r' below the orthorhombic→rhombohedral phase transition is greater for the fine-grained than for the coarse-grained material. This is supportive of the view that the higher domain-wall content of the fine-grained ceramic, compared with that of relatively coarse-grained material, is responsible for the grain-size dependence of ε_r'.

Measurements of the frequency-dependence of ε_r* in coarse-grained materials at temperatures close to T_c revealed a relaxation in ε_r* at intermediate-frequencies (~0.1 to 100 Hz), attributed to a domain-wall effect, and an increase in both ε_r' and ε_r'' at low frequencies associated with the presence of a barrier layer. The study is continuing.

Acknowledgements : We would like to thank the SERC for financial support and Mr. R. Holt, Mr. M. Daglish, and Dr. E. Gilbart for technical assistance.

REFERENCES

1. Jonker, G.H. and Noorlander, W., Science of Ceramics **1**, 1962, 255-264.
2. Buessem, W.R., Cross, L.E. and Goswami, A.K., J. Am. Cer. Soc. **49**, 1966, 33-36.
3. Bell, A.J. and Moulson, A.J., Ferroelectrics **54**, 1984, 147-150.
4. Martirena, H.T. and Burfoot, J.C., J. Phys. C : Solid State Phys. **7**, 1974, 3182-3192.
5. Arlt, G., Hennings, D. and de With, G., J. Appl. Phys. **58**, 1984, 1619-1625.
6. Arlt, G., Dederichs, H. and Herbiet, R., Ferroelectrics **74**, 1987, 37-53.
7. Bell, A.J. and Moulson, A.J., Brit. Ceram. Proc. **36**, 1985, 57-66.
8. Jonscher, A.K., J. Mat. Sci. **16**, 1981, 2037-2060.
9. Arlt, G. and Sasko, P., J. Appl. Phys. **51**, 1980, 4956-4960.
10. Poplavko, Y.M., Tsykalov, V.G. and Molchanov, V.I., Soviet Phys. Sol. State **10**, 1969, 2708-2709.

PIEZOELECTRIC AND PYROELECTRIC PROPERTIES OF 0-3 COMPOSITES CONTAINING $PbTiO_3$

C.E. MILLAR AND A.J. MOULSON
Division of Ceramics, School of Materials,
University of Leeds, Leeds LS2 9JT, UK

ABSTRACT

Piezoelectric-polymer composites offer improved properties over those of single phase materials for applications such as sonar and pressure sensitive devices. The present study concerns 0-3 composites comprising $PbTiO_3$ powder dispersed in a low density polymer matrix. Piezoelectric property data for corona poled composites is compared with that obtained by conventional poling. Data are presented covering a range of ceramic loadings and special attention is given to the effect of elastic modulus of the polymer phase on piezoelectric and pyroelectric properties of the composites.

INTRODUCTION

Lead zirconate titanate (PZT) is the most widely used of the piezoelectric materials because of its high piezoelectric activity. However for certain applications, such as hydrophones and other pressure-sensitive devices, PZT falls short of the ideal in several respects. Despite having a large piezoelectric coefficient d_{33}, its hydrostatic coefficient, d_h, is low. This is evident from the relationship:

$$d_h = d_{33} + 2d_{31}$$

in which, for PZT, d_{33} is approximately $-d_{31}/2$.

Furthermore, PZT is a relatively poor pressure-sensor because its high permittivity lowers the voltage coefficient, g_h (where $g_h = d_h/\varepsilon_o \varepsilon_r$). In addition, with a density of 7,900 kg m^{-3}, PZT is difficult to match acoustically to water, and its brittleness leads to poor mechanical shock-resistance limiting its usefulness in device applications. Many of these disadvantages can be overcome by using composite materials comprising PZT ceramic and a flexible polymer matrix. The polymer reduces permittivity, lowers density and increases mechanical shock resistance of the material. Despite the inevitable dilution of piezoelectric activity, the overall properties of the composites are superior to those of the ceramics alone. Over the past decade several investigators have examined composites of various structures, or 'connectivities'; reviews can be found elsewhere [1,2].

The most simple ceramic-polymer composite consists of a piezoelectric powder dispersed in a polymer matrix as shown in Fig. 1. The ceramic particles are considered to be unconnected and the polymer self-connected in 3 dimensions. Such a structure is described as (0-3), according to Newham's notation [3].

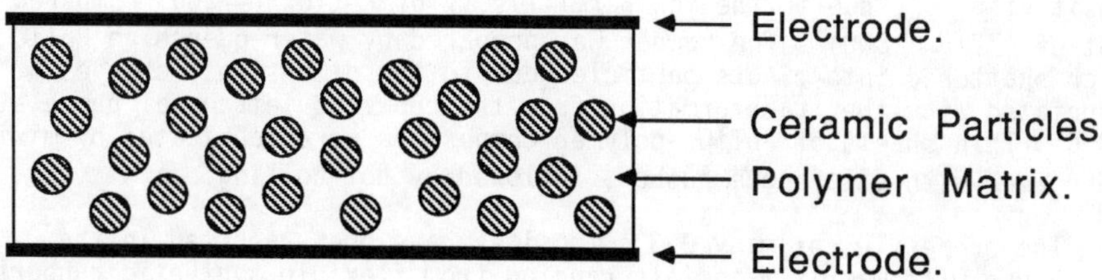

Figure 1. Schematic of a 0-3 composite containing discrete ceramic particles in a continuous polymer matrix.

To develop piezoelectric activity in piezoceramics and composites, they first have to be 'poled'. In this process an electric field (~ 10 MV m^{-1} for a 0-3 composite) is applied, which aligns the polar axis in crystallographic directions closest to the direction of the field. Conventional poling involves immersing the sample, often a disk, in heated oil and applying a DC electric field via electrodes on the samples' surfaces. An improved method, used successfully for the polymer, PVDF [4], is Corona-poling. The experimental arrangement is illustrated in Fig. 2.

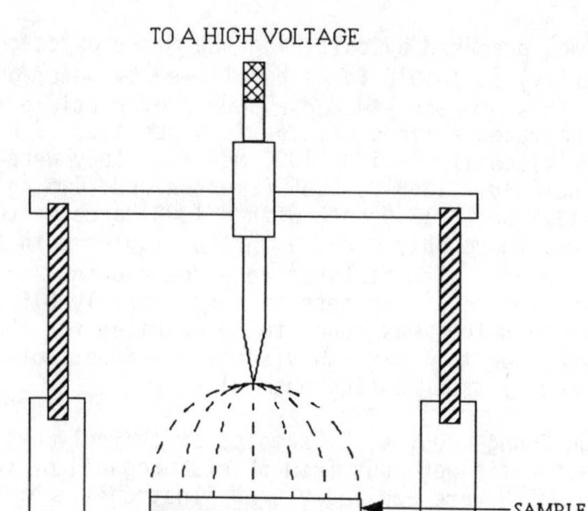

Figure 2. Experimental arrangement for the Corona poling technique.

A large DC potential is applied to a needle causing ionization of surrounding gas molecules. The charge is sprayed onto the unelectroded surface of a sample resting on an earthed plate, and develops the poling field between the sample faces.

In early piezoelectric composites, PZT comprised the ceramic phase, subsequently $PbTiO_3$ has been substituted leading to improved voltage sensitivity, g_h, due to the low permittivity of $PbTiO_3$ (~ 200) compared to that of PZT (~ 1800). The powder was prepared by water-quenching $PbTiO_3$, which shattered into minute particles due to the dimensional changes associated with the transformation from the cubic to tetragonal phase at 490°C. Thin sheets of $PbTiO_3$-polymer composites were fabricated by mixing the powder with chloropene rubber, followed by hot-rolling.

The properties of many 0-3 composites have been reported in the literature with matrix materials ranging from flexible synthetic rubbers to rigid epoxy resins [1]. However there has been little discussion about how properties of the matrix affect final piezoelectric and pyroelectric properties of the composite, although Sa-gang [6] briefly noted that the piezoelectric coefficients, d_h and d_{33}, increased with increasing flexibility.

In the present study 0-3 composites containing $PbTiO_3$ particles were examined. A fabrication technique for producing thin sheets is described, and the properties of samples poled by corona poling are compared with those poled by the conventional technique. Results are presented for composites containing a range of particle loadings and matrix materials of different elastic properties.

EXPERIMENTAL

The $PbTiO_3$ powder was prepared by calcining the mixed oxide components PbO and TiO_2 (99.9% purity) at 1000°C for 1 h followed by water quenching to room temperature. This process yielded a powder of particle size 1-5 µm. The polymers used embraced a range of elastic properties (2 MPa - 109 MPa) and were all of low viscosity (~ 100 - 1000 mPa s). They were based on epoxy resins and included: 1365-0, 1265 (Emerson and Cuming) and two Araldite resins CY1311 and CY1310 both with HY1300 hardener (Ciba-Geigy). The $PbTiO_3$ powder was thoroughly mixed with the liquid resin in amounts ranging from 10 to 60 v/o. The mixtures were deaired in a vacuum desiccator. Samples were cold-rolled to a thickness of approximately 0.5 mm and then cured. The rolling technique was found to be unsuited for the lower volume fractions (10-20 v/o) due to their low viscosity. Sheets of these volume fractions were formed by tape-casting onto glass.

To measure the Young modulus, E, samples of the polymers with dimensions of high aspect ratio were cut from sheets prepared by tape casting. Loads in the range 1-5 N were applied via an 'Instron' machine, the modulus being determined from the generated stress/strain curves.

After electroding with an air-drying silver paint, samples were Corona-poled. For comparison a few samples were poled conventionally. All the composites were poled at 80°C for 15 min with a field of 10 MV m^{-1} which was maintained until the composites cooled to 40°C. Composites were aged for at least 48 h before measurements were made.

The extent of poling was determined from the 90° domain orientation measured from the X-ray powder diffraction reflections from the 001 and 100 planes, before and after poling, and using the relationship:

$$\text{extent of poling} = \frac{\text{Area (001)}}{\text{Area (001)} + \text{Area (100)}}$$

The d_{31} coefficient was measured at 100 Hz using the technique described in the IRE standard [7]. To measure the hydrostatic coefficient, d_h, the electroded specimen was immersed in oil in a chamber and subjected to a pressure of 140 MPa, the generated charge was collected on a charge amplifier and measured by a technique similar to that described elsewhere [8].

RESULTS AND DISCUSSION

The conventional and corona poling techniques were both effective in poling the composites, as evidenced by the coefficients shown in Table 1.

TABLE 1

Comparison of piezoelectric and pyroelectric coefficients for 60 v/o $PbTiO_3$ / 40 v/o 1365-0 samples poled by conventional and corona poling

Poling technique	d_{31} /pC N^{-1}	d_h /pC N^{-1}	λ μC m^{-2} K^{-1}	$\frac{I(001)}{I(001) + I(100)}$
Conventional	12	36	106	76
Corona	14	37	109	79

Although both poling techniques seemed effective, Corona-poling offers several advantages over conventional poling. The absence of a conducting electrode means that charge cannot be quickly channeled to a weak spot. In the present study no electrical breakdown was encountered during Corona-poling; in contrast, some samples, particularly thin sections, failed during conventional poling. An additional advantage is that Corona poling can easily be adapted for the poling of large areas and for continuous poling. For the present study, where large poling fields were necessary, Corona poling was adopted.

The diluting effect of the polymer on the properties of the $PbTiO_3$ is shown in Fig. 3. 60 v/o was found to offer a good balance between properties and ease of fabrication. Composites containing high solid loadings were difficult to fabricate and resulting materials had poor mechanical properties.

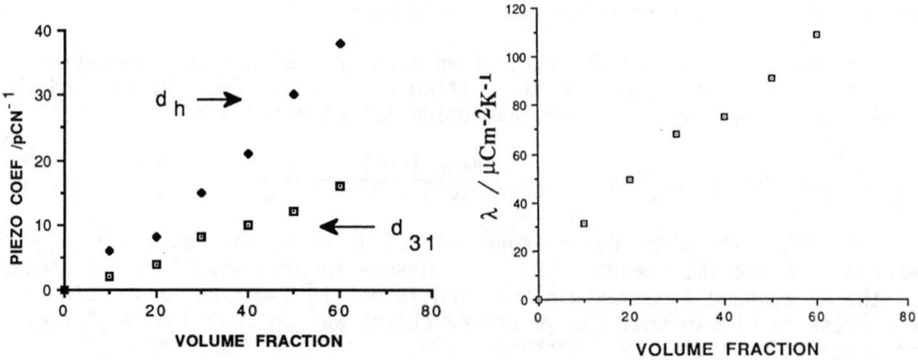

Figure 3. The dependence of piezoelectric coefficients d_{31} and d_h and pyroelectric coefficient λ on volume fraction of composites consisting of $PbTiO_3$ and matrix 1365-0.

TABLE 2

Properties of 60 v/o $PbTiO_3$ composites fabricated with polymer matrices embracing a range of elastic properties.

Sample	d_{31} /pC N^{-1}	d_h /pC N^{-1}	λ /μC m^{-2} K^{-1}	E /MPa	I(001) / [I(001) + I(100)]
$PbTiO_3$-1365-0	14.0	37	109	2	0.79
$PbTiO_3$-1265	10.0	32	89	10	0.68
$PbTiO_3$-CY1311	6.4	25	85	36	0.86
$PbTiO_3$-CY1301	1.0	10	12	111	0.85

The final column in Table 2 indicates that all samples were poled to a similar extent. The results show that as the Young modulus of the polymer phase decreases, the applied load is increasingly transferred to the ceramic particles leading to an increase in piezoelectric coefficient.

The pyroelectric coefficient also increases with increasing matrix flexibility, ranging from 12-109 μC m^{-2} K^{-1}. The measured pyroelectric coefficient is the sum of the primary and secondary pyroelectric co-efficients. The secondary coefficient is due to the mismatch in thermal expansion coefficients between the filler and the matrix, causing the ceramic particles to be stressed and so developing a piezoelectric charge. The magnitude of the secondary effect is dependent on the thermal expansion coefficients, the Young modulus of the polymer and the piezoelectric co-efficient. In the case of the composites the secondary coefficient is in the opposite sense to the primary coefficient therefore reducing the total pyroelectric coefficient with increasing Young modulus of the polymer phase.

In conclusion, optimum properties in the composites are obtained using a flexible polymer matrix and high volume fraction of lead titanate.

REFERENCES

1. Levinson, L.M., Electronic Ceramics, Properties, Devices and Applications, Marcel Dekker, New York, 1988, pp. 92-128.

2. Newnham, R.E., Browan, L.J., Klicker, K.A., Cross, L.E., Composite Piezoelectric Transducers, Mater. Eng., 1980, **112**, 93-106.

3. Newnham, R.E., Skinner, D.P., Cross, L.E., Connectivity and Piezo-electric-Pyroelectric Composites. Mat. Res. Bull., 1978, **13**, 525, 536.

4. Das Gupta, D.K., Doughty, K., Corona charging and piezoelectric effect in polyvinylidene fluoride. J. Appl. Phys., 1978, **49**, 4601-3.

5. Banno, H., Saito, S., Piezoelectric and Dielectric Properties of Composites of Synthetic Rubber and $PbTiO_3$ or PZT. Jap. J. Appl. Phys., 1983, **22** supp/22-2, 67-69.

6. Sa-gong, G., Safari, A., Jang, S.J., Newnham, R.E., Poling Flexible Piezoelectric Composites. Ferroelectrics, 1986, **5**, 131-142.

7. Mason, W.P., Jaffe, H., Methods for measuring piezoelectric, elastic and dielectric coefficients of crystals and ceramics. Proc. IRE, 1954, **42**, 921-930.

8. Hartley, N.P., Squire, P.T., Putley, E.H., A New Method of Measuring Pyroelectric Coefficients. J. Phys. E: Sci. Instrum., 1972, **5**, 787-789.

2.382

DEVELOPMENT OF CRYSTAL PHASES AND NONLINEAR PROPERTIES IN CHEMICALLY PREPARED VARISTOR CERAMICS

O.MILOŠEVIĆ, D.VASOVIĆ, D.POLETI, Lj.KARANOVIĆ, V.PETROVIĆ and D.USKOKOVIĆ
Institute of Technical Sciences of SASA, Belgrade, Yugoslavia

ABSTRACT

Advanced ceramic processing techniques were used to fabricate ZnO based varistor ceramics having the nonlinearity coefficient from 35 to 45, the breakdown field in the range from 400 to 1000 V/mm and the leakage current far bellow 1 $\mu A/cm^2$. Processing steps included chemical synthesis of powders with complex initial composition, mechanical activation of calcinated powders and sintering at 1473 and 1573 K. Various analytical methods, such as are lighting and scanning microscopy, X-ray, particle size analyzis and EDAX were used for the characterisation of powders and resulting nonlinear ceramics. The formation process of the main phases: ZnO, spinel and intergranular layer was investigated as well as the influence of the same on resulting nonohmic properties.

INTRODUCTION

Metal-oxide varistors based on ZnO represent multiphase ceramic semiconductive devices showing high nonlinear current-voltage characteristics (1). The non-Ohmic properties are generally described by the empirical relation $J=C\cdot K^{\alpha}$, where J and K are the current density and the applied field, respectively, C-constant and α is the nonlinearity coefficient, generally ranging from 5-50. Conventional processing of ZnO varistors involves searching for optimum phase content by means of high-temperature solid and liquid phase reactions of oxides as starting materials (2,3). Unfortunatelly, it is hard to avoid microstructural inhomogeneites,concerning nonuniform distribution of the main phases: ZnO, spinel and intergranular layer by this procedure. That can cause lowering of nonlinearity and enhancement of degradation phemonema (4).Also,both usefulness and the tendency to apply these devices in high-voltage transmission systems have generated the necessity for usage high-voltage capabilities in smaller volumes (5).That was the reason why such a preparation of fine-grained reactive, high purity powders with controlled composition by various chemical methods (6,7,8,9) has become the most important part in varistors production having high reliable ceramic and electrical properties.

The purpose of this paper is to describe the improved varistor production procedure that includes chemical preparation of powders and their mechanical activation. The conditions of these powders consolidation and varistor formation process were investigated in order that ceramics with high nonlinearity coefficient and a low leakage current can be obtained.

EXPERIMENTAL

To obtain varistor powders with complex initial composition containing 92.7 mol% ZnO and 7.2 mol% of addittives, the method of evaporation of ZnO suspensions with additive solutions was utilized. The following additives were used: $Co(NO_3)_2 \cdot 6H_2O$, $Ni(NO_3)_2 \cdot 6H_2O$, $Cr(NO_3)_2 \cdot 9H_2O$, $Bi(NO_3)_3 \cdot 5H_2O$, $Mn(CH_3COO)_2 \cdot 4H_2O$ and $SbCl_3$. The obtained powder was calcinated at 973 K. Details concerning powder preparation are given elsewhere (9). Several batches of this powder were mechanically treated by means of agate high-speed planetary ball mill with acceleration 12 g (Retsch, typ PM 4). The experiments were performed by basic sundiscs rotation of 50 and 250 rpm., during irreversibile rotating regime, while balls diameters were of 10 and 20 mm and powder to ball mass ratio, m_p/m_b, was 1/10 and 1/20. Treating time ranged from 10 to 120 min. (Table 1):

TABLE 1.
Conditions of the mechanical treatment in high-speed planetary ball mill

rpm	m_p/m_b	t(min)	batch
non treated			A.0
50	1/10	10	A.1
		60	A.2
		120	A.3
250	1/10	10	A.4
		60	A.5
	1/20	10	A.6
		60	A.7

Powders from batches A.0 and A.3-A.7 were compressed into pills 1-5 mm in high and diameter of 8 mm. Green samples (~50% theoretical density) were isothermal sintered in air at temperatures of 1473 and 1573 K, held for 5-480 min and after that quenched in air.Particle size distribution was performed at calcinated powders A.0-A.7. using laser diffraction analyser of "Malvern 3600" type. Development of crystal phases in powders was followed at "Phillips 1051" diffractometer under standard experimental conditions. Microstructural analysis of free surfaces of sintered samples, as well as polished and etched ones were performed at light microscope.Chemical etching was done during 10s using HNO_3 solution.The grain size was determinated from etching surfaces. Scannning electron microscopy was applied during observation of fractured surfaces. Energy dispersive x-ray analysis (EDAX) was performed in qualitative phase determination in samples sintered at 1573 K (for A.0). X-ray diffraction (XRD) was used in crystal phase detection of sintered ceramics,shaped in pills. Electrical properties were determinated at polished samples: the nonlinearity coefficient within the range from 0.1-1.0 $mA/cm^2 (\alpha_1)$, 1.0-10.0 mA/cm^2 (α_2), the breakdown field, K_C, at 1 mA/cm^2 and the leakage current, J_L, for the voltage of 0.8 K_C.

RESULTS AND DISSCUSION

The results of particle size distribution of powders A.0-A.7 are plotted as particle diameters against cumulative weight undersize in logarithm probability papers (Fig.1). The near linearity observed for all examinated powders implies the fact that distribution may be approximated by log-normal distribution. The mean particle sizes, D , were graphically determined for each distribution and are also presented at Fig.1. It is shown that a mechanical treatment lasting for 10-120 min (batches A.1-A.3) causes a gradual decreament in mean particle size to at least 0.55 m. In the same way, the percentage of submicron particles gradually increased from 3.1 (for referent batch A.0) to 96.5%, after 120 min. (A.3). By raising the basic sundiscs rotation to 250 rpm, the decreament in the mean particle sizes is forced with time when powder to ball mass ratio is 1/10 (A.4, A.5). In that case, the percentage of submicron particles after 10 min only is 64% with D =1,07 μm. When powder to ball mass ratio changes to 1/20, for the same experimental conditions and for balls of 20 mm in diameters (batch A.7), extremely monodispersive powder is obtained containing 92.5% of particles,the mean particle sizes of which was 1.9 μm.

Fig.2a. represents the XRD patterns of the powders obtained from A.0 and A.7 batches. It is evident that in powder A.0 a well crystallized ZnO phase and a poor crystallized spinel phase are present. No change in phase composition has been noticed in connection with the referent mixture,on A.1-A.6 XRD patterns.However,when powder A.7 is concerned,the disturbance of ZnO and spinel lattices occurred what has a widen diffraction maximum, of lower intensity,being characteristic for these two phases,as a result.

After isothermal sintering of compacting samples at 1573 K/5 min (A.0) X-ray spectra indicate the existance of a phase composition being typical for varistor ceramics (Fig.2b).The presence of a well crystallized ZnO, together with spinel and tetragonal β-Bi_2O_3, lattice parameters of a=7,7425 and c=5,6313 A (JCPDS 22-515) is evident. ZnO grains 9μm in diameter, contain Co, Mn and Ni ions as substituents (9) and vary linearly with sintering time according to the relation $D \sim t^{1/6}$ at 1473K and $D \sim t^{1/7}$ at 1573 K. The similar law is presented elsewhere (10).

The spinel phase with main composition Zn-Sb, contains Cr, Mn, Co and Ni as substituents. It appears in form of elongated chain-like structures along ZnO grain boundary, that during prolongation of sintering time are transformed into grain-like, one-axis structures.

Tetragonal β-Bi_2O_3 phase can be transformed or even dissapeared by prolongation of the sintering time. This is proved by dissapearance of the characteristic peak at 3.2 Å after 180 min at 1573 K and by appearance of a small peak at 3.12 A (Fig.2b).

In samples sintered at 1473 K, batches A.3-A.7, a more uniform distribution of particles size of spinel are noticed when compared to those in the referent mixture. The relative particles sizes of ZnO in those systems are as follows: 5 μm (A.3) and 3.5 μm (A.4, A.5, A.6, A.7). The most optimum microstructure, regarded from the viewpoint of distribution of all constituents, was noticed in sintered samples of A.7 powder. These systems have also shown a high chemical instability versus decomposition reagents.

Note that the average breakdown voltages of individual grain boundaries, given by the slope of the curve K=f(1/D)(10), plotted in Fig.3, are 5-6 V for the samples obtained from A.0, sintered at 1473 and 1573 K and from A.3-A.5 sintered at 1473 K. Much greater values of the breakdown voltage per grain boundary given in our paper and also for amine processed ZnO varistor ceramics (~5V)(5) than those reported so far

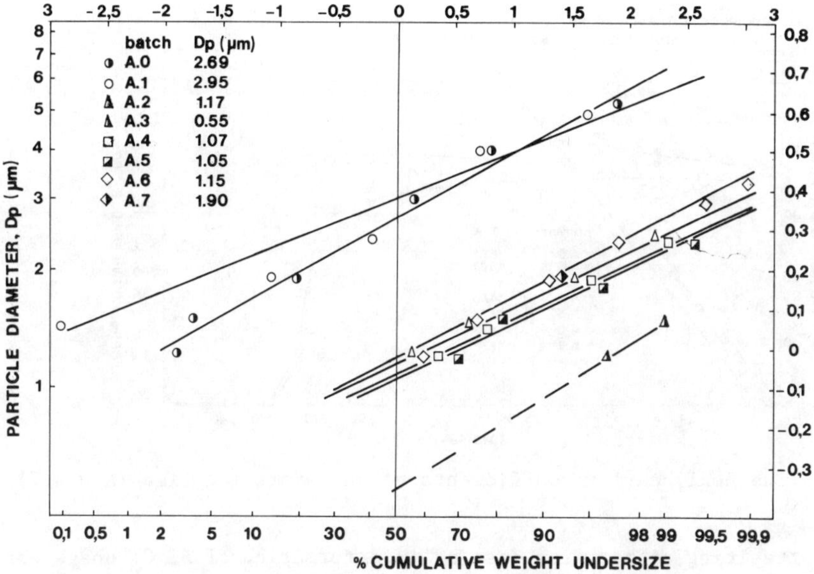

Figure 1. Log-normal probability plot of powders particles A.0-A.7

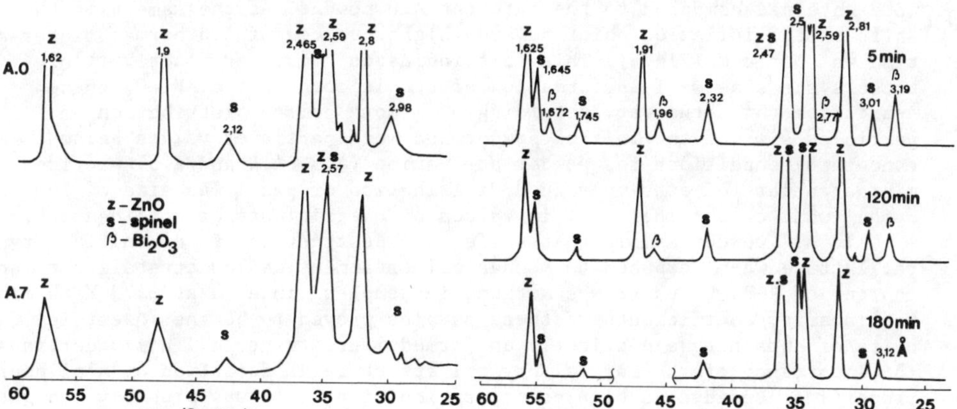

Figure 2. XRD patterns of A.0 and A.7 (a) and isorthermally sintered samples at 1573 K for 5,120 and 180 min. (batch A.0)(b)

(~2-3V) (10) for conventional procedures could be expressed by more narrow distribution of ZnO grain size when ceramic is processed from fine grained chemically prepared powders (5,11).

Coefficients of nonlinearity, α_1 and α_2, in function of isothermal sin ering time at 1473 i.e. 1573 K are given in Fig.3 for A.0 mixture, i.e. for A.3-A.7 ones. The corresponding values for powder samples of A.0 isothermally sintered at 1573 K range from 37-42 (α_1) i.e. from 32-35 (α_2), being characterized by leakage currents below 1 $\mu A/cm^2$, to 120 min.approximatively. After that time an abrupt fall of nonlinearity was

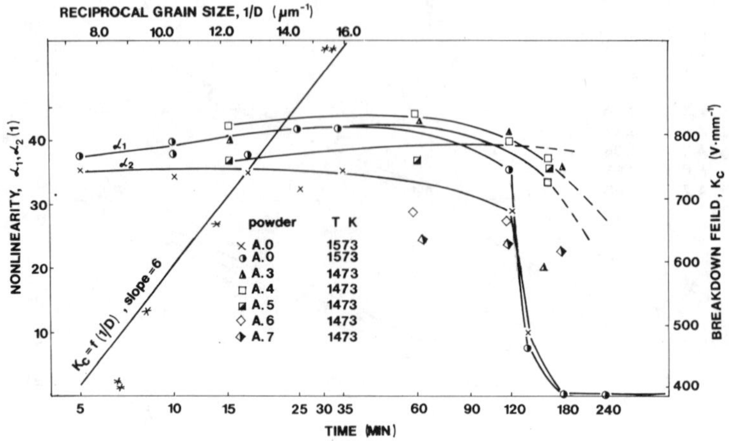

Figure 3. The nonlinearity coefficients versus sintering time (A.0-A.7)

noticed, resulting either from loss or transformation of Bi_2O_3 phase what was proved by the change of the relative diffraction maxima of tetragonal Bi_2O_3 phase in function of isothermal sintering time (Fig.2b).

During sintering of mechanically activated A.3 and A.4 powders at 1473 K, higher corresponding values for coefficients of nonlinearity (~45) were obtained compared to the referent A.0 powder. At the same time the shift of the inflexion point towards higher values of isothermal sintering time was noticed (Fig.3). This fact indicates a greater reactivity of these systems as well as formation of stable compounds of Bi_2O_3 phase. Apart from this, the very obtaining of a more narrow distribution of particles size of the initial powder and mean particles values below 1 μm concerning conditions for powder activation (A.3, A.4 and A.5) provides microstructure of extremely good distribution of particles size of ZnO and spinel what causes increment in values of coefficients of nonlinearity.

In A.7 powder sample that suffered a destruction of the initial crystal lattice when exposed to mechanical treatment, a low crystallinity and content of β-Bi_2O_3 phase was noticed in samples sintered at 1473 K/60 min. Nonlinearity coefficients of these samples proved to be the lowest in spite of obtaining a distinctly uniformed microstructure. This underlines the importance of the initial crystal structure in formation of nonlinear properties. Besides, a high concentration of defects was probably brought into this system during mechanical grinding process carried out under conditions that correspond to A.7 batch. This could cause a volume alloying of ZnO lattice and frustrated formation of a heavy potential barrier along grain boundary. The above presumptions should be experimentally proved.

CONCLUSION

In this paper we have investigated the influence of mechanical activation of powder obtained by chemical synthesis on structural changes and electrical properties. The following properties of powder were analysed: particle size distribution, crystallographic composition and characteristics of sintered samples: microstructure,phases development and electrical properties. It was proved that under activation conditions in planetary ball mill, where no change occurs in connection with phase composition of powder, the obtaining of more narrow distribution of powder particle size of 1 μm approximately leads to obtaining of more narrow distribution of ZnO and spinel particles size in sintered samples as well as to higher values of nonlinearity coefficients (35-45) and barrier voltage of 5-6 V. Temperature of nonlinear properties formation is lower for approximately 100 K than is the corresponding value of the referent non-activated powder.Under conditions when a destruction of the crystal lattice of initial phases occurred during mechanical activation of powder,the resulting values of nonlinearity are very low even apart from the fact that practically monodispersed powder was obtained. This points to the importance the initial crystal structure of powder plays in formation of nonlinear properties.

ACKNOWLEDGEMENT

The paper was financially supported by the Republic Committee for Sciences of Serbia through the Project "Physical Chemistry of Condensed Systems".

REFERENCES

1. Matsuoka, M., Nonohmic properties of zinc oxide ceramics, J. J. Appl. Phys., 1971, 10(6), 736

2. Inada, M., Crystal phases of nonohmic zinc oxide ceramics, J. J. Appl. Phys., 1978, 17(1), 1

3. Santhanam, A.T., Gupta, T.K. and Carlson, W.G., Microstructural evaluation of multicomponent ZnO ceramics, J. Appl. Phys., 1979, 50(2), 852

4. Bruckner, W., Bather, K.H., Moldenhauer, W., Wolf, M. and Lange, F., Inhomogeneites and single barriers in ZnO varistor ceramics, Phys. Stat. Sol. (a), 1980, 59, K1

5. Hishita, S., Yao, Y. and Shirasaki, S., Zinc oxide varistors made from powders prepared by amine processing, J. Am. Ceram. Soc.,1989,72(2),338

6. Lauf, R.J. and Bond, W.D., Fabrication of high field zinc oxide varistors by sol-gel processing, Ceram. Bull., 1984, 63(2), 278

7. Sonder, E., Quinby, T.C. and Kinser, D.L., ZnO varistors made from powders produced using a urea process, Ceram. Bull., 1986, 65(4), 665

8. Dosch, R.G., The effects of processing chemistry on electrical properties of high-field ZnO varistors, In Science of Ceramics Chemical Processing, ed.Hench and Ulrich D.R., Willey & Sons, 1986, pp.311.

9. Milošević, O., Vasović, D., Poleti, D., Karanović, Lj., Petrović,V. and Uskoković, D., Microstructural and electrical properties of ZnO varistors prepared by coprecipitation and evaporation of suspensions and solutions". Second Varistor Conference, Schenectady, USA, December 1988

10. Wong, J., Barrier voltage measurement in metal oxide varistors, J. Appl. Phys., 1976, 47(11), 4971

11. Emtage, P.R.,Statistics and grain size in zinc oxide varistors, J. Appl. Phys., 1979, 50(11), 6833

THE CRYSTAL STRUCTURE OF HIGH-Tc Bi-BASED SUPERCONDUCTORS

Shigeo HORIUCHI, Kaoru SHODA* and Yoshio MATSUI

National Institute for Research in Inorganic Materials, Tsukuba, Ibaraki 305, Japan
*Permanent address: Ube Industries Ltd, Ube, Yamaguchi 755, Japan

ABSTRACT

The crystal structures of high-Tc Bi-based superconductive oxides, $Bi_2Sr_2Ca_{n-1}Cu_nO_{2n+4+x}$ (n=1~3), are examined mainly by means of high-resolution electron microscopy. They are fundamentally consisted of perovskite-like layers interleaved with Bi-O planes. The Cu-O planes, which are responsible for the superconductivity, increase in number with n. Characteristic features are 1) a prominent lattice bending, 2) Bi-concentrated bands, 3) a structural modulation and 4) a frequent intergrowth. These are compared with those so far reported for other high-Tc superconductors like $YBa_2Cu_3O_{7-y}$, with the discussion on the relationship between microstructures and the critical temperature Tc.

INTRODUCTION

The analysis of crystal structures is important from the viewpoint not only of the interpretation on the origin of the physical and chemical properties but also of finding the guideline for exploring further new materials. We have examined the recently found Bi-based superconductors, $Bi_2Sr_2Ca_{n-1}Cu_nO_{2n+4+x}$ [1-3], to find out a prominent structural modulation.[4,5] Because of this the conventional x-ray or neutron diffraction method are not so useful for these superconductive materials and, therefore, we have tried to analyze the crystal structure mainly by means of high-resolution transmission electron microscopy (HRTEM).

EXPERIMENTAL

Specimens of (2201) phase (n=1), (2212) phase (n=2) and (2223) phase (n=3), whose critical transition temperature Tc are 20, 80 and 110 K respectively, were obtained by different heating processes. Powder x-ray diffraction and electrical resistivity were measured. Small blocks of the crystals were crushed in an agate mortar. Fine fragments created were mounted on the microgrids and observed in the electron microscopes of JEM2000EX (200kV), JEM4000EX (400kV) or H-9000 (300kV) type.

EXPERIMENTAL RESULTS

From a series of HRTEM images taken for different orientations, it has been made clear that each crystal is composed of layers, interleaved by the lines of darkest spots.[4-6] This suggests that the structure is similar to that of so-called Aurivillius-type.[7] The layers interleaved by the Bi-O planes have the perovskite-like structure.

In electron diffraction patterns the array of weak spots indicates the formation of a superstructure. They appear at non-integral positions so that it must be of incommensurate-type.

Fig.1(a) is an HRTEM image of the (2212) phase (orthorhombic, a=5.4, b=27 and c= 30.8Å), where electrons are incident along the a axis. The (020) crystal lattice planes are prominently bended, by about 20 % at maximum. Moreover, the darkest spots closely adjoin in small regions. From the theory of the image formation we have considered that these spots are due to the concentration of Bi atoms and we have named these as Bi-concentrated bands.[5] This has been proved true based on the computer simulation of image intensity.[6] The calculated image, which well fits to the real one, is shown in the micrograph with same magnification. A model of the structure used for the calculation is shown in Fig.1(b). It is consisted mostly of 5 subcells, each of which is identical with that by powder X-ray or neutron diffraction[8-11] as well as by HRTEM observation.[12-14] Some important conclusions obtained from the model are; 1) Bi atoms are partly replaced by Sr in the regions between adjoining Bi-concentrated bands. 2) In the perovskite-like layers metal planes are stacked in the sequence of Sr-Cu-Ca-Cu-Sr. 3) A part of Sr are substituted by Bi. 4) In the Ca planes some are replaced by Sr. 5) The positions of oxygens can be speculated based on the crystal symmetry (S.G.Amma) and the atomic distance. It may be said that CuO_5 square piramids are formed.

The (2223) phase (a=27, b=5.4 and c=37 Å) grows from the (2212) phase.[15] The crystals are both plate-like and the diameter of (2223) plates is much larger than that of (2212) ones. However such impurity phases as CuO, Ca_2CuO_3 and $Sr_3Cu_5O_z$ are left even after long time heating.

Another structural interest of this phase is the frequent intergrowth of layers with different thickness.[16-18] Any specimens intended for getting the (2223) phase have always included the intergrowth for the system Bi-Sr-Ca-Cu-O. The prominent bending of (020) lattice planes and the Bi-concentrated bands are noticed also in the layers of the (2223) phase.[17] The diffraction patterns indicated the formation of the structural modulation similar to that for the (2212) phase.

We have examined the effect of replacing oxygens partly by fluorines.[19] The specimens fired in air were composed of very large grains of the (2223) phase, in which only a trace of fluorines were doped. However they showed a slight increase in Tc; Tc=113K, Tc_{onset}=120K and Tc_{end}=106K. Fig.2 shows an example of the microstructures. In (a) frequent intergrowth is seen. A local interface arises. The structure is schematically shown in (b). Cu-O planes of both sides must partly be linked at the interface.

The characteristic features with the lattice bending, Bi-concentrated bands and the structure modulation are essentially the same for the (2201) phase (a=26, b=5.4 and c=24Å).[20]

The (2212) phase was examined by the low-temperature electron microscopy, where the specimen is kept under 10 K.[21] The results show that the crystal structure with the modulation along the b axis is not substantially changed on the transition to the superconducting state.

DISCUSSION

The diffraction spots at non-integral positions are described by the four-dimensional vector, $h\underline{a}^*+(k+m/s)\underline{b}^*+l\underline{c}^*$, where h,k,l and m are integers.[9] $s/\underline{b}^*$ means the wave length of the modulation. The value s is measured to be 4.75 from a diffraction pattern. We may say that the structural modulation is originated from the lattice bending, which are related to the fluctuation in the concentration of Bi atoms. Although the intensity of the reflections should generally be interpreted based on the four-dimensional space theory[22], the geometry can be explained intuitively based on the Fujiwara's method.[23] It is seen in the HRTEM image like Fig.1 that major larger supercells and minor smaller ones are mixed in an almost regular way with the ratio of 3:1. The value of s can then be understood from the viewpoint of mixing.

It has been known that almost all the high-Tc superconductors are common in following two secifications, 1) two-dimensional Cu-O plane and 2) hole doping into the parent crystals, which otherwise are anti-ferromagnetic. From the structural viewpoint, it seems important to understand the role of the Cu-O planes. Fig.3(a) is to show the atomic distances, r_1 (Cu-sqr.O), r_2 (Cu-apex O) and r_3 (M-apex O), in and around the CuO_5 square pyramids. (b) is a diagram, which summarizes a relation r_1 and r_2/r_3 vs Tc, for $La_2SrCu_2O_6$ (Tc=0K), $La_{1.85}Sr_{0.15}CuO_4$ (40K), $Pb_2Sr_{2.24}Nd_{0.76}Cu_3O_x$ (70K), $Bi_2(SrCa)_3Cu_2O_x$ (80K), $YBa_2Cu_3O_7$ (90K) and $Tl_2Ba_2CaCu_2O_8$ (110K) (broken lines)[24]. It is seen that the value of Tc increases with r_2/r_3, which must be related to the tendency of the preferential coordination in the square plane. On the other hand, the fluctuation in the value due to the bending of lattice planes in the Bi-based superconductors is in the range shown by the solid lines in the figure. The fluctuation in r_1 suggests the valency change and, further, the variation of the hole concentration in the square planes. Therefore, we may assume that the structural modulation is related to the mechanism of the superconductivity.

Acknowledgement; The authors would like to express their thanks to Director N.Setaka of NIRIM and Prof.S.Saito for continuous support, and Drs.H.Nozaki, Y.Onoda, M.Ohta, E.Takayama-Muromachi, A.Ono, S.Takekawa, F.Izumi, A.Yamamoto, F.P.Okamura and M.Onoda of NIRIM, Drs.H.Maeda and Y.Tanaka of NRIM, Prof.S.Sueno of Tsukuba Univ. and Drs.H.Daimon, F.Sakai, K.Uchida for valuable discussion.

REFERENCES

1) J.Akimitsu, A.Yamazaki, H.Sawa and H.Fujiki: Jpn.J.Appl.Phys. 26 (1987) L2080.
2) C.Michel, M.Hervieu, ,M.M.Borel A.Grandin, F.Deslands, J.Provost and B.Raveau: Z.Phys. B68 (1987) 421.
3) H.Maeda, Y.Tanaka, M.Fukutomi and T.Asano: Jpn.J.Appl.Phys. 27 (1988) L209.
4) Y.Matsui, H.Maeda, Y.Tanaka and S.Horiuchi: Jpn.J.Appl.Phys. 27 (1988) L361.
5) Y.Matsui, H.Maeda, Y.Tanaka and S.Horiuchi: Jpn.J.Appl.Phys. 27 (1988) L372.
6) S.Horiuchi, H.Maeda, Y.Tanaka and Y.Matsui: Jpn.J.Appl.Phys. 27 (1988) L1172.
7) B.Aurivillius: Ark.Kem. 1 (1949) 463.
8) E.Takayama-Muromachi, Y.Uchida, Y.Matsui, M.Onoda and K.Kato: Jpn.J.Appl.Phys. 27 (1988) L556.
9) M.Onoda, A.Yamamoto, E.Takayama-Muromachi and S.Takekawa: Jpn.J.Appl.Phys. 27 (1988) L587.
10) J.M.Tarascon, Y.LePage, P.Barboux, B.G.Bagley, L.H.Greene, W.R.McKinnon, G.W.Hull, M.Giroud and D.M.Hwang: Phys.Rev.B. 37 (1988) 9382.
11) M.A.Subramanian, C.C.Torardi, J.C.Calabress, J.Gopalakrishnan, K.J.Morrissey, T.R.Askew, R.B.Flippen, U.Chowdhry and A.W.Sleight: Science, 239 (1988) 1015.
12) D.Shindo, K.Hiraga, M.Hirabayashi, M.Kikuchi and Y.Shono: Jpn.J.Appl.Phys. 27 (1988) L1018.
13) Y.Hirotsu, O.Tomioka, T.Ohkubo, N.Yamamoto, Y.Nakamura, S.Nagakura, T.Komatsu and K.Matsushita: Jpn.J.Appl.Phys. 27 (1988) L1869.
14) H.W.Zandbergen, W.A.Goen, F.C.Mijhoff, G.vanTendeloo and S.Amelinckx: Physica C, 156 (1988) 325.
15) S.Horiuchi, K.Shoda, M.Tsutsumi, K.Kosuda and H.Nozaki: ISIJ Internat. 29 (1989) 209.
16) Y.Matsui, H.Maeda, Y.Tanaka, E.Takayama-Muromachi, S.Takekawa and S.Horiuchi: Jpn.J.Appl.Phys. 27 (1988) L827.

17) Y. Matsui, S. Takekawa, H. Nozaki, A. Umezono, E. Takayama-Muromachi and S. Horiuchi: Jpn. J. Appl. Phys. 27 (1088) L1241.
18) S. Ikeda, H. Ichinose, T. Kimura, T. Matsumoto, H. Maeda, Y. Ishida and K. Ogawa: Jpn. J. Appl. Phys. 27 (1988) L999.
19) S. Horiuchi, K. Shoda, H. Nozaki, Y. Onoda and Y. Matsui: Jpn. J. Appl. Phys. 28 (1989) L621.
20) Y. Matsui, S. Takekawa, S. Horiuchi and A. Umezono: Jpn. J. Appl. Phys. 27 (1988) L1873.
21) S. Horiuchi. K. Shoda, M. Iwatsuki, Y. Harada and Y. Matsui: Jpn. J. Appl. Phys. 28 (1989) L386.
22) P. W. deWolff, T. Junssen and A. Janner: Acta Crystallogr. A37 (1981) 625.
23) S. Horiuchi, X. H. Zhang, K. Shoda, M. Ohta and Y. Matsui: to be published.
24) S. Sueno: J. Cryst. Soc. Jpn. 31 (1989) 119.

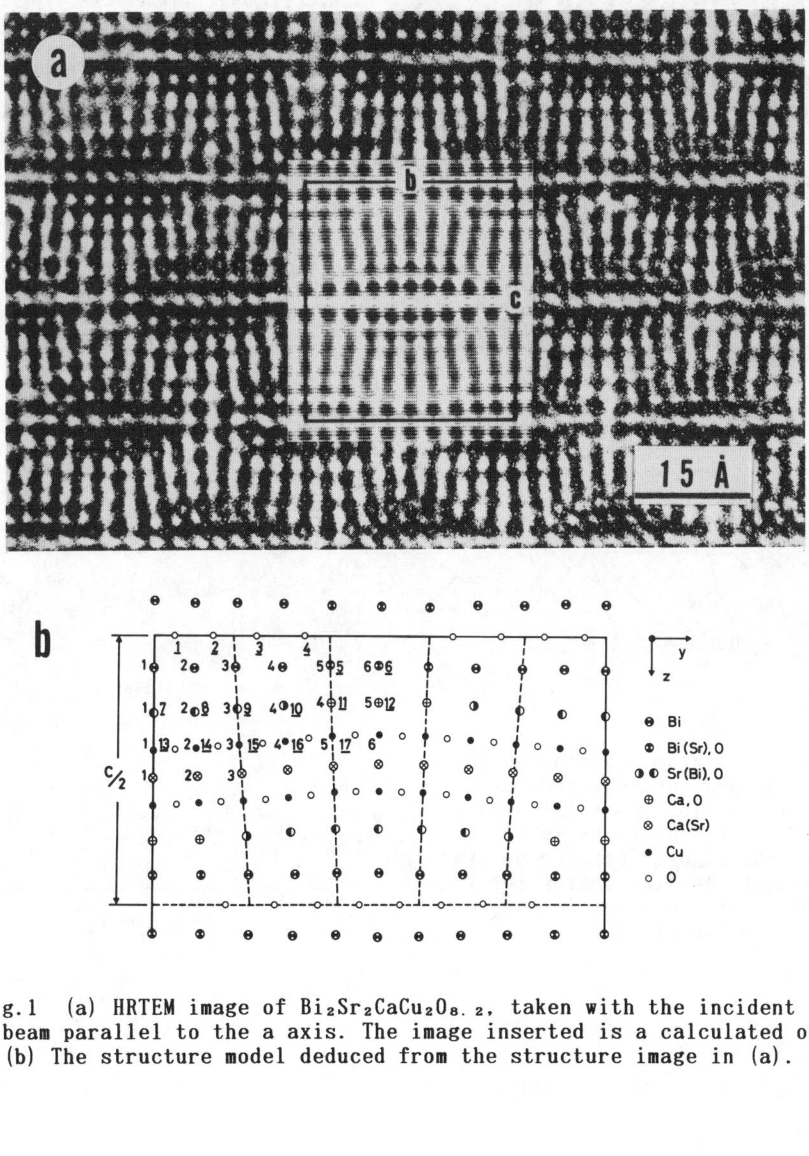

Fig.1 (a) HRTEM image of $Bi_2Sr_2CaCu_2O_{8.2}$, taken with the incident beam parallel to the a axis. The image inserted is a calculated one. (b) The structure model deduced from the structure image in (a).

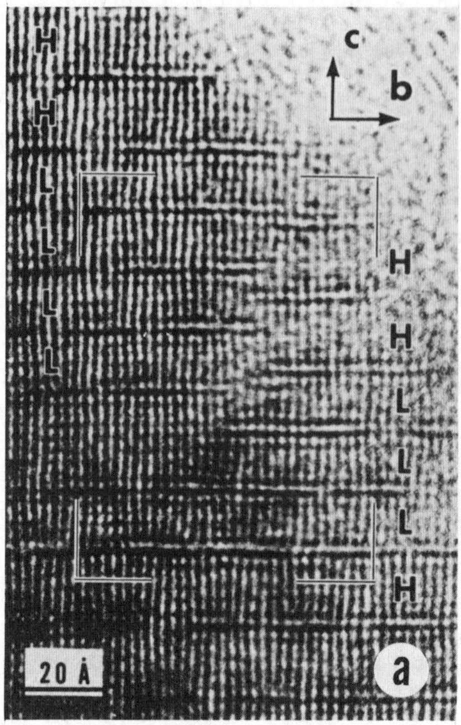

Fig.2 (a) A microstructure of the Bi-based superconductors prepared by doping fluorines. H:(2223) phase, L:(2212) phase (b) Schematic representation of the lattice planes in the area outlined in (a).

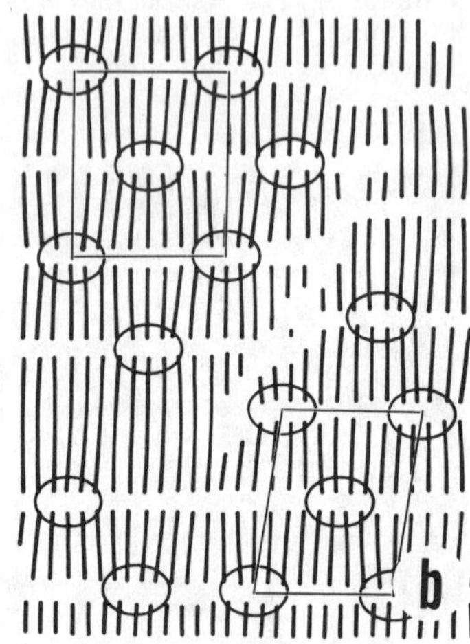

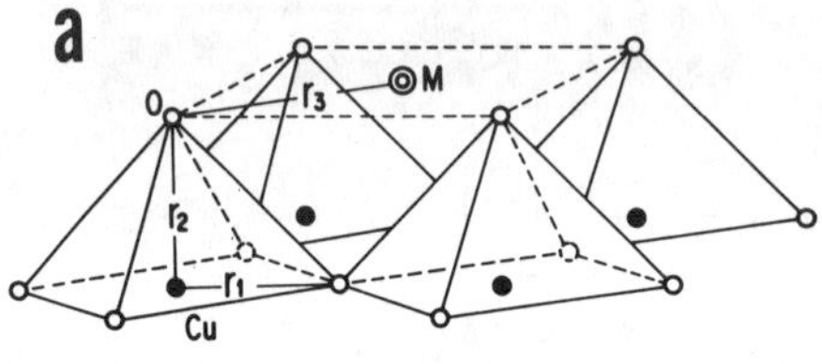

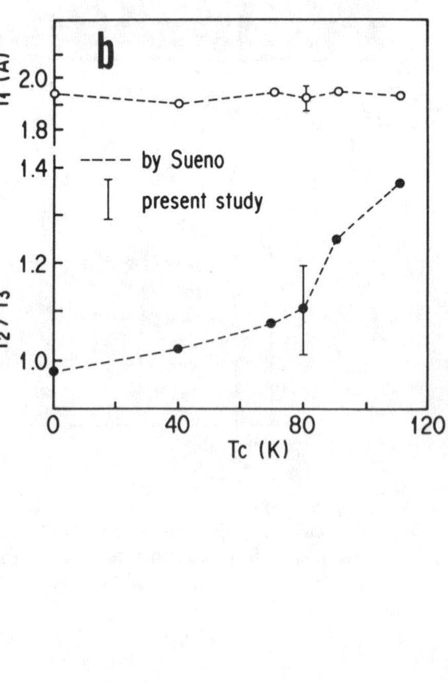

Fig.3 (a) Atomic distances in and around the CuO_5 square pyramids. (b) The relationship between r_1 and r_2/r_3 vs Tc for various high-Tc superconductive oxides. The fluctuation for Tc=80K ($Bi_2Sr_2CaCu_2O_8$) is due to the bending of lattice planes.

PHASE RELATIONSHIPS IN AIR IN THE SYSTEM $BaO-CuO-Cu_2O$, RELEVANT IN THE DEVELOPMENT OF CERAMIC SUPERCONDUCTORS

ANTÓNIO JOSÉ LAU and ANA MARIA SEGADÃES
Depº Engª Cerâmica e do Vidro, Universidade de Aveiro
3800 AVEIRO, PORTUGAL

ABSTRACT

Ceramic materials based on $YBa_2Cu_3O_7$ are the best superconductors known so far. It is known that the oxygen stoichiometry, very sensitive to the processing conditions, plays a fundamental role in the superconducting properties of these oxides. Their development depends on the control of processing variables and the knowledge of phase compatibility relationships in the non-condensed system $Y_2O_3-BaO-CuO-Cu_2O$ is, therefore, of utmost importance. The present work was aimed at determining the relevant part of the isobaric phase diagram in air of the sub-system $BaO-CuO-Cu_2O$, up to 1200°C, using mainly thermogravimetric analysis, as a first approach to the system $Y_2O_3-BaO-CuO-Cu_2O$ into which this work will progress. Melting in the CuO-rich part of the system was found to begin at about 925°C, BaO.CuO appears to melt incongruently at about 1150°C, and considerable extent of oxygen non-stoichiometry was detected in the solid phases.

INTRODUCTION

The report of superconductivity above 90K in yttrium-barium-cuprates by Wu *et al.* and subsequent identification of the superconducting phase as $YBa_2Cu_3O_x$ ($6<x<7$), has generated a tremendous amount of interest in these materials. Their superconducting properties are extremely dependent on the oxygen stoichiometry [1] which, in turn, is very sensitive to the processing conditions. In reducing atmospheres, the final products contain metallic Cu and no ternary oxides [2].

Typically, these materials are processed by solid state reaction between Y_2O_3, $BaCO_3$ or BaO or BaO_2 or $Ba(OH)_2$, and CuO at temperatures between 900 and 1000°C, followed by an oxygen annealing treatment, either at 500°C, or at 900°C followed by slow cooling. It has been shown [2] that the reaction mechanism depends on the starting materials used. $BaCuO_2$ is the first crystalline phase formed when BaO_2 is used, followed by the formation, above 800°C, of $YBa_2Cu_3O_x$ from the reaction between Y_2O_3 and $BaCuO_2$. Pure $YBa_2Cu_3O_x$ can be obtained only above 950°C, if the heating time is longer than 2h. A different reaction mechanism was found when $BaCO_3$ was used. The formation of $YBa_2Cu_3O_x$ occurs at a slightly higher temperature and the reaction rate is controlled by the decomposition of the carbonate. Thus, samples prepared below 950°C always contain unreacted starting materials.

It is not surprising to find in the literature available many disagreements among authors and even contradictory results. Experimental effort has been carried out to map isothermal regions, compatible phase assemblages and initial melting temperatures in the system Y-Ba-Cu-O and, to assess the phase equilibrium in the overal system, it is important to know the phase relations, including the development of liquid phases in the sub-system BaO-CuO-Cu_2O.

Plotting and interpreting phase equilibrium relationships in systems containing oxides of variable valency elements is somewhat more complicate than in systems of stable oxides. In the former, the stability fields of the various phase assemblages are not only functions of temperature and composition but also of oxygen partial pressure in the atmosphere.

The dissociation curve of an unstable oxide at a given oxygen partial pressure is an isobaric binary diagram in which the equilibrium of two condensed phases (solids or liquids) is isothermal (*i.e.* invariant). When a second, stable, oxide is added to this system, the compositions of initially fully oxidized mixtures will move, as the temperature rises and mixtures lose oxygen, into the ternary diagram along oxygen reaction lines. In a ternary non-condensed system, complete melting (*i.e.* equilibrium of one solid+liquid+gas) is no longer isothermal, even at a fixed oxygen pressure. In other words, the liquidus oxygen isobars are univariant curves on the Liquidus surface of the system and the temperature of complete melting of a given mixture, and the composition of the resulting liquid phase, is given by the intersection of the particular isobar with the mixture's oxygen reaction line.

However, in a ternary isobaric non-condensed system, the equilibrium of three condensed phases (*e.g.* two solids+liquid) is invariant and, therefore, melting will begin at a temperature corresponding to the intersection of the particular isobar with the univariant boundary curve in that part of the system, and not at the corresponding ternary invariant point.

When such a ternary isobaric diagram is projected back along the oxygen reaction lines onto the fully oxidized edge of the composition triangle, a binary-looking phase diagram is obtained, in which horizontal lines represent the invariant, isothermal, equilibrium of three condensed phases. Such a diagram, determined by single-crystal X-ray diffraction, was repported by Roth *et al.* [3] for the system $BaO-CuO_x$ in air, and is reproduced in Figure 1.

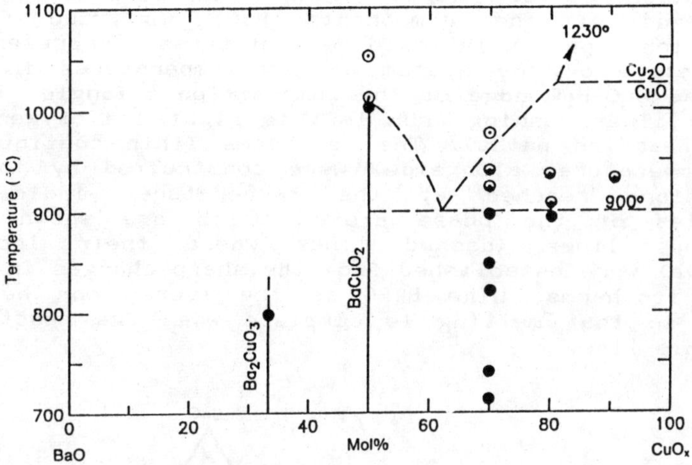

Figure 1. Preliminary experimental phase diagram of the system $BaO-CuO_x$ in air [3].

MATERIALS AND METHODS

All experiments were carried out in air, data being recorded as thermogravimetric (weight loss) curves obtained for selected fully oxydized compositions heated up to 1200°C, at 2°C/min. BDH copper oxide was used. As for the barium oxide, due to its strong tendency to hydrate in air, other compounds were tested, namely $Ba(OH)_2$ and M&B $BaCO_3$. However, when mixed with CuO, both barium compounds decompose over a range of temperatures overlapping the decomposition range of CuO, wich renders it impossible to separate the weight loss due to the Ba source from that due to the evolution of oxygen as the copper oxide changes valency.

The compound CuO.3BaO was then prepared from CuO and $BaCO_3$ by isostatically pressing the appropriate amounts of the reagents at 180 MPa and sintering at 880°C (to avoid liquid formation) for 6h. Pellets were then ground and X-rayed, and the whole procedure was repeated twice, when only traces of carbonate were detected in the X-ray diffraction patterns.

TGA was then carried out on 6 mm cylindrical pellets prepared with CuO and CuO.3BaO, unidirectionally pressed in a hard-metal die at 141.5 MPa, without any binder, wrapped in platinum foil. The weight loss recorded was converted into dissociation curves (O/Cu ratios versus temperature) and from these curves the isobaric diagram shown in Figure 2 was constructed.

RESULTS AND DISCUSSION

Figure 2 shows the isobaric diagram constructed from the dissociation curves, compositions being expressed as molar percentages of the components BaO, CuO and Cu_2O. The compositions of fully oxidized mixtures, representing the stable state of the system at low temperatures in air, lie along the CuO-BaO edge of the composition triangle. The dashed straight lines running from left to right are oxygen reaction lines (reaction paths). The isotherms (thin continuous lines with temperatures alongside) were constructed by joining the compositions reached at the temperatures indicated. The boundaries of the phase areas, which are shown as thick continuous lines (dashed lines where their location is tentative) were established from the sharp changes in direction of the isotherms. Line HCFG is the isobar on the Liquidus surface so that melting is complete when the reaction paths cross this line.

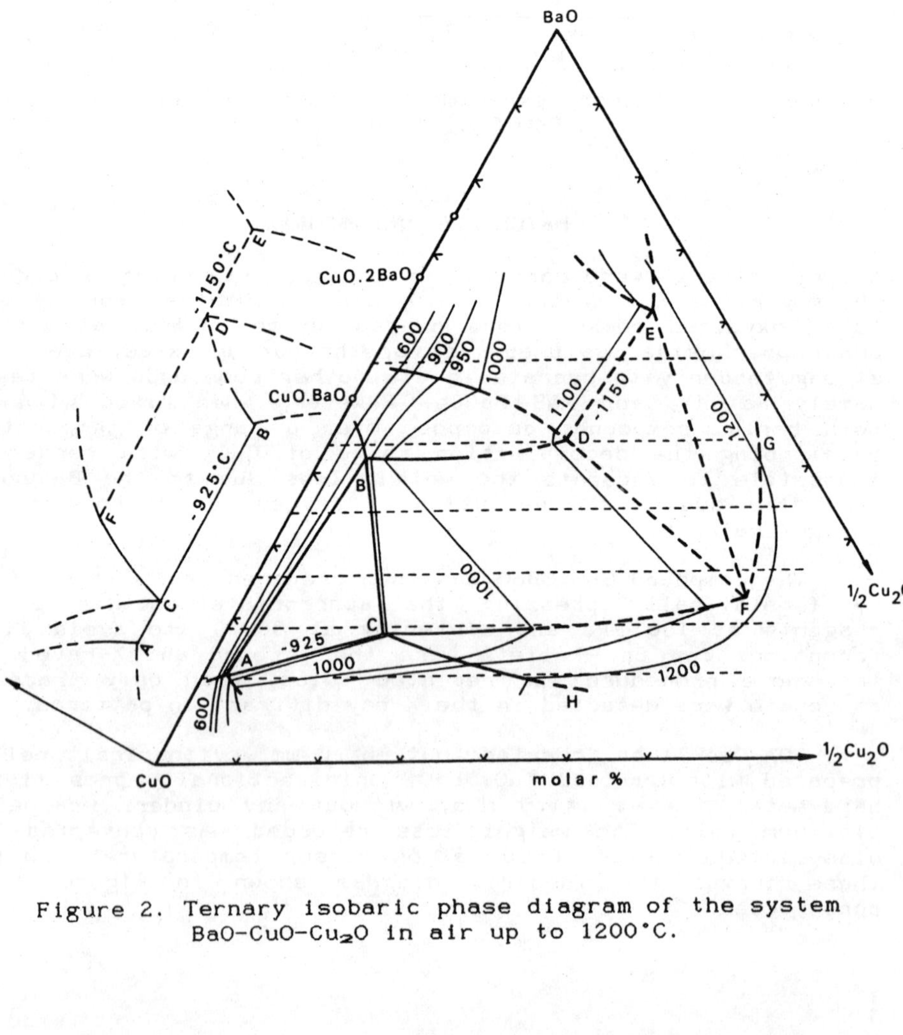

Figure 2. Ternary isobaric phase diagram of the system $BaO-CuO-Cu_2O$ in air up to 1200°C.

On the side of the composition triangle of Figure 2 is shown the quasi-binary isobaric temperature/composition diagram obtained by plotting the temperatures along the various phase boundaries against the composition of the mixtures expressed in terms of their initial BaO and CuO contents.

CONCLUSIONS

The temperature of initial liquid formation in the CuO-rich part of the diagram, around 925°C, is in good agreement with that determined by Roth *et al.* (Figure 1). However, the results now obtained, within the limits of accuracy of the experimental technique, suggest that the compound BaO.CuO melts incongruently and that there is an appreciable extent of oxygen non-stoichiometry in all the solid phases. Quenching-experiments will be carried out to check this reported results.

REFERENCES

1. Rao, C.N.R., Ganguly, P., Gopalakrishnan, J., and Sarma, D.D., Mechanism of High-Temperature Superconductivity in $YBa_2Cu_3O_{7-\delta}$: Crucial Role of Oxygen. Mat. Res. Bull., 1987, 22, 1159-63.

2. Leskelä, M., Mueller, C.H., Truman, J.K., and Holloway, P.H., Preparation of Y-Ba-Cu-O High Transition Temperature Superconductors Using BaO_2 and $BaCO_3$. Mat. Res. Bull., 1988, 23, 1469-77.

3. Roth, R.S., Davis, K.L., and Dennis, J.R., Phase Equilibria and Crystal Chemistry in the System Ba-Y-Cu-O. Adv. Ceram. Mater., 1987, 2[3B], 303-12.

SYNTHESIS AND SINTERING OF DIFFERENT $YBa_2Cu_3O_{7-x}$ COMPOUNDS

J. WANG*, F. BOTEREL**, G. DESGARDIN*, J.M. HAUSSONNE**, J. LOSTEC**
and B. RAVEAU*
* ISMRa, Laboratoire CRISMAT, Bd du Maréchal Juin, 14032 Caen Cedex, France
** CNET, , Lannion B, OCM/TAC, 22301 Lannion, France

ABSTRACT

In addition to the classical orthorhombic "O" and tetragonal "T" $YBa_2Cu_3O_{7-x}$ phases, we prepared a tetragonal (c = 3a) phase, so called β. Sintering of these different phases has been studied. Sintering of the β phase is possible at 920°C leading to a ceramic similar to the one prepared using a powder obtained by a chemical process. The influence of the different sintering parameters on the supraconducting properties is discussed.

INTRODUCTION

When studying the synthesis of $YBa_2Cu_3O_{7-x}$ by the mixing and calcining powder technology, we have been able to obtain a pure tetragonal phase with c = 3a so called β appearing previously to the formation of the orthorhombic (O) and tetragonal (T) phases classicaly reported ‹1›..
Table 1 gives the cristallographic data for the different $YBa_2Cu_3O_{7-x}$ phases.
These results point out that the calcined powders which can be used as starting material to sinter $YBa_2Cu_3O_{7-x}$ can be quite different and may lead to different ceramics.

Table 1. Cristallographic data for the different $YBa_2Cu_3O_{7-x}$ phases

	β phase ‹4› x = 0.2	"O" phase ‹2› x = 0.1	"T" phase ‹3› x = 0.8
Parameters	a = 3,866 c = 11,599 non supra	a = 3,825 b = 3,893 c = 11,676 T_C › 90K	a = 3,857 c = 11,971 non supra

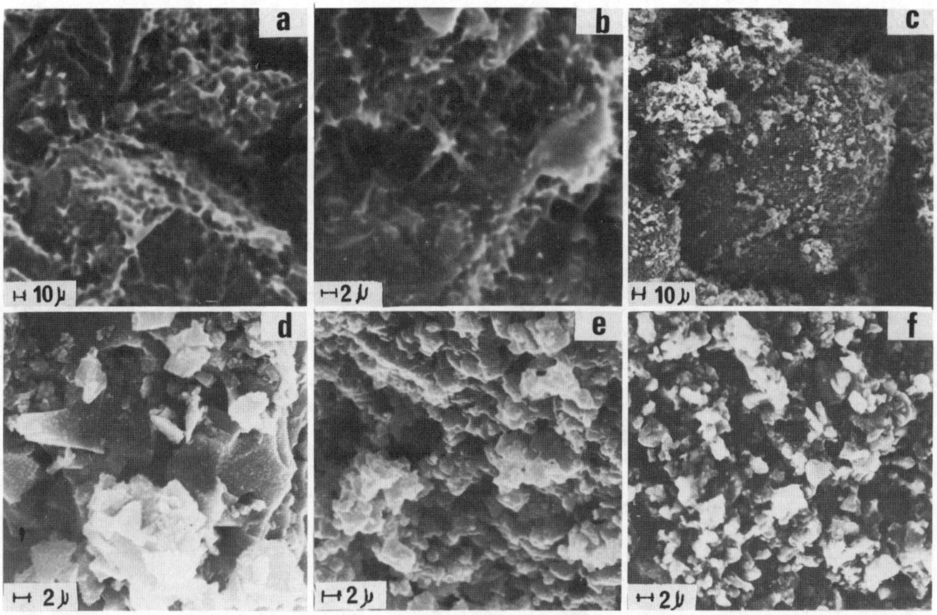

Figure 1. Microstructure of the differents powders : a) "T" G = 1000, b) "T" G = 5000, c) "O" G = 1000 ; d) "O" G = 5000 ; e) "β" G = 5000 ; f) "RP" G = 5000.

Figure I shows the different morphologies of the powders obtained after the calcining schedules resumed in table 2. We shall discuss also about a chemically processed powder*.
Tetragonal phase T shows huge grains (> 50 µ) surrounded by small ones. Careful examination shows that big grains are in fact aggregates, and that small grains present fibrous growth.
The orthorhombic phase "O" looks like big spheres which are composed mainly of fine grains with a narrow size distribution owing irregular shapes, and a flat form.

Table 2. Different process used for the synthesis of four $YBa_2Cu_3O_{7-x}$ powders

"β" phase	"O" phase	"T" phase	Orth. RP phase
Heating rate 400°C/h in O_2, 930°C, 0 h Quenched in O_2	Heating and cooling rate 80°/h 900°C, 6 h in air	Heating rate 80°/h 930°C, 10 h in air Quenched in air	Precursor (chemical process)

---

* We thank Rhône Poulenc Society who kindly supplied us this powder.

Phase β and "Rhône Poulenc (RP), in spite they are obtained from quite different ways present very similar microstructures, characterized by fine grains (<3 microns) and a quite narrow size distribution.

Dilatometric analysis

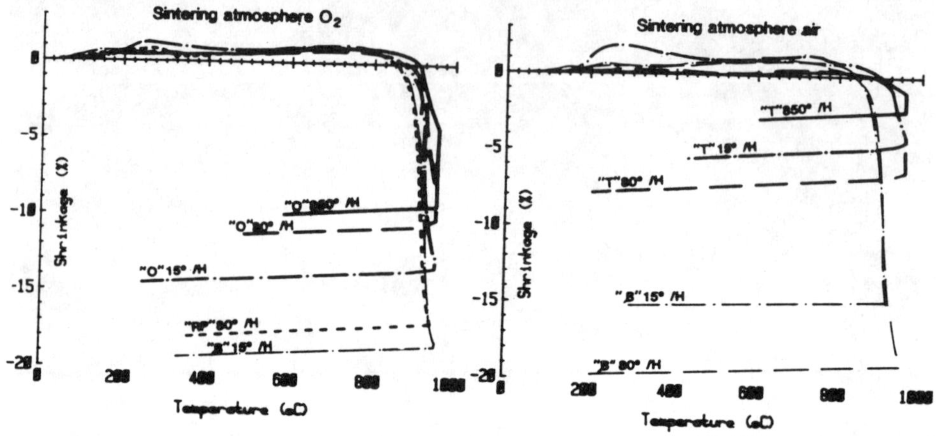

Figure 2. Dilatometric curves for the different YBaCuO powders sintered in air or in oxygen.

Fig. 2 show as an example the obtained dilatometric curves when sintering these four powders with different sintering schedules, either in air or in pure oxygen atmosphere. Both "RP" and "β" phases densifie at low temperature ; material seems full densed at 950°C and can be sintered at 920°C. On the contrary, when sintered in air or in oxygen at 950°C, phase "T" leads to a very small shrinkage and phase "O" to a poor shrinkage. One cannot say today if these differences are due to the morphologic characteristics of powders or to the cristallographic characteristics, or both.

We can see the very sensitive effect of heating rate, and particularly with low heating rate an abnormal expansion of sample around 300°C that can be attributed to a partial reoxydation of the material.

Influence of the atmosphere is noticeable only by the fact that shrinkage is slighly more important when sintering phases "O" or "T" in pure oxygen, although the temperature of beginning of shrinkage is higher.

Microstructures observed after sintering

Generaly speaking, the atmosphere of sintering seems to have practically no influence on the microstructures observed on the different materials;

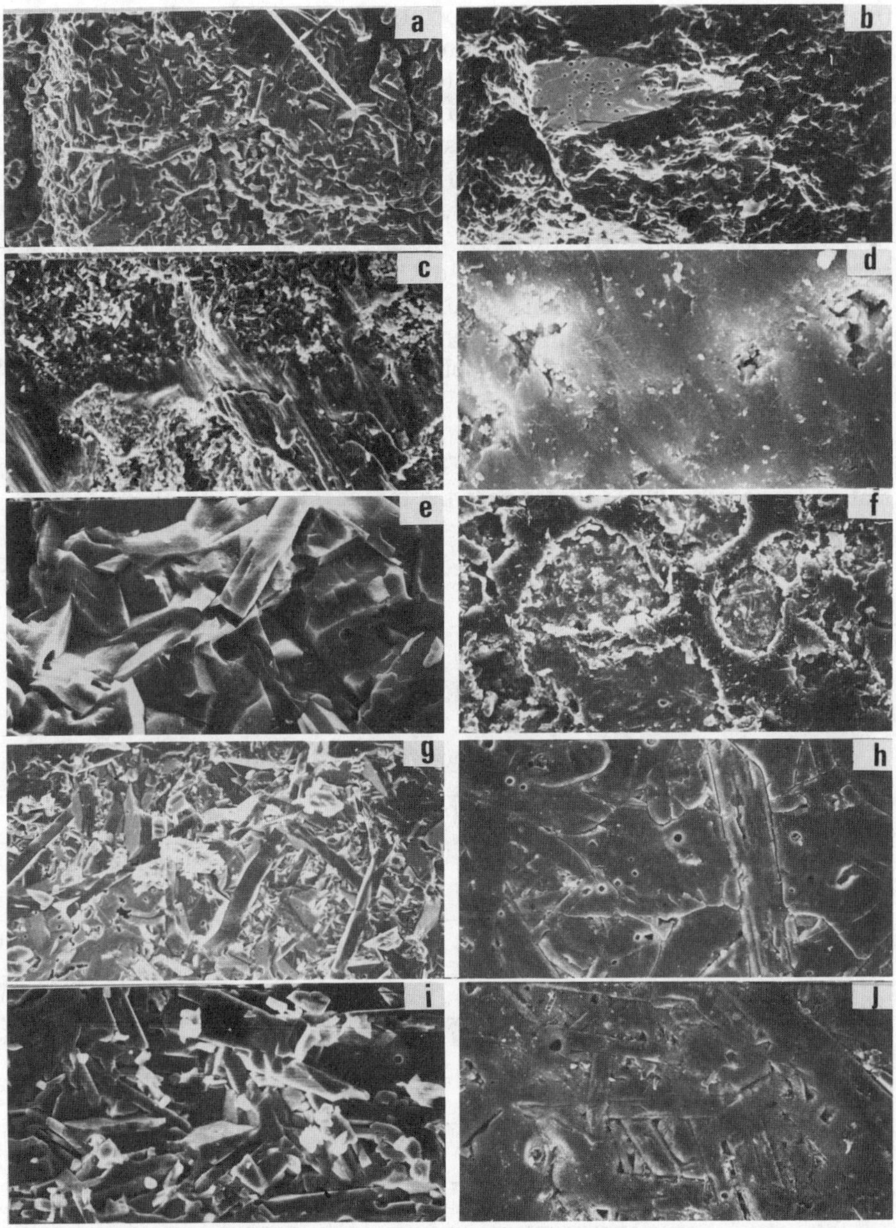

Figure 3. Microstructures after sintering at 950°C, 2 h. a) "T" heating rate 15°/h, b) "T" heating rate 80°C/h, c) "T" heating rate 850°/h, d) "T" heating rate 80°/h after chemical etching, e) "O" heating rate 80°/h, f) "O" after chemical etching, g) "β", h) "β" after chemical etching, i) RP, j) RP after chemical etching.

10μ

On the contrary, Figure 3a to 3c point out, when considering sintering of tetragonal phase "T", the main importance of the sintering kinetic on microstructure. A rapid heating rate lead to an inhomogeneous microstructure resulting from a partial melt (Fig. 3a). When using a 80°C/h heating rate the microstructure is more homogeneous in spite of the presence of some large blocks among small grains (Fig. 3b). These blocs disappear for low heating rate (15°C/h). The ceramic is then composed of small, partially faceted grains, owing irregular shapes (Fig. 3c). Chemical etching shows a small porosity and the presence of a secondary phase (Fig. 3d). Heating rate has less influence on the microstructures of the ceramics sintered using the orthorhombic "O" phase. Faceted and elongated grains can then be observed on fractures. (Fig. 3e). Polished and etched samples point out an important inhomogeneity and poor contrasted grain-boundaries (Fig. 3f). Ceramics sintered from β phase powder show large rectangular grains (Fig. 3g) with grains boundaries irregularely etched by chemical attack (Fig. 3h). This can be observed also on the ceramics sintered using the "RP" powder showing moreover a second phase present (Fig. 3i). Polished and etched samples seem to have in this case a lack of matter between parallel edges of grains (Fig. 3j). This lack of matter can be attributed to an uncontrolled dissolution either of $YBa_2Cu_3O_{7-x}$ grain boundaries or of the second phase.

Superconducting properties

Table 3. Superconducting characteristics after sintering at 950°C 2 h.

		Precursor									
Rate of sintering	Sintering atmosphère	β phase phase			Orthorhombic phase			Tetragonal phase		RP phase	
		Jc	% supra	J'c	Jc	% supra	J'c	Jc	% supra	Jc	% supra
15°C/h	air	60			53	50		50	60	3	47
	O_2	81	50		163	72/46		116	58		58
80°C/h	air	2	63		5		360	25		10	42
	O_2	62			173	75/42		90	55	100	58
850°C/h	air	2		310	24	50/32		35	48	3	
	O_2	3	67/24	410	148	44/40		75	50	40	

Jc : critical current (A/cm^2) (resistivity measurement).
% supra : deduced from FONER measurements : Weak Field/Field increasing.
J'c : critical current after annealing in oxygen (450°C/16 h).

All sintered materials own superconducting properties even when sintered in air, with T_C located between 90 and 92K. However, critical currents measured with the resistive method are mainly dependent on the sintering atmosphere and on the used powder precursor.
Figure 4 shows these critical currents versus shrinkage for ceramics sintered using various parameters. It confirms the main importance of the sintering atmosphere and shows that the critical current cannot be correlated directly to the microstructure. It also point out the influence of the heating rate on the critical current. Although, the ceramics issued from the "O" phase are far to be the best densed, they own better critical currents than the others. So, this figure demonstrates clearely that densification is not the parameter to be considered to forecast the value of J_C as both the best and the less densified samples own the worth critical currents.

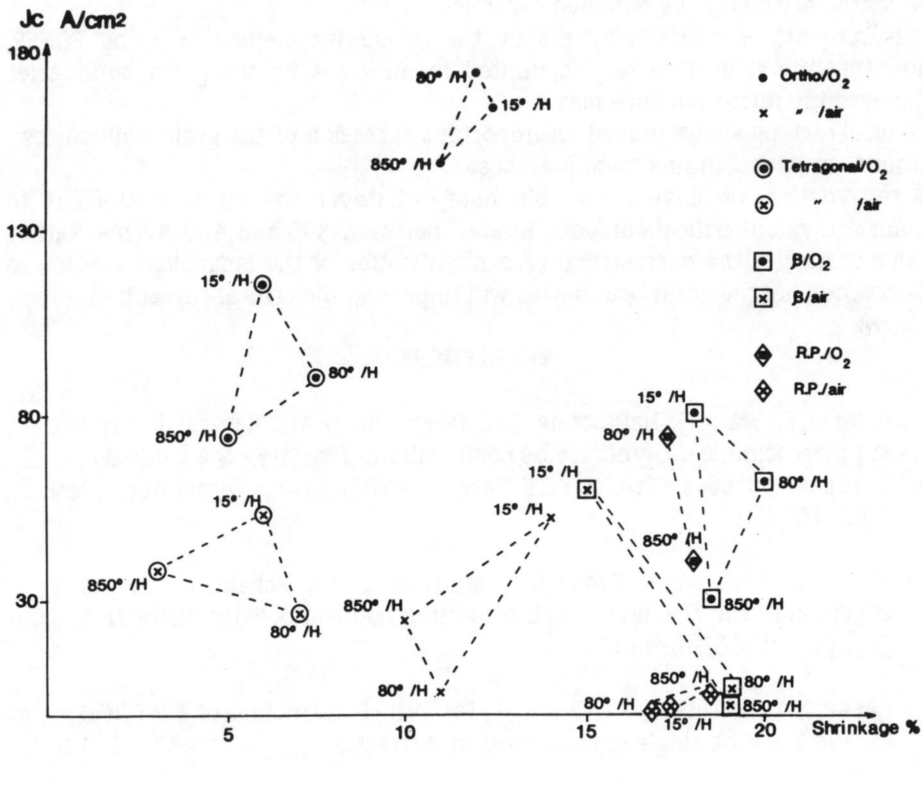

Figure 4. Critical current versus shrinkage for ceramics $YBa_2Cu_3O_{7-x}$ sintered using various precursors and various sintering process.

Foner measurements give both indications concerning critical current and volume ratio of superconducting phase in the bulk. It is noteworthy that the values of critical currents are then very high (10^4-10^5A/cm^2) and that in this case the ceramic sintered using tetragonal phase T seems to own the higher value.
Ceramics sintered from "O" and "β" phases own in a weak magnetic field the higher ratio of superconducting phase up to 75 % but this ratio fall down when the field increases (Table 3). This point out the influence of the characteristics of the grain boundary on the critical current. When reoxydising the ceramics at 450°C during 16 h, whatever the ceramic is made from, and whatever the sintering schedule is, the resistive critical current grows to 300-400 A/cm^2.

CONCLUSION

This study has shown that it is possible by using powders such as phases "β" or "RP" to sinter.,YBaCuO at temperature lower to 950° and consequently to modify the characteristics of the obtained ceramics.
Measurements of critical currents by the resistivity method or using FONER, show that the critical current is limited in all cases by the grain boundaries whatever the microstructure may be.
Chemical etching shows indeed an important agression of the grain boundaries : rather than joined, grains seem juxtaposed.
By reoxydation we have been able here, whatever the microstructure is to obtain equivalent critical currents, located between 300 and 400 A/cm^2. Rather than a change of the microstructure, a modification of the technology leading to the creation of true grain boundaries will improved the critical current.

REFERENCES

1. Boterel, F., Wang, J., Haussonne, J.M., Desgardin, G. and Raveau, B., Etude de la physicochimie de synthèse de compositions $YBa_2Cu_3O_{7-x}$ à l'aide de trempes. Journées d'études S.E.E. Caen, 1988, Industrie Ceramique, 1989, **231-10**, 716.

2. François, M., Junod, A., Yvons, K., A study of the Cu-O chains in the high T_c superconductor YBaCuO by high resolution neutron powder diffraction, Solid Stat. Com., 1988, **66**, 1117.

3. Renault, A., McIntyre, G.J., Collin, G., Pouget, J.P. Structure of $Ba_2YCu_3O_{6.26}$ at 20 and 298K by single crystal neutron diffraction. J. Phys., 1988, **48**, 1407.

4. Caignaert, V., Wang, J., Desgardin, G., Raveau, B., Boterel, F. and Haussonne, J.M., The low temperature tetragonal form $YBa_2Cu_3O_{6.7}$.Neutron diffraction structure determination. To be published.

A VERSATILE SOL-GEL METHOD FOR SYNTHESIS OF CERAMIC SUPERCONDUCTORS

O. Van der Biest, J. Fransaer, T. Eggermont, O. Arkens,
Department of Metallurgy and Materials Engineering
K.U. Leuven, Belgium

ABSTRACT

A sol-gel method is described based on the use of ethylene-diamine-tetra-acetic acid (EDTA) as a complexing agent in the base solution. An important advantage of the method is its versatility due to the fact that EDTA forms stable complexes for a wide range of elements. $YBa_2Cu_3O_{7-x}$ superconductors as well as a range of compositions in the Bi-Pb-Ca-Sr-Cu-O system have been prepared using this method. Precise control of stoichiometry is possible.
Powders prepared via this technique allow the use of reduced calcining temperatures and times compared with preparation via the solid state. Thermal analysis of the calcination process and some electrical properties of the resulting materials will also be presented.

1. INTRODUCTION

Superconducting ceramics are complex multi-component oxides, in which any chemical inhomogeneity is strongly reflected in the electrical properties. Initially most of these compounds were prepared by mixing the elementary oxides or their pre-cursors in the solid state, then calcine and grind. The last two steps need to be repeated several times in order to obtain a reasonably homogeneous product. The danger of segregation and deviations of stoichiometry is always present when manipulating powders, e.g. due to the loss of fines or the formation of undesirable reaction products, which do not dissociate readily.
The advantages of wet chemical synthesis routes were quickly realised and a number of them have been published. Most of these methods however cannot easily be transferred from one superconducting compound to another.It is the goal of this paper is to present a method which is applicable to a wide range of superconducting oxides.

2. USE OF PM'- PH DIAGRAMS

When designing wet chemical methods based on complex formation with cations in aqueous solutions the use of pM'-pH diagrams published by Kragten [1] is useful. These diagrams delineate the boundaries of solution and precipitation as function of the pH of the solution and metal concentration (pM'=-log[M']). For many cations the hydroxides will be the compounds precipitated at high pH. For others, specifically Ca, Ba and Sr wich are of interest here, the precipitate is a carbonate. Curve a in Figure 1 represents such a boundary for $BaCO_3$.
The effect of the presence of a complexing agent is presented in the diagrams by lines, which delineate the solution/precipitation boundary as function of the excess ligand added to the system. Curve b in Figure 1 shows the boundary for an excess citrate concentration of 0.01 M whereas curve c shows the same for an excess EDTA of 0.01 M. It is clear that EDTA provides a much wider range of pH and metal concentrations for which the solution is stable.

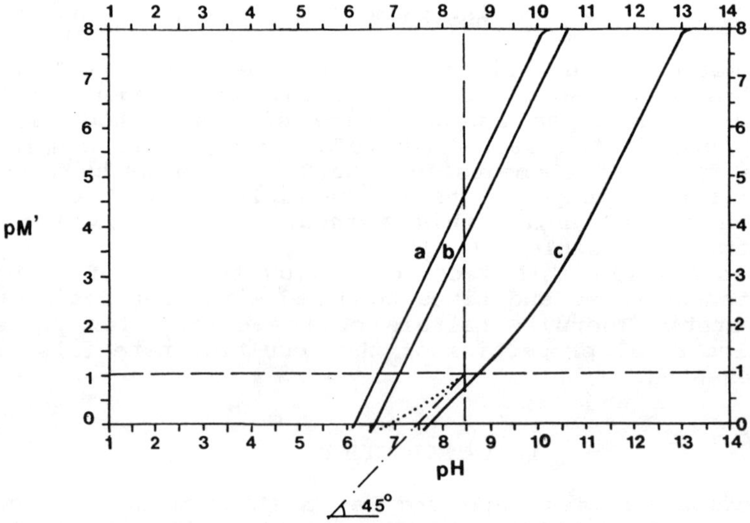

Figure 1 : pM'-pH diagram for Barium [1].

The gellation of the solutions by evaporation of water can also be represented on the diagrams : The solution follows a straight line at 45° to the pH axis and in the design of the method the starting condition should be such that during this evolution the precipitation regions in the respective diagrams are avoided. For the cations of interest here, Ba represents the most difficult case, because the evaporation line is close to the precipitation region. Because of the presence of NH_3 in our solution which also evaporates during gellation, the actual path followed is thought to be the dotted line in Figure 1.

3. PREPARATION METHODS AND RESULTS

3.1.The Y-Ba-Cu-O system

Preparation

Solutions of the three cations are prepared separately, starting from Y_2O_3, $BaCO_3$ and Cu metal using nitric acid. Instead of $BaCO_3$, $Ba(NO_3)_2$, which dissolves easier, can also be used, if its stoichiometry is controlled. EDTA is added to the three solutions in a 1 : 1 molar ratio with a slight excess added as needed (derivable from the pM'-pH diagrams). The pH is adjusted to a value of about 8 by adding ammonium (NH_3, 25 % by weight in water). This is necessary, because the weak EDTA acid does not fully dissociate at lower pH values. Again the pM'-pH diagrams serve as a guide for the precise value. The three solutions are mixed together with the most alkaline one added last.

The liquid solution is heated in two stages to evolve water. Upon slow heating at temperatures up to 80°C the liquid turns into a transparant blue glassy resin. Subsequent heating expands the resin into a dark organic foam, that combines with oxygen in the air at about 220 °C. This solid pre-cursor, which is largely amorphous, is subsequently calcined at high temperature (900-940°C) to obtain the superconducting phase.

Properties

The total carbon content of one of the powders, calcined for 8 hrs at 920°C is 0.1 wt%. This compares favourably with other values found in literature[2].Superconducting properties of the $YBa_2Cu_3O_x$ oxide prepared by this method, have been reported [3].

3.2. The Bi-Pb-Ca-Sr-Cu-O system

Preparation.

Both the $Bi_2Sr_2Ca_1Cu_2O_y$ and the $Bi_2Sr_2Ca_2Cu_3O_y$ stoichiometries have been synthesized, with various levels of Pb substituting for Bi. Again separate solutions of the cations are prepared. Bi_2O_3 and PbO are dissolved in HNO_3. It is useful to limit the amount of HNO_3 in this system, because $BiONO_3$ can precipitate and this is not taken into account in the diagrams from Kragten [1]. Hence, the Ca and Sr solutions are prepared by dissolving the carbonates in an EDTA solution. Ammonium is also used here to bring the solutions up to the required pH.

After mixing solutions, a transparant gel is obtained which can be converted to the powder as before.

Thermal Analysis.

Thermal analysis is used to provide information on the temperatures at which volatile matter is lost from the gel and the key temperature at which the decomposition is complete. Figure 2 shows the thermogravimetric analyses for the precursors. The sample is oven dried at 75°C for 24 hours, cooled in an exsiccator and measured immediately thereafter since the material is hygroscopic in this state. The scan is taken at a heating rate of 10°C/min and under a flow of oxygen

(4 l/h) to remove the decomposition products. After the initial loss of water, the $Bi_2Sr_2Ca_1Cu_2$-oxide precursor exhibits three distinct stages of decomposition. Most EDTA-compounds decompose between 200 and 250°C. The first major weight loss in the sample is thus due to the break up of the EDTA-compounds probably into carbonates and nitrates.
Volatilization of NH_4NO_3 and the elimination of residual CO_3^{2-}, NO_3^- and free EDTA-acid also contribute to the first weight loss. The last weight loss in the TGA curve is the decomposition into the oxide.

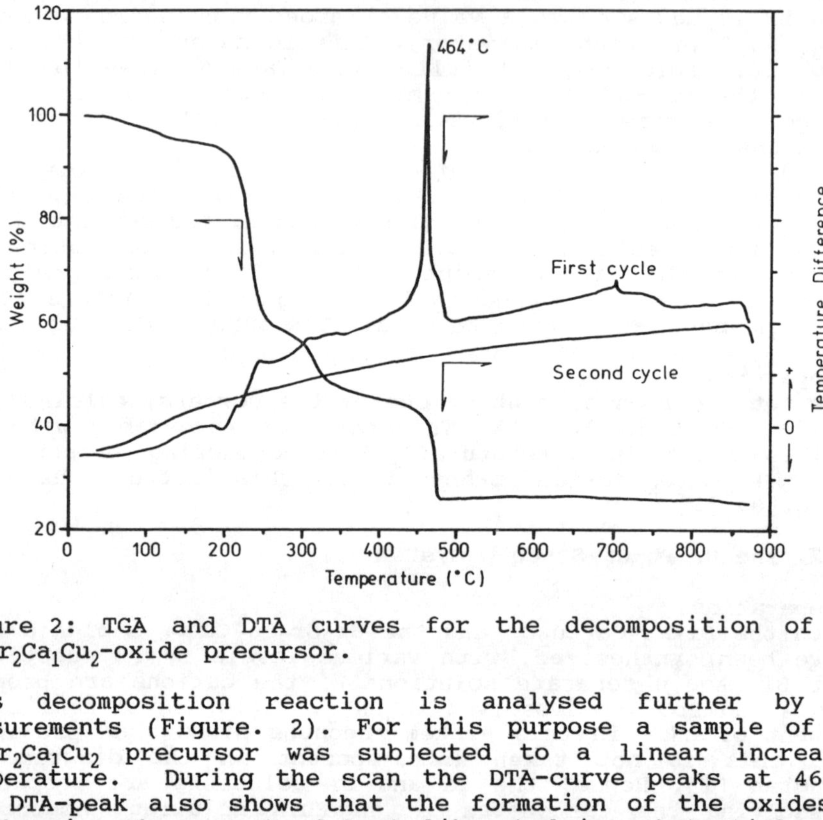

Figure 2: TGA and DTA curves for the decomposition of the $Bi_2Sr_2Ca_1Cu_2$-oxide precursor.

This decomposition reaction is analysed further by DTA measurements (Figure. 2). For this purpose a sample of the $Bi_2Sr_2Ca_1Cu_2$ precursor was subjected to a linear increasing temperature. During the scan the DTA-curve peaks at 464°C. The DTA-peak also shows that the formation of the oxides is exothermic. The amount of heat liberated is substantial since the reaction heats up the sample and sample holder.
After dissociation, the material consists of oxides since the weight remains constant even after prolonged heating at temperatures up to the melting point. This shows that the formation of oxides with sol-gel precursors is complete at temperatures far below the temperatures needed when processing these materials by solid-state reactions.

Electrical Properties.
As an example, Figure 3 shows the electrical resistivity as function of temperature of a $Bi_{1.6}Pb_{0.4}Sr_2Ca_2Cu_3O_y$ specimen. The measurements are done using the standard four-probe technique and evaporated-on gold contacts. There is a strong

drop in the resistivity at 115K, but the resistivity does not become zero till 98K. The electrical properties of these materials will be reported elsewhere.

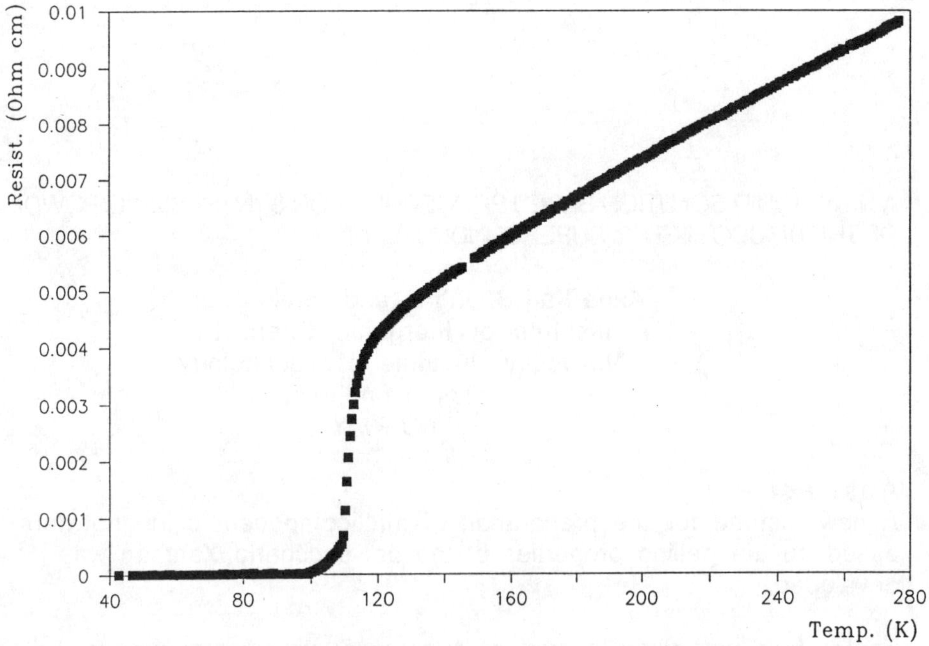

Figure. 3: Four-probe resistivity measurements of $Bi_{1.6}Pb_{0.4}Sr_2Ca_2Cu_3O_y$ sample sintered at 845°C for 240 hours and cooled slowly to room temperature.

4.DISCUSSION

The versatility of the synthesis method presented here, stems from the large stability constant of most EDTA-complexes and the large number of elements with which EDTA forms complexes. The method is thought to be easily extendable to other compounds e.g the Tl-Ba-Ca-Cu-oxide, since EDTA also forms a stable complex with thallium.

J.F. wishes to thank the IWONL for financial support.

REFERENCES

1 KRAGTEN J. , Atlas of Metal-Ligand Equilibria in Aqueous Solution, Ellis Horwood Limited, John Wiley & Sons, 1978.

2 UNO.N., ENOMOTO N.,TANAKA Y., Jap. J. Apl. Phys., 1988, 27, L1003-L1006.

3 FRANSAER J., ROOS J.R., DELAEY L., VAN DER BIEST O., ARKENS O. and CELIS J.P., J. Appl. Phys., 1989, 65, 3277-3279.

A NEW LIQUID SOLUTION BASED PROCEDURE FOR SYNTHESIZING POWDERS OF THE BISCCO HIGH Tc SUPERCONDUCTORS.

Anne Kari B. Jahre and Stein Julsrud
Institute of Inorganic Chemistry
Norwegian Institute of Technology
7034 Trondheim
NORWAY

Abstract

A new method for the preparation of multicomponent oxide powders based on the gelling properties of the polysaccharid Xanthan was developed.

The method was used to produce superconducting compounds in the system Bi_2O_3-SrO-CaO-CuO(-PbO). The results were evaluated in terms of phase-purity and superconducting properties.

Even though single phase materials were not achieved, it was concluded that the method was promising in it´s simplicity.

Introduction

During the last couple of years the number of members in the family of high Tc oxide compounds has been growing rapidly. What characterizes this development is the increasing complexity of the compounds with higher Tc. The highest Tc´s are now found in systems with at least five components (1). To prepare phase-pure samples in systems with this many components is often a demanding task.This led us to consider solution based methods to ensure homogeneous mixing of the component oxides on a microscopic level.

The polysaccharid Xanthan (fig.1) has for several years been known to form gels with divalent and trivalent ions (2). As far as the authors are aware of, this property has not been utilized for the production of ceramic materials. In this paper the first attemps to prepare

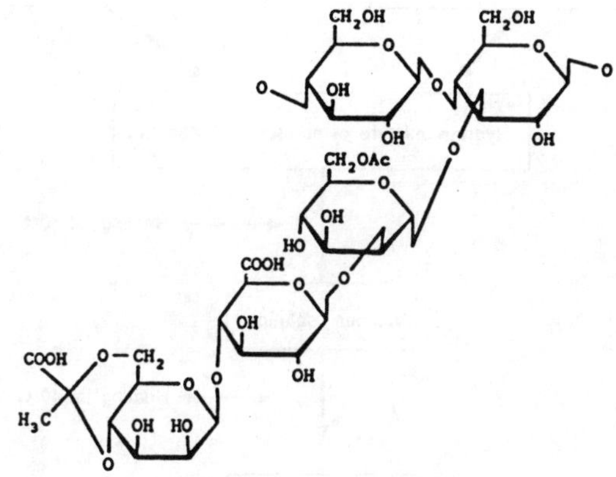

Fig 1 The structure of one repeating unit of Xanthan (4).

superconductors in the system Bi_2O_3-CaO-SrO-CuO(-PbO) via a gel-forming step due to Xanthan, are reported.

Method.

Chemicals

the chemicals used (PbO, Bi_2O_3, $SrCO_3$, $CaCO_3$ and CuO) were all from Merck - grade p.a. The Bi_2O_3 was dried in air at 600°C prior to use.

Preparation of the precursor.

The steps in the process is shown in fig.2.

Nitrate solutions of the metals were made by dissolving stoichiometric amounts of the oxides and carbonates in nitric acid. The 1% solution of Xanthan was prepared by adding 1g of Xanthan to 100ml distilled water, stirring the solution for 12h at room temperature.

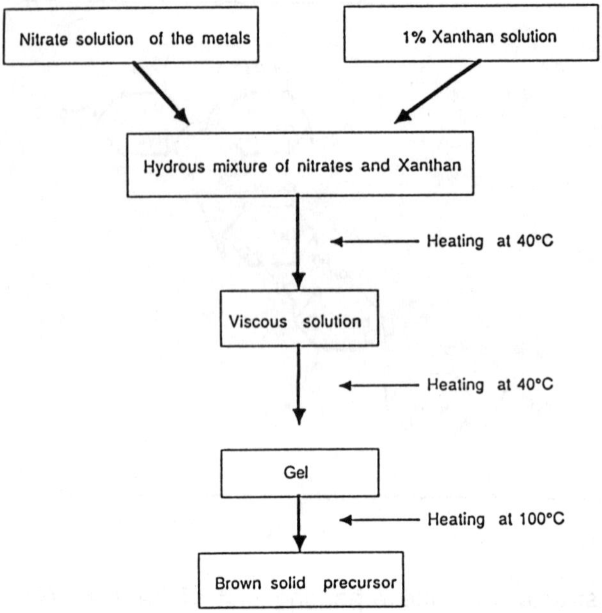

Fig 2 Flowchart for the preparation of precursors via a Xanthan-gel.

Thermal treatment of the precursor.

The precursors were decomposed at 650°C in air for 24h. The ensuing thermal treatment of pellets pressed from this powder is summarised in the following table:

Composition:	$Bi_2Sr_2Ca_1Cu_2O_x$			
Sample	Heat Rate	Temp./time	Cool.rate	atm.
AK1	200°/h	840°/12h	300°/h	air

Composition:	$Bi_7Pb_3Sr_{10}Ca_{10}Cu_{18}O_x$			
Sample	Heat Rate	Temp./time	Cool.rate	atm.
AKB1b	200°/h	840°/180h	100°/h	
		450°/4h	very high	air
AKB1d	200°/h	840°/90h	200°/h	O_2
	200°/h	840°/10h	100°/h	
		450°/4h	very high	air

Susceptibility measurements.
The temperature of the superconducting transitions were found using AC-susceptibility measurements.

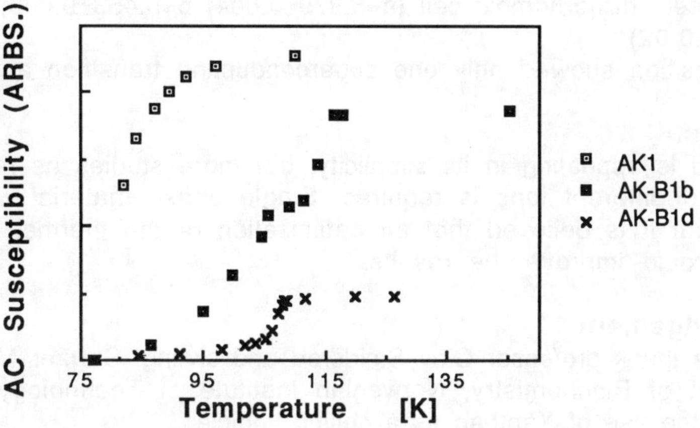

Fig 3 The results from the susceptibility measurements

X-Ray studies.
The samples were characterized using a Phillips PW 1710 diffractometer with $Cu_{K\alpha}$-radiation. The scan-rate used was 0.5° 2θ per minute. The diagrams were indiced and the lattice parameters refined using a version of the Appleman-Evans least square procedure (3).

Results.
The analysis of the X-ray diffraction patterns showed that none of the compositions were single phased. The diagrams could however be indexed assuming two phases in all cases :

AK1:
Abundant phase: tetragonal cell (a=b=5.399±0.01; c=30.785±0.008)

Minor phase: orthorhombic cell (a=5.333±0.009; b=5.460±0.007; c=30.96±0.005)
The superconducting transition for this composition was at 90K (see fig. 3).

AK-B1b:
Abundant phase: tetragonal cell (a=b=5.419±0.02; c=30.14±0.02)
Abundant phase: tetragonal cell (a=b=5.417±0.002; c=37.03±0.02)

This composition showed two superconducting transitions; at 108K and 99K (fig.3).

AK-B1d:

Abundant phase: tetragonal cell (a=b=5.405±0.001; c=37.07±0.01)
Minor phase: orthorhombic cell (a=5.379±0.004; b=5.453±0.003; c=30.045±0.02)

This composition showed only one superconducting transition at 108K.

Conclusion.

The method is appealing in its simplicity, but more studies as regards selectivity to different ions is required. Single phase material was not achieved, but it is believed that an optimization of the thermal treatment could improve the results.

Acknowledgement

The authors thank professor Olav Smidsrød and siv.ing. Størker Moe (Department of Biochemistry, Norwegian Institute of Technology) for suggesting the use of Xanthan as a gelling vehicle.

References.

1. Sleight,A.W.,Subramanian,M.A.,and Torardi,C.C., MRS Bull.,14(1989)45.

2. Cottrell,I.W.,Kang,K.S., and Kovacs,P. in "Handbook of water-soluble gums and resins."(Ed. R.L.Davidson) McGraw-Hill Book Company :PP24-1 - 24-31.

3. Benoit,P.H., American Mineralogist, 72 (1987) 1018

4. Jansson,P.E.,Kenne,L. and Lindberg,B. Carbohydr. Res. 45 (1975) 275

DENSIFICATION OF $Y_1Ba_2Cu_3O_x$ SUPERCONDUCTING CERAMICS

C.MOURE, J.F.FERNANDEZ, P.RECIO and P.DURAN
Instituto de cerámica y Vidrio (CSIC). Electroceramics Department. Arganda del Rey. 28500 Madrid, Spain.

ABSTRACT

The study of sintering and densification of $Y_1Ba_2Cu_3O_x$ powders, obtained from several synthesis processes was carried out. Density and microstructure were studied. The crystalline structures were determined by means of XRD. The solid state reaction between oxides and barium carbonate leads to synthesized powders which densified at high temperatures, with liquid phase formation. Better results were obtained using a chemical process of oxalate coprecipitation. A solid state reaction between intermediate barium copper and barium yttrium compounds gave a good sinterable powders at low temperatures.

INTRODUCTION

One of the most common procedure for synthesis of superconducting $Y_1Ba_2Cu_3O_x$ material consists of mixing CuO, Y_2O_3 and $BaCO_3$ powders, repeatedly calcined at high temperatures and grounded until complete reaction (1). The use of alternative methods as oxalate, nitrate, citrate coprecipitation or sol-gel technique allows the obtention of more sinterable powders (2-3). Nevertheless, all the powders are sintered above 930ºC, in presence of a liquid phase (4), which causing compositional changes. In the present work a comparative study of the sintering behaviour of powder, synthesized at low temperatures, from barium-yttrium and barium-copper binary compounds previously prepared, with the two $Y_1Ba_2Cu_3O_x$ prepared by more usual methods was made.

EXPERIMENTAL METHODS

In view of their different sintering behaviour three kind of $Y_1Ba_2Cu_3O_x$ powders, obtained from alternative routes, have been studied:

a) A classical two-step procedure of calcining three-component oxide and carbonate mixtures (Y_2O_3+$BaCO_3$+CuO), (Powder I).
b) An oxalate coprecipitation procedure carried out from a nitrate solution, (Powder II).
c) A bypass procedure of reacting intermediate phases previously synthesized. (Powder III).

The powder I was obtained from stoichiometric mixtures of the corresponding oxides. The raw materials were throughly blended by ball milling with isopropanol. The blend was calcined at 860ºC in air for 16 hours. The calcined powder was ground, ball milled and calcined at 950ºC for 16 hours in air, and cooled to R.T. with a cooling rate of 1ºC/min. The obtained material was reground and attrition milled to give a small-grained powder.

The powder II was prepared by a method described elsewhere (3). The synthesis was carried out at 875ºC for 8 hours.

The bypass procedure was the following: Two intermediate compounds Y_2BaO_4 and $BaCu_2O_3$ were prepared. The corresponding powders were mixed in the adequated amounts, and heat-treated at 900ºC for 4 hours to give a pure $Y_1Ba_2Cu_3O_x$.

The calcined powders II and III were also attrition milled. All the powders were isopressed at 200 MPa. The pressed bodies were sintered at temperatures ranging between 920 and 960ºC with times up to 8 hours. The heating rate was 3ºC/min and the cooling rate 1ºC/min. Annealing treatment were not given afterwards. The apparent density was measured by an Archimedes method. The crystalline structure was determined by XRD analysis and the microstructures were observed by SEM.

RESULTS AND DISCUSSION

The synthesis procedures by means of which powder I and II have been prepared, are well known processes, and are not necessary to explain here. The third process is a variation of a new process described elsewhere (5).

The Y_2BaO_4 compound is obtained from an intimate blend of Y_2O_3 and $BaCO_3$ by calcining at 1350ºC for 2 hours. According to Kwestroo et al (6) this compound is a stable phase belonging to the Y_2O_3-BaO system. The compound is strongly attrition milled to achieve a mean grain size of 1 μm. The $BaCu_2O_3$ oxide is an intimate oxide mixture of $BaCuO_2$ and free CuO obtained by calcining at 920ºC for 8 hours a $BaCO_3$ and very fine CuO blend. The partially melted product was also attrition milled. The calcining time must be sufficiently long for achieving the fully carbonate decomposition. Adequate mixture of both synthesized phases were heat-treated at several temperatures and times. The best schedule for synthesizing $Y_1Ba_2Cu_3O_x$ was 900ºC for 4 hours in air with rapid cooling in the furnace. The obtained powder had orthorhombic structure. The detailed study of this synthesis process will be exposed elsewhere.

Figure 1 shows the densification curves against temperature for the three powders. The temperature of maximun density for a constant time of 2 hours lowers from 960ºC for powder I to 930ºC for powder III. The densification level also increases according the same behaviour. Figure 2 shows the density

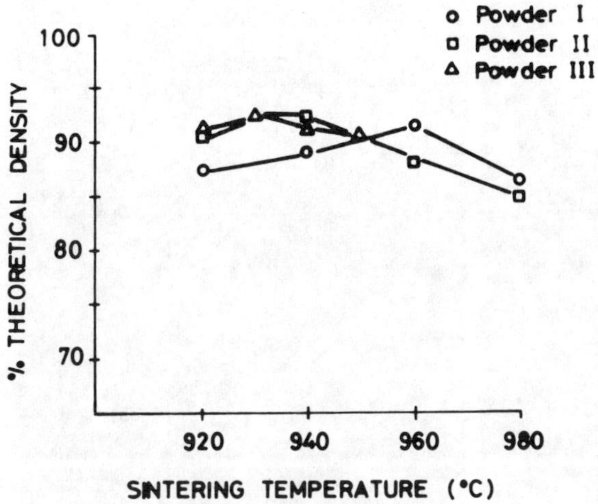

Figure 1. Densification curves versus temperature.

time curves for the optimal density temperature. It can be seen that the powder III densifies at very high density values for short times (1h).

The weight losses during the sintering of the several samples have different behaviour. The powder I, shows very low weight losses (~1%), as it correspond to its higher synthesis temperature, whereas the oxalate powders show the highest weight losses (~5%), owing to the imcomplete reaction caused by the low synthesis temperature, which leading to an incomplete oxalate decomposition. The powder III shows also low weight losses (~1%).

The figure 3 shows the microstructure of the best densified samples. It can be observed that those materials which have been sintered above 940ºC develop a coarse-grained microstructure with crystals higher than 20 μm. These microstructures are porous and weak, and show large regions with pulling out of grains caused by polishing. The samples corresponding to the powder II show a bimodal microstructure with large grains near the sample surface and small grains in the core zone. The samples sintered from the bypass powders show a fine-grained microstructure, with low porosity, mainly located at the grain boundaries and the triple points.

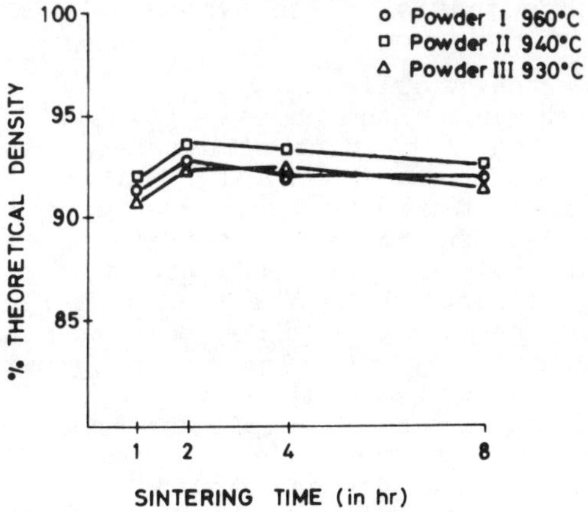

Figure 2. Densification curves vs time for optimal density temperature.

The XRD analysis of sample surfaces indicates the existence of strong differences in the final crystalline structures. Whereas the lest densified bodies have an orthorhombic type I structure, (samples I and II), the samples with highest density (III) have developed a tetragonal symmetry. Along with the tetragonal structure some amounts of green phase Y_2BaCuO_5 were obseved. (Fig.4).

The presence of orthorhombic I superconducting phase is strongly affected by several processing parameters. The sintering temperature, the oxygen partial

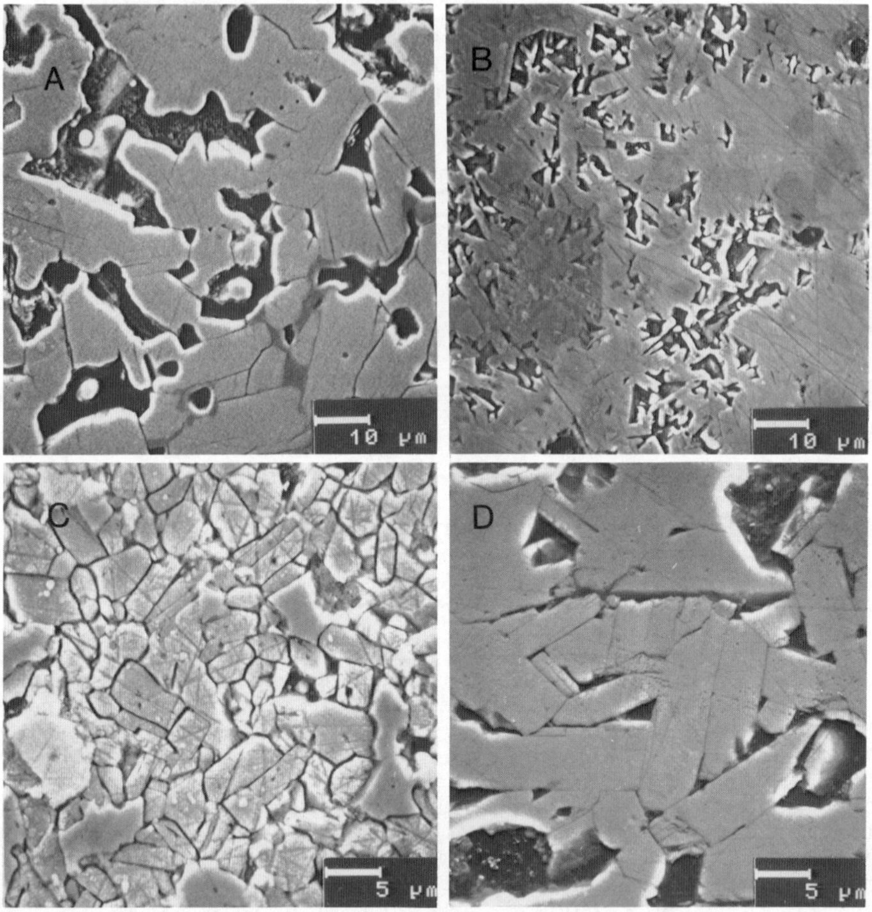

Figure 3. SEM micrographs of maximum density samples: a)Powder I b)Powder II and, Powder III c)1th synthesis, d)2th synthesis.

pressure, and particularly the cooling cycle, play an important role on the tetragonal-orthorhombic transition. Futhermore the density and ceramic microstructures also strongly influence the oxigen-diffusion path. In that sense high densities and small-grained microstructures seems to be harmful for complete tetragonal-orthorhombic transition.

The developing of small-grained microstructure for the III powder samples could be attributed to the low sintering temperatures, the small powder particle size, or to the presence of the "green phase", which playing a growth controlling-rate role. The appearing of this green phase, after sintering could be caused by an incomplete formation reaction of the intermediate phases. The $BaCu_2O_3$ oxide melt partially at low temperatures and the formed liquid can stop the carbonate decompositon. Small deffects in barium content leads to rapid fomation of the Y_2BaCuO_5 phase at 920-930ºC.

More carefully processed powders of $BaCu_2O_3$ oxide were

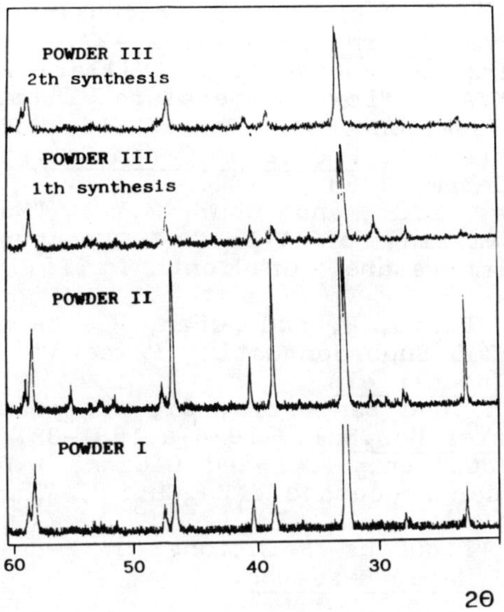

Figure 4. XRD of samples surface.

mixed with freshly prepared Y_2BaO_4, and pure $Y_1Ba_2Cu_3O_x$ was obtained by heat treatment at 910ºC for 2 hours. The sintering of these new powders at 930ºC for 2 hours leads to sintered bodies with a high density. The weight losses are somewhat smaller. The XRD shows the presence of orthorhombic $Y_1Ba_2Cu_3O_x$ without traces of Y_2BaCuO_5. The microstructure is a fine-grained by comparing to those of the samples I and II.

The results of this second sintering series confirm that the small-grained microstructures developed in the first series are originated by the presence of the green phase, according to the recent conclusions of No et al (7). The growth of coarse grains at temperatures as low as 930ºC could be produced by the formation of a transient liquid from very small amounts of unreacted CuO and $BaCuO_2$, which showing an invariant point at 920ºC (8).

CONCLUSIONS

Powders synthesized at high temperatures and/or long times are little reactives. They must be sintered at temperatures as high as the synthesis temperature and lead to sintered bodies with poor microstructures. The density shows relatively good values. The processing via oxalates allows the obtaining of better densities but the microstructures are still poor.

The use of powders processed at low temperatures from intermediate barium compounds give sintered bodies with the best density values. Small-grained microstructures seems to hinder the oxigen diffusion and leads to tetragonal samples. The density plays a role less important in that oxygen diffusion control. More work is being given to have a better knowledge of this density-microstucture-thermal schedule-crystalline symmetry relationships.

REFERENCES

1. Itoh, T., Uzawa, M. and Uchikawa, H., "Formation of Superconducting Ceramics $Y_1Ba_2Cu_3Ox$ from a Mixture of Y_2O_3,

$BaCO_3$ and CuO", J.Mater.Sc.Letters. 1988,7,130-32.

2. Sale, F.R. and Mahloojchi, F." Citrate Gel Processing of Oxide Superconductors". Ceram.Intern. 1988,14,229-37.
3. Moure, C., Duran, P., Vieira, S. Bourgeal, S. Villar, R., Aguilo, A. and Ramos, M.A., "Low Temperature Thermal Expansion of a High Tc Superconductor Phase Y1Ba2Cu3Ox Prepared from Oxalate Precursors". Proc.Brit.Ceram.Soc. Vol 40, ed. R.Freer, Inst. of Ceram. 1988,237-42.
4. Rahaman, M.N., de Jonghe, L.C. and Chu, M.Y., "Mass Transport and Chemical Demixing of Y-Ba-Cu-O Compounds During Sintering in Temperature Gradients". To be published.
5. Moure, C., Fernandez, J.F., Recio, P. and Duran, P., "A new Route for Synthesizing YBaCuO Superconducting Powders", to be published in Silicates Industriels.
6. Kwestroo, W, van Hal, H.A.M. and Langereis, C., "Compounds in the System $BaO-Y_2O_3$", Mater.Res.Bull. 1974,9,1631-38.
7. No, K., Verhoeven, J.D., McCallun, R.W. and Gibson, E.D., "Grain Size Control in Powder Processed $Y_1Ba_2Cu_3O_x$". To be published.
8. Aselage, T., Keefer, K., "Liquidus Relations in Y-Ba-Cu Oxides". Submitted to the J.Mater Research.

PREPARATION AND CHARACTERIZATION OF HIGHLY ORIENTED CERAMICS AND LARGE CRYSTALS OF $YBa_2Cu_3O_x$ SUPERCONDUCTOR

P. Murugaraj, J. Maier* and A. Rabenau
Max-Planck-Institut für Festkörperforschung,
Heisenbergstr. 1, 7000 Stuttgart 80, F.R.G.
*Affiliated to Max-Planck-Institut für Metallforschung,
Institut für Werkstoffwissenschaft, Heisenbergstr. 5,
7000 Stuttgart 80, F.R.G.

ABSTRACT

We successfully prepared BaO-deficient Y-Ba-Cu-O ceramics with high degree of orientation and extended the same technique to prepare large crystals (typically 3x3x3.5mm). The critical temperature for these materials lies between 90 and 94K. By wet chemical and EDX analysis of these materials, Ba deficiency and a possible Y- and Ba- sublattice disorder is suggested. Detailed SEM investigations on these materials are presented. Electrochemical and defect chemistry investigations are briefly discussed.

INTRODUCTION

Using K_2CO_3 as an 'intermediate dopant' we found that it is possible to prepare oriented ceramics [1,2] as well as large crystals [3]. Due to the anisotropy of the transport properties, there is a considerable need for textured samples or single crystals for basic research as well as for applications. In this paper we present our recent results on the processing aspects of these materials, on their characterization and also briefly on transport properties (oxygen diffusion, high temperature defect model and critical current densities).

1. Preparation of highly oriented agglomerates:

In our preparation procedure, BaO-deficient (≈5 mole %) composition (Y_2O_3, $BaCO_3$, CuO) in conjunction with K_2CO_3 is used as a starting material [1]. After the first calcination at T=950°C, we obtain a K-free porous agglomerate consisting of several highly oriented sub-regions [1]. These regions consist of several well formed platelets arranged in an ordered manner. X-ray diffraction analysis shows highly intense (00ℓ) peaks [1].

Without K-addition this effect was not observed. Using a stoichiometric composition does not result in such a pronounced texture effect. It is suggested that K is incorporated in the '123' lattice intermediately. During the course of the reaction at 930-950°C, K vanishes (partly due to volatlization and partly by reaction with the crucible material), leaving behind pure '123' compound with a slight Ba-nonstoichiometry ($YBa_{2-\beta}Cu_3O_x$, $\beta \approx 0.05$) ($T_c \approx 90K$).

With respect to Y-Ba-Cu-O processing, currently there are numerous reports available in the literature. K has been used as mineralizer [4], as sintering aid [5], and also as flux for crystal growth (in carbonate and chloride forms) [6]. There are also reports in which K is suggested to substitute Ba in the '123' lattice [7-9].

The presence of K_2CO_3 may affect the '123' material in different ways: 1) acting as a crystal habit modifier, 2) depressing the melting point due to intermediate doping and 3) aiding sintering. The products which are obtained from K-added reactant mixtures, contain morphologically well formed grains (Fig.1a) when compared to that obtained from K-free reactant mixtures (Fig.1b).

2. Preparation of oriented ceramics:

The porous agglomerate thus obtained, is ball milled using a tungsten carbide ball mill. The particles are sieved and categorized according to their sizes. These powders are then pelletised under uniaxial pressure (7 tons cm^{-2}) and sintered at 950°C in air [1]. It is found that the degree of orientation obtained in the final product depends critically on the initial particle size and morphology. In order to achieve high degree of orientation (a-b plane normal to the pressure axis), following criteria are found to be important: 1) The particles should have platelet morphology ($\perp$ to c-axis) with high aspect ratio; 2) The particles should have sizes between 10-60μm.

3. Preparation of large crystals:

The formation of highly oriented sub-regions indicated that this 'K_2CO_3-method' can be used for growing crystals also. The reaction mixture is heated at 1000°C for 40 h and slow-cooled. This results in the formation of a dense mass well separated from the walls of the container, consisting of several crystal regions. The upper portion of the charge is relatively flux free with relatively low amount ($\leqslant$5%) of second phase. The crystal regions are typically of size 3mmx3mmx1.5mm. Each region is made up of thin crystal platelets (20-100μm thick) (Fig.2) [3]. The platelets are arranged in an highly ordered fashion with less than 5° tilt. Thus the whole crystal region can be considered as a mosaic crystal. This 'K_2CO_3-technique' is found to work in the synthesis of rare earth barium cuprate crystals also. Moreover these materials are more stable at high temperatures when compared to Y-Ba-Cu-O system. Crystals prepared by this technique are found to undergo clean fracture on the a-b surface (Fig.3) indicating ordering of the platelets. Microstructure of the free standing crystals of $GdBa_{1.95}Cu_3O_x$ clearly reveal the formation of well defined crystallographic faces that are made up of edges of platelets (Fig.4a and 4b). The chemical analysis on these crystal samples

Figure 1. SEM micrograph of a)typical $YBa_{1.95}Cu_3O_x$ grains obtained during ceramic synthesis by 'K_2CO_3-method' and b) microstructure of a conventional ceramic.

Figure 2. SEM micrograph of the $(Y_{\alpha}Ba_{1-\alpha})_3Cu_3O_x$ crystal showing the platelet growth pattern.

Figure 3. SEM micrograph of the surface containing c-axis obtained during the fracture of a-b surface of a $(Y_{\alpha}Ba_{1-\alpha})_3Cu_3O_x$ crystal.

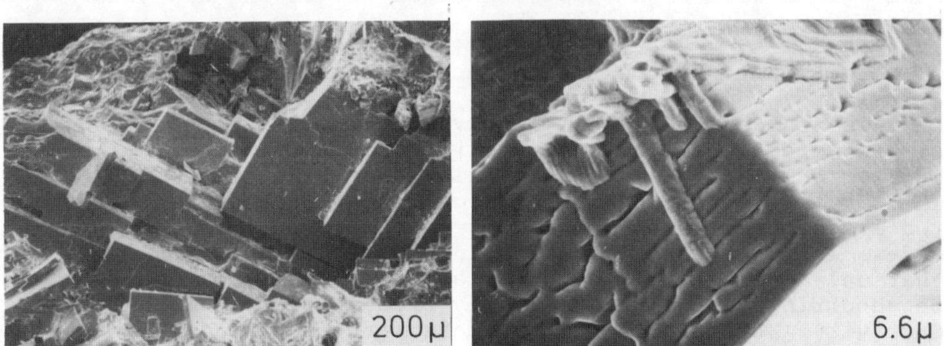

Figure 4. SEM micrograph of: a)free standing crystals of $(Gd_{\alpha}Ba_{1-\alpha})_3Cu_3O_x$ and b) magnified picture of one crystal.

suggests mutual disorder in the Y(or Gd)- and Ba-sublattices. The chemical formulae for Y(or Gd)-Ba-Cu-O crystals, are close to $(Y_{\alpha}Ba_{1-\alpha})_3Cu_3O_x$ with $0.33 \leq \alpha \leq 0.45$ (Y(or Gd)- and Ba positions are crystallographically equivalent positions in the ideal perovskite lattice). Such a possibility of disorder is demonstrated in the formation of $La_1(Ba_{1-\beta}La_{\beta})_2Cu_3O_x$ [10]. In any case, spectroscopic studies (IR, Raman, ellipsometric etc.) [11] show that there is no observable change in the lattice vibrational mode frequencies with respect to the values calculated assuming an ideal '123' structure. Moreover the transition to the superconducting state (resistive and magnetic) is quite sharp suggesting the absence of detoriation in superconducting properties [2,3].

4. Anisotropy of the electrical conduction:

Conductivity measurements on the mosaic crystals show the room temperature resistivity anisotropy factor (ratio of the electrical conductivity perpendicular and parallel to the c-axis) to be ≈ 50 [12]. Similar values are measured for the oriented ceramics [12].

In consideration of their size, the superconducting blocks containing crystal regions are found to tolerate considerably high currents (current density: > 250 A cm^{-2} in 500 G field at 77 K) [13,3,12].

5. Chemical diffusion and ionic conductivity:

In order to study the oxygen diffusion process in the Y-Ba-Cu-O system, electrochemical investigations were carried out. The samples in pellet form were sandwiched between electronically blocking electrodes ($ZrO_2(Y_2O_3)$). Galvanostatic measurements were made on the glass encapsulated cell [12]. From the steady state potential and its transient behaviour, the ionic conductivity (σ_{ion}) and oxygen diffusion coefficient ($\bar{D}$) can be calculated [12]. E.g., on weakly oriented ceramics, $\sigma_{ion} = 4.1 \times 10^{-7} \Omega^{-1} cm^{-1}$ and $\bar{D} \approx 2.5 \times 10^{-7} cm^2 s^{-1}$ at $P \approx 0.2$ bar O_2 and $T = 420°C$. The activation enthalpy for $_{ion}$ is identical with tracer diffusion measurements [14]. D obtained from the conductivity relaxation induced by a sudden change in the oxygen potential agrees well with electrochemical values and also with those reported in literature [12,15].

6. Defect model in the semiconductive region:

In the semiconductive ($>650°C$) region, the conductivity changes from n- to p- type at oxygen composition around 6-6.3. A simple defect model given by us [12] explains the slope in the $\lg \sigma - \lg P_{O_2}$ diagrams in a simple manner. In contrast to the literature models, where the large concentration changes involved are tackled as in simple defect chemistry, this model considers ionic-electronic defect interaction. It basically assumes, that only a small portion of holes formally introduced by the neighbouring gas phase, is really free and responsible for conductivity. For more details of this model the reader is referred to Ref.[12]. Further work is in progress to elucidate the contributions of various factors such as valence changes of defects on the conduction mechanism.

Acknowledgements:
We thank M. Erz, B Reichert and L. Viczian for technical help.

References

1. P. Murugaraj, J. Maier and A. Rabenau, Preparation of highly oriented superconducting ceramics. Solid State Commun., 1988, **66**, 735-738; German Patent P 38 05954. 1-45.
2. P. Murugaraj, J. Maier and A. Rabenau, Y-Ba-Cu-O superconductor: Preparation of highly oriented ceramics, improvement of durability and precise determination of redox state. Solid State Ionics, 1989, **32/33**, 1183-1187.
3. P. Murugaraj, J. Maier and A. Rabenau, Preparation of large crystals of Y-Ba-Cu-O superconductor. Solid State Commun., 1989 (in print).
4. S.-G. Eriksson, L.-G. Johansson and C. Olsson, Doping effects on the 123-system. Physica C, 1988, **153/154**, 902-903.
5. I.W. Chen, S. Keating, and C.Y. Keating, Superconductivity and the tailoring of lattice parameters of the Y-Ba-Cu-O compound. Advanced Ceramic Mater., 1987, **2**, 457-460.
6. S. Bosi, T. Puzzer, G.J. Russell, S.L. Town and K.N.R. Taylor, Large single crystals of $YBa_2Cu_3O_{7-\delta}$ superconductors from chloride fluxes. J. Mater. Sci. Letts., 1989, **8**, 497-500.
7. S.X. Dou, A.J. Bourdillon, X.Y. Sun, H.K. Liu, J.P. Zhou, N. Savvides, C.C. Sorrell and K.E. Easterling, Dependence of the superconducting transition temperature on the radui of alkali and alkaline earth dopants in $YBa_2Cu_3O_{7-x}$. Phys. Stat. Sol. (b), 1988, **147**, K153-157.
8. I. Kontoulis, X. Turrillas, R. Meisels, S. Bungre and B.C.H. Steele, Increase of the O-T transition temperature by Potassium doping of $YBa_2Cu_3O_{7-\delta}$. J. Less. Com. Metals, 1989, **150**, 139-145.
9. N.-L. Wu, E. Ruckenstein L.-Q. Wang, and P. Mattocks, The effects of alkali impurities on the superconductivity and powder characteristics of $YBa_2Cu_3O_7$. Mater. Letts., 1988, **6**, 211-215.
10. G. Collin, A.C. Audier, P.A. Albouy, S. Senoussi, R. Comes, M. Konczykowski and F. R.-Albenque, The $La(Ba_{2-x}La_x)Cu_{3-y}O_z$ single crystals: strucțures and conductivity. J. Phy. France, 1989, **50**, 77-90.
11. I.B. Ferreira, M. Bauer, L. Genzel, P. Murugaraj and J. Maier, FIR study of $YBa_2Cu_3O_6$ phonon anisotropies. (to be presented at the "Phonons '89" III Int. Conf. on phonon physics, Aug. 21-25 '89, Heidelberg; 13a) P. Knoll, C. Thomsen, M. Cardona, P. Murugaraj and J. Maier (in preparation); 13b) J. Kircher. M. Aluani, M. Garriga, P. Murugaraj, J. Maier, C. Thomsen, M. Cardona, O.K. Anderson and O. Jepsen, Anisotropy of the dielectric function in $YBa_2Cu_3O_6$. (submitted to Phy. Rev. B).
12. J. Maier, P. Murugaraj. G. Pfundtner and W. Sitte, Defect chemical investigations of definded $YBa_2Cu_3O_x$-superconductors. Ber. Bunsengesw. Phys. Chem. (to be published).
13. W. Becker, S. Elschner, P. Murugaraj and J. Maier (to be published).
14. Y. Ikuma and S. Akiyoshi, Diffusion of oxygen in $YBa_2Cu_3O_{7-\delta}$. J. Appl. Phys., 1988, **64**, 3915-3917.
15. J.-H. Park, P. Kostic, J.P. Singh, Electrical conductivity and chemical diffusion in sintered $YBa_2Cu_3O_y$. Mater. Letts., 1988, 6, 393-397.

MICROSTRUCTURE AND PROPERTIES OF CITRATE-GEL PROCESSED Bi-CONTAINING Cu-OXIDE SUPERCONDUCTORS

M. RAJABI and F.R. SALE
Manchester Materials Science Centre,
University of Manchester/UMIST,
Grosvenor Street, Manchester M1 7HS.

ABSTRACT

Citrate gel processing has been used to prepare superconducting oxides in the Bi-Ca-Sr-Cu system. This paper reports data for a composition of $Bi_2Ca_2Sr_2Cu_3O_x$. DTA has been used to study the decomposition of the organic precursor and the reactions of the oxide products of pyrolysis. $Bi_2Ca_2Sr_2Cu_3O_x$ was shown to melt at 910°C. The products of precursor decomposition at 750°C underwent reactions at 800°C and 860°C prior to fusion. AC magnetic susceptibility measurements have been made upon sintered materials to determine T_c as a function of sintering atmosphere. SEM with EDAX analysis, optical microscopy and XRD have been used to identify at least three phases in the superconducting pellets. Sharp transitions to a superconducting state were observed in the range 80-85K for pellets sintered in air at 865°C. The initial onset of superconductivity occurred at 105-110K.

INTRODUCTION

Since the discovery of superconductivity in the Bi-Sr-Cu-O system (1), many investigations have been carried out of Bi-containing Cu-oxide superconductors (2-5). In the Bi-Sr-Ca-Cu-O system at least two superconducting phases have been detected (2) and within compacts of any given composition it has been usual to detect three or more phases. In an attempt to determine whether the multitude of phases is a result of either heterogeneity in the pellets or the basic phase relationships which exist within this complex system, citrate gel processing has been used in an attempt to overcome heterogeneity problems. Previously, such processing has been used successfully for a number of complex oxides and superconductors (6,7) and has been shown to produce oxide powders of high homogeneity. This technique obviates the several cycles of firing and grinding of oxides or carbonates which have been shown to be required in conventional processing. Typically in the production of Bi-Sr-Ca-Cu-O Bi_2O_3, $SrCO_3$, $CaCO_3$ and CuO are mixed, calcined at temperatures in the mixture of oxides and carbonates that may have been produced by a variety of other physical and chemical routes.

Figure 1 shows a DTA trace obtained for the powder produced by pyrolysis at 300°C and 750°C. It is clear that there are 3 prominent endotherms, the highest temperature one of which has a small shoulder. It was anticipated that the endotherms indicated temperature regions where changes in sintered product could be expected. Accordingly, pellets were sintered at 800°C, 865°C, 885°C and 906°C (i.e. before and after each endotherm) in air for 1 hour and then air-quenched. Hot-stage microscopy and visual examination of the pellets showed that the high temperature endotherm was associated with fusion of the material. In addition it is clear from published data that the shoulder on the endotherm is caused by the decomposition of calcium carbonate.

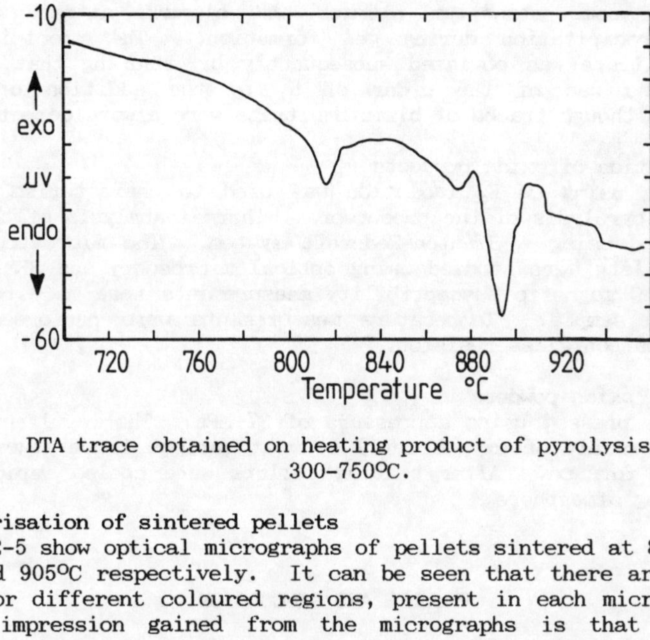

Figure 1. DTA trace obtained on heating product of pyrolysis at 300-750°C.

Characterisation of sintered pellets

Figures 2-5 show optical micrographs of pellets sintered at 800°C, 865°C, 885°C and 905°C respectively. It can be seen that there are at least 3 phases, or different coloured regions, present in each micrograph. The initial impression gained from the micrographs is that the regions coarsen as temperature increases. However, this is not the complete explanation of the effect of sintering temperature as was seen by EDAX analysis of the regions. In addition, XRD showed the presence of CuO in samples sintered at 800°C and Cu_2O in the samples sintered at 906°C. The full 4 regions which have been identified by colour; white, light grey, dark grey and black, are only apparent in material sintered at 885°C. The compositions of some of the regions appeared to be constant from one sintering temperature to another and then change with further changes in temperature. For example, the light grey region approximates to 2:1.1:2:1.4 (Bi:Ca:Sr:Cu) for samples sintered at 800, 865 and 885°C but changes to 2:0.8:1.8:1.1 at 906°C i.e. becomes either enriched with Bi or depleted in Ca and Cu. Similarly, the white region has a very similar overall composition of approximately 2:1:2:1.5 at 800°C and 865°C but becomes enriched in Bi and Sr at 885°C and 906°C. Both these regions have compositions close to 2:1:2:2, which has a reported T_c of 80-83K (2-5), and seem to be Cu-deficient variants on this phase. Further TEM work is in progress to study these regions. It is interesting to note range 750-850°C for periods of 5 to 15 hours, reground, calcined, reground again and eventually pressed and sintered (2). Citrate processing may replace this sequence by gel formation and a single or two-stage firing sequence to give high homogeneity powders. Recently, it has been shown that some mechanical treatment of powders may be necessary to remove agglomerates. However, this treatment is not necessary from the point of view of chemical homogeneity.

EXPERIMENTAL PROCEDURE

Preparation of precursors

Initial precursors were produced in the manner described previously for $YBa_2Cu_3O_x$ superconductors (7). XRD of these precursors showed the presence of some strontium nitrate and bismuth nitrate and hence indicated precipitation during gel formation. The precipitation of strontium nitrate was obviated subsequently by ensuring that the pH of the solutions was of the order of 6, by the addition of ammonium hydroxide, although traces of bismuth nitrate were always detected.

Characterisation of oxide products

XRD analysis using Cu K_α radiation was used to characterise the oxide products of pyrolysis of the precursors. Thermal analysis of the powders was performed using a Stanton-Redcroft system. The microstructures of sintered pellets were studied using optical microscopy and SEM with EDAX analysis. AC magnetic susceptibility measurements were made relative to a pure lead sample. Temperature measurements were performed using a calibrated carbon/glass resistor.

Sintering of oxide powders

Powders were pressed using a pressure of 37 MPa. The resultant compacts were heated in air or oxygen for 1 hour at various temperatures using a simple tube furnace. After heating pellets were cooled rapidly within the sintering atmosphere.

RESULTS AND DISCUSSION

Precursor pyrolysis

Because the decomposition/oxidation of citrate nitrate gel precursors is very exothermic, and may lead to loss of homogeneity by phase separation or volatilisation as overheating occurs, a low temperature decomposition of 300°C was always performed prior to higher temperature processing. This is particularly necessary when ammonium hydroxide has been used because of the possibility of forming small amounts of ammonium nitrate. For the products discussed here a second stage pyrolysis at 750°C was always used. XRD data obtained for the product of pyrolysis of a precursor at 750°C showed that citrate gel processing had disappointingly failed to yield a single homogeneous powder, in contrast to that obtained for $YBa_2Cu_2O_{7-x}$ (7). It was clear that during pyrolysis reaction had occurred with CO_2, a product of pyrolysis, to produce the stable carbonate $CaCO_3$ along with traces of $SrCO_3$. In addition traces of Bi_2O_3 and the mixed oxide Bi_2CuO_4 were detected. SEM and EDAX analysis of the pyrolysis product failed to indicate any regions of gross segregation. Consequently, the product was treated as if it were a fine intimate that samples sintered at 865°C were the only ones to show superconductivity. Also EDAX analyses of the dark regions in the samples sintered at 865°C gave a composition approximating to $Bi_{2.1}(CaSr)_{3.9}Cu_{3.4}$ which is very close to $Bi_{2.1}(CaSr)_{n+1}Cu_n$, where n = 3, for which a T_c of 105-110K has been determined (1,5). Hence these regions are likely to be the ones which cause the onset of superconductivity at approximately 107K.

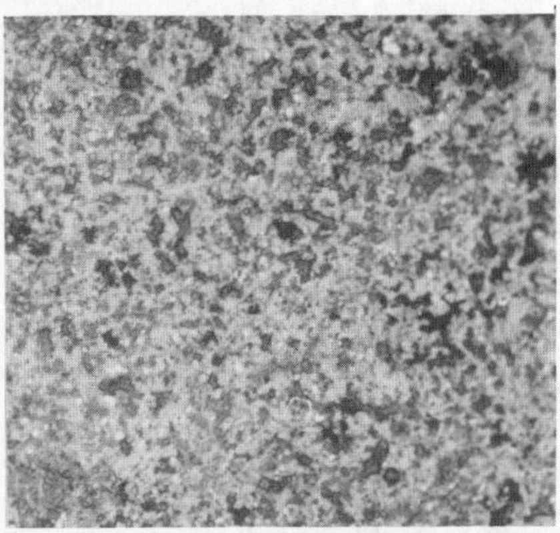

Figure 2. Product sintered at 800°C.

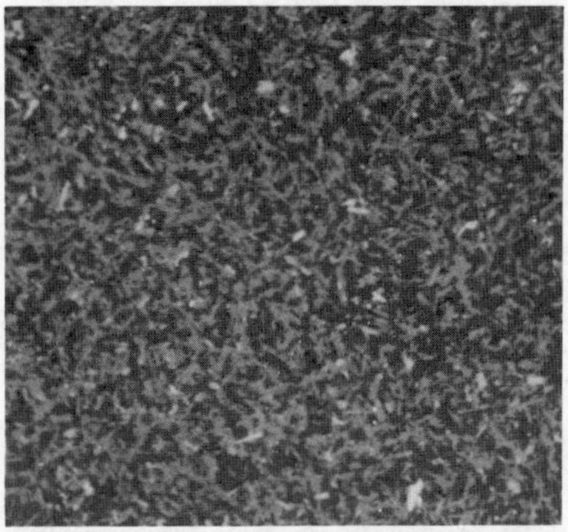

Figure 3. Product sintered at 865°C.

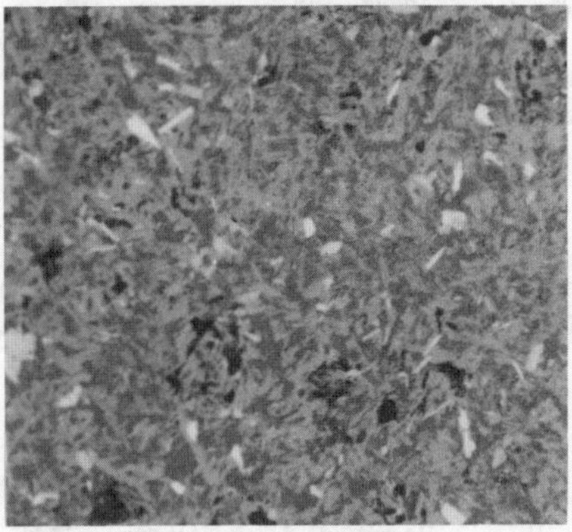

Figure 4. Product sintered at 885°C.

10 µm

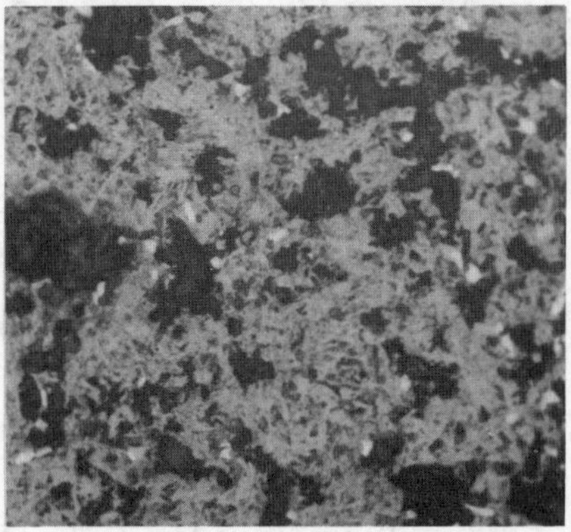

Figure 5. Product sintered at 906°C.

Figure 6 shows the relationship between AC magnetic susceptibility and temperature for two samples sintered at 865°C. Sample A was sintered and quenched in air whilst sample B was sintered and quenched in oxygen. Both samples were made from the same batch of powder. Dou et al (4) have previously reported that sintering in oxygen did not significantly affect the superconducting properties of material relative to air treatments. However, the present data show a positively detrimental effect of the use of oxygen. Further work including the TEM characterisation of samples prepared in oxygen and air are presently in hand to explain this observation.

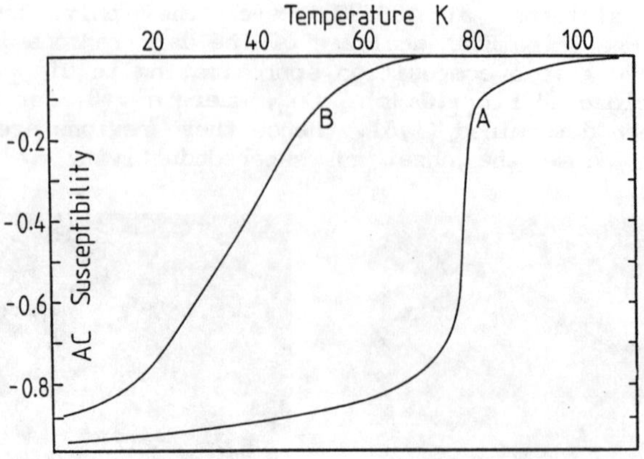

Figure 6. Relationship between AC magnetic susceptibility and temperature for products sintered at 865°C. A in air, B in oxygen.

ACKNOWLEDGEMENTS

The authors would like to thank Dr. J.W. Ross of the Physics Department, Manchester University for the AC magnetic susceptibility measurements and SERC for partial funding of this project.

REFERENCES

1. Michel, C., Hervieu, M., Borel, M.M., Grandin, A., Deslandes, F., Provost, J. and Raveau, B., Z. Phys. B., 1987, 68, 421.

2. Maeda , H., Tanaka, Y., Fukutomi, M. and Asano, T., Jpn. J. Appl. Phys., 1988, 27, 1209.

3. Hazen, R.M., Prewitt, C.T., Angel, R.J., Ross, N.L., Finger, L.W., Hadidiacos, C.G., Veblen, D.R., Heaney, P.J., Hor, P.H., Meng, R.L., Sun, Y.Y., Wang, Y.Q., Xue, Y.Y., Huang, Z.J., Gao, L., Bechtold, J. and Chu, C.W., Phys. Rev. Lett., 1988, 60(12), 1174.

4. Dou, Z.X., Liu, H.K., Bourdillon, A.J., Tan. N.X., Savvides, N., Zhou, J.P. and Sorrell, C.C., Supercond. Sci. Technol., 1988, 1, 78.

5. Tallon, J.L., Buckley, R.G., Gilberd, P.W., Presland, M.R., Brown, I.W.M., Bowden, M.E., Christian, L.A. and Gognel, R., Nature, 1988, 333, 153.

6. Anderton, D.J. and Sale, F.R., Powder Metall., 1979, 1, 8.

7. Mahloojchi, F. and Sale, F.R., Ceramics Int., 1988, 14(4), 229.

DEVELOPMENT OF Bi-Sr-Ca-Cu-O CERAMIC MATERIALS

M. PEUCKERT, J. BOCK, B. HETTICH, M. SCHWARZ
Hoechst AG,
Postf. 800320, D-6230 Frankfurt/M. 80, F.R.G.

ABSTRACT

Bi-Sr-Ca-Cu-O superconductors have been prepared with single, double, and triple CuO_2-layers and corresponding Tc-values of 60-75 K, 80-90 K, and 100-110 K. While the single and triple layer compounds may be prepared by reacting solid mixtures of the oxides of Bi, Sr, Ca, and Cu a process starting with a stoichiometric oxide melt was found to be advantageous for the double layer compound. Using PbO as an additive the triple layer compound can be prepared as a pure phase according to X-ray diffraction. Nevertheless, an almost identical superconductive behaviour has been observed with a nonstoichiometric, multiphase sample.

INTRODUCTION

Since the discovery of the Bi family of high Tc superconductors by von Schnering (1) and Maeda (2) numerous attempts have been made to further increase the critical temperature. Especially, the preparation of the so-called "110 K-phase" has been tried and achieved by several research groups (3-6). Basically, these studies succeded by reacting nonstoichiometric mixtures of Bi, Sr, Ca, and Cu oxide close to the melting point (3) and by adding PbO as a flux (4-6). The goal of the present investigation was to develop synthesis routes for single phase material of all the 3 first members of $Bi_2Sr_{2-x}Ca_{n+x}Cu_{n+1}O_y$ with n=0, 1, and 2 and to establish a relation between critical temperature and the index n.

MATERIALS AND METHODS

The raw materials used were Bi_2O_3 (99.0%), SrO (99.0%), CaO (99.0%), CuO (99.0%), and PbO (99.0%) all by Riedel-deHaen.

Single layer compounds $Bi_2Sr_{2-x}Ca_xCuO_{6+y}$ were prepared by reacting the appropriate oxide mixtures at 800 to 820 ^oC in air.

The double layer compound $Bi_2Sr_2CaCu_2O_{8+y}$ was prepared by completely melting a mixture of the parent oxides at about 1100 ^oC, casting this melt into Cu moulds, and slowly cooling to room temperature. Afterwards, the ingots were subjected to annealing in air according to the following program: heating to 815 ^oC, keeping at 815 ^oC for 12 to 60 h depending on the size of the ingot and slow furnace cooling at a rate of 20 K h^{-1} or less.

The triple layer compound was prepared following the procedure described by Endo et al. (6). A Pb-free material with the same superconductivity characteristics was prepared by reaction of a 1112 mixture of Bi_2O_3, SrO, CaO, and CuO for 80 h at 870 °C and subsequent temperature cycling between 600 and 870 °C.

The samples were characterized by SQUID measurements at 1.5 mT using a S.H.E. susceptometer, by DC resistance and X-ray diffraction.

RESULTS AND DISCUSSION

The quasiquaternary system Bi-Sr-Ca-Cu-O depicted in figure 1 shows the ranges of existence of the superconductors $Bi_2Sr_{2-x}Ca_{n+x}Cu_{n+1}O_y$ with n=0, 1, and 2 and also the not superconducting $(Sr,Ca)CuO_2$. The miscibility ranges have been determined in this study as about $0 \leq x < 0.7$ for n=0 and $-0.3 \leq x \leq 1.0$ for n=1.

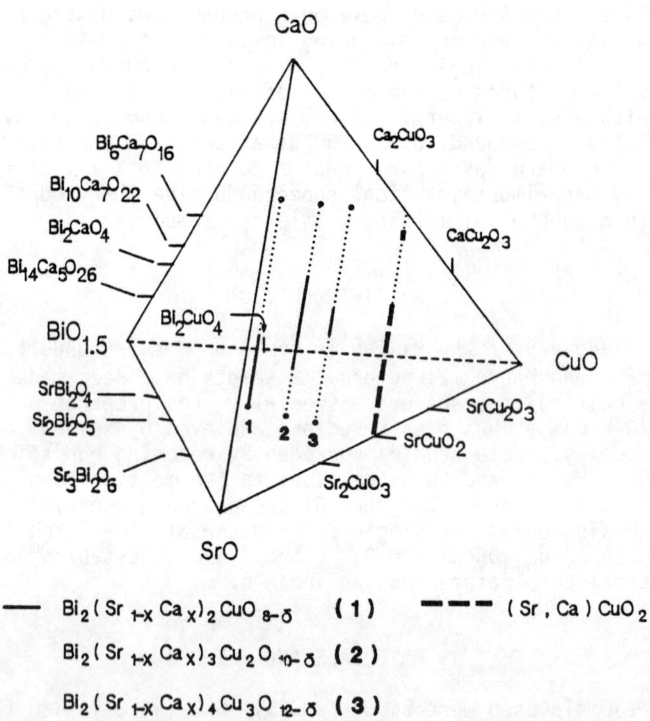

Figure 1. The quasiquaternary system Bi-Sr-Ca-Cu-O.

In figure 2 the phase compositions obtained in our experiments on the preparation of the single CuO_2-layer superconductor are shown as determined by XRD. Up to Ca/Sr ratios of 1/3 over 85% is $Bi(Sr,Ca)_2CuO_y$ with less than 5% of the double layer compound and the rest being Bi-Sr-O, CuO and other cuprates. For $Bi_2Sr_{1.3}Ca_{0.7}CuO_{6+y}$ a Tc of 75 K was measured by susceptibility. The SQUID data lead to around 70 vol.-% superconducting material at 60 K indicating that the high Tc value has to be associated with the

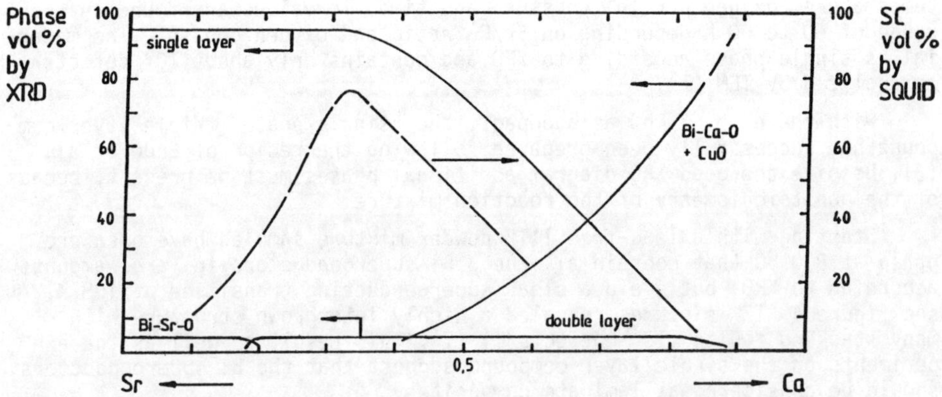

Figure 2. Formation of single layer $Bi_2(Sr,Ca)_2CuO_{6+y}$ with various Sr/Ca ratios as determined by XRD and SQUID measurements.

single layer compound. TEM pictures, on the other hand, showed that this material is strongly intergrown with lamellas of the double layer type (7). Therefore, as long as there is no single phase material as proved by XRD and TEM this superconductive ceramic should better be described as a laminate composite on a nanometer scale.

As already outlined in the experimental section the double layer compound $Bi_2Sr_2CaCu_2O_{8+y}$ can be prepared as truely single phase powder and bulk material by a special melt process. The superconductor crystallizes from the solidified melt following the complicated sequence of phase transformations listed in figure 3. The phase transformations strongly depend on

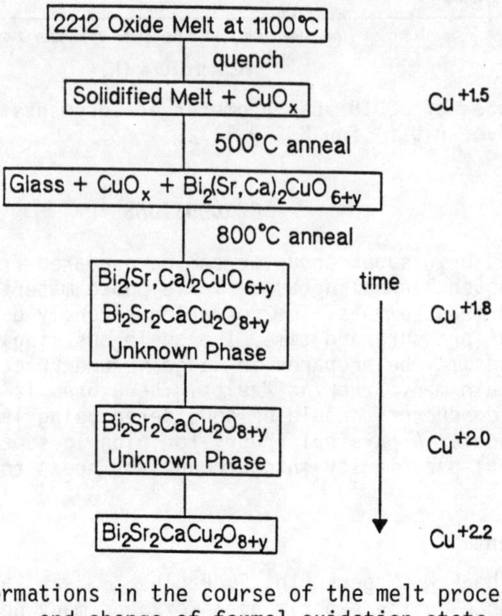

Figure 3. Phase transformations in the course of the melt process for $Bi_2Sr_2CaCu_2O_{8+y}$ and change of formal oxidation state of Cu.

temperature, oxygen partial pressure and time. Tc values are found in the range of 80 to 90 K depending on Sr/Ca ratio and oxygen content. The material is single phase according to XRD and contains only about 10^7 defects/m as analyzed by TEM (7).

With the help of PbO as a dopant, the "single phase" triple layer compound has successfully been prepared following the recipe of Endo et al. (6). Despite the good XRD diagram additional phases must be present, because of the nonstoichiometry of the reaction mixture.

Starting with a lead-free 1112 powder mixture samples have been prepared at 870 ^oC that contain all the 3 Bi superconductors in large amounts according to XRD, but yield a clear superconducting transition at 108 K, see figure 4. TEM pictures revealed a highly intergrown structure with many stacking faults (10^{10} defects/m) (7). This result as well as the experiments on the single layer compound suggest that the Bi superconductors should be considered as laminate composites.

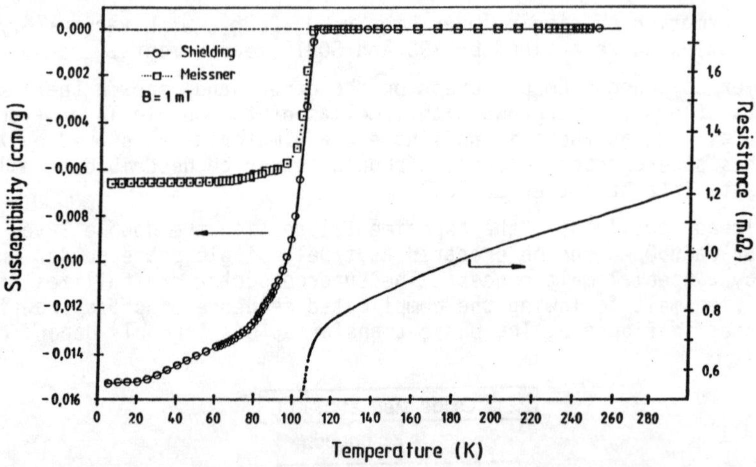

Figure 4. Traces of SQUID susceptometer measurements (left) and resistance curve (right) for BSCCO-1112.

CONCLUSIONS

The $Bi_2Sr_2CaCu_2O_{8+y}$ superconductor can be prepared from an oxide melt process by subsequent annealing as a single phase material according to XRD, TEM and SQUID measurements. The formation strongly depends on temperature, oxygen partial pressure and time. The analogous single and triple CuO_2-layer compounds may be prepared in large volume fractions according to XRD, but they contain many stacking faults. Therefore, it is suggested that the Bi-type superconductors should be considered being laminate composites on a nanometer scale. A physical theory for high Tc superconductors which take this structural peculiarity into account yet needs to be developed.

Acknowledgement

The authors thank K. Magerkurth, E. Paulus, F. Rat, and R. Reichling for their assistance in the experimental work and many helpful discussions.

REFERENCES

1. Von Schnering, H. G., Walz, L., Schwarz, M., Becker, W., Hartweg, M., Popp, T., Hettich, B., Müller, P. and Kämpf, G., Angew. Chem., 1988, **100**, 604-7; Angew. Chem. Int. Ed. Engl., 1988, **27**, 574-6.

2. Maeda, H., FED HiTcSc-ED Workshop, Miyagi-Zao, 1988, June 2-4, pp. 3-10.

3. Tarascon, J. M., McKinnon, W.R., Barboux, P., Hwang, D. M., Bagley, B. G., Green, L. H., Hull, G., LePage, Y., Stoffel, N. and Giroud, M., Phys. Rev. B, 1988, **38**, 8885-92.

4. Green, S. M., Jiang, C., Yu Mei, Luo, H. L. and Politis, C., Phys. Rev. B, 1988, **38**, 5016-8.

5. Takano, M., Takada, J., Oda, K., Kitaguchi, H., Miura, Y., Ikeda, Y., Tomii, Y. and Mazaki, H., Jap. J. Appl. Phys., 1988, **27**, L1041-3.

6. Endo, U., Koyama, S. and Kawai, T., Jap. J. Appl. Phys.,1988, **27**, L1476-9.

7. Eibl, O., private communication.

2.436

Bi-CONTAINING SUPERCONDUCTORS PREPARED FROM RAPIDLY-QUENCHED AMORPHOUS FILMS

Masahiro Yoshimura, Tae-Hyun Sung, Nobuo Ishizawa
and Zenbe-e Nakagawa
Research Laboratory of Engineering Materials,
Tokyo Institute of Technology,
4259 Nagatsuta, Midori, Yokohama 227, JAPAN.

ABSTRACT

Amorphous films with $Bi_2Sr_2CaCu_2O_z$ and $Bi_2Sr_2Ca_{2-x}Cd_xCu_3O_y$ compositions were prepared by the rapid quenching after rapid melting technique without any containers. The amorphous materials were crystallized in the $Bi_2Sr_2CaCu_2O_z$ phase above 800°C through the $Bi_2Sr_2CuO_z$ phase crystallized above 460°C. In $Bi_2Sr_2Ca_{2-x}Cd_xCu_3O_y$ composition, the zero resistivity temperature (T_{ce}) increased and tailing of transition vanished with increased doping of Cd. The film with the composition $Bi_2Sr_2Ca_{1.6}Cd_{0.4}Cu_3O_y$ exhibited a T_{ce}=87 K without tail. Further substitution of Cd for Ca brought about lowering in the T_c, because a semiconducting phase with c_o=2.5 nm formed instead of a c_o=3.0 nm phase.

INTRODUCTION

Bi-containing superconductors have a great significance in the research for higher T_c than 100 K and in easiness to obtain amorphous materials. These systems could produce amorphous materials by a rapid-quenching-after-rapid-melting (RQRM) technique[1-8]. Melting techniques are potentially very interesting because they have following merits; (1) they can prepare complex geometries, such as films, wires and odd-shaped articles. (2)They can provide great flexibility for microstructral control, as well as for enhancing properties through chemical homogeneity and the selectivity of composition[3]. They also have the advantage of the crystallization process which undergoes from amorphous phase through metastable phase. We succeeded in preparing superconducting $Bi_2Sr_2CaCu_2O_z$[1,8] and series of $Bi_2Sr_2Ca_{2-x}Cd_xCu_3O_y$[4-6] using this technique. In attempt to synthesize Bi-containing amorphous materials, one has met with difficulty due to reaction with Pt and alumina crucibles employed[9,10]. These problems could be solved in the RQRM process without any containers. The RQRM technique has another merit to rapidly heat and quench the sample using a Xe arc imaging furnace, that is, the compositional changes were negligible[1]. The main object of the present study is to explore the possibility of obtaining amorphous materials in the Bi-Sr-Ca-Cd-Cu-O system. A second object is to verify the crystallization process of superconducting phases from amorphous materials and to explore the effects of the substitution of Cd for Ca.

MATERIALS AND METHODS

Series of $Bi_2Sr_2CaCu_2O_z$ and $Bi_2Sr_2Ca_{2-x}Cd_xCu_3O_y$ samples (x=0.0, 0.1, 0.2, 0.3, 0.4, 0.5, 0.6) were prepared from mixtures of 99.99% Bi_2O_3, $CaCO_3$ and CuO, and of 99.9% $SrCO_3$ and CdO powders. They were mixed thoroughly in an agate mortar and pestle using ethanol for 4h and calcined at 800°C for 12 h. After crushing, sample bars, 2.5mmϕ x 150 mml were prepared by cold isostatic pressing at about 294MPa for 3 minutes. The bar was suspended in a Xe arc-imaging furnace(SC-5D,Nichiden Kikai Inc., Tokyo, Japan). The bottom of the bar was melted within a few seconds to avoid excessive evaporation by applying 1kW power in air. The molten droplet with a diameter of 3 ϕ mm dropped onto the twin roller rotating at 2000 rpm. to yield the films with 20 μm thick and several cm^2 areas. The cooling rate was estimated to exceed 10^5°C/sec[11]. The films thus obtained were analyzed on the crystallization process of the amorphous materials using a thermoanalyzer, TG-DTA (Regaku, 8078G2) coupled to a computerized data acquisition system. They were annealed at various temperatures on the (100) cleavage surface of the MgO single crystal in air. All samples were investigated using an X-ray powder diffractometer(Regaku Co. RU-200). The electrical resistivity was measured from room temperature to electrical resistivity was measured from room temperature to 20K on the annealed films of using a D.C. four-probe method using a constant current of 0.5mA. The electrical contacts were made with fine silver wire and silver paste.

RESULTS AND DISCUSSION

The film obtained had about 20 μm in thickness, which verified an amorphous material by X-ray powder diffraction. The surface was very clean and smooth[1]. The X-ray powder diffraction sometimes suggested the presence of a small amount of CaO in addition to the amorphous phases after the RQRM technique, particularly in Ca-rich samples, as reported previously[3-5].

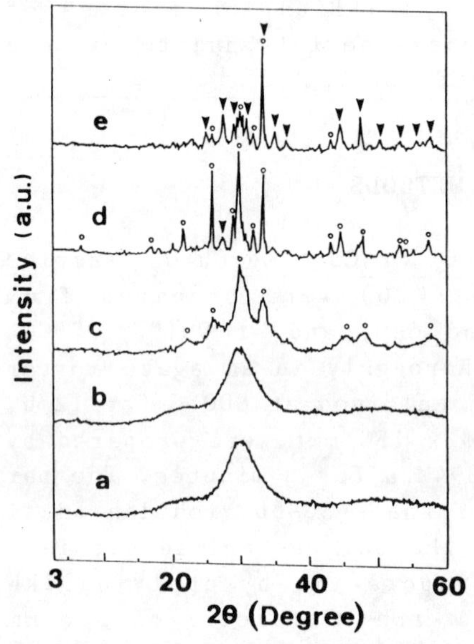

Figure 1. The X-ray powder diffracton patterns of the $Bi_2Sr_2CaCu_2O_z$ films.
(a)as-prepared amorphous film,
(b)at 460°C quenched,
(c)at 490°C quenched,
(d)at 790°C quenched,
(e)at 800°C quenched.
▼ : c_o=3.0 nm phase,
o : c_o=2.5 nm phase.

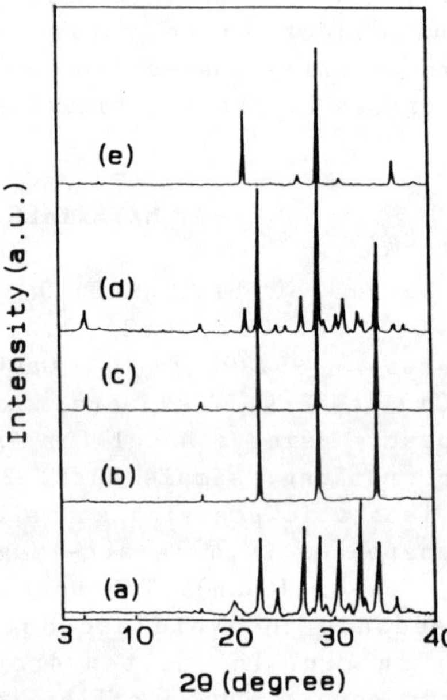

Figure 2. The X-ray powder diffraction patterns of the $Bi_2Sr_2Ca_{2-x}Cd_xCu_3O_y$ films after annealing at 870°C for 5h:
(a)x=0.0 full scale 0.8 kcps,
(b)x=0.2 full scale 10 kcps,
(c)x=0.4 full scale 40 kcps,
(d)x=0.5 full scale 4 kcps,
(e)x=0.6 full scale 10 kcps.

Since the vaporized materials were identified to be α- and β-Bi_2O_3 by X-ray powder diffraction, the compositional change in molten $Bi_2Sr_2CaCu_2O_z$ is calculated to be 4.5×10^{-3} mol (in maximum) of Bi_2O_3/sec[1].

Figure 1 shows the crystallization process[1] from the amorphous materials to $Bi_2Sr_2CaCu_2O_z$ above 800°C through the $Bi_2Sr_2CuO_z$ phase crystallized above 460°C. Reheating of the complete amorphous materials yielded amorphous films at 460°C(Fig.1(b)), the c_o=2.5 nm phase at 490°C(Fig.1(c)) or above up to 790°C(Fig.1(d)), and then the c_o=3.0 nm phase above 800°C, as seen in Fig.1(e).

In $Bi_2Sr_2Ca_{2-x}Cd_xCu_3O_y$ composition[3-6], annealed samples yielded the formation of the c_o=3.0 nm phase until x=0.4 and emerged the c_o=2.5 nm phase from x=0.5, and then the sample with x>0.6 changed into the c_o=2.5 nm phase(Fig.2(e)). Figure 2 shows the extent of orientation improved with increasing Cd content. The sample with x=0.4 shows the best-oriented structure[4]. The $Bi_2Sr_2Ca_2Cu_3O_y$(x=0.0) sample has the lattice parameters of a_o=0.54 nm and c_o=3.08 nm as a pseudo-tetragonal lattice. With increasing Cd content, the structure changed from a pseudo-tetragonal to the orthorhombic and the cell volume increased as plotted in Fig.3.

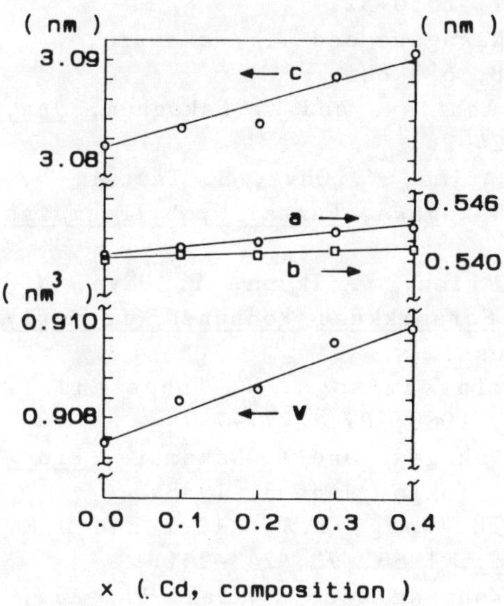

Figure 3. Lattice parameters of a, b and c axes and volume of c_o=3.0 nm phase as a function of Cd content.

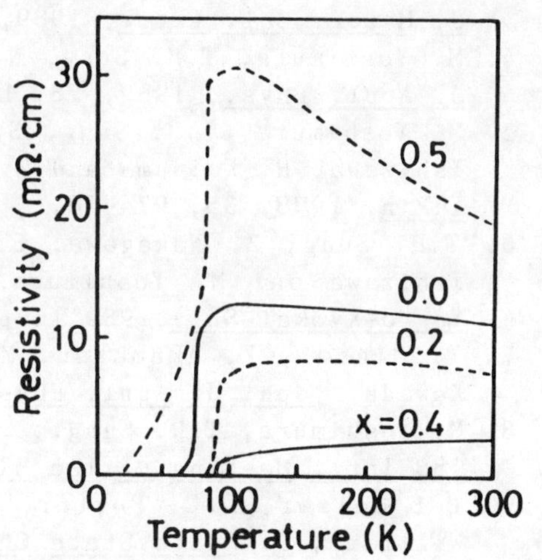

Figure 4. Temperature dependence of the $Bi_2Sr_2Ca_{2-x}Cd_xCu_3O_y$ films after annealing at 870°C for 5h.

These results suggest that Ca was substituted by Cd or the certain alteration took place in cation distribution of Bi, Sr and Ca[4].

Figure 4 shows the resistivity-temperature curves for the samples with x=0.0 to 0.5. With increasing Cd content from x=0.0 to 0.4, the resistivity at room temperature decreased and tailing of transition vanished. $Bi_2Sr_2Ca_2Cu_3O_y$(x=0.0) film showed Tc(endpoint)=67 K with a tail. Tc(endpoint) gradually increased and exhibited the highest Tc(endpoint)=87 K without a tail on the sample with x=0.4. However, on the behaviors of the sample with x=0.5, the resistivity at room temperature increased and Tc(onset) decreased as well as Tc(endpoint) with tail. These results clearly indicate that Bi-containing superconductor improved the superconductivity , such as to be raised up to T_{ce}=87K with c_o=3.0 nm phase as well as to be well oriented by doping of Cd in $Bi_2Sr_2Ca_2Cu_3O_y$.

REFERENCES

1. M. Yoshimura, T.H. Sung, Z. Nakagawa and T. Nakamura, Jpn. J. Appl. Phys.,1988, **27**, L1877-79.
2. T. Komatsu, R. Sato, K. Imai, K. Matusita and T. Yamashita, Jpn. J. Appl. Phys., 1988, **27**, L550-52.
3. M. Yoshimura, T.H. Sung, Z. Nakagawa and T. Nakamura, J. Mater. Sci. Lett., 1989, **8**, 687-88.
4. M. Yoshimura, T.H. Sung, N. Ishizawa and Z. Nakagawa, Jpn. J. Appl. Phys., 1989, **28**, L424-26.
5. M. Yoshimura, T.H. Sung, S. Arima, Y. Ohya, S. Tsutai, N. Ishizawa, H. Koinuma and Z. Nakagawa, Funtai Oyobi Funmatsu Yakin, 1988, **35**, 974-78.
6. T.H. Sung, Z. Nakagawa, S. Arima, Y. Ikuma, Y. Ohya, N. Ishizawa and M. Yoshimura, Seramikkusu Ronbunshi(formerly Yogyo-Kyokai-Shi),1989 in press.
7. T. Minami, Y. Akamatsu, M. Tatsumisago, N. Tohge and Y. Kowada, Jpn. J. Appl. Phys., 1988, **27**, L777-78.
8. M. Yoshimura, T.H. Sung, Z. Nakagawa and T. Nakamura, Proc. '88 Intl. Meeting of The MRS, Tokyo, June 3, 1988.
9. URS Geiser, L.C. Porter, H.H. Wang, T.A. Allen and J.M. Williams, J. Solid State Chem., 1988, **73**, 243-251.
10. L.P. Cook, C.K. Chiang, W, Wong-Ng and J. Blendell, Adv. Ceram. Mater., 1987, **2** [3B], 656-659.
11. M. Yoshimura, M. Kaneko and S. Somiya, J. Mater. Sci. Lett., 1985, **4**, 1082-84.

CRITICAL CURRENT DENSITY OF YBCO CYLINDRICAL SAMPLES. EFFECT OF THE RADIUS AND THE MICROSTRUCTURE

René KORMANN, Francois LAINEE, Jean-Pierre GANNE, Bruno LLORET*
Ceramics Department, Central Research Laboratory
Thomson-CSF, Domaine de Corbeville, 91404 Orsay, France

*LCC, 74 Route de Savigny, 21201 Beaune, France

ABSTRACT

The critical current density J_c was measured on samples having different radius and microstructures. J_c decreases as the radius increases and is dependent on microstructure defects such as second phases and micro-cracks. This behaviour is explained using a model of weak links.

INTRODUCTION

New high T_c superconducting ceramics behave as granular supraconductors. In order to know the intrinsic limits of the transport properties of ceramics and the effect of the microstructure features on those limits, we have measured the transport critical current on different samples :
1. Rods having the same microstructure but different radius
2. Rods having the same shape but different microstructures.

SAMPLE PREPARATION

These samples were prepared using two kinds of YBCO powders.
C : a chemical route powder provided by Rhône Poulenc
MO : a mixed oxide route powder : prepared in Thomson Laboratory by mixing the oxides in stoechiometric proportions, firing and grinding.

The first set of samples was shaped using the chemical route powder. This powder was isostatically cold pressed and machined to the desired radius before firing. The firing temperature was 940°C for 16 h. The second set was made using the two kinds of powder : bars of 5 x 5 x 50 mm were cold pressed at the same pressure. Three sintering temperatures were tried 910, 925 and 950°C. We obtained 6 samples of different microstructures (C 910, C 925, C 950, MO 910, MO 925, MO 950). The samples were cut into 1.2 x 1.2 x 34 mm. bars. The microstructure was characterized by SEM and optical microscope and is reported elsewhere (1).

CRITICAL CURRENT DENSITY MEASUREMENT TECHNIQUES

Four point critical current measurements were made using 10 ms current pulses (fig.1).

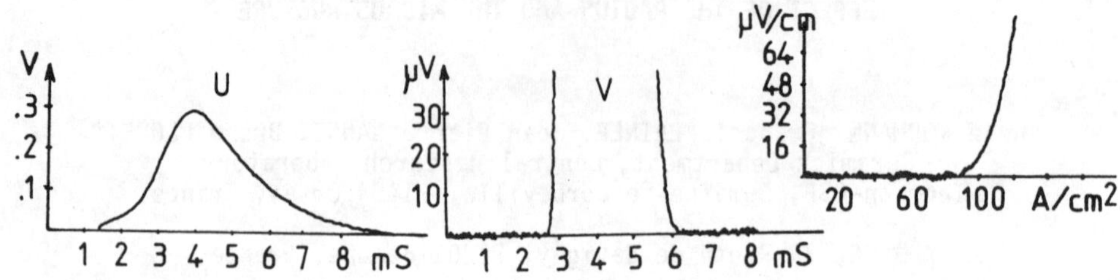

Figure 1.

The voltage appearing across the reference resistance and the superconductor when a current pulse is flowing through them is recorded with a numerical oscilloscope. These two sets of values allow us to know the E = f (J) relation. A control of the absence of heating was made by comparing the two values of J_c obtained in each experiment, one at the rising side of the pulse, the other at the falling side. This test always proves satisfactory for the samples tested. Four indium contacts were ultrasonically soldered on each sample, providing contact resistance less than 10 mΩ.

Due to various causes of noise, the critical current is defined as the current for which the electrical field measured across the sample is less than 5 µV/cm.

RELATION BETWEEN THE CRITICAL CURRENT AND THE RADIUS OF THE SAMPLE

The critical current was measured on all samples of the first set. After the measurements they were cut in order to check the homogeneity of the microstructure. The results of critical transport current are reported fig. 2.

The J_c value decrease when the radius r of the sample increases. A plot of the product J_c x r / 2 vs : r (J_c xr/2 is the calculated magnetic field at the surface of the rod) show a constant value for large radius and a decreasing to smaller one (fig.3). This behaviour can be fairly well explained by using a model of weak links granular superconductor. In this approach the material is represented as superconducting grains coupled by non superconducting grain boundaries. Each of these weak links behaves like a Josephson junction.

The behaviour of an array of junctions has already been studied by Peterson and Ekin (2, 3) and a law between J and H for the new

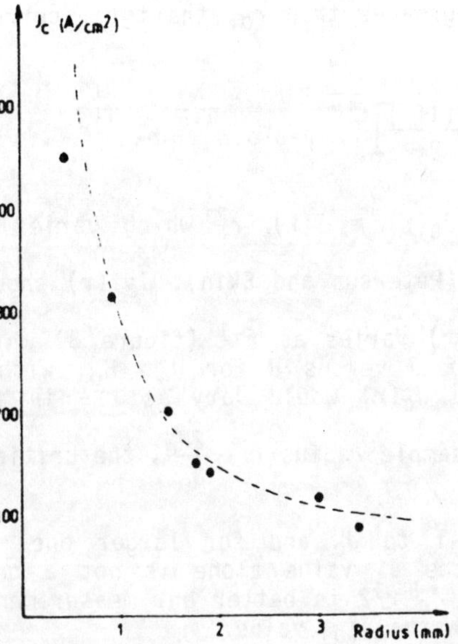

Figure 2.

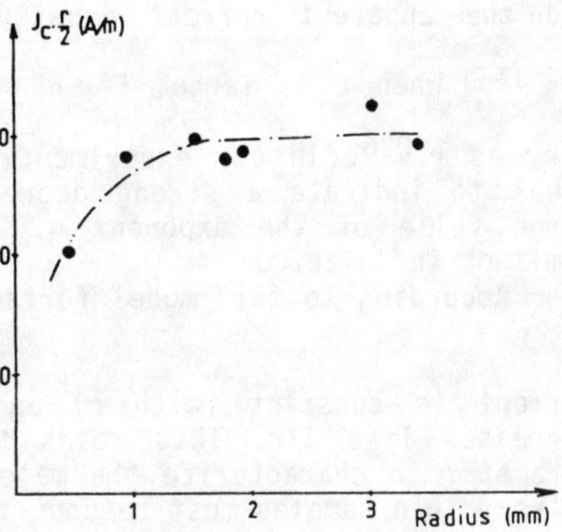

Figure 3.

superconductors has been found.

$$J_c\ (H) = J_c\ (0) \text{ for } H < H_c = 0.3\ H_0$$

$$J_c\ (H) = J_c\ (0) \times 0.15\ (H/Ho)^{-1.5} \text{ for } H > 0.3\ Ho$$

If we represent the J(H) relations by the following formulae, which are similar to Peterson and Ekin's results :

$J(H) = J_0$ for $H < H_0$

$J(H) = J_0\ (H/H_0)^{-\alpha}$ where α is a positive real number for $H > H_0$.

We can calculate the current distribution $J(x)$ inside the sample, the total current $I(r)$ in a sample of radius r, and the apparent critical current density $J_c(r) = I(r)/\ r^2$, by solving the equation given by Ampere's theorem :

$$rJ(r) = \frac{d}{dr}\ (rH(r))$$

The results are :

$J(x) = J_0$ if $x < r_0$, where r_0 is the radius where H becomes equal to H_0 : $r_0 = 2\ H_0/J_0$

$$J(x) = J_0\ \left[2(1+\alpha)/\ (2+\alpha)\ \right]^{-\frac{\alpha}{1+\alpha}}\ x^{-\frac{\alpha}{1+\alpha}} \text{ if } x > r_0$$

If the radius r of the sample is greater than r_o, the total current in the sample is :

$$I(r) = \pi r_o^2 J_o \left[1 - \frac{2(1-\alpha)}{2+\alpha}\right]^{\frac{1}{1+\alpha}} + \left[\frac{2(1+\alpha)}{2+\alpha}\right]^{\frac{1}{1+\alpha}} \pi J_o r_o^{\frac{\alpha}{1+\alpha}} r^{\frac{2+\alpha}{1+\alpha}}$$

and the apparent current density is $J_c(r) = I(r)/r^2$ which varies as $r^{-\frac{\alpha}{1+\alpha}}$ when r is large. For $\alpha \sim 1.5$ (Peterson and Ekin), $J_c(r)$ should vary as $r^{-0.6}$. In our experiment, $j_c(r)$ varies as r^{-1} (figure 3) which tends to indicate a strong decrease of J versus H for $H > H_o$, with a large value of the exponent α. Indeed $J_c(r)$ would vary as r^{-1} in the limit of infinite α.

According to this model for small sample radius $r < \frac{2H_o}{J_o}$, the critical current is constant (with r) and equal to J_o and for larger one, it decreases like 1/r. This means that the J_c value alone is not a good parameter to characterize the material, J_c r/2 is better but measurements on very thin samples must be done to know the $J_{(o)}$ value.

RELATION BETWEEN THE CRITICAL CURRENT AND THE MICROSTRUCTURE

The results for the six samples at the second set are reported figure 4.

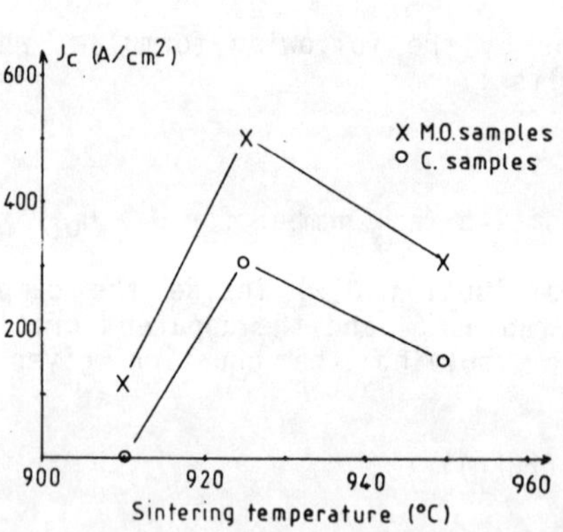

Figure 4.

All the samples but C 910 are superconducting in the bulk at 77 K and all have the same section which, as we saw, is important to compare the intrinsic electrical properties. The results can be interpreted in the following way : in ceramics the coupling between grains is determined by the sintering and the oxygenation, as grains boundaries take their

final characteristics at this stage of the process. According to the theory of weak coupling, various microstructural characteristics can improve or degrade the quality of coupling between grains. These characteristics may be classified in two groups :
- those which decrease the measured value of J_c by decreasing the effective of conducting volume : porosity and microcrack
- those which influence the value of H_0 or $J_{(0)}$ ie the grain size, the oxygen content, the cristallographic relations, the intragranular films...

The lower value obtained for C 910 and MO 910 is probably due to the amount of unreacted materials which decrease for higher sintering temperature (ie for C 925 and MO 925). Microcracks appearing at grains boundaries of big grains and due to the temperature expansion mismatch (4) are undoubtely responsible for the lower J_c value of C 950 and MO 950 (more details on these microstructures are reported in 1).

CONCLUSION

The critical current was measured on different samples varying only by their diameter or their microstructure. The results show that :
- the critical current density decreases as the inverse of the radius for thick enough samples.
- the highest values of critical current density were obtained with a sintering temperature at 925°C.
- the lower values obtained for lower sintering temperature may be due to unreacted phase and for higher sintering temperature to the presence of microcracks.

All these results can be interpreted with the weak links model. Accordingly the measured critical current depends on three parameters :
- the radius of the sample
- the critical magnetic field H_0
- the critical current $J_{(0)}$.

REFERENCES

1. Lainée F., Kormann R., Lloret B., Labeyrie M. and Ganne J.P., To appear in proceedings of the ECS Meeting, Maastricht June 1989.

2. Peterson R.L. and Ekin J.W., Physica C. 157, 325-333 (1989).

3. Peterson R.L. and Ekin J.W., Phys. Rev. B, 37, 9848, (1988).

EFFECT OF SCANDIUM ON SUPERCONDUCTIVE PROPERTIES OF THE Y(Sc)-Ba-Cu^{-} OXIDES

SHATALOV V.V., BUCHIKHIN E.P., KOSYNKIN V.D., KUSJ V.E., SYCHEVA V.J., CHERNOPLEKOV S.M., IONOV R.A., BASHUN V.N., SINCHENKO A.A., EVSTIGNEEV V.V., ZHABREV G.I.

All-Union Research Institute of Chemical Technology, Kashirskoe shosse, 33, 115409 Moscow, USSR

ABSTRACT

Using the method of solid phase synthesis we have succeeded in creating Sc-containing volumetrical superconductive ceramics with composition approaching "1-2-3". The structure and phase composition of the products with various Y:Sc ratio have been studied. The values of T_c and critical current densities, obtained at various sintering temperatures, have been measured.

The present information on Y-Sc-containing ceramics does not supply any clear conception about the effect of the rare earth element substitute - scandium - on such important parameters as the critical current density or transition range into the superconductive state. The phases formed by raising Sc-contents in $Y_{0.4-x}Sc_xBa_{0.6}CuO_y$ samples at x>0.325 have not been identified, the influence of the synthesis conditions - temperature, the rate of heating and cooling, etc. - on the superconductive properties of samples has not yet been investigated.

The paper presented aims at a systematized study considering the effect of the synthesis conditions of the ceramics with $(Y_{1-x}Sc_x)_yBa_{1-y}CuO_z$ (y=0.4) composition on their structure and superconductive properties. A special task has been set to investigate the region with a small scandium content.

METHODS

The oxides with $Y_{0.4-x}Sc_xBa_{0.6}CuO_z$ composition were obtained via a solid phase reaction. The x values varied from 0 to 0.2. The initial compounds Y_2O_3 (99.99%), $BaCO_3$ (99.5%), CuO (99.11%) and Sc_2O_3 (99.9%) were taken accordingly to the stoichiometry of the final product. The ground mixture was sintered at temperature 900°C for 4-4.5 hours with subsequent gradual cooling in an oxygen current.

The sintered material was reground and pressed into pellets 9 x (0.9-3) mm under pressure 10^3 kg/cm^2.

The pellets were sintered at temperature 900-950°C for 2-4 hours. The cooling up to the room temperature proceeded with the rate 6-7°C per min. in an oxygen current. The sintered pellets were used to measure the critical temperatures (T_c) individually. The j_c value was measured by a standard four-contact method.

X-RAY STUDIES

The X-ray studies were conducted on a DRON-2 diffractometer using the filtered copper emission with voltage 30 kV and current 20 mA. The travel speed of the counter with a continuous recording of the diffractional picture on a diagram band within the angle interval 2θ = 20-30 grades was 1 grade per minute. In view to raise sensitivity the diffractogram parts 2θ = 27.2-31.2° were recorded three times for each product in the regime of a step-wise motion with a step Δ2θ = 0.05° and an exposition for 20 seconds for every point.

The data obtained for the pellets after the annealing at 950°C for 2 hours and grinding are presented in Table 1 and Figure 1 . The pellets compositions were: $Y_{0.4-x}Sc_xBa_{0.6}CuO_y$ where x = 0.02; 0.04; 0.06; 0.08; 0.4. In accordance to the X-ray study the phase approaching $YBa_2Cu_3O_x$ with an ortho-

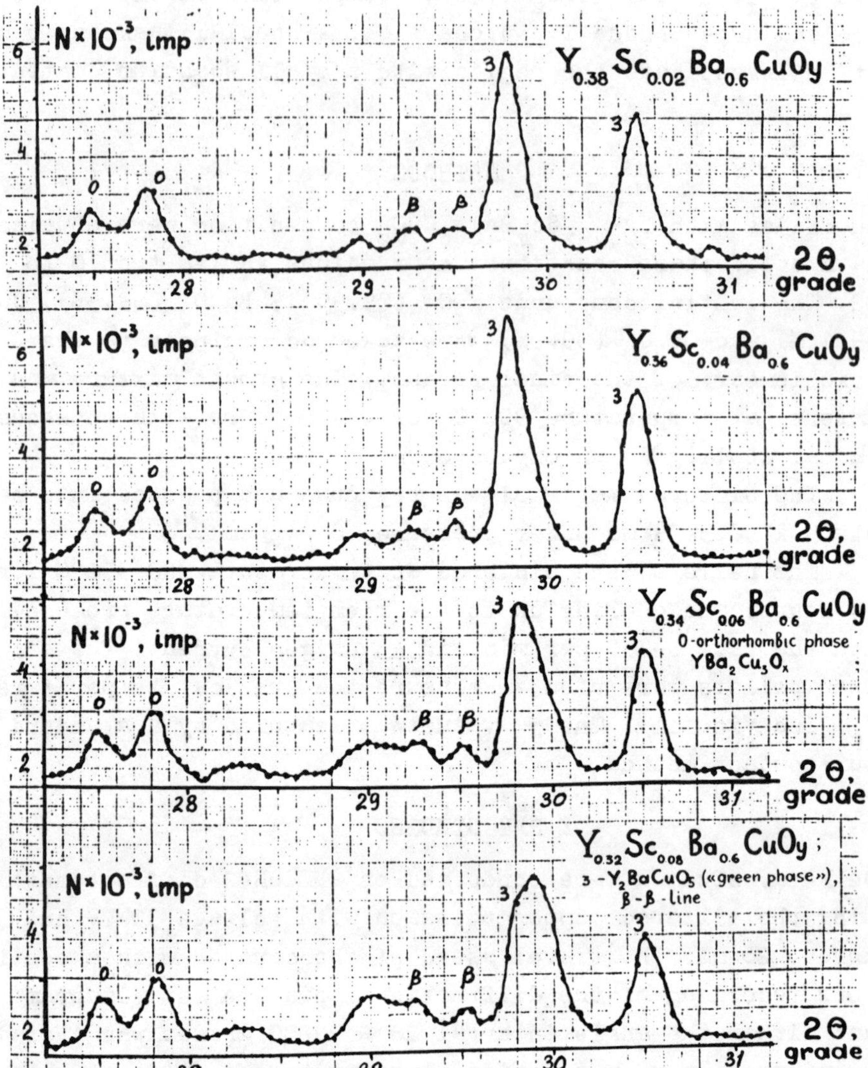

Figure 1. Diffractograms fractions of $Y_{0.4-x}Sc_xBa_{0.6}CuO_y$ pellets

rhombic structure comprises he base of the products with small Sc contents.

The parameters of an elementary cell of this phase, whithin the limits of experimental error, do not differ from the parameters of a compound without Sc. When changing the Sc contents within the interval Sc/(Y+Sc) = 0.05-0.20 the intensities of the diffraction lines did not change (Figure 1).

Besides the basic phase there are certain amount of Y_2BaCuO_5 and CuO present in the samples. A slite shift of diffraction lines of Y_2BaCuO_5 seems to depend on a partial replacement of Y by Sc, but it has to be specially proved. The intensities of diffractional lines of Y_2BaCuO_5 decrease with the raising contents of Sc in the product up to x = 0.06-0.08. At the same time the weak lines grow stronger, they might belong to the composition $Sc_{0.4}Ba_{0.6}CuO_y$ (Table 1).

TABLE 1

X-ray study data of the product with $Sc_{0.4}Ba_{0.6}CuO_y$ composit.

2θ, grade	I	d x 10, nm	2θ, grade	I	d x 10, nm
21.13	20	4.205	36.61	2	2.455
21.60	3	4.114	37.05	9	2.426
22.76	4	3.907	37.40	4	2.404
23.40	3	3.802	38.75	22	2.324
23.69	14	3.756	39.39	8	2.287
24.00	2	3.708	40.11	11	2.248
24.71	3	3.603	40.42	5	2.232
25.96	10	3.432	40.97	3	2.203
28.38	21	3.145	41.80	8	2.161
29.16	44	3.062	42.39	11	2.132
30.04	100	2.975	43.04	62	2.102
31.96	40	2.800	44.44	10	2.039
32.50	6	2.755	45.15	7	2.008
32.83	21	2.728	47.08	4	1.930
33.90	7	2.644	47.75	10	1.905
34.17	5	2.624	48.50	6	1.877
34.68	5	2.587	48.89	6	1.863
35.11	5	2.556	49.19	5	1.852
35.57	18	2.524	49.71	10	1.834
35.96	6	2.497			

The annealed products without Y of the above composition have quite different diffraction picture as that of the compound without Sc or with a small amount of it.

MEASURING T_c AND j_c VALUES

The data obtained when measuring the temperature of the Sc-containing ceramics transiting into the superconductive state as well as the densities of the critical currents (77K) are given in Table 2. The samples have been manufactured by annealing the pellets under various conditions.

TABLE 2

Properties of $Y_{0.4-x}Sc_xBa_{0.6}CuO_y$ samples conditioned by annealing temperatures of pellets

No.	Annealing temperatures, °C	Parameter	X values						
			0	0.02	0.04	0.06	0.08	0.12	0.20
1	2	3	4	5	6	7	8	9	10
1.	900	$T_{c\ in}$, K	87.0	non-conductive phases					
		$T_{c\ R=0}$, K	63.0						
		j_c, A/cm^2	8.4						
2.	930	$T_{c\ in}$, K	87.9	88.2	88.2	88.2	-	-	-
		$T_{c\ R=0}$, K	65.8	82.7	80.0	76.0	-	-	-
		j_c, A/cm^2	-	-	-	-	-	-	-
3.	950	$T_{c\ in}$, K	89.0	90.7	90.7	90.7	90.7	89.0	89.0
		$T_{c\ R=0}$, K	80.2	88.7	88.6	88.3	88.9	85.6	85.7
		j_c, A/cm^2	39	-	37	47	49	54	8.9

Thus, introduction of small amounts of scandium into ceramics of $Y_{0.4}Ba_{0.6}CuO_y$ composition considerably increases $T_{c\ R=0}$ value and has little effect on the initial temperature of transition into the superconductive state and the density of the critical current. Improvement of the superconductive properties of Y-Sc-containing ceramics depends on the optimal conditions for the mixture and pellets annealing. The research along these lines is still going on.

DOPANT EFFECTS ON THE SUPERCONDUCTIVITY OF THE Bi-Sr-Ca-Cu-O SYSTEM

TSUNEYUKI KANAI, TOMOICHI KAMO and SHIN-PEI MATSUDA
Hitachi Research Laboratory, Hitachi Ltd., Hitachi, Ibaraki 317, JAPAN

ABSTRACT

The effects of 34 different dopants on the $BiSrCaCu_2O_y$ ceramics were investigated. The dopants can be classified into four categories. Dopants in the first group, i.e. Fe, Co, Ni and Zn, dissolve and substitute into the copper site and can significantly reduce the transition temperatures (T_c's) of the high- and low-T_c phase. The second group of dopants, i.e. rare-earth elements, substitute into the Ca sites and cause the disappearance of the high-T_c phase. The dopants in the third group, i.e. alkaline metals and phosphorus, have a strong tendency to decompose the superconducting phase. The remainder of the dopants investigated were classified as the fourth group. Superconductivity of the high- and low-T_c phase of the group remains almost uneffected.

INTRODUCTION

Many studies have reported on the high-T_c superconducting oxides since the discovery of the 40K-class superconductor of the La-Ba(Sr)-Cu-O system [1,2] and 90K-class superconductor of the Y-Ba-Cu-O system [3]. This was followed by the discovery of superconductivity in the Bi-Sr-Ca-Cu-O system [4] and Tl-Ba-Ca-Cu-O system [5] with T_c's above 100K.

The Bi-Sr-Ca-Cu-O system contains two main superconducting phases, i.e. the high-T_c phase (105K) with the composition of $Bi_2Sr_2Ca_2Cu_3O_y$ and the low-T_c phase (75K) with the composition of $Bi_2Sr_2Ca_1Cu_2O_y$. It has been reported that incorporation of Pb ions [6] into the system causes a higher volume fraction of the high-T_c phase. This suggests that the addition of foreign element to the Bi-Sr-Ca-Cu-O system could change the free energy of formation of the high- and low-T_c phase.

In this paper, the effects of 34 different dopants on the $BiSrCaCu_2O_y$ ceramics were investigated to clarify the phase stability effect by the dopants on the high- and low-T_c phase and to explore possible improvements in T_c.

EXPERIMENTAL PROCEDURE

Thirty-four different doped samples were prepared with the nominal composition of $BiSrCaCu_2M_xO_y$ by a solid-state reaction in air. The starting materials were 3N purity powders of Bi_2O_3, $SrCO_3$, $CaCO_3$, CuO and the dopants. The powders were mixed and calcined at 800°C for 5h in air. The calcined powders were pulverized, pressed into pellets and sintered at 860°C for 5-10h in air. Crystal phases of the samples were examined by X-ray diffraction analysis using Cu-Kα radiation. Electrical properties

TABLE I. Effects of the dopants on the T_c's of $BiSrCaCu_2M_{0.05}O_y$ ceramics sintered at 860°C for 5 h in air. The T_c^H and T_c^L represent the onset of superconductivity in the high- and low-T_c phases, respectively.

Group	Dopant	Ionic	Radius(Å)	Tc^H (K)	Tc^L (K)
	undoped			116	78
	Co	3+	0.63	103	64
	Fe	3+	0.64	107	59
I	Ni	2+	0.68	110	74
	Zn	2+	0.74	112	73
	Yb	3+	0.86	—	90
	Ce	4+	0.90	—	84
II	Y	3+	0.92	—	93
	Sm	3+	1.00	—	89
	La	3+	1.14	—	85
	P	5+	0.35	—	90
	Li	1+	0.68	—	89
III	K	1+	1.33	—	86
	Rb	1+	1.47	—	86
	B	3+	0.23	116	82
	Be	2+	0.35	118	82
IV	Si	4+	0.42	115	79
	Ge	4+	0.53	116	79
	V	5+	0.59	116	77
	Mo	6+	0.62	117	78
	W	6+	0.62	114	77
	Ga	3+	0.62	115	83
	Cr	3+	0.63	116	78
	Mg	2+	0.66	119	74
	Ta	5+	0.68	115	80
	Ti	4+	0.68	116	79
IV	Nb	5+	0.69	112	73
	Sn	4+	0.71	114	77
	Sb	3+	0.76	113	79
	Hf	4+	0.78	114	78
	Zr	4+	0.79	116	78
	Mn	2+	0.80	116	78
	Sc	3+	0.81	113	75
	Cd	2+	0.97	117	81
	Ba	2+	1.34	114	75

were measured by a standard four-probe method.

RESULTS AND DISCUSSION

The effects of 34 different dopants on the superconductivity of the Bi-Sr-Ca-Cu-O system were investigated. Table I summarizes the dopant effects on the T_c's of the high- and low-T_c phase in the $BiSrCaCu_2M_{0.05}O_y$ ceramics. The dopants were classified into four groups.

Dopants in the first group are Co, Fe, Ni and Zn. Figure 1 shows the effect of the dopant concentration of the first group on the T_c onset. The dopants belonging to this group significantly reduced the T_c's of the low-T_c phase, but the manner and extent of the T_c reduction in each dopant differed. The addition of Fe or Co reduced the T_c's of the low-T_c phase to a greater extent than that of Ni or Zn. The EDS analysis revealed that the actual dopant level of the low-T_c pahse was far less than that of nominal compositions, particulary in the Ni or Zn samples. This may be one of the reasons why Ni or Zn doping has less of an effect on the T_c's compared to Fe or Co doping.

Figure 2 shows the temperature dependence of the resistivity of the Fe-doped samples. The curves show drastic changes on increasing the doping level of Fe ions; i.e., the disappearance of the high-T_c phase occurs over 4 mol%, while the disappearance of the semiconducting behavior occurs above 10 mol%. In the case of Co doping, the disappearance of the high-T_c phase is observed above 5mol% of doping. X-ray analysis revealed that the diffraction patterns change significantly by increasing the amount of Fe doped; i.e., the 2212 phase occurs under a 10 mol% dopant level, the appearance of the 2201 phase together with the 2212 phase

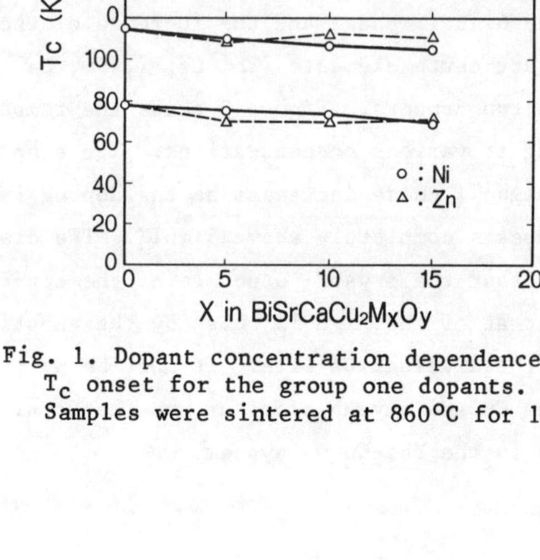

Fig. 1. Dopant concentration dependence of T_c onset for the group one dopants. Samples were sintered at 860°C for 10 h.

occurs around the 10 mol% level and the disappearance of the 2212 phase occurs at the 20 mol% level. A similar tendency is observed in the Co-doped samples.

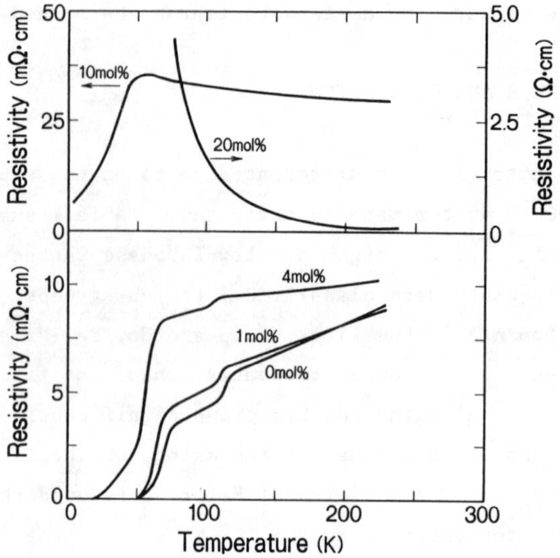

Fig. 2. Temperature dependence of the resistivity for the various Fe concentrations. Samples were sintered at 860°C for 10 h in air.

There are three crystal structures in Bi-Sr-Ca-Cu-O system. The configuration of the oxygen atoms around a Cu atom differs in each crystal, i.e. a 6 coordination number in the 2201 phase, a 5 coordination number in the 2212 phase, a 4 and 5 coordination number in the 2223 phase. The Fe or Co atoms are considered to substitute into the Cu site and prefer the octahedral sites. Thus, the substitution by these atoms into the Cu sites is more probable in the crystal structures containing higher coordination numbers in the Cu sites. This is considered to be one reason why the resistivity-temperature curves show the disappearance of the high- and low-T_c phase followed by the semiconducting behavior with the Fe and Co addition.

The dopants comprising the second group give rise to the disappearance of the high-T_c phase and the increase of the T_c's in the low-T_c phase. All the rare-earth elements (Yb, Ce, Y, Sm, La) investigated are classified as group two dopants. Figure 3 shows the temperature-resistivity curves for Y doping at various concentrations. The step wise change in the curves of the high-T_c phase decreases as the doping level increases, and the change disappears completely above 5 mol%. The disappearance of the high-T_c phase means that the crystal stucture of the low-T_c phase becomes more stable than that of the high-T_c phase by the substitution of the trivalent Y atoms into the divalent Ca site. It must be pointed out that the configuration of the Cu and oxygen atoms around the Ca site is identical to that of the Y sites in the $YBa_2Cu_3O_7$ system.

The dopants in the third group (P, Li, K and Rb) were characterized by the disappearance of the high-T_c phase and the increase of the T_c's of the low-T_c phase by 8-10 K. The dopants in this group have a tendency to decompose the superconducting phases, especially the high-T_c phase.

The remainder of the dopants investigated were classified as the fourth group. The group was characterized by the limited solubility and shows little reactivity in the Bi-Sr-Ca-Cu-O ceramics.

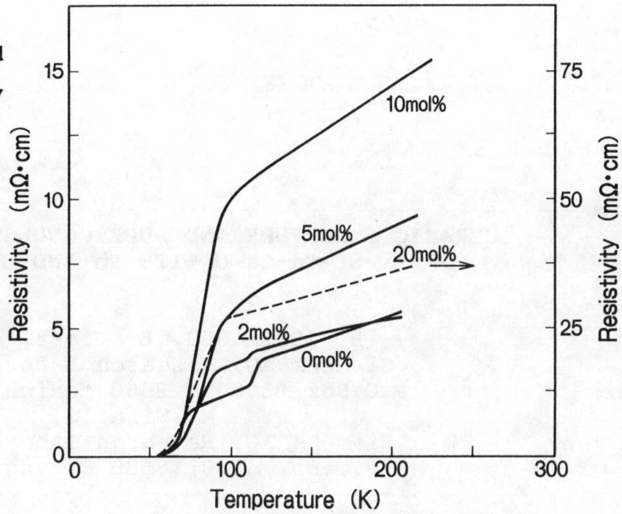

Fig. 3. Temperature dependence of the resistivity for the various Y concentrations. Samples are sintered at 860°C for 10 h.

REFERENCES

1. J. G. Bednorz and K. A. Muller: Z. Phys. B64(1986) 189.
2. S. Uchida, H. Takagi, S. Tanaka and K Kitazawa: Jpn. J. Appl. Phys. 26(1987) L1.
3. M. K. Wu, J. R. Ashburn, C. J. Torng, P. H. Hor, R. L. Meng, L. Gao, Z. J. Haung, Y. Q. Wang and C. W. Chu: Phys. Rev. Lett. 58(1987) 909.
4. H. Maeda, Y. Tanaka, M. Fukutomi and T. Asano: Jpn. J. Appl. Phys. 27(1988) L209.
5. R. M. Hazen, L. W. Finger, R. J. Angle, C. T. Prewitt, N. L. Ross, C. G. Hadidacos, P. J. Heaney, D. R. Veblen, Z. Z. Sheng, A. El Ali and A. M. Hermann: Phys. Rev. Lett. 60(1988) 1657.
6. M. Takano, J. Takada, K. Oda, H. Kitaguchi, Y. Miura, Y. Ikeda, Y. Tomii and H. Mazaki: Jpn. J. Appl. Phys. 27(1988) L1041.

CERAMIC STRUCTURE AND SUPERCONDUCTING PROPERTIES OF Bi-Sr-Ca-Cu-O WITH Pb AND Sb SUBSTITUTION

B. SEEBACHER, B. Jobst, G. Zorn
Siemens AG, Research Laboratories,
P.O.Box 830953, 8000 Munich 83, FRG
M. Wilhelm,
Siemens AG, Research Laboratories,
P.O.Box 3240, 8520 Erlangen, FRG

ABSTRACT

Ceramic specimens of Bi-Pb-Sb-Sr-Ca-Cu-O show superconductivity with zero resistivity between 30 K and 106 K, depending on sintering procedure and chemical composition.
For the nominal composition $Bi_{1.7}Pb_{0.4}Sr_{1.7}Ca_{2.1}Cu_3O_x$ almost single phase material with a structure corresponding to $Bi_2Sr_2CaCu_2O_8$ is obtained at low reaction temperatures as shown by XRD. With increasing reaction temperature phases corresponding to $Bi_2Sr_2CuO_6$ and $Bi_2Sr_2Ca_2Cu_3O_{10}$ are formed additionally. Diffractometry, resistance and a.c. inductance data demonstrate that a significant portion of the so-called 110 K phase is formed. For a substitution of Bi by .3 at.% Pb and .1 at.% Sb zero resistivity is obtained at 106 K. Microphotographs show that the occurrence of the 110 K phase is associated with a plate-like porous structure without shrinkage.

INTRODUCTION

Since the discovery of superconductivity in the system Bi-Sr-Ca-Cu-O three phases with the compositions $Bi_2Sr_2CuO_6$ (2201), $Bi_2Sr_2CaCu_2O_8$ (2212) and $Bi_2Sr_2Ca_2Cu_3O_{10}$ (2223) have been identified. They are characterized by the transition temperatures of 10 K, 80 K and 110 K, respectively [1]. The preparation of single phase material with zero resistivity at 110 K has been shown to be extremely complicated. Generally all three phases are present in bulk material. Investigations of substitutions for Bi have shown, that Pb and Sb facilitate the preparation of 2223 material [2,3].

We investigated the influence of various annealing procedures on superconducting properties, phase composition and microstructure of Pb and Sb substituted samples.

EXPERIMENTAL PROCEDURE

Samples with various starting compositions were prepared from oxide and carbonate powder reagents by conventional ceramic procedure or by co-precipitation. Pellets were sintered at 800 to 880 °C for 2 h to 100 h in an atmosphere with varied oxygen partial pressure. The samples were characterized by X-ray diffractometry (XRD), SEM microscopy, standard four-probe d.c. resistance and a.c. susceptibility measurements. Electrical contacts were made by sputtering of Au.

RESULTS

For Pb substitution almost single-phase material was obtained for the starting composition $Bi_{1.7}Pb_{0.4}Sr_{1.7}Ca_{2.1}Cu_3O_x$. For low annealing temperature XRD patterns revealed only the 2212 phase with a minor amount of CuO and of an unidentified phase. Zero resistivity was obtained at 62 K. With increasing annealing temperature phases corresponding to 2201 and 2223 were formed, additionally. At 843 °C for PO_2 = 0.075 bar or 860 °C for PO_2 = 0.2 bar almost single phase material with a composition corresponding to 2223 was formed (fig. 1). Nearly 16 % of the Bi atoms were replaced by Pb.

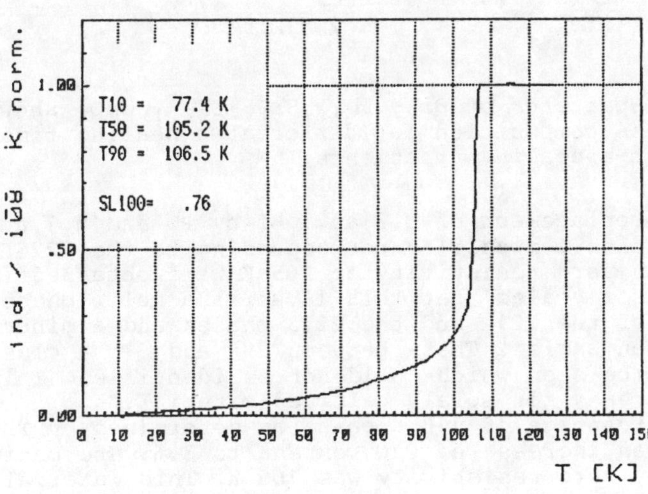

Fig. 1 Temp. dependence of susceptibility of $Bi_{1.7}Pb_{0.4}Sr_{1.7}Ca_{2.1}Cu_3O_x$.

Fig. 2 shows the forming of the 110 K phase dependent on annealing time. Post-annealing at 830 °C caused a decomposition whereas a subsequent post-annealing at 845 °C restored the starting XRD-pattern (fig. 2c-e). The occurrence of the 110 K phase was associated with a plate-like porous structure without shrinkage during annealing (fig. 3). A more dense structure was obtained at higher reaction temperatures. However, extensive decomposition of the superconducting phases, especially the 110 K phase, was then occurring.

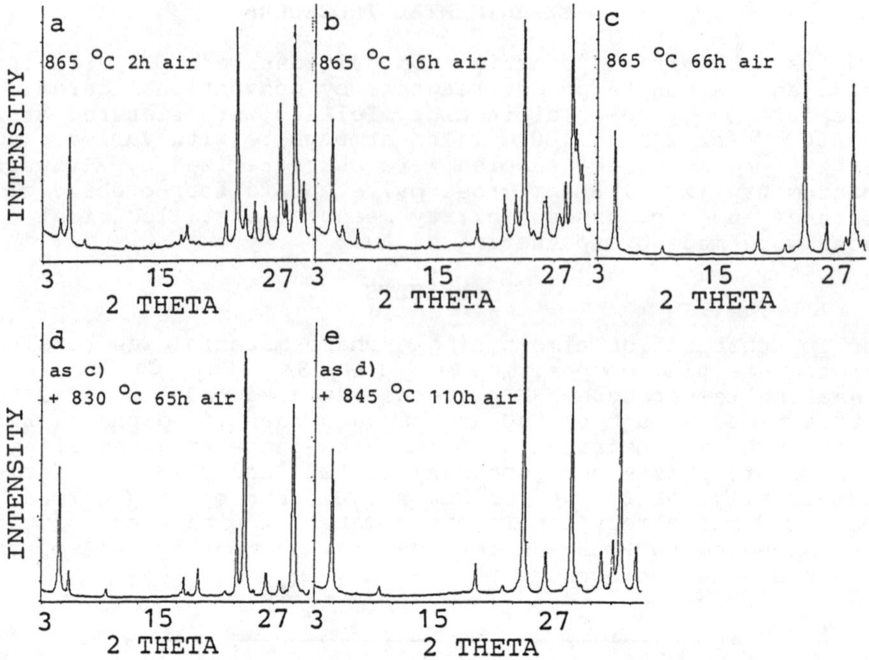

Fig. 2 XRD-patterns of $Bi_{1.7}Pb_{0.4}Sr_{1.7}Ca_{2.1}Cu_3O_x$ showing the phase composition for different annealing times and post-annealing treatments.

For a replacement of 0.3 at.% Bi by Pb and 0.1 at.% by Sb in the starting composition corresponding to the 2223 phase samples with zero resistivity at 105 K were obtained (fig. 4). XRD patterns revealed that this is still a multi-phase material consisting of the 2212 and the 2223 phases and a minor amount of CuO. At an angle 2 Theta between 29° and 31° a cluster of 4 peaks is to be seen which could not be identified. This cluster appeared already for solely calcined material.

The percentage of 110 K phase as determined by XRD is increased by an increase of CuO content to 3.6. The maximum temperature for zero resistivity was 106 K. This material contained only a minor portion of 2212 phase (about 10 %). Because of the increased Cu content the amount of CuO second phase was increased, too. An increase of Pb content lowered the temperature for zero resistivity to 93 K. The atmosphere during reaction was air for the stoichiometrical compositions whereas a reduced oxygen partial pressure of 0.1 bar was necessary for nonstoichiometrical compositions.

The ceramic structure of these materials is similar to that of materials with Pb substitution, only. The samples consisted of a network of loosely connected plate-like crystals.

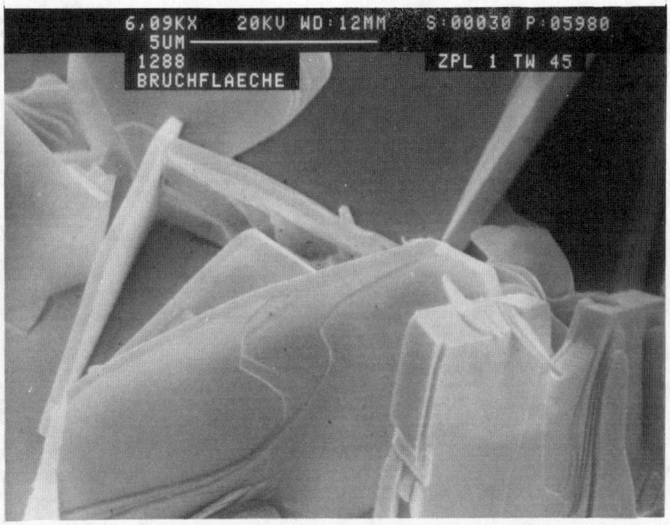

Fig. 3 SEM photograph showing the plate-like structure of $Bi_{1.7}Pb_{0.4}Sr_{1.7}Ca_{2.1}Cu_3O_x$ (U. SCHEITHAUER)

The volume of the samples was expanded by 50 % compared to that of the starting material. But the dimension of the crystalline plates were only half of those of Sb-free compositions (fig. 5)

The width of the temperature range for max. percentage of 110 K phase or max. temperature of zero resistivity amounted to ± 2 °C, only. An increase of annealing temperature by 10 °C more than the most favourable temperature caused extensive

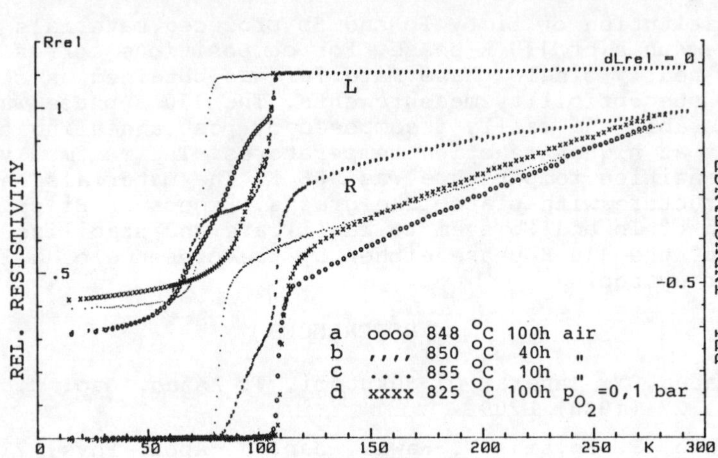

Fig. 4 Temp. dependence of resistivity and diamagnetic signal of $Bi_{1.7}Pb_{0.3}Sb_{0.1}Sr_{1.7}Ca_{2.1}Cu_3O_x$ (a-c) and $Bi_{1.7}Pb_{0.3}Sb_{0.1}Sr_{1.7}Ca_{2.1}Cu_{3.6}O_x$ (d) for different reaction procedures

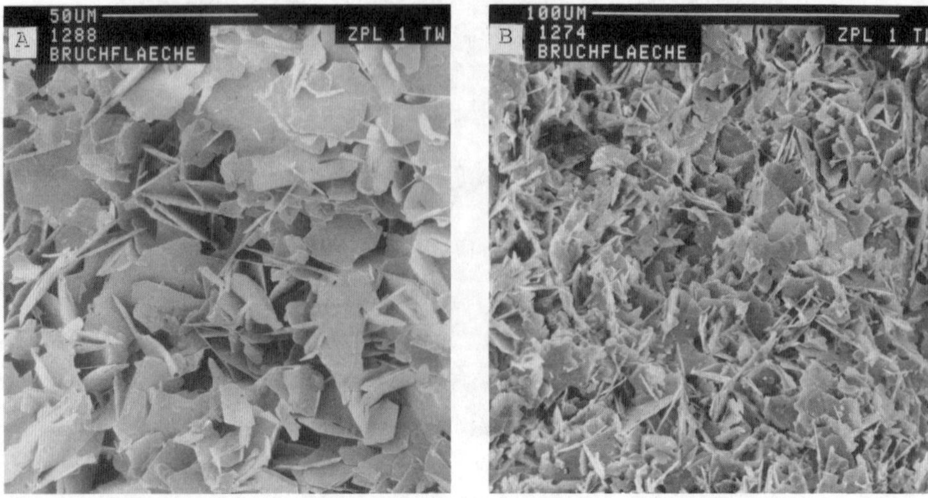

$Bi_{1.7}Pb_{0.4}Sr_{1.7}Ca_{2.1}Cu_3O_x$ $Bi_{1.7}Pb_{0.3}Sb_{0.1}Sr_{1.7}Ca_{2.1}Cu_3O_x$

Fig. 5 SEM photographs of fractured cross sections (U. SCHEITHAUER)

decomposition of the 110 K phase. The temperature for zero resistivity was lowered to nearly 80 K. The material was partially melted and showed an added content of second phase.

CONCLUSIONS

The substitution of Bi by Pb and Sb produced materials with higher amounts of 110 K phase. For compositions corresponding to 2223 nearly single phase material was obtained as shown by XRD and susceptibility measurements. The 110 K phase was thermally unstable and easily decomposed by post-annealing treatments or at higher reaction temperatures. The maximum value for the transition temperature was 106 K. The materials had a porous structure with plate-like grains. Excess of Bi or Cu and addition of Pb and Sb seem to facilitate and stabilize the formation of the 110 K phase either by the presence of a liquid phase or by doping.

REFERENCES

1. H. Maeda, Y. Tanaka, M. Fukutomi, T. Asano, Jap. J. Appl. Phys. 27 (1988) L209
2. U. Endo, S. Koyama, T. Kawai, Jap. J. Appl. Phys. 27 (1988) L1476
3. Liu Hongbao, Cao Liezhao, Zhou Ling, Mao Zhiqiang, Zhang Weijie, et.al, preprint
4. B. Seebacher, G. Schindler, Physica C 153-155 (1988) 615-6

POLYCRYSTALLINE SUBSTRATES FOR $YBa_2Cu_3O_{7-x}$ /YBaCuO FILMS

Hrnčíř K., Janda M., Skácel V., Koller A., Fiedlerová J.,
Research Institute of Electrotechnical Ceramics,
Hradec Králové, Czechoslovakia

Vlček M., Horák J., Institue of Chemical Technology,
Pardubice, Czechoslovakia

ABSTRACT

X-ray diffractometry /XRD/, optical and scanning electron microscopy /OM, SEM/, energy disspersive X-ray analysis /EDXA/ and measurements of thermal dilatation were used in order to characterize polycrystalline ceramic substrates of $SrTiO_3$, YSZ, Y_2BaCuO_5, Ba_2SiO_4, $BaZrO_3$ and $BaAl_2O_4$. Reactions of these compounds with YBaCuO powder between 400 °C and 900 °C were studied using measurements of the ac-susceptibility changes. It was found that barium containing substrate materials did not react so strongly with YBaCuO as $SrTiO_3$ and YSZ materials did. The most promissing substrate material for YBaCuO films seems to be Y_2BaCuO_5.

KEYWORDS : High-T_c superconductors, polycrystalline substrate, $YBa_2Cu_3O_{7-x}$, Y_2BaCuO_5

INTRODUCTION

It has been generally accepted that high-T_c YBaCuO superconductors might find their first practical applications in the form of thin or thick films. Properties of YBaCuO films, unlike bulk samples, are modified through their interactions with substrates on which superconducting films were deposited.

In this paper some basic facts about mechanical and chemical interactions between YBaCuO films and their substrates are summarized. Preliminary results on the preparation and characterization of different ceramic substrates are introduced. Some future tasks are outlined.

THEORETICAL

A film-substrate system is usually in the state of mechanical stress even if no external forces act on this system. Let us suppose for simplicity that the thickness of the film is much smaller than the substrate thickness and at the same time that the elasticity moduli of the film and the substrate are equal. The internal macrostress in the film is given by the equation.

$$\sigma = \sigma_i + \sigma_T \qquad (1)$$

where σ_i is the so called growth stress and σ_T is the thermal stress component.

The growth stress is usually, but not necessarity, independant on the material of the polycrystalline substrate.

On the contrary, thermal stress arises as the consequence of the difference between thermal expansivities of the film and the substrate, according to the equation

$$\sigma_T = E_F(\alpha_f - \alpha_s)\,(T_1 - T_0) \qquad (2)$$

where E_F is the thin film elasticity modulus $/E_F = E(1-\nu)^{-1}/$, α_f, α_s are thermal expansion coefficients of the film and substrate respectively, T_1 is temperature during the film deposition and T_0 is a service temperature. For YBaCuO films the T_0 value may be considered as 77 K.

Chemical interactions between YBaCuO films and substrates were found for all substrate materials studied until now. Papers were published in which practically identical substrates were used for YBaCuO films deposited by various mathods and different physical properties of superconducting films were found. From these findings one may conclude that the YBaCuO reactivity with the substrate is not only governed by the material combination (film-substrate) but also by the microchemistry of YBaCuO clusters.

Experiments have proved that Ba, most probably in the form of BaO, interacts with substrates most strongly /1/. There is also evidence of the presence of BaO on the surfaces of YBaCuO films /2/. From the view point of reactivity one might consider YBaCuO as a mixture oxide material consisting of Y_2O_3, BaO and CuO_{1+x}, in which the BaO component is mainly responsible for the enormous chemical activity of YBaCuO.

These ideas are in good agreement with the results of the paper /3/ which revealed some details on interactions between powders of YBaCuO and some materials considered as possible candidates for making ceramic substrates. Analyzing the reaction products the above-mentioned autors identified newly formed phases that had always contained BaO. The smallest reactivities with YBaCuO showed materials containing strongly electropositive component, such as Ba, Sr and Mg.

EXPERIMENTAL

The main aim of this work was to prepare suitable dielectric substrates for the deposition of YBaCuO films. A usual ceramic technology was used to prepare polycrystalline materials of $SrTiO_3$, yttrium stabilized zirconia (YSZ), and barium containing compounds : $BaZrO_3$, Y_2BaCuO_5, $BaSiO_4$ and $BaAl_2O_4$ respectively. The starting mixtures of oxides and carbonates with purities better than 99,2 % were calcinated at temperatures which are given in Tab, 1. After calcination powders of substrate materials were prepared by wet milling. These powders were used for the reactivity studies as well as for shaping substrates.

To estimate the reactivity of the substrate materials with YBaCuO the mixtures of powders in 1:2 mole ratio (substrate: YBaCuO) were annealed for 30 minutes in oxygen at 400 ^oC, 600 ^oC, 800 ^oC and 900 ^oC respectively. After each annealing step the ac-susceptibility change $\Delta\chi$ was computed from the measurements performed at 77 K and 300 K respectively. A pure YBaCuO powder was used as reference sample. The value relates to the volume decrease of the supercondunctinq phase in the studied mixture of powders.

The calcinated powders of substrate material were pressed (200 MPa) into the form of rectangular substrates of 16x5x2 mm^3, and 50x5x5 mm^3 respectively. The sintering temperatures are given in Tab. 1.

Thermal expansion coefficients of these substrates were measured up to 1000^oC. The OM, SEM, XRD and EDXA were performed to gain deeper understanding of processes which modified qualities of substrates under study.

RESULTS AND DISCUSSION

The XRD analysis had shown that no foreign phases were present in any substrates material. Chemical homogenity studies by EDXA indicated a weak aluminium based contamination of Y_2BaCuO_5 and $BaZrO_3$ substrates. No other chemical inhomogenities were found.

From the view-point of substrate porosity studied substrates might be classified in two groups. Highly dense almost pore-free substrates were prepared from $SrTiO_3$ and YSZ. On the contrary, ceramic substrates of barium containing materials displayed a higher concentration of pores. In the second group the sintering process advanced farthest onto $BaZrO_3$ where the ceramic surface is pracitcally pore-free, but on fractured surfaces closed pores were found /Fig 1/.

The worst macrostructure exhibited $BaAl_2O_4$ having enormous porosity and only weak mechanical joints among adjacent grains. The results of thermal expansivities and reactivities of substrate materials with YBaCuO are summarized in Tab. 1.

TABLE 1
Calcination and sintering temperatures, thermal expansion coefficients and relative susceptibility changes

	T^{+} /^{o}C/	T^{++} /^{o}C/	$\alpha \cdot 10^{6}$ / K^{-1}/	$\Delta\chi$ /%/
$YBa_2Cu_3O_{7-x}$	900	930	13,7	100,0
$SrTiO_3$	1200	1380	10,8	52,5
YSZ	1360	1680	9,9	73,1
Y_2BaCuO_5	940	1130	12,5	93,7
$BaZrO_3$	1250	1680	7,0	79,7
Ba_2SiO_4	1300	1450	-	60,0
$BaAl_2O_4$	1250	1680	8,4	-

T^{+} = calcination temperature, T^{++} = sintering temperature
α = linear thermal expansion coeffient between 20 ^{o}C and 600°C,
$\Delta\chi$ = ac - susceptibility changes related to YBaCuO after annealing at 900 ^{o}C.

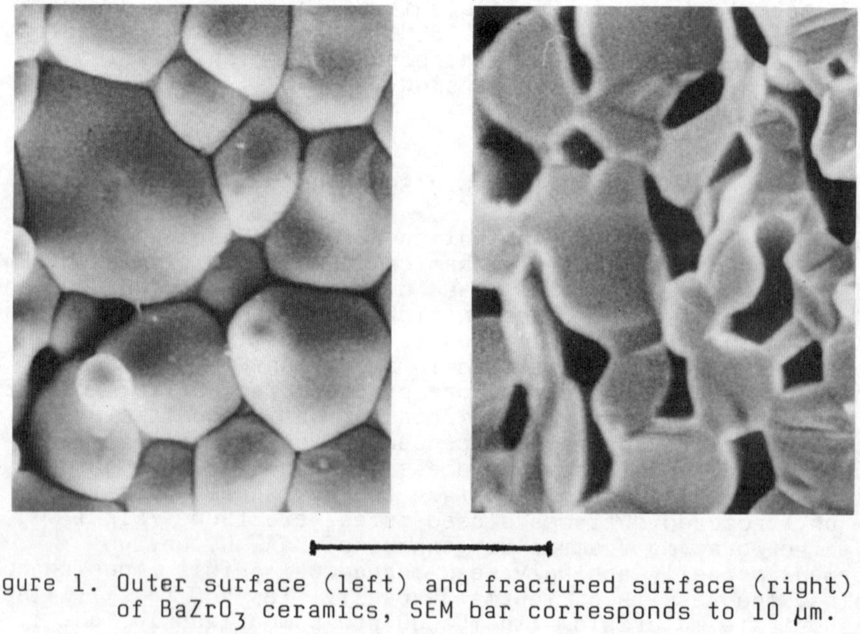

Figure 1. Outer surface (left) and fractured surface (right) of $BaZrO_3$ ceramics, SEM bar corresponds to 10 μm.

The insulating substrate materials studied in this work might be classified as follows. The most promising candidate for polycrystalline insulating substrates is Y_2BaCuO_5 because of its good match of α coefficient and low reactivity with YBaCuO even after annealing at 900 ^oC in oxygen. Good chemical stability with respect to YBaCuO showed $BaZrO_3$ but unfortunately its low thermal expansion would create problems when mastering mechanical properties of superconducting films.

OUTLOOK

More detailed studies of physical properties of Y_2BaCuO_5 polycrystalline material are needed before its responsible use as a substrate material for YBaCuO films. Some new results gained on this material will be published in the near future/4/. Experiments focused on the control of the macrostructure of other barium containing substrates are in progress now.

REFERENCES

1. Cima, M.J.,Schneider,J.S., Peterson, S.C. and Coblenz, W., Reaction of $Ba_2YCu_3O_{6,9}$ films with yttria-stabilized zirconia substrates. Appl.Phys.Lett.,1988, 53, 710-2

2. Bursill, L.A. and Fan, Xu Dong, Surface structure of the ceramic superconductor $YBa_2Cu_3O_{7-x}$. Phys.Stat.Sol.,1988,107, 503-510

3. Koinuma H.,Fukuda K.,Hashimoto T., Fueki K., Chemical interaction between $YBa_2Cu_3O_{7-x}$ and substrate materials in the solid state. Jap. J. Appl. Phys., 1988, 27, L1216 - L1218

4. Jansa, Z., Hrnčíř, K., Janda, M.and Horák, J., Electrical conductivity of the polycrystalline Y_2BaCuO_5. Phys.Stat.Sol.,(to be published)

PRODUCTION OF A $YBa_2Cu_3O_7$ CYLINDER FOR MAGNETIC FIELD SHIELDING

L.P.L.M. RABOU
Netherlands Energy Research Foundation, ECN
P.O. Box 1, 1755 ZG Petten, the Netherlands

Z. TARNAWSKY* and H.P. VAN DER MEULEN
Natuurkundig Laboratorium, Universiteit van Amsterdam
Valckenierstraat 65, 1018 XE Amsterdam, the Netherlands
* on leave from Solid State Physics Department
Academy of Mining and Metallurgy
al Mickiewicza 30, 30-059 Krakow, Poland

ABSTRACT

A ø 19 x 50 mm cylinder has been produced from a high-quality $YBa_2Cu_3O_7$ powder prepared by a citrate synthesis route. The cylinder is to be used as a shield against the magnetic field ripple of a high-field magnet. Data are presented on the shielding capacity of cylinders produced by isostatic pressing. At increasing field values above 0.5 T the maximum ripple field which can be shielded is virtually constant at 2 mT (peak to peak). A pronounced hysteresis effect is observed at decreasing field values. The addition of AgO to $YBa_2Cu_3O_7$ powder produced no measurable effect.

INTRODUCTION

The discovery of materials with superconducting properties at temperatures above 77 K, the boiling point of nitrogen, has led to wild speculations about their impact on society. Applications have been considered from small-scale electronic components and magnetic field sensors to large-scale magnets for energy storage. However, most applications require at least several years of development. High current applications at considerable fields will further require orders of magnitude improvement of the critical current density.

A well-known property of superconducting materials, often used to demonstrate superconductivity, is their ability to exclude magnetic fields from their interior. Shielding of magnetic fields by high-temperature superconductors has been studied by several groups both for discs [1, 2] and for cylinders [3-5]. The magnetic field which can be shielded from the centre of a superconducting cylinder is proportional to the critical transport current density [2]. For typical critical current densities realized at present this means that only fields of the order of 1 to 10 mT can be shielded. Still, this may be sufficient, e.g. to allow SQUID operation at 77 K [5].

The high upper critical field of $YBa_2Cu_3O_7$ makes it potentially useful at very high magnetic field strengths. Tarnawski et al. [6] have proposed to use a $YBa_2Cu_3O_7$ cylinder as a shield against the magnetic field ripples of large hybrid or Bitter-type magnets. A cylinder consisting of seven separate rings has been used at the Nijmegen High Field Magnet Laboratory to perform specific-heat measurements on the heavy-fermion compound UPt_3 in magnetic fields up to 24.5 T [7]. It was concluded that a considerable reduction of the noise level could be realized upto the highest fields. In an attempt to improve the ripple shielding, three cylinders have been produced by isostatic pressing. The present article reports test results and discusses the effect of AgO addition to the $YBa_2Cu_3O_7$ powder.

EXPERIMENTAL PROCEDURE

Cylinders were produced by isostatic pressing of powder to the approximate shape. The correct outer dimensions were obtained after sintering by dry processing with diamond tools. Powder for cylinder A was produced by the solid-state reaction method. Powder for the other two cylinders was prepared by citrate pyrolysis [8]. For cylinder C 15% by weight AgO was mixed to the calcined powder. Cylinder A was sintered in air for 12 hours at 920°C, followed by 6 hours at 940°C and cooling to room temperature in 30 hours. Cylinders B and C were sintered at 940°C in air for 12 hours, followed by cooling at 60°C per hour with a 6 hours stop at 450°C. An annealing treatment at 425°C in pure O_2 for 45 hours (cylinder B) or 63 hours (cylinder C) increased the weight of the cylinders by an amount equivalent to 0.04 (B) or 0.05 (C) oxygen atoms per formula unit. The density of cylinders A, B and C was approximately 60%, 85% and 85%, respectively, of the theoretical density.

Measurement technique

Measurement of the shielding capacity for magnetic ripple fields has been performed in the bore of an 8T magnet with the cylinder suspended in helium vapour at a temperature close to 4.2 K. A sinusoidal ripple field with a frequency of 100 Hz was generated by a coil surrounding the cylinder. The signal of a pick-up coil inside the cylinder was detected with the use of a lock-in amplifier and visually inspected on an oscilloscope. Cylinder A was cooled in zero (i.e. the earth magnetic) field, B and C were cooled in the remanent field of the magnet.

RESULTS

In a previous experiment the shielding effect has been found to be independent of the frequency of the varying field in the range between 50 Hz and 20 kHz [6]. This observation was confirmed by the present experiments. Figure 1 demonstrates the shielding effect for a varying external magnetic field of triangular shape. At low field levels the cylinder shields the field completely and no signal is observed by the sensing coil. Just above the threshold shielding field, the signal of the sensing coil deviates from zero when the external field reaches its extreme values. The sharp drop to zero at reversal of the field change indicates that the entered flux is trapped (Figure 1b). When stronger fields are applied the signal is zero

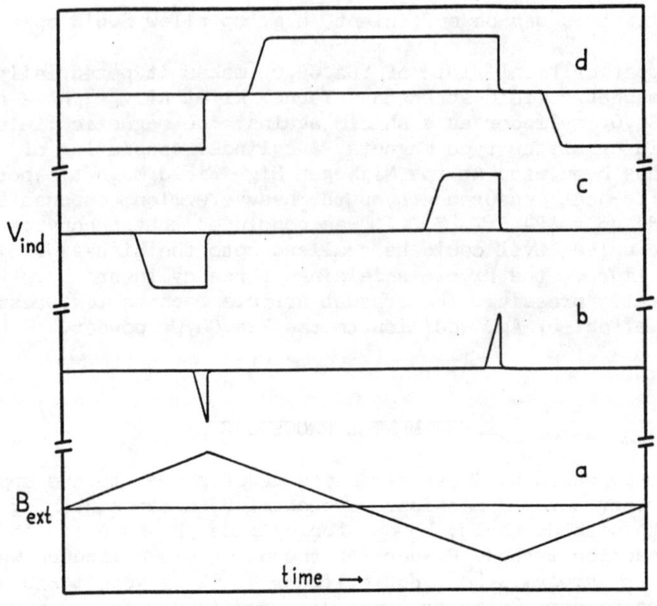

Figure 1. Magnetic field shielding by a $YBa_2Cu_3O_7$ cylinder.

a. Externally applied field
b. Induced voltage in the sensing coil at peak fields slightly above the maximum field which is shielded
c. Same as b at 1.3 times threshold field
d. Same as b at 6 times threshold field

(the vertical scale differs for b, c and d).

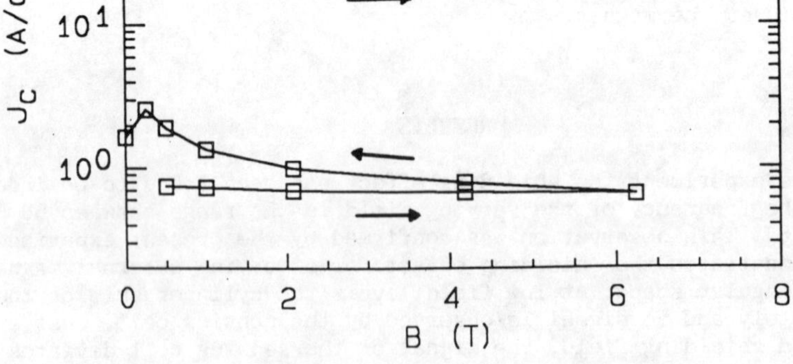

Figure 2. Magnetic field dependence of J_c for increasing and decreasing fields (indicated by arrows).

□ cylinder A X cylinder B O cylinder C

for shorter periods (Figure 1c). Finally, at large values of the external field hardly any shielding is observed (Figure 1d).

The presence of a constant external field influences the magnitude of the varying field which can be shielded by its effect on the critical current density. Results of measurements of the shielding capacity of the three cylinders, translated into a critical current density are shown in figure 2. At large fields the critical current density reaches almost constant levels of 0.7 A/cm² for cylinder A and 28 A/cm² for cylinders B and C. These values correspond to field levels (peak to peak) of 0.05 mT and 2 mT, respectively.

The above results were obtained for measurements performed at increasing values of the constant field. At decreasing field much higher values are obtained. This hysteresis effect has been observed in previous experiments [6] and in measurements of the transport current density [9, 10]. It is related to the magnetization of the grains which reduces the effective field at the weak links [9]. The hysteresis effect is negligible for the field values at which the cylinder is to be used [10].

DISCUSSION

The critical transport current density J_c at 4.2 K is found to be almost independent of the magnetic field above 0.5 T. This confirms earlier observations by two of the present authors [6] and other groups who performed measurements upto 20 T e.g. [10]. A large difference in J_c is observed for the three cylinders. A similar effect has been observed by Macfarlane et al. [1]. According to Alford et al. [11] the difference in porosity accounts for a factor of 3 difference in J_c, much smaller than the factor of 40 observed. Hence, it is concluded that the quality of the superconducting material used is essential.

The addition of AgO is found to have a negligible effect on J_c. This result is in accordance with the conclusions of Tiefel et al. [12] that the strong magnetic hysteresis effect observed for silver or silver-oxide doped material is not due to an improved J_c but to silver-enhanced grain growth. The authors plan to use cylinder B (without silver addition) in future work on the specific heat in high magnetic fields of UPt_3 and similar compounds.

Acknowledgements
This work is sponsored by the Dutch Ministry of Ecomonic Affairs through the National Program on High Temperature Superconductors. It is also part of the research program of the Stichting FOM (Foundation for Fundamental Research on Matter). The assistance of Messrs. Roskam, Berkeveld and Kroone is gratefully acknowledged.

REFERENCES

1. Macfarlane, J.C., Driver, R., Roberts, R.B. and Horrigan, E.C., Electromagnetic shielding properties of yttrium barium cuprate superconductor. Cryogenics 28 (1988) 303-5.

2. Okada, T., Takahata, K., Nishijima, S., Yoshida, S. and Hanasaka, T., Applicability of oxide superconductor to magnetic shielding. IEEE Trans. Magn. 25 (1989) 2270-2.

3. Tjukanov, E., Cline, R.W., Krahn, R., Hayden, M., Reynolds, M.W., Hardy, W.N., Carolan, J.F. and Thompson, R.C., Current persistance and magnetic shielding properties of $Y_1Ba_2Cu_3O_x$ tubes. Phys. Rev. B36 (1987) 7244-7.

4. Eberhardt, F.J., Hibbs, A.D. and Campbell, A.M., Flux trapping and magnetization of hollow superconducting cylinders. Cryogenics 28 (1988) 681-4.

5. Symko, O.G., Yeh, W.J., Zheng, D.J. and Kulkarni, S., Magnetic shielding and relaxation characteristics of superconducting $YBa_2Cu_3O_7$ tubes. J. Appl. Phys. 65 (1989) 2142-4.

6. Tarnawski, Z., van der Meulen, H.P., Franse, J.J.M., Kadowaki, K., Veenhuizen, P.A. and Klaasse, J.C.P., New possibility of magnetic ripple shielding for specific heat measurements in hybrid magnets. Cryogenics 28 (1988) 614-6.

7. Van der Meulen, H.P., Tarnawski, Z., Franse, J.J.M., Perenboom, J.A.A.J., Althof, D. and van Kempen, H., Specific heat measurements of UPt_3 in magnetic fields upto 24.5 T. Submitted to Phys. Rev. B.

8. Blank, D.H.A., Kruidhof, H. and Flokstra, J., Preparation of $YBa_2Cu_3O_{7-\delta}$ by citrate synthesis and pyrolysis. J. Phys. D 21 (1988) 226-7.

9. Qian, Y.J., Tang, Z.M., Chen, K.Y., Zhou, B., Qiu, J.W., Miao, B.C. and Cai, Y.M., Transport hysteresis of the oxide superconductor $YBa_2Cu_3O_{7-x}$ in applied fields. Phys. Rev. B 39 (1989) 4701-3.

10. Ikeno, Y., Gotoh, K Kume, A, Kanagawa, M. and Kohno, O., High field properties and characteristics of oxide superconductors. Physica B 155 (1989) 164-6.

11. Alford, N. McN., Clegg, W.J., Harmer, M.A. Birchall, J.D., Kendall, K. and Jones, D.H., The effect of density on critical current and oxygen stoichiometry of $YBa_2Cu_3O_x$ superconductors. Nature 332 (1988) 58-9.

12. Tiefel, T.H., Jin, S., Sherwood, R.C., Davis, M.E., Kammlott, G.W, Gallagher, P.K. Johnson Jr, D.W., Fastnacht, R.A. and Rhodes, W.W., Grain-growth enhancement in the $YBa_2Cu_3O_{7-\delta}$ superconductor by silver-oxide doping. Mat. Lett. 7 (1989) 363-6.

SUPERCONDUCTIVE LAYERS ON COPPER RIBBONS

VLADIMIR HAJKO Jr.,
STEFAN MOLOKAC, MILAN TIMKO, ANTON ZENTKO, ONDREJ HUDAK
Institut of Experimental Physics, Slovak Acad. of Sci.
043 53 Kosice, CZECHOSLOVAKIA
Yu.N. IVASHCHENKO, Yu.V. MILMAN, S.V. PAN, V.V. KOVYLYAYEV
Institut for Problems of Material Science
Ukrainian Acad. of Sci. 252180 Kiev, USSR

ABSTRACT

Microcrystalline ribbons were prepared from copper alloy doped with Y and Ba by melt spinning method. This method is one of the perspective ways for preparing high temperature superconductors in the form of ribbons, wires or foils. The structure, susceptibility and resistance of thermooxidized Y-Ba-Cu ribbons were studied. The surface oxide layers of the ribbons have a granular microcrystalline structure of a multiphase nature. The critical current $I_C \sim 10^{-4}$ A and resistance jump of $\sim 10^7$ were obtained at room temperature. The critical fields H_{C1} and H_{C2} were estimated to be equal to zero for $T_C \approx 330$ K. On the base of experimental data the samples seem to display the presence of regions with very low resistance forming a fairly unstable " Josephson medium ".

INTRODUCTION

Experimental work (1, 2, 3) and others, as well as our previous work (4, 5) rank among the first in the field of preparing superconductors by oxidation of alloys and ribbons.

Our access differs from all others in the fact that we chose for the base an alloy the substantial part of which is copper (95 wt.%). The remaining 5 wt.% are Y and Ba, with the ratio Y : Ba = 1 : 2 in at.%.

Microcrystalline ribbons can be prepared by rapid quenching of the melt on a rotating metal wheel. By this procedure we can achieve an increase in the solid state solubility. The higher concentration of defects in the structure leads to a more rapid segregation an diffusion of Y and Ba. In consequence of that surface oxide layers get formed with a much greater volume concentration of Y and Ba than in the basic metal matrix. It is caused by the fact that Y and Ba have a considerably higher affinity to oxygen than copper in the temperature range from 500 to 900 K .

MATERIALS AND METHODICS

The alloy ribbons were prepared by melt spinning method. The dimensions of such quenched ribbons were $\sim 50\ \mu m$ in thickness, 2 - 3 mm in width and some metres in length. The 12 mm long samples were statistically selected from different sites of the ribbon and afterwards were annealed in the range of temperature from 625 to 775 K in air.

The structure of the samples was observed both by optical microscopy and by SEM. Microanalyses were realized using x-ray spectrometer SUPERPROBE JCXA - 733. Metallographically polished cross sections of the ribbons were observed.

The electric resistance as a function of temperature was measured by the four-point probe method. A silver paint was used to improve these contacts. Magnetic susceptibility was measured in an alternating field with a frequency of 15 kHz.

RESULTS AND DISCUSSION

An oxide layer $\sim 6\ \mu m$ thick is formed on the surface of the ribbon after oxidation at 625 K/2 h + 775 K/3 h in air.

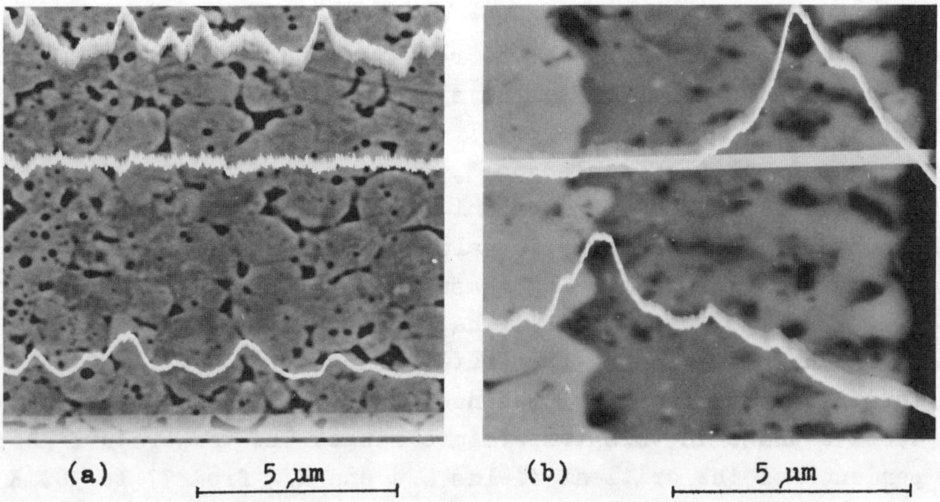

Figure 1. SEM micrographs obtained by SUPERPROBE JCXA - 733. Y-Ba-Cu ribbon before annealing (a), the lines for Ba, Cu, Y are begining from the top. The bottom line = analysed line. The concentration of Ba and Y across the surface oxide layer is shown in (b). Ba - upper line, Y - lower line.

From x-ray spectrometry microanalyses follows that distribution of Y and Ba across the ribbon thickness is approximately equable - Fig. 1(a). Their volume contents in the surface oxide layer is much higher than in the metal matrix. It is evident that Ba has a greater tendency to diffuse in the direction of the free surface than Y - Fig. 1(b).

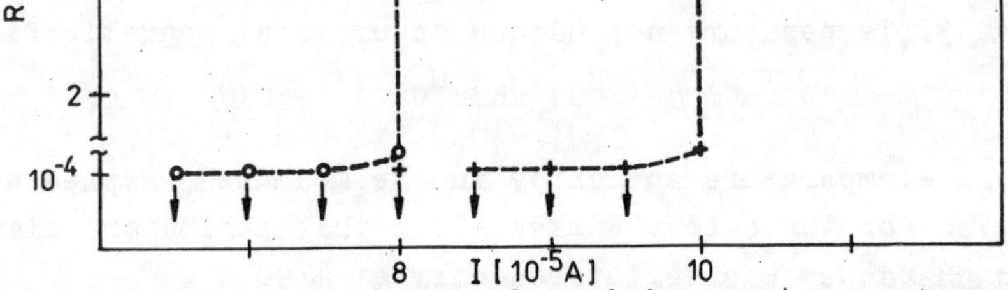

Figure 2. Current dependence of resistance at room temperature. + increase in current, o decrease in current.

Current dependence of the electric resistance at room temperature is in Fig. 2. The resistance jump from $R_L < 10^{-4}\Omega$ (sensitive limit of measuring system) to $R_H \sim 8.10^2\,\Omega$ was registered at critical current $I_C \sim 10^{-4}$ A. The sample returned to a state of "low resistance", showing some hysteresis, after decreasing of the current $I_C \sim 7.10^{-5}$ A. Minimal deviations of the characteristic values I_C, R_L and R_H were registered during the following measuring cycles.

Magnetic susceptibility has a diamagnetic character with a few anomalies in the temperature range from 4,2 to 300 K. This fact testifies the presence of more than one superconductive phase in this temperature range. The temperature dependence of the critical fields H_{C1} and H_{C2} from 77 to 300 K can be seen in Fig. 3. By extrapolation of these values to zero magnetic field we can obtain the critical temperature value of the assumed high-temperature superconductive phase $T_C \approx 330$ K.

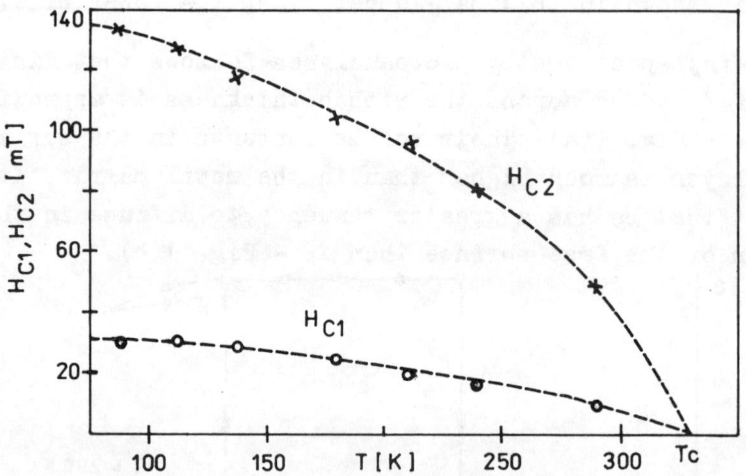

Figure 3. Temperature dependence of critical magnetic fields H_{C1} and H_{C2} .

The high-temperature anomalies in the measured samples were observed for three-four weeks. After this period the diamagnetic character was registered only at ~ 90 K .

CONCLUSIONS

The presence of superconductivity above room temperature is probably proved by the achieved results in the studied samples of microcrystalline Y-Ba-Cu ribbons. The approximate volume contents (about 0.5 %) of the superconductive phase (marked X) with $T_c \approx 330$ K in the surface oxide layer may be determined from the results of magnetic susceptibility measuring. The relatively low stability of " X phase ", about 3 - 4 weeks, is probably caused by increased diffusion of oxygen from places enriched by oxygen to the places impoverished by it, i.e. also into unoxidized metal matrix.

REFERENCES

1. Yurek, G.J., Vander Sande, J.B., Wang, W.X., Rudman, D.A., Zhang, Y. and Matthiesen, M.M., Synthesis of a Superconducting Oxide by Oxidation of a Metallic Precursor. Met. Trans. A, 1987, 18A, October, 1813-7.

2. Yurek, G.J., Vander Sande, J.B., Wang,W.X. and Rudman, D.A. Direct Synthesis of a Metal/Superconducting Oxide Composite by Oxidation of a Metallic Precursor. J. Electrochem. Soc., 1987, Vol.134, 10, 2635-6.

3. Matsuzaki, K., Inoue, A., Kimura, H., Moroishi, K. and Masumoto, T., Preparation of a High T_c Superconductor by Oxidation of an Amorphous $La_{1.8}Sr_{0.2}Cu$ Alloy Ribbon in Air. Japanese J. of Appl. Phys., 1987, 26, April, L 334-6.

4. Hajko, V.,Jr., Diko, P., Csach, K. and Molokac, S., High Temperature Superconductivity in Rapidly Quenched La-Ba-Cu Oxidized Ribbons. Czech. J. Phys.,1987, B37, Oct., 1203-6.

5. Hajko, V.,Jr., Diko, P., Csach, K. and Molokac, S., Czechoslov. Patent No: PV 4699/87, filed June 25, 1987.

TEXTURED SUPERCONDUCTING THICK FILMS OF $YBa_2Cu_3O_{7-x}$

J.S. ABELL, F. WELLHOFER, T.C. SHIELDS,
School of Metallurgy, University of Birmingham, UK
D. HOLLAND, L. BLUNT, M.E. YAKINCI,
Department of Physics, University of Warwick, UK
K.N.R. TAYLOR
Department of Physics, University of New South Wales, Australia

ABSTRACT

The fabrication of superconducting thick films of YBCO on both alumina and zirconia substrates is described. The surface morphologies and cross-sectional microstructures are compared and the observation of a spherulitic-type morphology with a high degree of grain alignment is reported for tracks on zirconia substrates. These tracks exhibit enhanced superconducting properties with critical currents measured of several hundred A/cm^2.

INTRODUCTION

Screen printing is a well established technology which at present is widely used in the production of hybrid electronic devices with a large variety of materials combinations, including metal oxides. Hence the development of screen printed circuits with the new superconducting oxides is of considerable attraction, providing that the particular problems associated with these ceramics can be overcome. One such problem is the reactivity of these oxides with most commercially available substrate materials at the high sintering temperatures required to establish good grain inter-connectivity. Several reports have appeared of the preparation of thick films on various substrate materials [1-5]but none with superconducting properties approaching those found in bulk ceramics.

Earlier experiments in our own laboratories have demonstrated the significance of sinter temperature and time on the adhesion of superconducting tracks of YBCO to alumina substrates[6]. Short anneals at low temperature (850°C) resulted in poor adhesion while longer anneals at higher temperatures (950°C) gave good adhesion but the tracks contained several degradation products and reduced electrical properties. The establishment of a well-defined interfacial reaction layer to provide good adhesion without severe track degradation is considered to be essential to the fabrication of viable superconducting tracks. Track thickness is clearly an important parameter in this respect and subsequent experiments confirmed this to be the case[6]. A change from semiconducting to metallic behaviour above Tc occurs for track thicknesses between 80 and 160μm on alumina with a corresponding increase in Tc. Shorter times at higher temperatures (1050°C) also improved the metallic behaviour but still led to a reduced transition temperatures compared with bulk material. Evidence for partial melting of the surface was found on these tracks which, if properly controlled, could lead to enhanced current densities in the manner established for bulk material[7]. In this paper we report the fabrication and properties of textured thick films of YBCO on zirconia substrates with enhanced superconducting properties by a short high temperature sinter.

MATERIALS AND METHODS

YBCO powders were prepared by solid state synthesis of commercially available Y_2O_3, CuO and $BaCO_3$ at 950°C in air for 24h as described previously[6], and slowly cooled (2°/min). The powders were subsequently ball milled for 2h. Other material used in this study was supplied by ICI Advanced Materials. Both batches of powder had a normal distribution of particle sizes centred on 2μm. Suitable inks were prepared by ultra-sonic mixing of a constant weight ratio of powder with Blythe 63/2 organic binder. Screen printing was performed through a 325-mesh stainless steel screen with a simple pattern consisting of tracks of width 1-5mm. onto substrates of alumina or yttrium-stabilised zirconia (YSZ).The track thickness was varied by multiple screen printing passes with intermediate drying stages at 110°C. Controlled heating rates of 5°C/min., designed to burn off the carrier, were followed by sintering at 975°C or 1025°C for 6min in flowing oxygen with a subsequent slow cool at 2°C/min. The fired track thicknesses were in the range 25-40μm and were assessed by means of optical and SEM microscopy, XRD and X-ray microanalysis, and resistive transition measurement. Current carrying capacities were measured using a pulsed, four point probe method[8].

RESULTS AND DISCUSSION

The preparation conditions and superconducting properties of the thick films are summarised in Table 1. The best characteristics were found for the tracks on YSZ substrates, particularly those sintered at the higher temperature. The room temperature resisitivities measured on these tracks were consistently an order of magnitude lower than those on alumina. The ratio R_{300}/R_{100} indicating the normal state behaviour is a useful parameter in a comparative assessment of these tracks. A high value indicates good metallic conductivity and is often associated with good superconducting properties. This ratio has been used to calculate the resisitivity at 100K and it can be seen that there is a significant difference when tracks sintered at 975°C and 1025°C on YSZ substrates are compared. The values are approximately three times lower for the higher sinter temperature.

This normal state metallic behaviour together with a sharp resistive transition with To=90.6 K for the 3mm track on YSZ is shown in Fig.1.A measurement of current carrying capacity was taken on this track at three different temperatures giving results which correspond to J_c values of several hundred A/cm^2. On the assumption that the whole of the track is carrying current and is of uniform cross-section, the values imply a J_c of ~750 A /cm^2 at 77K. However the actual values are likely to be higher since the track cross section is by no means uniform.

TABLE 1

Preparation conditions and superconducting properties of the films.

Substrate	Temperature (°C)	Width (mm)	Thickness (μm)	To (K)	ρ_{300K} (mΩcm)	ρ_{300K}/ρ_{100K}	ρ_{100K} (mΩcm)
YSZ	1025	1	40	90.6	2.5	2.5	1.0
YSZ	1025	3	40	89.8	2.4	2.6	0.9
YSZ	1025	5	40	89.5	2.2	2.3	1.0
YSZ	975	1	40	88.1	4.4	1.6	2.8
YSZ	975	3	40	89.8	5.6	1.5	3.8
YSZ	975	5	40	88.0	3.9	1.5	2.6
Alumina	1025	5	40	56	36	1.3	27.7
Alumina	975	5	40	40	31	Semiconductor	
Alumina	1025	5	30	44	119	Semiconductor	
Alumina	975	5	30	40	18	Semiconductor	

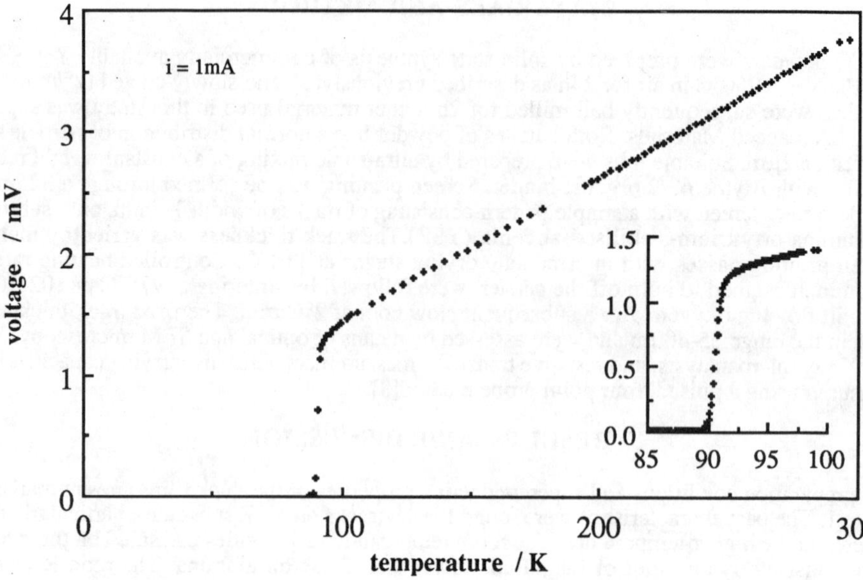

Figure 1. Resistance vs temperature for a film sintered at 1025°C on YSZ substrate.

The surface morphologies observed on the tracks were significantly different. At 975°C the material on both alumina and YSZ substrates appears well sintered but with little evidence of any significant grain growth. The biggest difference was observed in the tracks sintered at the higher temperature. While the microstructure of the alumina based tracks, shown in Fig.2(a), were composed of randomly oriented elongated grains similar in nature to that normally observed in ceramic samples, the morphology of the tracks on YSZ substrates was significantly different as seen in Fig.2(b). Low magnification observation of the tracks revealed a characteristic spherulitic-type morphology with discrete areas of the spherulite exhibiting an apparent radial distribution of grains. Closer observation, however, shows that many of the rosettes actually have a predominant set of aligned grains giving a linear or bipolar dimension to the structure (Fig.2c). This grain structure is very similar to that seen in melt textured bulk material [7] and in view of the sinter temperature involved we believe that partial melting has occured in these tracks. Areas of several impurity phases were also found including (211), CuO and $BaCuO_2$ all of which are consistent with partial melting. Needles of CuO were also located primarily towards the edge of the tracks.

Polarised optical microscopy of these surfaces revealed the characteristic domain structure associated with the twinned orthorhombic phase, and more particularly the orthogonal patterns of domains clearly indicated that the majority of grains were oriented with the c-axis normal to the track and substrate (Fig.2d). However, the bipolar regions in the spherulites did not exhibit orthogonal arrays of domains indicating that these highly directional grains were not of the same orientation as the remainder of the rosette. The overall c-axis preferred orientation was confirmed from the XRD patterns recorded from the surface of the high temperature tracks. A comparison of such patterns taken from tracks sintered at 975°C and 1025°C on YSZ is shown in Fig. 3 where the enhancement of the 00l lines associated with preferred orientation of the c-axis normal to the track surface is clearly seen.

Since this characteristic textured morphology was confined to the YSZ substrate tracks, the cross-sectional microstructure was studied in order to determine whether the development of a particular interfacial reaction layer was playing a role in the nucleation and growth of these spherulitic regions. An SEM back scattered electron micrograph of a cross section taken from a

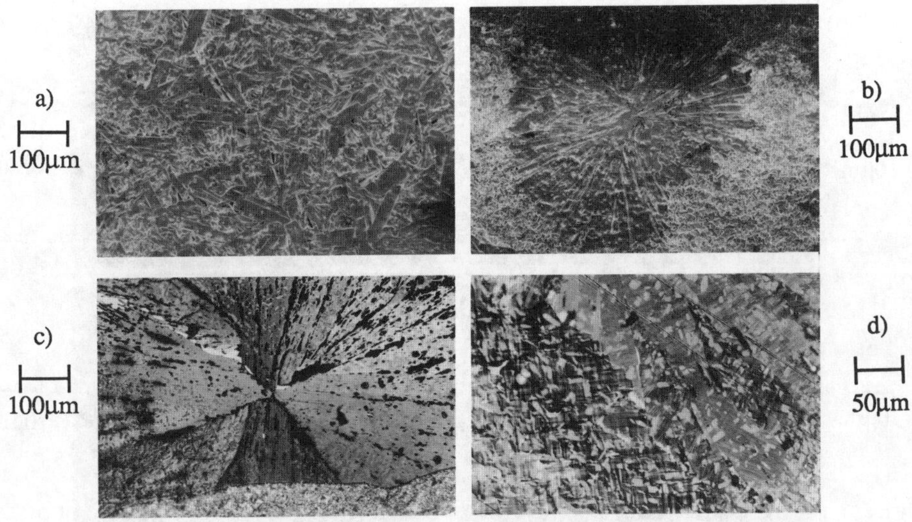

Figure 2. Surface morphologies of tracks sintered at 1025°C on (a) alumina, (b,c,d) on YSZ.

track sintered at 1025°C is shown in Fig.4(a). The elongated (a,b) grains lying approximately parallel to the plane of the substrate are evident from this micrograph, although this was by no means the only orientation seen on these cross-sections. The orientations of these grains and the geometry of the observed domain structures reveal that all orientations are present along the length of a track, but there is a preponderence of grains with the c-axis normal or nearly normal to the plane of the substrate in accord with the surface observations. A well-defined interfacial reaction layer is clearly seen, together with other inclusions located between the track and the substrate. The rest of Fig.4 shows X-ray maps of the same area, tuned to the elements Y,Ba and Cu. These maps and spot EDX analyses indicate that the interfacial layer is $BaZrO_3$ and the other inclusions are CuO. The formation of this layer may well be connected with the observed texturing by providing preferred nucleation sites for the solidification of the 123 phase as suggested previously [9], but the presence of the low-melting point CuO phase at the interface seems inconsistent with nucleation at the interface, and implies a solidification nucleated at the surface and proceeding down through the track with the CuO being the last phase to solidify.

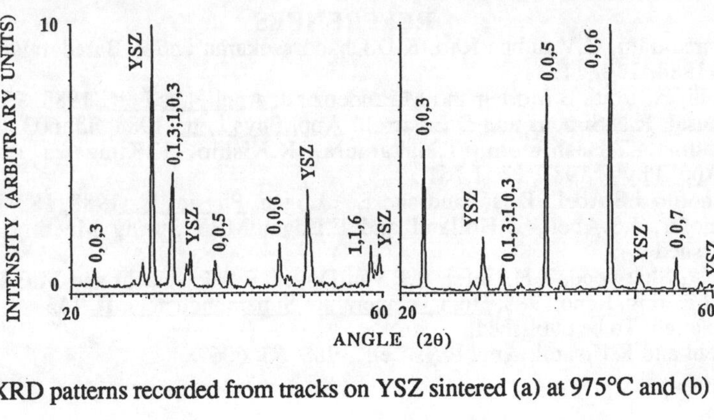

Figure 3. XRD patterns recorded from tracks on YSZ sintered (a) at 975°C and (b) at 1025°C.

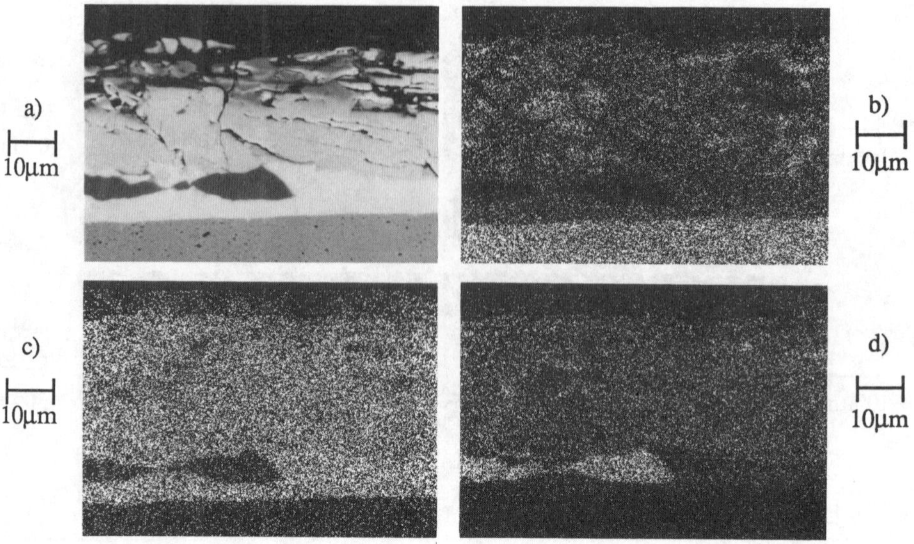

Figure 4. SEM micrographs on cross-section of YSZ track sintered at 1025°C showing interface reaction layer in (a) BSE mode, (b) Yttrium (c) Barium and (d) Copper x-ray map.

The high current densities measured on one of these YSZ tracks is associated with the high degree of inter-grain connectivity achieved due to the partial melting and to the preferred orientation resulting from the nucleation and growth of the 123 phase under these conditions. The interface studies show, however, that the degree of orientation is by no means optimised in these tracks; an understanding of the precise mechanism of nucleation is required before complete grain alignment can be achieved and associated higher critical currents can be supported.

ACKNOWLEDGEMENTS

The Lucas Advanced Engineering Centre and ICI Advanced Materials are gratefully recognised for their continued support . Special thanks to MS Colclough for measurements of the critical currents. One of us (KNRT) wishes to thank the SERC for support during an extended visit to the High Temperature Superconductivity Group, Birmingham University.

REFERENCES

1. U.V.Varadaraju, G.V.Subba Rao, K.D.Chandrasekaran and A.Baradarajan: Thin Solid Films, 1988, **164**, 119.
2. M.Sacchi, F.Sirotti, B.Morten and M.Prudenziati: Appl.Phys.Lett., 1988, **53**, 1110.
3. N.P.Bansal, R.N.Simons and D.E.Farrell: Appl.Phys.Lett., 1988, **53**, 603.
4. H.Koinuma, T.Hashimoto, T.Nakamura, K.Kishio, K.Kitazawa and K.Fueki: Jap.J.Appl.Phys., 1987, **26**, L761..
5. F.Wellhofer, J.S.Abell, D.Holland and E.A.Logan, Physica C., 1988, **153-155**, 399.
6. F.Wellhofer, J.S.Abell, D.Holland and E.Logan:MRS Spring Meeting, Reno.1988, Unpublished
7. S.Jin, R.C.Sherwood, T.H.Tiefel, R.B.van Dover, R.A.Fastnacht and M.E.Davies: MRS Spring Meeting, Reno.1988, High Temperature Superconductors, II, 153.
8. M.Colclough. To be published.
9. J.Tabuchi and K.Utsumi: Appl Phys Lett, 1988, **53**, 606.

SPECTRAL, SUPERCONDUCTING AND MAGNETIC PROPERTIES OF $Ba_2YCu_{3-x}M_xO_{7-\delta}$

N. Brničević, Lj. Tušek-Božić, P. Planinić*, A. Turković, S. Popović, B. Šantić, Ž. Toman, Rudjer Bošković Institute, 41000 Zagreb, Yugoslavia
G. Leising, E. Schweiger, V. Wippel, Institut für Festkörperphysik, Technische Universität Graz, A-8010 Graz, Austria

ABSTRACT

Partial replacement of copper atoms in the superconducting orthorhombic phase $Ba_2YCu_{3-x}M_xO_{7-\delta}$ (M=V, Nb, Ta) with the 5a group of transition metal elements stabilized this phase in a wide range of dopant concentrations. As the dopant amounts increased, the superconducting properties gradually decreased due to the formation of $BaVO_3$, $Ba_3V_2O_8$, Ba_2YNbO_6 or Ba_2YTaO_6 and CuO. X-ray powder diffraction, EPR spectra, electrical resistivity and magnetic susceptibility measurements of the doped samples were correlated.

INTRODUCTION

The influence of transition metal elements on structural and superconducting properties of the orthorhombic phase $Ba_2YCu_3O_{7-\delta}$ was extensively studied in many laboratories. It was shown that magnetic rare earth ions substituted for yttrium in $Ba_2YCu_3O_{7-\delta}$ has essentially no effect on Tc (1) indicating that the superconducting phenomena should be at the copper side. Substitution of copper might provide important information regarding the relationship between structure, chemistry and superconductivity. Special attention was paid to the partial substitution of copper by divalent (2) (Zn^{2+}, Ni^{2+}) or trivalent (3) (Fe^{3+}, Co^{3+}, Al^{3+}) transition metal elements in order to get information on the nature of different copper sites in the crystal lattice of the orthorhombic phase $Ba_2YCu_3O_{7-\delta}$. Considerably less attention was given to the substitution of copper atoms by the transition metal elements in the oxidation state higher than +3. We report here on the investigation of the influence of pentavalent transition metals (V, Nb, Ta) on structural, superconducting, magnetic and mechanical properties of the doped $Ba_2YCu_3O_{7-\delta}$ samples.

MATERIALS AND METHODS

A series of samples $Ba_2YCu_{3-x}M_xO_{7-\delta}$, M = V, Nb, Ta, were prepared with $\underline{x}$ ranging from zero to 1. The mixtures of Y_2O_3, $BaCO_3$, CuO and pentoxides of V, Nb or Ta were treated isothermally at 860° and 900°C with intermediate grinding and pelletizing. The pellets were finally fired at 950°C, and then slowly cooled to room temperature (cooling rate 80°C/h). Complete decomposition of $BaCO_3$ was verified by infrared spectroscopy. The X-ray powder diffraction patterns were taken using a Philips counter diffractometer (monochromatized Cu$\underline{K}$ α radiation). The resistivity was measured by the four probe geometry in a closed cycle refrigirator. The contacts were checked to be ohmic in the temperature range measured. The magnetic susceptibility was determined by an AC susceptometer working at 28 Hz.

RESULTS AND DISCUSSION

In order to study the phase composition by the X-ray powder diffraction the whole range of concentrations in each system was covered with a series of samples, in steps of $\Delta\underline{x}$ = 0.1 or 0.2. Characteristic powder diffraction patterns of some selected samples containing niobium are shown in Fig. 1. The sample without dopant, i.e. for $\underline{x}$ = 0, was the phase $Ba_2YCu_3O_{7-\delta}$, having orthorhombic symmetry. As the amount of dopants increased, the fraction of the orthorhombic phase gradually decreased, and the samples

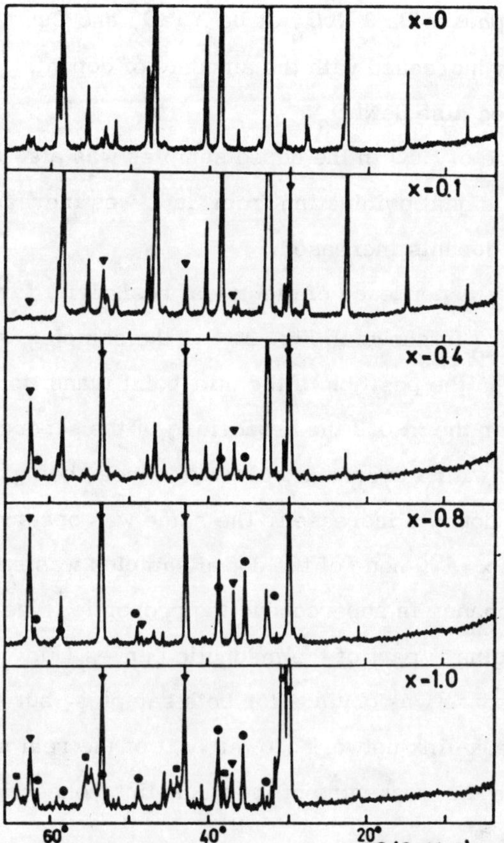

Figure 1. X-ray powder diffraction patterns of $Ba_2YCu_{3-x}Nb_xO_{7-\delta}$ for several values of x: (▼) Ba_2YNbO_6, (■)$BaNbO_3$, (●) CuO.

with high amount of dopants, for x≥0.8, did not contain this phase. No change to tetragonal symmetry of the orthorhombic phase was observed as the amount of dopants increased. The disappearance of diffraction lines of the orthorhombic phase coincided with the transition of the samples to the insulating state. In the samples with vanadium, besides the orthorhombic phase, the phases $BaVO_3$, $Ba_3V_2O_8$ and CuO were observed for x≥0.1, and their fraction increased with the amount of vanadium. In the samples containing niobium or tantalum, besides the orthorhombic phase, the nonsuperconducting phases Ba_2YNbO_6 or Ba_2YTaO_6 and CuO were observed, and their fraction increased with the amounts of dopants. The samples with x(Nb) = 1 contained also $BaNbO_3$.

The presence of CuO in the doped samples was also confirmed by the EPR spectra. The signal originating from Cu^{2+} was increasing in intensity as the amount of dopants increased.

Temperature dependence of electrical resistivity for the samples $Ba_2YCu_{3-x}Nb_xO_{7-\delta}$ is shown in Fig. 2. For the samples with $x \le 0.2$ no essential change in the position of the mid-point transition temperature was observed. For the $x > 0.2$ the broadening of the superconducting transition accompanied with a decrease in Tc took place, being more pronounced as the amount of dopants increased. The same was observed for vanadium and tantalum. At x = 1.0 none of the doped samples was superconducting. A more evident change in superconducting properties was obtained from the real and imaginary part of the magnetic curves (Fig. 3). The step-like diamagnetic response was obvious for both samples, but the step corresponding to the weak-link network (lower part of the real susceptibility curves) indicated the weakening of the weak-link connectivities for the sample with x = 0.2.

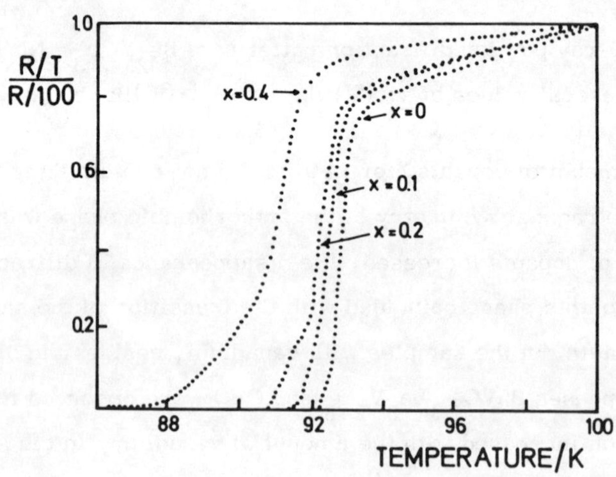

Figure 2. Electrical resistance (normalized at 100 K) versus temperature for $Ba_2YCu_{3-x}Nb_xO_{7-\delta}$

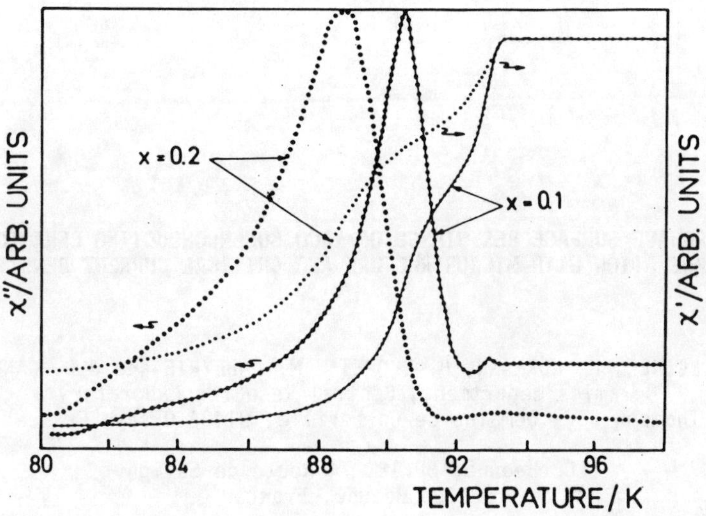

Figure 3. The real (χ') and imaginary (χ'') magnetic susceptibility versus temperature for $Ba_2YCu_{3-x}Nb_xO_{7-\delta}$

CONCLUSION

Transition metal elements vanadium, niobium and tantalum in their highest oxidation state when added to $Ba_2YCu_3O_{7-\delta}$ partly replacing copper stabilized the orthorhombic phase from transformation to the tetragonal symmetry. Depending on the dopants level the fraction of the nonsuperconducting phases $BaVO_3$, $Ba_3V_2O_8$, Ba_2YNbO_6 or Ba_2YTaO_6 and CuO was increasing and dominated for the x = 1.0. Consequently the superconducting properties were gradually destroyed and vanished at the same values of x.

REFERENCES

1. Tarascon, J.M., McKinnon, W.R., Greene, L.H., Hull, G.W. and Vogel, E.M., Phys. Rev. B, 1987, 36, 226-234.

2. Jones, R., Edwards, P.P., Harrison, M.R., Thanyasiri, T. and Sinn, E. J. Am. Chem. Soc., 1988, 110, 6716-6720.

3. Siegrist, T., Schneemeyer, L.F., Waszczak, J.V., Singh, N.P., Opila, R.L., Batlogg, B. and Rupp, L.W., Phys. Rev. B, 1987, 36, 8365-8368.

MICROWAVE SURFACE RESISTANCE OF YBCO SUPERCONDUCTING CERAMICS. CORRELATION WITH MICROSTRUCTURE AND CRITICAL CURRENT DENSITY

F. LAINEE, R. KORMANN, B. LLORET*, M. LABEYRIE and J.P. GANNE
Ceramics department, Central Research Laboratory
Thomson-CSF, Domaine de Corbeville, 91404 Orsay, France

*LCC Beaune, BP 18, 74 Route de Savigny
21200 Beaune, France

ABSTRACT

Ceramic samples of superconducting $YBa_2Cu_3O_{7-x}$ were produced at three sintering temperatures, using two different ceramic precursors. Microwave measurements show that critical transport current density is not simply correlated with microwave surface resistance. This result is discussed in terms of microstructure of the samples.

INTRODUCTION

The discovery of high-T_c ceramics superconductors by Bednorz and Müller |1| in 86 opened broad new perspectives for industrial use of superconductivity. In the range of VHF and microwave frequencies (100 MHz - 10 GHz) low T_c superconductors (Nb, Nb_3Sn) have proved useful to make low losses devices. On the contrary new superconducting ceramics have been found to be hardly better than copper at 77 K and a few GHz |2|. With such results the challenge is still open to improve R_s for superconducting ceramics and make them industrially useful.

The object of this study is to measure R_s for different YBCO ceramic samples obtained by controlled processing, and to correlate R_s with microstructural or other electric properties. Such properties as phase composition, porosity, grain-size and critical-current density will be considered.

MATERIAL PREPARATION

Ceramic samples of composition $YBa_2Cu_3O_{7-x}$ were prepared by classical solid state reaction techniques. Two kinds of ceramics precursors were

used to produce the samples :
1. a chemical-route powder (thereafter named C-powder). This powder was provided to us by Rhône-Poulenc (grains-size distribution centered around 1 μm, specific area : 2,6 m^2/g).
2. a mixed-oxide route powder (MO powder) prepared in Thomson LCC laboratory by classical solid-state reactions (grains size : 3-7 μm, specific area : 1 m^2/g).

MO and C powders were cold pressed in shapes of green samples 5 mm x 5 mm x 5 cm. Sintering of the samples was made according to the following cycle : 100°C/h in air to raise temperature to T_S, 12 h at T_S, 100°C/h from T_S to 500°C in O_2, 5 h at 500°C in O_2, slow cooling to room temperature.

Three sintering temperatures were used for both powders : T_S = 910°C, T_S = 925°C and T_S = 950°C. The six samples obtained (C 910, C 925, C 950, MO 910, MO 925, MO 950) were cut into rods of 1,2 mm x 1,2 mm x 3-4 cm for microwave and critical current measurements. Parts of samples were also used for X-ray analysis, optical microscope and scanning electron microscope observations.

ELECTROMAGNETIC MEASUREMENTS

Two electromagnetic measurements were performed to characterize the samples :
- 1. critical current density J_C at 77 K
- 2. surface resistance at 77 K, at frequencies around 3,6 and 9 GHz.

J_C measurement techniques (4 probe pulsed current technique) are reported in detail in reference |3|. Microwave surface resistance measurements were made using a coaxial resonator method. This method was first developped in our laboratory (ref. |4|).

RESULTS AND DISCUSSION

1. Microstructure

The density of the sintered samples is shown in Table 1. For all the sintering temperatures C powder provides more dense samples than the MO powder.

TABLE 1
Microstuctural data for C and Mo samples

sample	C 910	C 925	C 950	MO 910	MO 925	MO 950
Relative density (%)	87	91	93	80	86	90
grain size (μm)	< 2	5- 20	25-130	< 10	8-25	16-50

1

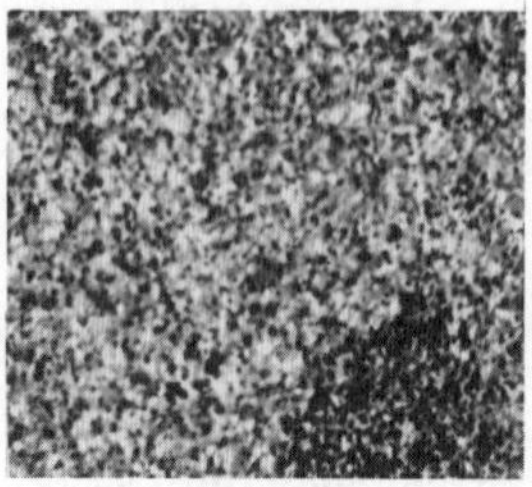

C 910 ⟝ 10 μm

2

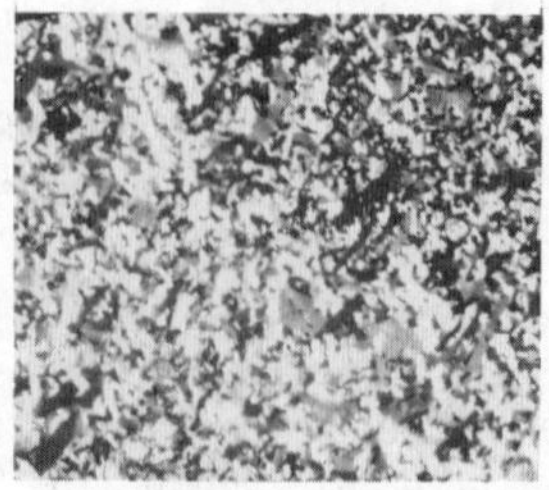

C 925 ⟝ 20 μm

3

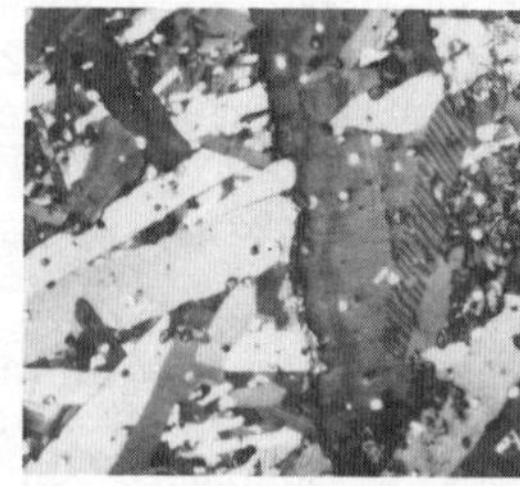

C 950 ⟝ 20 μm

MO 910 ⟝ 10 μm

4

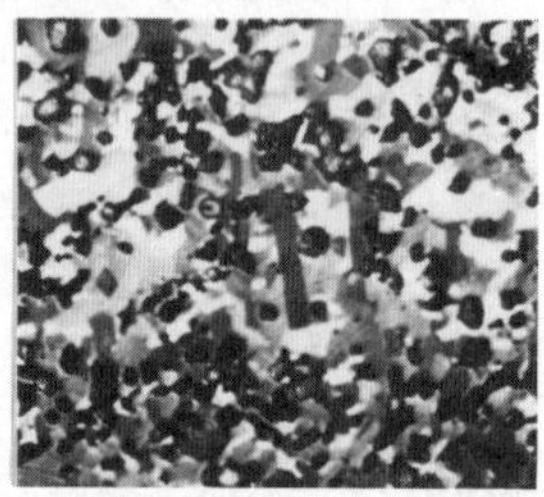

MO 925 ⟝ 20 μm

5

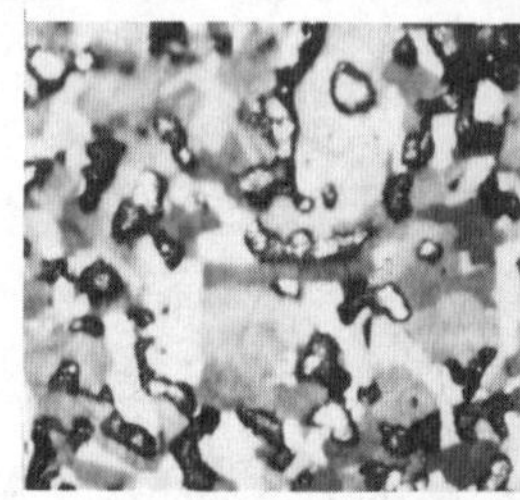

MO 950 ⟝ 20 μm

6

Microstructure of the samples

The higher reactivity of C powder is also apparent on microstructural characteristics. Photographs 1 to 6 are optical views of the microstructure showing the grain size increases with sintering temperature both for C and MO powders (see Table 1).

The pictures 1 to 6 also reveal a difference in the statistical distribution of grain shapes. C samples tend to have long and needle like grains when sintered at or above 925°C, while MO samples tend to have more isotropic grains. In samples C 925, C 950, MO 910, M 925, M 950 all the grains showed twins, which is a necessary result if samples are well oxygenated. In sample C 910 the grains were too small to permit an easy observation of the twins. As C 910 is only 87 % dense, while C 925 is 91 % dense, we feel it justified to suppose that C 910 is at least as well oxygenated as C 925. Another microstructural feature is the existence of microcracks within the grains or at the grains boundaries. A careful examination of the samples by optical or electron scanning microscopy shows that :

- C 925, MO 925 and MO 950 present a few microcracks mainly located at grain boundaries.
- C 950 shows many intragrain and intergrain microcracks.

X-ray analysis of all the samples reveals the presence of small amounts of unreacted phases $BaCO_3$, $BaCuO_2$, Y_2BaCuO_5 and CuO (at most a few percents).

2. Critical current density measurements

The measurements results are reported in figure 1.

3. Microwave surface resistance (R_S) measurements

For industrial applications the major interest of YBCO is the possibility to work at 77 K. This is why we focus here on microwave properties at this temperature. For all samples measurements of R_S were made around 3,5, 7 and 10,5 GHz. Around 3,5 GHz. The lowest values of R_S are obtained for MO 950, with 13 m / , still higher than values for copper (9 m /). An extrapolation of the results for MO 950 shows that it would have losses lower than copper below 2 GHz at 77 K. This clearly limits the industrial applications of YBCO ceramics.

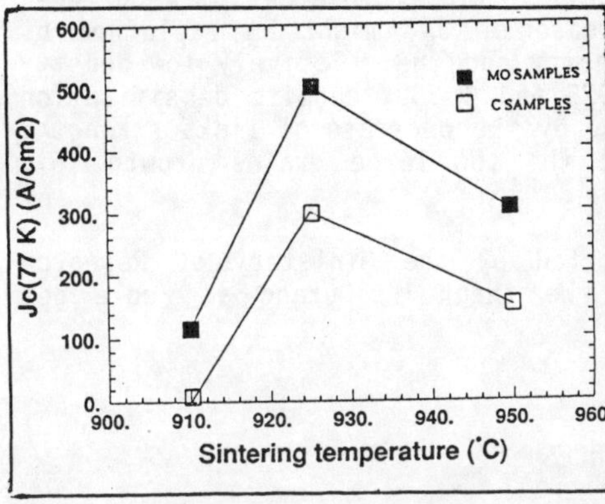

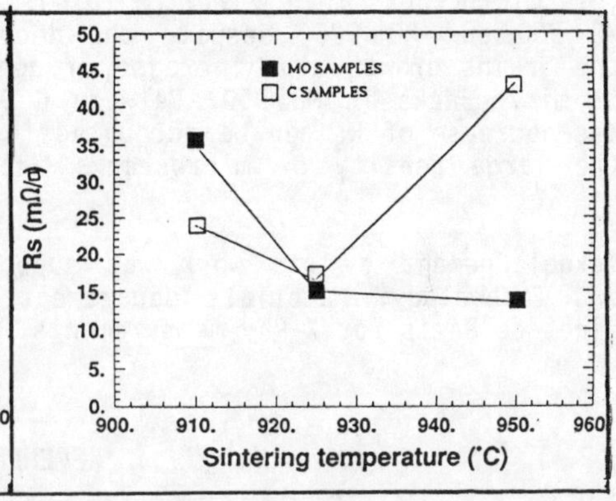

Figure 1. Critical current density at 77 K versus sintering temperature.

Figure 2. Surface resistance at 77 K and 3,5 GHz versus sintering temperature.

Figure 1 and 2 exhibit a striking difference between the behaviours of C and MO samples. While in the case of C samples the variations of R_S and J_c are correlated this is not the case for MO samples. In that last case between T_S = 925°C and T_S = 950°C a decrease of J_c occurs in parallel with a decrease of R_S. This clearly proves that the microstructural factors controlling J_c and R_S are not the same. We believe this is the first time that such an experimental observation has been made.

Various losses mechanisms have been proposed to account for measured surface resistances of YBCO ceramics. We believe our results are in support of ideas of granular superconductivity, where YBCO ceramics are assumed to be a collection of superconducting grains coupled by weak non superconducting links acting as Josephson junctions (critical current density J_c). According to us, in such an image J_c of the samples should be sensitive to the weakest links in the samples while R_S should receive contributions from all the juctions of the surface. Bearing this idea in mind we shall estimate the influence which the following microstructural elements may have on R_S :

- distribution of J_c of weak links (including anisotropy effects) :

an improvement in the distribution of J_c should in any model result in both improvements of J_c and decrease of R_S.

- grains size :

an increase of grains size should result in a decrease of R_S. Indeed if the average grains size is a and the average grain boundary thickness is d, the relative surface of grain boundaries goes like d/a. Such an influence of grain size has been observed in |5| on powder samples up to a grain size of 10 μm. For grains size above 50 μm internal boundaries are found to increase microwave losses.

- Porosity and microcracks : an increase in their density should result in an increase of R_S, through an increase of the penetration depth of the wave. An indirect effect of an increase of microcracks density consists in the degradation of links strength.

Those qualitative arguments help us to understand the improvement of R_S between C 910 (MO 910) and C 925 (MO 925). Indeed C 925 (MO 925) has higher J_C (links strength), higher density, larger grains and hardly more microcracks than C 910 (MO 910). The little improvement of R_S between MO 950 and MO 925, despite the decrease of J_C, might be explained by the grains growth, the increase of density and the relatively low density of microcracks in MO 950. Between C 925 and C 950, despite densification the decrease of R_S can be accounted for by the decrease of links strength, the large density of microcracks and the too large grains growth |6|.

Acknoledgement : This work was supported by the Ministery of Research and Technology, Materials department. We thank M. Durand at Ecole des Mines de Paris for X-Ray measurements.

REFERENCES

1. Bednorz G. and Müller K.A., Z. Phys. B64, 189 (1986).

2. Hein M., Klein N., Müller G., Piel H., Röth R.W., Wu B 87-22 (1987) Proceeding of the third workship on RF Superconductivity, Argonne, September 1987.

3. Kormann R., Lloret B., Lainée F. and Ganne J.P., To appear in proceedings of the E.C.S. Meeting at Maastricht, June 89.

4. Labeyrie M., Mage J.C., proceedings of the high T_C superconducting ceramics conference, Caen, September 1988.

5. Gould R., Jackson E.M., Renovard K., Grittenden R., Bhagat S.M., Spencer N.D., Dolherst L.E., Warmsbecker R.F.; Physic 156 (1988).

AMORPHOUS TO CRYSTALLINE TRANSFORMATION IN YBCO SUPERCONDUCTORS

R. DAL MASCHIO, A. MADDALENA*, S. DIRE', P. FERRERO
Dipartimento di Ingegneria, Università di Trento, 38050, Mesiano di Povo,Italy.
*Dipartimento di Ingegneria Meccanica, Sezione Materiali, Università di Padova, 35100, Padova, Italy.

ABSTRACT

Following recent tendencies in YBCO research, this work was undertaken to investigate the crystallization of amorphous samples prepared either by the sol-gel method or by quenching the molten mixture. The transformation from amorphous to crystalline was studied by simultaneous differential thermal and thermogravimetric analyses (DTA,TGA) and X-ray diffraction (XRD).

INTRODUCTION

The crystallization of amorphous materials may be a potential attractive route in order to obtain ceramic superconductors. Non-crystalline bulk materials are obtained either by melt-quenching or by sol-gel method. Melt-quenching allows production of samples with increased density and therefore improves electrical and mechanical properties. On the other side, the state of the art in the sol-gel method permits the production of films and fibers characterized by high homogeneity at relatively low sintering temperatures. The aim of this study is the investigation of the different crystallization behavior of amorphous YBCO samples prepared by these two techniques.

EXPERIMENTAL

$YBa_2Cu_3O_{7-x}$ samples were prepared by the conventional solid-state sintering method in air. In order to obtain amorphous YBCO, a rapid solidification method was used (1). This procedure involves melting of droplets from the $YBa_2Cu_3O_{7-x}$ sintered samples using a laboratory gas blow torch flame; the droplets fall between two counter-rotating copper rollers (60 mm diameter, 60 r.p.m.). The quenched flakes (50 µm thick and 10 mm wide) were almost amorphous although some crystallinity was still present. For the sol-gel method, standard techniques were employed in the manipulation of air-sensitive compounds and the solvents were distilled under Ar atmosphere. According to literature (2), a copper sol was obtained from anhydrous $Cu(CH_3COO)_2$ ($4.5 \cdot 10^{-2}$ moles) in 2-methoxyethanol in Ar atmosphere. In the same way, $Y(CH_3COO)_3 \cdot 4H_2O$ ($1.5 \cdot 10^{-2}$ moles) produced a clear sol. $Ba(OC_3H_7)_2$ ($3.0 \cdot 10^{-2}$ moles) was added to the yttrium-sol; then copper-sol was added to the mixture which assumed a dark brown color. The mixed sols were hydrolyzed, refluxing the solution for few hours, then the solvent was evaporated affording the final powder YBCO product in virtually quantitative yield. The powder was then calcined at 300°C to eliminate the organic compounds. DTA and TGA analyses were performed in static air with a heating rate of 10°C/min. XRD spectra were recorded using a Cu Kα radiation with a Ni filter. DTA patterns were utilized for the choice of thermal treatments; samples were heated in air with a heating rate of 200°C/h at different temperatures and subsequently quenched and XRD analyzed. The superconductivity at liquid N_2 temperature of samples was measured by the a.c. magnetic susceptibility method.

RESULTS AND DISCUSSION

In Figure 1 are represented DTA patterns of YBCO samples obtained respectively a) with a conventional solid state sintering, b) by melt-quenching method and c) by sol-gel method. In curve a) the DTA peaks can be probably ascribed to orthorhombic-tetragonal phase transformation (950°C) and to the 1-2-3 phase incongruent melting (1034°C) (3). These two endothermic peaks appear also in the melt-quenched sample pattern (curve b)). The peak at about 845°C, as results also from XRD 870°C heated sample analysis, may be attributed to 1-2-3 crystallization. The broad exothermic effect from 250 to 350°C, corresponding to a weight increase in TGA curve (not reported), is probably due to an oxidation . In curve c) the sol-gel pattern of a 60°C heated sample shows primarily a broad exothermic peak from 200 to 400°C corresponding to a wide weight decrease in TGA curve and ascribable to the reaction between the components in the primary gel (4). The endothermic effect at 820°C, owing to the increased reactivity of the fine sol-gel powder, may be attributed to 1-2-3 crystallization. As regards the peak at 978°C, it's reversible on cooling and it results absent on reheating where the system behaves like the sintered. This behavior may be attributed to the porous structure of the DTA sample, due to the bloating of gel at about 200°C (5).

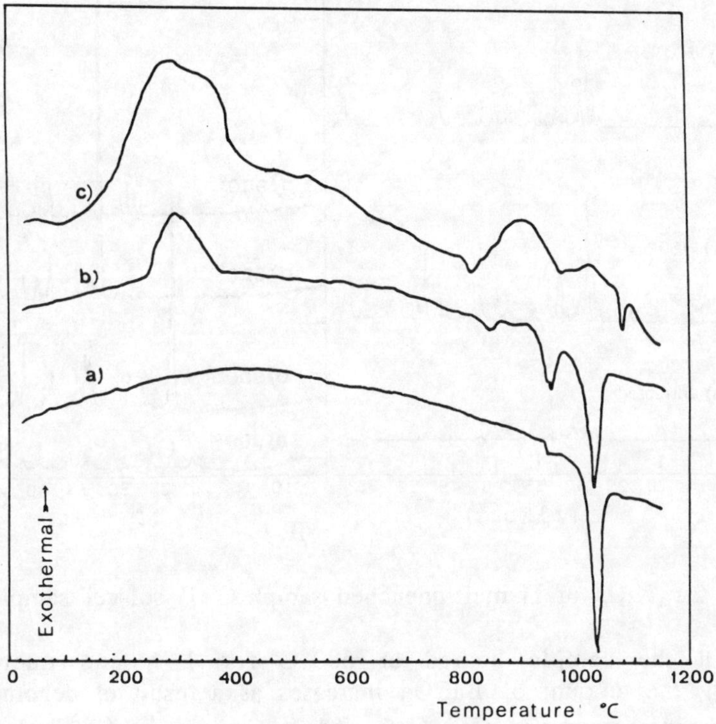

Figure 1. DTA patterns of YBCO: a)sintered, b)melt-quenched, c)sol-gel.

XRD of melt-quenched samples (Figure 2-I) are characterized in untreated specimen (-a) pattern) by Y_2O_3 in an almost amorphous matrix. On the contrary in a sample heated to 750°C for 1h and quenched (-b) pattern), it's possible to see peaks characteristic of $BaCuO_2$ and CuO. In c) pattern (870°C for 1h and quenched), it begins to appear 1-2-3 phase with the contemporary decreasing of the other phases. A prolonged heat treatment at 900°C for 16h followed by slow cooling in O_2, gives a sample with a typical superconducting behavior at 90 K. XRD of sol-gel samples are shown in Figure 2-II. The a) spectrum refers to the sample heated at 300°C for 6 h and shows, in an amorphous matrix, peaks attributable to $BaCO_3$ and acetate compounds as reported by Kozuka et al. (5).

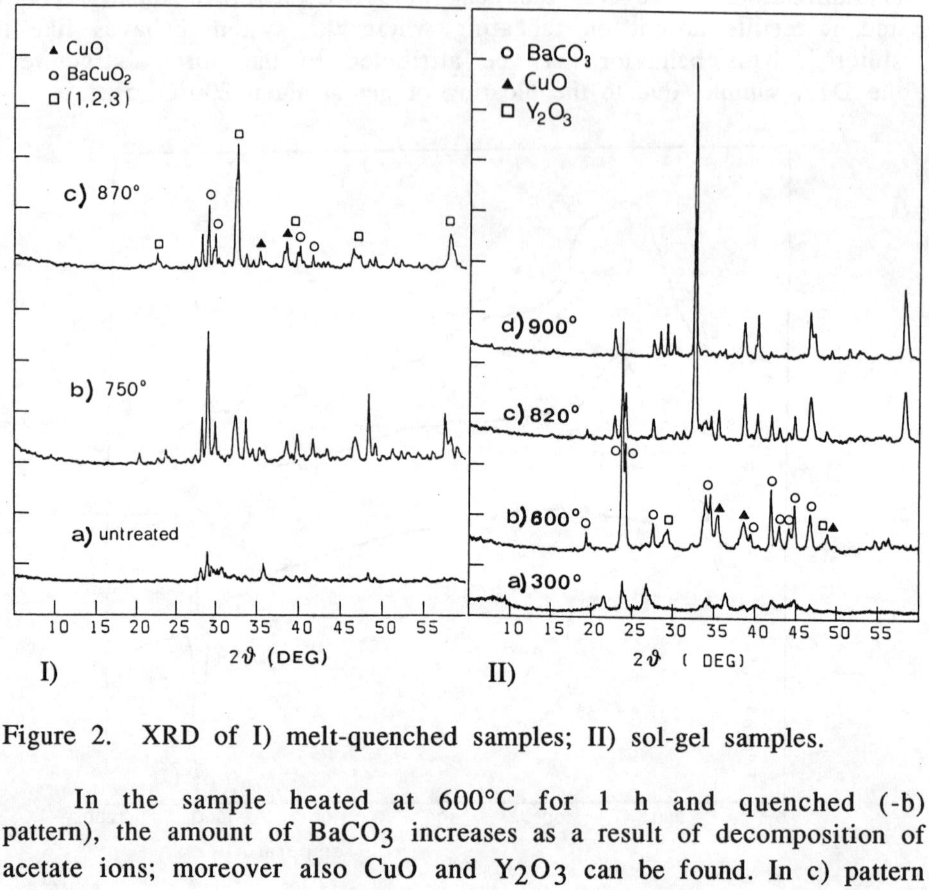

I) II)

Figure 2. XRD of I) melt-quenched samples; II) sol-gel samples.

In the sample heated at 600°C for 1 h and quenched (-b) pattern), the amount of $BaCO_3$ increases as a result of decomposition of acetate ions; moreover also CuO and Y_2O_3 can be found. In c) pattern (820°C, 1h and quenched) 1-2-3 phase begins to form and contemporary the other compounds decrease. Pattern d) (900°C, 1h and quenched) shows almost exclusively the 1-2-3 phase, although the splitting in the main diffraction peak at 32.8° cannot be seen. This effect may be ascribed to the too short heat treatment time and to quenching. In fact the same sample, treated in air at 900°C for 16h and subsequently cooled down to R.T. in O_2, shows the typical 1-2-3 orthorhombic XRD pattern and a good superconductivity transition, as evidenced in Figure 3. In conclusion, the crystallization of 1-2-3 phase, although through different intermediate compounds, is obtained from both melt-quenching route and sol-gel method at a lower temperature than traditional powder sintering method. In fact 1-2-3 phase starts to form at about 820°C in sol-gel powder and at about 845°C in melt-quenched samples.

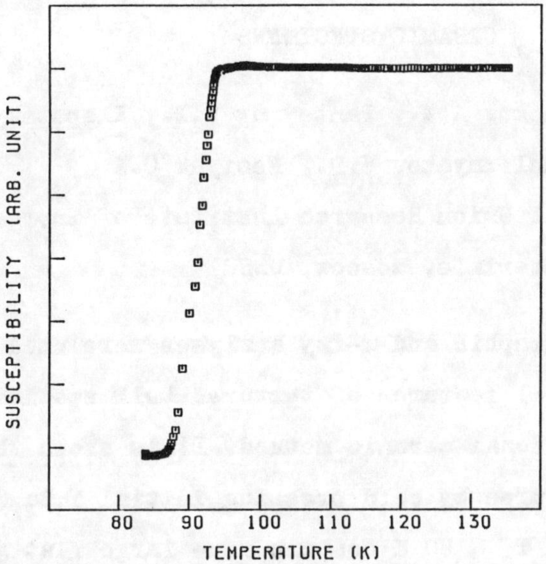

Figure 3. A.C. magnetic susceptibility of 900°C, 16h cooled in O_2 sol-gel sample.

REFERENCES

1. Kim, N.K., Drodzyk, L., Payne, D.A., Friedmann, T.A., Wright, W.H., Ginsberg, D.M., Mater. Lett., 1987, **5**, 387-90.

2. Ravindranathan, P., Komarneni, S., Bhalla, A., Roy, R., Cross, L.E., J. Mater. Res., 1988, **3**, 810-12.

3. Jiang, X.P., Zhang, J.S., Huang, J.G., Jiang, M., Qiao, G.W., Hu, Z.Q., Shi, C.X., Mater. Lett., 1988, **7**, 250-55.

4. Barboux, P., Tarascon, J.M., Greene, L.H., Hull, G.W., Bagley, B.G., J. Appl. Phys., 1988, **63**, 2725-29.

5. Kozuka, H. Umeda, T. jin, J.S., Sakka, S., in Better Ceramics Through Chemistry III, eds. C. J. Brinker, D.E. Clark, D.R. Ulrich, MRS, Pittsburgh, 1988, pp. 639-42.

INVESTIGATION INTO THE INFLUENCE OF HIGH TEMPERATURE HEAT TREATMENT ON THE STRUCTURE OF TEXTURED BULK $Y_1Ba_2Cu_3O_{7-x}$ CERAMIC SPECIMENS

Shikov A.K., Pantsyrny V.I., Klepatsky V.E., Zheleznyakov V.D., Nestyuk O.I.
All Union Research Institute of Inorganic Materials, Moscow, USSR

Metallographic and x-ray analyses were used to investigate structural features of textured bulk specimens produced by the traditional ceramic method. It is shown that specimens manufactured by cold pressing initial $Y_1Ba_2Cu_3O_{7-x}$ powder having T_c 90 K contain some large flat grains up to 30 mcm in size with the "C" axis normal to the pellet surface. It is established that with further high temperature sintering within 850-960°C the degree of the texture becomes less.

From the time the discovery of the superconductivity above the liquid nitrogen temperature was reported 1 a great number of investigations were published dealing with the $Y_1Ba_2Cu_3O_{7-x}$ structure and the interrelation between the structure and the synthesis parameters, primarily, heat treatment conditions. Due to the peculiar crystal structure of a superconductor a strong anisotropy is suggested in the superconductivity properties such as the critical current and critical magnetic field in different crystallographic directions 2 . To-day, a great quantity of the methods of $Y_1Ba_2Cu_3O_{7-x}$ synthesis have been developed and it is unambiguously confirmed that production of highly textured specimens leads to the maximum critical current densities up to 10^6 A/cm^2 at 77 K 3 . However, this high level of the critical parameters is reched in thin film small size specimens with the film thickness of some fractions of a micrometer. Some successful attempts are also known to prepare bulk textured specimens, in particular, using a highly sophisticated control method of molt texturing 4 ; these attempts attest that high current properties, viz., $1.75x10^4$ A/cm^2 at 77 K , have been attained. Along with the indicated method the work 5 shows the feasibility of preparing textured bulk specimens by the simplest solid state ceramics synthesis by varuing the synthesis temperature. To produce the texture at the intermediate synthesis stage $Y_1Ba_2Cu_3O_{7-x}$ powder having the tetragonal structure was produced by quenching from 950°C. The aim of our work was to investigate the structure of textured bulk $Y_1Ba_2Cu_3O_{7-x}$ specimens produced by the traditional ceramics synthesis and the mode of the influence of the heat treatmenttemperature on the specimen structure at the final sintering stage.

$Y_1Ba_2Cu_3O_{7-x}$ powder prepared by the solid state synthesis of Y_2O_3, $BaCO_3$ and CuO and having the onset temperature of the transition to the superconductivity 90 K was used as an initial material. The powder particle morphology is illustrated in fig.1 which shows that the average particle size of the powder was 3 mcm with the scatter in the individual particle sizes from 0.3 to 30 mcm. The mass contents of yttrium, barium and copper metal atoms in the initial powder were 12.9%, 38.6% and 27.8%, respectively. According to the data of x-ray and diffractometric analyses the content of phase 123 in the powder was in excess of 96%, fig.2a.

Morphology of the initial $Y_1Ba_2Cu_3O_{7-x}$ powder

Fig.1

The powder was cold compacted into pellets 18 mm dia 2-3 mm high. Then the pellets were heat treated at 700°C for 16 h in air to increase their strength. The x-ray analysis of the heat treated pellets showed a clear-cut texture with the "C" axis normal to the pellet surface (fig.2b).

Diffraction patterns of initial powder and pellets of $Y_1Ba_2Cu_3O_{7-x}$

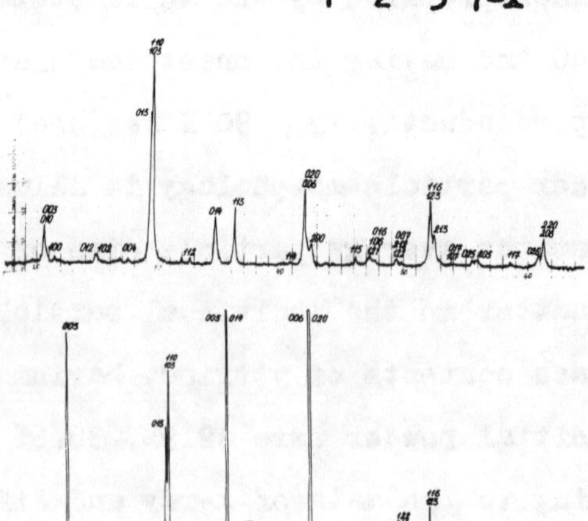

Fig.2

The investigation into the grain structure revealed some flat surface grains up to 30 mcm in size with the interstices between them filled with finer equiaxed crystallites, see fig.3.

Surface morphology of a pellet heat treated at 700°C, 16 h

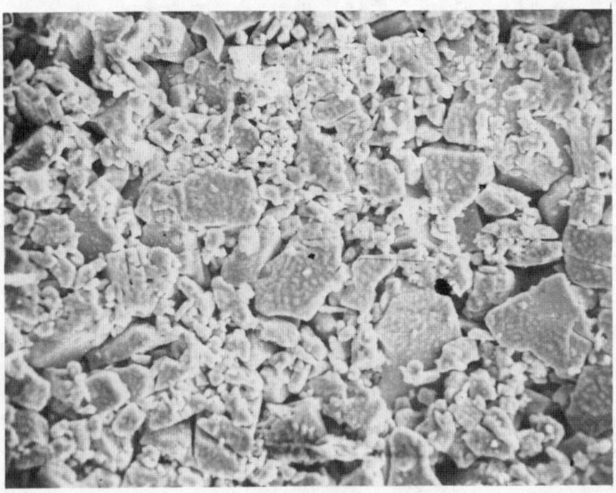

Fig.3

The subsequent heat treatment of textured specimens was conducted at 850-900°C for 5 h as it is known 6 that a further increase of a heat treatment duration does not significantly affect the structure and primarily the density of specimens. The data of the x-ray analysis show that with an increase of the sintering temperature the degree of the specimen texture is gradually reduced (fig.4).

With a heat treatment temperature rise the specimen grain structure changes mainly through varying the quantitative ratio between large flat grains and interstitial fine equiaxed ones the number of which diminishes (fig.5).

Diffraction patterns of high temperature treated specimens

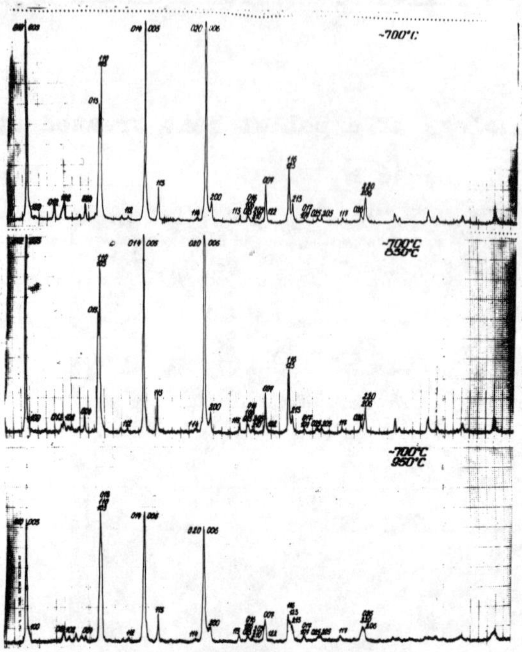

Fig.4

Surface morphology of high temperature treated specimens

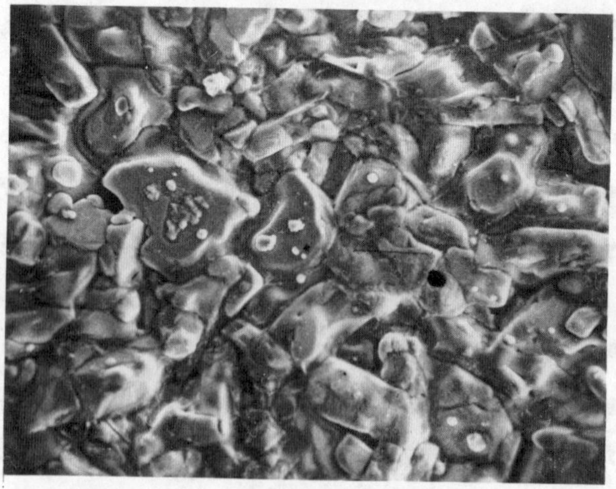

Fig.5

Thus, the work shows that texture $Y_1Ba_2Cu_3O_{7-x}$ specimens can be produced by the traditional ceramics synthesis process and the texture of ceramic specimens is reduced by a heat treatment at a higher temperature 850-960°C. Now investigation of the interrelation between the structure and current carrying capacity is in progress.

References

1. K.Wu, J.R.Ashburn, J.R.Torng et al. Phys.Rev.Letters 58 (1987) pp.908-910.
2. A.D.Nikulin, V.J.Filkin, I.I.Davydov et al. "Influence of heat treatment on $YBa_2Cu_3O_{7-x}$ critical properties, composition and structure "Presented to the I Pacific Ocean-Asia Conf. Singapure 24.06-02.07.88.
3. Y.Enamoto, T.Murakami, M.Suzuki, K.Morivaki. Jap. J.Appl. Phys. v.26 (1987) No 7, pp.L1248-L1250.
4. S.Jin, T.F.Tiefel et al. Appl.Phys.Lett. v.52 (1988) pp.2074-2076.
5. M.Omori, S.Tomiyoshi, H.Yamauchi et al. Jap. J.Appl.Phys. v.26 (1987) p.L1421-L1423.
6. B.V.Revdi, K.Jain, S.Singh et al. Supercond. Sci. and Technol. v.1 (1988) pp.148-152.

ASPECTS OF TRADITIONAL CERAMICS

J. ENRIQUE NAVARRO
ASOCIACION DE INVESTIGACION DE LAS INDUSTRIAS CERAMICAS
Ctra. Borriol km 0,7, 12004 CASTELLON (SPAIN)

ABSTRACT

In the last few years there has been a strong technological development in the traditional ceramic products such as tiles and sanitaryware.

The ceramic manufacturing processes evolution towards higher performance, more flexible production and lesser quality variation systems is tremendous. Because of that, we can find old fashioned process stages or even old fashioned whole processes whithin the implantation period which is sometimes smaller than five years.

Clear examples of those features in tile manufacturing are the development and evolution of single firing not only in floor tiles but also in wall tiles.

Refering to sanitaryware the progress has ocurred not only in the under pressure slip casting methods but also in the implantation of highly flexible manufacturing units which allow a great product variety.

In this work it is shown these evolution development in the last years and the possible trend not only in the technical field but also the market tendency in the future.

CERAMIC FLOOR AND WALL TILES

Today, the term "ceramic tile" covers a large variety of products distinguished by their primary composition, method of production, structural and physical characteristics, mechanical strength, surface finish, dimensions and colour.

In Europe, the classification established by the European Committee for Standardization (CEN) divides ceramic tiles into groups based on their methods of production (extrusion or dust pressing) and their degree of water absorption (an important factor in determining such characteristics as mechanical and frost resistance).

As noted above, ceramic tiles can be glazed or unglazed. Most of the tile manufactured is glazed. Glazes afford an almost infinite variety of colours and decoration. They can also provide a variety of finishes, from mat to satin to high gloss, depending on the desired effect.

The classification of ceramic tiles according to groups, specific product standards as well as to the characteristics required in their different applications, is listed in TABLE 1.

TABLE 1

Classification of ceramic tiles according to their groups and their specific product standards

WATER ABSORPTION	*GRUPO I* E<3%	*GRUPO IIa* 3%<E<6%	*GRUPO IIb* 6%<E<10%	*GRUPO III* E>10%
SHAPING				
A	*GRUPO AI* EN 121	*GRUPO AIIa* EN 186	*GRUPO AIIb* EN 187	*GRUPO AIII* EN 188
B	*GRUPO BI* EN 176	*GRUPO BIIa* EN 177	*GRUPO BIIb* EN 178	*GRUPO BIII* EN 159

The wide range of ceramic products to be found on the market is determined by their multiple uses, and depends upon the site where they are to be located. At present the most common uses to which these materials are put, are:

- *Interior cladding in the home (kitchens, bathrooms,etc)*
- *Interior floor coverings in the home*
- *Exterior coverings (facades)*
- *Outdoor paving (terraces)*
- *Floor coverings in public buildings (hospitals, schools, offices, etc*
- *Various types of floor and wall coverings (swimming pools, floors in industrial installations, etc.)*

As far as technological innovation is concerned, the trend to diversify ceramic floor and wall coverings as regards format, texture and technical characteristics, is to be emphasized.

The automation of most of the stages in the production process and the increasingly rational constant control of this process have led to a product with an ever greater quality and ever improving properties and performance. On the other hand, the introduction of sophisticated glazing methods, together with the continuous development of new glazes, frits and ceramic pigments, and the recovery of artisan decorating procedures have brought the product to the same or even higher levels of quality while maintaining the esthetic expression of traditional ceramics.

The consolidation of industrial design at the heart of the enterprises is the climax of a whole product strategy which is to keep it competitive against other alternative products, thus reaching levels of performance which are comparable, if not clearly superior, to those of other materials.

The future of the product is decided on the field of surface treatment and its ability to fulfil the esthetic and functional needs of architecture and present day interior decoration.

At present the great manufacturing groups to be found in modern technology, are:

- Plants processing raw materials for bodies, mainly involving spray driers with discontinuous wet grinding.
- Plants manufacturing twice-fired ceramic coverings with a fast second firing cycle in roller kilns or second firing glazing plants with a fully automated process.
- Plants manufacturing single-fired ceramic floor coverings which start their process at the pressing stage or also include their own spray drying plant.
- Plants producing single-fired ceramic coverings or twice-fired ceramic coverings with fast firing in the first and the second cycles, the former having the same characteristics as the latter.
- Plants manufacturing split tiles by the classical method.
- Plants manufacturing split tiles with the direct setting process, whereby the split tiles are loaded onto the kiln car directly after stiff extrusion.

As opposed to these manufacturing systems, survival of traditional double-firing techniques focussing exclusively on ceramic coverings, will depend upon their ability to offer an excellent quality product with a high added value due to the glazing.

In a sector characterized by constant technological evolution it is practically impossible to elucidate medium or long-term future technological trends. The following is to be foreseen within the short term:

- The complete link up of all the stages in the production cycle to form an integral line ranging from the shaping of the piece to the packaging and labelling of the final product, thus affording the possibility of complete production management.
- The incorporation of robotics at some stages of the process, either to take over certain handling operations or to carry out series of decorating operations.
- The optimization of the different stages in the processing of raw materials for bodies, both in the field of continuous wet grinding for large quantities of spray dried bodies as well as in that of dry grinding with subsequent granulation.
- The continuing research into new bodies and glazes to improve the physical-chemical characteristics above all in floor coverings and to increase the texture and chromatic possibilities of ceramic coverings.

The production capacity of the Spanish ceramic floor and wall tile industry is about 230 millions sq. m. per annum at present. This production is more than enough in the present situation with its intensive activity in residential construction and remodelling both in Spain as well as in the most important international markets.

A significant worldwide production overcapacity must be highlighted, which allows full employment even in the present state of affairs in the residential building industry. This overcapacity and the possibility of a negative change in the present situation render new production increases ill-advised whether these be carried out by new enterprises or by enlargement of existing ones unless these enterprises have both a firm position on the market as well as the size and installations to face a hardening competition.

Production follows the trends of demand, which proceeds more and more towards, higher quality and more differentiated products. As a result, the offer of common types and formats has diminished thus producing their revaluation as well as new products with an increased esthetic contribution. In TABLE 2, figures on ceramic floor and wall tile imports in the U.S.A. are listed, in which the average prices of the different countries stand out.

TABLE 2

U.S. TILE IMPORTS ACCORDING TO COUNTRY OF ORIGIN (1988)

	Million sq.ft	%	Million dollars	%	Average price
TOTAL	524.0	100	410.91	100	0.78
ITALY	248.0	47.3	210.04	51.1	0.85
MEXICO	61.9	11.8	30.34	7.4	0.49
SPAIN	58.8	11.2	43.3	10.5	0.74
JAPAN	36.5	7.0	55.8	13.6	1.53
BRAZIL	35.1	6.7	15.5	3.8	0.44
S.KOREA	17.2	3.3	10.7	2.6	0.63
THAILAND	12.7	2.4	7.0	1.7	0.55
TAIWAN	6.7	1.3	4.6	1.1	0.69
W.GERMANY	5.0	1.0	7.0	1.7	1.40
OTHERS	42.0	8.0	26.5	6.5	0.63

The Spanish ceramic floor and wall tile industry and the whole of all related activities, which range from the manufacture of frits and glazes to the manufacture of machinery may be considered to have finished their process of technological renovation.

In TABLE 3, the investment in this sector over the last few years, is to be found.

TABLE 3

Total investment in the ceramic floor and wall tile sector

YEAR	*TOTAL INVESTMENT 1000 Sp. pts*	*NEW ENTERPRISES*	*ENLARGEMENT OF EXISTING ENTERPRISES*
1983	*1.287.677*	*8*	*14*
1984	*2.310.922*	*6*	*28*
1985	*2.759.523*	*19*	*44*
1986	*3.823.051*	*15*	*33*
1987	*4.041.304*	*18*	*39*
1988	*9.294.246*	*26*	*52*

These figures reflect the renovation of the Spanish ceramic floor and wall covering plants.

At present, more than 85% of the production of ceramic floor and wall coverings is obtained through a fast firing cycle, incorporating the latest technology.

In the single firing field and especially with regard to single-fired products with a body which has a medium or high water absorption capacity, thus being particularly suitable for interior cladding, the effort has been great and well directed.

Export figures of the last few years are to be found in TABLE 4, as we can observed the export increase made up of high price, high quality product segments coinciding with the general trend of Spanish production, is to be observed.

TABLE 4

Export of Spanish wall and floor tile industry

AÑO	*10^3 sq m*	*Million Pts*
1985	*50.254.*	*29.338*
1986	*46.228*	*28.624*
1987	*55.798*	*38.288*
1988	*66.692*	*48.490*

SANITARYWARE

In vitreous china bathroom sanitaryware, representing almost all world-wide production, materials characteristics and product typology suggest no significant innovations are to be expected.

As far as production technology is concerned, we have an almost artisan situation at present, which is very manpower intensive and has a manual process or at most functions with the aid of simple mechanical systems, this with the exception of the large industrial groups.

In these processes, labour costs represent a more than 40% share of production costs, thus making clear the need to rationalize the use of manpower, which innumerable enterprises of this kind have started to bring about.

Here follows the description of the most important technological innovations in the manufacture of vitreous china:

- *Glazing with programmable robot.*
- *Firing in a roller kiln for large numbers of uniform products.*
- *Firing in intermittent kilns for limited numbers and diversified products.*
- *Semi-automatic bench casting and fast drying of the plaster moulds*
- *Mechanical medium-pressure bench casting.*
- *Casting with an automatic high-pressure press.*
- *Continuous drying in a fast dryer.*

It seems probable that world demand for vitreous china sanitaryware will continue its long term growth, though this will be interrupted from time to time by political and economic crises. Where alternative materials prove competitive in quality and price, manufacturers of vitreous china sanitaryware will be forced to make or market sanitaryware in the new materials, as was the case with everything needed for the bathroom some years ago.

The Spanish vitreous china sanitaryware sector is made up of six enterprises, production evolution being illustrated by TABLE 5. Production shows a positive evolution in the last few years in accordance with the development in residential construction and remodelling.

TABLE 5

Sanitaryware in Spain

	PRODUCTION		EXPORT	
AÑO	Tons	Million Pts	Tons	Millions Pts
1982	81.700	9.112	14.600	1.557
1983	70.338	8.270	17.159	2.368
1984	69.998	8.730	19.597	2.870
1985	71.386	10.179	21.792	2.984
1986	79.000	12.378	19.375	2.922
1987	82.000	13.487	18.678	2.913
1988	87.000	14.190	23.856	3.214

NEW WAYS IN THE FIRING TECHNOLOGY FOR WHITEWARE CERAMICS (Porcelain)

Dipl.-Chem.R.Sladek, Riedhammer GmbH & Co.KG,
Klingenhofstraße 72, 8500 Nürnberg

abstract:

The general development of technology - in some fields one can even talk of a "speedy" development - also influenced the process of modernization in the ceramic industry, though only gradually.
Reflections on how to obtain a better flexibility for the firing process resulted in corresponding new developments which guarantee a faster firing, a lower energy consumption and an optimum quality due to a more precise sintering process control. Parallelly to these achievements, endeavours are made to provide for an ecological relief and to bind and/or to dispose of the "classical" harmful substances released in the production of ceramics (fluorides and hydrocarbons) accordingly.
The realization of the ceramist's "dream" to develop a technology for a continuous production line even for whiteware seems to be quite near. The preservation and improvement of competitiveness can no longer be guaranteed without keeping up with the technical progress.
The use of a modern firing technology may create the necessary prerequisites such as a new technology for the production of hotel porcelain.
The firing units can be operated both continuously and periodically, whereas the latter operation type concerns mainly small-size production units.

\- - - - -

For putting the theory of fast firing respectively modern firing technique into practice, the following requirements had to be satisfied:

I. Further development of the firing technology, for example high-velocity burners, and thus a different geometry of the kiln cross-section (at present useful kiln widths of 2.5 to 4.5 m can be realized in continuous or intermittent kilns).

II. Introduction of microprocessor-controlled firing and regulation systems

III. Selection of high-quality insulating firebricks and high-temperature fibres for the temperature range of up to 1300°C, for example for sanitary ware and tableware such as earthenware, stoneware, vitreous china and bone china; moreover the appropriate kiln furniture and conveying means qualities (mullite/oxide ceramic rollers, SiSiC-rollers i.e. rollers of infiltrated SiSiC, kiln furniture on cordierite basis).

IV. Selection of high-quality insulating firebricks and high-temperature fibres for the temperature range of above 1300°C, for example for porcelain (fancy, household and hotel porcelain), with the appropriate kiln furniture and conveying means - mainly on non-oxide ceramic basis.
The fast firing in the high-temperature or porcelain range depends largely on the quality of these materials.

V. Better flexibility in kiln construction
- Prefabrication
- Light-weight construction
This also permits the exact application of the respective optimum firing curve with the corresponding energy balance.

- - - - -

For the changeover from production on the basis of a conventional technology to modern firing technology, the following has to be taken into consideration:

1. Possible change of the properties of the body
2. Constructive design of the kilns
3. Corresponding kiln furniture quality
4. Reduction of the kiln furniture
5. Increased use of mechanization and automation possibilities
6. Reduction of the specific energy consumption
7. Reduction of the environmental pollution by raw materials and binders
8. If required, arrangement of filter and fluoride absorption plants as well as postcombustion plants for hydrocarbons for reducing the fluoride and hydrocarbon content of the flue gases

The adaption of the firing units to the properties of the product to be fired leads to a "tailor-made" kiln design with an optimum regulation and firing technique.

The firing units can be operated both continuously and periodically, whereas the latter operation type concerns mainly smallsize production units. The modern passage kiln is certainly more economical in energy consumption than an intermittent kiln, however, there must be considered that the costs for personnel which represent a large portion of the total expenses exerts a great influence on the economic efficiency.

Kilns economical in energy are for example:

- On the sanitary sector:
  The fibre-lined fast-firing tunnel kiln with a specific energy consumption of less than 1200 kcal/kg of ware net and the roller kiln with less than 800 kcal/kg of ware net in comparison to the classical muffled tunnel kiln with 2500-3500 kcal/kg of ware net.

- For the glost firing of porcelain:
  The most modern roller kiln with 1800-2200 kcal/kg of ware net with a setting weight of 25 kg/m² in comparison to the known tunnel kilns with 5000-8000 kcal/kg of ware net.

Short kiln cycles - especially with intermittent kilns - have become possible due to the modern lining technique, namely the combined variant with insulating firebrick lining and fibre "wallpaper" glued on the kiln hot faces; or the all-fibre lining - a variant used mainly for temperature ranges of up to 1300°C with growing tendency. However, those linings are also used in tunnel kilns which can be operated periodically for the compensation of production peaks or which are kept at a lower temperature or shut down during the weekends.

The operation of all these kilns would be unthinkable without the use of electronic systems for the automatic regulation of the time/temperature/atmosphere profile and the use of heating-up and temperature reduction programs.

<u>Tendencies in the Sanitary Field</u>

The open-flame tunnel kiln has superseded the muffled tunnel kiln due to its enormous advantages.

- Roller kilns with cycles of 8 to 10 hours and outputs of 20-40 tons/24 hours are in operation with automatic circulation systems and, if desired, with storage units for the weekend (fig. 1).

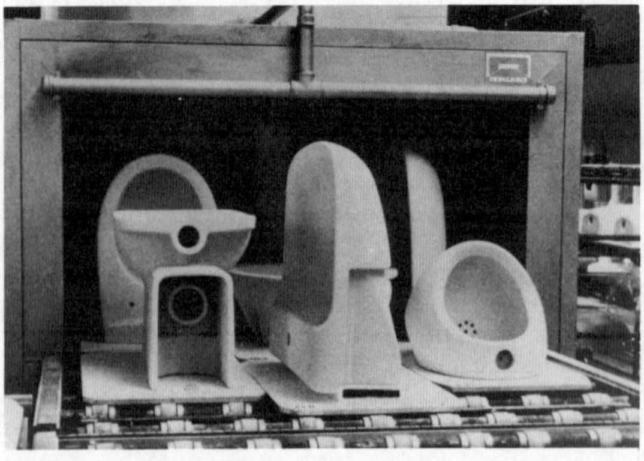

fig. 1

- All-fibre lined tunnel kilns with cycles of 12 to 18 hours which can also be operated intermittently, represent the latest technology (fig. 2)

fig. 2

- Shuttle kilns as an alternative for firing green ware, but mainly also for repair-firing, can no longer be dispensed with in rather big factories (fig. 3)

fig. 3

Tendencies in the Tableware Field

Here the introduction of the fast-firing technology - even though a little bit slower than for example in the sanitary field - is clearly foreseeable. A flat cross-section of both the fast-firing kiln with table conveyance (top and bottom burners) - see fig. 4 - and the roller kiln - see fig. 5 - will certainly become a part of the modern technology of the continuous production line system working with a certain rhythm.

A new technology for the production of hotel porcelain is under development, the process sequence of which (pressing with integrated decoration, glazing and once-firing) from the preparation to the finished product would take 10 to 16 hours only.

fig. 4 - fast-firing kiln with table conveyance for glost firing porcelain at 1400°C
kiln passage time: 4-6 h
output/24 h: 5-10 tons

fig. 5 - roller kiln for glost firing porcelain
kiln passage time: 4 h
output/24 h: 5 tons

An assortment separation in the sense of the technological "line conception" is imperative since the pressing factor is the wage cost portion. However, for smaller factories the alternative of using a shuttle kiln, that means intermittent operation, is absolutely reasonable, particularly with regard to saving personnel considering the 5- or 6-day working week.
The all-fibre lined tunnel kiln can be considered another alternative for biscuit-firing resepectively glost-firing.

Tendencies for the Future - General

On principle, there are three directions for the development of the manufacturing process:

a) The intermittent firing operation with a 5-day week rhythm

b) The continuous firing operation adopted in the early Sixties with the use of tunnel kilns.

c) The continuous firing operation in view of a flexibilization of the working time, i.e. a better utilization of the machinery and thus an output increase by 20 to 25 %.

All three directions are based on the common tendency to automize the working steps from the shaping to the finished product. The ceramic industry will surely show an increased interest in finding solutions that are beneficial for the economy, the ecology as well as their competitiveness.

NEW TRENDS AND DEVELOPMENTS OF CERAMIC TILES PRODUCTION TECHNOLOGIES

G. NASSETTI(*), M. POPPI(**)
(*)Centro Ceramico, Bologna, Italy; (**)Poppi S.p.A., Sassuolo, Italy

ABSTRACT

A review is made of the more important innovations relative to fabrication technology (in particular as regards the phases of raw materials preparation, drying and firing) recently introduced by the ceramic floor and wall tile industry. The description of the innovative solutions regarding machinery and equipment is accompanied by a discussion of the requirements leaded to such innovations and characterization of the advantages resulting from their introduction with regard to quality of the semi-worked pieces and final products as well as the optimization of energy consumption and management of the processes as compared with traditional technology. The picture that emerges provides a precise detailed outline of the evolutional trends in a sector, that of ceramic floor and wall tile, in which Italy holds an important and stable position of leadership, not only as regards production capacity but also in relation to progress in fabrication technology and the attainment of higher levels of quality.

INTRODUCTION

The purpose of this report is the analysis and the detailed explanation of the latest technological innovations in ceramic production processes, especially for floor and wall tiles.

Firstly, let's consider the previous history of technological innovation, by running through its most important steps and by evidencing its reasons: in this way we can understand that technological innovations originate from specific necessities.

As a matter of fact, technology isn't an abstract concept, but it's rather the way goods are produced: it's an extremely concrete operative way and it has to represent the instantaneous and latest knowledge degree in that particular field, in order not to become obsolete. It must therefore be evaluated according to the industrial result it's aiming at and, accordingly, from an economical stand point.

Tile technology was born in Italy in 1902 (table no. 1) when the first tile production started with 100 workers and proved to be interesting as far as the tile output. It was an extremely handicraft-technology and very similar to the one used by housewives cooking biscuits. It was all started by the presence and the utilization of local clays, for producing floor and wall tiles with good physical-mechanical features and long life, to be used mainly in those rooms of the house - such as bathrooms and kitchens - where cleaning easiness and hygiene were the most important requisites.

During the 50's two significant technological innovations were introduced: the automatic press and the tunnel kiln, that caused handicraft-activities to change into large industrial companies.
In this case the main target was the productivity increase inside the different companies, in order to satisfy the needs of a continuously growing market. As a matter of fact during these years the tile began to be regarded as a furnishing and interior decorating item, thus overcoming its use as a mainly technical material for bathrooms.

The 60's and the beginning of the 70's experienced the main ceramic industry development in Italy. The production cycle was completely automatized through the continuous production lines, the material handling automation and the spray-drying introduction and development. This process replaced the discontinuous working of the filter-pressing for body wet preparation.

Between 1974 and 1982 three events determined the sector crisis:

- the productivity increase due to the automatized lines;
- the single-layer fast firing introduction and development;
- the energy crisis that contributed to the fast firing spreading, as it allowed remarkable energy savings with respect to the traditional stack firing.

The 1982 crisis was caused right by the ceramic product offer overcoming the true market demand.

Between 1980 and 1986 many innovative technologies were introduced, thus allowing energy savings (energy costs peak was in 1975) and, at the same time guaranteeing the same output level even with a lower manpower. This resulted in the offer being readjusted to the market demand.

The trend of the technological innovation towards "energy" saving continued until the middle '80s and then, subsequent to the fuel cost remarkable decrease

Table 1

REFERENCE DATA	EVENTS	MAIN REQUIREMENTS
1902	Rise of the technological ceramic process in Sassuolo (Italy)	Exploitation of local clays
50's	Introduction of the automatic press and the tunnel kiln	Promotion from crafts factories to industries
60's & 70's start	- Carrying out of continuous processes - Automation of materials flows - Introduction of the spray drier	Increase of processing line productivity
1974-1982	- Increase of automatized line productivity - Introduction & development of single layer fast firing - Energetic crisis	More exploitation of plant productivity keeping the same manpower
1982	Big crisis of the sector	The offer of the product exceeded the requirement of the market
1985	Energy costs highest increase	
1980-1986	- Introduction of technological innovations with lower energy consumptions - Keeping of production quantities & decreasing of manpower	- Reduction of energy costs - Adaption of production quantities to real requirement of the market
After 1986	- Increase of production quantities and manpower - Introduction of technological innovations with environmental and product quality aims	- Reduction of environmental impact - Increase of product competitiveness

(fig. no. 1), innovative technologies were introduced showing environmental and/or product qualification trends, as well as utilizing energy in a more rational way.

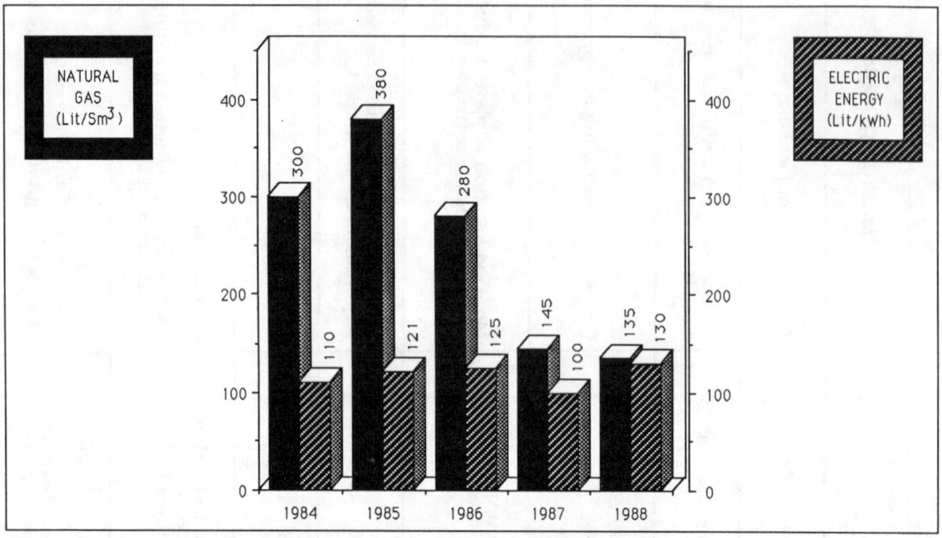

Figure 1. Trend of energy costs.

THE INNOVATIONS

While analyzing the innovative technologies brought about in the ceramic field after 1982, we want to follow the production steps sequence within the production process (table no. 2), so as to evidence its details more clearly. We put into evidence the starting year of the innovation and the main reasons that brought to it, subdivided between:

- **energy** reasons, that is consumption limiting and rational energy use;

- **operational** reasons, that is greater cycle automation and production times reduction;

- **environmental** reasons, that is reduction of the environment impact due to the production cycle;

- **product qualification** reasons, that is improvement of the tile aesthetic and technological requisites.

Table 2

PROCESSING PHASE	TECHNOLOGICAL INNOVATIONS	INTRODUCTION ON THE SECTOR	MAIN REQUIREMENTS
PREPARATION OF RAW MATERIALS (grinding and moisture content regulation)	Dry process: GRANULATION	1983	Energy
	Wet process: CONTINUOUS GRINDING	1985	Energy & Management
	DISCONTINUOUS MILLS OPERATED BY INVERTER	1988	Energy & Management
SHAPING	SINGLE PRESSING	first tests 1972-76 carried on 1983-88	Management & Environment
DRYING	SUPER FAST DRYING (hot air, I.R.)	1984	Management & Product improvement
GLAZING	DRY GLAZING	1988	Environment & Product improvement
FIRING	I.T.P. INTEGRAL THERMAL PROCESS	1986	Management & Product improvement

In the raw material preparation step the following processes were introduced:

- pressing powders granulation;
- continuous wet grinding;
- inverter installation on discontinuous wet mills.

The main purpose of these technologies was that of reducing energy consumptions during a production process step, implying high thermal consumptions, particularly during the wet cycle, due to spray-drying, as well as electric consumptions, due to the grinding in drum ball-mills.

Granulation supplies powders quite similar to the spray-dried product as for particle size and shape: it agglomerates mechanically, by moistening, the

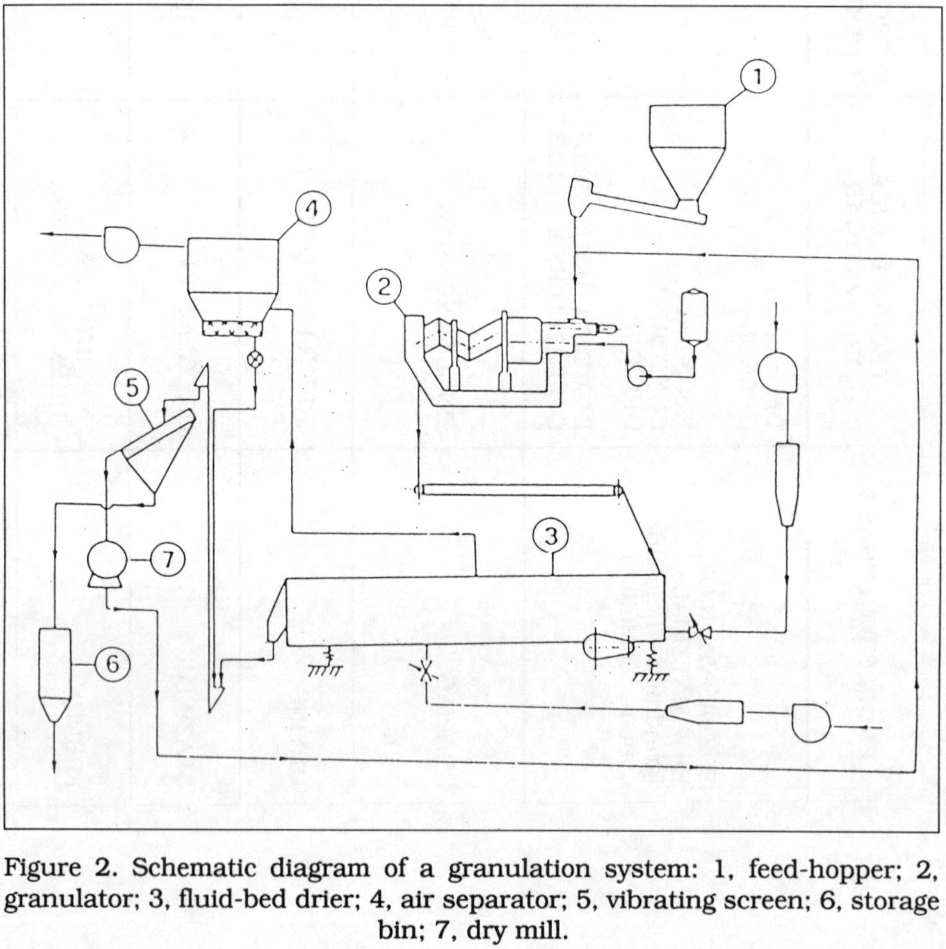

Figure 2. Schematic diagram of a granulation system: 1, feed-hopper; 2, granulator; 3, fluid-bed drier; 4, air separator; 5, vibrating screen; 6, storage bin; 7, dry mill.

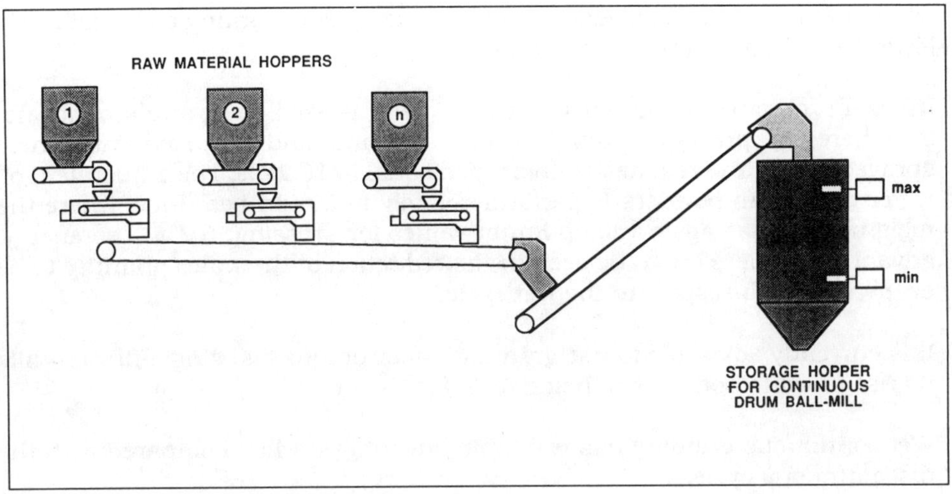

Figure 3. Automatic batch-proportioning plant.

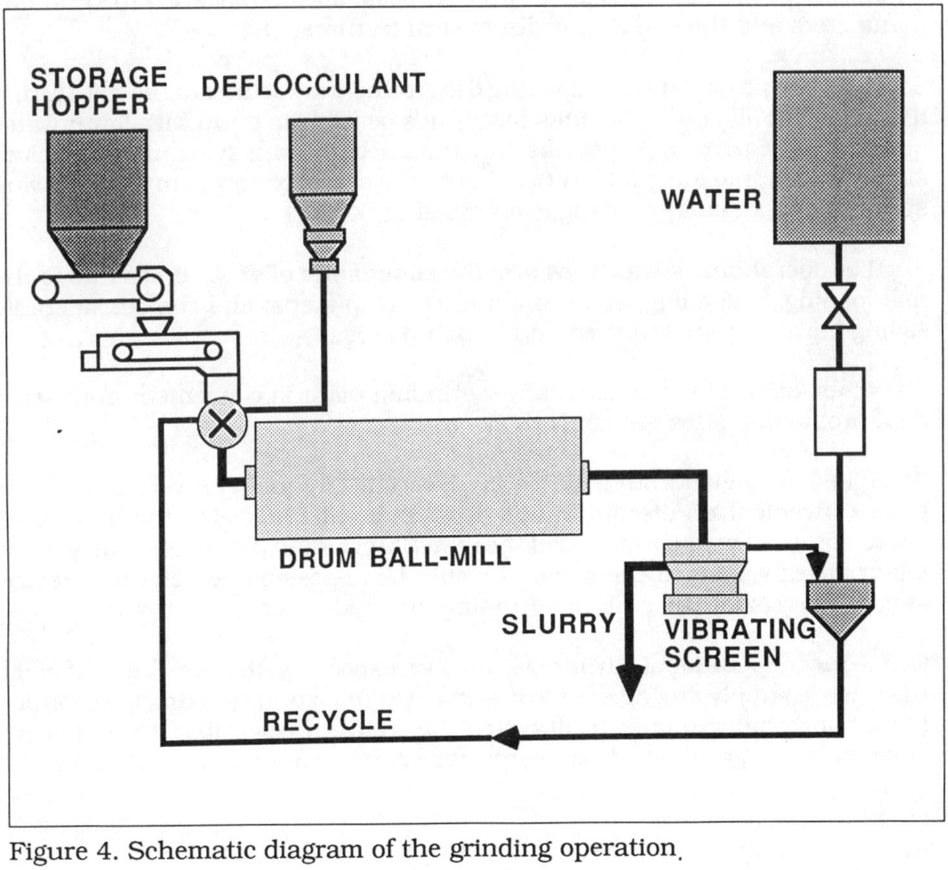

Figure 4. Schematic diagram of the grinding operation.

previously dry-ground powders and gives energy and working costs definitely lower than those of the wet process.

In order to obtain granulation a special machine is used, the granulator (fig. no. 2), where the previously dry-ground clay compound is mixed with finely sprayed water, and reaches an hydration equal to 10 12%. The adjustment of water content in powders is performed lately in a fluid bed drier, where the moisture reaches again the optimum values for pressing: 5 7%. The energy advantage is therefore to be seen in the reduction of the water quantity to be evaporated with respect to the wet cycle.

It is currently advisable to use granulation for porous fast single-firing, while its use in white body is still being tested.

Wet continuous grinding has two main advantages when compared with the discontinuous cycle.

Firstly the complete automation of the ceramic body preparation cycle (fig. no. 3), that ensures a better evenness and the possibility of repeating the grinding parameters and the slip chemical-physical features.

Secondly the possibility of unloading the ground products from the mill during its movement (fig. no. 4). Thanks to this, it's possible to grind with lower water quantities (5 6 percent points less than the discontinuous system) for a thicker slip, without unloading difficulties due to the slip thixotropy: in the following spray-drying an energy saving is obtained.

Further operational advantages are: the elimination of down times due to the mill loading, unloading, centering and start-up steps; and the considerable saving both in space required and in manpower.

The installation of a continuous wet grinding plant is convenient only when high production sizes are concerned.

When instead high output isn't required, variable frequency inverters turn out to be convenient on discontinuous drum ball-mills, allowing the mill motor speed adjustment as a function of the material grinding curve. Obviously, if the motor speed remained the same, it would be impossible to get such electric supply savings for the mill and grinding time reductions.

Consequent operational advantages involve, especially, the possibility of using the electric supply during the most convenient consumption times (according to the hourly differentiated tariff system), as well as the possibility of increasing once more the grinding department productivity without increasing the mill number.

As for the shaping step, the single-pressing system - better known as pressing-glazing process - has been tested for some years: it consists in the tile glazing simultaneously with their shaping, obtained by pressing the glaze, as spray-dried powder, in the mould already containing the powder for the support.

The advantages connected with this system are the elimination of the traditional glazing line on one side and the dry glaze application on the other, which cannot but give benefits to the surrounding environment.

There are substantially two innovative methods of the super-fast drying. The first is the utilization of the drying hot air flow by "pulses" in the transverse direction and not continuously in the material flow direction. The drying air temperature can therefore be kept higher, thus allowing a faster water transfer from the piece core to its surface.

The second makes use of the infrared rays technology and is based on the water capacity of absorbing radiations in heat transmission through irradiation.

Both systems ensure cycles 7-8 times lower than those of the traditional fast ones.

Glaze application as dry powder, by means of hot systems or cold ones, such

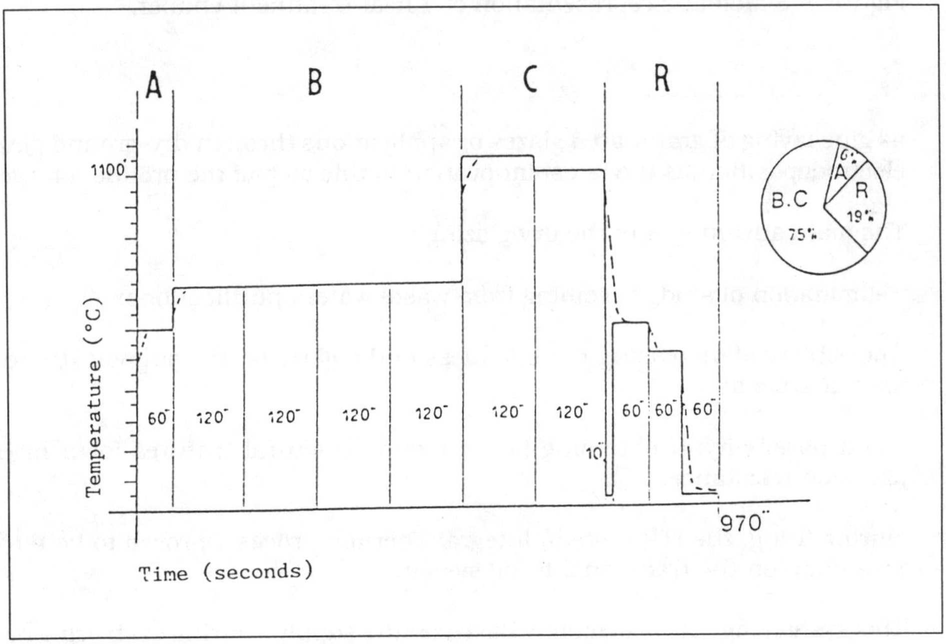

Figure 5. ITP histogram for one type of single fired redware.

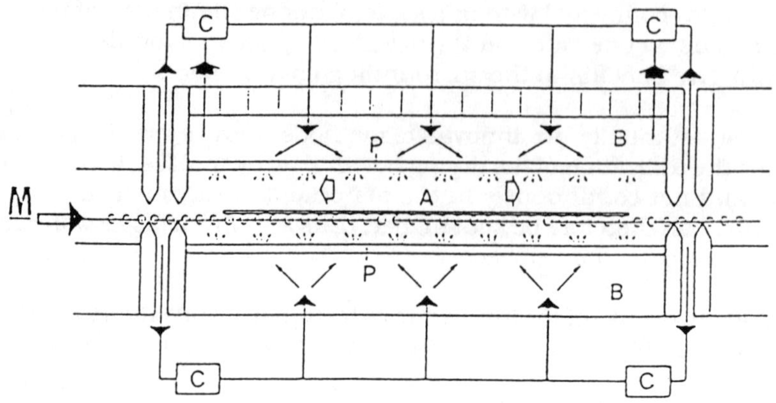

Figure 6. Schematic representation of a heat treatment chaber.

as glue laying of granulated glazes or applications through dry-ground glazes electrodeposition, is the latest innovation in this step of the production cycle.

The main advantages of the dry glazing are:

- elimination of sludges coming from waste waters purifications;

- possibility of increasing the thickness of the glaze on the support (through special systems);

- and possibility of obtaining better micro-structural features for a higher abrasion resistance.

During firing, the ITP system, Integral Thermal Process, proved to be a true revolution on the traditional firing system.

This system (fig. no. 5) is mainly based on the supply - during each firing cycle step - of the exact heat quantity for the completion of the exothermal and endothermal reactions generating the ceramic product. For each product type it has been possible to create specific ranges of temperature to be kept at, corresponding exactly to the best reaction kinetics.

The kiln (fig. no. 6) is formed by a set of insulated chambers placed one near the other, at different thermal levels, where the piece heating takes place through radiating surfaces and not through combustion fumes.

A technical consequence of this process is obviously the firing cycles reduction, as they go well under those of the traditional fast firing, and improvement in the technological features of the fired product.

CONCLUSIONS

Summing up, we can say that during the latest years ceramic technology has been able to reply to the new different requirements and that, once the energy crisis has been overcome, it is now directing towards solutions that could not but rationalize the production process and improve the product technological features, still preserving the environment.

IS FAST FIRING CONDUCIVE TO MOISTURE EXPANSION?

RICHARD BOWMAN
CSIRO Division of Building, Construction and Engineering
P.O. Box 56, Highett, Victoria 3190, Australia

ABSTRACT

While there is considerable moisture expansion literature, apparently only Fabbri and Venturi [1] have dealt specifically with the expansion of fast-fired tiles. They recommended that there should be an international standard for determining the moisture expansion of glazed ceramic tiles. Therefore one must question whether fast firing is conducive to moisture expansion, particularly in the light of a large number of recent pop-up (floor tile lifting) failures in Australia involving fast-fired tiles, some of which have unusually shaped natural expansion curves.

INTRODUCTION

Several European countries have resisted the introduction of an international standard for determining the moisture expansion of glazed ceramic tiles, despite evidence that some glazed tiles have moisture expansions in excess of those permitted for unglazed tiles [1]. Although several pop-up failures can be attributed to high concrete shrinkages or poor fixing practices, it must be acknowledged that when tiles have been correctly laid on old concrete slabs and a failure occurs, then tile expansion is likely to be a significant factor.

By considering the shape of moisture expansion curves, this paper questions not only whether fast firing is conducive to moisture expansion, but also the relative kinetics of tile expansion; and thus the role of irreversible tile expansion in differential movement failures, and the relevance of accelerated moisture expansion determinations.

DISCUSSION

General Shape of Moisture Expansion Curves

Bowman [2] has proposed that the measured in-service expansion of fired clay products is primarily due to moisture expansion and relief of thermal stresses. While Figure 1 depicts some typical moisture expansion curves of some British tile bodies [3], each curve is likely to represent a wide range of expansion behaviour as evidenced by the one highlighted point. Although bodies may be classified as being highly sensitive to moisture expansion if they have a high initial expansion, this should not be regarded as being critical if there is negligible further expansion, since little expansion will occur in service. Such behaviour may possibly be attributed to expansion due to the relief of

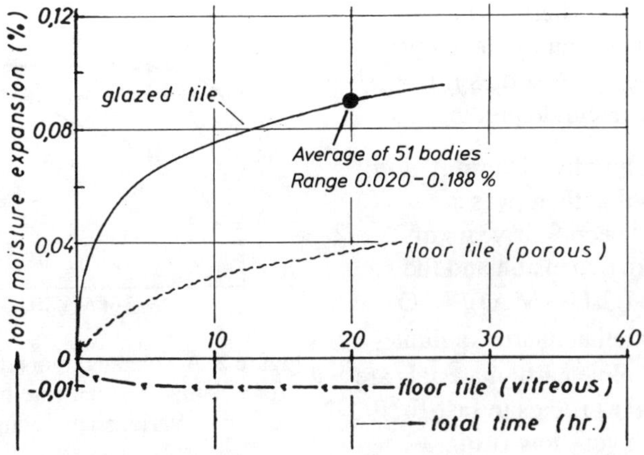

Figure 1. Typical accelerated moisture expansion curves for British tiles [3].

residual thermal stresses, rather than moisture expansion. A body which has a low initial expansion but then fails to reach a limiting value with time is likely to pose the most problems, since the in-service expansion is likely to be high.

Causes of Moisture Expansion

Capillary absorbed water tends to produce a contraction. The moisture expansion of a ceramic body is caused by the chemical and physical adsorption of water on its surfaces, which reduces its surface energy, thereby releasing some of the compressive force under which the body is held, and the body behaves elastically and expands [4]. The magnitude of the expansion will depend on the reduction of surface energy produced, the internal surface area of the body and its elastic modulus. Lehmann *et al.* [5] established a relationship between the moisture expansion and the decrease of surface energy of a ceramic body.

The difference in moisture expansion behaviour between various bodies can also be described in terms of their body constituents. Crystalline material, unless it is fractured, has a relatively small specific surface and a low surface energy. Consequently it does not contribute greatly to moisture expansion. Amorphous material has a much larger specific surface and a much higher surface energy, and therefore contributes very considerably to moisture expansion. Glass, although it has a small specific surface initially, may be attacked by water which leaches out network-modifying ions. This attack produces amorphous surfaces and may also open up pores that had previously been sealed. It is this alteration of glass that produces the slow steady rise in the final stages of the moisture expansion curves [4]. Norris *et al.* confirmed this behaviour [3], although the use of porous single phase compacts distorted the relative contribution of the glassy phase. Figure 2, however, confirms the role of glass in hard-fired bodies, where moisture expansion was induced in an earthenware type body by autoclaving at 1 MPa [4]. However, it is known that such a treatment may induce reactions that would not occur under ambient atmospheric conditions [3]. It is particularly worth stressing that since moisture expansion curves may cross one another, measurements of expansions after a fixed time or specified treatment are no indication of the relative size of bodies after some other period. There is thus no substitute for determining the natural moisture expansion of bodies if one wants to obtain accurate predictions.

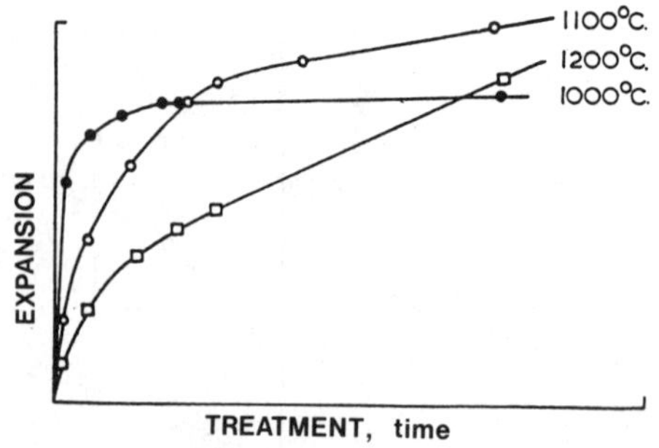

Figure 2. Accelerated moisture expansion curves for a British earthenware body after firing to various temperatures [4].

Influence of Chemical Composition
Cole [6] found that there was a relationship between four years of natural moisture expansion and the molecular ratio $(CaO + MgO)/(K_2O + Na_2O)$ for Australian quartz-kaolinite-mica clays. Alkaline earth materials are commonly used as fluxes in fast-fired bodies to obtain very low firing shrinkages. It is fortuitous that they also reduce the moisture expansion of bodies due to the formation of crystalline compounds. Boucher [7] found that Cole's relationship gave a poor correlation for some South African split tile bodies. A better fit was obtained when the accelerated long-term moisture expansion (the difference between 96 and 24 hours steaming at atmospheric pressure) was plotted against the alkali content $(Na_2O + K_2O)$.

Accelerated Moisture Expansion Tests
Accelerated moisture expansion tests generally rely on steam, either at or above atmospheric pressure, to induce expansion in the ceramic body, although a 24-hour boiling treatment is used in the European Norm for determining the moisture expansion of unglazed tiles. Unfortunately there is virtually no literature correlating the results of this accelerated test with natural moisture expansion measurements.

McDowall and Birtwistle [8] developed a relationship between an accelerated 4-hour atmospheric steam-induced moisture expansion treatment and the long-term natural expansion of several Australian brick types. The 5-year expansion of British and Australian bricks are respectively 1.6 [9] and 2 to 3 [10] times the expansion induced by a 4-hour steam treatment. The difference in behaviour indicates that the results obtained from a standard test cannot be universally used to predict potential expansion. Accelerated test methods further suffer from the limitation that they rely upon an extrapolation of a single point, implicitly assuming that all moisture expansion curves have a similar shape. Potential expansion can best be predicted by manufacturers determining the long-term expansion of their products and correlating these data with the results of accelerated tests. In the case of new products, manufacturers can determine the initial natural expansion behaviour and supplement this with a series of mild accelerated expansion treatments.

Estimation of Past Expansion
Where it is considered that a failure may be due to tile expansion, there is a need to estimate the past expansion, and if possible its kinetics, as the critical component of the expansion is the portion which occurs after the tiles are laid. Most workers have concluded that a standard reheating treatment is unlikely to completely recover the natural expansion of a body [11]. Boucher *et al.* [12] found good agreement between repeated

desorption/steam cycles, although the original expansion was not determined. Lomax and Ford [13] found that similar expansions were noted after the second and third cycles, but that these tended to be less than those obtained after the first cycle.

Bowman [2] has suggested that the contraction which is measured after reheating can be expressed as a function of contraction due to desorption, crack annealing and further sintering, and expansions due to thermal displacement and the recommencement of the expansion phenomenon.

Robinson [14] indicated that the reheat contraction associated with moisture expansion occurs primarily between 150 and 400°C. Smith [4] postulated that physically adsorbed water should be driven off by 446°C, and the chemically adsorbed water by 705°C. Norris *et al.* [3] found that most moisture expansion could be recovered at a reheating temperature just above 300°C, but the rapid initial expansion could only be recovered by heating to about 1000°C. This high temperature was probably required to anneal microcracks which had resulted from residual stresses. Reheating at or above 900°C will cause some fast-fired products to shrink, either due to sintering or to phase changes which result in a higher coefficient of thermal expansion. The European Norm uses a reheating temperature of 600°C, but this could result in anomalous behaviour due to thermal displacements associated with the quartz inversion [15]. Since moisture expansion commences within the kiln during cooling [16], and it is common practice to commence measurements at a fixed interval after the produce exits the kiln, it can be argued that one is not determining absolute expansion. Chemical adsorption is likely to predominantly occur during the conditioning period, and there appears little need to remove the chemically adsorbed water with a high temperature reheat. Bowman [15] has proposed a reheat temperature of 450°C which should not unduly affect glazed tiles, being less than the transition point of almost all glazes. Thus it may not be necessary to grind the glaze of the tile, as this appears to have little influence on the degree of moisture expansion [1,17].

Recent Australian Observations

Cole (pers. comm., 1988) has found that some bodies have a higher contraction when reheated at 860°C, than when reheated at 450°C. The ensuing natural expansion of the body reheated at 860°C was commensurately higher than that of the body reheated at 450°C due to an appreciable initial expansion, although the subsequent rates of expansion were very similar. Reheating to 860°C may cause the desorption of chemically bound water and the rehealing of microcracks, with subsequent chemical adsorption and micro-cracking occurring within the cooled body, although the difference in the magnitude of the movements could also be attributed to anomalies associated with the quartz inversion.

Cole and Banks (pers. comm., 1988) have observed some unusual natural expansion curves in some low porosity tile bodies reheated at 860°C, where there has been an initial contraction and, after a period of relative stability, there has then been a sudden onset of expansion. Learmonth and Stewart (pers. comm., 1989) have independently found similar behaviour in both Australian and imported tiles after a 460°C reheating treatment. Some typical curves are shown in Figure 3.

Such behaviour suggests that fast-fired tiles may have a microstructure where a thin glass layer coats potentially reactive components. The initial moisture contraction, due to absorbed water, has often been seen in products with less than 5% apparent porosity [3]. The subsequent expansion is consistent with progressive hydrolysis of the glassy phase, with the production of an amorphous phase. The onset of reaction and its rate would be determined by the chemical composition of the glassy phase and the level of residual stresses within the body, since these determine the resistance to stress corrosion. Stress corrosion will be a more significant factor in service due to the stresses applied by differential movements.

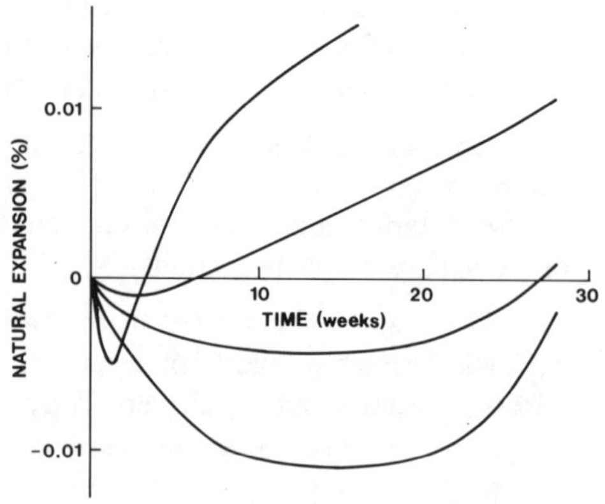

Figure 3. Observed behaviour of tiles which contract and later expand.

This behaviour is disconcerting because the potential expansion of a tile could be badly underestimated in an accelerated test, and particularly where it is calculated by multiplying the induced expansion by a factor. The variable time taken before expansion becomes positive in the above curves should be considered in the context that a 4-hour atmospheric steaming treatment has been found to correlate well with 6 months of natural exposure [8].

It thus appears that fast firing may be conducive to moisture expansion if the glass fraction has a chemical composition that is favourable to hydrolysis. This is unlikely to be the case when appropriate additions of alkaline earths are made.

Most of the past moisture expansion studies have relied on accelerated rather than natural expansion measurements, and this is probably why the behaviour of bodies which contract and subsequently expand has not been previously reported. In the past there has been virtually no correlation between the microstructure of a body, the chemical composition of the glass phase and the natural expansion of the body. Such studies are now warranted since the analytical tools are available, and it appears that fast firing of some compositions will result in glassy phases that have a high potential for expansion.

A Different Philosophy for a New Test Method

Tile manufacturers should benefit from a different philosophy for determining the moisture expansion of tiles, because there is insufficient published data to have sufficient faith in any test method being able to adequately characterise expansion behaviour, such that specifiers can be given a reliable estimate of the expected in-service expansion, or that investigators can later determine this expansion component.

Since manufacturers have the most intimate knowledge of their produce, it would appear preferable that they should be asked to classify their products as being of negligible (<0.02%), low (0.02 to 0.04%), medium (0.04 to 0.06%) or high (>0.06%) moisture expansion potential, based on the likely in-service expansion over a fixed time interval, perhaps a 5-year period commencing 2 weeks after leaving the kiln, since they are unlikely to be laid prior to this. If such classifications were adopted, there would be no need to have a compliance limit within the tile standard, nor a mandatory requirement to test to the standard. While it would allow for improved design of tiling systems, manufacturers of high expansion tiles would need to give the estimated potential expansion. There would, however, still be a need for a recognised standard test method for the purposes of quality assurance, settling disputes, and specification where a purchaser seeks to specify tiles in terms of a maximum permissible moisture expansion. Since glazed tiles expand in the same way as unglazed tiles, there is no logical reason to exclude them from any test method for determining moisture expansion. A series of mild expansion inducing treatments would be most likely to yield the most useful information.

CONCLUSIONS

Fast firing of some body compositions may be conducive to moisture expansion. This may be due to a thin glass layer coating potentially reactive components. Stress corrosion of the glass coating may expose more reactive components resulting in significant adsorption of moisture and accelerated rates of expansion.

Natural moisture expansion curves may have very different shapes, and accelerated test methods which use a single treatment to induce moisture expansion do not necessarily indicate the natural moisture expansion potential of a tile.

Manufacturers should determine the long-term natural moisture expansion behaviour of their products. In the case of a new product, the initial natural moisture expansion measurements should be supplemented by a series of accelerated moisture expansion measurements. Manufacturers should classify their products as being of negligible, low, medium or high moisture expansion potential, based on the likely expansion of the tiles in service over a fixed time interval. Since manufacturers would be relying on their intimate knowledge of their products to make such classifications, there need not be a compliance limit within the standard, nor a mandatory requirement to test to the standard.

There should be a recognised standard test method for settling disputes, for quality assurance purposes, and for specification purposes if a purchaser seeks to specify tiles in terms of the maximum permissible moisture expansion as determined by the test method. Glazed tiles expand in the same manner as unglazed tiles, and there is thus a need for an international standard test method for determining the moisture expansion of glazed ceramic tiles. Significant experimental work may be required to devise a test method for determining the potential and past moisture expansion of tiles, given the wide diversity of product. Any reheating treatment to determine past expansion or to condition samples for determining potential expansion, should be below 573°C to avoid dimensional changes associated with the quartz inversion. A reheating temperature of 450°C should be suitable for both glazed and unglazed tiles.

ACKNOWLEDGMENTS

The author is grateful to his colleagues Drs W.F. Cole and P.J. Banks and to Messrs D. Learmonth and N.C. Stewart (PGH Eureka Tile, Ballarat) for the provision of details of their experimental findings.

REFERENCES

1. Fabbri, B. and Venturi, I., Theoretical and experimental considerations on the standardization of moisture expansion for ceramic tiles. *Ceramurgia*, 1988, **18**, 10-5.
2. Bowman, R., A theoretical reconsideration of certain aspects of brick expansion. Proc. 7th Int. Brick Masonry Conf., BDRI, Melbourne, 1985, 297-304.
3. Norris, A.W., Vaughan, F., Harrison, R. and Seabridge, K.C.J., Size changes of porous ceramics caused by water and soluble salts. Proc. 6th Int. Ceramic Congress, Wiesbaden, West Germany, 1958, 63-76.
4. Smith, A.N., Investigations of the moisture expansion of porous ceramic bodies. *Trans. Br. Ceram. Soc.*, 1955, **54**, 300-18.
5. Lehmann, H., Abraham, D. and Farooq, M.K., Moisture expansion IV: kinetics and mechanism of the moisture expansion of porous ceramic bodies at low water vapour partial pressures. *Ber. Dtsch. Keram. Ges.*, 1969, **46**, 551 and 601.
6. Cole, W.F., Predictions on moisture expansion from chemical data. *J. Aust. Ceram. Soc.*, 1975, **11**, 20-1.
7. Boucher, P.S., Moisture expansion of ceramic products. Part 2: Unrestrained moisture expansion of split unglazed ceramic tiles. CSIR Research Report 627, CSIR, Pretoria, 1986.
8. McDowall, I.C. and Birtwistle, R., Predicting the long-term moisture expansion of fired clay products. Proc. 2nd Int. Brick Masonry Conf., BCRA, Stoke-on-Trent, 1971, 75-9.
9. Lomax, J. and Ford, R.W., A method for assessing the long term moisture expansion characteristics of clay bricks. Proc. 8th Int. Brick Masonry Conf., Dublin, 1988, 92-102.
10. Zsembery, S., Sharpe, K. and McDowall, I.C., Review of the experimental data on prediction of long-term natural expansion of fired clay bricks. Proc. 8th Aust. Ceram. Conf., Aust. Ceram. Soc., 1986, 327-34.
11. Waters, E.H., Hosking, J.S. and Hueber, H.V., Tests for potential and past moisture expansion of ceramic building units. ASTM Bulletin No. 245, 1960, 55-9.
12. Boucher P.S., Frost, L.J. and Howden, M.G., Moisture expansion of ceramic products. Part 1: Rapid method for determining the unrestrained moisture expansion of burnt clay bricks. CSIR Research Report 336, CSIR, Pretoria, 1977.
13. Lomax, J. and Ford, R.W., Moisture expansion: investigations into a method for assessing the long-term moisture expansion of clay bricks. *Trans. J. Br. Ceram. Soc.*, 1983, **82**, 79-82.
14. Robinson, G.C., The reversibility of moisture expansion. *Am. Ceram. Soc. Bull.*, 1985, **64**, 712-5.
15. Bowman, R., The design of an accelerated test method to estimate irreversible expansion. *Mater. Sci. Forum*, 1988, **34-36**, 997-1001.
16. Stevens, T.J. and Holland, A.E., Moisture expansion of fired clay: experiments in controlled environments. *J. Aust. Ceram. Soc.*, 1972, **8**, 1-5.
17. Ravaglioli, A., Fiori, C. and Vecchi, G., Ceramic tile expansion through imbibition of water and steam: a statistical study. *La Ceramica*, July-Aug. 1976, 17-26.

PORE STRUCTURE IN PLASTER MOULDS

SVEN KARLSSON AND ROGER CARLSSON
Swedish Ceramic Institute, Box 5403, S-402 29 Göteborg, Sweden

ABSTRACT

Test pieces of two plaster moulds used in production of wash basins were investigated. Test pieces were drilled from the moulds after 10, 20, 30 ... up to 98 and 151 castings, respectively. The open porosity, the pore size distribution (mercury) and the microstructure of mould surface layer as well as interior layers were examined. The pore size distributions show that the percentage of large pores ($1.7<\phi<3.3$ μm) increase with mould age but also with the distance from the casting surface into the mould. On the other hand, the percentage of medium pores ($0.7<\phi<1.7$ μm) decreases at the same time. The microstructure near the drying surface in a plaster mould used for approximately 100 castings is coarser compared to a not used one. The gypsum crystals change in shape and the "fibre-type" structure seen earlier has disappeared. Closer to the casting surface microstructural changes are not so clearly visible.

INTRODUCTION

The plaster mould structure, both the surface as well as the interior one, depends on a number of factors. According to Lach (1) the properties of a plaster mould are a function of material properties such as composition and plaster/water ratio but also of external properties such as temperature, grain size, and additives. The pore size distribution of plaster moulds are quite often similar to each other but with variation in the mean diameter depending on internal and external properties.

As gypsum has a certain solubility in water, about 0.2 g/100 cm^3, a small amount of gypsum crystals are removed from the mould surface at slip casting. This leads to an increased surface roughness and finally the mould must be replaced due to too rough mould surface. Together with the slip water small amounts of ions and dissolved gypsum are transported into the mould structure. At drying the salts will deposite on the gypsum crystals thus slightly reducing the pore volume. Calculations, which have been made, show that the porosity decreases about 1% after 100 repeated slip castings. Kiwinska (2) measured the Na^+-migration in the mould by adding ^{24}Na to the slip. That investigation showed that the pore system was "filled" with sodium after about 36 castings. Earlier investigations by the present authors (3) showed that the Na-content increases in the pores in dried moulds, but not to that extent reported by Kiwinska. Increased Na-content is especially found near the drying surface and to somewhat lower content also near the casting surface.

This paper reports on examinations of the pore size distribution and microstructure at various depths in plaster moulds which have been used for up to 150 repeated castings with a standard sanitary ware slip.

EXPERIMENTAL

Two plaster moulds in a battery casting line producing wash basins were chosen. The moulds were made of Supraduro(R) and Alfor 100(R) plaster, respectively. Both the moulds were cast in a factory plaster mould shop using a plaster/water ratio of 1.30/1. Other details of the mould casting and drying procedures are not available.

The two plaster moulds were after drying put up together with similar moulds to fill a normal battery casting line. From the two selected moulds cylindrical test pieces were drilled before the first filling with a standard sanitary ware slip and then continously after 10, 20, 30 ... casting cycles until it was decided to replace the moulds. The mould made of Supraduro(R) plaster were used for 98 casting cycles and the mould of Alfor 100(R) plaster for 151 casting cycles. The cylindrical test pieces had a diameter of approximately 4 cm and a depth of approximately 5 cm. After the test piece was drilled from the mould the remaining hole was filled with fresh plaster/water slurry and thus the mould could be used again. The moulds were dried between each casting cycle using the normal drying procedure. After some specific measurements performed at the factory the test pieces were dried in 55 °C and delivered to SCI for microstructural examinations. In order to follow up the change in the microstructure caused by repeated slip casting cycles in the moulds the open porosity, the pore size distribution and the microstructure were examined. The total pore volume and the pore size distribution were measured using a Hg-porosimeter (Pore Sizer 9305, Micromeritics). The microstructure was examined using a scanning electron microscope (SEM) (JSM-35, Jeol). Both the pore size distribution and the microstructure were investigated on smaller test specimens selected at various distances from casting surface. The extrusion of mercury from the pores was normally recorded at a controlled depressurization step. In the following, results of the examinations of the mould made of Alfor 100(R) plaster are presented.

RESULTS AND DISCUSSION

A general cumulative pore size distribution graph including depressurization to atmospheric pressure is shown in figure 1. From the figure it is seen that for the investigated plaster mould, the pore size distribution is quite narrow. Generally, more than 95 v/o of the pores have equivalent pore diameters within the range of $0.5<\phi<3$ μm. After complete depressurization about 30-35 v/o of the intruded mercury still remains in the pore structure. A second intrusion completed about 1 hour after the first extrusion gives a similar graph as for the first one. The subsequent extrusion follows exactly the first one as well. The amount of mercury left in the structure and the result of the second intrusion/extrusion cycle confirm that the plaster mould pore structure has ink bottle type pores or pores with irregular pore diameters which are slowly emptied at depressurization. Continous recording of the continued extrusion at atmospheric pressure shows that all the mercury once intruded will extrude from the pore structure within approximately 25 hours.

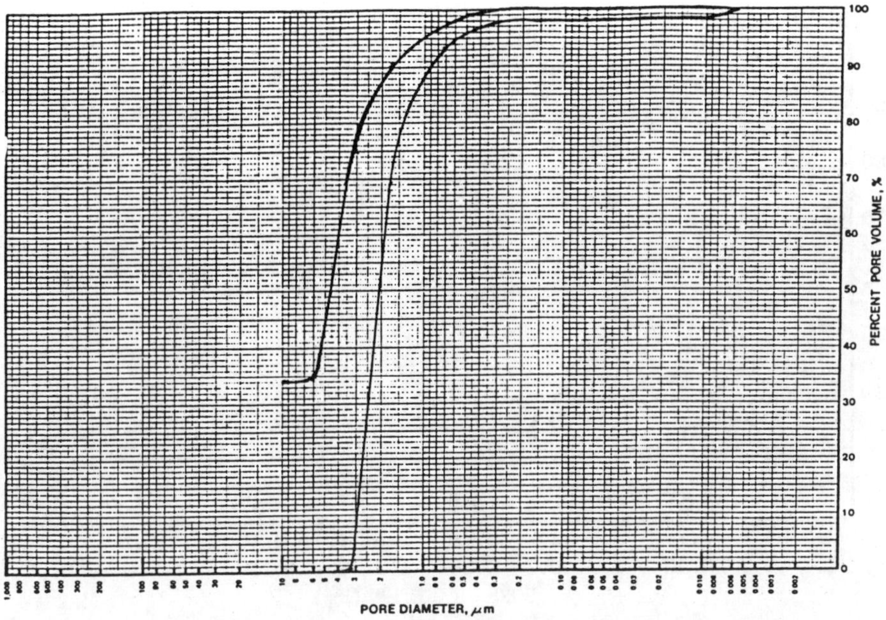

Figure 1. Cumulative pore size distribution of an unused mould. Alfor 100[(R)], plaster/water ratio 1.30/1. Mercury surface tension 485 mN/m and contact angle 130°.

The change in pore size distribution with repeated use of the moulds have been followed by examinations of test specimens taken from the mould after 0, 10, 80 and 151 casting cycles, respectively. From each test piece four smaller test specimens were selected at various depths from the casting surface. The examined test specimens were taken as 0-5, 6-12, 13-21 and 22-30 mm from the casting surface. This distance covers about 60% of the total mould thickness. The result of these four selected specimens are shown in figure 2. The pores within the pore size range of $0.36<\phi<3.3$ µm were divided into three size groups and the percentile pore volumes for each casting and each depth are shown in the figure. The total pore volume of the investigated specimens are typically 0.43-0.47 cm^3/g for the actual pore sizes that have been measured, that is, $0.006<\phi<150$ µm. The higher value is normally valid for the interior layers in the moulds. Looking at the pore volumes, there is a significant increase, approximately 5%, between the first and the second layer independent of the number of castings. The variation with mould age within each layer shows significant changes for the top and for the bottom layer. For the top layer (0-5 mm) the pore volume increases with about 4 % and for the bottom layer (22-30 mm) the pore volume decreases with about 5% from 0 to 151 castings. For the other depths the pore volumes are more or less independent of the number of castings.

The figure shows the percentage of the total pore volume divided into three classes of pore sizes. The percentile pore volumes for each class are shown at various depths for the four mould ages. It is seen that for each mould age the percentage of large pores, $1.7<\phi<3.3$ µm, slightly increases with depth in the mould. Accordingly, the percentage of medium pores, $0.7<\phi<1.7$ µm, decreases with depth. The variation for the finest pores, $\phi<0.7$ µm, is not so clearly shown but the tendency is the same as for the medium pores. Comparisons between mould ages show that for all the depths, the percentage of large pores increases with mould age. At the same time, the percentage of medium pores decreases with mould age.

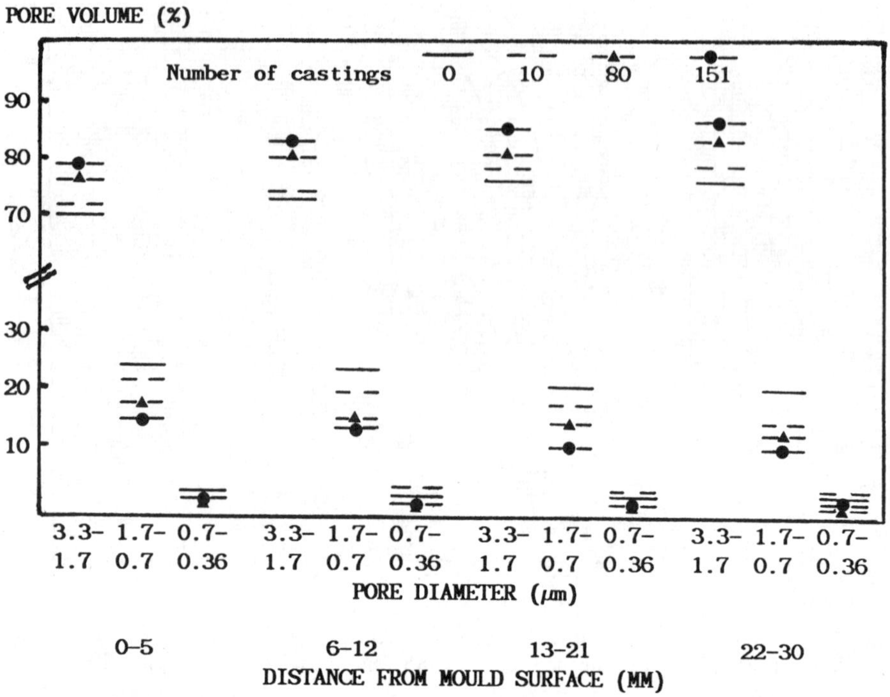

Figure 2. Pore size distribution at various depths in plaster moulds after specific number of casting cycles. Alfor 100(R), plaster/water ratio 1.30/1.

The increase of large pores in the top layer is caused by chemical corrosion of the mould surface. This opens up the surface, enlarges the pores and slightly increases the total porosity of this layer. The increase of the large pores in deeper layers is caused by a continued chemical corrosion and partially dissolving of gypsum crystals as well. Smaller pores are enlarged. The concurrent reaction where salts are formed and deposited on the crystals can probably to some extent be seen in the decrease of the medium and the fine pore sizes. The amount of added ions to the mould is low and measurement of the Si-content of used moulds show a higher but constant Si-content through the mould compared with not used moulds (3). The specimens examined in this investigation were not taken from the lower half of the mould thickness where deposition of salts are expected during the drying stage. Examinations of layers close to the drying surface might show more pronounced pore size reduction and decreased porosity.

The coarsening of the mould surface and the pore enlargement are confirmed by microscopic examinations of the mould surface. The from the beginning almost smooth surface have after a few castings become rougher. Continoued use creates still larger pores and some air bubbles are seen. Examinations of the microstructure from the mould surface through the mould to the drying surface have been made on both fractured as well as polished surfaces normal to the direction of slip water movement. The latter examination was done on Supraduro(R) plaster moulds where the pores were vacuum impregnated with a plastic filler.

A part from the coarsening of the mould surface no real differences in the microstructure can be seen on the fractured surfaces. The gypsum crystals seem to be almost uninfluenced of the 151 casting cycles in comparison with a not used mould. The variations in pore size distribution which have been measured by mercury intrusion are probably too small to be seen without image analysing of the microstructure. On the other hand examinations of the polished surfaces show more distinct influences of the mould age. The structure is more open and coarse in the used mould, especially near the drying surface. In this part of the mould most of the "fibre-type" gypsum crystals have disappeared. Figure 3 shows the microstructure of polished specimens near the casting surface and the drying surface, respectively.

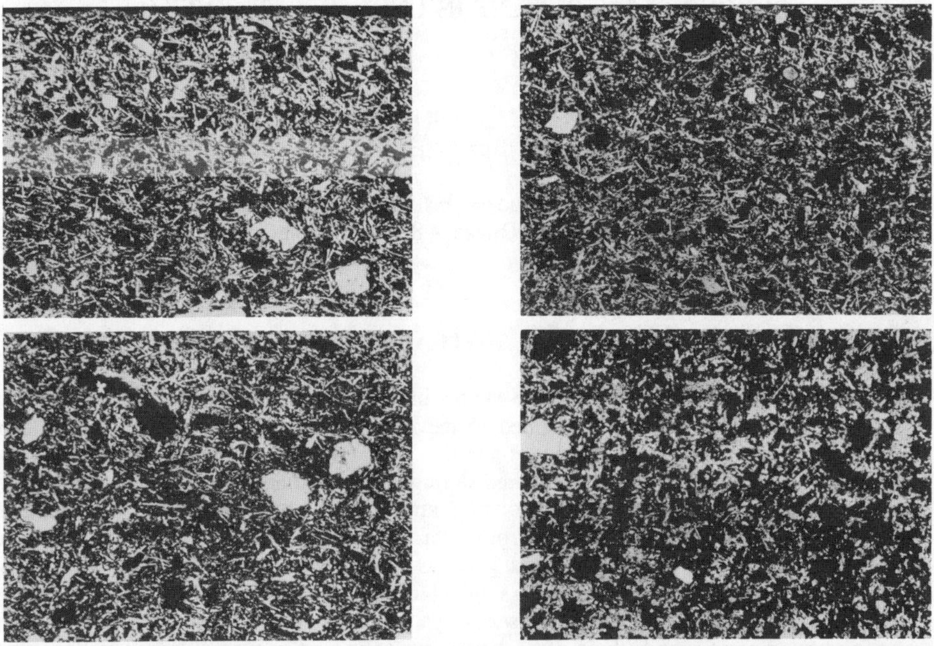

a) b)
Figure 3. Polished specimens near casting (top) and drying (bottom) surface. Supraduro(R) plaster/water ratio 1.30/1. ——— =100μm. a) no castings b) 98 castings.

ACKNOWLEDGEMENTS

Mr Bertil Holmström, Ifö Sanitär AB, is gratefully acknowledged for his permission to examine the test pieces and for his encouragement in this investigation. For the microscopic examinations Mr Lars Eklund is thanked. This work has been financially supported by STU.

REFERENCES

1. Lach, V., Microstructure of hydrated moulding plaster. Interceram, 1985, **34**, 31-34.

2. Kiwinska, M., Electrolyte migration in the gypsum moulds - isotope measurements. Szklo Ceram., 1973, **24**, 354,360.

3. Karlsson, S., Andersson, K. and Carlsson R., Electrolyte migration in plaster moulds. In Proc. Science of Ceramics 12, ed. P. Vincenzini, Faenza 1983, p95-99.

COLOUR DEVELOPMENT IN UNGLAZED FLOOR TILES

SVEN KARLSSON, ROGER CARLSSON AND BADIRUDDIN AHMAD*
Swedish Ceramic Institute, Box 5403, S-402 29 Göteborg, Sweden

*Present address: Bangladesh Institute of Glass and Ceramics
Tejgaon, Dhaka - 8, Bangladesh

ABSTRACT

The colour development and crystalline phases in 13 unglazed floor tiles have been studied. Glass compositions which are presumed to the real composition of the glass phase in the tiles were prepared by mixing raw materials and melting. The colour behaviour and crystallization of the glasses were studied during heat-treatment in a gradient furnace. The surface colour of tiles fired in a reducing atmosphere becomes darker with increased temperature. The change in colour is from light red to black. Phase analysis of the surface shows quartz, mullite, cristobalite, and glass as main phases. The dark tiles also show a fairly high content of iron-cordierite (solid solution), which probably causes the surface colour. The analysis also indicates phases which are not so easily identified. Crystallization of a glass composition which is close to the real one in the tiles shows a yellow-brown surface colour and needle-like Fe_2O_3 TiO_2 (pseudo-brookite) crystals. Crystallization of a glass composition without TiO_2 results in a dark-brown surface colour and hematite crystals. This development together with the clay can cause a reddish-brown colour of the fired tiles.

INTRODUCTION

In naturally occurring clays, iron oxides are the most potent colourants. Titanium and in some cases manganese exert a lesser influence. Since iron oxides are the dominant colourants in structural clay products, their behaviour and interaction with other body constituents in atmospheres other than air merit examination.

Harrel and Russel (1) postulate that not only the atmosphere but also the raw materials control the colour development. They summarize the controlling factors as: the chemical composition of the clay, the physical condition of the clay, the temperature of firing, the degree of reduction, the rate of cooling, and the rate of oxidation. After the firing, the iron oxides exist in the products as (1-2): pure oxides, solid solution in other structures, dissolved in the glass phase, colloidal in the glass phase.

According to Salmang and Scholze (3) an extreme reducing atmosphere can reduce SiO_2 to gaseous SiO. This compound is not stable and react as 2 SiO $\rightarrow$ SiO_2 + Si. The pure Si may give a black surface colour. In reducing atmospheres, solution of FeO in the glass phase results in a black glass. This colour will remain if no recrystallization appears during cooling in an oxidizing atmosphere (4). The preceived colour of a fired body is the result of interactions between the pigment and the glassy phase (5). Melting and crystallization of high-iron-content glasses have been reported by Pavlushkin (6) who compares the crystallization of a slag glass and a synthetic mixture.

The present paper describes investigations of colour development and crystallization behaviour of some unglazed floor tiles fired in a reducing - oxidizing atmosphere. Calculated glass compositions of the tiles were mixed and melted, heat treated and investigated for colour development and crystallization. These heat treatments were made in order to study the behaviour of the glass phase during firing. This can be of help in answering the question whether the glass phase affects the colour of the fired tiles.

EXPERIMENTAL

The colour development was studied in 13 unglazed floor tiles taken form an industrial firing in a periodical gas-heated kiln. At the top temperature of the kiln curve, the atmosphere was first reducing for 2 hours followed by 2 hours of oxidizing atmosphere before cooling down. The tiles represented the normal colour variation during such a firing. The tiles were investigated with an X-ray diffractometer directly on the top surface and with a scanning electron microscope (SEM) on a cut and polished surface perpendicular to the top surface. The FeO-content of some of the tiles was measured by chemical analysis.

Five glass compositions were prepared, table 1. Composition No. 1 was based on previous investigations of the glass content and its composition in various ceramic bodies (7). The glass content of the tiles in this investigation was between 19-36 w/o and had an approximate composition of (w/o): 72 SiO_2, 4.5 TiO_2, 9 Al_2O_3, 5 Fe_2O_3, 1 CaO, 2 MgO, 7 K_2O, 1 Na_2O. The glass compositions were prepared of high-purity raw materials: SiO_2, Al_2O_3, TiO_2, Fe_2O_3, $CaCO_3$, MgO and K_2CO_3. The raw materials were mixed in an agate mortar and melted twice in a platinum basin at 1550°C for a total time of 20-30 hours in an electrically heated furnace. After the firing they were quenched in cold water. From the glasses test samples were cut and heat treated for various times and under various atmospheres in an electrically heated gradient furnace with a top temperature of 1160°C. The atmospheres used were obtained by mixing pure gases from cylinders and water vapour. After the heat treatment the samples were investigated visually for colour and analyzed on the surface with an X-ray diffractometer for crystalline phases. The microstructure was observed by SEM. Some of the glasses were also heat treated together with floor tiles in a gas heated Bickley kiln using the normal temperature and atmosphere.

TABLE 1
Composition of investigated glasses

Glass	Composition (w/o)						
No.	SiO_2	TiO_2	Al_2O_3	Fe_2O_3	CaO	MgO	K_2O
1	71	5	8	6	1	-	9
2	74.4	-	8.5	6.5	1.1	-	9.5
3	73.2	2	8.2	6.2	1	-	9.3
4	73.2	5.2	8.2	3	1	-	9.3
5	71	5	8	5	1	2	8

RESULTS AND DISCUSSION

Colour and crystallization of unglazed floor tiles

The surface colours and identified phases with a semiquantitative analysis of the studied tiles are presented in table 2. Tile No. 1 is fired only in an oxidizing atmosphere. The top temperature during firing increases with darker surface colour. The darkest tile, No. 13, shows a black metallic surface.

TABLE 2
Colour development and phase analysis of floor tiles

Tile no	Surface colour	Q	M	Cr	Co	S	F	G	X
1	yellow-brown i)	1	2	4	-	-	-	5	3
2	light red i)	1	2	-	-	3	4	4	?
3-4	light red i)	1	2	6?	-	3	5	4	?
5	dark red i)	1	2	6	-	3	?	5	4
6	dark red i)	1	2	3	?	-	-	5	4
7-8	red-brown i)	1	2	3	6	-	-	5	4
9-10	dark red-brown ii)	1	2	3	5	-	-	4	6
11	dark red-brown ii)	1	2	3	6	-	-	4	5
12	almost black ii)	1	2	3	6	-	-	4	5
13	black ii)	1	2	3	5	-	-	4	-

Q = Quartz, M = Mullite, Cr = Cristobalite, Co = Cordierite, S = Spinel, F = Feldspar, G = Glass, X = Unknown phase, 1...6 = highest...lowest content, - = not appearent, ? = probably appearent in very small content, i) interior colour same as surface colour, ii) interior colour lighter than surface colour.

Phase analysis on the surface shows quartz, mullite, and cristobalite as dominant phases followed by the glass phase. In the darker tiles there is also iron-cordierite (ss). The surface layer of the lighter tiles probably also contains spinel and feldspar. In all of the tiles, except No. 13, diffraction peaks which have not been identified appeared. The intensities of these peaks decrease as the surface colour becomes darker. Analysis of a layer just below the surface also shows decreased intensities of these peaks. Further grinding to deeper layers does not change the intensities any more. The phase or phases which have not been identified are accordingly dominant in the top layer of the tiles. The strongest peaks on the unidentified phase(-s) are as follows (d-values with relative intensities within parentheses. 7.17 (100), 5.18 (10), 3.72 (30), 3.58 (40), 3.22 (30), 2.92 (40), 2.77 (40).Phase analysis of pulverized samples of the surface layer and at various depths in the tiles does not show any significant difference from the analysis of the surface layer or layers just below the surface. As before, the intensities of the unidentified peaks decrease from the surface layer to the inner parts of the tiles. This is especially true for tiles Nos. 1-7. The decrease is not so pronounced in the darker tiles. The content of iron-cordierite increases the darker the tiles are. The increase is the most pronounced in the surface layers, where the highest contents of iron-cordierite are found. The unidentified peaks do not refer to any composition in the system Fe_2O_3-TiO_2 or any other colourant agent that has been proposed in the literature (2).

X-ray fluorescence analysis of the surface layer shows that the contents of Fe, Mn and K increase the darker the tiles are. The higher content of iron depends on a higher content of iron-cordierite in the darker tiles, but probably also on an increased solid solution of

Fe_2O_3 in the glass phase at higher temperatures. Chemical analysis of the outer surface layer of some of the tiles shows a higher content of FeO the darker the tile is. This confirms the increased content of iron-cordierite in the dark tiles.

Samples etched in HF (1 v/o, 2-3 min) and analyzed by SEM show that the structure is more porous in the lighter tiles. The glass phase seems to be uniformly spread throughout the tile. No difference in structure can be seen between the surface layer and inner parts of the tile. The Fe-content is uniformly distributed, but also highly concentrated to distinct parts. Figures 1a and 1b show the microstructure of the surface layer of tile No. 2 (light) and tile No. 13 (dark), respectively.

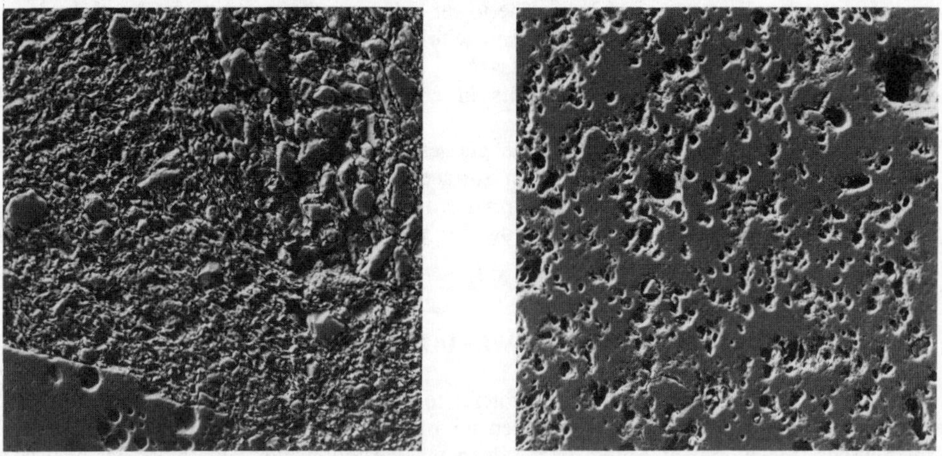

a) b)
Figure 1. Microstructure of surface layer a) tile No 2 b) tile No 13 —— =100μm.

Crystallization behaviour of the glass compositions

Before heat treatment glasses Nos. 1, 2 and 5 were black with a bright surface, especially glass No. 5. Glasses Nos. 3 and 4 were umbra coloured and partly transparent. Results of a heat treatment in a gradient furnace are presented in table 3 and explained below.

TABLE 3
Colour and crystallization of heat treated glass compositions

Glass No	Colour: Crystalline part	Colour: Glass phase	Crystalline part: Phase	Crystalline part: Shape	Crystalline part: Size (μm)	Crystalline part: Content (%)
1	yellow-brown	dark brown	Fe_2O_3 TiO_2	needle	5	10
2	dark brown	black	Fe_2O_3	6-8 edged	1-2	10
3	ivory	yellow	α-cristobalite	-	-	10
4	white	yellow	α-cristobalite	-	-	25
5	yellow-green	dark brown	Fe_2O_3 TiO_2	needle	4-5	10

The crystallization is for all five glasses located to the upper surface. For glasses Nos. 1 and 5 the needle shaped crystals of Fe_2O_3 TiO_2 (pseudo-brookite) are randomly oriented, in glass No. 5 they are located deeper in the glass body. The crystalline part is normally more porous than the glass phase. The contents of Fe and Ti are the same in both the glass phase and the crystallized phase. In glass No. 2 the crystalline part has a higher content of iron. Heat treatment in only a reducing atmosphere results also in crystallization of Fe_2O_3 but to a lower content. In glasses Nos. 3 and 4 the crystallization has caused cracks in the bodies. Glass No. 4 is more crystallized than glass No. 3. Glasses Nos. 1 and 3 were also heat treated in a Bickley kiln in a normal firing together with floor tiles. Colour development and crystallization are the same as for heat treatment in a laboratory gradient furnace. During heat treatment, temperature gradients of 20-30°C as well as various reducing atmospheres have not changed the colour and crystallization of the glasses.

The heat treatment shows that minor changes in the glass compositions (table 1) result in different crystallization behaviours. The lower TiO_2 content of glass No. 3 and the lower Fe_2O_3 content of glass No. 4 in comparison with glasses Nos. 1 and 5 are not sufficient to form pseudo-brookite (Fe_2O_3 TiO_2). Glass No. 2 without TiO_2 but with about the same content of Fe_2O_3 as glass No. 3 results in crystallization of hematite compared with cristobalite for glass No. 3.

The crystallization behaviour of the glasses indicates that hematite is the dominant colouring agent in the tiles. The brown surface colour of the heat treated glass No. 2 together with the clay ought to produce a red-brown colour of the tile. The greenish-yellow surface colour of glasses Nos. 1 and 5 together with clay cannot produce a reddish-brown colour but a light greyish colour instead.

ACKNOWLEDGEMENTS

One of the authors, Mr B. Ahmad, is grateful to the International Seminar in Physics and Chemistry, University of Uppsala, Sweden for offering the fellowship and to the Swedish International Development Authority for financing the fellowship. This work has also been supported by the STU. Gratitude is also expressed to Partek Höganäs Byggkeramik AB for providing ceramic bodies and kind information.

REFERENCES

1. Harrel, G.O. and Russel, R. Jr., Influence of ambient atmosphere in maturation of structural clay products. Ohio State Univ., Eng. Exp. Stat. Bull., 1967, 204.

2. Maggetti, M., Mineralogische Untersuchingen zur Farbe gebrannter Tone. Bull. Soc. Frib. Sc. Nat., 1975, **64**, 46-62.

3. Salmang, H. and Scholze, H., Die Keramik, Physikalische und Chemische Grundlagen. 5 ed., Berlin, 1968, p 279.

4. Houseman, J.E. and Koening, C.J., Influence of kiln atmospheres in firing structural clay products. J. Am. Ceram. Soc., 1971, **54**, 75-89.

5. Fischer, P., Some comments on the colour of fired clays. CFI/Ber DKG, 1983, **60**, 320-327.

6. Pavlushkin, N.M., Change in the phase composition of high-iron-content glass upon crystallization. Inorg. Mater., Cons. Bur. Transl., 1968, **4**, 1224-1235.

7. Carlsson, R., SILIKAT-RAPPORT 72-4, 1972, **7**, (in Swedish).

POSSIBLE EVALUATION OF ADHESION BETWEEN GLAZE AND SUBSTRATE IN TRADITIONAL CERAMICS.

R. Dal Maschio*, G. De Portu§, B. Fabbri§ and J. Shen+.
*Engineering Department of the University, Trento, 38050, Italy.
§Research Institute for Ceramic Technology C.N.R., Faenza, 48018, Italy.
+Shantong Institute of Industrial Ceramics, Zi-Bo, China.

ABSTRACT

In this work, tensile test results are compared with some data of interface fracture resistance by the introduction of a mechanically stable crack with usual indentation procedure perpendicularly to the interface, after sectioning normal to the glazed surface, such that one radial crack propagates along the interface.

INTRODUCTION

The ceramic glazed products are constituted by two different components, the substrate and the glaze, connected by an interface.

As a matter of fact the mutual partial compenetration gives rise to an interfacial zone with characteristics different from both the substrate and the glaze. In the industrial practice these products are well characterized from the point of view of both substrate and glaze; on the contrary the interface zone is completely ignored. But the quality of this type of products is influenced also by interface characteristics, strictly connected to the adhesion between glaze and substrate. Particularly the interface zone influences the resistance against chipping by impact, a typical practical situation on ceramic crockery and kitchen tiles.

Aim of this paper is to analyze the techniques potentially useful for this purpose, in order to evaluate experimentally the adhesion between the glaze and the ceramic substrate.

THEORETICAL APPROACH

A classical approach to the adhesion measurement consists in the tensile test, that is carried out by glueing a pull-off bar to the coating, and than subjecting the coating to a uniaxial tensile force.

Practically different experimental procedures can be used according to the specimen geometry and manner of failure analysis and reporting (1). Several different test methods have been recently proposed for evaluating adhesion in coated materials (2). Among them indentation on coatings seems one of the most promising and simple (Fig.1) (3). However this test has been developed and used mainly for thin (generally < 100 μm) and soft films on rigid bodies. This situation is different from the traditional ceramics systems where the glaze has a stiffness similar to the substrate and a thickness always higher than 100 μm. In fact preliminary experimental tests carried on according to this technique showed the impossibility to reach the condition for crack formation.

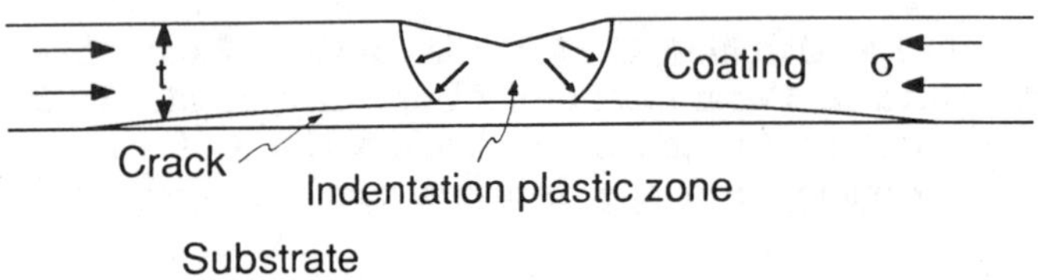

Figure 1. Schematic of surface indentation-induced delamination at the interface.

A different approach to the problem is the technique consisting in a Vickers indentation on the interface (Fig.2). In this way the median cracks start from the indentation site corners and propagate along the interface itself with a substantially semicircular front. A possible solution in term of strain energy release rate could be obtained adapting the theoretical model proposed by Willis (4) to the present situation. After mathematical manipulations and assumptions connected to our experimental configuration the critical strain energy release rate can be written as:

$$\mathcal{G}_C = \alpha\, \Phi(E_{1,2},\nu_{1,2})\, \Psi(E_{1,2},H_{1,2})\, a^4/c^3 \qquad (1)$$

where:

$$\Phi(E_1,E_2,\nu_1,\nu_2) = ((b^2-d^2)/b)(1+4k^2)$$
$$b = (1-\nu_1)/(2\pi G_1) + (1-\nu_2)/(2\pi G_2)$$
$$d = (1-2\nu_1)/(4\pi G_1) - (1-2\nu_2)/(4\pi G_2)$$
$$k = 1/(2\pi)\ \ln((b+d)/(b-d))$$
$$\Psi = 1/(1/E_1+1/E_2)^2/(\sqrt{E_1/H_1}+\sqrt{E_2/H_2})^2$$

with $E_{1,2}$, $G_{1,2}$, $\nu_{1,2}$ and $H_{1,2}$ Young modulus, shear modulus, Poisson ratio and hardness of material 1 and 2 respectively; a is the semidiagonal of Vickers impression, c is the median crack length and α is a constant equal to 0.0197 (5).

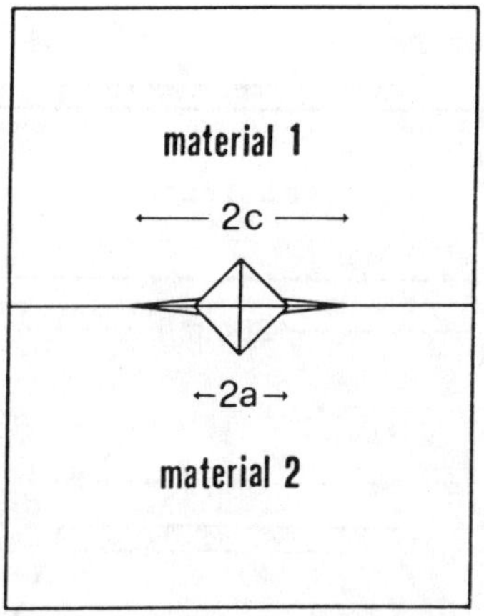

Figure 2. Schematic of Vickers indentation on the interface.

EXPERIMENTAL PROCEDURE

Two kinds of commercial glazed floor tiles were used for the present study. The first one was a porous body (11% water absorption), the second one a dense stoneware (1% water absorption). The glaze thickness was 450 μm and 400 μm respectively. The tiles were cut by diamond sawing into rectangular bars with proper sizes. Then these bars were fixed on a clamping apparatus and grinded to obtain a tapered body as showed in Figure 3. On both sides, 5 mm of the glaze were grinded away, so that its residual area was 10 mm x 10 mm (±0.2 mm). This reduced area was chosen in order to allow the fracture to propagate at the interface rather than in the substrate. To avoid the generation of small cracks, $280^{\#}$ diamond grinding wheel was used and the feeds were controlled (up to 0.1 mm for the porous tiles and up to 0.05 mm for the stoneware). During process, attention was paid to avoid tensile stress between the glaze and the body as much as possible.

Two samples having the same configuration and dimensions were sticked together in opposite position by heating them to a proper temperature (900 °C for the porous tiles and 1180 °C for the stoneware, respectively) with a soaking time of 15 min. At those temperature the glaze was just melted and no observable chemical and physical changes occurred in the samples. Tensile strength measurements were carried out using a universal testing machine with self aligning tapered grips. The cross head speed was 5 mm/min. If the fracture did not occur at the interface the datum was rejected.

The porosity seems to be a limit for this technique. In fact we tried to test a tile with higher porosity (18% water absorption) and the fracture always occurred in the body.

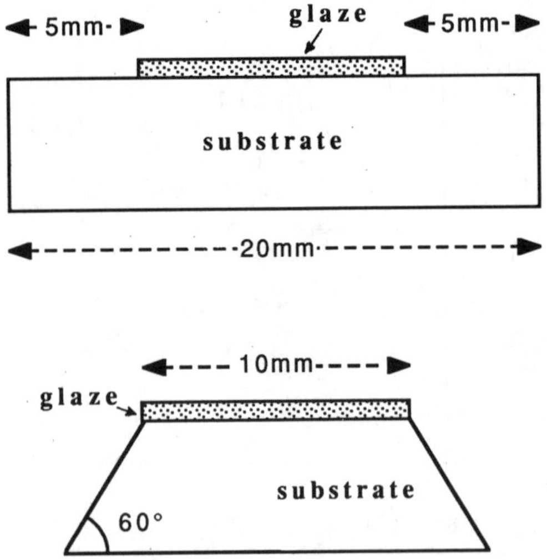

Figure 3. Schematic of samples preparation for tensile test.

RESULTS AND DISCUSSION

The fracture strength at the interface was 4.3±0.3 MPa for the porous tile and 6.3±0.4 MPa for the stoneware. The low standard deviations suggest that these values are typical of the materials. So this technique is suitable for measuring interfacial tensile strength of tiles. These preliminary results need further investigations in order to verify the reliability of this test.

From the point of view of the indentation test on the interface, the same two glazed floor tiles were sectioned and indented, the porous tile with 19.6 N and the dense stoneware with 29.4 N respectively. In the former sample the indentations were characterized by a diffuse cracking system without the typical radial cracks pattern, probably due to the porous ceramic substrate. On the contrary in the latter sample it has been possible to evaluate the interface cracks length (Fig.4); such measures were treated by means of eq.(1), so obtaining a critical strain energy release rate $\mathcal{G}_C$ of 2.5 J/m^2.

CONCLUSIONS

Different test methods were considered in order to evaluate the adhesion strength between glaze and substrate in ceramic tiles. Both modified indentation and tensile tests seem promising; however, further investigations are necessary in order to evaluate the quality of the data obtained and the reliability of the techniques.

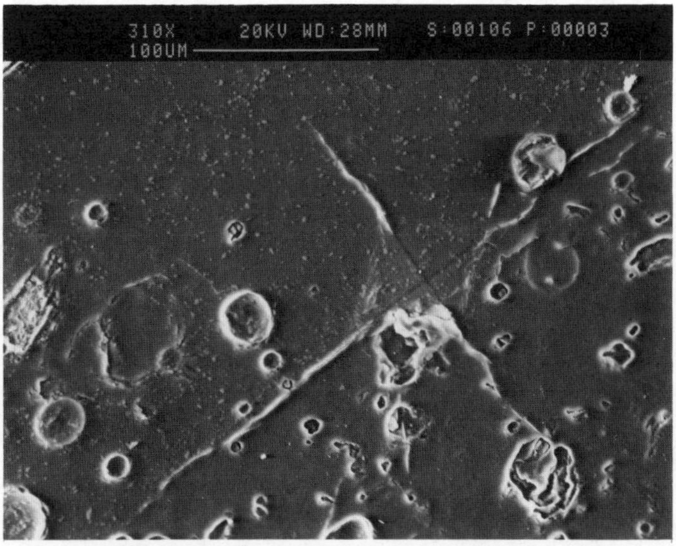

Figure 4. Typical interface Vickers indentation on a stoneware tile.

REFERENCES

1. Berndt, C.C. and Ostojic, P., Strength Testing of Plasma Sprayed Coatings. Proceedings of the International Symposium on Advanced Thermal Spraying Technology and Allied Coatings, 13-15 May 1988, Osaka, Japan.
2. Colin, C., Boussuge, M., Valentin, D. and Desplanches, G., Mechanical Testing of Plasma-Sprayed Coatings of Ceramics. J. Mater. Sci., 1988, **23**, 2121-28.
3. Marshall, D.B. and Evans, A.G., Measurement of Adherence of Residually Stressed Thin Films by Indentation. I. Mechanics of Interface Delamination. J. Appl. Phys., 1984, **56**, 2632-38.
4. Willis, J.R., Fracture Mechanics of Interfacial Cracks. J. Mech. Phys. Solids, 1971, **19**, 353-68.
5. Sglavo, V.M. and Dal Maschio, R., Adhesion Testing of Plasma Sprayed Ceramic Coatings on Metals by an Indentation Technique. European Applied Research Reports, 1989, to be published.

TECHCON DRYER

H.J. TIMMERS
Technisch Centrum voor de Keramische Industrie
De Steeg
The Netherlands

ABSTRACT

The article starts with a short historic review of the drying process of heavy clay products from ancient times to modern industrial practice, pointing to a number of drawbacks of modern dryers. In a further theoretical review attention is paid to moisture conduction, heat transfer and evaporation. With the aid of a model the minimum costs of gas(fuel) and electricity are calculated and at the end a description is given of the dryer.

INTRODUCTION

The brick making process has already been handed down to us by the Bible which in Ex. 5 says that the Jews, as slaves in Egypt, had to make mud bricks with the aid of wooden moulds. The bricks were put on the ground to dry by being exposed to sun and wind.

Since ancient times up to the Industrial Revolution (about 1850) the method of shaping and drying of the bricks is hardly changed. In the Netherlands the first "artificial" dryer was built in 1899. But it was not until the sixties that in the Netherlands all heavy clay materials were dried "artificially". Now, after about thirty years, several dryers need to be replaced.

A disadvantage of the present chamber dryer is that it is only loaded and unloaded if the pressing machine is in operation and that is, for practically all brick works in the Netherlands, for about 38 hours a week. This causes a peak demand of energy for the drying process around the middle of the week and a minimum demand over the weekend. If the kiln produces much "waste" heat, a surplus of heat may occur over the weekend.

Only continuously operating tunnel dryers have a constant energy demand during the week. One of the drawbacks of this type of dryer is that one cannot always anticipate a change in clay type or brick format as flexibly as one would wish.

Another drawback of the chamber dryer is that the air travel distance between supply and discharge is at least 2 m. This causes local differences in drying as a consequence of which longer drying times and more energy are required.
To solve these problems a new, so-called Techcon dryer has been developed in cooperation with the management and staff of the Zennewijnen brickyard. A number of theoretical aspects involved in the development of this dryer will be discussed below.

SOME THEORY

The following two important aspects must be noted:
1. What is the optimum drying rate during the drying process, and
2. Which drying conditions (temperature, relative humidity, air speed) must be chosen to achieve optimum results.

The drying systems applied in the heavy clay industry can be described in theory by using the so-called "agitator" model (fig. 1) [2].

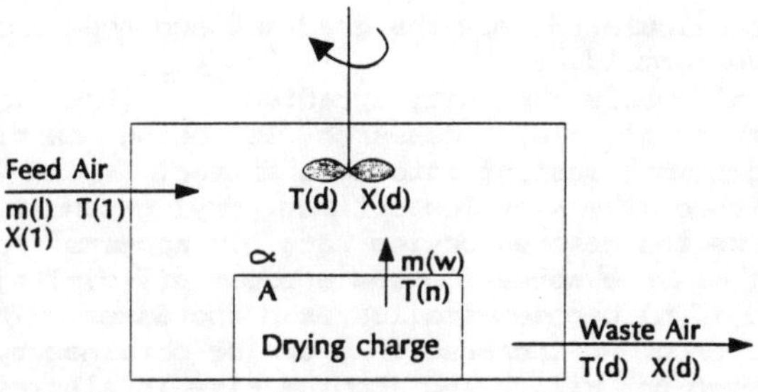

Figure 1. Agitator model with:

m(1)	=	supply air (kg/hour)
T(1)	=	inlet temperature (°C)
X(1)	=	inlet humidity (kg water/kg air)
T(d)	=	chamber temperature (dry bulb, °C)
X(d)	=	humidity in the chamber (kg/water kg/air)
T(n)	=	product surface temperature (wet bulb, °C)
m(w)	=	water evaporation (kg/hour)
A	=	surface area of the products (m^2)
α	=	average heat transfer coeff. (kJ/m^2.hour.°C)

The agitator model assumes that an ideal mixing process takes place in the drying chamber. The mixture, when discharged, is equal to the mixture in the drying room.
The heat balance can be described as follows [3]:

$$\alpha . A . (T(d) - T(n)) = m(w) . r \qquad (1)$$

in which:

α	=	average heat transfer coefficient (kJ/m^2.hour.°C)
A	=	product surface area (m^2)
T(d)	=	room temperature (dry bulb, °C)
T(n)	=	product surface area temperature (wet bulb, °C)

m(w) = water evaporation (kg/hour)
r = evaporation heat (circa 2500 kJ per kg of water)

The left part of the equation refers to temperature, relative humidity and air speed.
The right part refers to the allowable drying rate and thus, depends inter alia on the composition of the raw material and the shape of the product.
The moisture transport can be described as follows [4]:

$$\frac{Q}{A} = -K.\gamma.\frac{dW}{dy} \qquad (4)$$

in which:
Q = quantity of evaporated water (kg/s)
K = coefficient of moisture conduction (K-value, m^2/s)
A = drying surface (m^2)
γ = specific weight by volume (kg/m^3)
dW/dy = humidity gradient (m-1)

The K value is calculated from the gradient and the amount of evaporation, using the above formula.
The maximum allowable humidity gradient at the various drying phases differs per type of clay. Research is being carried out by TPD-TNO Eindhoven to get more insight into this matter.
Having established the K value optimum drying conditions will have to be found to realize the desired drying rate. It appears from formula (1) that at a constant water evaporation the product $\alpha(T(d)-T(n))$ is constant. This means that T(d)-T(n) becomes smaller as α increases. This allows the heat consumption to drop. An increase in α can be obtained by increasing the air speed. This, however, will bring about a rise in electricity consumption.
After drying tests with green waal format bricks were carried out [1], the following formula was developed empirically:

$$q' = 0.9\ (1 + vo).(T(d) - T(n)) \qquad (3)$$

in which:
q' = evaporated water per waal format (g/hour)
vo = speed across section of empty dryer (m/s).

The effect of the air speed on heat consumption and costs of gas and electricity can be calculated with the aid of formula (3) and the formulas in [6] which have been derived to set up the Mollier diagram.
In fig. 2 the results of a calculation model have been worked out for circulation fans of different capacities, a wet bulb of 35 °C, evaporation of 20 g/hour/wf and two gas prices.
It appears that, in this example, an optimum electricity consumption can be obtained at approx. 25 Wh/kg of water, without using waste heat. When using waste heat the optimum will move to lower values. For a rapid, uniform drying process good turbulence in the drying chamber is essential.
The above applies mainly to the first drying stage. The second drying stage is much more difficult to describe theoretically. As a matter of fact it is not possible to give a brief description of all aspects of the drying process. A number of important elements which are still under study have been described.

DESCRIPTION OF THE TECHCON DRYER

The pallets carrying the green bricks are collected into an elevator. The pallets are conveyed outof the elevator and into the car by an overhead finger car (fig. 3).

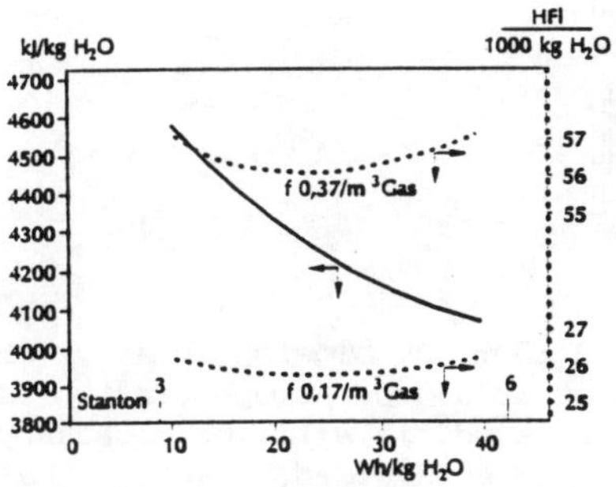

fig 2.

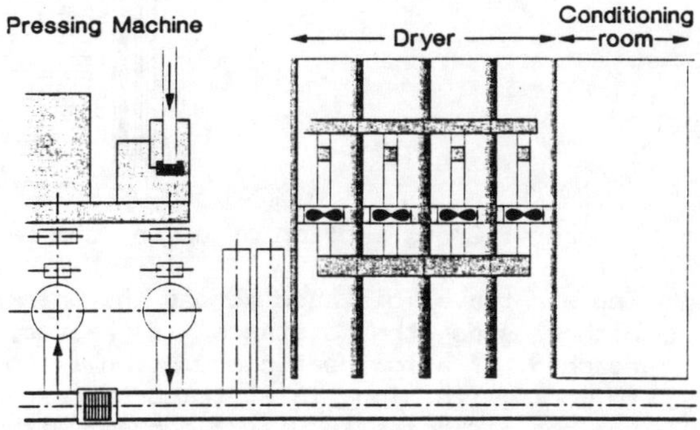

fig 3.

Figure 2. Heat consumption and costs to evaporate water from green bricks at increased electricity consumption of the fans (arrows point to the axis).

Figure 3. Techcon dryer, transport, dry and wet sides and press.

By turning the car through 180° another charge can be set. The filled car is transported automatically to the conditioning room concerned by a travers car. Subsequently the travers car removes a car with dried products from another conditioning room and takes this car to the unloading equipment. Here, the pallets carrying the products are removed from the car which is thereupon reloaded with wet products.
This automatic car transport system enables the operation ofthe pressing machine and the dryer to be disconnected. The dryer receives the products to be dried from the conditioning rooms which are also used to store the bricks which are already dry. This disconnection permits an optimum drying time to be selected and provides a continuous process during the whole week. The energy demand of the dryer is now adjusted to the energy supply of the kiln. In the drying chamber a number of cars are set behind each other. This is done in such a way that the air is blown against the "headers" of the bricks loaded on the car. On either side of the drying chamber movable columns are installed with fans (fig. 4).

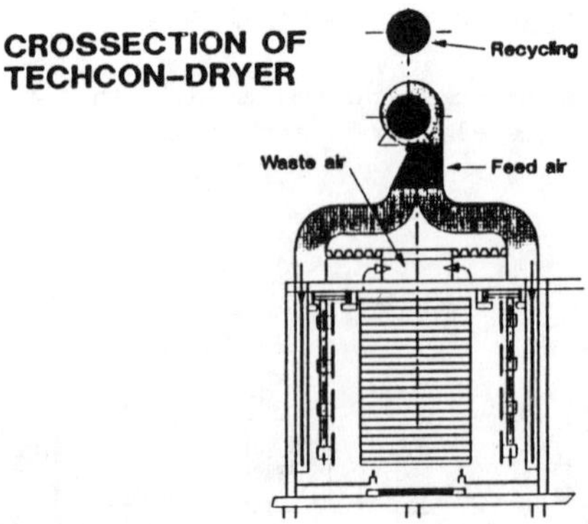

Figure 4. Cross section of drying chamber

The air travel distance across the bricks is 2.30 m. By blowing in air from either side the distance is halved, so that a good homogenization is reached. If a low heat consumption is to be achieved it will be necessary to work with high inlet temperatures. It should be avoided, however, that the hot air comes in direct contact with the wet product. To prevent this a system has been chosen consisting of vertical tubes spaced approx. 2 m apart fed by a horizontal air duct located on top of the dryer.
The vertical tubes are on each side provided with slots running over the entire tube length, through which the air is blown into the dryer, parallel to the chamber wall. The hot air is directly mixed with the air in the chamber. The wet air is discharged via a central duct and can be recirculated either partly or wholly.
It is, of course, also possible to reach the same result by applying other circulation systems in the chamber.
For a chamber dryer a heat consumption of 4200 kJ/kg of water (approx. 1000 kcal/kg) is feasible (without heat recovery from the discharged air).
This can be realized with high inlet temperatures and a low specific air consumption. If, however, this should lead to part of the waste heat of the kiln not being used, the costs might be higher than if all waste heat would be used with a lower efficiency of the dryer. Attempts are therefore made to reduce the amount of waste heat of the kiln.

REFERENCES

1. Van der Velden,J.H., Drogen en Keramiek, Rapportnr. 853.204/e, T.P.D./T.N.O. 17-11-1988.
2. Junge, K., Umwalzlufter fur Senkung de Enegieverbrauchs und zur Vergleichmassigung der Trocknung, Ziegelindustrie International, 1/85, 10, 4/85, 227.
3. Stichting Opleiding Keramiek, Syllabus Keramiek.
4. Timmers, H.J., Enige fysische aspecten van het drogen van keramische producten, Klei en keramiek, 1967, 166.
5. Keramisch Insttuut T.N.O. Bepaling van formaat factoren van verschillende grofkeramische vormlingen, Klei, 1959, 205.
6. Ham, Ph.J., Mollier-h/x-diagrammen voor vochtige lucht, geconstrueerd door middel van de computer, Klimaatbeheersing, 13(1984) nr 8.

INVENTORY OF RAW MATERIALS FOR THE DUTCH HEAVY CLAY INDUSTRIES

(and a calculation method for the mineralogical composition of a clay)

J.H. VAN WIJCK, H.J. TIMMERS, M. VAN MEERTEN
Technisch Centrum voor de Keramische Industrie,
Postbus 40, 6994 ZG De Steeg, The Netherlands

ABSTRACT

Considering the diminishing availability of clay in The Netherlands as a consequence of the tightening rules for environment protection, an inventory research has been carried out into twelve clay deposits. In this contribution a survey is given of clay deposits in The Netherlands, being exploited for the manufacturing of heavy clay products (building or paving bricks and roofing tiles) or may be so in the future. The way of geological origining is dealt with, as well as several properties of the clay sediments. Those properties of each clay deposit have been collected in a so-called "raw material sheet". With the help of an example the physical, chemical and thermal properties and mineralogical compositions are dealt with which are characteristic for Dutch clays. A calculation method for the mineralogical composition on the basis of a quantitative chemical, XRD and thermal analysis is explained.

INTRODUCTION

In The Netherlands, the production of surface minerals like sand, gravel and clay is coming into conflict with the protection of landscape, nature and environment. This results in tightening rules and limitation of the earth removing possibilities. Moreover, raw materials are getting scarce as a consequence of working out the existing deposits.

However, a modern heavy clay industry in The Netherlands demands well-defined raw materials of a constant quality. It is thinkable that in the future clays of less convenient qualities must be used. In order to record the properties of clay deposits which are in exploitation by this time or will be so in the future, the Technical Centre for the Ceramic Industries, in imitation of Van Amerongen and Van der Velden, carried out an inventory research into twelve clay deposits in The Netherlands.

IMPLEMENTATION

Of each clay deposit a so-called raw material sheet has been made. On that sheet the situation, the geological type of the sedimentary deposit and the present industrial exploitation have been indicated.
It concerns Quarternary (from 1 million years ago until now) fluviatile, marine and eolic deposits as indicated in figure 1.

A Holocène marine clay (extruded)
B Early-Holocene marine calcareous clay
C,D,E,F,G Holocene Rhine clays
H Early-Holocene Rhine clay
J Holocene Maas clay
K Early-Pleistocene Rhine clay
L Pleistocene loam
M Pleistocene loess-loam

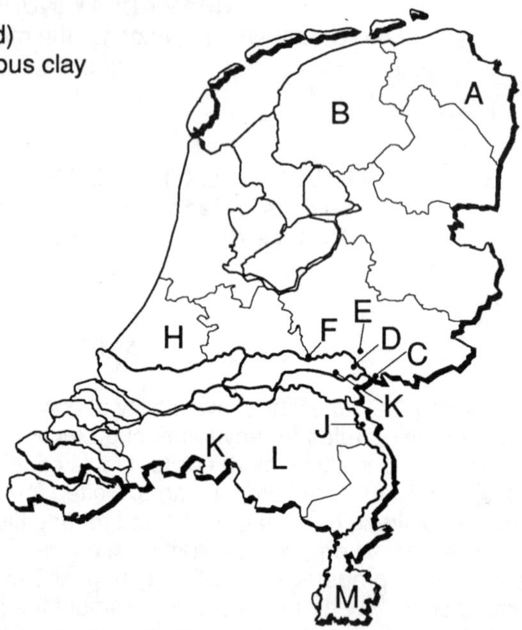

Figure 1. Survey of researched clay deposits in The Netherlands.

Further, of each clay deposit the physical and thermal properties and the chemical and mineralogical composition are indicated. The research of the physical properties includes the determination of the grain size distribution and its deduced grain size characteristic [2], the specific surface [3], the plastic behaviour [4], drying shrinkage and moisture conduction. The determined moisture conduction coefficient gives information about the evaporation rate from a moist clay if a certain moisture gradient is applied into the clay [5].
Thermal properties have been researched with the help of thermogravimetrical, differential-thermal and dilatometer analyses.
The chemical composition has been determined from different grain size fractions. The samples have been prepared in different ways. The chemical composition was determined using atomic absorption spectrophotometry (a.a.s.) as well as X-ray fluorescence. Not only main elements were determined, but also other elements which are important for the ceramic industries, such as F, Cl and S, as well as the content of organic material.

MINERALOGICAL COMPOSITION

In the Technical Centre for Ceramics a calculation method has been developed, especially for Dutch clays, approaching the mineralogical composition of a clay rather close.
Not only a usual chemical analysis and qualitative X-ray diffraction are used, but also the content of organic material, CO_2, S, solvable SO_4, the loss on ignition, particles smaller than 10 μm, the specific surface, optical analysis and a dilatometer record. The chemical analysis serves as a basis for the calculation. Refinements are introduced with the help of other data.

<u>EXAMPLE:</u> Clay J (for chemical analysis see table 1).

Rutile/anatase (TiO_2):
The percentage of TiO_2: 0.7%.
Organic material:
The percentage of organic material from the chemical analysis: 0.7%.
Pyrite (FeS_2):
If the content of sulphur (S) amounts to more than 0.1%, this will be reduced with 1/3 of the content of sulphate (SO_4). The remainder is multiplied by 2 (120/64) and gives the pyrite content. Generally, in Dutch clays this is less than 1% and usually nil.
Goethite ($Fe_2O_3.H_2O$):
From the Fe_2O_3 amount which is solvable in HCl (generally 90 to 95% of the total Fe_2O_3 percentage), 2/3 of the pyrite percentage is subtracted (160/240). Goethite is the remaining percentage Fe_2O_3 x 1.1 (178/160): 4.5%.
Calcite/dolomite (Ca/$MgCO_3$):
The content $CaCO_3$/$MgCO_3$ is 100/44 resp. 84/44 x the content of CO_2. If too little CaO is available for the calculation up to $CaCO_3$, also dolomite will be present. This can be proved by means of X-ray diffraction. 0.4% CO_2 in combination with 0.3% CaO is rounded off to 0.5% dolomite.
Quartz, clay minerals, feldspars:
The remainder R (93.6%) consists of quartz (Q), clay minerals (CM) and feldspars (F). In the Netherlands, clay minerals and feldspars consist average half of SiO_2.
Now we consider two comparisons:

$$Q + (CM + F) = R$$
$$Q + 0.5(CM + F) = SiO_2.$$

For R = 93.6 and SiO_2 = 78 counts: CM + F = 31 and Q = 62.
Illites and feldspars:
Illites and feldspars consist for about 1/8 part of K_2O. The percentage K_2O x 8 gives the percentage illites and feldspars: 1.6 x 8 = about 13%.
The ratio illites : feldspars is for Dutch clays about 2 : 1 [1]; 8% illites, 5% feldspars.
Kaolinite:
The content of kaolinite in Dutch clays is about 15% of the content of particles smaller than 10 μm [1]; 0.15 x 41 = 6%.
Chlorite:
Chlorite is present for at least 1%. If proved (optically, by X-ray diffraction) 3%.
Expanding clay minerals:
Expanding clay minerals are calculated by reducing (CM + F) by the calculated percentages of feldspars, illites, kaolinite and chlorite; 31 - (13 + 6 + 3) = 9%.

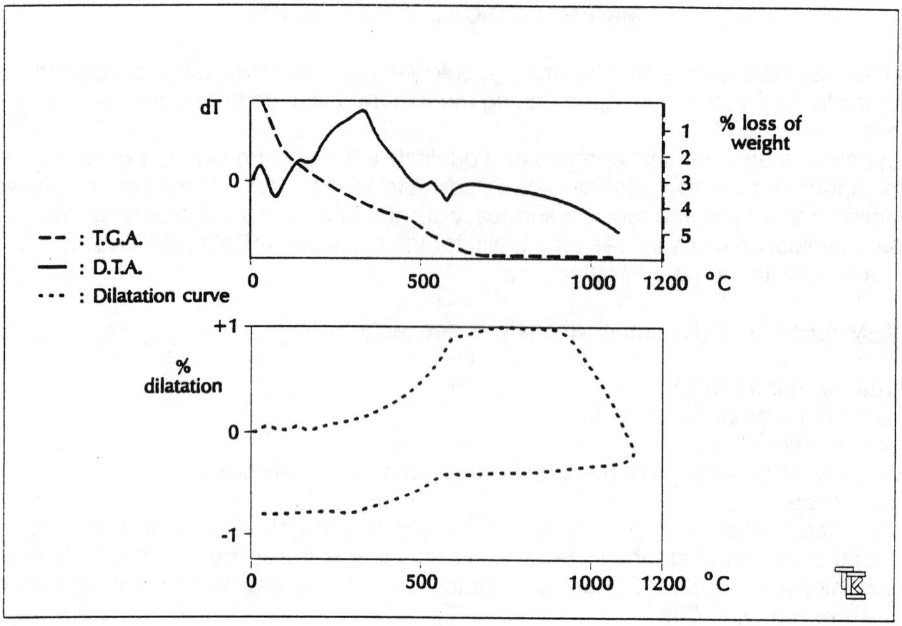

Figure 2. Thermal analysis of clay J.

Physical properties (% m/md)			Chemical composition (% m/md)			Mineralogical composition (%m/md)	
Grain size distribution			SiO_2	=	78	Rutile/Anatase	0,7
			Al_2O_3	=	9,9	Organic material	0,7
< 250 µm	=	96	TiO_2	=	0,7	Pyrite	-
< 125 µm	=	83	Fe_2O_3	=	4,5	Goethite	4,5
< 65 µm	=	76	CaO	=	0,3	Dolomite	0,5
< 45 µm	=	70	MgO	=	0,7	Quartz	61
< 20 µm	=	50	K_2O	=	1,6	Illite	8
< 10 µm	=	41	Na_2O	=	0,03	Feldspars	5
< 2 µm	=	27	L.O.I.	=	4,1	Chlorite	3
< 0.5 µm	=	12			99,8	Expanding clay minerals	10
specific surface = 85 m²/g			CO_2	=	0,4		
			O.M.	=	0,7		
			F	=	0,033		
			Cl	=	0,028		
			S	=	0,01		

Table 1. Some physical, chemical and recalculated mineralogical properties of clay J.

CHECKS AND REFINEMENTS:
Starting from the specific surface of the clay and from a specific surface of the distinct minerals kaolinite (50 m^2/g), illite (200 m^2/g), organic material (600 m^2/g) and expanding clay minerals (600-800 m^2/g) the percentage of expanding clay minerals can be calculated in a different way:

Total of clay	85 m^2/g
Kaolinite	3 m^2/g
Illite	16 m^2/g
Organic material	4 m^2/g
Expanding clay minerals	62 m^2/g

This contribution to the specific surface can be explained from a contribution of 10% expanding clay minerals. A correction is applied upon the quartz content: 62 - (10 - 9) = 61%.

Still some adjustments can be introduced with the help of data from the dilometer curve and X-ray diffractograms.
It is thinkable that the exposed calculation method will also be applicable to other than Dutch clay deposits. In that case several assumptions should possibly be adapted.

RESULTS

The results of the research will presently be collected and published titled "Raw material sheets for the Dutch ceramic industries". This edition enters into the geological origining of clay deposits, several properties and the connection between those properties. Besides, some determination methods will further be explained.

REFERENCES

1. Van Amerongen, H. and Van der Velden, J.H., Composition and properties of 31 types of clay, part I and part II, CTI - TNO 70 - 04032, 1970.

2. Bloemen, G.W., Calculation capillary perviosity from granular composition and content of humus. Landbouwkundig tijdschrift, 1979, 91, 329 - 335.

3. Timmers, H.J., Some physical aspects of drying ceramic products. Klei en Keramiek, 1967, 17, 166 - 182.

4. Velden, J.H. van der, Analysis of the Pfefferkorn test, Klei en Keramiek, 1979, 29, 42 - 56.

5. Timmers, H.J., Techcon dryer. Proceedings 1st European Ceramic Society Conference, including Science of Ceramics 15. Maastricht, 1989.

THE INFLUENCE OF PROCESSING CONDITIONS ON THE PORE SIZE DISTRIBUTION AND DURABILITY OF KEUPER MARL BRICKS

ROBERT WILLIAM FORD
Section Manager, Building Materials Division
British Ceramic Research Limited,
Queens Road, Penkhull, Stoke-on-Trent, ST4 7LQ, U.K.

ABSTRACT

The relation between pore size distribution and the resistance to freeze/thaw conditions has been examined for a range of bricks made from Mercian Mudstone [Keuper Marl]. Results show how raw material processing, the moisture content and the method used to form the bricks affect these characteristics.

INTRODUCTION

The influence of the pore structure of clay bricks on their resistance to freeze/thaw conditions has been described by several authors[1-4]. Unfortunately attempts to relate and compare results from different investigations are hindered by the fact that no commonly accepted criterion of durability is available. In addition, a simple general relation between pore size distribution and durability encompassing all types of clay brick is unlikely. In particular, the intrinsic strength of the material determined by the degree of vitrification must have a considerable influence on the ability of the product, when saturated with water, to withstand the stresses produced by freezing.

The present investigation was limited to one type of raw material. The Keuper Marls were laid down during the second half of the Triassic Period and are located mainly in the Midlands of England. The marls are illitic and have a narrow vitrification range with rapid shrinkage occurring at temperatures above about 1100°C. For this reason the firing of the bricks is terminated before the onset of this rapid vitrification and the resulting products are fairly porous [>10% water absorption] and relatively low in strength [<45N/mm²]. In the majority of cases they do not have a high resistance to freeze/thaw conditions. Because the bricks produced by different plants are fired to about the same level of vitrification [or more correctly level of sintering] any differences in durability are likely to be caused by factors other than the firing treatment. The present investigation was carried out to examine some of these factors.

PORE SIZE DISTRIBUTION AND FROST RESISTANCE OF BRICKS

Samples of eight types of Keuper Marl bricks were obtained from five plants. All the bricks were produced by extrusion and, with one exception, were fully deaired. The bricks were subjected to a number of tests and the results are given in Table 1.

Resistance to freeze/thaw conditions

A detailed description of the test procedure has been given elsewhere[5]. Each type of brick was built into a test panel [3-brick wide x 10-courses high] using 1:¼:3 mortar. After curing for at least 28 days the panel was immersed in water for 7 days before subjecting one face to alternate freeze/thaw conditions for 100 cycles. Damage occurring during the test was assessed after 50 cycles and 100 cycles.

TABLE 1
Properties of bricks

Sample No.	Freezing Test		Pore Size Distribution				Water Absorption	
	50 cycles	100 cycles	% within size range [μm] >10	1 - 10	0.1 - 1.0	<0.1	5hr boil	sat. coef.
B1*	none	none	7	71	21	1	15.1	0.73
B2	3	4	3	35	22	-	25.8	0.77
A2	2	15	11	56	33	-	-	0.78
A1	12	35	10	38	52	-	18.5	0.81
D3	13	35	2	5	92	1	16.7	0.95
G1	10	16	4	42	52	2	20.5	0.80
D1	25	35	3	4	88	5	16.2	0.94
C2	29	35	2	15	82	1	19.6	0.89

* Non-deaired

With the exception of brick B1 all samples showed some damage after 100 cycles. Bricks which fail in less than 100 cycles would probably not be classified as F [Frost Resistant] as defined in BS 3921[6], although there is at present no completely acceptable criterion for categorizing bricks O, M or F based on the results of this panel freezing test. Nevertheless, it is considered that the test is able to place bricks in a ranking order of durability.

The bricks have been arranged in descending order of durability in Table 1. Bricks A1, D3 and G1 and bricks D1 and C2 have been grouped together since the test results indicated no significant difference.

Pore size distribution

This was determined on a small sample taken from a brick using a mercury intrusion method [Micromeritics].

Water absorption by weight

Five bricks of each type were used to determine the mean water absorption after 24h immersion in cold water and 5h immersion in boiling water. The saturation coefficient:

$$\left[\frac{\text{24h soak value}}{\text{5h boil value}}\right]$$

is also reported in Table 1.

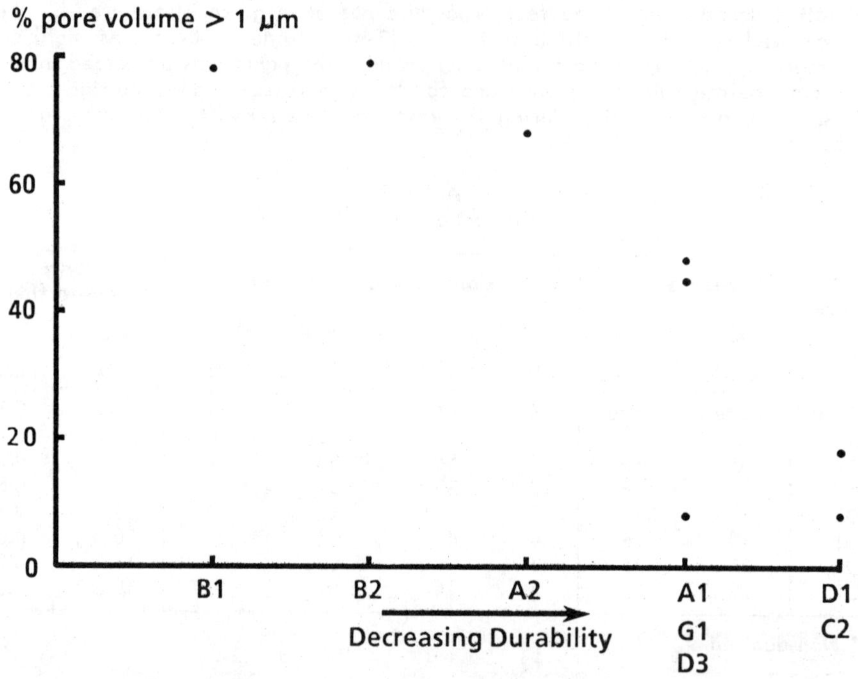

FIGURE 1: Pore size and durability of bricks

The general relation which exists between pore size distribution and resistance to freeze/thaw cycling is illustrated by Figure 1 where the brick samples have been arranged in a descending order of durability. This clearly demonstrates that bricks with a higher proportion of coarser pores are more durable. In this particular case a coarse pore is considered to have a diameter greater than 1.0µm. The criterion of what constitutes a coarse pore is faily arbitrary but a limiting size of 1.0µm more clearly differentiates between the various bricks than if, for example, a diameter of 10µm was chosen. There is no evidence from these results that the durability correlates with the water absorption values for the range of bricks examined. In general, as already mentioned, the low level of sintering during firing with Keuper Marl bricks means that the porosity of the fired brick is largely determined by the original porosity of the unfired brick. More importantly, the pore size distribution of the unfired and fired brick must be closely related so that changes in the texture of the unfired brick should be reflected in the durability of the fired product. The investigation was therefore extended to consider some of the factors which could influence this texture with particular emphasis on the composition and processing treatment of the raw material.

PORE SIZE DISTRIBUTION AND PARTICLE SIZE

Four samples of Keuper Marl obtained from different sites were analyzed for particle size distribution following complete dispersion and deflocculation. Previous results had demonstrated that bricks produced from these marls gave a range of durabilities and pore size distributions. The results in Table 2 indicate no correlation either between the total surface areas of the original raw material and the pore system of the fired material or between the particle diameter of the raw material and the pore diameter within the fired brick.

TABLE 2
Particle size and pore size

Keuper Marl	Particle Size [Raw material]		Pore Size [Fired brick]	
	Area [m²/g]	> 1μm [%]	Area [m²/g]	> 1μm [%]
D3	2.56	72	2.46	7
C1	2.98	67	2.31	17
B1	3.12	52	0.61	79
A2	3.13	72	0.63	67

Clearly, aggregated clay particles are only partially dispersed during processing and shaping in the presence of water; the extent of this dispersion must be a prime determinant of body texture and this is examined in the next Section.

Processing conditions and pore size distribution

Briquettes measuring 75x38x25mm were produced from a laboratory extruder from two different Keuper Marls. Prior to tempering, each marl was dry-ground thro' a 2.4mm aperture sieve. The processing and extrusion conditions are described in Table 3. The conditions were varied to examine the effect of:

Consistency [soft and hard] of the extruded body	[coded S or H in Table 3]
Deairing and non-deairing	[coded D or N in Table 3]
Plastic working of the tempered mix [1 pass or 10 passes through the extruder]	[coded 1 or 10 in Table 3]

Briquettes were dried carefully and fired to 1075°C at 50°C/h with a four hour soak at top temperature. Water absorption values [5h boil and 24h soak] and pore size distributions were determined and the results are given in Table 3.

TABLE 3
Processing conditions and fired properties of briquettes

Briquette No.	Processing Conditions		Water Absorption			Pore Size Distribution [μm]			
	Code	Extrusion Moisture	Boiled [5h]	Soaked [24h]	Sat. Coef.	>10μm	1 - 10	0.1 - 1.0	<0.1
	Marl A								
A1	S/1/N	24.2	18.2	13.7	0.75	1	60	38	-
A2	S/10/N	23.8	18.0	13.2	0.73	2	58	41	1
A3	H/1/N	19.7	16.8	13.7	0.81	2	46	50	1
A4	H/1/D	20.1	16.3	13.0	0.80	1	54	54	1
A5	H/10/N	16.1	16.7	13.8	0.83	7	25	68	1
A6	H/10/D	15.9	14.1	10.8	0.77	2	34	64	-
	Marl B								
B1	S/1/N	20.7	12.8	8.3	0.65	3	72	25	-
B2	S/10/N	19.2	11.6	7.4	0.64	3	71	26	-
B3	H/1/N	17.3	10.6	7.4	0.70	6	60	33	1
B4	H/1/D	17.7	9.6	6.2	0.64	2	64	34	1
B5	H/10/N	15.3	9.3	6.3	0.67	5	55	39	1
B6	H/10/D	15.2	8.8	5.7	0.65	2	57	39	1

A number of general effects are demonstrated by these results:

[a] Stiff extrusion produces an increase in the proportions of finer pores [less than 1μm] at the expense of pores greater than 1μm.

Compare briquettes A1 with A3, A2 with A5 and B1 with B3, B2 with B5

[b] Deairing has little effect on the pore size distribution. The main effect is to reduce the proportion of large pores [greater than 10μm].

Compare briquettes A3 with A4, A5 with A6 and B3 with B4, B5 with B6

[c] Plastic working [re-pugging] at a stiff consistency produces a marked increase in the proportion of fine pores. This effect is not nearly so great at softer consistencies.

Compare briquettes A3 with A5, A4 with A6 and B3 with B5, B4 with B6

The negative correlation between saturation coefficient and the proportion of pores bigger than 1.0μm is shown in Figure 2 for the works-produced bricks [see Table 1] and the laboratory briquettes [see Table 3].

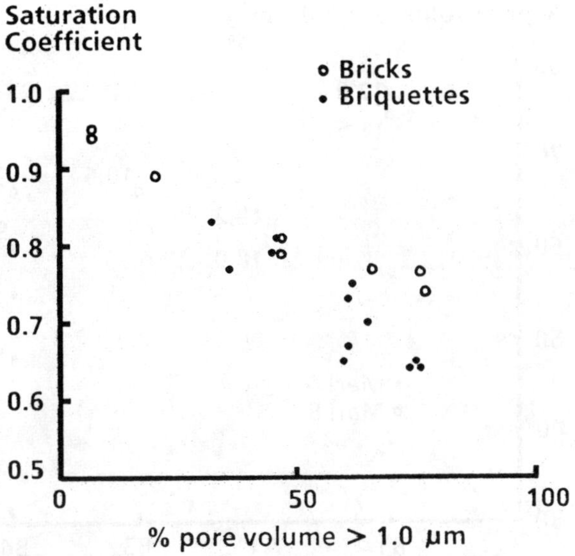

FIGURE 2: Pore size and saturation coefficient

It is important to emphasise that, although the relation demonstrated in Figure 2 indicates that saturation coefficient has some relevance when assessing durability indirectly, it should not be assumed that this parameter can be used to uniquely characterize the frost resistance of a clay brick irrespective of the amount of vitrification attained during firing.

FREEZE/THAW TESTS ON BRIQUETTES

The durability of the briquettes produced under the conditions described in Table 3 were examined using the following procedure. The briquettes were saturated by immersion in boiling water for 5 hours and were then arranged together in an insulated tray with one face only exposed to cyclic freeze/thaw conditions. These consisted of 2 hours freezing at -15°C, followed by 20 minutes thawing by spraying the exposed face with water. Progressive failure of the briquettes up to 200 cycles was noted, and it was possible to place the samples in a ranking order of durability. This has been done in Figure 3 which demonstrates how the relative durability is related to the proportion of coarser pores [>1.0μm]. The number against each point is the mean water absorption value of each sample. It is interesting to note that with both Keuper Marls, a reduction in water absorption resulting from stiffer extrusion, or plastic working or de-airing, is associated with a reduction in durability because of a concomitant increase in fineness of the texture of the fired body. This contrasts with the effect of any change in water absorption arising from changes in the firing treatment.

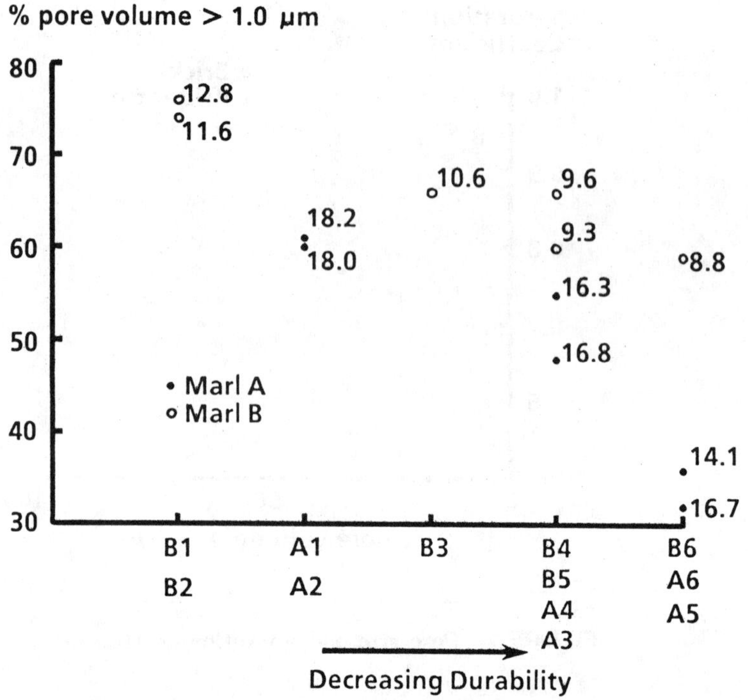

FIGURE 3: Pore size and durability of briquettes

SUMMARY AND CONCLUSIONS

A general relation between durability as assessed by laboratory testing and pore size distribution has been demonstrated, for Keuper Marl bricks, with better durability being associated with a higher proportion of pores larger than 1µm in diameter. A similar correlation has been established for a range of briquettes produced using various processing and extrusion conditions.

There is no relation between the pore size distribution of a fired Keuper Marl brick and the fully dispersed particle size of the raw material. However, the pore size distribution of the fired material is markedly influenced by the degree of processing and by the extrusion conditions used in manufacture.

In general terms, a raw material subjected to excessive plastic working and then extruded with de-airing at a stiff consistency contains a higher proportion of finer pores and, consequently, is more prone to damage by freeze/thaw conditions.

The negative correlation which has been demonstrated between the proportion of pores bigger than 1.0µm and the saturation coefficient of the fired product indicates that a low saturation coefficient is associated with good durability. It is not however, the sole determinant of this property.

REFERENCES

1. MARUSIN, S., The Effect of Variations in Pore Structure on the Frost Resistance of Porous Materials.
Cem. and Conc. Res. 11,115,1981.

2. ROBINSON, G.C., The Relationship between Pore Structure and Durability of Brick.
Bull.A.C.S. 63,[2],295,1984.

3. MAY, J.O. and BUTTERWORTH, B., Studies of Pore Size Distribution Part III.
Science of Ceramics I - Page 201.

4. MAAGE, M., Frost Resistance and Pore Size Distribution of Bricks.
Materiaux & Constructions 17,345,1984.

5. WEST, H.W.H., FORD, R.W. and PEAKE, F., A Panel Freezing Test for Brickwork.
Br.Ceram.Trans.J. 83,112,1984.

6. BRITISH STANDARDS INSTITUTION, Clay Bricks.
BS 3921 : 1985.

ACKNOWLEDGEMENT

The author thanks the brick manufacturers who sponsored this work and also supplied the samples.

INFORMATISATION AND AUTOMATION IN THE CERAMIC INDUSTRY

J.C. MARKS
Stichting Technisch Centrum voor de Keramische Industrie
P.O. Box 40, 6994 ZG De Steeg, The Netherlands

SUMMARY

A proposition is done in order to come to a CERAMICS EXCHANGE COMMUNICATION STANDARD for the benefit of Management Information Systems in the industry.

We base ourself on the sector research report "Informatisation and Automation in the Ceramic Industry" which was done in brick- and tile works in the Netherlands and with suppliers of means of productions in Europe.

INTRODUCTION

To manage or in other words to make decisions, means making a choice from alternatives all the time. When one wants to make such a responsible choice, one will need information about matters like why such a choice will have to be made and information about the possible consequences resulting from a certain choice. Someone who does not have this kind of information at all, cannot possibly make a decision he can be accounted for. If he chooses anyway, he will (in a free adaption of the well-known proverb) act "through his hat". And one can hardly claim that if one acts like that such a decision can be accounted for.

In order to prevent an information and automation chaos, within every company one will have to choose a system and an information infrastructure. Suppliers will have to anticipate to the demands of the users and in order to get that far, one will have to proceed to the delivery of standard automationsystems and/or standard communicationsystems, which will result in a greater availability of machinery and processes and add to this that the integration of new means of production in the existing structure will also be possible.

INFORMATISATION

Numerous data are often collected in order to registrate the consequences of an earlier taken decision. For example the registration of production- and salesresults or the bookkeeping of an organization. What now is the use of such information afterwards? Thus information after already a decision has been made. Those information are only relevant in accordance with new decisions having to be taken. Roughly taken one can distinguish two cases, being:

- feed-back information: one collects data about the consequences of an earlier taken decision with the purpose to learn from it for the future; this information serves also an improvement in processes how to take decisions in the future;
- accountinformation: one collects data about earlier decisions, on the basis of which the deciding person itself or someone else will have to judge, so deciding whether a certain task has been carried out properly.

When we are looking at any decisionsystem, it appears that such a system can be divided into three partial systems, namely:

- a managed system, being the partial system having to be operated;
- a managing organ, being the operating person or group of persons (aside we would like to make the remark that a machine can also be an operating organ);
- a management informationsystem, being the partial system producing information for the operating organ on the basis of data of the operated system itself (internal data) or on the basis of data about the surroundings (external data).

Computers and their applications form new constitutional elements on different levels of policy, management and administration in an organization.

When one wants to express that it is much more than:
doing the same as one used to -with new means- in a different way, we speak of informatisation of the informationsociety. Topmanagement realises that information is not only a tool for working efficiently and effectively, but also a strategic weapon in competitive struggle. This forces the organisation to a strategic information policy and -planning. Here fits the remark that an informationsystem is not only a tool of management, but also a proof of management.

AUTOMATION

When a machine is equipped with measuring possibilities and when the measuring results are feeded back to the operating organs without human intervention, the mechanization evolves to automation. Because, when in fact the machine takes over human mental work, we can then already speak of artificial intelligence (AI). We see practical application of this in machinery for Flexible Production Automation (FPA). The essence of automation is always some form of managing.

The difference between automation of primary productionprocesses and operatingprocesses can be rather diffuse, namely in organisations in which the productionprocess itself consists for the greater part of dataprocessing. The automation of the managing of primary productionprocesses and the automation of organisations are mostly refferred to as management informationsystems.

STATE OF AUTOMATION

In the first phase of the branche research "Information and Automation in the Ceramic Industry" we checked with the principals whether the models we used can be applied within the ceramic industry. These companies were examined in a detailed way and a report was made for each company. It appears that we can very well use the models and the display of the results.

Consequently we drew up an inventarisation on this basis among 30 companies. These companies model for approximately 50 companies and thus for approximately 70 % of the production capacity in the Netherlands.

Registration of production data
A good registration of production data is important for:
- the decrease of energy expenditure
- the planning of more maintenance
- to substruct investment decisions
- a dynamic costprice calculation
- the obtaining of reliable data in relation with causes of disturbances
- to limit the chance on human errors with taking in and processing production data.

CONCEPT MANAGEMENT INFORMATIONSYSTEM

The informationsystem is the whole of people, means, data and procedures aimed at producing appropriate information. The managing organ is often splitted up in hierarchial decision levels, because the information differs per level as well as the conditions under which the information has to be given.

As a concept a Management Information System can also exist without computers. But it is the power of the computer that makes a MIS practically executable. The Management Information System is an evoluating concept. One can consider the MIS-concept as an important addition to the concept of business economies, operational research and organization theories about management and decision making.

In the sector research report the limits (interfases) of the system are discussed further and it makes clear what information has to be obtained from productionsystems (the functions).
On the basis of the outputdata we set up a basismodel for a PMIS/EMIS-system, which is available on diskette.

INTEGRATION OF AUTOMATIONSYSTEMS

A recent important development is the fact that the end user, with the help of a personal computer and powerful standard software, himself can get acces to the data and development models. It is not necessary anymore to use only one databank, so external sources can be used.
The implementation of a single, highly integrated system is complex. The MIS-concept has to be constructed coöperatively from a federation of subsystems, which will be -when needed- developed, implemented and adjusted to changing insights and means of production. Herewith one should conform to a global plan, standards and procedures for the Management Information System. For the above mentioned a certain amount of being consequent or standardization among the suppliers will be necessary, namely for the accessability of external services and the physical possibilities of connection among the different types of apparatus.

STANDARDIZATION

Concerns within the ceramic industry will have to make a choice about the automationsystem to be applied and about an information- infrastructure. From the sector research it appears that there are three options when seeing the existing and offered systems within the ceramic industry.

1. Technologically seen is the standardization on the ISO-OSI model based MAP the best way to come to an open communication infrastructure within the companies. The present PLC market leaders will introduce their MAP version in the coming one or two years.
   The standardization of this system has some objections, being:
   - insufficient certainty whether the top three levels of the ISO model harmonize sufficiently with each other
   - the high costprice when we consider the purpose it is going to be used for.
2. The method to standardize in automationsystems as well as in communicationsystems, is to choose for one mark and type of process system for the entire company. This will -depending on the mark chosen- limit the number of potential suppliers, but it has great advantages in the area of usability, flexibility, optimalizability and serviceability. As communicationsystem a low-cost mark-innate connection can be chosen, which can surely cope with the demands made in this specification.
3. Within existing production companies it's usually not possible anymore to standardize automationsystems. This also counts when a maximum freedom of choice is desirable in suppliers of means of production, because then it will not be possible to choose a mark-innate communication system within the company. From the sector research it appears that one harmonizing physical connection for all systems can be delivered, being: RS 232/485.

With the application of this communication connection all automationsystems offered by the suppliers can be connected in a simple way with a central (personal) computer, independent of the mark, operatingsystem or software. When a standard communication protocol is agreed on, then we can speak of a Low Cost Multi Vendor Network for the benefit of Management Information Systems in the ceramic industry. For this we introduce the CERAMICS EXCHANGE COMMUNICATION STANDARD (CECS).

CERAMICS EXCHANGE COMMUNICATION STANDARD (CECS)

If we base ourselves on the at all offered automationsystems present RS 232 C communication gate, in that case there are still several possibilities to come to a joining among the systems. For this agreements will have to be made between suppliers and users. Standard connections, protocols and not mark-tied and freely tradable programmes have to be used. The data transmission happens on three levels.

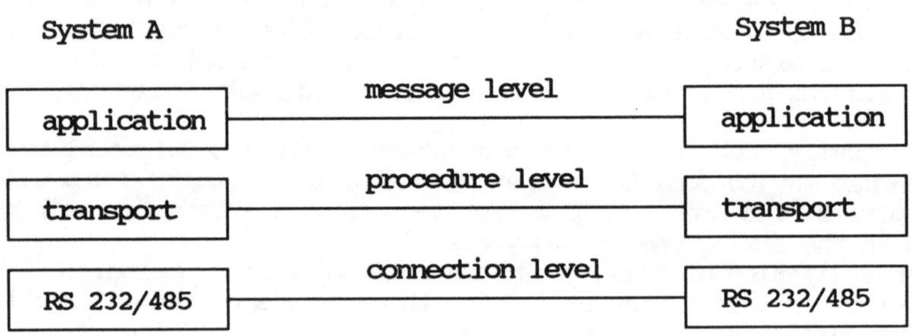

CERAMICS EXCHANGE COMMUNICATION STANDARD (CECS)

CONSULTED LITERATURE

- Bestuurlijke Informatiesystemen en automatisering
  T.M.A. Bemelmans, Stenfert Kroeze, ISBN 90 20715909

- Branche onderzoek Informatisering en Automatisering in de Keramische Industrie
- Stichting Technisch Centrum voor de Keramische Industrie, De Steeg
- Instituut Informatie Technologie Produktieautomatisering, TUE/TNO Eindhoven

- Electronica in de keramische industrie
  KleiGlasKeramiek nr. 1 1988
  Keramische Zeitschrift nr. 3 1988
  J.C. Marks

- Management informatiesystemen
  G.B. Davis & M.H. Olsen, Academic Service, ISBN 90 6233 1890

INFILTRATION IN STEEL REFRACTORIES

J.E. DAVIES^, P.M. MARQUIS^ & M. HANKIN*
^School of Metallurgy & Materials, University of Birmingham, Birmingham, U.K.;
*Foseco (FS), Tamworth, Staffordshire, U.K.

ABSTRACT

An improved rotating dip test has been developed for testing refractories used in steel casting. The erosion of 'Black' and 'White' refractories have been investigated and their relative performance in both test and service compared. The impact of plasma sprayed barrier coatings have also been evaluated.

INTRODUCTION

In the continuous casting of steel one of the most demanding roles for a refractory is the tundish to mould pouring tube, the submerged entry shroud (SES) or submerged entry nozzle (SEN). These components have to withstand severe thermal shock and be resistant against attack from the aggressive flux on the surface of the steel . Erosion and corrosion of the tubes in this area often causes their failure.

Originally fused silica was widely used in this application but advances in black refractory technology have produced shroud materials which combine good thermal shock behaviour and steel erosion resistance. As a result carbon bonded alumina-graphite has been accepted as the standard material for these tubes. Although the general performance of alumina graphite is acceptable, flux line erosion is still a major problem. Tube lifetimes have been increased by a factor of 2 to 3 by the use of carbon bonded zirconia graphite materials at the flux line. However, there is still a need for improved performance and research into new materials for pouring tubes. One of the problems faced during development of new materials for use in continuous casting is selecting a suitable simulative test. Even the most sophisticated experiment cannot fully reproduce the unique conditions in a continuous casting mould and more simplified tests have to be used.

Shaw et al [1] have described a laboratory scale ‘ rotating finger’ type of test but unfortunately these workers had difficulty in obtaining wear rates comparable to those observed in service, erosion depths mainly being greater in actual service than those observed in the dipping test. A major drawback of this method is the contamination of the mould flux and steel by the containing vessel and the samples. This factor can markedly influence wear rates particularly in small scale tests. Although the use of more exotic crucible materials can minimise these effects the optimum solution is to increase the scale of test to a level where contamination effects are less marked.

Work in Japan by Nakamura et al [2] has investigated the effect of mould powder chemical composition on the erosion of of SENs' in a static dip test using a high frequency induction furnace. Although this method proved a valuable comparative test there were still discrepancies between the results and those obtained in service. The work reported here describes the extension of this approach with the objective of obtaining an improved correlation with service results and clarifying the reasons for any discrepancies between test and service performance.

EXPERIMENTAL

In the experimental studies both carbon alumina-graphite bonded and zirconia graphite were investigated as widely used refractories in continuous casting. As comparators fused silica and zircon based refractories were selected, silica on the basis of its long term use and the zircon material as an example of a more porous material of a composition which is regarded as having potential in the shroud application. In addition the zircon based material was also plasma sprayed with a zirconia coating to evaluate the impact of a barrier coat on the test approach. The chemical compositions of the materials studied are given in Table 1. A standard mould flux of 0.95 basicity (CaO/SiO_2) was used, having a melting point of 1150°C and a viscosity of 2.5 poise at 1400°C. Two different grades of steel were used, a low carbon steel and a high manganese low alloy steel both fully killed with aluminium.

TABLE 1
Chemical composition of materials evaluated

CHEMICAL COMPOSITION (wt%)	Alumina Graphite	Zirconia Graphite	Fused Silica	Zircon	Plasma Sprayed Zirconia Coating
Alumina	56				
Zirconia		80		61.5	92
Silica	14	5	99	38	
Calcium Oxide		3.5			
Yttria					8
Carbon	28.5	10			
Others	1.5	1.5	1.0	0.5	

Testing Method
Rotary dip tests were undertaken in a 350Kg high frequency induction furnace. A diagram of the experimental layout is shown in Figure 1. Samples were preheated to red heat (approx. 900-1000°C) using a propane burner, placed over the induction furnace and then immersed in the molten steel. Mould powder was added to the surface of the molten metal, which was kept at 1550°C, to give a flux layer of between 10 and 20mm. The tube holder was rotated at 40 r.p.m. for one hour, the duration of the test. Mould powder was added at frequent intervals to maintain the flux cover. Samples were also obtained which had been used in actual service for known periods of time and for which both steel and flux compositions were known.

Analysis of Test Samples
The erosion depth at the flux line was measured at the outer edge of the sample using the portion above the flux and molten metal as a reference. Samples were then cut from the test materials at the flux line, sectioned, mounted in resin and then polished using standard metallographic techniques to a one micron diamond paste finish. Final preparation was undertaken with a combination of chemical and mechanical polishing with a colloidal silica sol in a caustic solution. Samples were then coated with a thin layer of carbon and examined in a JEOL 840 scanning electron microscope. Energy dispersive X-ray analysis was used to determine the chemical composition of the phases present and back scattered imaging to establish the phase distribution.

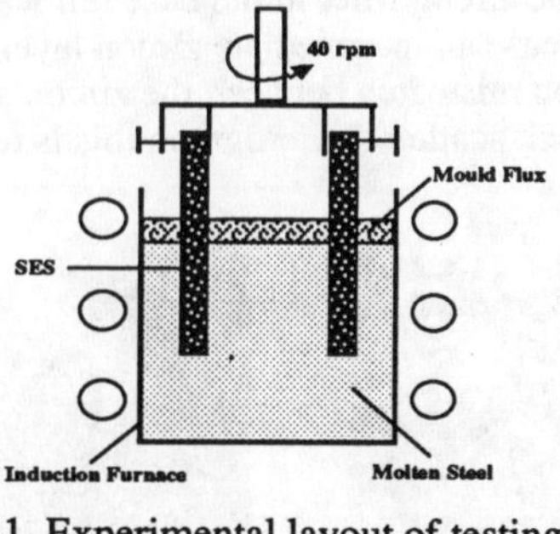

Figure 1. Experimental layout of testing method.

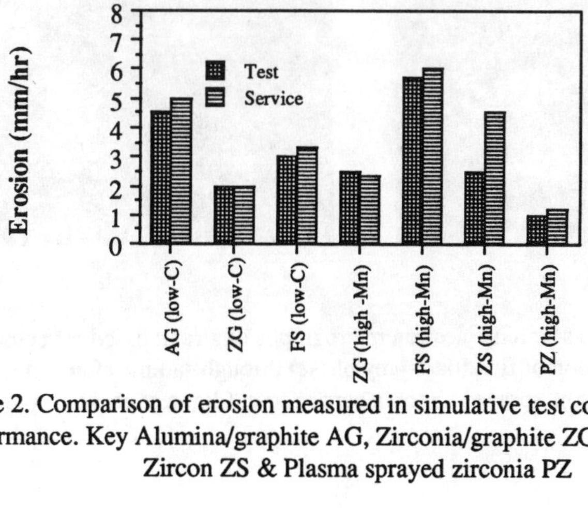

Figure 2. Comparison of erosion measured in simulative test compared to service performance. Key Alumina/graphite AG, Zirconia/graphite ZG, Fused silica FS, Zircon ZS & Plasma sprayed zirconia PZ

RESULTS & DISCUSSION

The maximum erosion depths recorded in the simulative test are compared with the actual service performance in Figure 2. It can be seen that the test has clearly indicated differences in performance and highlighted the impact of high manganese levels in the steel. The optimum performance was achieved using the plasma sprayed barrier coating which showed no sign of thermal spalling. In the case of the black refractories and fused silica, the patern of service performance is closely reflected by the simulative tests. However, in the case of the zircon material the service wear was significantly higher than the simulative test result.The different wear mechanisms of the materials were deduced from analysis in the scanning electron microscope. In the case of the black refractories and fused silica flux penetration is limited and the wear occurs by a process of attrition at the flux/refractory interface. In contrast the more porous zircon refractory shown in Figure 3a reveals almost total penetration of the flux throughout the sample. It can also be seen that the zircon filler particles are dissociating and as the zircon breaks down the refractory can be eroded by the smaller particles being washed into the flux. The difference in performance appears to be related to flux penetration and it is proposed that in actual service the flow of molten metal down the bore of the SES or SEN tube creates a Venturi effect drawing in the flux and accelerating the erosion above that in the simulative test where there is no comparative metal flow. Reinforcement of this mechanism is shown from the effect of the plasma zirconia coating shown in Figure 3b. The impact of this layer is to inhibit flux penetration and delay the dissociation of the zircon filler in the base refractory thus creating the improved performance.In this case an intermediate zircon layer has also been laid down to minimise the thermal expansion mismatch between the zirconia coating and zircon substrate and as can be seen from the back scattered micrograph this is retained even after plasma spraying.

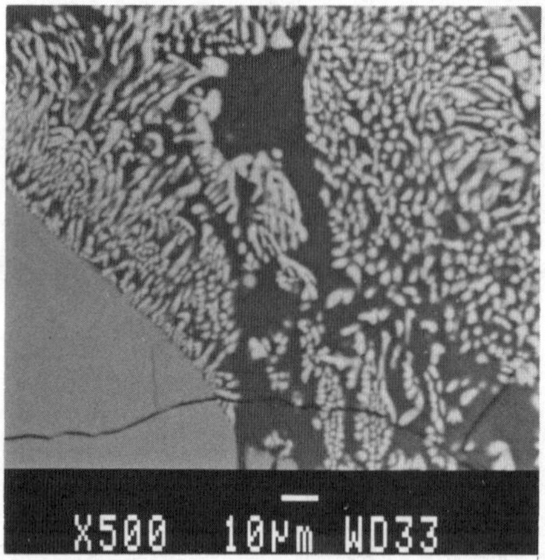

a.

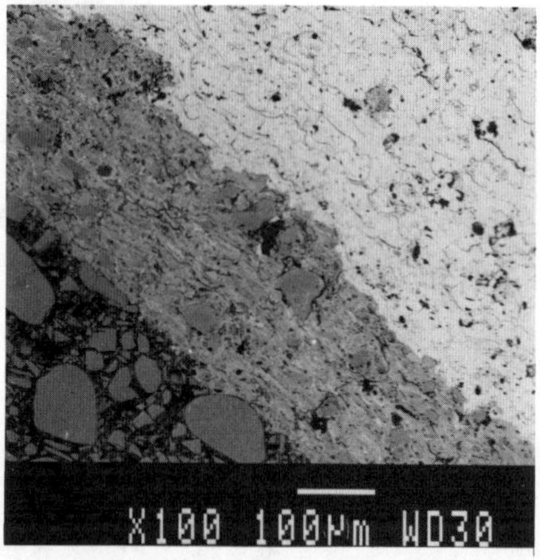

b.

Figure 3. Backscattered scanning electron micrographs of zircon based refractory showing
a) penetration of flux (dark grey phase) through sample after test
b) plasma sprayed zirconia barrier coat (lightest phase)

CONCLUSIONS

A simulative test configuration has been described which measures the erosion and corrosion of flux line materials for submerged entry shroud and nozzle applications, and gives improved correlation with in service performance. This is due in part to the larger scale of the test limiting the effects of contamination, but also rotation of the samples is important since it simulates the washing action of the flux over the refractories surface as the mould oscillates and the metal level fluctuates. Turbulent flow behaviour is also created in service by steel flow patterns in the mould and more localised behaviour at the steel/flux/refractory interface, caused by the 'Marangoni effect' [3]. Clearly the effects of contamination have not been totally eliminated and it is for this reason that the comparatively short test time was selected. The main source of contamination is from the induction furnace lining which progressively contaminated the flux with alumina during the test. As a result the viscosity of the flux increases reducing its penetrating power, notably for more porous refractories, although frequent additions of mould powder can offset this behaviour to some extent. There are still discrepancies with more porous materials (>20%) and it has been noted that the Venturi effect created by steel flow down the bore in service is not simulated in the test and is a potential explanation of the discrepancies.

The results have also shown that a plasma sprayed zirconia barrier coat can significantly extend the service life of the more porous refractories. When applied to a zircon refractory substrate these coatings have been found to show excellent thermal spalling resistance and very low wear rate at the flux line.

This route therefore offers potential for producing 'white' refractories with good flux line erosion and corrosion resistance whilst avoiding the problems associated with the carbonaceous material in the black refractories. In particular these types of coated white refractories would combine oxidation resistance, flexible pre heating regimes, impermeability and strength retention during casting, reduced alumina build up and most significantly insensitivity to low carbon, high oxygen level steel grades which are highly corrosive to current black refractories.

ACKNOWLEDGEMENTS

The work reported in this paper was undertaken with the support of a grant from the U.K. Science and Engineering Research Council and Department of Trade and Industry.

REFERENCES

1. G.B. Shaw, R.M. Taylor & M.D. Chambers, UK Inst. Ceram. Convention, 1988.

2. Y. Nakamura, T. Ando, K. Kurata & M. Ikeda, Trans. ISIJ (Japan), 1986, 26, 1052.

3. E. Hoffken et. al., Proc. 4 Int. Conf. Cont. Casting, Brussels, 1988, 461.

PREPARATION AND PROPERTIES OF ALUMINA-CHROMIA (CORUNDUM) REFRACTORIES CONTAINING MULLITE

A. HARABI* and T.J. DAVIES
Manchester Materials Science Centre
Grosvenor Street, Manchester M1 7HS UK

ABSTRACT

Mixtures of alumina and chromia powders have been heat treated to produce a modified corundum. The effect of varying the amount of chromia, sintering temperature and time, and addition of mullite on the densification and strength of the sintered bodies is reported.

INTRODUCTION

The addition of chromia to corundum-mullite refractories should improve certain high temperature properties of the product and several workers (1,2) have pointed out that the addition of chromia to mullite or corundum-mullite improves the resistance of mullite bricks (or mullite-corundum) to ferruginous slag reactions. Additionally, Stanislaw (3) showed that when corundum (alumina-chromia) was reinforced by mullite it had a good thermal shock resistance and Makarov et al (4) have recently reported that the strength of corundum ceramics is increased by the formation of $(Al,Cr)_2O_3$ solid solution in the presence of mullite. Of the few studies which have been reported on this system (2,5), the most detailed study was published by Roeder et al (6).

It was considered that the use of mullite as a second phase, uniformly distributed throughout a corundum matrix (in which alumina is partially replaced by chromia) could be effective in restricting crack propagation, thus strengthening and toughening the ceramic.

Because of the low melting point of the glassy phase in the pseudo-ternary system and the low limit of solubility of chromia in silica, the pseudo-subsystem mullite-Al_2O_3-Cr_2O_3 was chosen as the region of immediate interest. Only small amounts of chromia were used to avoid loss of chromia during firing; this was one of the major reasons for

---

* Solid State Physics Laboratory, Institute of Physics, University of Constantine 25000, Algeria. (After January 15, 1990)

choosing compositions near to the pseudo-binary mullite modified corundum system.

MATERIALS AND METHODS

Starting materials were chromic oxide (M100), α-alumina (AMS 9), and fused mullite. The average particle size of these materials were 0.40, 0.55, and 1.40 μm respectively. These materials were dried at about 170°C for 24 hours, weighed to the desired composition and the mixtures were then vibratory mixed.

Weighed samples of these mixtures were cold pressed at about 100 MPa and presintered at 1100°C overnight. Finally, sets of 3 samples, at least, from each composition were sintered at different temperatures (1600, 1650 and 1700°C) for different times (15, 60, 240, and 720 minutes). Both heating and cooling rates of samples were constant at 20°C/min.

The investigation consisted of two parts. The first part was a study of the matrix composition (alumina + different percentages of chromia) and in the second part, a selected composition of this 'modified' corundum was used as a matrix, to which additions of fused mullite were made (0-25 wt% fused mullite).

The microstructures and the fracture surfaces were investigated by optical microscopy and by a Philips 505 Scanning Electron Microscope (SEM), equipped with an EDAX facility. Phase identification of the sintered samples was by X-ray diffraction, using a Philips 1230 and Cu K_{α} radiation. The tensile strengths of the sintered samples were determined by diametral compression testing.

RESULTS AND DISCUSSION

Sintering

Figure 1a gives the effect of chromia addition on densification of samples sintered at 1600°C for different times. This figure shows that addition of more than 5.0 wt% chromia to alumina decreases densification of samples sintered at 1600°C for 60 minutes or less; density increased with increased sintering time and temperature. In contrast to this, the addition of fused mullite to the modified corundum (alumina + 2.5wt% chromia) gave a small increase in sintered density (Fig. 1b) and faster initial sintering rates for short sintering times. The variation in density at lower chromia addition is presumed to be due to the bimodal particle size distribution of the starting materials (7).

Grain Growth (Microstructure)

The average grain size versus chromia addition of samples sintered at 1600 and 1650°C for different times is illustrated in figures 2a and 2b; a decrease in grain size for modified corundum containing 5.0 wt% chromia occurs. This decrease is not evident at higher sintering temperatures and for longer sintering times (e.g. T = 1650°C, t = 12 hrs). The figures also show an increased grain growth rate in the modified

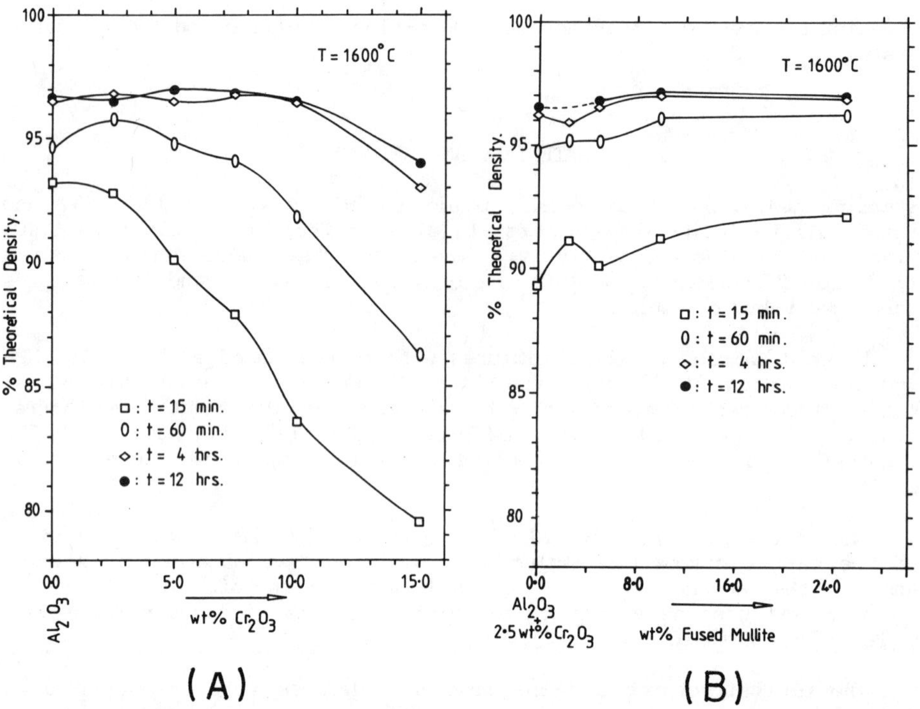

Figure 1. Effect of chromia (a) and fused mullite (b) addition on densification of samples sintered at 1600°C for different sintering times.

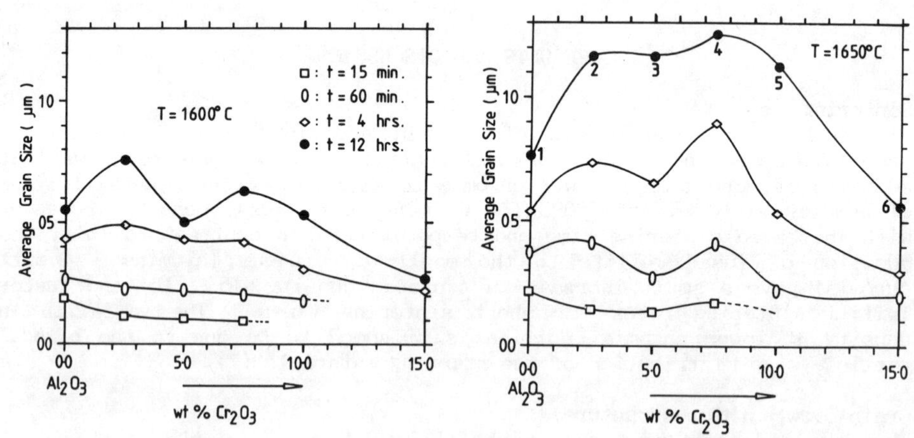

Figure 2. Effect of chromia addition on the average grain size of samples sintered at 1600°C (a) and 1650°C (b) for different sintering times.

corundum. Microstructurally, it is seen that chromia additions promote a change from polygonal grains (in pure alumina) to tabular grains (fig. 3). Tabular grains were also observed when fused mullite was added to the modified corundum (up to 10 wt% fused mullite). Further additions of fused mullite (up to 25 wt%) produced a duplex microstructure.

Tensile Strength

Addition of 2.5 wt% chromia to alumina doubled the tensile strength of samples sintered at 1600°C for 15 minutes. With further addition of chromia, this strength decreased gradually from 5.0 wt% chromia compositions to very low values at high concentrations of chromia. It was also found that the strength decreased at the highest sintering temperature and long sintering times. Correlation between the tensile strength, density, and average grain size has led to the conclusion that tensile strength was mainly controlled by sintered density, fraction of open porosity, and average grain size.

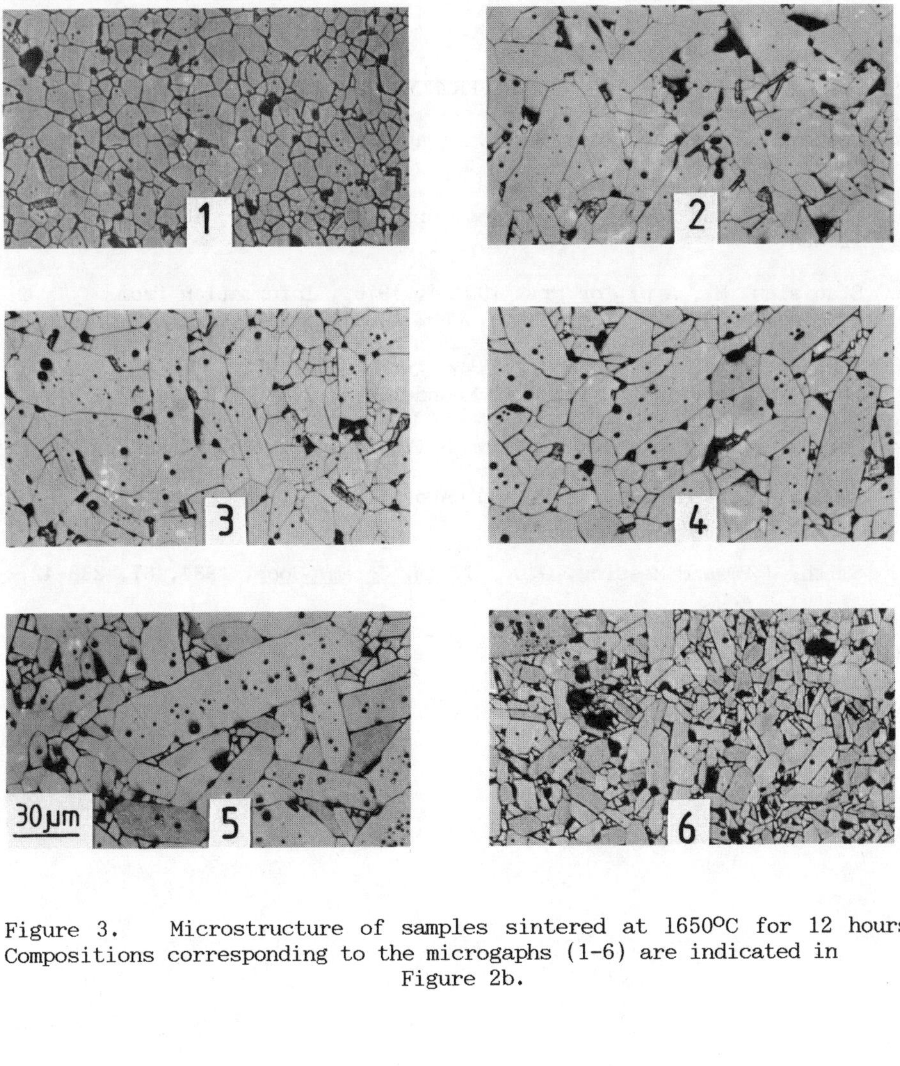

Figure 3. Microstructure of samples sintered at 1650°C for 12 hours. Compositions corresponding to the micrographs (1-6) are indicated in Figure 2b.

The strength of samples sintered at 1600°C and 1650°C for a short time (60 min.) increased by addition of up to 2.5 wt% fused mullite and then decreased with further additions. This decrease was not so significant for a medium time of sintering (4 hours). However, increasing the sintering time to 12 hours produced a substantial increase in strength, to about 500 MPa (TRS), occurring at 10 wt% fused mullite additions.

CONCLUSIONS

Mullite - modified corundum ceramics have been produced which show interesting microstructural states and mechanical strengths comparable (or better than) commercial high temperature ceramics; it is possible that these new compositions could be used as primary ceramics or, in a finely divided state, as binders for high alumina ceramics.

REFERENCES

1. Chadeyron, A.A. and Rees, W.J., Trans. Brit. Ceram. Soc., 1942, 42, 136-

2. Ford, W.F. and Rees, W.J., Trans. Brit. Ceram. Soc., 1946, 45, 125-36.

3. Stanislaw, M., Appl. or Pr. 190114, 1976. Information from Chemical Abstracts File Search, CA92 (18): 151949s.

4. Makarov, T.A., Alekseev, S.V., and Artamonova, M.V., Tr. Inst.-Mosk. Khim.-Tekhnol. Inst., Im. D.I. Medndeleeva, 1983, 128, 79-82.

5. Solacolu, S., Proc. 8th Conf. on Silicate Ind., 1965, 777-8D.

6. Roeder, P.L., Glasser, F.P. and Osborn, E.F. J. Am. Ceram. Soc., 1968, 51, 267-73.

7. Smith, J.P. and Messing, G.L., J. Am. Ceram. Soc., 1984, 67, 238-42.

SOL-GEL PROCESS OF PHOSPHATE ALUMINA SYSTEM.

G. Palavit, P. Vast, R. Chi, L.Montagne.
Laboratoire de Chimie Appliquée. Université des Sciences et Techniques de Lille Flandres-Artois. FRANCE.
B. Thierry, M. Descamps.
C.R.I.T.T. Zone Industrielle de la Petite Savatte Maubeuge FRANCE

ABSTRACT.

We present in this paper a study of the chemistry involved in the formation of aluminium metaphosphate gels and coacervates. ^{31}P and ^{27}Al NMR and FTIR have been used to bring out the mechanism of complexation of aluminium with metaphosphates. The gel results from an instantaneous and complete complexation of the aluminium with phosphates, while in the coacervate this complexation is in competition with the hydrolysis of phosphate chains. However, in both cases, the aluminium is hexacoordinated to water and phosphates groups. This can be explained by the existence of a more or less covalent bond between aluminium and metaphosphates.

INTRODUCTION.

Phosphates cements are generally derived from Al_2O_3-P_2O_5 mixes[1-6]. Our work in the realm of cement pastes is based on metaphosphate coacervates. These were first discribed by Kanazawa for the MgO-Polyphosphates-H2O system[7]. Recent work[8-9] has shown this system to be a powerful approach in the development of the formu-ation of phosphates cements.

We present here a structural study of aluminium metaphosphate coacervates and gels in the early stages of their formation using ^{31}P and ^{27}Al NMR coupled with FTIR.

RESULTS AND DISCUSSION.

Metaphosphate coacervates and gels are systems based on the complexation of a cation with phosphates chains[10]. The choice of the generative cation: Ca^{2+}, Mg^{2+}, Mn^{2+}, Al^{3+}, ... allows the formulation of glass and cements precursors[11].

Aluminium coacervates and gels are both prepared by the addition of an aluminium nitrate solution to a sodium metaphosphate solution (Graham's salt) under slow stirring. The aluminium coacervate is obtained by a rapid addition of aluminium nitrate, whereas the aluminium gel is obtained by a very slow addition of aluminium nitrate.

^{31}P NMR showed the following phosphates species in the aluminium coacervate and gel (**Fig.1a,b**): orthophosphates at 0.6 ppm (refered at 0 ppm for 85% H_3PO_4), chains-end groups and pyrophosphates at -10 ppm, intermediates PO_3 groups of shorts chains at ca. -17 ppm, rings and intermediates PO_3 groups of long chains at -21 ppm. A peak at -32 ppm is also observed, which is attributed to aluminium-phosphate bonds.

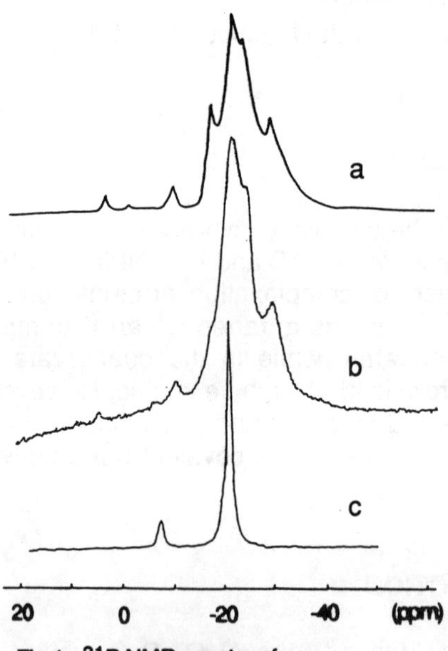

Fig.1: ^{31}P NMR spectra of
a: aluminium metaphosphate gel
b: aluminium metaphosphate coacervate
c: calcium metaphosphate coacervate

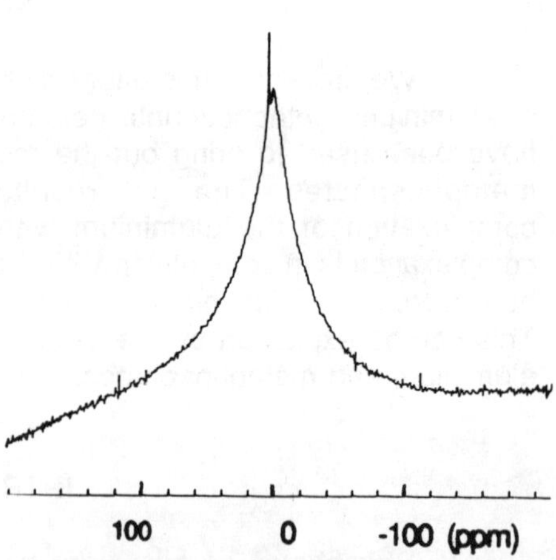

Fig.2: ^{27}Al NMR spectrum of
Aluminium metaphosphate coacervate

We can consider that a metal-phosphate bond is created first by an ionic bond between the cation (Al^{3+}, Ca^{2+}, Na^{+},...) and the phosphate anion, then by the polarisation of oxygen atoms by the metallic cation. During the first step, the greater the cation charge and the smaller the cation size, the greater the energy of the process. In the second step, trivalent cations have a greater polarising effect than alkalines ions. As a result, aluminium-oxygen-phosphorus bonds in the gel as well as the coacervate are more covalent than a classical phosphate-metal bond. Thus, the P-O-Al bond is shifted to -32 ppm compared to the -21 ppm peak of metaphosphates. Moreover, solid state NMR studies show that covalent $AlPO_4$ bond occurs at -30 ppm[12].

As a matter of comparison, **Fig.1c** shows the NMR spectrum of a calcium metaphosphate gel. Calcium-phosphate bonding is ionic, so the phosphorus resonance at -32 ppm is not observed. Furthermore, the shealding effect of the aluminium cation on the phosphorus observed in NMR, leads to a decrease of the hydrolysis activation energy[13].

^{27}Al NMR spectra were recorded for aluminium metaphosphate coacervates.The quadrupolar momentum of the Al nucleus is commonly used as a probe to obtain information about the Al coordination[14]. As can be seen in **Fig.2** , aluminium cations are present in two environments.The sharp peak at 0 ppm, refered to aluminium nitrate, is due to the free aluminium cation $Al(H_2O)_6^{3+}$ [15]. The wide peak at -4 ppm is attributed to phosphate-complexed aluminium. We note that the aluminium is octahedrically ligated, not tetrahedrically as can be observed in alumina gels[16]. The complexation of the hexahydrated Al^{3+} cation by metaphosphates can be schematized as follow:

$$Al(H_2O)_6^{3+} + (PO_3^-)_n \dashrightarrow Al(PO_3)_n(H_2O)_{6-n}^{+(3-n)} + nH_2O \quad n=1,2,3$$

Complexation of phosphates with aluminium has been shown by Raman spectroscopy[15]. P-O-Al bonding is found at 902 cm^{-1}. We compared aluminium metaphosphate coacervates and gels by infrared spectroscopy. The infrared spectra (**Fig.3**) show broad bands in the proton region between 2500 and 3500cm^{-1}, and a strong one at 1640 cm^{-1}. They are attributed to the several kinds of water present in the system[17]. 1350-1400 cm^{-1} bands are assigned to nitrates and in-plane deformation of P-(OH) groups. These bands are strong in the coacervate and very strong in the gel. The 1274 and 1157 cm^{-1} bands correspond respectively to the antisymmetrical vibrations of O-P-O and P-O(H) groups, their symmetrical vibrations occurring at 1090 and 1041 cm^{-1}. The O-P-O bands are strong in the coacervate, but less so in the gel. The 830 cm^{-1} peak is attributed to the nitrates groups. The 600 to 800 cm^{-1} range in the coacervate spectrum contains bands at 608, 667 cm^{-1} (deformations) and 737 cm^{-1} (symmetrical vibrations of P-O-P groups). These bands are hidden in the gel spectrum.

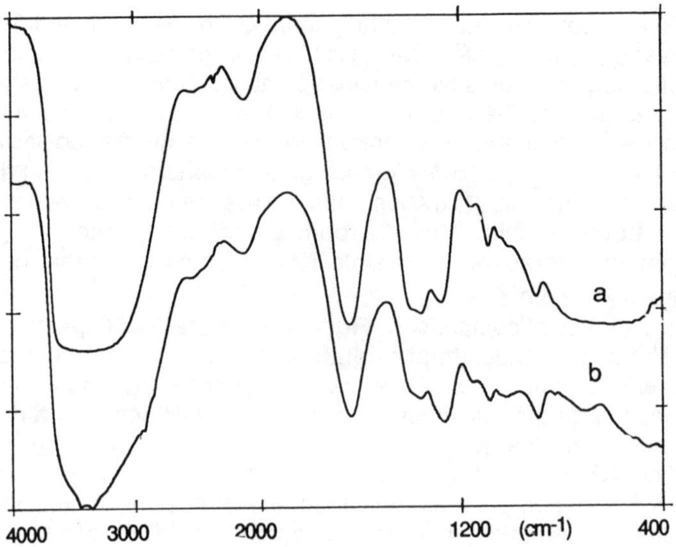

Fig 3: IR spectra of
a: aluminium metaphosphate gel
b: aluminium metaphosphate coacervate

The antisymmetrical vibration of P-O-P bonding, occuring at 890 cm^{-1}, is partially masked for the coacervate and completely for the gel. This confirms the NMR results: some of the Al^{3+} ions don't form complexes immediately with phosphates. Thus, all the P-O-P are not bonded with aluminium, which leads to a 10 cm^{-1} shift towards high frequencies. When Al^{3+} is completely complexed, the 890 cm^{-1} band is shifted to 900 cm^{-1}. In the case of the gel, one can see that the P-O-P band is at 900 cm^{-1}. Unlike the coacervate, all the Al^{3+} ions are complexed to phosphates in the early stage of the formation of the gel.

EXPERIMENTAL PROCEDURE.

Aluminium Metaphosphate coacervate is prepared by rapid addition of a 2M aluminium nitrate solution in a 1.25M solution of sodium metaphosphate (Graham'salt) under vigorous agitation. The coacervate is formed instantaneously and separates from the supernatant. Aluminium metaphosphate gel is synthesized with the same reagents and the same proportions, but the aluminium nitrate solution is added very slowly and under gentle stirring. The gel sets to a homogeneous and transparent mass in about 15 mn.

^{31}P and ^{27}Al NMR spectra were recorded on a Brüker WP 80 spectrometer operating at 32.4MHz and 20.8MHz for phosphorus and aluminium respectively.

A M1730 Perkin-Elmer Infrared Fourier Transform spectrometer was used to performed the infrared spectra over the range 4000-400 cm^{-1}. The samples were pressed between two AgCl windows.

CONCLUSION.

The NMR and infrared arguments showed that early stage aluminium metaphosphate gels and coacervates have a different behavior according to the complexation of aluminium cations with phosphates groups. The complexation is instantaneous for the gel while requiring a certain length of time to be complete for the coacervate. The slow addition of aluminium solution to the metaphosphate solution leads to a complete complexation of the phosphate groups. On the other hand, since the Al^{3+} ions are added rapidly, the coacervate appears almost immediately, and the tangled chains prevent the total incorporation of the aluminium cations. The further complexation of the free aluminium is in competition with aluminium induced hydrolysis of the phosphate chains. Contrary to the case of alumino-silicates, aluminium is never tetra-coordinated in the metaphosphate systems. Aluminium remains always hexa-coordinated, and therefore never takes the place of phosphates in the formation of network-like structures.

[1] S.J.Lukasiewicz, J.S.Reed; Am. Ceram. Soc. Bull., **66** [7] 1134-38 (1987).

[2] W.D.Kingery; J. Am. Ceram. Soc., **33** [8] 239-41 (1950).

[3] J.E.Cassidy; Am. Ceram. Soc. Bull., **56** [7] 640-43 (1977).

[4] F.J.Gonzalez-Aleu; Ph.D.Thesis, Pennsylvania State University, State college, PA, November(1980).

[5] I.V.Tananaev, V.P.Orlovskii; Russ. J. Inorg. Chem., **31** [8] (1986).

[6] A.A Chistyakova, V.A.Sivkina, V.I.Sadkov, A.P.Khashkovskaya, L.G.Povysheva; Izvestiya Akademii Nauk SSSR, Neorganicheskie Materialy, **5**, 9, 1573-80 (1969).

[7] T. Kanazawa, T. Umegaki.; Bull. Chem. Soc. Jpn., **46**, 3587 (1973).

[8] M.C.Wagner, W.G.Klemperer, J.F.Young; 91st annual meeting of the American Ceramic Society, Indianapolis, (1989).

[9] R.A.Steinke, D.K.Agrawal, R.Roy, D.M.Roy; 91st annual meeting of the American Ceramic Society, Indianapolis, (1989).

[10] M.Draoui, G.Palavit, P.Vast; Revue de chimie Minérale, **22**, 256 (1985).

[11] G.Palavit, P.Vast, L.Montagne; VII Int. Symposium on ceramics, Bologne, (1989).

[12] R.J.P.Williams,R.G.F.Giles,A.M.Posner; J.C.S. Chem. Comm. 1051 (1981).

[13] M. Watanabe; Bull. Chem. Soc. Jpn., **55**, 3766-69 (1982).

[14] D.Muller, G.Berger, I.Grunze,G.Ladwig, E.Hallas, U.Haubenraisser; Phys. and Chem. of Glasses,**24**,2,37-42 (1983).

[15] J.W.Akitt, N.N.Greenwood, G.D.Lester; J. Chem. Soc. (A), 2450 (1971).

[16] L.F.Nazar, L.C.Klein; J. Am. Ceram. Soc. **71**, 2, C-85, (1988).

[17] G.Palavit, A.M.Bera, P.Vast; Calorim. Anal. Therm., 1568 (1986).

THE INFLUENCE OF THE MOISTURE CONTENT ON THE COMPACTION BEHAVIOUR AND PRODUCT PROPERTIES OF A WALL TILE GRANULATE

J. VAN DER ZWAN
TNO Institute of Applied Physics, Ceramics Dept
P.O. Box 595, 5600 AN Eindhoven, The Netherlands

ABSTRACT

Results are given of a research project on the compaction behaviour of wall tile granulates with various moisture contents. The relationships between compaction pressure, density and porosity of the compact, and mechanical properties of the green product are presented. The water content of the pressing powder has a large effect on the compaction mechanisms of the spray dried granulate. However, a limiting maximum moisture content exists. Above 8-9 w % water the compact density will not increase with increasing pressure. Thus, compact density depends on compression technology and water content. The known compaction formulae can be extended with a moisture content depending factor.

INTRODUCTION

In earlier publications the densification of agglomerated material is described (1.2). The compaction of wall tile granulates is described as a four stage process:
(i) filling of the holes between the granules
(ii) fragmentation and plastic deformation of the granules
(iii) filling of the holes between the primary particles
(iv) fragmentation and plastic deformation of the primary particles
Since the moisture content (M.C) more than any other factor besides the compaction pressure of the granulate, largely controls the density and the resulting mechanical properties of the green (and final) product, a knowledge of the relationships between pressure-compaction and the material properties is very important.

In (1) and (2) the densification of agglomerated materials is described by a modified Kawakita equation. The relationship between pressure P and modulus of elasticity E or (tensile) strength σ can be described by equations which are analogous to the Kawakita equation. These equations have the form:

$$C_{zs} = C_{max} \cdot \frac{b.P}{1+b.P} \; ; \; E = E_{max} \cdot \frac{c.P}{1+c.P} \; ; \; \sigma = \sigma_{max} \cdot \frac{d.P}{1+d.P} \qquad [1a,1b,1c]$$

where C_{max}, E_{max} and σ_{max} are the (extrapolated) maximum obtainable relative volume compaction respectively max.obt. modulus of elasticity, resp. max. obt. strength (by cold pressing), and b, c, and d are constants. (see figure 1)

The relative volume compaction C_{zs} can also be defined as a function of volume order pressure (V_p), or as a function of compact density ρ_p or compact porosity e_p, according to:

$$C_{zs} = \frac{V_b - V_p}{V_b} = 1 - \frac{\rho_b}{\rho_p} = \frac{e_b - e_p}{1 - e_p} \qquad [2]$$

In our previous work we have pointed out that the initial compaction volume V_b (and related ρ_b and e_b) depends on specific behaviour and properties of the granulate c.q. granules.

In 1980 - 1981 the ceramics department of TPD-TNO carried out a research project with the main objective to establish the influence of the moisture content on densification behaviour and green product properties of an industrial wall tile granulate. Not only a mere description of the change in compaction mechanisms as a function of M.C. of the granulate, but also the quantification and incorporation of the M.C. in the equations mentioned above (when possible) was part of the research program. Most of all, industry was interested in determination of an optimum in M.C. for the compaction process - when present.

MATERIALS AND METHODS

A Dutch wall tile manufacturer delivered granulates with 6 different M.C.'s, from which by conditioning 10 granulates were prepared with a M.C. ranging from 0,8 to 8,7 w % (on dry mass). Each granulate was cold pressed to cylindrical samples of approximately 35 mm diameter and 4-7 mm thickness (depending on pressure), on a 10-ton Norton-press. The pressure diagrams were registered. Each granulate was compacted at 10 different pressures. The pressure ranges from 0 to 80 MPa. At each pressure 20 samples were produced. In total 1600 samples had to be characterized.

Characterization was done by measurement of dimensions and (dry) weight, from which the apparent density ρ_p, apparent porosity e_p and the relative volume compaction C_{zs} were calculated. The tensile strength σ_p and the modulus of elasticity E_p of the samples were calculated from the results of the diametral compression test, which was carried out on an Instron testing machine. For each of the series pressed at different pressures and M.C., the mean value and standard deviation of the pressure and compact properties were calculated. From all these data the values of the constants in the equations [1] and [2] can be computed for each moisture content.

RESULTS AND DISCUSSION

Both the density and the mechanical properties depend on pressure and M.C. of the pressing powder. In the equations [1a, 1b, 1c] the constants b, c, d, and C_{max} are clearly depending on M.C.; the relationship between E_{max} and M.C., and between σ_{max} and M.C. is less explicit. The relationship between the granulate properties ρ_0 (bulk density), ρ_t (tap density), ρ_b (initial compact density), ρ_{max} (max. obt. compact density), and moisture content is shown in figure 2. As both constant b, the densities ρ_b and ρ_p, and C_{max} are a function of M.C., we can deduce from equations [1] and [2]:

$$\rho_p = f(P,M.C.), \text{ resulting in: } \rho_p = \frac{\alpha+\beta.P}{1+\gamma.P} \qquad [3]$$

where α,β and γ are moisture depending constants.
The calculated values of α,β and γ are given in table 1 for several moisture contents.

TABLE 1
Values of constants α,β and γ as a function of moisture content for compacts produced from spray dried wall tile granulates.

CONSTANT	MOISTURE CONTENT (% m/m_d) 2	4	6	8	10
α	1,42	1,36	1,29	1,23	1,17
β	0,15	0,27	0,38	0,48	0,57
γ	0,07	0,13	0,18	0,23	0,28

Figure 3 illustrates the quantitave model of eq. [3] which describes the observed phenonema during compaction very well. Although it was more or less known from experience, also in industry, that a too dry or a too wet granulate results in a less good or even problematical compaction process, we can learn from the produced data and equation [3] that indeed a limiting maximum in moisture content of the pressing powder exists.

W.M. Price gave already in 1972 a plausible explanation for this phenomenon, although his densification experiments were carried out on other masses and with an entirely different laboratory pressing device (3). In figure 4 an overall view is given (based on earlier work of Frisch c.s.(4)) of the rel. change in compact density and strength as a function of M.C. and pressure. Zeropoint in this figure are the normal industrial pressing conditions at 15 MPa and 8% moisture content (on dry).

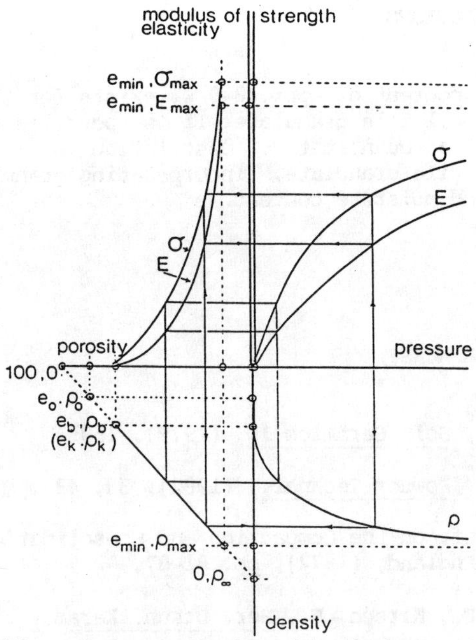

Figure 1. Relationship between pressure and compact properties.

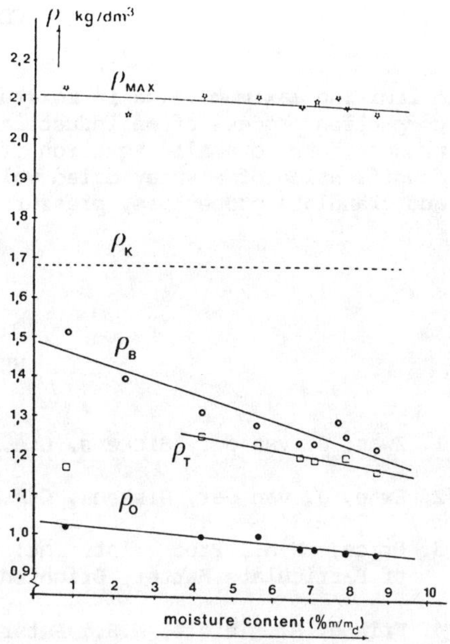

Figure 2. Granule properties as function of moisture content.

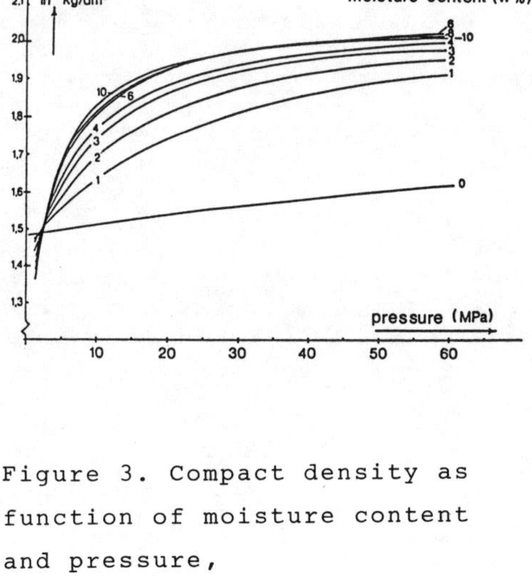

Figure 3. Compact density as function of moisture content and pressure,

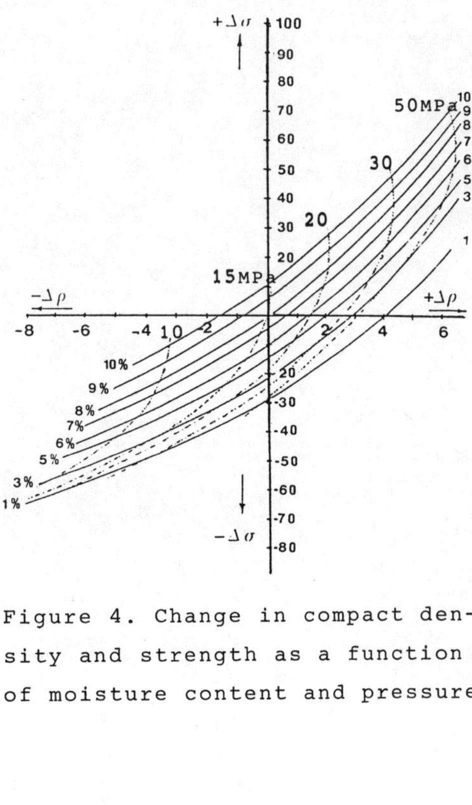

Figure 4. Change in compact density and strength as a function of moisture content and pressure.

CONCLUSIONS

A limiting maximum value in moisture content of appr. 8–9 w% exists for the compaction process of an industrial wall tile granulate. It is possible to produce one overall equation for a quantitative description of the densification of a spray dried wall tile granulate, incorporating granule and granulate properties, pressure and moisture content.

REFERENCES

1. Zwan, J. van der, Siskens, C.A.M., Sci. Ceramics **10**, (1979), 159

2. Zwan, J. van der, Siskens, C.A.M., Powder Technol., (1982), **33**, 43

3. Price, W.M., Proc. 1st. Int. Conf. on the Compaction and Consolidation of Particulate Matter, Brighton, England, (1972), pp. 81–87.

4. Frisch, B., Thiele, W.R., Peter, H., Kirsch, R., Ber. Dtsch. Keram. Ges., (1977), **54**, 224, 259.

ANALYTICAL INVESTIGATION ON VAPOUR CONDENSATIONS FROM SINGLE LAYER-ROLLER KILNS FAST FIRING

GIUSEPPE TOSI and PAOLO ZANNINI (*)
University of Modena, Chemistry Department,
Via G. Campi, 183   I 41100 - Modena, Italy

ABSTRACT

Analytical characterization of solid condensations inside fast firing kilns was made by means of XRF, XRD, AAS, SEM-EDS, TGA, DTA and classic wet analysis, on 48 samples collected from very different production technologies. Aim of the work was to localize critical zones and temperatures, and to assess the average chemical and mineralogical composition, then to relate got results with possible pollution starting materials.
It was found that:

a) HIGH temperature condensations (500 - 800 °C.) are mainly composed by $PbSO_4$ in Anglesite or glassy form, while
b) LOW temperature condensations (100 - 300 °C.) are as NH_4Cl, NH_4BF_4, $(NH_4)_2SO_4$ etc.

INTRODUCTION

The final step of all traditional ceramic production technology is firing, that induces irreversible chemical and physical modifications to the inorganic materials mixtures, everyway would they be formed. Thus, a great attention is always payed on studying and controlling all the parameters that have a very next importance on the characteristics of the final product, like temperature distribution, maximum firing temperature, cycle speed, pressure/depressure rates etc. .
On the contrary, very scarcely is usually investigated on kiln atmosphere, as vapours, oxygen, unburnt etc. content, forgetting that other sets of reaction occour on firing. Some attention to such arguments is given only when problems are found in different sections of the plant (e.g. at exausts depurator, or on air pollution controls).
the same, technical literature gives large relevance to evaporation reactions, pyrolysis, oxidations and reductions, but only respect with their influence on final products. Nevertheless a lot of problems connected to

condensation inside the kiln are very known by production people, from dropping on tiles to serious damages of kiln ceilings.

WHERE THE CONDENSATIONS ARE FOUND

A typical side-view of a single-layer roller kiln is shown in Figure 1. Ceramic materials are heated and fired in free flame from over- and under the rollers and, being cheminery at the beginning of the kiln, kiln atmosphere moves opposite to material advance direction, with spyral-like waving from maximum temperature zone, back to inlet of the kiln. Such a way, firing zone heat is used to help pre-heating of material and heat recover for burners is also possible.

The insulating materials and the small working section give a very good homogeneity to the system, and temperature distribution of the kiln ceiling respect with the material on the rollers one is very similar, but shifted, for the said reasons, backwords of some meter. So, every material, structural or on firing into the kiln, having enough vapour pressure at got temperature, will be vapourized, moved to the cheminery and will encounter a sharp temperature gradient, being induced to condensate at "cold gates" like viewing windows, refractory creeps etc., or a softer cooling on large, flat surfaces, like kiln ceilings. The more the material is volatile, the farest will be found from maximum temperature zone.

According to different changing parameters, as working temperature, pressure of exausts, roughness of ceiling etc., it was possible to localize two main ranges of condensation temperatures, like shown in Figure 1. Stalactitic products were found on examining more than 48 kilns, while working or on mantenaince.

It was tryed to evaluate the importance of single factors of vaporization, relating to: 1) glaze composition, 2) body mixture composition, 3) process additives, like water, deflocculants, glues ..., 4) structural materials, 5) fuels. While is possible to make general considerations for 4) and 5), it is not possible an analytical evaluation, on working plants, of 1), characterizing one, and only one, glaze (as it is too often changed), while a good accuracy was possible on 2) and 3), by collecting samples from:

* production kilns for red body, dry ground, unglazed (RD)
* production kilns for red body, dry ground, glazed (RDG)
* production kilns for red body, spray-dryed,glazed (RS)
* production kilns for white body, spray-dryed, glazed (WS)

EXPERIMENTAL AND RESULTS

For each sample were checked morphology, quantity, temperature of condensation (by means of an expecially arranged Cromel/Alumel thermocouple), than samples were dry ground at passing 100 microns sieve and stored in oven at 105°C. . Analytical techniques were selected trying to get the

FIGURE 1
Typical side-view of a roller kiln
(on evidence condensations location)

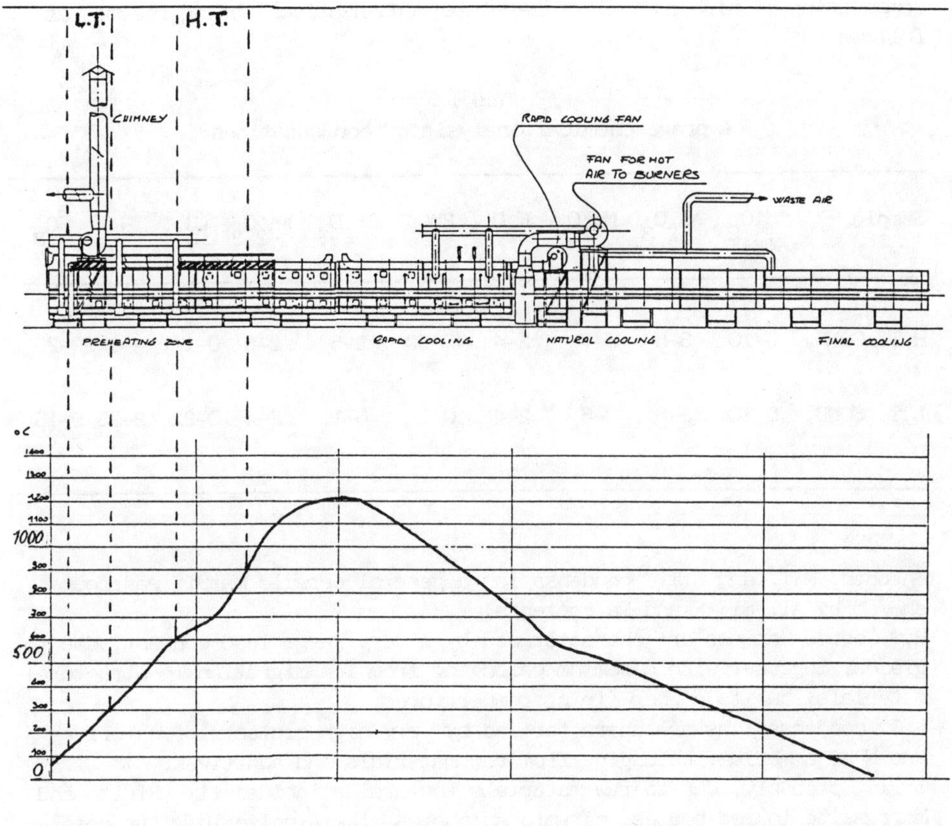

largest characterization without having any one privileged.
So, A.A.S., F.R.X., Elemental analysis, Colorimetry, D.R.X., S.E.M.-e.d.s. and Thermal analysis were collected.
Of course, got results indicate a substantial difference between chemical composition of high and low temperature condensations, being resumed as follows :

TABLE 1
Average chemical analysis of condensations

Sample	SiO_2 (%)	Al_2O_3	Na_2O	K_2O	PbO	Cr_2O_3	Fe_2O_3	Cl	N	SO_3
H.T. COND.	5-10	3-6	5-7	2-4	40-60	2-5	2-6	0-1	—	10-20
L.T. COND.	5-10	3-6	3-6	2-4	0-1	0-1	2-4	10-22	12-25	5-15

On both, H.T. and L.T., condensation a general ceramic matrix was found, showed by Al, Si, K and Na contents.
H.T. condensations are characterized by a very large lead content, due to glazes, together with chromium belonging from ion migration in structural materials, due to higher firing temperatures.
L.T. condensations are characterized by very high contents of chlorides and **N** as ammonium, belonging from raw materials and additives.
It is, probably, due to raw materials (natural sulphides like Pirite and Marcassite in red bodies, ceramic sulphates like Alunite in white ones) and fuels the quite high content of sulphur here checked.
Boron and fluorine, even if very important for exausts pollution, were not sistematically checked (but, anyway, not found) because too volatile to give condensations into the kiln.
From mineralogical analysis, got by powder Xray diffraction and D.T.A., also by comparison with standards, glassy matrix with $PbSO_4$ as Anglesite crystalline form was found for H.T. condensations. L.T. condensations have ammonium saline form, with appreciable contents of NH_4Cl, NH_4BF_4, $(NH_4)_2SO_4$ etc. .

CONCLUSIONS

Analytical results could, quite well, be fitted to the possible causes, by examining the condensation structural elements compared with their respective content in glazes, raw materials, waters, fuels etc. .
For instance, lead content in glazed tiles was analyzed by getting chemical analysis before and after firing:

TABLE 2

Oxide (%)	UNFIRED TILE ("base" + "flame" glazed)	FIRED TILE (cycle 50', T.max. 1160°C.)
Na_2O	2.9	1.3
K_2O	2.1	2.7
PbO_2	14.5	3.2
SO_3	1.2	—

Al, Si, Fe, Ti, Zr, Ca and Zn without any significant change.

Thus, very high volatility of lead was confirmed, and, more, H.T. condensations were not found into unglazed tiles production kilns.
A very good correlation between production processes (RD,RDG,RS,WS) and condensations composition was not always possible to obtain, as raw materials play a major role on determining condensations nature.

PRELIMINARY REPORT ON THE MIGRATION OF LEAD AND CADMIUM FROM CERAMIC ARTICLES SOLD IN ITALY AND INTENDED TO COME INTO CONTACT WITH FOODSTUFFS

BRUNO FABBRI and GIAMPAOLO ERCOLANI
C.N.R., Research Institute for Ceramic Technology
Via Granarolo 64, 48018 Faenza, Italy

ABSTRACT

A sampling of about 600 italian ceramic pieces for foodstuffs has been characterized in terms of lead and cadmium migration with reference to the European Economic Community norms. Furthermore, a qualitative analysis of the glazed surfaces has been performed by x-ray fluorescence technique. Only a few articles evidenced a too high lead release, while no article showed a significant cadmium release. Sometimes the lead release is due to an incorrect work, while in other cases the cause is the use of unsuitable products.

INTRODUCTION

The problem of lead migration from ceramic articles commonly used to contain food is still a current one. It is continuously nourished by discussions that every now and then stir up, taking occasional cases of product contests as a starting point. This seems to happen, at least as far as Italy is concerned, especially in relation to the American market, which is one of the most attentive to everything concerning the consumer defence. Taking one of these events as a starting point, our Institute organized a statistical investigation about the quality of products made and/or commercialized in Italy. To this purpose we availed ourselves of the cooperation of the artisans/ceramists association "Confartigianato", which involved an organization quite independent such as the National Consumers Union (UNC).

MATERIALS

Two different samplings were prepared: the first was made by the UNC to represent the industrial mass-production, and the second directly by us taking into consideration only the artisan products.

The UNC bought, from 16 different selling places in Rome (from pedlars to supermarkets), about 50 different articles, each of them in five copies. Almost all the pieces were made in Italy, except for three foreign articles, two from Yugoslavia and one from Brazil. The trade-mark imprinted on the italian products singles out 26 industrial factories accidentally distributed in various italian areas. From a technological point of view, they are mostly made of household porcelain or glazed earthenware, both white or coloured, often decorated, with extremely different shapes and sizes.

The material we collected directly from the same producers consisted of more than 100 articles, each of them in four copies, representing a significant part of the famous italian artistic/artisan production. The articles, mainly of majolica type, come from 25 different artisan shops, distributed in six different typical areas of production in the regions of Venetia, Liguria, Tuscany, Emilia-Romagna, Umbria and Sicily. They represent all the typologies normally commercialized; frequently, however, it is a question of articles that, even though they have morphological characteristics suitable to contain food, show aesthetical features qualifying them more as destined to ornamental purposes.

EXPERIMENTAL

All the collected sampling was subjected to two kinds of investigations concerning respectively:

1) lead and cadmium release;
2) chemical composition of the glaze.

Pb and Cd release has been analyzed according to the procedure prescribed by the present european norm (1). It provides for:

a) filling of the pieces with 4% acetic acid solution and taking of this after a permanence of 24 hours in the dark at room temperature;
b) measure of Pb and Cd concentrations in the attack solution by means of atomic absorption spectrophotometry;
c) expression of the migration results of each element in mg/l in the attack solution, as for fillable pieces are concerned (interior depth more than 25 mm), or in weight of removed element per unit of surface (mg/dm^2) as to not fillable pieces.

The migration limits provided by the same norm are: 0.8 and 0.07 mg/dm^2 of Pb and Cd, respectively, with regard to not fillable pieces; 4.0 and 0.3 mg/l of lead and cadmium, respectively, talking about the fillable ones.

The chemical composition of the glazes has been qualitatively investigated by means of X-ray fluorescence analysis with wavelength dispersion system (WD-XRF). To this purpose, from a sufficiently plane zone of each ceramic article a disk of 5 cm in diameter was extracted obtaining directly the test-piece.

Then, the most interesting elements were searched, and lead amount was approximatively evaluated by comparison with analogous investigations relative to known glaze samples.

RESULTS AND DISCUSSION

The industrial production

None of the considered articles showed Pb and Cd migration over the limits provided by the current regulations. The fillable objects originated attack solutions with a lead concentration as high as at maximum one tenth of those thresold values (for example UNC12 and UNC33 samples). As for the not fillable objects, some migration values approach more to the allowed limit (0.66 mg/dm^2 for UNC2 sample), but all the others are at least 20 times below.

The chemical analysis revealed the presence of lead in the glaze of all the objects, but the concentrations are highly different from a sample to another. Most glazes contain less than 5% PbO, and a lot of these less than 1% PbO, but there are also glazes containing more than 10% PbO. In spite of this, the lead migration is, as we said before, always very below the limit, as in UNC27, UNC30 and UNC33 samples.

As at least four copies of each article were always analyzed, we must point out also the great homogeneity of the considered production. Such homogeneity derives from the use of sufficiently repetitive devices, thanks moreover to the exceptional standardization and mechanization of the production process.

Also the cadmium migration analyses confirm the total safety of these products with reference to the utilizer health, being the found values always at least ten times lower than the limit provided by the norm. On the other hand, the x-ray fluorescence analyses revealed traces of cadmium only in the samples UNC37, UNC16 and UNC4, in decreasing order, in relation with their particular decoration.

The artisan/artistic production

Lead migration. About 35% of the considered articles has a glassy coating containing less than 1% PbO, while about 20% of them contains lead oxide in the range 1-5%. The remaining articles (about 45% of the total) are characterized by a plumbiferous or highly plumbiferous glassy coating. In fact, about 20% of the whole articles contains 10 to 20% PbO and the remaining ones 20 to 50% PbO.

But the articles not in agreement with the norm for the lead release are only eight; their characteristics are summarized in Table 1. All these have glazes with high lead content, with the only exception of CNR97; for this last, in fact, XRF analysis revealed only small traces of lead in the glaze. Lead migration from this article, therefore, can be justified only by an intense mobilization of the element, that could be produced by the very high quantity of copper added to colour the glaze itself. To support this hypothesis, it can be observed that a similar copper concentration has been found also in the lead releasing sample CNR39, while Cu traces have been revealed for CNR20, CNR32 and CNR56. On the other hand, it is well known that such colouring cation acts as a source of instability for Pb ion (2-7), as well as the proportion of sodium with respect to the sum of sodium and potassium (8).

Formal and decorative features of the mentioned incriminated pieces reveal that some of them (in particular the bowls CNR20 and CNR39 with iridescence effects after third firing) are produced more as objects for domestic ornamentation than as food containers. Furthermore, a significant

TABLE 1

Characteristics of the artisan/artistics ceramic articles that showed a high lead release

Article (N°)	Brief description	Pb release (mg/l)	PbO (%) (glaze)
CNR20	Bowls in majolica decorated with	23 - 66	10 - 20
" 39	iridescence effects in third firing	47 - 72	" "
CNR22	Cup in white glazed majolica without interior decorations	0.3 - 50	10 - 20
CNR32	Small plate with blue glaze and differently coloured decorations	1.0 - 1.6 (mg/dm^2)	30 - 50
CNR44	Bowl and jug in earthenware covered only	7 - 14	10 - 20
" 56	by lead containing transparent glaze	4 - 74	30 - 50
CNR66	Soup plate in white majolica decorated with small black lines irregularly disposed	5 - 8	10 - 20
CNR97	Small bowl in majolica internally green coloured	4 - 6	1

fact is that the unique two pieces simply covered with transparent glaze, among all those examined, both have an intolerable lead migration. But the most lead containing articles do not show any dangerous release, even if the glaze composition is quite variable mainly with reference to the contents of Na, K and Ca.

In conclusion, also in the artistic ceramic field the articles out of norm are rare and often previously identifiable with high probability. The chromatic showiness of the decoration (for example CNR89, CNR90, CNR91 and CNR92) is not included among the characteristics useful for this identification, as most people usually think. Sometimes different migration values were registered for the different samples of a same articles (CNR22 for example). This situation denotes a certain sensibility of lead to modify the intensity of its bonds in the glaze presumably in relation with probable differences of treatment introduced by the artisan/ceramist during some working phase.

Cadmium migration. None of the artisan ceramic articles revealed a cadmium release over the safety values indicated by the norms. This is mainly due to the little diffusion of the use of cadmium containing colouring materials. In fact, only six samples showed that element among the components of their glaze decoration, namely CNR86, CNR89, CNR90, CNR91, CNR92 and CNR76 in approximative order of decreasing abundance. Effectively, cadmium release from these articles is higher than the other ones, but in a different order of abundance. This means that the use of cadmium is possible without exceeding the allowed maximum release also when the corresponding colour covers a wide surface area of the ceramic piece, as in the article CNR91.

CONCLUSIONS

The examined sampling of ceramic articles destined to come into contact with foodstuffs is sufficiently wide to allow some general conclusions about their quality with reference to the problem of toxic metals release.

All the italian ceramic articles, of both artisan/artistic and industrial production, are free from dangerous cadmium migration. All the industrial ceramic production is completely free also from lead release.

With reference to the only artisan production, about 6% of the articles were found with an exceeding lead release. This can be due to either the use of unsuitable materials or an incorrect working, but the showiness of the decoration never constitutes an index for dangerous situations. Therefore it is necessary to introduce a generalized use of a symbol which can distinguish the artistic ceramic articles from those effectively destined to come into contact with foodstuffs.

REFERENCES

1. Counsil of the European Economic Community, Ceramics into contact with foodstuffs. Directory 84/500, CEE L277/12, 20-10-1984.
2. Ravaglioli, A., Maeda, T., Preparazione di vernici verdi brillanti al Pb-Cu non cedenti Pb. Ceramica Informazione, 1974, **9**, 5-9.
3. Eppler, R.A., Formulation of glazes for low Pb release. Am. Cer. Soc. Bull., 1975, **54**, 496-499.
4. Reynolds, C., Metal release from tableware. Br. Ceram. Review, 1979, **39**, 26-32.
5. Melpignano, A., Palmonari, C., Timellini, G., Lead solubility of raw materials for ceramic glazes. Interceram, 1984, **33**, 48-52.
6. Takashima, H., Relation between glaze structure and lead pollution. Interceram, 1985, **34**, 15-16.
7. Burke, F.M., Leachability of lead from commercial glazes. Ceram. Eng. Sci. Proc., 1985, **6**, 1394-1399.
8. Fabbri, B., Possibili problemi applicativi correlati alla normativa per la cessione di piombo e cadmio dagli oggetti ceramici per alimenti. Ceramica Informazione, 1987, **22**, 526-529.

COMPARISON OF TECHNOLOGICAL PROPERTIES OF BODIES FOR MAJOLICA AND EARTHENWARE PRODUCTS

CESARE FIORI
C.N.R.-IRTEC, Via Granarolo 64, 48018 Faenza, Italy
VIVIANO VENTURI
Geoceramic Researches, 40050 Monte S.Pietro (Bologna), Italy

ABSTRACT

Two bodies widely used in Italy for the production of tableware in majolica and earthenware respectively were examined in detail, with regard both to their chemical, mineralogical and particle size composition characteristics, and to their technological properties. The comparison of the two materials revealed considerable differences in mineralogical and particle size, even though the two bodies share a high carbonates content. On the other hand, the technological properties present less marked differences which may be related more to the different particle size distribution in the two bodies than to different chemical-mineralogical compositions.

INTRODUCTION

This paper addresses the chemico-mineralogical and technological characterization of two plastic bodies for the production of tableware in earthenware and majolica respectively. These bodies are marketed in the form of parallelepipeds or cylinders weighing some 20 kg, on the domestic Italian market. Basically, the earthenware body consists of mixtures of imported clays which are white or light grey in colour after firing (low iron oxide content) and of carbonates, while for the majolica body one or more calcareous clays of the Plio-Pleistocene age are used; these clays outcrop in various Appennine regions. The raw materials (the carbonates are previously ground to a very fine granulometry: "ventilated" type) are dispersed in water by means of turbopulpers; this is followed by sieving, so as to remove all the particles in excess of 100 µm, and then filterpressing and vacuum extrusion (with approximately 20% humidity) for the forming of a continuous column of material that is cut into pieces of standard length. These are immediately wrapped in a polythene film in order to retain the moisture which is essential for good formability, and palletized.

CHEMICAL, MINERALOGICAL AND PARTICLE SIZE COMPOSITION

Table 1 shows the chemical and mineralogical composition of the two bodies. The mineralogical composition was determined by computerized rational analysis starting from the chemical data, in order to obtain the percentage content of the minerals identified by X-ray analysis. Approximately 60% of the earthenware body consists of illitic-Kaolinitic clays, which in turn contain a considerable amount of quartz and small percentages of feldspar, and some 40% is made up of carbonates. In majolica, the clayey component is illitic-chloritic and the quartz-feldspar and carbonate fractions form part of these clays, thereby constituting true natural bodies ready for use after simple processing (wet homogenization, sieving). In addition, the majolica body contains iron oxides which account for the "red" colour after firing, and traces of substances of organic origin. It should be noted that the overall content of carbonate is much higher in the earthenware body (where dolomite greatly prevails over calcite) than in the majolica body (where the carbonates are almost purely calcite with only traces of dolomite). In terms of chemical composition, the earthenware body is much poorer in silica and alkalies, it has a really low iron content and is almost entirely without carbon of vegetable origin, while it is much richer in alkali-earth oxides (CaO+MgO) than the majolica. Ignition loss is much higher with the earthenware, chiefly as a result of the higher carbonates content.The thermodifferential curves (DTA) of the two bodies, given in Fig. 1, show generally similar patterns, but there are significant differences which reflect the different mineralogical composition. The earthenware curve is characterized by two very pronounced endothermic peaks: the first of these, between 500° and 600°C, is typical of a material with a high kaolinite content; the second, "split" peak, between 750° and 850°C, corresponds to the decomposition of the dolomite. The majolica curve shows a broad exothermic effect in the 320-450°C range, due to the presence of substances of organic origin; the endothermic effects are less marked than is the case of the earthenware as a result both of the type of clayey minerals and the lower carbonates content. The thermogravimetric analysis curves (TG), Fig.2, confirm the observations already made vis-à-vis the more pronounced overall loss shown by the earthenware body; only at low temperatures does the TG curve for majolica reach greater partial loss values, as a consequence of the higher illite content and the presence of substances of organic origin. The two dilatometric curves, Fig.3, have a similar pattern. In the heating phase we note: a sharp increase in expansion between 550° and 600°C, due to the allotropic transformation of the quartz; the onset of shrinkage starting

Table 1 - Chemical and mineralogical compositions, (1) earthenware, (2) majolica.

	(1)	(2)		(1)	(2)
I.L.	21.31	12.94	quartz	24	24
SiO_2	41.34	48.58	Na-feldspar	3	9
Al_2O_3	13.08	13.93	calcite	4	18
TiO_2	0.80	0.83	dolomite	34	1
Fe_2O_3	0.55	5.79	illite+muscovite	13	27
CaO	14.34	10.13	chlorite	-	11
MgO	7.56	3.10	kaolinite	20	4
K_2O	1.24	2.70	hematite	-	4
Na_2O	0.38	1.05	accessories	2	2
organic C	0.04	0.56			

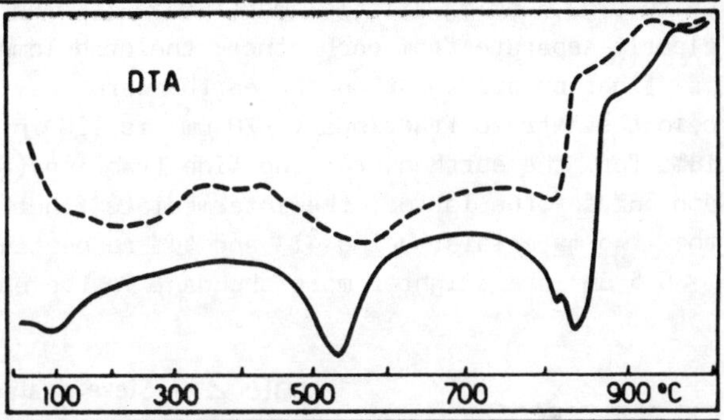

Figure 1
Differential thermal analysis

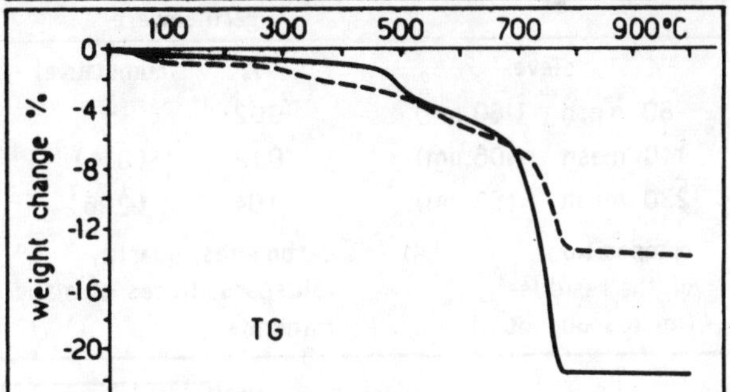

Figure 2
Thermogravimetric analysis

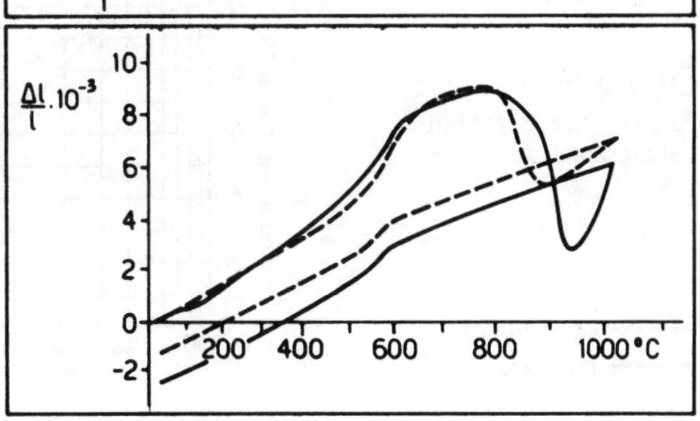

Figure 3
Dilatometric curves

—— earthenware
---- majolica

at approximately 800°C, due to the decomposition of the carbonates, which terminates rapidly as new crystalline phases begin to develop (gehlenite, diopside, anorthite, etc.) characterized by a high thermal expansion coefficient, which cause further expansion. The return dilatometric (cooling) curve, starting from relatively high final expansion values, terminates near the starting point, revealing a quite low dimensional variation. This is important since it means that firing curves can be designed for both the bodies which will generate products with dimensions more or less identical to those of the green pieces (no shrinkage or expansion after firing). With regard to particle size, the results of the sieve analysis are given in Table 2, while the cumulative particle size distribution curves, obtained using Sedigraph equipment, are given in Fig.4. As may be seen, residue in the 230 mesh sieve (particles greater than 63 µm in size) is quite low, being little more than 2%. The two particle size curves are clearly separate from each other; the granulometry of the majolica body is finer on average than the earthenware body. The differences mainly concern the extreme fractions (> 20 µm) is 12% in the case of majolica and 18% for the earthenware; the fine fraction (< 2 µm) is 47% for the former and 38% for the latter; the intermediate fraction (2-20 µm) is similar in the two materials, being 41% and 44% respectively. The very fine fractions (< 0.5 µm) are slighter more abundant in the earthenware body.

Table 2 - Sieve analysis

	earthenware		majolica	
sieve	%	(cumulative)	%	(cumulative)
80 mesh (180 µm)	0.02	–	0.38	–
140 mesh (106 µm)	0.12	(0.14)	1.90	(2.28)
230 mesh (63 µm)	1.94	(2.06)	0.20	(2.48)
composition of the residues (microscope obs.)	carbonates, quartz, feldspars, traces of mafic minerals.		quartz, feldspars, carbonates, mafic minerals, traces of sulphides and microfossils.	

Figure 4
Particle size distribution (Sedigraph)

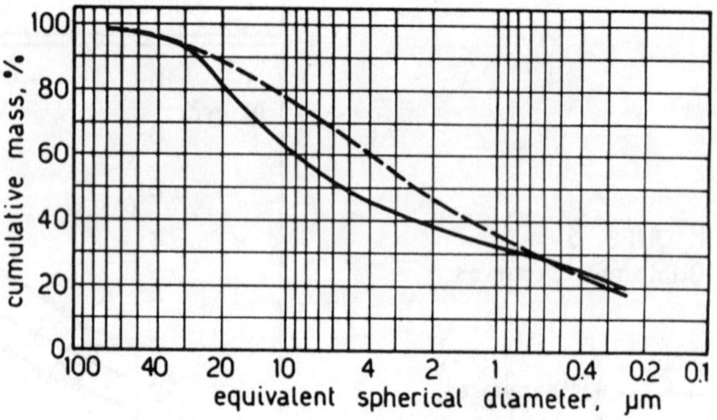

—— earthenware
---- majolica

Table 3 - Technological properties

Unfired bodies			earthenware	majolica
- Atterberg's limits:				
upper plastic limit		(water%)	37.8	34.7
lower plastic limit		(water%)	25.5	17.8
plasticity index			12.3	16.9
- water for extrusion		%	18.5	18.5
- drying shrinkage (linear)		%	5.3	5.8
- bending strength after drying		kg/cm^2	42	92
Bodies fired at 1020 °C				
- dimensional variation (from dried to fired)		%	+0.2	-0.7
- water absorption		%	20.4	16.1
- bending strength		kg/cm^2	216	302
- weight loss		%	21.4	12.6
- thermal expansion coefficient:				
(50-400°C)	linear	$.10^{-7}$°C^{-1}	76.0	77.1
	cubic	$.10^{-7}$°C^{-1}	228.0	231.3

TECHNOLOGICAL CHARACTERIZATION

The results of the technological tests are given in Table 3. The plasticity index is slightly higher for the majolica, this being related to the generally finer particle size of the latter. Samples of the two bodies were prepared using laboratory extrusion equipment; for both the materials, the amount of water in the body was 18.5%. The drying shrinkage in these samples was slightly higher in the case of majolica (5.8% as against 5.3%). Worthy of note is the much higher bending strength (more than double) of the dried majolica samples compared to the earthenware specimens; this is mainly due to the better granulometric assortment of the former. The mechanical strength of the earthenware after drying is nevertheless good. Firing tests at 1020ºC generated not very dissimilar results for the two bodies. Dimensional variation after firing is almost non-existent in the earthenware, while the majolica shows slight shrinkage (less than 1%). Open porosity, as may be inferred from the water absorption values, is slightly lower in the fired majolica; as a result the majolica's mechanical strength is greater than that of the earthenware, but in absolute terms the bending strength of the latter is decidedly good. The linear and cubic expansion coefficients of the two fired bodies are almost identical and are suitable for the most widely used glazes without the risk of crazing.

MARKET PROSPECT FOR CERAMIC TABLEWARES IN NIGERIA

EGUAKHIDE P. OAIKHINAN
Bendel State University, Ekpoma, Nigeria

ABSTRACT

The Nigerian consumption of ceramic tablewares remains extremely high despite strong and continuous import restrictions enacted by government over the last few years. This has made the Nigerian market increasingly attractive to foreign producers who have problems of over capacity and and static markets in their domestic sectors. About 53 million tonnes of various types of ceramic tablewares are currently being imported into Nigeria from Great Britain, Korea, China and Japan. Local production first approximates to 5.8 million tonnes per annum despite the availability of manpower and essential resources that are of major interest to the tableware industry.

INTRODUCTION

Traditionally, ceramic tablewares include cups, saucers, tea/coffee pots, plates and dishes. Nigeria has sufficient quantities of raw materials that are needed for the production of these products (1,2). Prior to 1960, ceramic tablewares were only found in the homes of high-income earners in Nigeria. The oil boom of the early seventies brought ceramic tablewares within the reach of the common man in the street. Today, Nigeria imports large quantities of ceramic tablewares from technologically developed countries, particularly England, Germany and Korea. Sometimes delivery promises often take more than twelve months because of government restrictions on importation and other foreign exchange problems. Shortages of ceramic tablewares now threaten the development of Nigerian tourist and hotel industries. The present paper examines the market and problems of investment in ceramic tableware production in Nigeria.

CONSUMPTION PATTERN

As previously reported (3), the growth rate of importation of ceramic tablewares in Nigeria between 1979 and 1985 increased by over 100 percent (Table 1).

TABLE 1
Imports of ceramic tablewares in Nigeria

Year	Quantity (tonnes '000)	Estimated Value (₦'000)
1979	1,644	12,137
1980	799	7,777
1981	2,843	26,706
1982	1,737	19,410
1983	1,037	12,840
1984	338	6,270
1985	3,545	9,193

The fluctuations in the importation pattern is due to economic constraints and has no distinct urban and rural segmentation.

The steady increase in population, and changes in the taste of an average Nigerian for metallic or plastic tablewares is creating a substantial demand for ceramic tablewares. Unless significant steps are taken to manufacture them locally from available mineral resources, importation will remain a major source of supply for a very long period of time in Nigeria.

LOCAL SUPPLY

The Ceramic Manufacturers' Association of Nigeria has been striving hard to ensure adequate supply of ceramic products to the Nigerian market. At the moment, there are nine whiteware industries in Nigeria, out of which only three produce tablewares for local consumption (Table 2).

TABLE 2
Whiteware Industries in Nigeria

Industry	Location	Capacity (tonnes '000)	Products
Arewa Ceramics	Missan	2,000	Sanitarywares
Ceramic Manufacturers	Kano	4,000	Sanitarywares
Modern Ceramics	Umuahia	850	Sanitarywares
Nigerian-Hungarian Ceramics	Ifon	3,000	Tiles/Tablewares
Ocean Five	Abeokuta	900	Tablewares
Porcelainware	Abeokuta	2,000	Tiles
Quality Ceramics	Uyo	2,000	Tiles/Sanitarywares
Richware Ceramics	Ilupeju-Lagos	780	Tablewares
Superceramics Manufacturers	Ijebu-Ode	2,000	Tiles

The level of operation of these industries has been very low because of raw materials shortages. The highest the Nigerian tableware industry has been able to achieve in capacity utilisation under the present depressed economic situation is just 47 percent on the average (3).

PROBLEMS OF INVESTMENT

The ceramic tableware industry usually uses a number of basic raw materials (e.g. clays, feldspar and quartz) whose properties are known precisely. In developed countries such as Britain and Germany, there are industries solely devoted to the processing of good quality raw materials for use in other manufacturing industries. For example, English China Clays (ECC) Devon, processed China and ball clays for the British Ceramic Industry. Here in Nigeria, ceramic manufacturers have to quarry and refine their own raw materials. Due to shortages of equipment, technical manpower and financial constraints, locally processed raw materials are often unable to match imported materials in terms of quality, consistency and price. All efforts so far made by the Nigerian Mining Corporation to mine and process quartz, feldspar and kaolin for the ceramics industry have also been faced with similar problems (5). However, a few industries that managed to process their own raw materials have had problems with glaze formulations and production. Hence their inability to manufacture high quality products that can resist foreign domination of the Nigerian market.

Other problems relate to government fiscal measures (6) and the structure of educational system (7) in Nigeria. For example, the kind of courses offer in Nigerian Polytechnics and Universities does not cater for the technological needs of the rapidly changing ceramics industry. Only two out of the twenty nine fully established Universities in Nigeria offer courses in ceramics engineering and technology. These Universities (i.e. ABU, Zaria and FUT, Owerri) lack basic facilities for developmental work in ceramics science and technology.

MARKET POTENTIALITIES

The market potentialities for ceramic tablewares in Nigeria can be seen at a glance from the demand-supply gap (Table 3).

TABLE 3
Demand-Supply Gap Projections for Ceramic Tablewares

Year	Aggregate demand (tonnes '000)	Gross Supply (tonnes '000)	Supply Gap (tonnes '000)
1986	50,220	5,364	44,856
1987	52,480	5,604	46,876
1988	54,900	5,856	49,044
1989	57,400	6,108	51,292
1990	60,040	6,384	53,656
1991	62,720	6,660	54,060
1992	65,500	6,948	58,552
1993	68,500	7,248	61,252
1994	71,620	7,632	63,988
1995	74,860	7,944	66,916
1996	78,300	8,292	70,008
1997	81,860	8,628	73,232
1998	85,600	9,060	76,540
1999	89,560	9,404	80,708
2000	93,680	9,972	83,708

There is excess demand over supply for ceramic tableware in Nigeria as local supply just represents only a paltry 10.5 percent of total demand.

With the present government stand on importation of ceramic products, indigenous ceramic industries at their current level of production, will not be able to provide a lasting solution to the demand for ceramic tablewares at the turn of the century. In order to meet future demands, the Nigerian ceramic industry and local entrepreneurs need the help of developed countries, preferrably on a joint-venture partnership, in the development of available resources in Nigeria for use in the tableware industry.

CONCLUSION

Nigerians need more than 55 million tonnes of ceramic tablewares at the moment for essential services in their domestic and tourist industries. Local supply is quite low (i.e. approximately 6 million tonnes) and despite government restrictions, importation has remained so high. To reduce the ever increasing demand-supply gap, there is an urgent need for joint-ventures between Nigerian entrepreneurs and foreign partners in areas solely devoted to the developments of ceramics in Nigeria.

ACKNOWLEDGEMENTS

The author wishes to acknowledge the assistance received from the Federal Ministry of Statics, Lagos during data collection. He also wishes to thank Professor Awele Maduemezia, Vice-Chancellor of the Bendel State University, for the financial support and encouragement.

REFERENCES

1. Oaikhinan, E.P., Characterisation of Certain West African Clays for use in the ceramics industry, M.Phil. Thesis, Univ. of Leeds, Great Britain, 1981.

2. Oaikhinan, E.P., Refractories in the Chemical industry in Nigeria, Interceram 37, 1988 (Aachen Proceedings 1987), 72-75.

3. Oaikhinan, E.P. Ceramic Developments in Nigeria, Materials Science Forum, 1988, Vols. 34-36, 1151-1156.

4. Umoru, M.S., Paper presented at the 2nd AGM of Ceramic Manufacturers Association of Nigeria, 1988.

5. Ojo, O.M. Exploration of Minerals, The NMC Magazine, Edition 3, 1987, 3-4.

6. Chigbo, G.O. Ceramics and Industrial Development in Nigeria, Paper presented at the inaugural meeting of N.I. Ceram., 1988.

7. Oaikhinan, E.P., Demand and Processing of ceramic sanitarywares in Nigeria, Proc. 2nd NCB Int. Seminar, New Delhi, 1989, 84-87.

THE CLAY-WATER RELATIONSHIP - A DIFFERENT APPROACH: MOISTURE RETENTION

M.B.G. Ketelaars, J. van der Zwan
TNO Institute of Applied Physics, Ceramics Dept.
P.O. Box 595, 5600 AN EINDHOVEN, The Netherlands.

ABSTRACT

In basic science on ceramic materials the drying process is in most cases described from a physical-technological point of view. We present a new approach for the description and interpretation of the drying behaviour of (green) ceramic products. This approach exists already 50 years in soil science and deals with the energy relationship between the solid and its water. The clay moisture retention (pF-)curves can be obtained by a set of measuring devices. From the pF-curves the pore-size distribution can be calculated. The obtained curves show the moisture retention or suction as function of the equilibrium moisture content of the clay mass. The curves characterize uniquely each clay, sand or mixture. Variations in deflocculation result in different structures and equilibrium moisture contents in the green products.

INTRODUCTION

The final properties of ceramics are the sum of many production steps, e.g. shaping, drying, glazing and firing. This study has centered on factors governing the drying process and their dependebility on the clay material properties. Moisture retention data of the clays depend on the structure of the clay aggregrates which are developed during the shaping sequence. This lead to three lines of research:

1. Study of the structure of clay suspensions by rhelogical and sedimentation measurements, and measurements of zeta-potentials (1, 10).
2. Study of the structure of clay in a plastic state by measurements of filtering properties under pressure and on plaster (1, 10), and measurements of moisture potentials (pF) at high moisture contents.
3. Study of the rigid clay during and after drying by measurement of the moisture potential at low moisture contents.

MOISTURE POTENTIAL

Porous materials such as soils, but also all unfired ceramic (clay) bodies are capable of retaining within their structure 'held' water, which does not drain under action of gravity. Water is held in such materials by

surface tension and absorption forces. The moisture potential relates the degree of water retention of a porous mass to its pore size distribution and to the forces withdrawing or supplying water. Moisture potental is defined as the work done per unit mass of water, by the water, when a small quantity of it is transported from the clay-water system to a free water surface at the same temperature and height as the clay (5). On the pF-scale, introduced by Schofield (2) the suction or potential is expressed as the common logarithm of the height in cm of the water column which the suction support.
Table 1 shows the relationship between the pF-scale, the equivalent negative hydraulic head, the suction in atm and the relative humidity. For more details of backgrounds, measuring methods and use, the reader is referred to literature (2, 3, 4, 5).
The curve which shows the relationship between the negative pressure (suction, hydraulic head) and the equilibrium water content of the investigated material (soil, clay) is a very important physical characteristic, and is known as the (soil) moisture retention curve. In figure 1 some typical pF curves are shown from literature (sand, sandy loam and clay loam) together with two curves from own experiments on two different clays (A: Arvor 1C kaolin; WB: highly plastic, illitic clay from Westerwald).
The curves demonstrate the effect of pore size distribution on the shape of the pF curve. In sandy soils most of the pores are relatively large and once these pores are emptied at a given pressure, only a small amount of liquid remains. The greater the clay content and the amount of fine pores, the greater the moisture content at any particular pressure and the more gradual the shape of the curve.

TABLE 1
pF scale, moisture tension and pore size of a (clay) mass

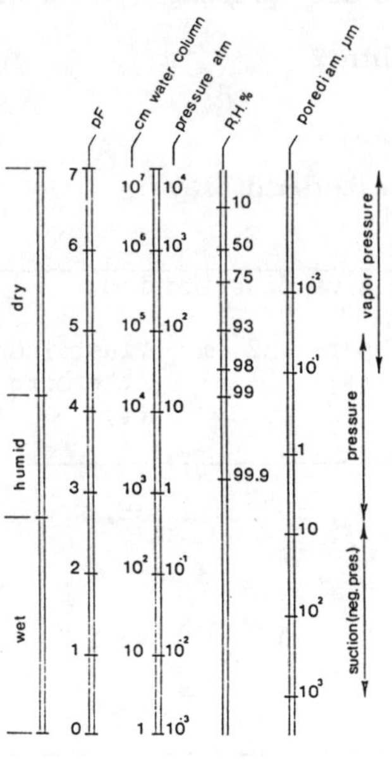

FIGURE 1
pF curves of some soils and two investigated clays

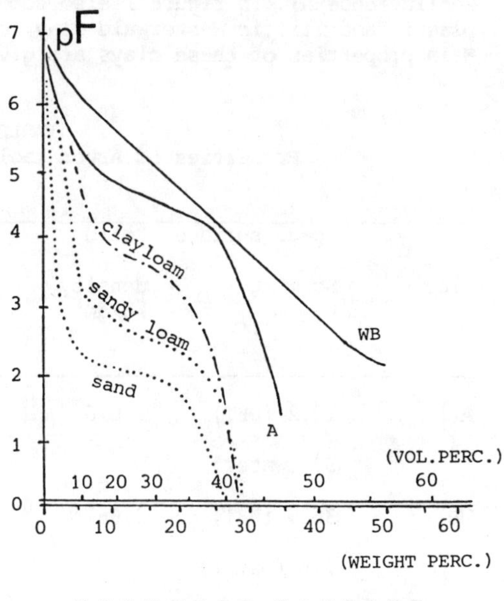

EXPERIMENTS

When porous materials are oven-dried, the suction of the moisture which is not removed rises to approximately 10^7 cm of water;pF equals than ^{10}Log h = 7 (h in cm water column).
The suction range pF 0 to pF 7 is traversed when the investigated clay is dried from a condition of saturation to oven dryness. Unfortunately no single measuring method can be used to cover the whole pF range, and in practice at least three methods have to be adopted.
We have used two direct methods, suction plates for the range pF 0 - pF 3 and pressure membrane apparatus for pF 3 - pF 5, and a so-called indirect method, a vacuum desiccator for pF 4,5 - pF 7.
The suction plate method consists of a sand-box apparatus and a kaolin-sand-box apparatus (Eykelkamp B.V., Giesbeek, NL). The pressure membrane apparatus (0 - 25 bar and 0 - 100 bar) were also delivered by Eykelkamp.(6)
The pF range 4,5 - 7 is covered by controlling the relative humidity in a desiccator. Such a controlled atmosphere can be obtained by using saturated salt solutions in a (vacuum) desiccator.
Samples are prepared by filter pressing as described in (1) and (10), from deflocculated clay suspensions with a density of 1,60 kg/dm^3; Na_2SiO_3 as deflocculant in concentrations ranging from 0,0244 to 0,244 molar (= 0,10 to 1,0 w% m/m_d) is used; filtering pressure 5 ± 0,2 bar. Samples, filtercakes, are contained in brass rings, height 20 mm, diameter 40 mm for suction measurements pF 0 - pF 2,7, and in teflon rings height 10 mm, diameter 35 mm for pressure membrane apparatus and desiccators.

MATERIALS

Although in the total research program 6 core clays were investigated, we focus in this publication on a highly kaolinitic clay from France: Arvor 1C kaolin (code A). In figure 1 also some results are presented of a highly plastic and illitic Westerwald clay, code WB.
Main properties of these clays are given in table 2.

TABLE 2
Properties of Arvor kaolin and Westerwald clay

clay	Spec. surface Sg m^2/g	Solid density kg/dm^3	Particle size distribution d_{50} (μm)	perc < 2 μm (%)	Plast.Index Atterberg (%)
A	12,2 (BET)	2,616	2,0	50	35,1
	37 (water)				
WB	35.9 (BET)	2,667	0,26	83	47,1
	145 (water)				

RESULTS AND DISCUSSION

As mentioned before, figure 1 shows the pF curves for Arvor kaolin (A) and a plastic, Westerwald clay 306 (WB) and 3 typical soils. The moisture retention curves of Arvor as a function of deflocculation are shown in figure 2. As can be seen from the curves, the equilibrium moisture content of the clay depends on the amount of Na_2SiO_3 in the slip, and shows a minimum at 25 mmolar Na_2SiO_3. At this concentration of deflocculant also the apparant viscosity, the cake thickness and several other properties show a minimum (1, 10).

D'Hollander (8) and Hartge (9) described a method to calculate a pore size distribution from a complete pF curve, and vice versa to calculate the moisture tension of a porous mass from a known pore size distribution. Because of some problems during the measurements at the low pF values (pF 0-2), we did not succeed in measuring the whole curve. The main problem was the unexpected long time required for reaching an equilibrium moisture content in the clay, especially in the wet and almost saturated clay samples (several months for each of the 10 installed moisture potentials). So in figure 3 only a pore size frequency and not a complete quantitative relationship between pore diameter and number of pores is presented for the Arvor kaolin.

Figure 4 shows the different pF curves of Arvor as a function of deflocculation. In the same figure the moisture content of the wet filtercake is given directly after filterpressing. The resemblence of the curves is very plausible as both the pF curves (an equilibrium moisture content!) and the "filtercake curve" reflect a moisture content after drainage by some form of suction or pressure.

Figure 2
pF curves of (de-)flocculated
Arvor kaolin

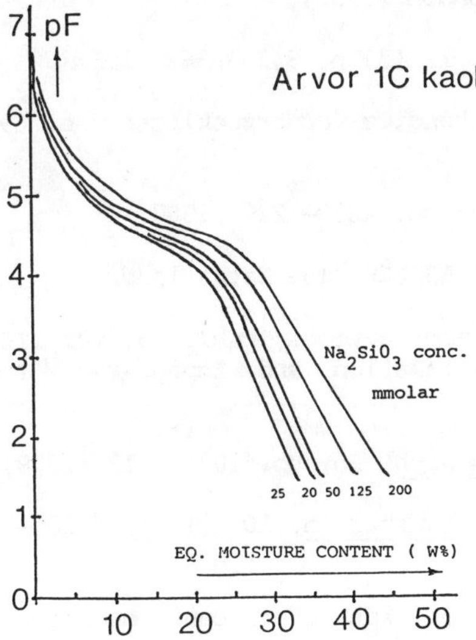

Figure 3
Pore volume frequency of
(de-)flocculated Arvor kaolin

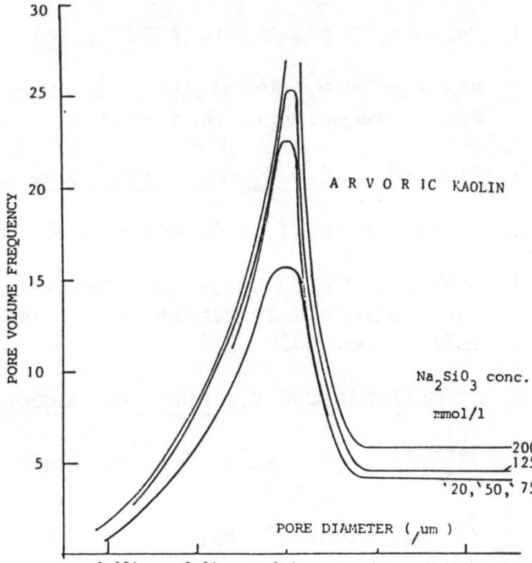

Figure 4: Moisture tension in Arvor filtercakes as function of deflocculation

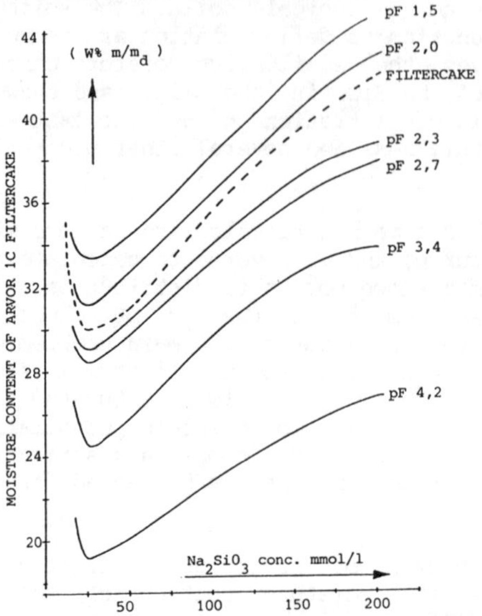

CONCLUSIONS

The pF curve is characteristic for each clay. Moisture stress depends on structure and therefore on degree of deflocculation of the clay mass. From pF data the pore-size distribution can be calculated. Moisture stress provides a powerful unifying concept for studying unfired ceramic bodies. It is the fundamental parameter governing their water movement, drying properties, workability and also dry strength (5).

REFERENCES

1. Siskens, C.A.M., Van der Zwan, J., Jacobs, J.C.H., Characterization of clays by the correlation of physical-chemical properties, in: Clay based Materials for the Ceramics Industry, Ed. H. Nosbusch an I.V. Mitchell, Elsevier Applied Science, London, 1988, pp. 57-63.

2. Croney, D., Coleman, J.D., Bridge, P.M., Road Research Techn. Paper no. 24, Her Majesty's Stationary Office, London (1952),p. 42

3. Coleman, J.D., Marsh, A.D.J., Soil Sci. **12** (2) p. 343 - 362 (1961)

4. Stakman, W.P., Verslagen van Landbouwkundige onderzoekingen no. 693, Pudoc, Wageningen, NL (1968).

5. Packard, R.Q., J. Am. Ceram. Soc. **50** (5) pp. 223 - 229 (1967)

6. Eijkelkamp B.V., P.O. Box 4, 6987 ZG GIESBEEK, The Netherlands

7. Stakman, W.P., Drainage priciples and applications, publ. 16, Vol III, International Institute for Land Reclamation and Improvement, ILRI 1974, pp. 221 - 251

8. D' Hollander, E.H., Water Resources Research, **15**, pp. 107 - 112 (1979)

9. Hartge, K.H., Z.f. Kulturtechnik und Flurberein, **10** (1) pp. 20 - 27 (1960)

10. Van der Zwan, J., Ketelaars, M.B.G., Proc. 1st ECerS Conf. Maastricht, (1989) p. ....

THE SPECIFIC SURFACE EXPOSED.

JAN VAN DER VELDEN
Ceramics Department TPD-TNO
P.O.Box:513,5600 MB Eindhoven
The Netherlands.

ABSTRACT

The intention of the paper is to reveal the usefulness of the total specific surface,measured by ethylene-glycol retention, for assessing the shaping and drying behaviour of plastic clay bodies for clay building products.
Data of a number of relevant properties of Dutch and other European clays are plotted versus the total specific surface of the samples.
The results of the investigation demonstrate the importance of the parameter in question.

INTRODUCTION

With regard to the shaping and drying behaviour the particle size distribution of a clay body is still considered one of the main parameters,although its value turns out to be restricted to clay bodies of similar mineralogical association and geological evolution.
This paper deals with the usefulness of the total specific surface,measured by the retention of ethylene-glycol.[1] [2] [3]

RESEARCH METHOD.

The determination of the ethylene-glycol specific surface has been developed by Dyal and Hendricks.[1]
In order to shorten the period of time needed for the analysis,the research in question has been carried out by a slightly modified method.[2] [3]
The investigation relates to a selection of 20 clay samples (15 from the Netherlands and 5 from other European countries) with strongly varying compositions as indicated in table 1. [4]

Data of large number of properties were gathered and reviewed with respect to possible relationships.

TABLE 1.
General characteristics of 20 clay samples.

CHARACTERISTICS	UNITS	RANGE	AVERAGE
KAOLINITE	%	4 - 26	12
ILLITE + CHLORITE	%	8 - 36	16
DEGRADED ILLITE	%	0 - 46	14
Σ- CLAY-MINERALS	%	18 - 72	42
Σ- FILLERS + FLUXES	%	28 - 82	58
SiO_2	%	53 - 81	65
Al_2O_3 + TiO_2	%	8 - 23	15
Σ- FLUXES	%	4 - 19	11
LOSS ON IGNITION	%	2 - 13	7
< 2 μm	%	12-80	38
< 20 μm	%	30 - 93	65
> 200 μm	%	0 - 17	4
TOTAL SPECIFIC SURFACE	m^2/g	34-219	106

RESULTS.

The properties giving a fair correlation with the total specific surface are represented in figure 1.
These properties are:

- The equilibrium water content for adsorption and desorption at a temperature of 20 C and relative humidities (R.H.) ranging from 0.2 to 0.8. Only one air condition is repre- sented here.
- The ion-exchange capacity for sodium acetate.
- The crushing strength of small dry cylindrical test speci-mens.
- The plasticity index according to Atterberg.
- The tempering water according to Pfefferkorn at 16 mm PRH (Pfeff.Residual Height).

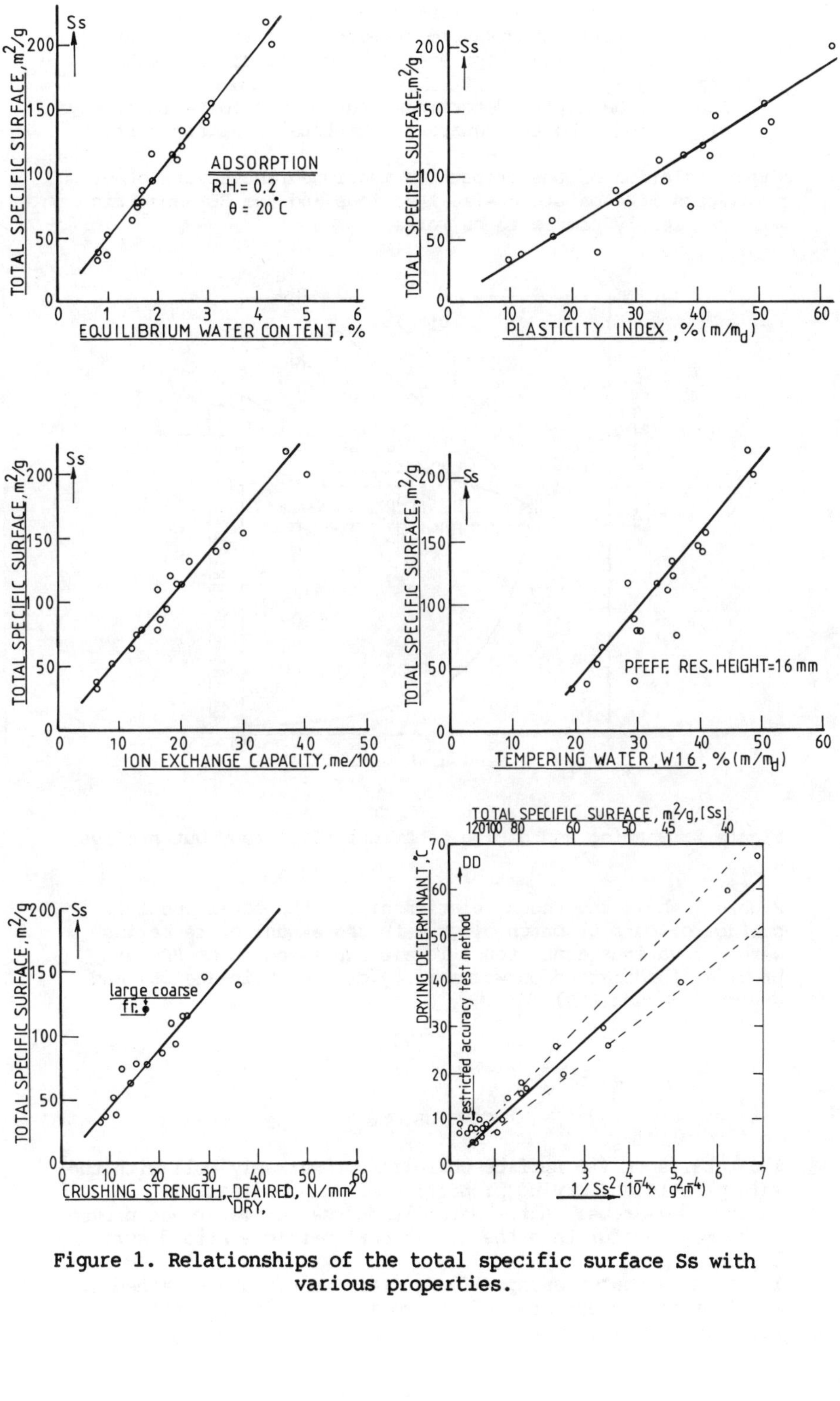

Figure 1. Relationships of the total specific surface Ss with various properties.

- The drying determinant (DD),being the maximum safe psychro-metric temperature difference of air,with which small cy-lindrical test specimens can be dried without cracking.
  Roughly,the drying determinant turned out to be inversely proportional to the square of the total specific surface.

The correlation of the properties mentioned above with other parameters,like particle size fractions and the BET-specific surface clearly proved to be worse.

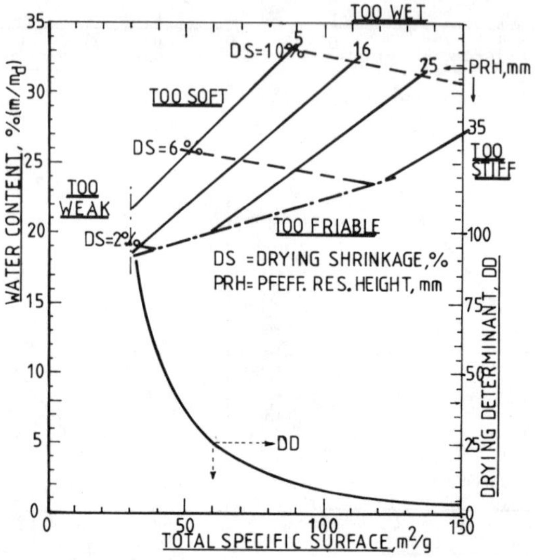

Figure 2. Shaping and drying behaviour of current Dutch clays

Figure 2 shows the rough relationship of the total specific surface of current Dutch clays with the amount of tempering water at various consistency levels,expressed in mm PRH and provides information on workability,drying shrinkage(DS) and drying behaviour(DD).

CONCLUSION.

The total specific surface correlates remarkably well with the main plastic and drying properties of clay bodies.
In its 20-year-use in the Netherlands,the parameter has proved to be more useful than the traditional particle size distri-bution.
In works laboratories,specific surface values are obtained by a calibrated measurement of the equilibrium water content at a relative humidity of 0.75.

REFERENCES

1. Dyal,R.S. and Hendricks ,S.B.,The measurement of the specific surface.
Soil Science,1950,69,421-432

2. Amerongen,H van, Measurement of the specific surface of clays.
Science of Ceramics,vol 3.

3. Douma,G.H.,De bepaling en betekenis van het specifiek oppervlak van klei voor de grofkeramiek.
Klei en Keramiek,1969,19,226.

4. Amerongen,H.van and Velden,J.H. van der, Samenstelling en eigenschappen van 31 kleisoorten.
Report No 70-04032, Ceramics Department MT-TNO.
Apeldoorn,The Netherlands

THE RHEOLOGICAL AND FILTRATION PROPERTIES OF CLAY

J. van der Zwan, M.B.G. Ketelaars
TNO Institute of Applied Physics, Ceramics Department
P.O. Box 595, 5600 AN Eindhoven, The Netherlands

ABSTRACT

The filtering properties of a clay slip and its rheological behaviour are quantitatively related. Casting rate, filtering rate and rheological parameters conform to the classical description of the deflocculation mechanism. A high filter cake resistance and a very low hydraulic conductivity of the cakes and casts can be explained with the high tortuosity ratio and compressibility of the (partly) flocculated cast and cake. In general, several properties of the clay slips and casts were influenced by the degree of flocculation, such that minimum c.q. maximum values of the properties were found at roughly the same state of flocculation. There are indications that this behaviour is influenced by the occurrence of orientated clay particles in the cast or cake and most probably, in the clay slips.

INTRODUCTION

The shaping process in the ceramics industry is very often the basis of the whole manufacturing process. Shaping has a substantial influence on the quality of the product and the efficiency of a production line. This study has centered on factors governing the shaping process and its dependence on the properties of the clay materials. Shaping of fine ceramic products often proceeds from "clay suspended in water", via "water retained in capillaries" to "rigid dry clay" (1). Therefore in this work the development of the structure of clay agglomerates along the shaping sequence is studied. This led to three research phases:

1. Study of the "structure" of clay suspensions
2. Study of the "structure" of clay in a plastic state as produced by filtering methods.
3. Study of the "structure" of a rigid clay during and after drying.

The state of agglomeration during the whole process is influenced by parameters such as grain-size distribution, mineralogy and colloidal performance. Therefore, the three research phases were carried out by varying the degree of flocculation as a function of the amount of electrolytes (Na_2SiO_3).

CHARACTERIZATION OF RAW MATERIALS

Six core clays have been investigated in our part of the E.C. program on clay based materials.(1) For the characterization of these clays the following properties have been investigated:

- chemical composition by X-ray fluorescence
- mineralogical composition by X-ray diffraction
- particle size distribution by wet sieving, X-ray absorption/sedimentation by sedigraph, and pipette centrifuge method.
- specific surface by one-point BET, glycerol-absorption and water-absorption method ("glycol-surface")
- true density by pyknometer method.

We focus in this publication on two clays, a French more or less pure kaolin, Arvor 1 C (code A) and a highly plastic illite containing clay from Westerwald (code WB). The main properties of these clays are given in table 1.

TABLE 1
Properties of Arvor kaolin and Westerwald clay

clay	Spec. surface Sg m^2/g	Solid density kg/dm^3	Particle size distribution d_{50} (μm)	perc < 2 μm (%)	Plast.Index Atterberg (%)
A	12,2 (BET) 37 (water)	2,616	2,0	50	35,1
WB	35.9 (BET) 145 (water)	2,667	0,26	83	47,1

DETERMINATION OF THE STRUCTURE IN CLAY SUSPENSIONS

The study on the structure of clay suspensions is done by rheological measurements, sedimentation measurements and measurements of zeta-potentials. The rheological properties are measured on clay slips with a standard spec. weight of 1600 g/cm^3, and as a function of the added amount of deflocculant (Na_2SiO_3) with the aid of a Haake RV3 viscosimeter. Casting properties were established by a well-defined method by casting small slabs on plaster from which the filter-time and water content of the casts could be measured.(2) On the same clay slip filtering experiments were carried out by filterpressing at 5 ± 0,2 bar (top of the filtercell). The determination of the zeta-potential of the clays is done by electro-osmosis method, i.e. not on seperate clay particles (as by electrophoresis), but on the whole bulk of the clay. Zeta-potential measurements give a good impression of the degree of flocculation in clay-water systems.

Sedimentation and transmission data give information on the different behaviour between flocculated and deflocculated slips.(3) From the calculated floc volume coefficients and sedimentation data the average size of the clay agglomerates can be estimated. However, the calculations of the average floc size formed from the agglomerates were not accurate enough for conclusions to be drawn.

DETERMINATION OF THE STRUCTURE IN CAST AND CAKE

From the data obtained by filterpressing (time, filtrate weight, weight and dimensions of the wet and dry filtercake) permeability of the cake and hydraulic conductivity can be calculated. The cakes produced by filter pressing are used for further experiments to determine the structure of the clay in a plastic state. The wet filtercakes are used for moisture retention (pF-)measurements.(4) From a complete pF-curve (pF 0-7) a pore size distribution can be calculated, and reverse, from a pore size or particle size distribution a water retention curve. After determination of the complete pF-curve the cakes are used for measurements of the clay-particle orientation in the product. By means of special X-ray techniques, using the reflection peaks [020] and [002] of kaolinite, it is possible to determine the rate of orientation of kaolinitic clay minerals in rigid products. The oven-dry cakes are completely impregnated with a polyglycolether (Hüls Polywachs 6000). Carefully polished products together with a "randomly orientated standard" and a "100 % orientated standard" were used for this experiment.(5)

RESULTS AND CONCLUSIONS

The rheological behaviour of the 6 clays were very different, but showed as a function of deflocculation the normal pattern with a more or less wide range of optimal deflocculation. Within this range the maximum deflocculation occured only in a rather small region. Especially the parameter "n" in the Ostwald equation shows the typical Newtonian behaviour of the kaolinitic clays Arvor, the pseudoplastic behaviour of the more illitic clays and even the dilatant behaviour of the Schmidt 1001 clay. Casting properties depend on deflocculation of the slip. The occurance of an extreme filtertime (in this case a maximum) on plaster can be explained by the classic theory of deflocculation of clay slips with "card houses", "card packs" and fully dispersed particles in slip and cast. The filtercake resistance is a property of the cake which can be compared with the filtertime of a cast on plaster. The agreement between filter cake resistance and the filtertime data was especially good for practically all 6 clays. The dependence of the zeta-potential of the clays as a function of Na_2SiO_3 concentration is in good agreement with the measured rheological properties of the slips, at least as far as the general shapes of the curves are concerned. The clays showed a clear relationship of the "Peak Ratio" [002]/[020] to the degree of flocculation. The minimum values were found at the low end (smallest percentage of Na_2O) of the deflocculation curve (see figure 1 for both clays).

From the described observations we can conclude that there is a clear relation between the amount of added deflocculant and the viscosity of the slip, filtertime on plaster, water content of the cast and filtercake, cake resistance, equilibrium moisture content at different pF-values, zeta-potential and the orientation of the kaolinite particles in the cake. Although it has not yet been proven, it is very likely that the agglomeration process is already going on in the clay slip and depends on the degree of deflocculation. Minima and maxima values in the relationships between deflocculation and slip and product properties are always found at the optimum deflocculation at the low end of the deflocculation curve.

Figure 1: Several properties of clay slips, filtercakes and casts as a function of deflocculation.
A=Arvor 1C kaolin; WB=Westerwald 306; CaRe=filtercake resistance (10^{16} m^{-2}); Ft=filtertime on plaster (min); PR=peakratio [002]/[020]; VIS=apparent viscosity (mPa.s); MC= moisture content filtercake (W% m/md); ZePo= zeta potential of clay solids (mV).

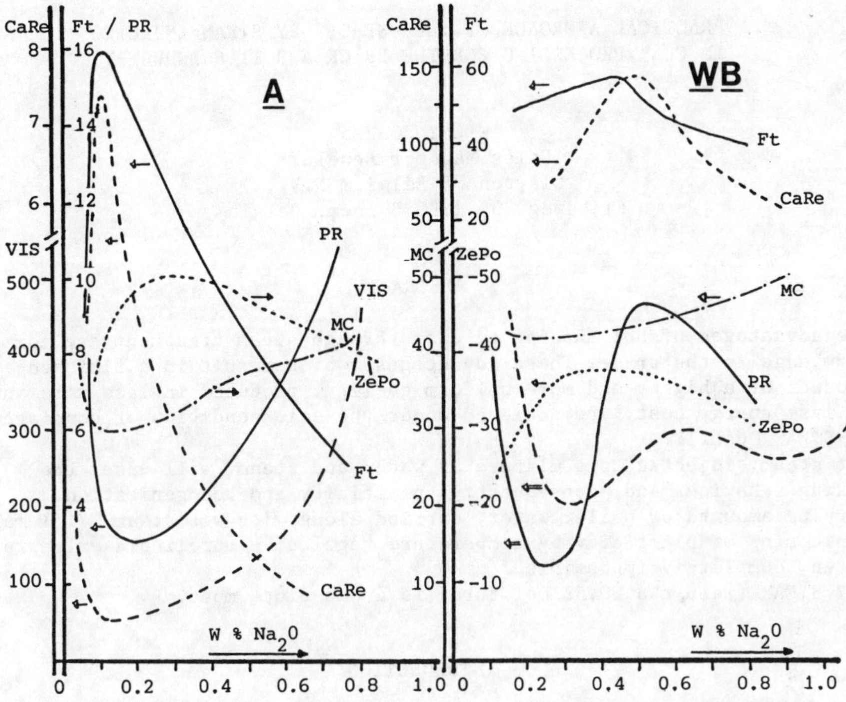

REFERENCES

1. Siskens, C.A.M., Van de Zwan, J., Jacobs, J.C.H., Characterization of clays by the correlation of physical-chemical properties, in: Clay based Materials for the Ceramics Industry, Ed. H. Nosbusch and I.V. Mitchell, Elsevier Applied Science, London, 1988, pp. 57-63

2. Siskens, C.A.M., Stevels, J.M., Welzen, J.Th.A.M., Van der Zwan, J., Science of Ceramics 9, Noordwijkerhout, 1977, pp. 272-280

3. Welzen, J.Th.A.M., The influence of surface-active agents on kaolinite, thesis, University of Technology, Eindhoven, 1979

4. Ketelaars, M.B.G., Van der Zwan, J., Proc.1st.ECerS Conf., Maastricht, (1989), pp.....

5. Martin, R.T., Ladd, C.C., Clay & Clay Minerals, 1975, 23, 17-25

A PRACTICAL APPROACH OF THE USE OF DRY STEAM INJECTION IN CLAY PROCESSING FOR THE BRICK AND TILE INDUSTRY

WILLY SOMERS
Sales Manager Benelux
Clayton of Belgium N.V.,
Rijksweg 30, 2680 Bornem, Belgium

ABSTRACT

The advantages of hot shaping of clay through steam treatment are common knowledge in the trade. These advantages, which result in a finished product of a higher and more uniform quality, produced in less time and at less energy cost, are achieved under the sole condition of consistency in STEAM QUALITY.
Wet steam, injected as a mixture of water and steam, will upset the mixing behaviour and, consequently, plasticity and homogenization. Varying amounts of boiler water, carried along with wet steam, will make monitoring of plasticity by temperature regulation unreliable or, more often, completely impossible.
DRY STEAM (less than 0,5% moisture) is an absolute must.

INTRODUCTION

In its natural state, clay is rarely suited to being pressed into molds from which bricks, tubes or tiles can be fired.
Ideally, the clay should be treated to guarantee its modelling consistency (plasticity) and this can be done by the adding of water, an auxiliary substance, plasticizer or heat.
When the clay is ready for production, it is processed into molds using either a stuffing press or mold burning press. These molds will then be dried before firing.
During the production cycle from clay to brick, the drying process is critical to ensure output and good quality. It is the clay temperature at which the mold is pressed that determines its drying behaviour.
The molding temperature (40°C - 70°C) is obtained by direct injection of "DRY" saturated steam into the clay, prior to pressing.

The temperature increase of the clay has the following advantages :

- better plasticity of the clay
- reduced absorbed electric power in the presses
- reduced sensitivity to "cracking"
- reduced energy consumption in the drying process
- higher maximum admissible drying velocity

PARAMETERS

Modelling Consistency or Plasticity : Dependent upon the production method used, the clay will have a given modelling consistency which is usually obtained by the addition of a specific quantity of water. In principle, the richer the clay, the more water that should be added. Within certain limits, the plasticity can also be influenced by temperature. As a result of increased molding temperature, 3% to 4% less mixing water can be used to obtain the same level of plasticity. Steam supply fluctuates between 30 jg. and 50 kg. of steam per ton of dry clay. This is dependent upon the type of production process and the type of clay being used. (see Table 1).

Homogenization : Under certain operating conditions, particularly with cold modelling, untreated "pits" can remain which, during the pressing and drying process, will cause cracking. These non-homogenized clay "pits" completely disappear when the clay is heated prior to pressing with direct DRY steam injection.

Absorbed power : Due to direct steam injection, the increased molding temperature reduces the required amount of mixing water to be added to ensure vital plasticity. The absorbed electric power of the presses drops by 10% to 20%.

Sensitivity to Cracking : Some cracks may develop in the molds both at the beginning and during the drying process. These cracks are known as "starting and ending cracks".
The cracking sensitivity of the mold is dependent upon the drying velocity. This velocity is in turn dependent on the starting temperature of the mold and the wet bulb temperature of the drying tunnel or

drying chamber.
Starting cracks result from the fact that the mold is heated too quickly at the commencement of the drying process. Ending cracks can be found in the center of the mold when a certain humidity gradient is exceeded. Both starting and ending cracks cause fractures during firing and, therefore, losses in production. Based on comparable patterns, drying time can be shortened by approximately 40% when the clay is heated.

<u>Drying time</u> : Due to the 40% higher maximum admissible drying velocity, 40% less drying time is necessary. Thermally-treated clay has an additional advantage in that no slow warm-up period is required, as is the case when placing cold molds in the drying plant.
In brick factories where the molding and firing process is carried out, the capacity of the ovens exceeds the capacity of the drying chambers and, therefore, the drying of the molds becomes a "bottle neck" within the production cycle.
Heating up the clay can increase the drying velocity and so the total production time can be considerably shortened.

<u>Drying costs</u> : A high molding temperature is excellent to combine with high wet bulb temperatures in the drying plant. With a wet bulb temperature TN = 70°C, the specific heat output of the drying air is around 800 kcal/kg. of water to evaporate at a T of 10°C.
At TN = 25°C, the required heat is around 1.800 kcal/kg. of water to evaporate, making a difference of 1.000 kcal/kg. of water.
Drying costs per ton with cold shaped molds having a water content of 35% (350 kg. per ton) and a "rest" humidity of 6% after drying, at TN = 25°C.

1.800 x (350 - 60) = 522.000 kcal per ton

Cost saving from the pre-heating of the clay :

Cold molds	:	522.000 kcal per ton
Hot molds	:	200.000 kcal per ton
Heat saving at drying	:	322.000 kcal per ton

Heat consumption using DRY steam injection in the presses at 20 - 50 kg. per ton (13.000 - 33.000 kcal per ton) results in a profit of appr. 300.000 kcal per ton.

GRAPH I

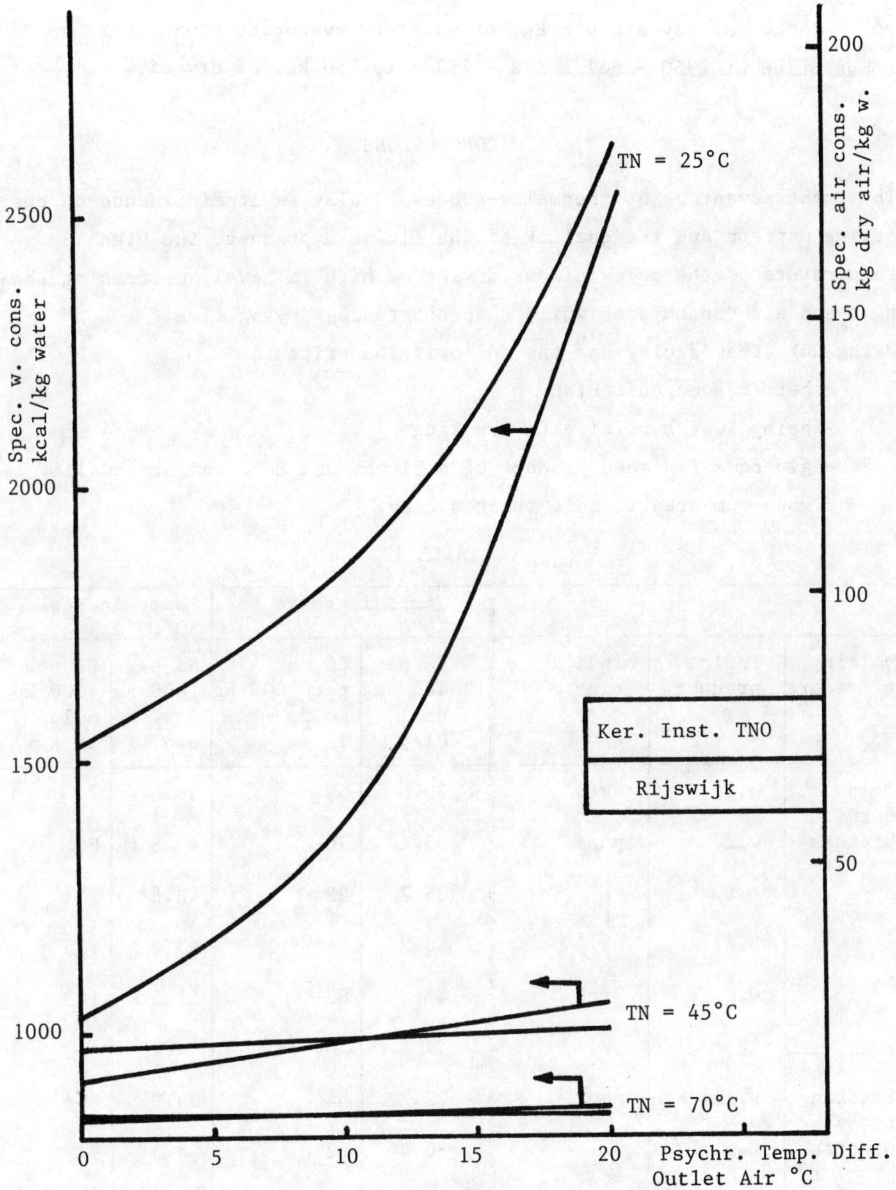

Spec. w. cons. kcal/kg water
2500
2000
1500
1000
Spec. air cons. kg dry air/kg w.
200
150
100
50
TN = 25°C
TN = 45°C
TN = 70°C
Ker. Inst. TNO
Rijswijk
0
5
10
15
20
Psychr. Temp. Diff. Outlet Air °C

Graph 1 illustrates (simultaneously) that the air consumption drops from 70 to 15 kg. of dry air per kg. of water to evaporate or a lower consumption of (350 - 60) x (70 - 15) = 15.950 kg. of dry air.

CONCLUSIONS

The great advantage of thermally-processed clay is its influence on the drying pattern and the quality of the finished product. The high temperature of the molds allows drying at high TN level, decreasing the heat and air consumption while also shortening drying time.
Using DRY steamed clay has the following benefits :

- better homogenization
- being less sensitive to cracking
- giving a finished product of a higher and more uniform quality when compared to cold-treated clay.

TABLE I

			Temp. increase 40°C		Temp. increase 50°C	
Molding procedure	Indic. product	Indic. clay	SS per 1000 kg dry clay	SS per per 1000 kg molds in kg	SS per 1000 kg dry clay	SS per 1000 kg molds in kg
Mold burn process	w.f.	very	29.6	59	36.8	74
	w.f.	poor	35.2	70	43.8	88
	c.f.	↓	35.2	92	43.8	114
	b.f.	↓	35.2	141	43.8	175
	w.f.	rich	41	82	51.5	103
Stuffing press process	w.f.	poor	27.2	54	33.9	68
	w.f.	↓	30.9	62	38.3	77
	w.f.	rich	34.5	69	43	86

SS = Steam Supply.

CURRENT RESEARCH INTO THE DRYING PROCESS OF DUTCH BRICKS

J. de Jong, A.H. de Vries
TPD-TNO Ceramics Department
P.O. BOX 513, 5600 MB Eindhoven, the Netherlands

ABSTRACT

A special laboratory air-conditioned drying apparatus was developed by the TNO working group "structural ceramics". The specifications of this device under full climatization conditions are: temp. 5-90 °C, relative humidity 10-95%, heating/cooling 0.3 °C/min.. Without humidity control: 0-200 °C and heating/cooling 1 °C/min.. Both temperature and rel. humidity trajectories are controlled by computer. Output consists of: air flow, temperature and rel. humidity in testing room and linear shrinkage and weight change of the object. Data are imported by a microcomputer, calculated, and printed as colour plots. The possibilities and limitations of this experimental device will be discussed as well as drying characteristics of some Dutch bricks.

INTRODUCTION

As a result of various developments drying clay has become the subject of renewed interest in the Netherlands. It is expected that in future, clays of a significantly different composition will have to be used, such as sedimentary clays and mixtures with residues.

When designing new hot air driers the optimum drying conditions have to be selected on the basis of knowledge about the drying behaviour of the relevant type of clay. In many factories, by optimizing the firing process, the amount of residual heat has been reduced to such an extent that additional fuel is required for drying.
It is therefore worth investigating whether these increased energy costs can be reduced. Given these developments it is important that more knowledge is built up about the drying behaviour of clays. In view of the wide variety of clays available, the relationship between clay composition and drying behaviour will also be investigated.

The research, which will be carried out by the Ceramics department together with other institutes, can be divided into the following segments.

1 Research into how to describe the drying behaviour of clays,
2 Research into improving the drying behaviour of clays by additives and/or by reducing the mixing water,
3 Modelling the water and heat transport in hot air drying plants.

The drying equipment described in this report was designed for these investigations. This equipment made it possible to study the optimum drying behaviour and sensitivity to cracking of green bricks during the drying process.

THE DRYING EQUIPMENT

The Ceramics department has a Grenco dry-climate chamber developed especially for drying experiments to study the drying behaviour of bricks.

The temperature and relative humidity can be controlled by a programmable curve. The following parameters were measured and stored in a file:
- air temperature in the test chamber,
- relative humidity of the air in the test chamber,
- change in weight of the sample,
- longitudinal linear drying shrinkage of the sample.

The measuring time depends on the duration of the experiment. The wet and dry bulb method is used for climate control. This is done using the psychrometric temperature differential, the difference in temperature between a wet bulb and a dry bulb. The relative humidity controller, which measures the psychrometric temperature differential through an interface, directs dry air or humidified air to the test chamber, as required.
The change in weight was measured by a built-in Sartorius balance. The fan switched off automatically while the weight was recorded to avoid the influence of air flows during measurement. The linear drying shrinkage along the length of the sample was measured with a digital Mitutoyo measuring device. The measuring equipment was directly connected by two rods placed in the sample. The drying shrinkage of the sample was indirectly calculated by measuring the movement of the rods relative to each other.

The measurements taken were processed by a programme. The following values were calculated:

The water content of the sample during drying,
Longitudinal linear shrinkage of the sample,
Drying rate,
Enthalpy of the air used for drying.

The following curves could be plotted automatically on the basis of the values measured and calculated:

Temperature/relative humidity profile,
Sample water content versus time,
Sample drying rate versus time,
Longitudinal linear shrinkage of the sample versus time,
Linear shrinkage of the sample versus water content: the Bigot curve
Sample drying rate versus sample water content.

STUDY OF THE DRYING OF A DUTCH RIVER CLAY

This section describes a study of the drying behaviour of a Dutch clay, carried out using the drying equipment described above. Many years practical experience have shown that this clay is difficult to dry. The study clearly shows that the drying equipment developed and the measuring method provided valuable information for studying the drying of clays for coarse ceramics.
The drying process for coarse ceramic green bricks should be such that the bricks do not distort or crack. The water content after the drying process should be so low that defects do not occur in the material during sintering . These criteria largely determine the drying process and the drying time.

The Bigot curve

The Bigot curve of a Waal size brick cut in half along its length was measured. The brick was dried using a linear temperature and humidity curve. The brick was dried a second time at 105 °C to determine its initial water content
The Bigot curve was used to determine the total drying shrinkage, the critical water content and the after-shrinkage.

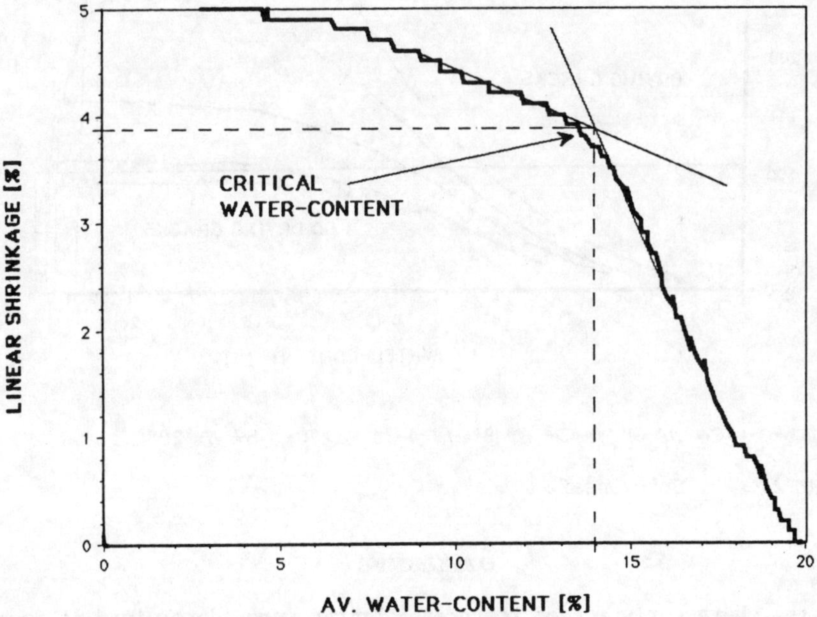

FIGURE 1 Bigot curve

- Total linear drying shrinkage : 5.0 %
- After-shrinkage : 1.2 %
- Critical water content : 14.2 %

Determining the safe and unsafe drying area

The drying rates were measured close to the initial water content (W(i)) and the critical water content (W(k)), at various psychrometric temperature differentials (temperature difference between the dry bulb and the wet bulb). Then the drying rates were measured at several constant psychrometric temperature differentials. These measurements were used to make a graph, see figure 2.
The optimum drying rate for these green bricks was 110 g/h.m^2. Figure 2 can be used to construct an initial drying curve. The results of this test are shown in figure 3 and 4.

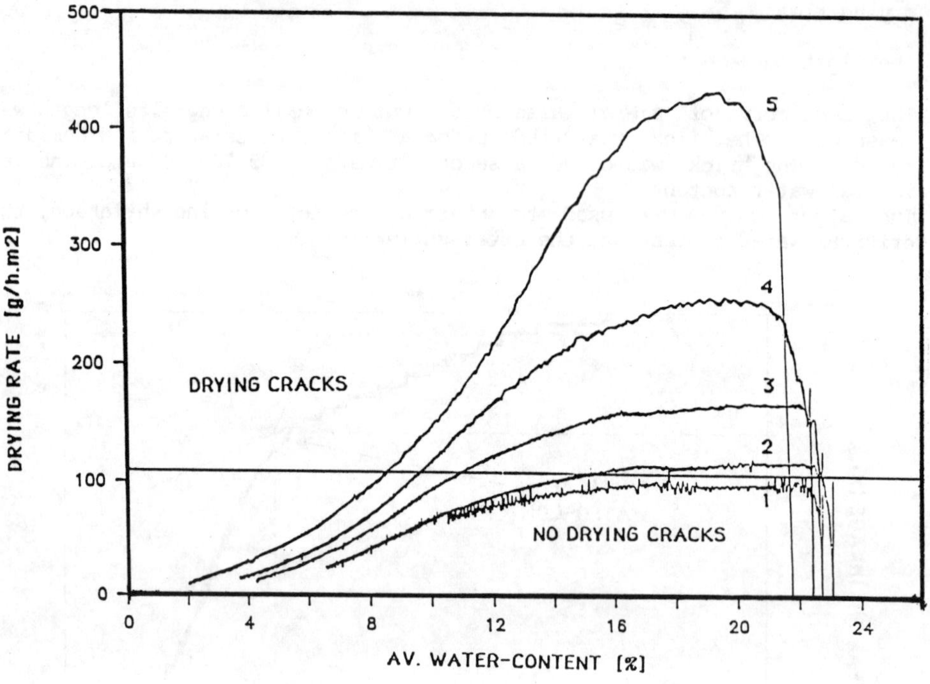

1= ΔT 5°C; 2= ΔT 6°C; 3= ΔT 8°C; 4= ΔT 12°C; 5= ΔT 20°C.

FIGURE 2 Safe/Unsafe

CONCLUSIONS

When the characteristics of the green bricks were determined it appeared that the Bigot curve as measured with the test equipment, provided more information than is found in the literature. A major difference compared with Bigot curves as found in the literature was the occurrence of two discontinuities (see figure 1). However, the literature does mention two discontinuities in the drying rate versus water content graph; these discontinuities probably correspond with the discontinuities in the Bigot

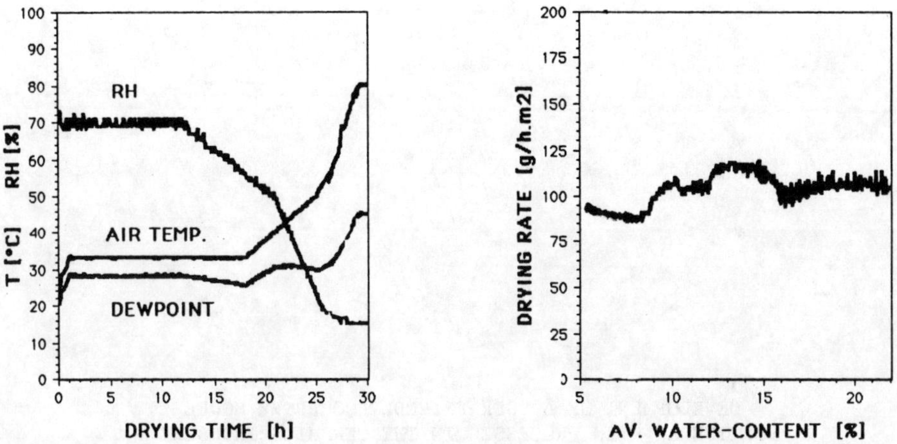

FIGURE 3 and 4 Initial drying curve and its measurement results

curve. It has been shown once again that the after- shrinkage specific to Dutch clays, forms an important parameter in determining the drying process. Increasing the drying rate after the critical water content has been reached can still cause drying cracks.

The drying equipment can be used to determine the maximum drying rate for a given mass. The optimum drying rate for the green bricks studied was 110 g/h.m^2. The drying conditions such as temperature, relative humidity and air flow speed, required to reach the drying rate will differ for each hot air drying plant. For this case, it was possible to meet the prerequisite that the green bricks be dried without cracks within 30 hours.
For the measurements at a constant psychrometric temperature differential a wet bulb temperature of 30 °C was arbitrarily chosen. It may be expected that at a higher wet bulb temperature the green bricks will be less sensitive to cracking during the constant rate period. This means that the optimum drying rate will then be higher resulting in a shorter drying time.

The experiments described above have provided valuable information about the behaviour of green bricks. The measuring equipment can be used to measure the drying behavior of various clay compositions which will contribute to solve drying problems in hot air drying plants.

DEVELOPMENT OF A USER-FRIENDLY COMPUTER MODEL FOR TUNNEL KILN PROCESSES IN THE CERAMICS INDUSTRY.

J.B. de Wit
TNO/MT Heat and Refrigeration Department
P.O. Box 342, 7300 AH Apeldoorn
The Netherlands

SUMMARY

Computer models are being used more and more to support field research. For this purpose the TNO Heat and Refrigeration Department has developed two computer models for the ceramics industry: Q-KILN, describing the static behaviour of the entire tunnel kiln and Q-CAR, describing the dynamic behaviour of tunnel kiln cars. A great deal of attention was paid to the userfriendlyness of these models in order to make them accessible to a wide range of users.

1. INTRODUCTION

An optimal design and optimal operation of industrial processes demand understanding of the relationships between all the relevant factors and even if, as the saying goes, experience is the mother of wisdom, this is usually gained at the cost of (unacceptably) expensive failures.

But the risks can be limited by using model development and simulation techniques. Moreover, these tools offer the possibility of optimization: these models can assist designers and users of industrial production facilities in finding optimal process conditions.

Computer models are being used more and more to support research. This also applies to the TNO Heat and Refrigeration Department's discipline. In the last three years two models have been developed for the ceramics sector, including a static overall model Q-KILN and a tunnel kiln car model Q-CAR, describing the dynamic behaviour of tunnel kiln cars.

Q-Kiln - stationary overall model of a tunnel kiln

Inputs : - Clay composition
- Geometry of kiln and product
- Proces variables

Outputs : - Energy consumption (MJ/ton product)
- Flue gas composition
- Drying air conditions volumes from cooling zone

Purpose : - Evaluation of energy consumption
- Optimisation of design of operation.

Figure 1 Summarized Q-KILN input and output

These models, and particularly Q-KILN, are very much different from the models developed 10 years ago; right from the start of development the main aspect was that the course of communication with the user should be simple and clear. Apart from that, it is undesirable that the user of these models is faced with submodels which do not easily link up with each other. In that context, the Q-KILN program acts as an "umbrella" under which new and existing programs perform. Q-KILN is developed for PC use.

Q-Car - Dynamic model of tunnel kiln cars

Inputs : - Time/temperature profile of tunnel kiln
- Geometry and materials

Outputs : - Thermal accumulation and transmission

Purpose : - Optimisation of design of tunnel kiln cars

Fig. 2 Summarized Q-CAR input and output

2. THE Q-CAR MODEL

The Q-CAR program allows analysis of the thermodynamic behaviour of the kiln car. Prior to use of Q-CAR for the analysis of kiln car designs, a study was carried out into whether 1D, 2D or 3D calculations are required for this type of design.
With regard to the objective, 1D calculations proved to be sufficiently accurate for almost any kiln car configuration used in practice. For this purpose the car was divided into thermally and geometrically identical kiln car parts. The basis for calculation of the non-stationary temperature distribution in any kiln car part is the following parabolic partial differential equation:

3. THE Q-KILN MODEL

Allthough the behaviour of the kiln parts is interesting, industry is obviously primarily concerned with the performance of the kiln as a whole. A static overall model has been developed on the basis of the research by ing. J.H. van der Velden [2].

As stated, the userfriendlyness was the main aspect in this developement. The user has three input screens available:
- the physical and chemical raw material characteristics;
- the shape characteristics (kiln packing, dimensions);
- the thermodynamic characteristics of the kiln and the kiln cars;
- the process variables (product temperatures, product rates, cooling air quantities, etc.).

For every variable, Q-KILN has a so-called default (= assumed) value which is representative of the values occurring in practice. The advantage of this is that users are not forced to enter very large numbers of parameters and the concomitant searching in manuals. So the user enters only the values to be changed. In conformity with the model by Van der Velden [2], Q-KILN is broken down into eight sections:
- preheating zone : 3 sections
- firing zone : 2 sections
- cooling zone : 3 sections

For each section the user enters the desired temperature profile and the desired production rate (with the kiln length as the resulting value) or the kiln length (with the production rate as the resulting value). On this basis Q-KILN calculates the power consumption per section, the required kiln length or production rate, the flue gas composition and the available drying air flows and temperatures. All the relevant heat flows are also calculated for each section. See figures 4 and 5.

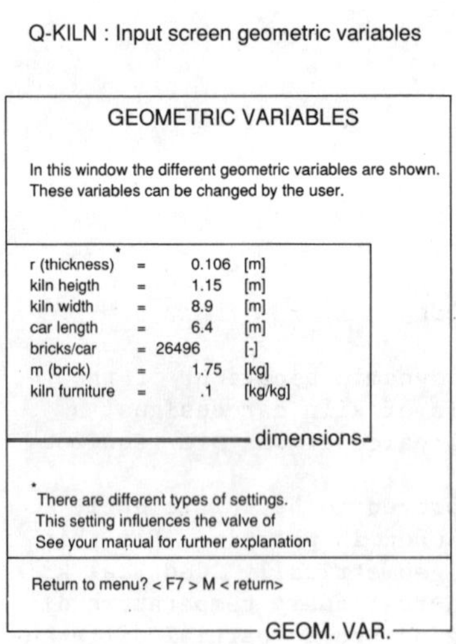

Figure 3 Shape variables input screen

$$\frac{\partial T}{\partial t} = \frac{\lambda}{\varrho . Cp} \cdot \frac{\partial^2 T}{\partial x^2}$$

λ = thermal conductivity (W/m K)
ϱ = density (kg/m³)
C_p = specific heat capacity (J/kg K)
$\frac{\partial T}{\partial t}$ = 1st derivative of the temperature with regard to time (K/s)
$\frac{\partial T}{\partial x^2}$ = 2nd derivative of the temperature with regard to location (K/m²)

As the thermal conductivity cannot be assumed to be constant in this process but, due to the large temperature range passed through, must be regarded as a function of the temperature, the above equation becomes nonlinear. The finite difference approach was used succesfully in solving this partial differential equation. For this purpose the equation was made discrete with regard to location and time. There is an initial value problem. As an explicit method was applied, the stability does not always remain stable for all values of the selected time increment. We have a stable solution if the following condition is satisfied:

$$\Delta t \leq \frac{(\Delta x)^2}{\sigma} \; ; \; \sigma = \frac{\lambda}{\varrho . C_p}$$

If kiln car materials with very different values are involved, it is possible that a very small time increment (dt) must be chosen, resulting in long computer times. Two boundary conditions and an initial condition are required to solve the above equation. Development of the boundary condition on the hot side is equal to the time/temperature profile of the product.

A constant temperature of 75 °C is assumed with regard to development of the boundary condition on the cold side of the kiln car if its time/temperature profile is unknown. A homogeneous kiln car temperature equal to the ambient temperature is assumed as the initial condition. Then the maximum allowable time increment (dt) is calculated on the basis of the above formula, allowing for the most unfavourable thermal conductivity (λ) value. Using the given time increment, the time is is incremented from time t = 0 until the specified final time is reached. The boundary conditions applicable to each time are calculated. The temperature dependent thermal resistances are calculated at these new boundary conditions. From these, together with the element temperatures, the heat flows are calculated, from which the new element temperatures are then calculated. The input and output heat flows and the accumulated heat in the kiln car are integrated. As the input of the physical characteristics of the kiln car and the car configuration can be carried out rapidly and in a userfriendly manner, a quick comparison between various kiln car designs is quite possible.

		flue gas composition			
		nm^3/sec.	kg/sec.	vol.% 'wet'	vol.% 'dry'
Nitrogen	(N_2)	3.59	4.94	70.86	76.69
Oxigen	(O_2)	0.85	1.22	15.32	16.58
Carbon Dioxide	(CO_2)	0.29	0.58	5.26	5.70
Water	(H_2O)	0.42	0.34	7.59	0.00
Argon	(Ar)	0.05	0.08	0.8	0.91
Methane	(CH_4)	0.01	0.01	0.12	0.13
Total		5.57	7.17	100.00	100.00

Figure 4 Output of flue gas composition

4. FUTURE DEVELOPMENTS

Allthough Q-KILN and Q-CAR can be used quite well at present, a number of expansions and adjustments are desirable:

- The drying process is not included in Q-KILN. Inclusion of a drying process model "under" Q-KILN is desirable as kiln and drier can then be optimally tuned to each other.
- Q-KILN and Q-CAR perform at the moment as individual program units. The desired integration has not yet been realized.
- At present Q-KILN only allows the user to optimize by trial and error, i.e. by varying the input, he can effect major output values such as the energy consumption to become minimal.

But it is desirable that the user is able to optimize quickly and easily, i.e. is able to define an optimization criterion by which Q-KILN calculates the optimum values for production rate, cooling air quantity, packing, etc. In this optimization criterion, the energy costs and other capital charges will be balanced against the production benefits.

5. REFERENCES

[1] Vemuri, V. and Karplus, W.J.,
"Digital Computer Treatment of Partial Differential Equations".
Prentice Hall, 1981.

[2] Velden, J.H. van der,
"Firing Process Model for Clay Ceramics in Tunnel Kilns",
TNO Report 86204.

RHEOLOGY OF REFRACTORY CONCRETE,

H. Gorter, J. v.d. Zwan
TNO Institute of Applied Physics,
Ceramics Department P.O. Box 513,
5600 MB Eindhoven, the Netherlands

ABSTRACT

The rheological properties of two low cement castables have been studied. One of these concretes contained a deflocculant. We studied the influence of the water content on the apparent bulk density before and after vibration, consistency, cold crushing strength and modulus of rupture. The consistency was measured with a Pilny-Martienson funnel. The results show a relationship between the water content of the mortar and the various material properties. By comparing the two concretes the influence of the deflocculant on the material properties becomes evident. Water content of the mortar is critical with respect to strength, consistency and the densification process. Doubt exists about the reliability of the consistency test. The results show that a watercontent close to standard PRE/R26 in fact is too high.

Introduction

In the last decade there has been progress in the production of refractory concrete. So called low- and ultra low cement castables have been developed. The use of ultra fine gel-forming powders and use of deflocculants made it possible to reduce the amount of cement in the concrete. This change in composition has several effects: advantageous effects are a higher refractoriness, a better corrosion resistance (1,3), lower water demand and a higher strength over the complete temperature range (2,3). A disadvantage of these concretes is the more difficult handling.

To achieve these qualities it is necessary to mix the concrete with the right amount of water. The water has two purposes: in the first place as a lubricant between the particles so the concrete can achieve a high apparent bulk density; secondly the water reacts with the cement. For both of these processes to be effective there should be enough water available. Too much water on the other hand has negative effects on the quality of the concrete.

This study aims to find whether consistency measurements with a Pilny-Martienson funnel (PRE/R26, 4) are sufficiently reliable to determine the correct amount of water. In addition the influence of a deflocculant is measured. To achieve these goals the influence of the water content of a mortar is measured on the consistency, the apparent bulk density before and after vibration, the modulus of rupture and the cold crushing strength.

Experimental procedure

The concrete employed in this study was a commercially grade alumina-based low-cement castable with 7 w.% cement. The bulk density was 3.78 kg/dm^3 as measured with a gas pyknometer method using the Stereopyknometer SPY-3 from Quantachrome Co. The amount of water absorbed by the concrete as received was 0.05 w.% and the surface area according to BET was 0.87 m^2/gr. The particle size distribution was measured with wet sieving and is given in figure 1. Two batches of this concrete were studied. The concrete A7 contained a deflocculant, while the concrete B7 did not contain this deflocculant. For every experiment a sample was taken from the concrete with a riffle box. Each sample was mixed with a predetermined amount of water in a Hobart planetary mixer in accordance with standard PRE/R26.
A Pilny-Martienson funnel was used to determine the consistency as discribed in standard PRE/R26. The funnel is mounted on a vibratory table that operates at 50 Hz and with the vertical amplitude adjusted to 0.75 mm. Run-out time was determined according to standard PRE/R26. A concrete has the correct consistency according to this standard if the run-out time is 15 s.

The apparent bulk density of the concrete has been determined as it was poured out and after 10 minutes of vibration. Apparent bulk density is based on the weight of the dry material. Six cylinders of 5 cm in diameter and 5 cm high were filled with mortar and the weight was determined. After vibration the resulting height of the mortar in the cylinders was determined. The results are shown as two curves in figure 3a for concrete A7 and in figure 3b for B7. The third curve shown in these pictures is the difference between the two apparent bulk densities called the densification curve.

fig. 1 Particlesize distribution by wet sieving for concrete A7 and B7

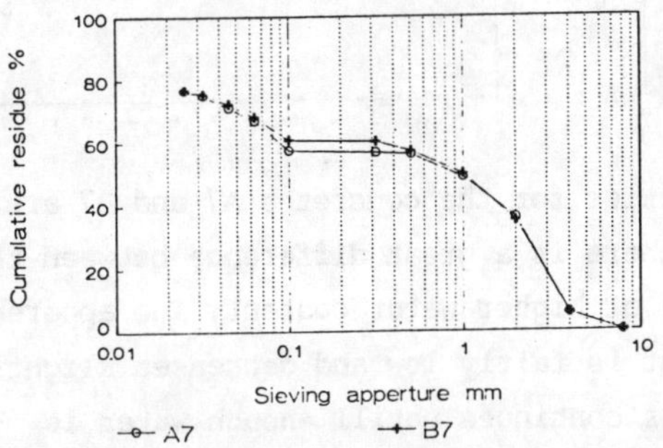

From the mortar each time six bars were cast with dimensions of 4 x 4 x 16 cm, according to standard PRE/R26, but the dimensions of the bars were in accordance with DIN 1164. The standard DIN 1164 was used to determine the modulus of rupture (MOR) and the cold crushing strength (CCS).

Results

In figures 2a and 2b the results of the consistency tests are shown. At low water content the concrete has no cohesive forces between the particles and the run-out time is low. With higher water content the cohesion between the particles increases and so does the run-out time, up to a level where the concrete just sticks to the funnel. At a further increase of the water content, the concrete starts to flow and the run-out time decreases. Both concretes A7 and B7 act in the same way, but there is a large difference between the water contents at which they start to flow. The correct water content according to PRE/R26 for concrete A7 is 6 w.% but for B7 this is 11 w.%. In addition there is a very broad range of water contents for which B7 will not flow out of the funnel. We conclude that the deflocculant is one of the most important parameters in controlling the necessery amount of water.

fig. 2 Run-out time vs. watercontent for (a) concrete A7 and (b) concrete B7

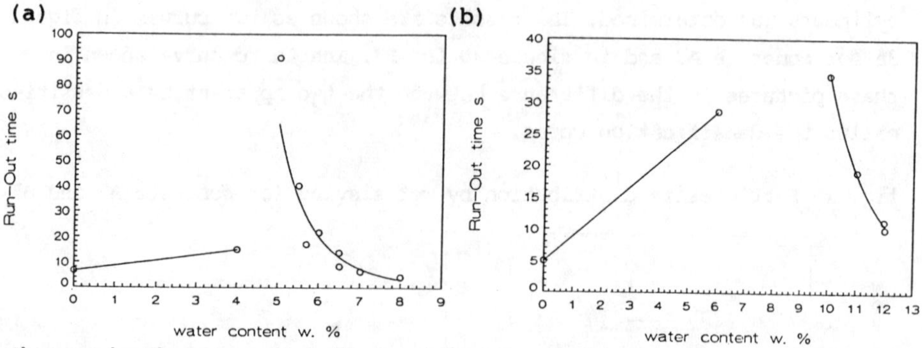

Changes in the apparent bulk densities for the concretes A7 and B7 are shown in figure 3a and 3b. Again there is a great difference between the behaviour of A7 and B7, especially at higher water content. The apparent bulk density at a low water content is fairly low and decreases slightly with increasing water content. This continues untill enough water is available between the particles so they can slide along each other and at then the apparent bulk density rises fast. This densification process starts with vibration at a lower water content than without vibration. The vibration helpes to overcome the relative strong cohesive forces between the particles. These forces decrease as the water content rises (6). Concrete A7 obtains an apparent bulk density of 85 % (relative to the bulk density of 3.78 kg/dm^3), but B7 only 65 %. B7 can not flow and can not reach a high apparent bulk density. The difference between the density before and after vibration, called the densification curve, is also shown in figures 3a and 3b. These curves show a distinct maximum, but the water content at that maximum does not coincide with 15 seconds run-out time (figs. 2a and 2b) but with a run-out time of approximatly 35 seconds.

The results of the measurements of CCS (cold crushing strenth) and MOR (modulus of rupture) are shown in figures 4 and 5. For concrete A7 the measurements show a maximum in strength for both CCS and MOR. This maximum appears at the same water content as the maximum for the densification curve in figure 3a.The measurements of CCS and MOR for concrete B7 (fig. 4b and 5b) show the behaviour that could be expected from the measurements of the apparent bulk density (fig. 3b). The

strength of concrete B7 is low compared to A7 and there is a large scatter of data points.

fig. 3 Relative density vs. watercontent for (a) concrete A7 and (b) concrete B7

(a) **(b)**

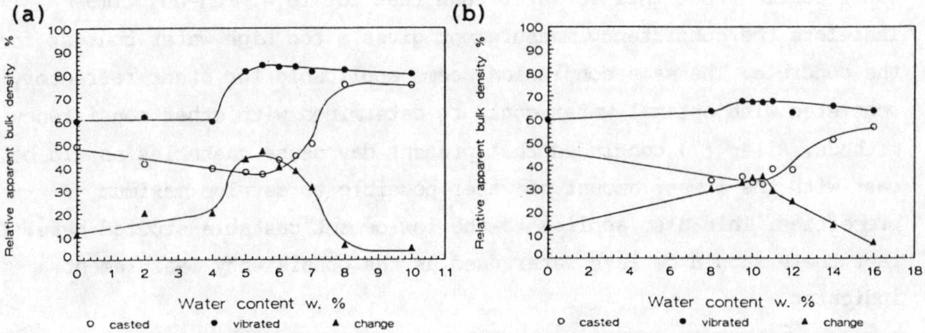

fig. 4 Cold crushing strength vs. watercontent for (a) concrete A7 and (b) concrete B7

(a) **(b)**

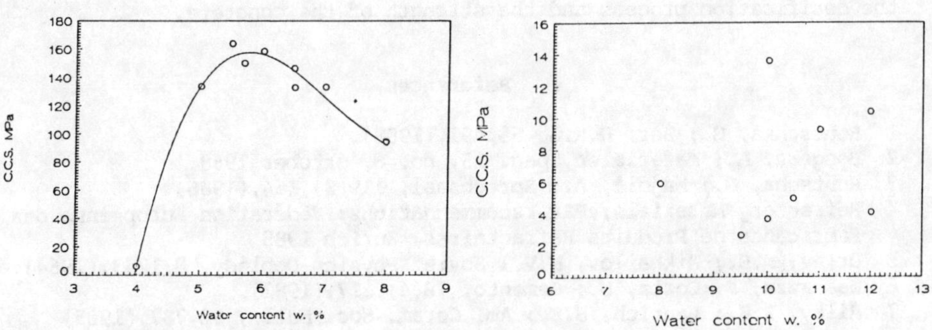

fig. 5 Modulus of rupture vs. watercontent for (a) concrete A7 and (b) concrete B7

(a) **(b)**

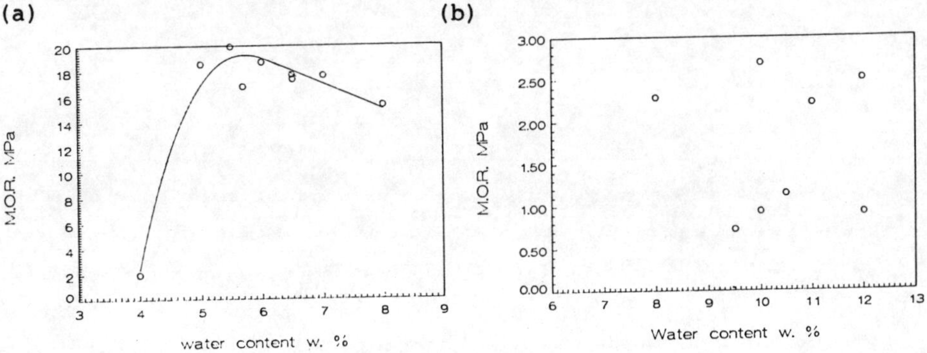

Conclusions

The optimal water content as measured with the Pilny Martienson funnel (15 s. run-out time) does not coincide with maximal MOR, CCS and maximal densification. Maximal MOR, CCS and densification appear at the same water content, but this is lower than that for 15 s. run-out time. Therefore the consistency measurement gives a too high water content for the concrete. The same conclusion seems applicable for other refractory concretes with optimal water contents determined with other consistency methods. Riley (7) concluded that present day dense castables should be cast with the lowest amount of water possible to develop maximum properties. This also applies to the low cement castable studied here. In fact there should be less water used as the consistency measurement indicates.

Comparing concrete A7 with B7 it is evident that the deflocculant in A7 has a major influence on the optimal water content, the flow behaviour, the desification process and the strength of the concrete.

References

1 Routschka, G.; Ber. D.K.G, **65**,491,(1988).
2 Bongers, A.; Materialen, pag. 13, no. **8**, oktober 1988.
3 Routscha, G.; Majdič, A.; Sprechsaal, **119**(3),164,(1986).
4 Refractory Materials; PRE recommendations; Fédération Européenne des Fabricants de Produits Réfractaires, Zurich 1985.
5 Ur'ev, N.B.; Mikhailov, N.V.; Soviet Physics-Doblady, **8**,1243,(1964).
6 Massazza, F.;Costa, U.; Cemento, **78**(4),217,(1982).
7 Riley, R.R.; Laurich, S.E.; Am. Ceram. Soc. Bull., **62**,787,(1983)

REFRACTORY BRICKS IN HIGHEST QUALITY PRODUCED WITH SGP-HYDRAULIC PRESSES

GERHARD ZOTTL
Head of Sales Dept. Presses
Simmering-Graz-Pauker AG
Brehmstraβe 16, 1110 Vienna, Austria

ABSTRACT

Report on the presses and setting equipment made by SGP for the refractory industry.

INTRODUCTION

Simmering-Graz-Pauker AG (SGP) is one of the leading Austrian enterprises in the capital goods and finished products industries. Its field of activity includes the branches of mechanical engineering, power and environmental protection technology as well as rail vehicles. The company's high quality of workmanship and its application of in-house high-tech research results in production have won world-wide recognition.

The SGP production programme for the refractories industry ranges from simple presses with manual mould filling to computer-assisted, freely programmable and fully automatic production lines with setting and unloading devices.

SGP-PRESSES

SGP presses are suitable for use with all free-flowing dry materials such as **magnesite, dolomite, compounds with tar, pitch, resin or carbon as binder, fire-clay or silica.**

SGP has two different press-designs in its programme:

The HSP oil-hydraulic floating table press is especially suitable for the manufacture of products with top and bottom surfaces of equal area, such as feather-edged bricks, slide gates, etc.

To suit differing situations, SGP offers its floating table moulding press in two versions, with bottom or top location of the piston.

The HSDP oil-hydraulic double-plunger press is in particular suitable for pressing parts which have unequally sized surfaces or which are of considerable height, such as spouts or chimney linings.
Standard and wedge-edged bricks, slide plates etc., can of course also be made economically and to highest quality standards on the HSDP double-plunger press.

The perfect double-plunger design and the fact the mould is fixed in the press frame permit separate preselection of pressure, stroke and speed for each plunger, thus ensuring maximum dimensional accuracy and uniformity of density of even tall workpieces or parts with unequally sized surfaces.
Moreover, compared with conventional press techniques, bricks of more uniform density and higher dimensional accuracy are made with a lower reject rate and less wear.

As a result of the systematic development, both systems of SGP presses have important **technological** and **technicoeconomic** advantages:

° **Universal adaptability**

° **Maximum reproducibility, precision and uniformity of brick dimensions and density**
Whatever its stroke, the differential piston is guided by means of bushes set at a constant maximum distance. Even with eccentric forces, for instance, in the manufacture of feathered-edge bricks. This design ensures maximum uniformity of dimensions and density. The control systems offered, which range from conventional controllers to freely programmable microprocessors can incorporate devices for automatic adjustment and correction of the filling depth, pressure release and venting to suit individual requirements.

As the brick discharge and the feeder box directions are at the right angles, the bottom plunger can also be cleaned automatically.

° **Up to 40 % energy savings**
The required power is infinitely adjustable by means of several axial piston pumps. Power and coolant consumption are almost zero during no-load operation.

° **Minimum maintenance and servicing costs**
All guide surfaces of the piston and the uprights are protected from dirt by bellows.
As the piston guide system is independent of the stroke, wear on guide bushes, packing and moulding tools is kept to a minimum.

- **Minimum downtime**
  The brick discharge and the feeder box directions are at right angles to one another.
  Therefore moulds can be changed very quickly without having to remove and readjust conveyor belts, racks and auxiliary equipment.
  The mould is automatically slid out and back by means of a hydraulic cylinder. No auxiliary appliances are needed and manual work is kept to a minimum.

- **Minimum installation times**
  At HSDP-presses the press frame is welded and stressfree annealed from a single piece without any loose parts.

To meet the different individual requirements, the SGP-presses can be supplied with fully developed auxiliary equipment. SGP has perfected appliances for improving quality and ease of operation as well as for reducing retooling times:

- **Oil leakage compensation**
  The pressing speed selected remains constant, independently of counterpressure, pump wear, oil temperature and viscosity, and oil leakage.

- **Spraying appliance**
  For automatic spraying of the top and bottom plunger with oil or water.

- **Electronic control systems**
  These are available with optional screen for easy storage and retrieval of production data on up to 150 brick formats.

- **Mixer systems**
  alternative with mixing shafts in the feeder box or above the feeder box with an own box.

- **Rapid clamping of moulds and quick-change devices for plungers**
  no screw connections are involved in changing moulds and plungers, so that total change-over times below 15 minutes are possible.

- **Feeder box filling**
  SGP delivers vibrator chute loaders or reversible conveyors for the various materials used, avoiding segregation.

- **Core-drawing equipment for spouts, Cowper-stone bricks etc.**

Further extras:

- **Gravimetric filling systems**
- **Special grippers for brick discharge**
- **Supplementary pressing units for glide plates**
- **Vacuum equipment for carbon bricks**
- **Press moulds and tools are also included in SGP's production programme.**

SGP-SETTING EQUIPMENT

Connected to the press for following heat treatment, SGP brick setting and discharging equipments have helped meet an urgent requirement of the present, namely automate many procedures which formerly had to be carried out manually for the most part, such as:

- **Removing of bricks from the press, and setting onto kiln cars, pallets, etc.**
- **Picking up of bricks from kiln cars, pallets, shelved cars and battens, and setting onto pallets fit for dispatch.**
- **Setting of bricks according to quality onto different locations**
- **Singling-out of reject bricks**
- **Utmost maintenance of customerspecific setting patterns**
- **Mastery of most complex setting patterns**
- **Minimal time spent changing brick size**
- **Discharging of misaligned brick stacks by sensor-controlled seeker set-head**
- **Close (jointless) setting of bricks fit for dispatch**

SGP brick setting and discharging equipment is built in modular design, thus ensuring that specific requirements can be fulfilled easily and individually at the same time.
One essential feature is the setting gripper, which is interposed between the press and the actual setting portal, and is designed to transfer the bricks leaving the press according to the setting pattern.

Among others SGP also designs

- **pallet setting equipment without portal unit**
- **gripper systems of all sorts for the most varied brick numbers**
- **brick weighing equipment**
- **brick measuring equipment**
- **brick marking and labelling equipment**
- **sanding equipment**
- **devices for automatic gripper change**

CO-OPERATION WITH THE CUSTOMER

The SGP package includes far more than the delivery of the product itself; it also includes training for operating staff and machine maintenance by an efficient customer service.
All steps, from the design of the package to manufacture, assembly and start-up, are carried out in co-operation with the customer. This is the only way to find an optimum solution to fit special individual requirements.

Presses and setting equipment of Simmering-Graz-Pauker AG,

the comprehensive solution.

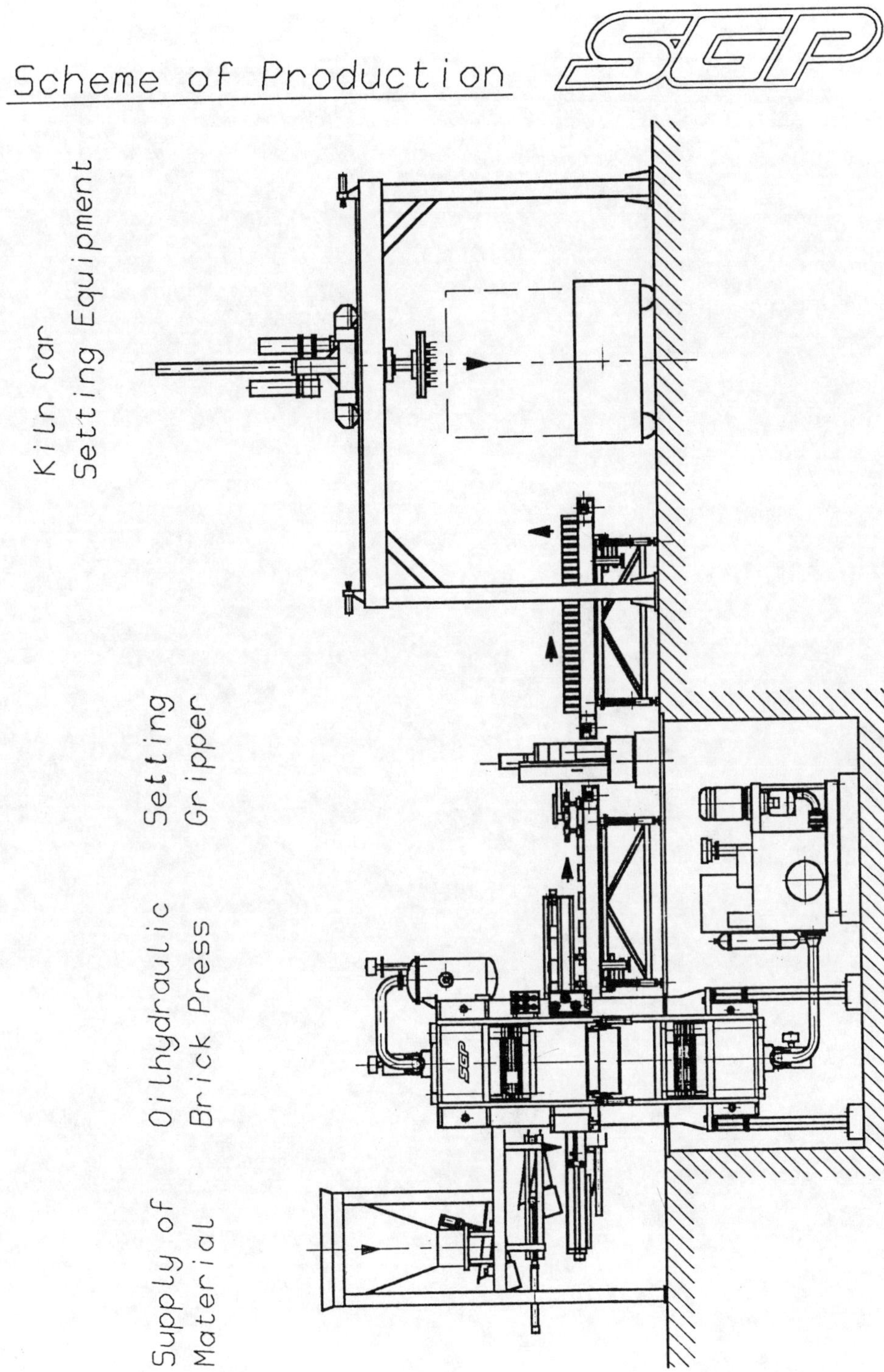
Scheme of Production
Kiln Car
Setting Equipment
Setting
Gripper
Oilhydraulic
Brick Press
Supply of
Material

LIST OF CONTRIBUTORS

INDEX OF FORMULAE

SUBJECT INDEX